# Mid-Infrared and Terahertz Quantum Cascade Lasers

Discover how mid-infrared and terahertz photonics has been revolutionized in this comprehensive overview of state-of-the-art quantum cascade lasers (QCLs).

Combining real-world examples with expert guidance, it provides a thorough treatment of practical applications, including high-power continuous-wave QCLs, frequency-comb devices, quantum-electronic transport and thermal transport modeling, and beam shaping in QCLs. With a focus on recent developments, such as frequency noise and frequency stabilization of QCLs, grating-outcoupled surface-emitting mid-infrared QCLs, coherent-power scaling of mid-IR and THz QCLs, metasurface-based surface-emitting THz QCLs, self-mixing in QCLs, and THz QCL sources based on difference-frequency generation, it also features detailed theoretical explanations of means for efficiency maximization, design criteria for high-power continuous-wave operation of QCLs, and QCL thermal modeling, enabling you to improve the performance of current and future devices.

Paving the way for new applications and further advancements, this is an invaluable resource for academics, researchers, and practitioners in electrical, opto-electronic, and photonic engineering.

**Dan Botez** is a Philip Dunham Reed Professor of Electrical and Computer Engineering at the University of Wisconsin–Madison. He is the recipient of the 2010 Optica Nick Holonyak Jr. Award, and a Fellow of Optica and the IEEE.

**Mikhail A. Belkin** is the Chair of Semiconductor Technology at the Walter Schottky Institute, Technical University of Munich. He is a recipient of the 2015 Friedrich Wilhelm Bessel research award among others. He is a Fellow of Optica and SPIE.

# Mid-Infrared and Terahertz Quantum Cascade Lasers

Edited by

DAN BOTEZ
University of Wisconsin–Madison

MIKHAIL A. BELKIN
Technische Universität München

CAMBRIDGE
UNIVERSITY PRESS

Shaftesbury Road, Cambridge CB2 8EA, United Kingdom

One Liberty Plaza, 20th Floor, New York, NY 10006, USA

477 Williamstown Road, Port Melbourne, VIC 3207, Australia

314–321, 3rd Floor, Plot 3, Splendor Forum, Jasola District Centre, New Delhi – 110025, India

103 Penang Road, #05–06/07, Visioncrest Commercial, Singapore 238467

Cambridge University Press is part of Cambridge University Press & Assessment, a department of the University of Cambridge.

We share the University's mission to contribute to society through the pursuit of education, learning and research at the highest international levels of excellence.

www.cambridge.org
Information on this title: www.cambridge.org/9781108427937
DOI: 10.1017/9781108552066

First published 2023

Printed in the United Kingdom by TJ Books Limited, Padstow Cornwall

*A catalogue record for this publication is available from the British Library.*

*A Cataloging-in-Publication data record for this book is available from the Library of Congress.*

ISBN 978-1-108-42793-7 Hardback

It is hard to believe that only about three decades have elapsed since the first conference in intersubband transitions took place in Cargese, Corsica, in the Fall of 1991 when the science of intersubband transitions was still in its infancy and the intersubband laser was but a vague dream. That conference and the second one, which followed it in 1993 in Whistler, Canada, brought together researchers excited by the newly open field, and the ensuing vigorous scientific exchange helped to usher in a number of significant developments, of which quantum cascade laser was by far the most momentous.

It is even harder for us to come to terms with the fact that the organizers of these two early gatherings, Emmanuel Rosencher (1952–2013) and H.-C. Liu (1960–2013), are no longer with us. Both of them were true visionaries and leaders who have made seminal contributions to the intersubband field on every level, and, tragically, both of them have left us way too early, in the same year of 2013. Emmanuel and H.-C. were well-known, respected and well-loved in the mid-infrared and THz communities, both professionally and as friends of many, including many contributors to this book. Emmanuel and H.-C. would have been pleased to see all the enormous progress in the QCL field attained just in the last few years, as certified by this book. It is only fitting therefore to dedicate this book to the memory of our friends and colleagues, Emmanuel Rosencher and H.-C. Liu.

# Contents

# Preface

It has been nearly 30 years since the first mid-infrared (mid-IR) quantum cascade laser (QCL) was demonstrated at Bell Laboratories in 1994. The first devices could only be operated in pulsed mode with cryogenic cooling, and it was far from clear at that time if lasers based on intersubband transitions could provide any practical levels of permanence. However, over the next eight years, the performance of QCLs improved dramatically: pulsed room-temperature operation was achieved in 1997 and the first continuous-wave (CW) room-temperature operating device was demonstrated in 2002. The wavelength coverage of QCLs has also expanded tremendously from the very first device operating at a wavelength of 4.2 μm. Modern mid-IR QCLs provide room-temperature CW operation across the entire 3–18 μm wavelength range with watts of CW output power over the 4–11 μm range. Several companies now offer mid-IR QCLs as commercial products that are used in both the defense and civilian sectors for applications such as infrared countermeasures, chemical and environmental gas sensing, free-space communications, high-resolution spectroscopy and microscopy, breath analysis, and other biomedical sensing.

Mid-IR QCLs operate at energies above the optical-phonon energies for semiconductors. Even before the first mid-IR QCLs were demonstrated, efforts were ongoing to also develop intersubband lasers that would operate at photon energies below the Reststrahlen band. These efforts resulted in the first demonstration of terahertz (THz) QCLs in 2001. Initially, THz QCLs could only be operated in pulsed mode with liquid helium cooling. However, over the years their performance has improved significantly, and they can now operate in pulsed mode on thermoelectric coolers. Active research efforts are ongoing towards developing devices capable of operation at room temperature. The progress in the temperature performance of THz QCLs is hindered by thermally activated carrier scattering in the upper laser levels and by the difficulty of achieving selective carrier injection into the upper laser level. In parallel, by using intra-cavity difference-frequency generation (DFG) in mid-IR QCLs, room-temperature THz emission was demonstrated in 2008 and the performance of these THz DFG-QCL devices has steadily been improved to reach mW-level peak output powers and CW powers of approximately over 10 microwatts at room temperature. While their output powers and the wall-plug efficiencies are lower than those of THz QCLs, THz DFG-QCLs offer a very broad tuning range (1–6 THz tuning was demonstrated) in addition to room-temperature operation. THz QCLs are now commercially available from at least two vendors in the USA and Europe. They find applications as local-oscillator sources for heterodyne detection and are used as light sources for high-resolution spectroscopy, imaging, microscopy, and nanospectroscopy.

Since the only comprehensive book on QCLs was published [1], there have been many exciting new developments in the research areas of mid-IR and THz QCLs. In 2012, frequency comb generation was discovered in mid-IR QCLs and frequency comb QCL sources were quickly developed for both the mid-IR and THz spectral regions with applications to spectroscopy and metrology demonstrated. In addition, key developments in mid-IR QCLs have been: the derivation, by taking into account carrier leakage, of fundamental limits in wall-plug efficiency significantly higher than previously estimated [1]; the discovery that carrier leakage triggered by interface-roughness scattering (IFR) generally dominates carrier leakage, and that its consideration fully explains experimental internal-efficiency values; and the achievement of a generalized IFR formalism that takes into account graded well/barrier interfaces and its use to reproduce the electro-optical characteristics of record-high performance short- and long-wave mid-IR QCLs. As for THz QCLs, the key developments have been: the achievement of THz QCLs that can operate at thermoelectric-cooler temperatures and THz DFG-QCLs with CW room-temperature operation; the demonstration of vertical-external-cavity QCL based on the concepts of optical metasurfaces; and the realization that QCLs can operate as self-pumped homodyne and heterodyne detectors that may be used for imaging and nanospectroscopy. In addition, various new beam-shaping and coherent-power scaling methods have been developed for both mid-IR and THz QCLs. This book addresses these important developments by having dedicated chapters written by experts on these aspects of QCL technology.

This book also covers crucial device features and applications not treated in the previous comprehensive monograph on QCLs [1], such as, regarding mid-IR QCLs, carrier leakage, and active-region designs for its suppression, the mechanisms behind the temperature dependence of the electro-optical characteristics of QCLs, how elastic scattering impacts the upper- and lower-laser level lifetimes at room temperature, and the means for achieving CW high-power ($>1.5$ W) and CW high wall-plug-efficiency operation; and, regarding THz QCLs, frequency stabilization of QCLs as well as the application of QCLs for chemical sensing using dual-comb spectroscopy, and other examples of QCL applications.

The book is aimed at both graduate students at the Master and PhD levels and practitioners in industry. In addition, it is intended to serve as reference material for QCL research, development, and applications.

The book starts with a chapter that focuses on a general treatment of intersubband transitions between quantum confined states in semiconductor heterostructures. The origin of these states is briefly described with the linear combination of atomic orbitals and then proceeds to k.P theory. The relationship between the interband and intersubband transitions, including their oscillator strength and selection rules, is established. Aside from radiative intersubband transitions, nonradiative transitions play important roles, and hence the most relevant of these scattering processes – electron phonon, interface roughness, and alloy disorder – are described in detail.

Chapter 2 covers state-of-the-art mid-IR QCLs emitting in the 4–10 μm wavelength range with emphasis on key issues not covered in previous books on QCLs as well as on important developments over the last decade. The foremost issue that is addressed – what does it take to achieve CW operation to multi-watt powers in a highly efficient manner? – is of interest to a wide range of applications. An updated review of the temperature dependence of the electro-optical characteristics of QCLs is presented by including elastic scattering and carrier-leakage triggered both by elastic and inelastic scattering, thus accounting for all mechanisms behind the internal efficiency. The significant impact of using graded-interface modeling is also addressed. Maximizing the CW wall-plug efficiency via band and elastic-scattering engineering as well as via photon-induced carrier transport is treated in detail. Then coherent-power scaling is discussed for both one- and two-dimensional (2-D) structures with emphasis on the optimal solution: high-index-contrast (HC) photonic-crystal (PC) lasers. Grating-coupled surface-emitting lasers are also treated with emphasis on those needed for 2-D HC-PC lasers; that is, devices holding potential to operate in diffraction-limited, single-lobe beam patterns to multi-watt CW output powers.

Chapter 3 reviews mid-IR QCLs emitting in the 15–28 μm range. Historically, 15 μm was a "barrier" wavelength above which QCL performance dramatically degraded, which was in large part due to increases in waveguide optical loss due to approaching the Reststrahlen band. This intrinsic limitation caused by multi-phonon absorption set forbidden or favorable spectral areas depending on the employed materials. However, over the past decade significant progress has been made by using material systems such as InAs/GaSb. The chapter considers specific properties of long-wavelength mid-IR QCLs based on different materials, and general issues related to QCL design. The results are presented in chronological order for each QCL material system.

Chapter 4 provides an overview of the state-of-the-art THz QCLs. Both active-region configurations and waveguide configurations of modern THz QCLs are discussed, and their specific properties are emphasized. The chapter also reviews the progress of THz QCLs towards room-temperature operation.

Chapter 5 presents a review of key challenges to QCL theory: (i) the choice of basis states when energy eigenstates are badly defined if the full periodic structure is considered; (ii) tunneling through barriers requires a treatment of quantum coherences: the advantages and disadvantages of approaches such as rate equations, Monte-Carlo simulations, density matrix, and Green's functions methods are addressed; (iii) gain evaluation is detailed, where broadening is of utmost importance; and (iv) electrical instabilities of the extended structure due to domain formation, which strongly affect the overall performance.

Chapter 6 gives an overview of a simulation framework capable of capturing the highly nonequilibrium physics of strongly coupled electron and phonon systems in QCLs. In mid-IR devices, both electronic and optical-phonon systems are described by coupled Boltzmann transport equations, which are solved using a technique known as ensemble Monte Carlo. The optical-phonon system is strongly coupled to acoustic phonons, whose dynamics and thermal transport throughout the whole device are described via a global heat-diffusion solver. Nonequilibrium optical phonons and anisotropic thermal transport of acoustic phonons are also presented.

Chapter 7 provides a comprehensive overview of the active new research topic of frequency comb generation in mid-IR and THz QCLs. It covers the key concepts and applications of optical frequency combs, gives the principles of frequency comb generation in QCL fast gain media, discusses methods of controlling the gain spectrum and attaining low group velocity dispersion in QCLs for broadband frequency comb generation, and describes measurement techniques for assessing QCL waveguide dispersion and performing comb characterization. The effect of radio-frequency injection on QCL frequency combs is also discussed in detail for the case of both frequency-modulated and amplitude-modulated frequency combs.

Chapter 8 discusses the topic of frequency noise and frequency stabilization of QCLs. High laser-frequency stability is needed to perform high-resolution and high-precision spectroscopy as well as frequency metrology with QCLs. The chapter reviews the origin of frequency noise in QCLs and describes measurements of the frequency noise power-spectral density in mid-IR and THz QCLs as well as techniques for frequency stabilization and frequency locking of QCLs.

Chapter 9 treats high-power, single-mode THz QCLs. Best performance is obtained with metallic cavities due to strong plasmonic-mode confinement of the optical mode. However, such lasers suffer from poor beam shapes, low output power, and multimode spectral behavior. The development of distributed-feedback (DFB) techniques to improve the spectral and modal properties thus becomes imperative. The theory, design methodologies, and key results from a sampling of a wide variety of DFB techniques that have been implemented for monolithic THz QCLs, in both edge-emitting and surface-emitting configurations, are presented.

Chapter 10 introduces a new class of THz QCLs based upon amplifying electromagnetic metasurfaces made from two-dimensional arrays of sub-wavelength antennas loaded with QCL gain material. Several types of device are described: vertical-external-cavity surface-emitting-lasers in which the amplifying metasurface is paired with external optics to form a laser cavity; monolithic metasurface lasers in which the metasurface array self-oscillates in a coherent array mode; and metasurfaces designed as free-space THz amplifiers. The metasurface THz QCLs approach allows the realization of lasers with high-quality beam patterns, scalable output powers, and broadly tunable single-mode emission.

Chapter 11 describes THz QCL sources based on intra-cavity DFG in mid-IR QCLs. These devices, commonly known as THz DFG-QCLs, are currently the only monolithic electrically pumped semiconductor laser sources of 1–6 THz radiation that can be operated at room temperature. They can now produce nearly 2 mW of peak terahertz power and up to 14 microwatts of continuous-wave terahertz power at room temperature. THz DFG-QCLs can provide widely tunable single-frequency THz output spanning 1–6 THz as well as THz frequency combs. The chapter gives an overview of the current state of the art and discusses approaches to improve the performance of these devices further.

Chapter 12 provides an insider's perspective and historical context to the present and rapid trajectory in widespread QCL deployment across multiple civilian- and defense-sector applications and markets, which have been building over the past two decades and will likely continue to benefit the quality of life and standard of living over the next hundred years.

Chapter 13 provides a detailed description of one of the most common QCL-based gas sensing methods – quartz-enhanced photoacoustic spectroscopy (QEPAS). The QEPAS technique is first described starting from basic physics principles. This is followed by a detailed discussion of the influence of quartz tuning forks, which is the core of a QEPAS sensor, on sensing performance. Different applications of the QEPAS technique and the minimum detection limits reached for the most performant QEPAS sensors are described for all gas species detected so far.

Chapter 14 addresses sensing and applications of mid-IR and THz QCL optical-frequency combs, which have emerged as the most promising sources for high-resolution spectrometers with broadband spectral coverage. Recent advancements in QCL and interband-cascade-laser frequency combs are overviewed as key steps towards field applications with truly integrated and scalable frequency-comb technology. Also discussed are dual-comb spectroscopy techniques that offer fast chemical sensing without the need for optomechanical tuning or dispersive spectrometers, and an overview is provided of the spectroscopic capabilities of dual-comb spectrometers. Measurement approaches and recent experimental implementations of mid-IR and THz dual-comb spectroscopy of chemicals using frequency-comb technology are treated as well.

Chapter 15 provides an overview of the effect of self-mixing in QCLs, where the feedback of a fraction of the light emitted from a QCL back into the QCL cavity changes the optical field inside the laser cavity and affects device bias voltage and other laser parameters. The coherent nature of this effect opens opportunities across a broad range of applications, particularly in the THz spectral region where it allows the reliance on THz detectors to be circumvented. The chapter gives a detailed overview of the physics of self-mixing and laser feedback interferometry in QCLs and provides examples of applications including measurements of the emission spectrum and linewidth enhancement factor of THz QCLs, fast coherent real-time THz imaging, THz nanoimaging, and high-resolution THz gas spectroscopy.

Finally, Chapter 16 presents an overview of THz QCL applications from the perspective of a THz QCL commercial manufacturer. Although THz QCLs still require cooling for operation, THz QCL applications have been multiplying steadily over the past decade, helped by the availability of turn-key Stirling engine cooler-based compact commercial THz QCL systems.

## Reference

[1]   Faist, J., *Quantum Cascade Lasers*, Oxford: Oxford University Press, 2013.

# Part I

## Bandstructure Engineering, Modeling and State-of-the-Art QCLs

# 1 Basic Physics of Intersubband Radiative and Nonradiative Processes

Jacob B. Khurgin

Johns Hopkins University

Intersubband (ISB) transitions were first observed experimentally in the 1980s [1] and immediately attracted attention due to their unique properties, of which the ability to be designed for a specific wavelength in the infrared (IR) range of the spectrum is perhaps the most significant one. Soon thereafter IR detectors based on the ISB transitions – quantum well intersubband detectors [2, 3] as well as incoherent emitters [4] – were realized, and in the following decade the Intersubband lasers, QCLs [5–7], which are the subject of this book, were developed. In addition to the flexibility of design, ISB transitions have a large oscillator strength, which allows one to create efficient sources and detectors that are only a few micrometers (i.e., a fraction of wavelength) thick. The concept of intersubband transition is easily understood using a simple potential-well model from any physics textbook, but this model does not provide an explanation of how the ISB transition is related to the interband transitions and how their strengths compare [8]. In this chapter we shall establish this connection, explain the selection rules, and point out that for the same frequency the interband and ISB transitions are just about equally strong, which explains why the interband cascade lasers (ICLs) that have been in development since the late 1990s [9, 10] provide strong competition to QCLs in the shorter wavelength range of the IR spectrum. The physics of ISB transitions [6, 11] is the subject of the first two sections of this chapter.

In Section 1.3 we describe the nonradiative ISB transitions which play an important role in the operation of QCLs. Specifically, we explain how the strong interaction between electrons and polar longitudinal-optical phonons, leading to picosecond-scale transitions within and between subbands, enables population inversion and lasing. We also discuss other important processes such as electron-electron scattering that thermalizes electrons within a given subband, interface-roughness (IFR) scattering that causes both transition broadening and transition-lifetime shortening, and alloy-disorder (AD) scattering, which also causes transition-lifetime shortening (this fact, for both IFR and AD scattering, being often underappreciated). The short background provided by this chapter is by no means a substitute for in-depth knowledge of the physics of semiconductor quantum wells and superlattices, which can be gained from a number of excellent texts on this subject [11–14]. Nevertheless, it aims to provide background sufficient for understanding the operation of various QCL designs that are covered in the rest of this book.

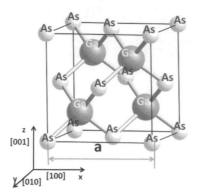

**Fig. 1.1** Zinc-blende crystal lattice of a typical III-V semiconductor (GaAs).

## 1.1 Electronic States in the III-V Semiconductor Quantum Wells

### 1.1.1 The Basis States and the Bond Orbital Picture

Although there have been attempts to generate infrared radiation using intersubband transitions in a wide range of semiconductors, at this time all practical QCLs are based on covalent III-V semiconductors, consisting of a group III (Al, In, or Ga) cations and group V (P, As, or Sb) anions arranged in a zinc-blende (ZB) cubic lattice. If only one species of anion and cation constitute the material, then it is called a binary alloy, such as, for instance, GaAs and InP. Otherwise, materials can be ternary (e.g., $Al_xGa_{1-x}As$ or $InAs_xP_{1-x}$) or quaternary (e.g., $In_yGa_{1-y}As_xP_{1-x}$) alloys. The composition of ternary- and quaternary-alloy materials is usually designed in such a way that their lattice constant is identical (or close) to that of a binary-alloy substrate, usually InP or GaAs, although alternative substrates have also been considered. It is the ability to adjust the composition of quantum wells (QWs) and their geometry that allows one to design the structures with ISB transitions over a wide range of IR and THz frequencies.

In the ZB lattice (Fig. 1.1) each group III cation atom is surrounded by four group V anion atoms in a tetragonal configuration and vice versa each group V ion is surrounded by four group III ions, thus, there are four tetragonal bonds directed along four diagonals of the cube – the <111> crystallographic directions. The bonds are formed by hybridization. Hybridization is a process by which first the four valence orbitals of each ion – one S-type $|S_{a,c}\rangle$ and three P-type: $|X_{a,c}\rangle$, $|Y_{a,c}\rangle$, $|Z_{a,c}\rangle$, all shown in Fig. 1.2, form four hybrid orbitals (Fig. 1.3) [15, 16],

$$\left|H_{a,c}^{(1)}\right\rangle = \frac{1}{2}\left|S_{a,c}\right\rangle + \frac{1}{2}\left|X_{a,c}\right\rangle + \frac{1}{2}\left|Y_{a,c}\right\rangle + \frac{1}{2}\left|Z_{a,c}\right\rangle,$$

$$\left|H_{a,c}^{(2)}\right\rangle = \frac{1}{2}\left|S_{a,c}\right\rangle + \frac{1}{2}\left|X_{a,c}\right\rangle - \frac{1}{2}\left|Y_{a,c}\right\rangle - \frac{1}{2}\left|Z_{a,c}\right\rangle,$$

$$\left|H_{a,c}^{(3)}\right\rangle = \frac{1}{2}\left|S_{a,c}\right\rangle - \frac{1}{2}\left|X_{a,c}\right\rangle - \frac{1}{2}\left|Y_{a,c}\right\rangle + \frac{1}{2}\left|Z_{a,c}\right\rangle,$$

$$\left|H_{a,c}^{(4)}\right\rangle = \frac{1}{2}\left|S_{a,c}\right\rangle - \frac{1}{2}\left|X_{a,c}\right\rangle + \frac{1}{2}\left|Y_{a,c}\right\rangle - \frac{1}{2}\left|Z_{a,c}\right\rangle,$$

$$(1.1)$$

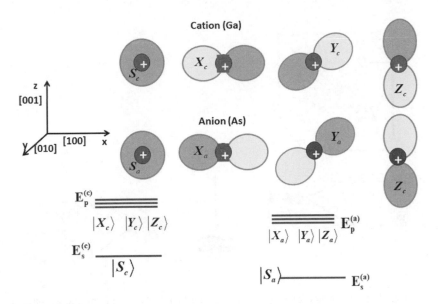

**Fig. 1.2** Basis states of four valence orbitals for cation and anion and their energies.

where indices a and c indicate anion and cation, respectively, from the four hybrid orbitals, and then each pair of hybrid orbitals directed towards each other form bonding,

$$\left| H_+^{(i)} \right\rangle = \alpha \left| H_a^{(i)} \right\rangle + \beta \left| H_c^{(i)} \right\rangle,$$  (1.2)

and anti-bonding,

$$\left| H_-^{(i)} \right\rangle = \beta \left| H_a^{(i)} \right\rangle - \alpha \left| H_c^{(i)} \right\rangle,$$  (1.3)

orbitals separated by the energy gap (which is **not** the fundamental direct energy gap related to the absorption edge). Four bonding orbitals are all filled with eight valence electrons shared by each anion–cation pair and thus constitute a fully occupied valence band. Four unfilled anti-bonding orbitals form the empty conduction band, and the energy difference between two bands represents the binding energy holding the crystal together. Since the bond is polar $\alpha > 1/\sqrt{2} > \beta$ indicating that electrons in the conduction band tend to be located near the group V anion (which is therefore negatively charged) and the electrons in the conduction band are to be found closer to the group III cation.

The above "bond orbitals" picture adequately describes the general features of ZB covalent semiconductors: the presence of a bandgap, mechanical hardness (due to the rigid "skeleton" formed by bonds), a relatively large dielectric constant (the latter due to long polarizable bonds), piezoelectricity, and optical nonlinearity (due to lack of central symmetry), as well as other properties. However, in order to understand the most relevant QCL electrical and optical properties one must consider the band structure near the center of the Brillouin zone, i.e., the states

$$u_{c,v}(\boldsymbol{k})e^{i\boldsymbol{k}\cdot\boldsymbol{r}},$$  (1.4)

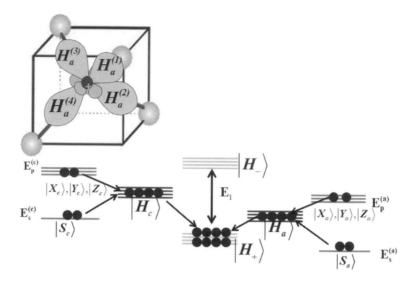

**Fig. 1.3** Four hybrid orbitals that bind the zinc-blende lattice and the energy diagram of hybrid bonding and anti-bonding schemes.

where $\mathbf{k}$ is the wavevector and $u_{c,v}(\mathbf{k})$ is the periodic wave Bloch function. At the center of BZ, where $\mathbf{k}=0$, the conduction and valence states split into triple-degenerate bonding and anti-bonding P-type states, and nondegenerate bonding and anti-bonding S-type states as shown in Fig. 1.4.

We thus obtain (see Fig. 1.5) the following states:

- The lowest valence-band (VB) state with energy $E_{s+} \sim -10$ eV,

$$u_{v1}(0) = \left| S_+ \right\rangle = \alpha \left| S_a \right\rangle + \beta \left| S_c \right\rangle, \tag{1.5}$$

which is a bonding combination of S-like orbitals of cations and anions. This band plays virtually no role in the interactions relevant to QCL operation and will from now on be neglected.

- Three degenerate valence band states with energy $E_{p+}$, which we take to be equal to zero:

$$u_{v2}(0) = \left| X_+ \right\rangle = \alpha \left| X_a \right\rangle + \beta \left| X_c \right\rangle,$$
$$u_{v3}(0) = \left| Y_+ \right\rangle = \alpha \left| Y_a \right\rangle + \beta \left| Y_c \right\rangle, \tag{1.6}$$
$$u_{v4}(0) = \left| Z_+ \right\rangle = \alpha \left| Z_a \right\rangle + \beta \left| Z_c \right\rangle.$$

These are the bonding combinations of P-like orbitals. These states are the ones in which the holes reside and are also the ones that mix with conduction-band (CB) states in QWs and are therefore extremely important for ISVB transitions.

- The lowest CB state with energy $E_{s-} = E_g$,

$$u_{c1}(0) = \left| iS_- \right\rangle = i\beta \left| S_a \right\rangle - i\alpha \left| S_c \right\rangle, \tag{1.7}$$

is the anti-bonding combination of S-like orbitals of anion and cation and is the most important state for the ISB transitions and thus plays the paramount role in

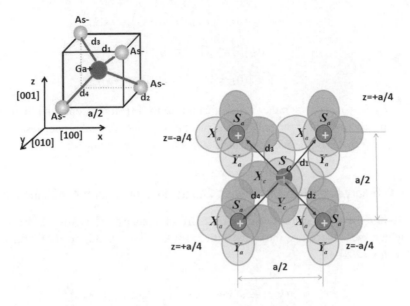

**Fig. 1.4** Binding between basis states in the zinc-blende lattice near the center of the Brillouin zone.

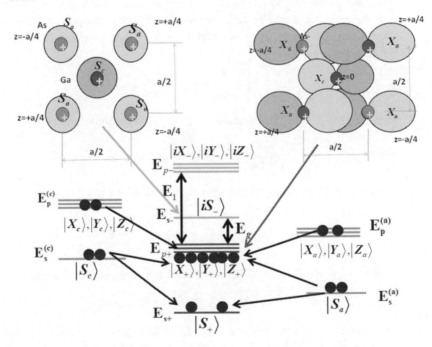

**Fig. 1.5** Energy and composition of states in zinc-blende lattice near the center of the Brillouin zone.

QCL operation. The factor of $i$ has been introduced for convenience as shown below.

- Finally, there are three higher lying degenerate CB states with energy $E_{p+} = E_1 \sim 5 - 7\,\text{eV}$:

$$u_{c2}(0) = \left|iX_-\right\rangle = i\beta\left|X_a\right\rangle - i\alpha\left|X_c\right\rangle,$$

$$u_{c3}(0) = \left|iY_-\right\rangle = i\beta\left|Y_a\right\rangle - i\alpha\left|Y_c\right\rangle, \qquad (1.8)$$

$$u_{c4}(0) = \left|iZ_-\right\rangle = i\beta\left|Z_a\right\rangle - i\alpha\left|Z_c\right\rangle.$$

These states play a rather insignificant role for ISB transitions, and their impact is mostly relevant to the bulk dielectric and nonlinear optical properties. Nevertheless, we shall keep these states for the time being.

## 1.1.2      k.P Theory Applied to Zinc Blende Semiconductors near the Brillouin Zone Center

To find the states near the center of the Brillouin zone one invokes **k.P** theory [17, 18], which seeks to express the states with small **k** as a linear combination of known states with **k = 0**, i.e.,

$$\Psi(k, r) = u(k)e^{ik\cdot r} = \sum_{m=1}^{M} C_m(k)u_m(0, r)e^{ik\cdot r}, \qquad (1.9)$$

where the summation index goes through all the bands, but in practice is always limited only to the states whose energies are not far from the energies of band edges, in our case to just the seven states described above. If one now substitutes (1.9) into the Schrödinger equation

$$\left[-\frac{\hbar^2}{2m_0}\nabla^2 - U(r)\right]\Psi = E\Psi, \qquad (1.10)$$

then uses the fact that the functions $u_m(0, r)$ are orthogonal, and that

$$\left[-\frac{\hbar^2}{2m_0}\nabla^2 + U(r)\right]u_m(0, r) = E_m u_m(0, r), \qquad (1.11)$$

one obtains a system of equations for the coefficients $C_m(k)$:

$$C_m(E_m - E + \frac{\hbar^2}{2m_0}) + \sum_{n \neq m}^{M} C_n \frac{i\hbar}{m_0}k \cdot P_{mn} = 0, \qquad (1.12)$$

where the matrix element of the momentum between two states $m$ and $n$ is

$$P_{mn} = -i\hbar \langle u_m(0)| \nabla |u_m(0)\rangle. \qquad (1.13)$$

The main matrix element of interest to ISB transitions is that between the lowest anti-bonding state in the conduction band (Fig. 1.6) and one of the top states in the valence band:

$$i\hbar \langle iS_-| \nabla |X_+\rangle = P\hat{x}, \qquad (1.14)$$

where $P$ is a real number and $\hat{x}$ is a unity vector along the [100] direction, with similar expressions for two other directions. The value of this matrix element is roughly the

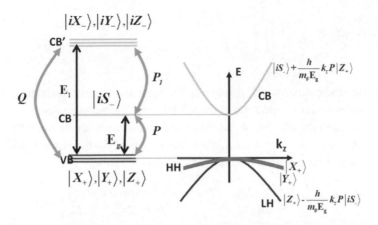

**Fig. 1.6** Energy and dispersion near the center of the Brillouin zone in the k.P model without spin-orbit interaction.

same for all III-V materials, which simply follows from the fact that the absolute value of the derivative should be on the scale of the ionic radii of cations and anions (i.e., somewhat on the order of $(\text{Å})^{-1}$ for most of the ions). Indeed, the values of this matrix element are usually quoted for the sake of convenience as follows:

$$E_P = 2P^2/m_0, \tag{1.15}$$

where $E_P$ ranges from 20 to 30 eV for all relevant semiconductors [19].

Another important matrix element connects the bonding and anti-bonding P-like states, for instance:

$$i\hbar \langle iY_- | \nabla | Z_+ \rangle = Q\hat{x}. \tag{1.16}$$

This matrix element $Q$, usually given as $E_Q = 2Q^2/m_0$, is of the same order as $P$ and is responsible for most of the dielectric properties of III-V semiconductors. But since the upper conduction bands are separated from the states in the lowest conduction band by as much as a few eV, their impact on ISB transitions is limited – all these states are doing is providing effective mass for heavy-hole states in the valence band. In addition, due to the lack of central symmetry there exists an additional matrix element between nominally anti-bonding S-like and P-like states,

$$i\hbar \langle iS_+ | \nabla | iX_+ \rangle = iP_1\hat{x}, \tag{1.17}$$

where $P_1 \approx (\alpha^2 - \beta^2)P$, which is responsible for all the second-order nonlinear effects, including second-harmonic generation and Pockels effect, but plays very limited role for intersubband transitions, mostly by slightly affecting all the effective masses of the electrons. If we now write the full k.P Hamiltonian for the energies that are close to the upper edge of the valence band, we obtain the following matrix in the basis of states $|iS_-\rangle, |Z_+\rangle, |X_+\rangle, |Y_+\rangle$:

$$\begin{pmatrix} E_g + \hbar^2 k^2/2m_0 & (\hbar/m_0)\,k_z P & 0 & 0 \\ (\hbar/m_0)\,k_z P & \hbar^2 k^2/2m_0 & 0 & 0 \\ 0 & 0 & \hbar^2 k^2/2m_0 - \hbar^2 k_z^2 Q^2/m_0^2 E_1 & 0 \\ 0 & 0 & 0 & \hbar^2 k^2/2m_0 - \hbar^2 k_z^2 Q^2/m_0^2 E_1 \end{pmatrix}.$$

(1.18)

In this approximation, which disregards the spin, the lowest conduction band state $|iS_-\rangle$ couples with the light state in the VB $|Z_+\rangle$, while two other states in the valence band do not couple with the conduction band and are considered to be "heavy" with effective mass $m_h^{-1} \sim m_0^{-1}(1 - EQ/E_1)$. One can then write the characteristic equation for the conduction and light-hole states as follows:

$$\begin{vmatrix} E_g + \hbar^2 k^2/2m_0 - E(k) & (\hbar/m_0)\,k_z P \\ (\hbar/m_0)\,k_z P & \hbar^2 k^2/2m_0 - E(k) \end{vmatrix} = 0.$$

(1.19)

The equation has two solutions – for the CB:

$$E_c(k) \approx E_g + \frac{\hbar^2 k^2}{2m_0} + \frac{\hbar^2 k^2}{2m_0}\frac{E_P}{E_g} = E_g + \frac{\hbar^2 k^2}{2m_c},$$

(1.20)

and for the light valence band:

$$E_1(k) \approx \frac{\hbar^2 k^2}{2m_0} - \frac{\hbar^2 k^2}{2m_0}\frac{E_P}{E_g} = -\frac{\hbar^2 k^2}{2m_l},$$

(1.21)

where we have introduced the effective masses of the electron and light hole as

$$\begin{aligned} m_c^{-1} &= m_0^{-1}\left(E_P/E_g + 1\right), \\ m_l^{-1} &= m_0^{-1}\left(E_P/E_g - 1\right), \end{aligned}$$

(1.22)

which can be substantially smaller than the effective mass of the free electron. For instance, the effective mass of the electron for GaAs is $m_c = 0.067m_0$ and for the light hole $m_c = 0.086m_0$. At the same time, since energy $E_1 \gg E_g$, the effective mass of the heavy hole is much larger, $m_h = 0.44m_0$.

Substituting solution (1.20) into (1.19) immediately gives us the coefficient $C$ that describes the admixing of the light VB state $|Z_+\rangle$ into the conduction band, and thus we obtain the overall expression for the wavefunction in the CB for small wave vectors:

$$\Psi_c(k_z, r) = \left[|iS_-\rangle + \frac{\hbar}{m_0 E_g}k_z P\,|Z_+\rangle\right] e^{ik_z z}.$$

(1.23)

It is instructive here to estimate the momentum of the state in the conduction band:

$$\langle \Psi_c^*(k) | \, p \, | \Psi_c(k)\rangle = \hbar k + \frac{2\hbar k P^2}{m_0 E_g} = \hbar k \frac{m_0}{m_c}.$$

(1.24)

This result explains why $\hbar k$ is called quasi-momentum and only represents a small part of the total momentum coming from the propagation term $\exp(ik_z z)$, while most of the total momentum originates from the band-mixing.

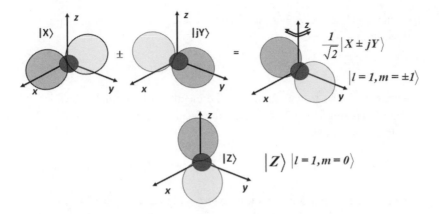

**Fig. 1.7** Basis states of the valence band with spin-orbit interaction.

## 1.1.3     Spin-Orbit Interaction and Band-Mixing in Bulk Semiconductors

The results obtained below give a good physical picture of the origin of the states in the CB with small wavevectors – there are mostly anti-bonding S-type states with the admixture of the bonding P-like states oriented along the direction of the wavevector. But in order to obtain a full picture, one must consider spin $S$ and orbital $L$ angular momenta which add up to make up a total angular momentum $J = L + S$ that must be conserved, i.e., each state must be an eigenstates with well-defined $J$. The orbital momentum of all P-like states is $l = 1$ and since spin angular momentum of 1/2 can be either parallel or anti-parallel to orbital momentum $J$ can take values of $1/2$ or 3/2. One can re-arrange six (including spin) wavefunctions in the VB and label them by their orbital angular momentum, its projection on the $z$-axis, $m$ (see Fig 1.7), the spin projection onto the same axis, $m_s$, and the projection of total angular momentum, $m_J$ [12, 18]:

$$
\frac{1}{\sqrt{2}}|X_+ + iY_+\rangle \uparrow = \left| m = 1, m_s = \frac{1}{2}, m_J = \frac{3}{2} \right\rangle,
$$

$$
|Z_+\rangle \uparrow = \left| m = 0, m_s = \frac{1}{2}, m_J = \frac{1}{2} \right\rangle,
$$

$$
\frac{1}{\sqrt{2}}|X_+ + iY_+\rangle \downarrow = \left| m = 1, m_s = -\frac{1}{2}, m_J = \frac{1}{2} \right\rangle,
$$

$$
\frac{1}{\sqrt{2}}|X_+ - iY_+\rangle \uparrow = \left| m = -1, m_s = \frac{1}{2}, m_J = -\frac{1}{2} \right\rangle, \tag{1.25}
$$

$$
|Z_+\rangle \downarrow = \left| m = 0, m_s = -\frac{1}{2}, m_J = -\frac{1}{2} \right\rangle,
$$

$$
\frac{1}{\sqrt{2}}|X_+ - iY_+\rangle \downarrow = \left| m = -1, m_s = -\frac{1}{2}, m_J = -\frac{3}{2} \right\rangle.
$$

Two states with projection of total angular momentum $m_J = \pm 3/2$ are already the eigenstates of $J$ with $J = 3/2$, while the four states with $m_J = \pm 1/2$ can be states with either $J = 1/2$ or $J = 3/2$; and therefore they get mixed by the spin-orbit interaction Hamiltonian:

$$H_{so} = \frac{eE_r}{2m_0^2 c^2 r} \boldsymbol{L} \cdot \boldsymbol{S}, \qquad (1.26)$$

where $E_r$ is the field that attracts the electron to the nucleus, which increases with increasing atomic number. Therefore, the energies of states with parallel angular and spin momenta increase by

$$\left\langle m = \pm 1, m_s = \pm \frac{1}{2} \middle| H_{so} \middle| m = \pm 1, m_s = \pm \frac{1}{2} \right\rangle = \frac{\Delta}{3}, \qquad (1.27)$$

where $\Delta$ is a spin-orbit splitting energy (for GaAs $\Delta = 0.4\,\text{eV}$), while energies of states with anti-parallel angular and spin momenta decrease by the same amount:

$$\left\langle m = \mp 1, m_s = \pm \frac{1}{2} \middle| H_{so} \middle| m = \mp 1, m_s = \pm \frac{1}{2} \right\rangle = -\frac{\Delta}{3}, \qquad (1.28)$$

and the states with the same projection of total angular momentum $m_J$ get mixed:

$$\left\langle m = 0, m_s = \mp \frac{1}{2} \middle| H_{so} \middle| m = \mp 1, m_s = \pm \frac{1}{2} \right\rangle = \frac{\sqrt{2}\Delta}{3}. \qquad (1.29)$$

Hence in the basis of $\left| m = \pm 1, m_s = \pm \frac{1}{2} \right\rangle$, $\left| m = 0, m_s = \pm \frac{1}{2} \right\rangle$, and $\left| m = \pm 1, m_s = \mp \frac{1}{2} \right\rangle$ the spin-orbit Hamiltonian can be written in matrix form as follows:

$$H_{so} = \begin{pmatrix} \Delta/3 & 0 & 0 \\ 0 & 0 & \sqrt{2}\Delta/3 \\ 0 & \sqrt{2}\Delta/3 & -\Delta/3 \end{pmatrix}, \qquad (1.30)$$

and the characteristic equation for energy becomes

$$\begin{vmatrix} \Delta/3 - E & 0 & 0 \\ 0 & -E & \sqrt{2}\Delta/3 \\ 0 & \sqrt{2}\Delta/3 & -\Delta/3 - E \end{vmatrix} = 0. \qquad (1.31)$$

This equation has three solutions, as shown in Fig. 1.8(a).
The original heavy-hole state with energy $\Delta/3$:

$$\left| J = \frac{3}{2}, m_J = \pm \frac{3}{2} \right\rangle = \frac{1}{\sqrt{2}} |X_+ \pm iY_+\rangle \uparrow\downarrow. \qquad (1.32)$$

The light-hole state also with energy $\Delta/3$:

$$\left| J = \frac{3}{2}, m_J = \pm \frac{1}{2} \right\rangle = \sqrt{\frac{2}{3}} |Z_+\rangle \uparrow\downarrow - \frac{1}{\sqrt{6}} |X_+ \pm iY_+\rangle \downarrow\uparrow, \qquad (1.33)$$

which is technically 2/3 light and 1/3 heavy state, and a split-off band with energy $-2\Delta/3$:

$$\left| J = \frac{1}{2}, m_J = \pm \frac{1}{2} \right\rangle = \frac{1}{\sqrt{3}} |Z_+\rangle \uparrow\downarrow + \frac{1}{\sqrt{3}} |X_+ \pm iY_+\rangle \downarrow\uparrow, \qquad (1.34)$$

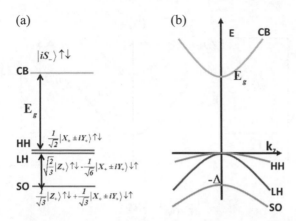

**Fig. 1.8** Energy band composition and dispersion near the center of the Brillouin zone in the **k.P** model with spin-orbit interaction.

which is technically 1/3 light and 2/3 heavy state. The heavy-hole state is not being mixed into the conduction-band states, but both light-hole and split-off hole states do, so we can rewrite the characteristic equation (1.19), for the three-band k.P Hamiltonian, in the basis $u_1 = |iS_- \uparrow\downarrow\rangle$, $u_2 = |Z_+ \uparrow\downarrow\rangle$, and $u_3 = 2^{-1/2}|(\pm X_+ + iY_+) \downarrow\uparrow\rangle$ as follows:

$$
\begin{vmatrix}
E_g + \dfrac{\hbar^2 k^2}{2m_0} - E(k) & (\hbar/m_0)\, k_z P & 0 \\[2ex]
(\hbar/m_0)\, k_z P & \dfrac{\hbar^2 k^2}{2m_0} - \dfrac{1}{3}\Delta - E(k) & \dfrac{\sqrt{2}}{3}\Delta \\[2ex]
0 & \dfrac{\sqrt{2}}{3}\Delta & \dfrac{\hbar^2 k^2}{2m_0} - \dfrac{2}{3}\Delta - E(k)
\end{vmatrix} = 0. \quad (1.35)
$$

If we now consider the state in the CB with a small wavevector, and substitute $E(k) \approx E_g$ into the last two lines, we immediately obtain

$$
E_c(k) \approx E_g + \frac{\hbar^2 k^2}{2m_0} + \frac{\hbar^2 k^2}{2m_0}\frac{E_P(E_g + 2\Delta/3)}{E_g(E_g + \Delta)} = E_g + \frac{\hbar^2 k^2}{2m_c}, \quad (1.36)
$$

which is shown in Fig. 1.8(b), where the effective mass is

$$
m_c^{-1} = m_0^{-1}\left(\frac{E_P(E_g + 2\Delta/3)}{E_g(E_g + \Delta)} + 1\right). \quad (1.37)
$$

In comparison to (1.22) the effective mass is changed by spin-orbit interaction rather insignificantly, but as we shall soon see the spin-orbit interaction becomes important when it comes to optical properties.

## 1.1.4 States in Quantum Wells in the Envelope Approximation

We now consider a semiconductor QW consisting of a narrow bandgap material layer, such as GaAs or InGaAs surrounded by higher bandgap material (AlGaAs or InP)

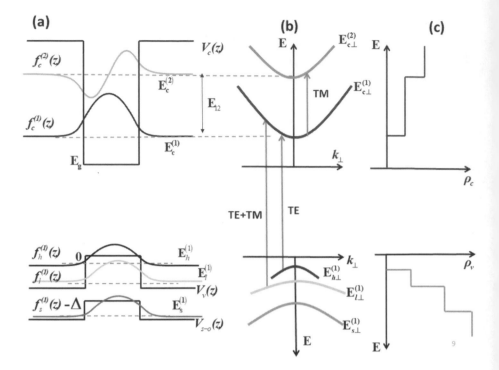

**Fig. 1.9** (a) QW subband energies and envelope wavefunctions, (b) in-plane dispersion with optical transitions for TE and TM polarized light, and (c) density of states.

as shown in Fig. 1.9(a). The confining potential can be described as $V_c(z)$ and the wavefunction of the $m$th state (subband) in the CB can be written as follows [12]:

$$\Psi_c^m(\mathbf{k}_\perp, \mathbf{r}) = \sum_{n=1}^{3} f_n^{(m)}(z) u_n(\mathbf{r}) e^{i\mathbf{k}_\perp \cdot \mathbf{r}_\perp}, \qquad (1.38)$$

where $f_n^m$ is the envelope function describing the contribution of the $n$th basis state in (1.35); i.e., $n = c, l, s$ for the conduction, light-hole, and split-off bands, respectively, to the state in the $m$th subband in the QW, and $\mathbf{k}_\perp$ and $\mathbf{r}_\perp$ are the wave vector and coordinate in the QW plane $xy$. The wavefunctions are shown in Fig. 1.9(a). Substituting this into the original wave equation we obtain

$$\left[ -\frac{\hbar^2}{2m_0} \nabla^2 + U(\mathbf{r}) + V_c(z) \right] \Psi_c^m(\mathbf{k}_\perp, \mathbf{r}) = E_c^m(\mathbf{k}_\perp) \Psi_c^m(\mathbf{k}_\perp, \mathbf{r}). \qquad (1.39)$$

One can separate the variables into the in-plane motion with energy

$$E_{c\perp}^{(m)} = \hbar^2 k_\perp^2 / 2m_c, \qquad (1.40)$$

as shown in Fig. 1.9(b), and the motion in the $z$-direction for which we substitute $\partial/\partial z$ operator in place of $ik_z$ in (1.35) and obtain the equation:

$$
\begin{pmatrix}
-\dfrac{\hbar^2}{2m_0}\dfrac{\partial}{\partial z} - E_c^{(m)} + V_c(z) & i\dfrac{\hbar}{m_0}\dfrac{\partial}{\partial z}P & 0 \\[2ex]
i\dfrac{\hbar}{m_0}P\dfrac{\partial}{\partial z} & -\dfrac{\hbar^2}{2m_0}\dfrac{\partial^2}{\partial z} - \dfrac{1}{3}\Delta - E_g - E_c^{(m)} + V_v(z) & \dfrac{\sqrt{2}}{3}\Delta \\[2ex]
0 & \dfrac{\sqrt{2}}{3}\Delta & -\dfrac{\hbar^2}{2m_0}\dfrac{\partial^2}{\partial z} - \dfrac{2}{3}\Delta - E_g - E_c^{(m)} + V_s(z)
\end{pmatrix}
$$

$$
\times \begin{pmatrix} f_c^{(m)} \\[1ex] f_l^{(m)} \\[1ex] f_s^{(m)} \end{pmatrix} = 0,
\tag{1.41}
$$

where $E_c^{(m)}$ is the energy of the $m$th conduction band, relative to the bandgap band at $k_\perp = 0$, and $V_v(z)$ and $V_s(z)$ are the potentials in the light-hole and split-off bands, respectively. Now for the state not far from the bottom of the CB one can re-write (1.41) as such:

$$
\begin{pmatrix}
-\dfrac{\hbar^2}{2m_0}\dfrac{\partial^2}{\partial z} + E_c^{(m)} + V_c(z) & i\dfrac{\hbar}{m_0}\dfrac{\partial}{\partial z}P & 0 \\[2ex]
i\dfrac{\hbar}{m_0}P\dfrac{\partial}{\partial z} & -\dfrac{1}{3}\Delta - E_g - E_c^{(m)} & \dfrac{\sqrt{2}}{3}\Delta \\[2ex]
0 & \dfrac{\sqrt{2}}{3}\Delta & -\dfrac{2}{3}\Delta - E_g - E_c^{(m)}
\end{pmatrix}
$$

$$
\times \begin{pmatrix} f_c^{(m)} \\[1ex] f_l^{(m)} \\[1ex] f_s^{(m)} \end{pmatrix} = 0.
\tag{1.42}
$$

The last equation in (1.42) readily yields

$$
f_s^{(m)}(z) = f_{l(m)}(z)\frac{\frac{\sqrt{2}}{3}\Delta}{\frac{2}{3}\Delta + E_g + E_c^{(m)}},
\tag{1.43}
$$

and substitution of (1.43) into the second equation of (1.42) results in

$$
f_l^{(m)}(z) = i\frac{\hbar}{m_0}P\frac{\frac{2}{3}\Delta + E_g + E_c^{(m)}}{(E_c^{(m)} + E_g)(E_g + E_c^{(m)} + \Delta)}\frac{\partial f_c^{(m)}(z)}{\partial z} = \frac{i\hbar}{2P}\left(\frac{m_0}{m_c(E_c^{(m)})} - 1\right)\frac{\partial f_c^{(m)}(z)}{\partial z},
\tag{1.44}
$$

where we have introduced the **energy-dependent** effective mass of the conduction band,

$$
m_c^{-1}(E_c^{(m)}) = m_0^{-1}\left(\frac{E_P(E_g + E_c^{(m)} + 2\Delta/3)}{(E_g + E_c^{(m)})(E_g + E_c^{(m)} + \Delta)} + 1\right).
\tag{1.45}
$$

Since the bandgap energy changes as a function of the coordinate, the effective mass is not only energy-dependent but also coordinate dependent. Now, by substituting (1.44) into the first equation of (1.42) we obtain a Schrödinger equation,

$$
-\frac{\hbar^2}{2m_c\left(E_c^{(m)}\right)}\frac{\partial^2 f_c^{(m)}}{\partial z^2} + V_c(z)f_c^{(m)} = E_c^{(m)}f_c^{(m)},
\tag{1.46}
$$

for the CB envelope function in the energy-dependent effective-mass approximation. This equation is solved using boundary conditions requiring continuity of $f_c^{(m)}(z)$ and

$m_c^{-1} \partial f_c^{(m)}(z)/\partial z$ and the values of subband energies $E_c^m$, shown in Fig. 1.9(b), can be determined. We can then write out the expression for the overall CB wavefunction (1.38) as follows:

$$\Psi_c^m(\boldsymbol{k}_\perp, \boldsymbol{r}) \uparrow\downarrow = \left[ f_1^m(z) |iS_- \uparrow\downarrow\rangle + \frac{i\hbar}{2P} \left( \frac{m_0}{m_c(E_c^m)} - 1 \right) \right.$$
$$\left. \times \frac{\partial f_1^m(z)}{\partial z} \left[ |Z_+ \uparrow\downarrow\rangle + \frac{r_\Delta}{\sqrt{2}} |(\pm X_+ + iY_+) \downarrow\uparrow\rangle \right] \right] e^{i\boldsymbol{k}_\perp \cdot \boldsymbol{r}_\perp}, \quad (1.47)$$

where

$$r_\Delta = \frac{\frac{\sqrt{2}}{3}\Delta}{\frac{2}{3}\Delta + E_g + E_c^{(m)}}. \quad (1.48)$$

Therefore the VB states get admixed into the CB in such a way that the parity of their envelope functions, $f_l^m(z)$ and $f_s^m(z)$, is opposite from the parity of the CB envelope function $f_c^m(z)$. The mixing is relatively small and can be evaluated by noting that the order of magnitude of the derivative of the envelope function is roughly $(1/d)$, where $d$ is its spatial extent. Then the relative weight of mixed-in VB states is roughly $(r_{\text{ion}}/d)^2$, where $r_{\text{ion}}$ is the ionic radius of the order of 1 Å. Therefore, the relative weight of the VB wavefunction is typically less than 0.1%. Nevertheless, as we shall see below, it is this mixing that engenders strong intersubband transitions.

## 1.2          Intersubband Transitions – Their Origin and Relative Strength

### 1.2.1          Interband Absorption in Quantum Wells

Let us evaluate the absorption coefficient between the two subbands, one in the valence band $\Psi_v^m$ (where $v$ can be either heavy (h)- or light (l)-hole subband) and one in the conduction band $\Psi_c^n$. The Hamiltonian of the interaction between electro-magnetic wave $\boldsymbol{E}(t) = \frac{1}{2}\boldsymbol{E}(\omega)\exp(i\omega t) + c.c.$ is [16]

$$H_E = \frac{e}{m_0} \boldsymbol{p} \cdot \boldsymbol{A}, \quad (1.49)$$

where $A$ is a vector potential, related to the electric field as $\boldsymbol{E}(t) = \partial A/\partial t$ and $\boldsymbol{p} = -i\hbar\nabla$ is a momentum operator. We shall start with Fermi's golden rule to evaluate the rate of change of the 2D carrier density due to the absorption of light as such:

$$\frac{dn_{2D}}{dt} = \frac{2\pi}{\hbar} \frac{e^2 p_{cv}^2}{m_0^2 \omega^2} \rho_{cv} \frac{E^2(\omega)}{4} F_{mn}, \quad (1.50)$$

where $p_{cv}$ is the matrix element of momentum operator between the Bloch functions of the conduction and valence bands; the $F_{mn}$ term,

$$F_{mn} = \left| \int f_c^{(n)}(z) f_v^{(m)}(z) dz \right|^2, \quad (1.51)$$

is the overlap of envelope wavefunctions, the factor of $1/4$ is related to the fact that only the positive frequencies of the electric field contribute to the absorption, and the joint density of states (Fig. 1.9(c)) is

$$\rho_{cv}^{(mn)}(\hbar\omega) = \frac{\mu_{cv}}{\pi\hbar^2} H(\hbar\omega - E_g - E_c^{(n)} - E_v^{(m)}), \qquad (1.52)$$

where $H$ is the Heaviside step function, and $\mu_{cv} = \left(m_c^{-1} + m_{v,|\perp}^{-1}\right)^{-1}$ is the reduced mass. Before continuing it is important to note that the in-plane valence band effective masses in QWs are different from their bulk values, namely $m_{hh,|\perp}^{-1} = \frac{3}{4}m_{lh}^{-1} + \frac{1}{4}m_{hh}^{-1}$ and $m_{lh,|\perp}^{-1} = \frac{3}{4}m_{hh}^{-1} + \frac{1}{4}m_{lh}^{-1}$ .

Let us first consider the interband transition between the heavy-hole subband and the conduction band, for which the matrix element can be found as follows:

$$\boldsymbol{p}_{12} = \langle -iS \uparrow\downarrow |\boldsymbol{p}| \frac{1}{\sqrt{2}} |(\pm X_+ + iY_+) \uparrow\downarrow\rangle = \frac{P}{\sqrt{2}}(\pm\hat{\boldsymbol{x}} + i\hat{\boldsymbol{y}}). \qquad (1.53)$$

For linearly polarized light we obtain $p_{cv}^2 = P^2/2 = m_0 E_P/4$. At the same time, we can pay attention to (1.45) and note that $m_c^{-1} \sim m_0^{-1} + 2P^2/m_0^2 E_g$. Since the effective mass of a hole is always higher than that of an electron, the reduced mass should be of the order of $\mu_{cv} \sim m_0 E_g/E_P$, and one can introduce a dimensionless parameter:

$$R_{cv} = \frac{E_P}{m_0 E_g}\mu_{cv}, \qquad (1.54)$$

which is on the order of unity for most III-V semiconductors (for instance, for GaAs $m_c = 0.67m_0$, in-plane $m_{hh,\perp} = 0.11m_0$, $\mu_{cv} = 0.041m_0$, and $E_P = 28.8\,\text{eV}$ we get $R_P = 0.82$). Therefore, from (1.52) $\rho_{cv} = R_{cv}m_0 E_g/E_P\pi\hbar^2$ and under the assumption that $\hbar\omega \sim E_{gap}$ and by introducing the power density of light propagating in the direction normal to the plane of QWs (Fig. 1.10(a)):

$$I_\omega = \frac{n_r}{2\eta_0}E(\omega)^2, \qquad (1.55)$$

where $n_r$ is the refractive index and $\eta_0 = 377\Omega$ is the vacuum impedance, one obtains from (1.50) the expression for the rate of increase of the energy density inside QWs as such:

$$\frac{du_{2D}}{dt} = \hbar\omega\frac{dn_{2D}}{dt} = \hbar\omega\frac{2\pi}{\hbar}\frac{e^2}{m_0^2\omega^2}\frac{m_0 E_P}{4}R_{ch}\frac{m_0 E_g}{E_P\pi\hbar^2}\frac{\eta_0 I_\omega}{2n_r}F_{mn}$$

$$= \frac{\eta_0 e^2}{4\hbar n_r}R_{ch}F_{mn}I_\omega = \alpha_{ch}^{(mn)}(\omega)I_\omega, \qquad (1.56)$$

where the absorption coefficient per one QW between the $m$th heavy-hole subband and $n$th conduction band subband is a remarkably simple expression:

$$\alpha_{ch}^{(mn)} = \frac{\pi\alpha_0}{n_r}R_{ch}F_{mn}, \qquad (1.57)$$

where $\alpha_0 = 1/137$ is the fine structure constant, indicating that practically any semiconductor 2D structure has an absorption coefficient per one layer on the order of 1%.

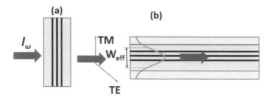

**Fig. 1.10** (a) Absorption for normal incidence light; (b) TE and TM absorption for the waveguide propagation.

If, on the other hand, one considers the propagation of the light in the dielectric waveguide mode (Fig. 1.10(b)) defined by the effective width $W_{\text{eff}}$ and containing $N_{\text{QW}}$, the number of quantum wells, the absorption will take place only for the TE (in-plane) polarized light and the absorption coefficient (now per unit of length) will be

$$\alpha_{\text{ch}}^{\text{TE}} = N_{QW}\alpha_{\text{ch}}^{(mn)} = N_{\text{QW}}\frac{\pi\alpha_0}{n_{\text{r}}W_{\text{eff}}}R_{\text{ch}}F_{mn}, \tag{1.58}$$

while the absorption of the TM waves will be zero.

For the transition between the light-hole VB and CB the momentum matrix element is given by

$$p_{cl} = \langle iS \uparrow\downarrow |\boldsymbol{p}|\sqrt{\frac{2}{3}}|Z_+ \uparrow\downarrow\rangle - \langle iS \uparrow\downarrow |\boldsymbol{p}|\frac{1}{\sqrt{6}}|(\pm X_+ + iY_+)\downarrow\uparrow\rangle$$

$$= \sqrt{\frac{2}{3}}P\hat{z} - \frac{P}{\sqrt{6}}(\pm\hat{x} + i\hat{y}). \tag{1.59}$$

Therefore, for the in-plane polarization the absorption coefficient of the light hole to the conduction band will be

$$\alpha_{cl}^{\text{TE}} = \frac{1}{3}N_{\text{QW}}\frac{\pi\alpha_0}{n_{\text{r}}W_{\text{eff}}}R_{cl}F_{mn} \approx \frac{1}{3}\alpha_{\text{ch}}^{\text{TE}}, \tag{1.60}$$

while for the TM polarization, normal to the QW plane, the light-hole state is expected to absorb very strongly:

$$\alpha_{cl}^{\text{TM}} = \frac{4}{3}N_{\text{QW}}\frac{\pi\alpha_0}{n_{\text{r}}W_{\text{eff}}}R_{cl}F_{mn} \approx \frac{4}{3}\alpha_{\text{ch}}^{\text{TE}}. \tag{1.61}$$

Of course, absorption of TM polarized light can only be observed if the incident light impinges onto a surface at an oblique angle, or, better in the waveguide geometry of Fig. 1.10(b).

## 1.2.2    Band-mixing Origin of ISB Transitions

Let us now turn our attention to our main subject – intersubband absorption. We now must evaluate the matrix element between two states described by (1.47) with $m = 1, 2$. First, let us consider the matrix element for the light that is TM (or normal to the QW plane) polarized:

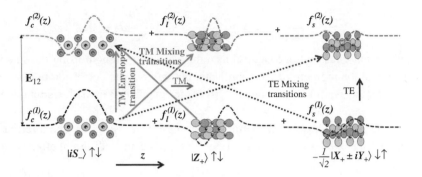

**Fig. 1.11** Composition of the states in the subbands and the origin of TM and TE transitions between the subbands.

$$p_{12,z} = -i\hbar \left[ \left\langle -iS_- \uparrow \Big| f_c^{(1)} - \frac{i\hbar}{2P} \left( \frac{m_0}{m_c(E_c^{(1)})} - 1 \right) \frac{\partial f_c^{(1)}}{\partial z} \langle Z \uparrow \Big| \right] \right.$$

$$\times \left[ \Big| iS_- \uparrow \rangle \frac{\partial f_c^{(2)}}{\partial z} + f_c^{(2)} \frac{\partial}{\partial z} \Big| iS_- \uparrow \rangle + \frac{i\hbar}{2P} \left( \frac{m_0}{m_c(E_c^{(2)})} - 1 \right) \frac{\partial f_c^{(2)}}{\partial z} \Big| Z \uparrow \rangle \right]$$

$$= -i\hbar \left[ \int f_c^{(1)} \frac{\partial f_c^{(2)}}{\partial z} dz - \frac{1}{2} \left( \frac{m_0}{m_c(E_c^{(1)})} - 1 \right) \int f_c^{(2)} \frac{\partial f_c^{(1)}}{\partial z} dz \right.$$

$$\left. + \frac{1}{2} \left( \frac{m_0}{m_c(E_c^{(2)})} - 1 \right) \int f_c^{(1)} \frac{\partial f_c^{(2)}}{\partial z} dz \right]. \tag{1.62}$$

The first term is an "envelope contribution" while the second and third terms are the contributions due to band-mixing, which turns out to be the dominant contribution to the strength of transition. This is explained in Fig. 1.11 where it is shown how k.P interaction mixes the light and split-off states into conduction subbands. For the two bands that are relatively close to each other, one can assume some average effective mass $\overline{m}_c^* = m_c(E_c^{(2)} + E_c^{(1)})/2$ and substitute it into (1.62) to obtain

$$p_{12,z} = -i\hbar \frac{m_0}{m_c} \int f_c^{(1)} \frac{\partial f_c^{(2)}}{\partial z} dz = \frac{m_0}{m_c} p_{12,z}^*, \tag{1.63}$$

where

$$p_{12,z}^* = -i\hbar \int f_c^{(1)} \frac{\partial f_c^{(2)}}{\partial z} dz \tag{1.64}$$

is a "quasi-momentum" matrix element of envelope wavefunctions, estimated without taking into account the increase caused by band-mixing. As one can see, band-mixing leads to a very large increase in the matrix element of the momentum for the ISB transition, similar to the one described by the expression (1.24) for the momentum of the electron in the conduction band. We can also find the matrix element of the coordinate of the intersubband transition,

$$z_{12} = \frac{p_{12}}{m_0 \omega_{21}} = -i\hbar \frac{1}{m_c \omega_{21}} \int f_c^{(1)} \frac{\partial f_c^{(2)}}{\partial z} dz = \int f_c^{(1)} z f_c^{(2)} dz, \qquad (1.65)$$

which indicates that one can estimate the transition strength using the standard definition of the moment of the coordinate, but without the derivation performed in this chapter the band-mixing origin [6, 20] of the "giant" ISB transition strength would not have been revealed.

The per-well absorption coefficient for the ISB transition can be inferred from the one for the interband transitions (1.56) by simply using a different effective density of states,

$$\rho_{12}(\hbar\omega) = \frac{1}{\pi} \frac{\Gamma}{(E_{21} - \hbar\omega)^2 + \Gamma^2}, \qquad (1.66)$$

where $N_{2D}$ is two-dimensional density of carriers and $\Gamma$ is broadening of the ISB transition to obtain

$$\alpha_{12}^{TM}(\omega) = \frac{\alpha}{W_{eff}} N_{QW} \hbar\omega \frac{2\pi}{\hbar} \frac{e^2 p_{12}^2}{m_0^2 \omega^2} \frac{N_{2D}\Gamma}{(E_{21} - \hbar\omega)^2 + \Gamma^2} \frac{\eta_0}{2n_r}, \qquad (1.67)$$

where $N_{2D}$ is two-dimensional carrier density. At resonance $\omega = \omega_{21}$; thus, the absorption amounts to

$$\alpha_{12}^{TM}(\omega_{12}) = \frac{N_{QW}}{W_{eff}} \frac{4\pi\alpha_0}{n_r} \frac{\hbar N_{2D} |p_{12}^*|^2}{\omega_{21} m_0^2 \Gamma} = \frac{4\pi\alpha_0}{n_r} \frac{\hbar\omega_{12}}{\Gamma} \frac{N_{QW} N_{2D}}{W_{eff}} z_{12}^2. \qquad (1.68)$$

Next, we can evaluate the strength of the transition for TE (or in-plane) polarized light. Since the heavy-hole like state $|2^{-1/2}(X_+ \pm iY_+)\downarrow\uparrow\rangle$ gets mixed into the conduction band with the opposite spin, the matrix element of in-plane momentum becomes [20, 21]:

$$
\begin{aligned}
p_{12,x} &= -i\hbar \left[ \langle -iS_- \uparrow | f_c^{(1)} - \frac{i\hbar}{2P} \left( \frac{m_0}{m_c(E_c^{(1)})} - 1 \right) \frac{r_{\Delta 1}}{\sqrt{2}} \frac{\partial f_c^{(1)}}{\partial z} \langle (X - jY)\downarrow | \right] \\
&\times \left[ f_c^{(2)} \frac{\partial}{\partial z} |iS_- \downarrow\rangle + \frac{i\hbar}{2P} \left( \frac{m_0}{m_c(E_c^{(2)})} - 1 \right) \frac{r_{\Delta 2}}{\sqrt{2}} \frac{\partial f_c^{(2)}}{\partial z} |(-X + jY)\uparrow\rangle \right] \\
&= -i\hbar \left[ -\frac{r_{\Delta 1}}{2^{3/2}} \left( \frac{m_0}{m_c(E_c^{(1)})} - 1 \right) \int f_c^{(2)} \frac{\partial f_c^{(1)}}{\partial z} dz \right. \\
&\left. \quad - \frac{r_{\Delta 2}}{2^{3/2}} \left( \frac{m_0}{m_c(E_c^{(2)})} - 1 \right) \int f_c^{(1)} \frac{\partial f_c^{(2)}}{\partial z} dz \right] \\
&= -\frac{i\hbar}{2^{3/2}} \int f_c^{(2)} \frac{\partial f_c^{(1)}}{\partial z} dz \left[ r_{\Delta 2} \left( \frac{m_0}{m_c(E_c^2)} - 1 \right) - r_{\Delta 1} \left( \frac{m_0}{m_c(E_c^{(1)})} - 1 \right) \right].
\end{aligned}
$$
$$(1.69)$$

As one can see, the matrix element for intersubband in-plane transition is significantly smaller than that for normal-to-plane polarization. This happens precisely because the transition must involve spin flip and would have been entirely forbidden if not for spin-orbit interaction. Using (1.48) one can obtain an approximate relation between the two:

$$\frac{p_{12,x}}{p_{12,z}} \approx \frac{\sqrt{2}\Delta\left(E_c^{(2)} - E_c^{(1)}\right)}{3E_g^2}. \tag{1.70}$$

Therefore, TE-polarized wave absorption between subbands is very weak in most relevant semiconductor QWs, with the possible exception of antimonides, in which case the bandgap energy and spin-orbit splitting energy are of the same magnitude. However, that does not mean that doped QWs do not absorb the TE-polarized light – this absorption is called free-carrier absorption and shall be considered next.

## 1.2.3    Free-Carrier Absorption and Its Relation to ISB Transitions

Free-carrier absorption occurs for the light polarized in the plane of the QWs (Fig. 1.12). For the absorption to take place between two states, one near the bottom of subband with $k_{\perp 1} = 0$ and one with $k_{\perp 2} > 0$, it is necessary to conserve both energy and momentum. Momentum conservation usually involves scattering by either a phonon or an imperfection, such as surface roughness. The matrix element of the scattering Hamiltonian $H_{scat}(k_{\perp 2})$ in general depends on the absolute value of in-plane wavevector $k_\perp$ ; i.e., on the energy:

$$E_2 - E_1 = \hbar^2 k_\perp^2 / 2m_c = \hbar\omega. \tag{1.71}$$

Essentially free-carrier absorption is a two-step process in which the electron from the bottom of subband scatters into the virtual state with wavevector $k_{\perp 2}$ and with energy $\pm E_{ph}$, which can be positive or negative, depending on whether the phonon is absorbed or emitted, or zero for all the elastic-scattering processes. Usually $E_{ph} << \hbar\omega$ and can be neglected, unless, of course, one is dealing with nitrides for which the LO–phonon energy is 90 meV, and thus comparable to the energy of the ISB transition itself. The scattering process is followed by the absorption of the phonon with energy $\hbar\omega$, and

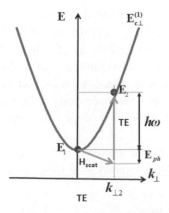

**Fig. 1.12** Origin of free-carrier absorption in doped semiconductor QWs.

the interaction Hamiltonian $ep \cdot A/m_0$, where the momentum of the electron in state $k_{\perp 2}$ is given by (1.23). Therefore $\langle ep \cdot A/m_0 \rangle_{22} = e\hbar k_{\perp 2} \cdot E/m_c\omega$, and one can write for the rate of absorption from a given state near $k_{\perp 1} = 0$, by using the Fermi golden rule,

$$R_{\text{abs}}(\omega) = \frac{2\pi}{\hbar} \frac{|H_{\text{scat}}(\omega)|^2}{\hbar^2\omega^2} \frac{e^2 E_\perp^2}{4\omega^2} \frac{\hbar^2 k_{\perp 2}^2 \langle \cos^2\theta \rangle}{m_c^2} \frac{1}{2}\rho_c, \tag{1.72}$$

where $E_\perp$ is the in-plane (TE) component of the electric field, and $\rho_c = m_c/\pi\hbar^2$ is the density of states in the CB (the factor $1/2$ in front of it indicates that the scattering processes typically preserve the spin) and $\theta$ is the angle between $k_{\perp 2}$ and the polarization of the electric field – obviously the mean value $\langle \cos^2\theta \rangle = 1/2$. Then, by using (1.71) and introducing the light intensity according to (1.55), we obtain

$$R_{\text{abs}}(\omega) = \frac{2\pi}{\hbar} |H_{\text{scat}}(\omega)|^2 \frac{1}{4}\rho_c \frac{1}{\hbar\omega} \frac{e^2\eta_0}{n_r m_c\omega^2}. \tag{1.73}$$

If we introduce the scattering rate inside the band as follows:

$$[\tau_{\text{scat}}(\omega)]^{-1} = \frac{2\pi}{\hbar} |H_{\text{scat}}(\omega)|^2 \frac{1}{4}\rho_c, \tag{1.74}$$

we obtain the rate of change of the two-dimensional energy density in QWs,

$$\frac{dU}{dt} = \hbar\omega N_{2D} R_{\text{abs}}(\omega) = \frac{e^2\eta_0}{n_r\tau_{\text{scat}}} \frac{N_{2D}I}{m_c[\omega^2 + \tau_{\text{scat}}^{-2}]} = \alpha_{fc}(\omega)I, \tag{1.75}$$

where we have added the additional term $\tau_{\text{scat}}^{-2}$ into the denominator in order to phenomenologically take into account the "uncertainty" in frequency caused by the scattering, but this can be rigorously derived if one follows full second-order perturbation theory rather than the simple Golden rule. Finally, for the per-well free-carrier absorption we obtain

$$\alpha_{fc}(\omega) = \frac{4\pi\alpha_0}{n_r} \frac{N_{2D}\hbar\tau_{\text{scat}}^{-1}}{m_c[\omega^2 + \tau_{\text{scat}}^{-2}]}. \tag{1.76}$$

Numerically, the typical scattering time is determined by the interaction with longitudinal-optical (LO) phonons, considered in greater detailed in the Section 1.4, but a typical value of this scattering rate is about $(1/200)\,\text{fs}^{-1}$ for the InGaAs, which for the doping density of $N_{2D} \sim 5 \times 10^{11}\,\text{cm}^{-2}$ results in free-carrier absorption per QW, as shown in Fig. 1.13. For a waveguide with effective three-dimensional doping density on the scale between $10^{17}$ and $10^{18}\,\text{cm}^{-3}$ the absorption coefficient is between 1 and $10\,\text{cm}^{-1}$ depending on the wavelength, which are values that are typically being measured experimentally.

If we neglect the frequency dependence of the scattering time, this expression (1.76) is exactly the one that follows from the Drude expression for the Drude theory for free-carrier complex conductivity:

$$\sigma(\omega) = \frac{e^2 N_{2D}\tau_{\text{scat}}}{m_c(1 + j\omega\tau_{\text{scat}})}. \tag{1.77}$$

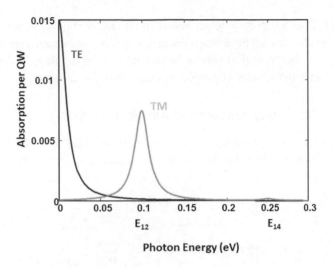

**Fig. 1.13** Spectra of free-carrier (TE) and intersubband (TM) absorption in doped QW. Note that the areas under the curves are equal.

If we now integrate (1.76) over the frequencies, we obtain

$$\int_0^\infty \alpha_{\text{fc}}(\omega)d\omega = \frac{2\pi^2\alpha_0\hbar N_{2\text{D}}}{n_{\text{r}}m_{\text{c}}}. \tag{1.78}$$

Then, if we integrate over the frequencies of the ISB absorption as in (1.68), we obtain

$$\int \alpha_{12}(\omega)d\omega = \frac{4\pi^2\alpha_0}{n_{\text{r}}}\omega_{12}N_{2\text{D}}z_{12}^2. \tag{1.79}$$

Now we can invoke the oscillator sum rule for all ISB transitions originating from the subband 1,

$$\sum_{m=2}^\infty \omega_{1m}z_{1m}^2 = \hbar/2m_{\text{c}}, \tag{1.80}$$

and obtain a very important result:

$$\int_0^\infty \alpha_{\text{fc}}(\omega)d\omega = \int_0^\infty \sum_{m=2}^\infty \alpha_{1m}(\omega)d\omega = \frac{2\pi^2\alpha_0\hbar N_{2\text{D}}}{n_{\text{r}}m_{\text{c}}}. \tag{1.81}$$

This result shows that ISB transitions are nothing but the quantized free-carrier transitions with total oscillator strength (integrated absorption) conserved as it is transferred from the broad-band continuous free-carrier absorption to the discrete ISB transitions. The quantization (one can think of it as reflections from the QW walls) relaxes momentum conservation rules, hence in ISB transitions there is no need for the "recoil" provided by phonon or impurity – the walls do the job perfectly well. Therefore, for a TM wave the free-carrier absorption is absent as long as all the carriers remain in the lowest subband. Of course, in QCLs the carriers can get excited to the upper subbands from where they do get absorbed into the continuum states [22]. This absorption typically does not have sharp features and sometimes is referred to as free-carrier absorption, but in truth it cannot be described by a simple $\omega^{-2}$ dependence (1.76), hence it is better to refer to it as nonresonant ISB absorption [22].

## 1.2.4    Comparison of the Band-to-Band and ISB Transition Strength

Let us now compare the absorption in a single QW for interband (1.58) and ISB transitions (1.68). If the transition is allowed, then, according to the oscillation sum rule,

$$\omega_{12} z_{12}^2 \approx \hbar/2m_c \tag{1.82}$$

and

$$\alpha_{12}^{TM}(\omega_{12}) = \frac{2\pi\alpha_0}{n_r} \frac{\hbar^2}{m_c} \frac{N_{2D}}{\Gamma} \frac{N_{QW}}{W_{eff}}. \tag{1.83}$$

Then we obtain for the ratio of ISB absorption (1.83) to the band-to-band absorption [8] (1.58),

$$\frac{\alpha_{12,\parallel}}{\alpha_{hh,\perp}} = \frac{2N_{2D}}{\pi\Gamma} \frac{\pi\hbar^2}{m_c} \approx \frac{2}{\pi} \frac{N_{2D}}{\rho_{cv}\Gamma}. \tag{1.84}$$

This is essentially the relation of effective densities of states for ISB and band-to-band transitions. The oscillator strengths of the transitions are for all practical purposes identical, and the "giant dipole moment" of the ISB transition is simply the consequence of a typical ISB transition having lower energy than a typical band-to-band transition. This can be shown in a more direct way by estimating the dipole moment of the band-to-band transition as follows:

$$z_{cv} \sim P_{cv}\hbar/m_0 E_g \sim \left(\frac{\hbar^2}{2m_c E_g}\right)^{1/2}, \tag{1.85}$$

which for GaAs turns out to be respectable 7Å – much larger than the bond length or even lattice constant. The result does not contradict the Bloch theorem, which only states that the electron wavefunction is periodic with a lattice constant and not that it is confined on the scale of lattice constant. Now we can re-write (1.82) by using the definition of effective mass (1.37) as follows:

$$z_{12} = \sqrt{\frac{\hbar}{2m_c\omega_{12}}} = \sqrt{\frac{P^2\hbar}{m_0^2 E_g\omega_{12}}} = \frac{P\hbar}{m_0 E_g}\sqrt{\frac{E_g}{\hbar\omega_{12}}} = z_{cv}\sqrt{\frac{E_g}{E_{12}}}. \tag{1.86}$$

So, to repeat, the one and only reason why an ISB transition has a relatively large dipole is simply the fact that it occurs typically in the mid- and far-IR where the transition energy is low. If, however, one compares the strength of an ISB transition in GaN and a band-to-band transition in InGaAs for the same wavelength, say 1.55μm, the dipole moments will be essentially identical. Thus, the main advantage of ISB transitions lies not as much in their strength, but in their flexibility – ability to obtain transitions at any mid- or far-IR wavelength by using standard medium-gap material systems like GaAs or InP, rather than venturing to more exotic narrow-gap III-V and II-VI materials.

## 1.3 Intersubband Scattering

Scattering plays an extremely important role in QCLs, as it is an inextricable part of both carrier transport through the QCL as well as of the lasing itself. One should distinguish between intrasubband and intersubband scattering. Intrasubband scattering is responsible for carrier thermalization as well as for level broadening. Intersubband scattering, in addition to contributing to electroluminescence (EL) linewidth broadening and carrier thermalization, is also responsible for the population and depopulation of laser levels, and is thus the key mechanism that determines whether population inversion can actually be attained between the upper and lower laser levels. In general, intrasubband scattering occurs on a significantly shorter time scale than intersubband scattering, mostly due to fact that the interaction strength does not involve the overlap between two different envelope wavefunctions. Therefore, often one can analyze the dynamics of populations in a QCL assuming that within each subband a thermal equilibrium has been reached. At any rate, intrasubband relaxation is just there and cannot be changed significantly by design; hence, we shall not consider it in this short introduction. Intersubband relaxation, on the other hand, depends on the overlap between the envelope wavefunction and the momentum change involved in scattering, hence it can be engineered to achieve favorable relationships between the lifetimes of upper- and lower-laser levels and simultaneously suppress leakage from the upper level, thus improving QCL characteristics.

There are many different mechanisms that contribute to scattering, but clearly, three of them are the dominant ones: scattering by the longitudinal-optical (LO) phonons, interface-roughness (IFR) scattering, and alloy-disorder (AD) scattering. The other important processes – acoustic-phonon scattering and electron-electron scattering – are always present as well, but their influence is not nearly as critical as those of the first three scattering processes.

### 1.3.1 Intersubband Scattering by Longitudinal-Optical Phonons

We start with the longitudinal-optical (LO) phonon scattering because it is this process that largely allows the population inversion to be achieved in QCLs [23]. That is why we shall derive all the relevant expressions rigorously. An LO phonon [24, 25] is an elastic wave propagating in the material with lattice basis consisting of two atoms, such as in zinc-blende lattice. In the optical mode, two basis atoms (cation and anion) move in the opposite directions, as shown in Fig. 1.14(a). Now, since two atoms comprising the basis are actually charged ions with effective charge of $+e^*$ for cation (Ga) and negative charge $-e^*$ for the anion (As), where $e^* = fe$ and a coefficient $0 < f < 1$ describes the polarity of the bond. Thus, if one introduces a relative displacement of the two ions as $u = u_c - u_a$, then one can write the differential equation of motion for the ions in the optical-phonon mode in the presence of an external harmonic electric field:

$$\frac{d^2 u}{dt^2} + \omega_{TO}^2 u = \frac{1}{2} \frac{e^*}{M_r} E_0 e^{i(q \cdot r - \omega t)} + c.c.,$$ (1.87)

where the frequency $\omega_{TO}$ at small wave vectors $q$ can be found as follows:

**(a)**          **(b)**

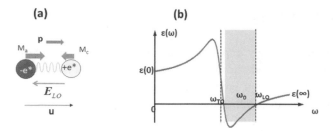

**Fig. 1.14** (a) Optical vibrations in a diatomic lattice of III-V semiconductor; (b) dispersion of the dielectric constant of polar semiconductor.

$$\omega_{TO}^2 \approx 2K/M_r, \tag{1.88}$$

where $K$ is the restoring force acting upon the ions and

$$M_r = M_c M_a (M_c + M_a) \tag{1.89}$$

is the reduced mass of the cation and anion. The ion displacement can naturally be written as a harmonic wave $\boldsymbol{u}(\boldsymbol{r}, t) = \boldsymbol{u}_0 e^{i(\boldsymbol{q}\cdot\boldsymbol{r}-\omega t)} + c.c.$ and, due to the presence of ion charge, each bond becomes polarized with a time-dependent dipole moment given by

$$\boldsymbol{p}(\boldsymbol{r}, t) = \frac{1}{2} e^* \boldsymbol{u}_0 e^{i(\boldsymbol{q}\cdot\boldsymbol{r}-\omega_0 t)} + c.c. \tag{1.90}$$

Hence, there exists a polarization wave in the lattice,

$$\boldsymbol{P}_L(\boldsymbol{r}, t) = \frac{1}{2} N e^* \boldsymbol{u}_0 e^{i(\boldsymbol{q}\cdot\boldsymbol{r}-\omega_0 t)} + c.c., \tag{1.91}$$

where $N$ is the density of bonds. The total polarization in the semiconductor must also include the electronic polarization, i.e., the polarization of the electrons inside the valence bonds, described by the electronic susceptibility in the presence of an external electric field

$$\boldsymbol{P}_{el}(\boldsymbol{r}, t) = \frac{1}{2} \varepsilon_0 \chi_{el} \boldsymbol{E}_0 e^{i(\boldsymbol{q}\cdot\boldsymbol{r}-\omega_0 t)} + c.c., \tag{1.92}$$

where the electronic susceptibility $\chi_{el}$ can be considered frequency independent as long as the frequency is far from the absorption edge in the semiconductor; thus it can be designated as

$$\chi_{el} = \varepsilon(\infty) - 1, \tag{1.93}$$

where $\varepsilon(\infty)$ is the dielectric constant in the optical range; i.e., at frequencies much higher than $\omega_{TO}$. Solving (1.87) and using (1.91) we immediately obtain

$$\boldsymbol{P}_L(\boldsymbol{r}, t) = \frac{N(e^*)^2 \boldsymbol{E}_0 e^{j(\boldsymbol{q}\cdot\boldsymbol{r}-\omega_0 t)}}{M_r(\omega_{TO}^2 - \omega^2)} + c.c., \tag{1.94}$$

and the total electric displacement can be found as follows:

$$D(r, t) = \varepsilon_0 E(r, t) + P_L(r, t) + P_{el}(r, t)$$

$$= \frac{1}{2}\varepsilon_0\varepsilon(\infty)E_0 e^{j(q\cdot r - \omega_0 t)} + c.c. + P_L(r, t) = \frac{1}{2}\varepsilon_0\varepsilon(\omega)E_0 e^{j(q\cdot r - \omega_0 t)} + c.c.$$

$$(1.95)$$

From (1.94) and (1.95) we immediately obtain the expression for the frequency-dependent dielectric constant,

$$\varepsilon(\omega) = \varepsilon(\infty) + \frac{N(e^*)^2}{\varepsilon_0 M_r(\omega_{TO}^2 - \omega^2 - i\omega\gamma)}, \tag{1.96}$$

where $\gamma$ is the phonon-scattering rate. The dispersion of the real part of dielectric constant is shown in Fig. 1.14(b). Near $\omega = \omega_{TO}$ the dielectric constant becomes very large, indicating that displacement can be not-zero even with a near-zero electric field. This is the resonance frequency of the transverse optical phonon which engenders no electric field. But at a frequency $\omega_{LO}$, defined as

$$\omega_{LO}^2 = \omega_{TO}^2 + \frac{N(e^*)^2}{\varepsilon_0\varepsilon(\infty)M_r}, \tag{1.97}$$

the dielectric constant is zero and that means that a longitudinal wave can propagate at this frequency. Indeed, if we assume that no external electric field is applied and the LO phonon is longitudinal, i.e., $\nabla \cdot u = q \cdot u = qu$, one obtains from the Maxwell equation for the displacement,

$$\nabla \cdot D = \varepsilon_0\varepsilon(\infty)q \cdot E_{LO} + q \cdot P_L = \varepsilon_0\varepsilon(\infty)qE_{LO} + qNe^*u = 0, \tag{1.98}$$

where the intrinsic field induced by the LO phonon is $E_{LO} = -Ne^*u/\varepsilon_0\varepsilon(\infty)$, now adds the restoring force into the equation of motion (1.87),

$$\frac{d^2 u}{dt^2} + \omega_{TO}^2 u = \frac{-N(e^*)^2}{\varepsilon_0\varepsilon(\infty)M_r}u, \tag{1.99}$$

from which the expression for eigenfrequency (1.97) readily follows. One can also obtain the expression for the dielectric constant at low frequencies from (1.96):

$$\varepsilon(0) = \varepsilon(\infty) + \frac{N(e^*)^2}{\varepsilon_0 M_r \omega_{TO}^2}. \tag{1.100}$$

From (1.97) and (1.100) one obtains the Lyddane–Sachs–Teller relation [26]:

$$\frac{\omega_{LO}^2}{\omega_{TO}^2} = \frac{\varepsilon(0)}{\varepsilon(\infty)}. \tag{1.101}$$

Now, also from (1.97), we can find the effective ion charge:

$$e^* = [\varepsilon_0\varepsilon(\infty)M_r\omega_{LO}^2[1 - \varepsilon(\infty)/\varepsilon(0)]/N]^{1/2}, \tag{1.102}$$

and the expression for the electric field:

$$E_{LO} = -\frac{1}{2}\sqrt{\frac{NM_r}{\varepsilon'\varepsilon_0}}\,\omega_{LO}^2 u_0 e^{i(q \cdot r - \omega_{LO} t)} + c.c., \tag{1.103}$$

where

$$\frac{1}{\varepsilon'} = \frac{1}{\varepsilon(\infty)} - \frac{1}{\varepsilon(0)}. \tag{1.104}$$

For both GaAs and lattice-matched to InP substrate $Ga_{0.47}In_{0.53}As$ $1/\varepsilon' \sim 0.014$. The potential of this electric field produced by LO phonon is then

$$\Phi_{LO} = \frac{1}{2q}\sqrt{\frac{NM_r}{\varepsilon_0\varepsilon'}}\,\omega_{LO}^2 u_0 e^{i(q \cdot r - \omega_{LO} t)} + c.c. \tag{1.105}$$

The $q$-vector dependence of this scattering potential is of utmost importance to the QCL. As we shall see, it is this dependence that facilitates achieving population inversion. Now, the energy of the LO-phonon mode is $NVM_r\omega_{LO}^2 u_0^2/2$, where $V$ is the quantization volume that can be quantized in increments of $\hbar\omega_{LO}$. Hence the ion displacement in the mode with $n_{LO}$ phonons in it is expressed by

$$NM_r\omega_{LO}^2 u_0^2/2 = (n_{LO} + 1/2)\hbar\omega_{LO}/V, \tag{1.106}$$

where

$$n_{LO} = \frac{1}{\exp(\hbar\omega_{LO}/k_B T) - 1}. \tag{1.107}$$

The Hamiltonian of interaction between the phonons and electrons is then given by [27]

$$H_{LO}^2 = e^2\Phi_{LO}^2 = \frac{e^2\hbar\omega_{LO}}{2\varepsilon_0\varepsilon' q^2 V}\left(n_{LO} + \frac{1}{2} \pm \frac{1}{2}\right), \tag{1.108}$$

where the $\pm$ signs correspond to the emission (absorption) of LO phonons. For the most part we are interested in the phonon-emission processes.

Let us now consider LO-phonon-assisted scattering between two subbands, $m$ and $n$, separated by an energy $E_{mn}$, as shown in Fig. 1.15 [28–32]. We assume that the electron resides near the bottom of the upper subband $m$ and its in-plane wavevector $k_{\perp m} \approx 0$. To conserve the energy, the electrons in the final state in subband 2 must all have a kinetic energy as follows:

$$\hbar^2 k_\perp^2/2m_c = E_{mn} - \hbar\omega_{LO}. \tag{1.109}$$

Now, we can use the Fermi Golden rule to write the expression for the scattering rate as such:

$$R_{LOm \to n} = \frac{2\pi}{\hbar}\sum_q \langle m| H_{LO}^2 |n\rangle\,\delta(E_{mn} - \hbar\omega_{LO} - \hbar^2 k_\perp^2/2m_c). \tag{1.110}$$

Next, we go from the summation over the phonon wavevector $q$ to integration and represent this wavevector as the sum of normal and in-plane wavevectors:

$$q = q_z + q_\perp, \tag{1.111}$$

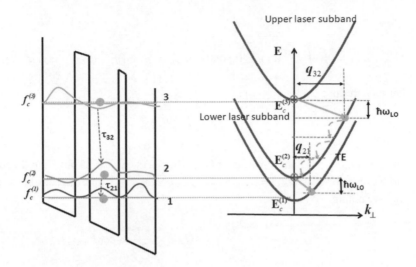

**Fig. 1.15** Energies and in-plane wavevectors involved in LO-phonon scattering in QCLs. Larger in-plane wavevectors are involved in the transition $3 \rightarrow 1$ than in transition $2 \rightarrow 1$, which causes the lifetime of level 3 to be longer than that of level 2.

where

$$q_\perp \approx k_\perp. \tag{1.112}$$

The result is as follows:

$$R_{\mathrm{LO}m \rightarrow n} = \frac{1}{8\pi^3} \frac{2\pi}{\hbar} \int\limits_{-\infty}^{\infty} \int\limits_{0}^{\infty} \frac{e^2 \hbar \omega_{\mathrm{LO}}}{2\varepsilon_0 \varepsilon' \left(q_\perp^2 + q_z^2\right)} G_{mn}^2(q_z) \frac{2m_c}{\hbar^2} \delta(q_\perp^2 - q_0^2) \pi \, dq_\perp^2 \, dq_z \, (n_{\mathrm{LO}} + 1)$$

$$= \frac{e^2 \omega_{\mathrm{LO}} m_c}{4\pi \varepsilon_0 \varepsilon' \hbar^2} \int\limits_{-\infty}^{\infty} \frac{G_{mn}^2(q_z)}{\left(q_0^2 + q_z^2\right)} dq_z, \tag{1.113}$$

where

$$G_{mn}(q_z) = \int f_c^{(m)}(z) e^{iq_z z} f_c^{(n)}(z) dz \tag{1.114}$$

and

$$q_{mn}^2 = \frac{2m_c}{\hbar^2}(E_{mn} - \hbar \omega_{\mathrm{LO}}). \tag{1.115}$$

The double integral in (1.113) can be evaluated as

$$\int\limits_{-\infty}^{\infty} \frac{e^{iq_z(z_1 - z_2)}}{q_z^2 + q_{mn}^2} dq_z = \pi \exp(-q_{mn}|z_1 - z_2|)/q_{mn}, \tag{1.116}$$

and one finally obtains the expression

$$R_{\mathrm{LO}m \rightarrow n} = \frac{e^2 \omega_{\mathrm{LO}} m_c}{4\varepsilon_0 \varepsilon' \hbar^2} \frac{F_{mn}(q_{mn})}{q_{mn}}, \tag{1.117}$$

where

$$F_{mn}(q_{mn}) = \int \int f_c^{(m)}(z_1)f_c^{(n)}(z_1)\exp(-q_{mn}|z_1 - z_2|)f_c^{(m)}(z_2)f_c^{(n)}(z_2)dz_1 dz_2. \quad (1.118)$$

Let us introduce

$$q_0 = \sqrt{\frac{2m_c\omega_{LO}}{\hbar}}. \quad (1.119)$$

For GaAs $q_0 \sim 0.25$ nm$^{-1}$ while for Ga$_{0.47}$In$_{0.53}$As $q_0 \sim 0.25$ nm$^{-1}$, and normalizing the wavevector $q$ as $Q_{mn} = q_{mn}/q_0 = \sqrt{E_{mn}/\hbar\omega_{LO} - 1}$ and the coordinate $z$ as $Z = zq_0$, we obtain

$$R_{LOm\rightarrow n} = \frac{e^2 q_0}{8\varepsilon_0\varepsilon'\hbar}F_{mn}(Q_{mn}) = \frac{\pi}{2\varepsilon'}\alpha_0 cq_0 F_{mn}(Q_{mn}), \quad (1.120)$$

where $\alpha_0$ is the fine structure constant and

$$F_{mn}(Q_{mn}) = Q_{12}^{-1}\int \int f_c^{(m)}(Z_1)f_c^{(n)}(Z_1)\exp(-Q_{mn}|Z_1 - Z_2|)f_c^{(m)}(Z_2)f_c^{(n)}(z_2)dZ_1 dZ_2. \quad (1.121)$$

The term in front of the $F_{mn}(Q_{mn})$ in (1.120) for GaAs is $1.2 \times 10^{13}s^{-1}$, while for Ga$_{0.47}$In$_{0.53}$As it is $0.9 \times 10^{13}s^{-1}$. What is left to calculate is $F_{12}(Q_{12})$. To do it, we consider an example of a square QW with infinitely high walls, having envelope wavefunctions:

$$f_c^{(m)}(Z) = \sqrt{\frac{2}{aq_0}}\sin(m\pi Z/aq_0), \quad (1.122)$$

where $a$ is the QW width. The results are shown in Fig. 1.16(a) for the allowed transitions in the QW with the width ranging from 5 to 20 nm. Of course, the sideband separation $E_{mn}$ cannot be easily varied for a given well thickness, and the actual shape of QCL's active regions is quite different from the square QW, but the general trend shown in Fig. 1.16(a) is correct, as the ISB scattering becomes strongest right when the ISB is resonant with the phonon energy and can lead to lifetimes as short as 200–250 fs. At the same time when the transition energy is in the range of 120–250 meV (which is the case for lasing transitions in the mid-IR), the scattering strength is reduced and lifetimes approach 1 ps. Using this fact one can achieve the difference between relaxation rates, $R_{LO3\rightarrow 2}$ and $R_{LO2\rightarrow 1}$, in Fig. 1.15, and make sure that the condition $\tau_{32} > \tau_{21}$ necessary for achieving population inversion in QCL is satisfied. One can further reduce the LO-scattering rates by considering transitions with reduced oscillator strength. For instance, as shown in Fig. 1.16(b), the ISB LO scattering for the dipole-forbidden transition $R_{LO3\rightarrow 1}$ is substantially weaker than for dipole-allowed transition $R_{LO2\rightarrow 1}$. This method is used for the case when the transition energy is small, such as in the THz region [33, 34]. But of course, the increase in lifetime is accompanied by a reduction in gain in this case, which makes attainment of the laser threshold in the THz range more difficult.

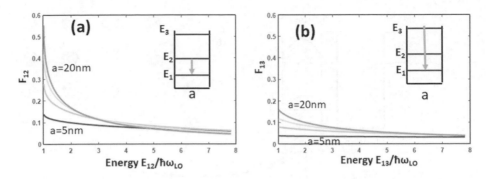

**Fig. 1.16** Relative strength of intersubband LO-phonon scattering vs. the intersubband separation energy for: (a) dipole-allowed and (b) dipole-forbidden transitions.

### 1.3.2  Alloy-Disorder Scattering

The active region of a QCL consists of a large number of wells and barriers having different compositions which can be binary, ternary, and sometimes quaternary alloys. The alloy scattering occurs in the ternary and quaternary materials such as InGaAs, InAlAs, or InGaAsP [31, 32, 35–38]. Consider a simple model with coupled QWs as shown in Fig. 1.17(a) [27]. At least one, and more often both QW and barrier materials are at least ternary or sometimes quaternary alloys. We consider a most simple well material, $In_xGa_{1-x}As$, that is often used in mid-IR QCLs. In Fig. 1.17(b) we show the arrangement of cation atoms In or Ga in the cation plane – and this arrangement is clearly aperiodic. Since the states in the conduction band are most strongly associated with S-states of cations, these electrons will see different effective potential near In and Ga ions – we should refer to them for generality as (A) and (B) ions – because In (A) and Ga (B) have different bandgaps and electron affinities. Let us now introduce the aperiodic lattice potential $U_{lat}(r)$, shown in Fig. 1.17(c), which has a constant value $E_{A(B)}$ within the distance $r_0$, commensurate with the size of the unit cell, from the cation A(B). This aperiodic potential differs from the mean lattice potential $\langle U_{lat}\rangle (r)$, which is periodic with value of $\bar{E} = xE_A + (1 - x)E_B$ at each unit cell, as shown by the dashed line in Fig. 1.17(c). The alloy-scattering potential is then simply the difference between the local and average lattice potentials seen by the conduction electron [12, 27]; i.e.,

$$\delta U(r) = \begin{cases} \delta E_A = E_A - \bar{E} = (1 - x)(E_A - E_B) = (1 - x)\delta E_{AB} & \text{on A site,} \\ \delta E_B = E_A - \bar{E} = x(E_B - E_A) = -x\delta E_{AB} & \text{on B site,} \end{cases}$$

$$(1.123)$$

where $\delta E_{AB} = E_A - E_B$ as shown in Fig. 1.17(d).

We can now calculate the matrix element of the alloy perturbation, at one particular A(B) site with co-ordinate $r_l$, between a given state in the subband $m$ of in-plane wavevector $k_{\perp m}$ and one in the subband $n$ of wavevector $k_{\perp n}$, as shown in Fig. 1.18:

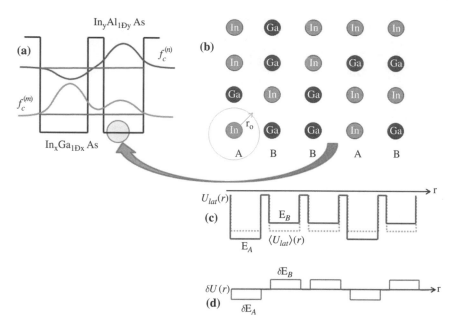

**Fig. 1.17** (a) Electron states in the QW comprising alloys. (b) Aperiodic arrangement of cations in an alloy. (c) Aperiodic lattice potential $U_{lat}(\mathbf{r})$ and its average value $< U_{lat}>(\mathbf{r})$. (d) Alloyed perturbation potential $\delta U_{lat}(\mathbf{r})$.

$$\langle m| \, \delta U(z_l) \, |n\rangle = S^{-1} \int_{|r-r_l|<r_0} \delta E_{A(B)} f_c^{(m)}(z) f_c^{(n)}(z) e^{i(k_{\perp m}-k_{\perp n})\cdot r_\perp} dr_\perp dz$$

$$\approx V_0 S^{-1} \delta E_{A(B)} f_c^{(m)}(z_l) f_c^{(n)}(z_l), \tag{1.124}$$

where $V_0 = \frac{4}{3}\pi r_0^3$, $S$ is the quantization area in the plane of QWs, and we have made the reasonable assumptions that $k_{\perp m,n} << 1/r_0$ and that the envelope wavefunctions do not change much over one unit cell – hence the AD scattering does not depend on the wavevector – which is dramatically different from the LO-phonon scattering process of Fig. 1.15. Therefore, the alloy-scattering due to one particular lattice site $l$ can be found with a Fermi Golden rule as follows:

$$R_{mn}^{A(B)}(z_l) = \frac{2\pi}{\hbar} V_0^2 S^{-1} \left| \delta E_{A(B)} \right|^2 \left| f_c^{(m)}(z_l) f_c^{(n)}(z_l) \right|^2 \rho_c, \tag{1.125}$$

where $\rho_c = m_c/2\pi \hbar^2$ is the density of states for one spin.

Now we need to perform the summation over all the lattice sites within a region occupied by an alloy – in our example a QW or a barrier region defined as $z_1 < z < z_2$. The volume of a unit cell in f.c.c. lattice is $a_0^3/4$; therefore the density of A atoms is $N_A = 4x/a_0^3$ and the density of B atoms is $N_B = 4(1-x)/a_0^3$. Therefore, performing integration over these densities, we obtain

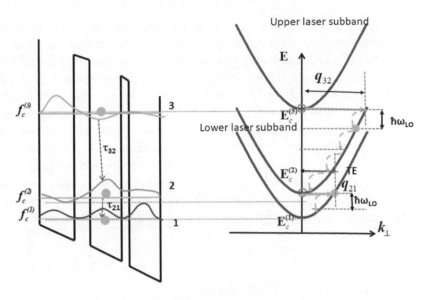

**Fig. 1.18** Energies and in-plane wavevectors involved in elastic intersubband scattering: alloy disorder, interface roughness, or ionized impurity.

$$R_{AD,m \to n} = \int_S \int_{z_1}^{z_2} N_A R_{mn}^A(z_l) dz_l dr_\perp + \int_S \int_{z_1}^{z_2} N_B R_{mn}^B(z_l) dz_l dr_\perp$$

$$= \frac{64\pi^3 r_0^6}{9\hbar a_0^3} (\delta E_{AB})^2 x(1-x) \int_{z_1}^{z_2} \left| f_c^{(m)}(z_l) f_c^{(n)}(z_l) \right|^2 dz_l \rho_c. \tag{1.126}$$

The size of $r_0$ is taken as half of the separation from the next cation in the f.c.c. lattice, i.e., $r_0 = a_0/2\sqrt{2}$, hence:

$$R_{AD,m \to n} = \frac{\pi^2 a_0^3 m_c}{72\hbar^3} (\delta E_{AB})^2 x(1-x) \int_{z_1}^{z_2} \left| f_c^{(m)}(z_l) f_c^{(n)}(z_l) \right|^2 dz_l. \tag{1.127}$$

Let us consider an example of a $Ga_{0.47}In_{0.53}$ As with $\delta E_{AB} \sim 0.6$ eV– the term outside the integral is equal to $0.82 \times 10^{13}$ nm$^{-1}$ s$^{-1}$ and the wavefunctions under the integral can be normalized to the QW width as $Z = z_l/a$ so that

$$R_{AD,m \to n} \sim \frac{0.82 \times 10^{13}}{a} F_{AD} \text{ s}^{-1}, \tag{1.128}$$

where the QW width $a$ is in nanometers, and $F_{AD} = \int_{Z_1}^{Z_2} \left| f_c^{(m)}(Z) f_c^{(n)}(Z) \right|^2 dZ < 1$ is the dimensionless overlap between the electron densities of the two states in the alloy region. Therefore, for a 20 nm active region the alloy-scattering time can be a few picoseconds, which is typically longer than LO-phonon scattering, but definitely cannot be neglected. Furthermore, the upper laser states are located, while using high conduction-band offset ternary materials where the lower states are located, on the one hand, the upper-state lifetime is enhanced due to reduced alloy-disorder scattering, and, on the other hand, the lower-state lifetime is decreased due to enhanced interface-roughness scattering (see 1.3.3). Thus, one can significantly improve the key lasing parameter – the lifetime ratio $\tau_{32}/\tau_{21}$, as shown in [35]. Of course, in THz QCLs with reduced LO-phonon scattering, the influence of the alloy scattering is even more pronounced. It should also be noted that the numerical factor in the result (1.127) strongly depends on the choice of the effective distance $r_0$ over which alloy disorder is "felt" by the electrons hence it should not be treated as an "exact" result. Note that [32, 37, 39] use a different approach that treats alloy scattering as a special case of interface-roughness scattering and obtain a result that differs from (1.127) by only approximately 10%. No matter what model is assumed, however, this order-of-magnitude estimate shows that alloy-disorder scattering is very important [35].

## 1.3.3    Interface-Roughness (IFR) Scattering

Consider another mechanism of ISB elastic scattering – due to the interface roughness [32, 36, 37, 40–42]. As shown in Fig. 1.19(a), the interface between QW and barrier is not smooth but changes as a function of the in-plane coordinate $r = (x, y)$. The interface profile of the $i$th interface can be described as a random function $z_i(r)$ of Fig. 1.19(b), which nevertheless can be characterized by the average position of the interface, $\bar{z}_i$ and a correlation function

$$C(r_1) = \int [z_i(r) - \bar{z}_i][z_i(r - r_1) - \bar{z}_i]\,dr = \Delta_i^2 e^{-r_1^2/\Lambda^2}, \qquad (1.129)$$

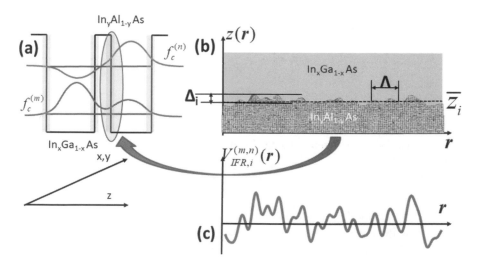

Fig. 1.19  (a) Electron states in the QW with interface roughness. (b) Interface with roughness. (c) Interface roughness potential $V_{\text{IFR},i}^{(m,n)}$.

where $\Delta_i = \langle (z_i - \bar{z}_i)^2 \rangle^{1/2}$ is the root-mean-square height and $\Lambda$ is the in-plane correlation length.

The perturbation Hamiltonian due to the $i$th interface roughness can be estimated as

$$V_{\text{IFR},i}^{(m,n)}(r) = \Delta E_{\text{c},i} f_{\text{c}}^{(m)}(\bar{z}_i) f_{\text{c}}^{(n)}(\bar{z}_i) [z_i(r) - \bar{z}_i], \tag{1.130}$$

where $\Delta E_{\text{c},i}$ is the conduction band offset at the $i$th interface. Now, the matrix element for the transition between two states in the conduction band – one near the bottom of the band $m$ with $k_{\perp m} \approx 0$ and the other in the band $n$ with the same energy and the in-plane wavevector $k_{\perp} \sim \sqrt{2 m_{\text{c}} E_{mn}/\hbar^2}$ can be evaluated as

$$V_{\text{IFR}}^{(mn)}(k_{\perp}) = \sum_i \int V_{\text{IFR},i}^{(mn)}(r) e^{ik_{\perp} \cdot r} dr = \sum_i \Delta E_{\text{c},i} f_{\text{c}}^{(m)}(\bar{z}_i) f_{\text{c}}^{(n)}(\bar{z}_i) \int [z_i(r) - \bar{z}_i] e^{ik_{\perp} \cdot r} dr, \tag{1.131}$$

which is essentially a Fourier transform of the random function $z_i(r) - \bar{z}$. Using the Wiener–Khinchin relation between the correlation function (1.129) and power spectrum of the function, we then obtain

$$\left| V_{\text{IFR}}^{(mn)}(k_{\perp}) \right|^2 = \sum_i \Delta_i^2 \Delta E_{\text{c},i}^2 \left| f_{\text{c}}^{(m)}(\bar{z}_i) f_{\text{c}}^{(n)}(\bar{z}_i) \right|^2 \times \pi \Lambda^2 e^{-\Lambda^2 k_{\perp}^2/4}. \tag{1.132}$$

Assuming that the band offsets, in-plane correlation lengths, and root-mean-square heights are the same for each interface, the IFR scattering rate then can be found [43] using

$$R_{\text{IFR},m \to n} = \frac{\pi m_{\text{c}}}{\hbar^3} \Delta E_{\text{c}}^2 \Delta_i^2 \Lambda^2 F_{\text{IFR}} e^{-\Lambda^2 k_{\perp}^2/4}, \tag{1.133}$$

where $F_{\text{IFR}} = \sum_i |f_{\text{c}}^{(m)}(\bar{z}_i) f_{\text{c}}^{(n)}(\bar{z}_i)|^2 < N_i/a^2$, $N_i$ is the total number of interfaces, and $a$ is the extent of the active region. Using $k_{\perp} \sim \sqrt{2 m_{\text{c}} E_{mn}/\hbar^2}$ one can obtain an order-of-magnitude estimate of the maximum possible IFR scattering rate as follows:

$$R_{\text{IFR},m \to n} = \pi N_i \frac{\Delta E_{\text{c}}^2}{\hbar E_{mn}} \frac{\Delta_i^2}{a^2} G_{\text{IFR}}(-\Lambda k_{\perp}/2), \tag{1.134}$$

where $G_{IFR}(x) = 4x^2 \exp(-x^2)$ is the wavefunction whose maximum value of 1.47 is achieved for $x = 1$, i.e., for $\Lambda = \sqrt{2\hbar^2/m_{\text{c}} E_{mn}}$, which is on the scale of a few nanometers. Typical values of correlation length $\Lambda$ reported in literature are indeed about 5–10 nm, while for mean value of $\Delta_i$ it is 0.1–0.15 nm. Therefore, for $N_i = 6$, $a = 15$ nm, $E_{mn} \sim 250$ meV, and $\Delta E_{\text{c}} = 500$ meV one can obtain transitions rates due to IFR as large as $3 \times 10^{12}$ s$^{-1}$. Note that the presence of the energy $E_{mn}$ in the denominator of (1.134) indicates that one can shorten the scattering time $\tau_{21}$ to accelerate the depopulation of the lower laser level in Fig. 1.18. In Fig. 1.20 the values of $G_{IFR}$ are plotted as a function of intersubband energy $E_{mn}$ for three different values of correlation length $\Lambda$. As one can see, for the larger values of $\Lambda$ IFR scattering is larger for the lower values of energy, i.e., it leads to the favorable result $\tau_{21}, \tau_{32}$, but for smaller $\Lambda$ value the situation can be reversed. The best way to engineer the lifetimes is

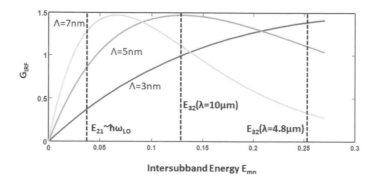

**Fig. 1.20** Interface-roughness scattering dependence on the ISB-transition energy.

to strategically place the interfaces where the two wavefunctions involved in the transitions are large which, in turn, makes $F_{IFR}$ large. Examples of this IFR engineering of nonradiative times can be found in [35, 43] where the lower-level lifetime $\tau_{21}$ was been reduced to less than 0.1 ps without affecting the upper-laser lifetime.

In addition to influencing the lifetimes, IFR also affects the linewidth of the laser transition since both the upper and lower levels are broadened due to intersubband and intrasubband transitions [32, 41, 42]. The intrasubband broadening can be evaluated, for vertical-transition devices, using essentially the same way as the intersubband scattering rate in (1.133), as discussed in [42, 44]:

$$\Delta\omega_{\text{IFR},m\to n} = \frac{\pi m_c}{\hbar^3} \Delta E_c^2 \Delta_i^2 \Lambda^2 \sum_i \left( \left|f_c^{(m)}(\bar{z}_i)\right|^2 - \left|f_c^{(n)}(\bar{z}_i)\right|^2 \right)^2 e^{-\Lambda^2 k_\perp^2/4}. \quad (1.135)$$

It should be noted that the in-plane wavevectors $k_\perp$ involved in the optical transitions are small, so the exponential term in (1.135) is close to unity and can be neglected. Furthermore, in most mid-IR QCL designs the electron injection from the reservoir into the active region occurs by tunneling [6] and thus can be strongly affected by IFR scattering [44]. Once the IFR scattering is taken into account, the barrier thickness needs to be reduced to obtain optimal performance, as has been proven in [45] where over 50% wall-plug efficiency has been attained at a heatsink temperature of 77 K. It is therefore quite reasonable to conclude that IFR broadening can easily be as important as that due to LO phonons as shown in [44]. One should also notice that intraband IFR scattering can be dealt with in entirely different fashion by treating it as time-independent localization in real space, i.e., as inhomogeneous broadening [46], still getting the same results in terms of its impact on the linewidth broadening.

## 1.3.4    Other Scattering Processes

The other ISB-scattering processes that can affect the intersubband lifetimes are acoustic phonons scattering (including piezoelectric) [24, 25, 27], and ionized-impurity scattering [47]. But neither one of these processes can be easily engineered to design active regions with favorable arrangement of lifetimes $\tau_{32}, \tau_{21}$. The relatively low doping density of the active regions of less than $10^{17}$ cm$^{-3}$ makes scattering rates due to ionized donor scattering less than $10^{11}$ s$^{-1}$, and since unlike IFR and AD scattering Coulomb forces are long-ranged, the selective doping of certain wells and barriers does not affect the rates strongly [48, 49] with the notable exception of THz QCLs [33] where the doping is all in the active region. And unlike LO-phonon scattering, the acoustic-phonon Hamiltonian is not wavevector dependent, hence the acoustic scattering rates for the upper- and lower-laser levels are roughly equal and can only be changed by reducing the overlap between the upper- and lower-state wavefunctions (so-called "diagonal transitions"), which is a strategy that can be used for any scattering mechanism and usually leads to reduced transitions' strength and gain.

## 1.4    Conclusions

While far from being a complete description of all the fundamental processes involved in the operation of QCLs, this introductory chapter attempts to answer a few important questions: what is the origin of strong optical intersubband transitions, how does their strength depend on the physical properties of semiconductors and geometry, and how can one reach population inversion enabling QCL operation? This information should help the reader to follow the subsequent chapters of this book. For deeper understanding one should turn attention to a number of excellent books on QCLs [6] and the physics of intersubband transitions [7, 11, 50] in general.

   Nevertheless, a short summary of the most important points made in this chapter can be made, specifically:

- ISB transitions have precisely the same origin as interband transitions, namely, on the microscopic level, the former and latter are both the transitions between the bonding and anti-bonding covalent orbitals in III-V compounds. The strength of ISB transitions is large not due to their unique character, but entirely due to the long wavelengths of the mid-IR or THz radiation.
- The main advantage of ISB transitions is that they can be engineered almost at will (in terms of transition energy and oscillator strength) entirely by changing the geometry of QWs using the same limited and well-tried set of III-V materials, arsenides and phosphides (with the recent addition of nitrides).
- The ISB absorption can also be treated as quantized free-carrier absorption with total oscillator strength of ISB and free-carrier absorption conserved.
- LO-phonon scattering plays the key role in the dynamics of QCLs. Due to fortuitous dependence of ISB LO-scattering rates on the subband energies separation, in the case of mid-IR QCLs, it is precisely the process that makes population inversion possible by assuring that the lower laser level gets depopulated faster than the upper level. This is difficult to attain in THz QCLs where the upper and lower levels are closely spaced and more ingenious methods must be used, as shown in following chapters.
- Interface-roughness and alloy-disorder scattering are both essential to engineering the lifetimes of the energy levels of QCL as well as to tunneling rates,

luminescence linewidth, and injection efficiency, as discussed in detail in Chapter 2.

With these basics placed at the readers' disposal, all I have left is to direct them to the exciting subsequent chapters of this book dealing with more practical aspects of QCLs.

## References

[1]   L. C. West and S. J. Eglash, "First observation of an extremely large-dipole infrared transition within the conduction band of a GaAs quantum well," *Appl. Phys. Lett.*, vol. 46, no. 12, pp. 1156–1158, 1985.

[2]   B. F. Levine, R. J. Malik, J. Walker, K. K. Choi, C. G. Bethea, D. A. Kleinman, and J. M. Vandenberg, "Strong 8.2 μm infrared intersubband absorption in doped GaAs/AlAs quantum well waveguides". *Appl. Phys. Lett.*, vol. 50, no. 5, pp. 273–275, 1987.

[3]   B. F. Levine, K. K. Choi, C. G. Bethea, J. Walker, and R. J. Malik, "New 10 μm infrared detector using intersubband absorption in resonant tunneling GaAlAs superlattices," *Appl. Phys. Lett.*, vol. 50, no. 16, pp. 1092–1094, 1987.

[4]   M. Helm, P. England, E. Colas, F. DeRosa, and S. J. Allen, Jr. "Intersubband emission from semiconductor superlattices excited by sequential resonant tunneling," *Phys. Rev. Lett.*, vol. 63, no. 1, pp. 74–77, 1989.

[5]   J. Faist, F. Capasso, D. L. Sivco, C. Sirtori, A. L. Hutchinson and A. Y. Cho, "Quantum cascade laser," *Science*, vol. 264, no. 5158, pp. 553–556, 1994.

[6]   J. Faist, *Quantum Cascade Lasers*. Oxford: Oxford University Press, 2013.

[7]   M. Razeghi, *Technology of Quantum Devices*. London; New York: Springer, 2010.

[8]   J. Khurgin, "Comparative analysis of the intersubband versus band-to-band transitions in quantum wells," *Appl. Phys. Lett.*, vol. 62, no. 12, pp. 1390–1392, 1993.

[9]   C-H, Lin, R. Q. Yang, D. Zhang, S. J. Murry, S. S. Pei, A. A. Allerman, and S. R. Kurtz, "Type-II interband quantum cascade laser at 3.8 μm," *Electron. Lett.*, vol 33, no.7, pp. 598–599, 1997.

[10]  I. Vurgaftmanm, R. Weih, M. Kamp, J. R. Meyer, C. L. Canedy, C S. Kim, M. Kim, W. W. Bewley, C. D. Merritt, and J. Abell, "Interband cascade lasers," *J. Phys. D: Appl. Phys.*, vol. 48, no.12, 123001, 2015.

[11]  R. Paiella, *Intersubband Transitions in Quantum Structures*. New York: McGraw-Hill, 2006.

[12]  G. Bastard, *Wave Mechanics Applied to Semiconductor Heterostructures*. Les Editions de Physique; Les Ulis Cedex, France, New York, NY: Halsted Press, 1988.

[13]  P. Bhattacharya, *Properties of III-V Quantum Wells and Superlattices*. London: The Institution of Engineering and Technology, 2007.

[14]  E. L. Ivchenko, and G. E. Pikus, *Superlattices and Other Heterostructures: Symmetry and Optical Phenomena*. 2nd ed. Berlin; New York: Springer: Berlin, 1997.

[15]  W. A. Harrison, *Solid State Theory*. New York: Dover Publications, 1979.

[16]  P. Y. Yu and M. Cardona, *Fundamentals of Semiconductors: Physics and Materials Properties*. 4th ed. Berlin; New York: Springer, 2010.

[17] E. O. Kane, "Band structure of indium antimonide," *J. Phys. Chem. Solids*, vol. 1, no. 4, pp. 249–261, 1957

[18] G. L. Bir, G. E Pikus, *Symmetry and Strain-induced Effects in Semiconductors*. New York: Wiley, 1974.

[19] I. Vurgaftman, J. R. Meyer, and L. R. Ram-Mohan, "Band parameters for III–V compound semiconductors and their alloys," *J. Appl. Phys.*, vol. 89, no. 11, pp. 5815–5875, 2001.

[20] R. Q. Yang, J. M. Xu, and M. Sweeny, "Selection rules of intersubband transitions in conduction-band quantum wells," *Phys. Rev. B*, vol. 50, no.11, pp. 7474–7482, 1994,

[21] J. B. Khurgin, "Intersubband spin pump," *Appl. Phys. Lett.*, vol. 88, no. 12, 123511, 2006.

[22] A. Wittmann, T. Gresch, E. Gini, L. Hvozdara, N. Hoyler, M. Giovannini, and J. Faist, "High-performance bound-to-continuum quantum-cascade lasers for broad-gain applications," *IEEE J. Quantum Electron.*, vol. 44, no. 1, pp. 36–40, 2008.

[23] G. Sun and J. B Khurgin, "Optically pumped four-level infrared laser based on intersubband transitions in multiple quantum wells: feasibility study," *IEEE J. Quantum Electron*, vol. 29, no. 4, pp. 1104–1111, 1993.

[24] B. K. Ridley, *Electrons and Phonons in Semiconductor Multilayers*. 2nd ed.; Cambridge: Cambridge University Press, 2009.

[25] B. K. Ridley, *Quantum Processes in Semiconductors*. 5th ed.; Oxford: Oxford University Press, 2013.

[26] C. Kittel, *Introduction to Solid State Physics*. 8th ed.; Hoboken, NJ: Wiley, 2005.

[27] J. Singh, *Physics of Semiconductors and Their Heterostructures*. New York: McGraw-Hill, 1993.

[28] U. Bockelmann and G. Bastard, "Phonon scattering and energy relaxation in two-, one-, and zero-dimensional electron gases," *Phys. Rev. B*, vol. 42, no. 14, pp. 8947–8951, 1990.

[29] F. Chevoir and B. Vinter, "Calculation of phonon-assisted tunneling and valley current in a double-barrier diode," *Appl. Phys. Lett.*, vol. 55, no. 18, 1859–1861, 1989.

[30] J. Faist, C. Sirtori, F. Capasso, L. Pfeiffer, and K. W. West, "Phonon limited intersubband lifetimes and linewidths in a two-dimensional electron gas," *Appl. Phys. Lett.*, vol. 64, no.7, pp. 872–874, 1994.

[31] Y. Chen, N. Regnault, R. Ferreira, B. F. Zhu, and G. Bastard, "Optical phonon scattering in quantum cascade laser in a magnetic field," *Physics of Semiconductors, AIP Conf, Proc.* vol. 1199, pp. 221–222, 2009.

[32] T. Unuma, M. Yoshita, T. Noda, H. Sakaki, and H. Akiyama, "Intersubband absorption linewidth in GaAs quantum wells due to scattering by interface roughness, phonons, alloy disorder, and impurities," *J. Appl. Phys.*, vol. 93, no. 3, pp. 1586–1597, 2003.

[33] B. S. Williams, H. Callebaut, S. Kumar, and Q. Hu, "3.4-THz quantum cascade laser based on longitudinal-optical-phonon scattering for depopulation," *Appl. Phys. Lett.*, vol. 82, no. 7, pp. 1015–1017, 2003.

[34] Q. Hu, B. S. Williams, S. Kumar, H. Callebaut, and J. L. Reno, "Terahertz quantum cascade lasers based on resonant phonon scattering for depopulation," *Philos. Trans. Roy. Soc. A*, vol. 362, no. 1815, pp. 233–247, 2004.

[35] D. Botez, J. D. Kirch, C. Boyle, K. M. Oresick, C. Sigler, H. Kim, B. B. Knipfer, J. H. Ryu, D. Lindberg III, T. Earles, L. J. Mawst, and Y. V. Flores, "High-efficiency, high-power mid-infrared quantum cascade lasers," *Opt. Mater. Express*, vol. 8, no. 5, pp. 1378–1398, 2018; Erratum: vol. 11, no. 7, p. 1970, 2021.

[36] F. Chevoir and B. Vinter, "Scattering-assisted tunneling in double-barrier diodes: Scattering rates and valley current," *Phys. Rev. B*, vol. 47, no. 12, pp. 7260–7274, 1993.

[37] T. Unuma M. Yoshita, T. Noda, H. Sakaki M. Baba and H. Akiyama, "Sensitivity of intersubband absorption linewidth and transport mobility to interface roughness scattering in GaAs quantum wells," Inst. Phys. Conf. Ser. No 174 (*Proc. of 29th Int. Symp. on Comp. Semicond.*), pp. 379–383, 2003.

[38] A . Vasanelli, A. Leuliet, C. Sirtori, A. Wade, G. Fedorov, D. Smirnov, G. Bastard, B. Vinter, M. Giovannini, and J. Faist, "Role of elastic scattering mechanisms in GaInAs/AlInAs quantum cascade lasers," *Appl. Phys. Lett.*, vol. 89, no. 17, 172120, 2006.

[39] R. Terazzi, "Transport in quantum cascade lasers," Ph.D. dissertation, ETH Zurich, pp. 103–106, 2011.

[40] A. Leuliet, A. Vasanelli, A. Wade, G. Fedorov, D. Smirnov, G. Bastard, and C. Sirtori, "Electron scattering spectroscopy by a high magnetic field in quantum cascade lasers," *Phys. Rev. B*, vol. 73, no. 8, 085311, 2006.

[41] S. Tsujino, A. Borak, E. Müller, M. Scheinert, C. V. Falub, H. Sigg, and D. Grützmacher, "Interface-roughness-induced broadening of intersubband electroluminescence in *p*-SiGe and *n*-GaInAs/AlInAs quantum-cascade structure," *Appl. Phys. Lett.*, vol. 86, no. 6, 062113, 2005.

[42] A. Wittmann, Y. Bonetti, J. Faist, E. Gini, and M. Giovannini, "Intersubband linewidths in quantum cascade laser designs," *Appl. Phys. Lett.*, vol. 93, no. 14, 141103, 2008.

[43] Y. T. Chiu, Y. Dikmelik, P. Q. Liu, N. L. Aung, J. B. Khurgin, and C. F. Gmachl, "Importance of interface roughness induced intersubband scattering in mid-infrared quantum cascade lasers," *Appl. Phys. Lett.*, vol. 101, no. 17, 171117, 2012.

[44] J. B. Khurgin, Y. Dikmelik, P. Q. Liu, A. J. Hoffman, M. D. Escarra, K. J. Franz, and C. F. Gmachl, "Role of interface roughness in the transport and lasing characteristics of quantum-cascade lasers," *Appl. Phys. Lett.*, vol. 94, no. 9, 091101, 2009.

[45] P. Q. Liu, A. J. Hoffman, M. D. Escarra, K. J. Franz, J. B. Khurgin, Y. Dikmelik, X. Wang, J.-Y. Fan, and C. F. Gmachl, "Highly power-efficient quantum cascade lasers," *Nat. Photonics*, vol. 4, pp. 95–98, 2010.

[46] J. B. Khurgin, "Inhomogeneous origin of the interface roughness broadening of intersubband transitions," *Appl. Phys. Lett.*, vol. 93, no. 9, 091104, 2008.

[47] W. Ted Masselink, "Ionized-impurity scattering of quasi-two-dimensional quantum-confined carriers," *Phys. Rev. Lett.*, vol. 66, no. 11, pp. 1513–1516, 1991.

[48] J. V. D. Veliadis and J. B. Khurgin, "Engineering of the nonradiative transition rates in nonpolar modulation-doped multiple quantum wells," *J. Opt. Soc. Am. B.*, vol. 14, no. 5, pp. 1043–1047, 1997.

[49] J. V. D. Veliadis, J. B. Khurgin, and Y. J. Ding, "Engineering of the nonradiative transition rates in modulation-doped multiple-quantum wells," *IEEE J. Quantum Electron.*, vol. 32, no. 7, pp. 1155–1160, 1996.

[50] M. O. Manasreh, *Semiconductor Heterojunctions and Nanostructures*. New York: McGraw-Hill, 2005.

# 2 State-of-the-Art Mid-Infrared QCLs: Elastic Scattering, High CW Power, and Coherent-Power Scaling

Dan Botez and Luke J. Mawst

University of Wisconsin–Madison

## 2.1 Introduction

The achievement in 1994 [1] of implementing the intersubband-transition laser concept [2], together with the achievement in 2002 of continuous-wave (CW) room-temperature (RT) lasing, opened the way for the development of quantum cascade lasers (QCLs) for a vast array of applications in the mid-infrared (IR) wavelength range. Since 2002 research efforts have unveiled highly relevant mechanisms behind QCL operation such as elastic scattering and carrier leakage, and devices have been developed with specific characteristics such as high CW power, high-speed modulation, low-power dissipation and coherent-power scaling. This chapter treats QCLs emitting in the 4–10 μm wavelength range with emphasis on what it takes to achieve high-CW-power ($\geq 2$ W), highly efficient devices. In particular, we update our review of the temperature dependence of the electro-optical characteristics of mid-IR QCLs [3] by including elastic-scattering mechanisms, the three-level rate-equations-based QCL model corrected for carrier leakage, and carrier leakage triggered by elastic scattering. As a result, we account for all mechanisms behind the QCL internal-efficiency value. Then, we cover the means for maximizing the pulsed and CW wall-plug efficiency via conduction-band and elastic-scattering engineering as well as via photon-induced carrier transport. Finally, we review coherent-power scaling for both one- and two-dimensional (2-D) photonic-crystal (PC) structures, and discuss grating-coupled surface-emitting (GCSE) lasers with emphasis on those needed for implementing high-index-contrast, 2-D PC lasers.

Sections 2.2 and 2.5 treat the electro-optical characteristics under pulsed and CW operation, respectively. Elastic scattering and carrier leakage are discussed in Sections 2.3 and 2.4. Finally, coherent-power scaling and GCSE QCLs are covered in Sections 2.6 and 2.7, respectively.

## 2.2 Electro-Optical Characteristics under Pulsed Operating Conditions

### 2.2.1 Threshold-Current Density

**Definitions**

As long as the tunneling-injection efficiency $\eta_{inj}^{tun}$ is close to unity, the threshold-current density $J_{th}$ is well approximated by [4]:

$$J_{th} = \frac{\alpha_m + \alpha_w + \alpha_{bf}}{\eta_{inj}\Gamma g} = \frac{\alpha_m + \alpha_w + \alpha_{bf}}{\eta_{inj}^{tun}\eta_p\Gamma g}, \tag{2.1a}$$

where $\alpha_m$ and $\alpha_w$ are the mirror and waveguide loss, respectively; $\alpha_{bf}$ is the equivalent loss corresponding to thermal backfilling of the lower-laser (*ll*) level [3]; $\Gamma$ is the optical-mode confinement factor; and $g$ is the differential gain in the case of unity $\eta_{inj}^{tun}$ value and no carrier leakage, and can be obtained by using the expression for the gain cross-section $g_c$ [5] divided by $e\Gamma$ and multiplied by the global "effective" upper-level lifetime $\tau_{up,g}$ [6] due to inelastic and elastic scattering [4, 5]:

$$g = \tau_{up,g}\frac{4\pi e}{\varepsilon_0 n_{refr}\lambda}\frac{z_{ul,ll}^2}{2\gamma L_p}, \tag{2.1b}$$

where $z_{ul,ll}$ is the dipole-matrix element between the upper-laser (*ul*) and *ll* level(s), $\lambda$ is the vacuum wavelength, $2\gamma$ is the full-width-at-half-maximum (FWHM) broadening of the transition [7, 8], and $L_p$ and $n_{refr}$ are the length and the effective refractive index of one cascading stage of the core region, respectively, Finally, $\eta_{inj}$ is the *total* injection efficiency [4] which is the product of the $\eta_{inj}^{tun}$ and $\eta_p$, the so-called pumping efficiency, which represents the degree of carrier-leakage suppression [4, 9]. Carrier leakage occurs from both injector-region states and the *ul* level, mostly within the active region (AR) [4, 9]. $\eta_p$ is well approximated by [9]:

$$\eta_p \cong 1 - J_{leak,ul}^{net}/J_{th} - J_{leak,inj}^{net}/J_{th} = 1 - J_{leak,tot}/J_{th}, \tag{2.1c}$$

where $J_{leak,ul}^{net}$ and $J_{leak,inj}^{net}$ stand for the net leakage-current densities, at threshold, from the *ul* level and injector states, respectively, triggered by LO-phonon scattering and interface-roughness (IFR) scattering [9]. These carrier-leakage mechanisms are discussed in Sections 2.4.2 and 2.4.3.

While the $\eta_{inj}^{tun}$ value is in general very close to unity (i.e., in the 0.97–0.98 range); thus, is usually taken to be unity, the $\eta_p$ value for conventional 4.6 μm-emitting QCLs, when considering only LO-phonon scattering [10], is ~0.85 at RT and ~0.77 at 360 K heatsink temperature. When considering elastic scattering as well, $\eta_p$ at RT has values of ~0.70 for conventional 8.2 μm-emitting QCLs [11], and can be as low as ~0.75 for record-high performance QCLs [9, 30, 46]. Thus, $\eta_p$ is a key factor in determining the $J_{th}$ value, especially under CW operating conditions [6] since the leakage currents strongly increase with increasing temperature [9, 10]. For short-barrier devices, that is, QCLs for which the *ul* level is the highest energy state in the AR, leakage does mostly occur to the continuum, in which case it can be considered a scattering event of a rate $1/\tau_{esc}$ [3, 5] (Fig. 2.1(a)). Then leakage can be incorporated as part of a modified *ul*-level lifetime [3, 5] $\tau_{ul,g}' = 1/(1/\tau_{ul,g} + 1/\tau_{esc})$, where $\tau_{ul,g}$ is the global *ul*-level lifetime [3, 6]. That is, for that specific case carrier leakage may be simulated as part of the *ul*-level lifetime rather than part of the injection efficiency. However, in tall-barrier devices, that is, for QCLs for which there are AR energy states above the *ul* level, which constitute the vast majority of state-of-the-art QCLs [3], carrier leakage mostly consists of carrier scattering to the next higher AR energy states followed by relaxation to low-energy AR states (Fig. 2.1(b)) [3]. Furthermore, leakage occurs not

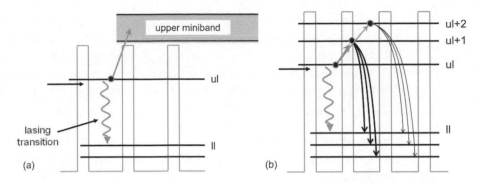

**Fig. 2.1** Schematic representation of the carrier leakage in: (a) short-barrier, and (b) tall-barrier QCLs. Only leakage from the upper-laser (*ul*) level is shown. Leakages from injector states are similar.

only from the *ul* level, but from injector states as well [9]. Therefore, for state-of-the-art QCLs leakage is mostly a *shunt-type* current through the next higher AR energy states (e.g., via the *ul* + 1 level in Fig. 2.1(b)), *not* leakage to the continuum. That is why the $\eta_p$ term has to be considered part of the injection efficiency (Eq. (2.1a)), just as for interband-transition devices [12]. One consequence is that if Eq. (2.1a) is rewritten with the definitions for $\alpha_{bf}$ [4] and for $\eta_p$ (Eq. (2.1c)), one obtains:

$$J_{th} = \frac{\alpha_m + \alpha_w}{\eta_{inj}^{tun}\Gamma g} + J_{bf} + J_{leak,tot} \simeq \frac{\alpha_m + \alpha_w}{\Gamma g} + J_{bf} + J_{leak,tot}, \qquad (2.1d)$$

from which it is clear that was has been called transparency-current density $J_{tr}$ [13, 14] is the sum of thermal-backfilling and carrier-leakage currents, not just thermal back-filling. This explains, for instance, why for devices of carrier-leakage current density ~20% of $J_{th}$ [3, 13] the extracted $J_{tr}/J_{th}$ value was found to be rather large (~45%) [13]. The other significant consequence is that the rate equations for a three-level QCL of unity $\eta_{inj}^{tun}$ value [5] have to be corrected accordingly:

$$\frac{\delta n_3}{\delta t} = \frac{J - J_{leak,tot}}{e} - \frac{n_3}{\tau_3} - Sg_c(n_3 - n_2) = \eta_p\frac{J}{e} - \frac{n_3}{\tau_3} - Sg_c(n_3 - n_2),$$

$$\frac{\delta n_2}{\delta t} = \frac{n_3}{\tau_{32}} + Sg_c(n_3 - n_2) - \frac{n_2 - n_2^{therm}}{\tau_2}, \qquad (2.1e)$$

$$\frac{\delta S}{\delta t} = \frac{c}{n_{refr}}\left[(g_c(n_3 - n_2) - (\alpha_m + \alpha_w))S - \beta\frac{n_3}{\tau_{sp}}\right],$$

where $J_{leak,tot}$ is total leakage at a current density $J$; $\tau_3$ is the state-3 lifetime *without* leakage considered; $n_3$, $n_2$, and $n_1$ are the sheet-carrier densities in the three states; $S$ is the photon-flux density; $\beta$ is the spontaneous emission factor; and $\tau_{sp}$ is the lifetime due to spontaneous emission from state 3. The gain cross-section $g_c$ is given by $g$ (Eq. (2.1b)) multiplied by $e\Gamma$ and divided by $\tau_{up,g}$, with the *ul* and *ll* levels being states 3 and 2. Setting $\delta/\delta t = 0$ and $S = 0$ (i.e., below threshold) and neglecting spontaneous emission, the solution for the $J_{th}$ value is simply given by Eq. (2.1d).

Other sources of carrier leakage are: (a) scattering to satellite valleys, such as the X or L valleys of InGaAs quantum wells, which is a significant factor only for InP-based QCLs emitting in the $\lambda \leq 3.5\,\mu m$ range and InAs-based QCLs emitting in the $\lambda \leq 3.2\,\mu m$ range [3]; and (b) transitions from injector states to low-energy AR states, which lower $\eta_{inj}^{tun}$ in two-quantum-well (QW) AR devices (e.g., [15]) or in InAs/AlSb QCLs at wavelengths $>15\,\mu m$ (Chapter 3).

## Temperature Dependence

The $J_{th}$ temperature dependence is defined by a characteristic temperature coefficient $T_0$ as such:

$$J_{th}\,(T_{ref} + \Delta T) = J_{th}(T_{ref})\exp(\Delta T/T_0), \tag{2.2}$$

where $T_{ref} + \Delta T$ is the heatsink temperature, $T_{ref}$ is the reference heatsink temperature, and $\Delta T$ is the range in temperature over which the $J_{th}$ increase can be approximated by an exponential function. For a given $\Delta T$ value, $T_0$ is determined by the terms in Eq. (2.1a) that are strong functions of temperature: $\eta_p$ and $\alpha_{bf}$. As far as $\eta_p$, it has been shown [10, 16] that, as long as the energy difference between the $ul + 1$ and $ul$ levels, $E_{ul+1,ul}$, $\geq 50$ meV, carrier leakage primarily occurs through the $ul + 1$ level. Then, for LO- [10, 16] and IFR-scattering- [9, 17]-triggered carrier leakage:

$$J_{leak,i}^{LO} \propto \exp\left(-\frac{E_{ul+1,i} + \hbar\omega_{LO}\left[(T_{ei}/T_\ell) - 1\right]}{kT_{ei}}\right) \quad \text{and} \quad J_{leak,i}^{IFR} \propto \left(-\frac{E_{ul+1,i}}{kT_{ei}}\right), \tag{2.3a}$$

where $i$ stands either for the $ul$ level or for an injector-region energy state, such as the ground state $g$. $T_{ei}$ is the electron temperature in the $i$th energy state, which, at threshold, has been defined [18, 19], for the injector miniband, as such:

$$T_{e,inj} = T_\ell + \alpha_{e-1} \cdot J_{th}, \tag{2.3b}$$

where $T_\ell$ is the lattice temperature and $\alpha_{e-1}$ is the electron-lattice coupling constant. It has been generally assumed that $T_{e,inj} \approx T_{e,ul}$ [10, 18]. Consequently as the $E_{ul+1,ul}$ value increases the leakage currents sharply decrease with increasing temperature and, in turn, the $J_{th}$ variation with heatsink temperature decreases; that is, $T_0$ increases. The fact that, for a given injector-doping level, the $T_0$ value increases with increasing $E_{ul+1,ul}$ value was experimentally demonstrated as early as 2009 [20]. Then, in 2010 it was shown that in order to obtain reasonably good estimates for the $T_0$ values of 4.5–5.0 $\mu m$-emitting QCLs, for both a conventional design and a high-$E_{ul+1,ul}$ design, one has to consider 'hot' electrons in the $ul$ level [10]. Significantly increasing the $T_0$ value by increasing the $E_{ul+1,ul}$ value was later demonstrated for 3.9 $\mu m$-emitting QCLs [21] as well as for 8–9 $\mu m$-emitting QCLs [3]. Note that when increasing the $E_{ul+1,ul}$ value by design, the $E_{ul+1,inj}$ value also increases, thus suppressing carrier leakage from injector states as well [9].

As for $\alpha_{bf}$, it is the product of $\Gamma g$ and the so-called backfilling-current density, $J_{bf}$ [5]:

$$J_{bf} = \frac{en_{\parallel}^{therm}}{\tau_{up,g}}, \tag{2.4}$$

where $n_{ll}^{therm}$ is the sheet carrier density in the $ll$ level that is thermally excited from energy states in the injector region. If the injector miniband is a single subband, then a rough estimate [5] is: $n_{ll}^{therm} \approx n_s \exp(-\Delta_{inj}/kT_{e,inj})$, where $n_s$ is the sheet-carrier doping density in the injector, $\Delta_{inj}$ is the energy difference, at resonance, between the $ll$ level and the ground state $g$ in the next injector miniband, and $T_{e,inj}$ is the electron temperature in that miniband. If electrons are considered to be in equilibrium with the lattice (i.e., $T_{e,inj} = T_l$), the calculated $J_{bf}$ values, when $\Delta_{inj} = 130$–$150$ meV, were found to be significantly smaller than derived values [3], which tended to confirm that hot electrons significantly increase backfilling and/or that leakage currents were not considered. For lower $\Delta_{inj}$ values ($\sim 100$ meV) the single-subband $n_{ll}^{therm}$ expression was found to strongly overestimate backfilling [22] and then, considering the injector miniband to have several subbands, a different expression was derived, which for 7.1 $\mu$m-emitting QCLs [22] was found to be in relatively better agreement with experiment. Nonetheless, in general, calculated $J_{bf}$ values are difficult to reconcile with experimental results. On the one hand, as pointed above, the transparency-current density is the sum of $J_{bf}$ and $J_{leak,tot}$. On the other hand, in devices of ultrafast, IFR-driven $ll$-level depopulation (i.e., $ll$-level lifetimes in the 0.04–0.10 ps range) [5, 9, 14] relatively low $\alpha_{bf}$ values are derived since, the system not being in a thermally steady state, the simple rate-equations thermal backfilling of the $ll$ level fails to apply [14]. Even though there is no accurate expression for calculating the $\alpha_{bf}$ value, the trends in measured $T_0$ values are as expected from the rough estimates for $n_{ll}^{therm}$: (a) $T_0$ increases with decreasing $n_s$; and (b) $T_0$ increases with increasing $\Delta_{inj}$ [3]. Historically, increasing the $\Delta_{inj}$ value to $\geq 100$ meV has allowed breakthroughs in QCL performance [3]: CW operation at 77 K in 1995, and RT pulsed operation in 1996. After the achievement of RT CW operation in 2002 [23] it steadily became apparent that what by and large affects the $T_0$ value, at and above RT, is carrier leakage from the active regions of QCL structures. Starting in 2009 carrier-leakage suppression via conduction-band (CB) engineering of the active region [10, 20, 24, 25] has led to large increases in the $T_0$ value above RT (e.g., from $\sim 140$ K to $\sim 250$ K for moderately doped, 4.5–4.9 $\mu$m-emitting devices, and from $\sim 160$ K to $\sim 240$ K for moderately doped, 8–9 $\mu$m-emitting devices) [3].

High $T_0$ values are relevant for: (a) maximizing the CW operating temperature of devices such as low-power, distributed feedback (DFB) lasers of broad tuning range [26]; (b) minimizing the threshold-dissipated powers for low-dissipation-power ($\leq 0.5$ W) devices [27]; and (c) stable CW operation of low-power ($\sim 100$ mW) devices for broadband-tuning applications [28]. However, for high-CW-power ($\geq 1.5$ W) operation the $T_0$ value is of secondary importance [3].

## 2.2.2 External Differential Efficiency

### Definitions and Experimental Data

The external differential efficiency $\eta_d$ for a QCL having $N_p$ stages is given by [4]:

$$\eta_d = \eta_i \frac{\alpha_m}{\alpha_m + \alpha_w} N_p, \tag{2.5a}$$

where $\eta_i$ is the internal differential efficiency per stage, which is well approximated by [4]:

$$\eta_i \cong \eta_{\text{inj}}^{\text{tun}} \eta_p \eta_{\text{tr}} = \eta_{\text{inj}} \eta_{\text{tr}}, \tag{2.5b}$$

where $\eta_{\text{inj}}$ is defined as above, for $J_{\text{th}}$, and $\eta_{\text{tr}}$ is the differential laser-transition efficiency [5]:

$$\eta_{\text{tr}} = \frac{\tau_{\text{up,g}}}{\tau_{\text{up,g}} + \tau_{\text{ll,g}}} = \frac{\tau_{\text{ul,g}}\left[1 - (\tau_{\text{ll,g}}/\tau_{\text{ul-ll,g}})\right]}{\tau_{\text{ul,g}}\left[1 - (\tau_{\text{ll,g}}/\tau_{\text{ul-ll,g}})\right] + \tau_{\text{ll,g}}}, \tag{2.5c}$$

where $\tau_{\text{ul,g}}$, $\tau_{\text{ll,g}}$ and $\tau_{\text{ul-ll,g}}$ are the global $ul$-level, $ll$-level and lasing-transition ($ul$-$ll$) lifetimes [6], respectively. The use of the same $\eta_{\text{inj}}$ expression for the external differential efficiency and the threshold-current density is justified by the fact that the pumping efficiency at threshold is basically the same as the differential pumping efficiency at and slightly above threshold [3, 4, 12].

The $\eta_d$ value can be expressed in terms of the slope efficiency $\eta_{\text{sl}} = \Delta P / \Delta I$, the rate of optical-power increase with drive-current increase above threshold, as such: $\eta_d = (e/h\nu)\eta_{\text{sl}}$. Taking $\eta_{\text{inj}}^{\text{tun}}$ to be unity, it has be shown [29], by defining for a four-level AR a differential pumping efficiency above threshold $\eta_p = \Delta J_4 / \Delta J$ where $J$ is the total current density and $J_4$ is the 'useful' current density flowing through the upper level (state 4), that, for small variations in $J$ above threshold, $\eta_p$ is identical with the pumping efficiency at threshold (i.e., $1 - J_{\text{leak,tot}}/J_{\text{th}}$), and that: $\Delta P / \Delta J = \eta_p \times \Delta P / \Delta J_4$. The fact that the slope efficiency includes the $\eta_p$ factor can also be derived from the corrected three-level rate equations (2.1e) by considering the useful current density flowing through the upper level, state $3$, $J_3 = J - J_{\text{leak,tot}} = \eta_p J$. Then, one can easily show that $\eta_{\text{sl}} \propto dS/dJ = \eta_p \times dS/dJ_3$, with $S$ being the photon-flux density. Thus, the fact that the internal efficiency has to take into account the degree of carrier leakage (i.e., $\eta_p$) is self-evident, just as for interband-transition lasers [12]. This also explains why the temperature dependence of the $\eta_d$ value for QCLs became to be understood only after carrier leakage was taken into account [10].

To obtain high output power and high wall-plug efficiency the $\eta_d$ value needs to be maximized. While $\alpha_w$ has been reduced to values as low as $0.5\,\text{cm}^{-1}$ at $\lambda = 4.6$–$4.9\,\mu\text{m}$ and $1.6\,\text{cm}^{-1}$ at $\lambda = 9\,\mu\text{m}$ [3] for moderate- to high-doped injector regions, and $N_p$ has been raised to 40–45 stages, the $\eta_i$ values were found to be rather low for conventional-type QCLs: 50–60% in the 4.5–6.0 $\mu$m wavelength range [3] and 57–58% in the 7–11 $\mu$m range [4] with, until recently, no explanation why that was the case. First of all, to increase $\eta_{\text{inj}}$ the carrier leakage was strongly suppressed by using CB engineering (see Section 2.4.4) such as step-tapering the AR barrier heights and well bottoms [3, 4] which has resulted in extracted $\eta_{\text{inj}}$ values as high as 84.5% [9]. Linear tapering of the AR barrier heights in the so-called shallow-well QCL [30], which has been shown [6] to be a linear tapered-active (TA)-type QCL, has resulted in an extracted $\eta_{\text{inj}}$ value of $\sim$75% [9].

Second, effective miniband-like carrier extraction from the AR was employed [4] by extracting carriers from the $ll$ level and one energy level below it, so-called resonant extraction (RE) [4]. Taking into account elastic scattering for 5 $\mu$m-emitting, step-tapered AR (STA)-RE [4] and 4.9 μm-emitting TA-RE QCLs [30] has resulted in enhancing $\eta_{tr}$ by 9–10% to values in the 91–94% range [9, 46]. In turn, for ∼5 μm-emitting STA-RE QCLs $\eta_i$ values as high as 77% were obtained [4]; that is, 30–50% higher than for conventional QCLs emitting in the 4.5–6.0 μm wavelength range. More recently, a 45-stage, ∼4.9 μm-emitting TA-RE QCL [31] has resulted in an $\eta_i$ value of ∼76%. Therefore, by combining carrier-leakage suppression with effective carrier extraction $\eta_i$ values have increased towards upper limits of ∼90% for mid-IR QCLs [4, 9]. In contrast, for diode lasers, that is devices for which the upper-levels lifetimes are of the order of nsecs, the $\eta_i$ upper limit is 100% since $\eta_{tr}$ is ∼100% [12].

Figure 2.2 shows a comparison of experimentally obtained $\eta_i$ values, for various QCL types, over the 4–11 μm wavelength range. Devices with both carrier-leakage suppression *and* miniband-like extraction have significantly higher $\eta_i$ values than conventional QCLs. More specifically, over the 4.5–6.0 μm range STA- and TA-RE QCLs have reached $\eta_i$ values of ∼77% at $\lambda = 5$ μm; and ∼70% and ∼76% at $\lambda = 4.9$ μm, respectively. Over the 7.0–9.0 μm range STA-RE QCLs reached extracted values of ∼74% at $\lambda = 7.8$ μm [11] and experimental values of ∼86% at both $\lambda = 8.4$ μm and $\lambda = 8.8$ μm [32]; that is, 30–50% higher than for conventional QCLs emitting in the 7–11 μm wavelength range. The relatively low $\eta_i$-value cases (i.e., 57–67%) correspond to devices of: (a) carrier-leakage suppression, but low $\eta_{inj}^{tun}$ and $\eta_{tr}$ values (e.g., ∼62% at $\lambda = 5.6$ μm for a two-QW design [15]); (b) miniband-like extraction, but without carrier-leakage suppression (i.e., ∼63.5% at $\lambda = 7.1$ μm [22, 32], 66% at $\lambda \sim 8.25$ μm [33], ∼67% at $\lambda = 9$ μm [32, 34] and ∼63.5% at $\lambda = 10.7$ μm [32, 35]). An ∼55% value was extracted [11] from experimental data of a conventional $\lambda = 8.2$ μm QCL [36] with strong carrier leakage. The upper-limit $\eta_i$ value of 90% is obtained by considering an ideal case when the lasing transition occurs via LO-phonon-relaxation to the top of a single-subband miniband, instead to a discrete energy state, thus, providing very short (∼0.1 ps) lower-level lifetimes [37], and a generic 1.0 ps effective upper-level lifetime (i.e., $\eta_{tr} \sim 90\%$), no carrier leakage and unity tunneling-injection efficiency. As we shall see, when considering both inelastic and elastic scattering for 4.0–5.5 μm-emitting devices of strong diagonal transition, the upper limit for $\eta_i$ can reach values of 95% when considering either abrupt or graded well/barrier interfaces.

## Temperature Dependence

The $\eta_d$ temperature dependence is defined by a characteristic temperature coefficient $T_1$ [3]:

$$\eta_d \left(T_{ref} + \Delta T\right) = \eta_d \left(T_{ref}\right) \exp(-\Delta T / T_1), \tag{2.6}$$

where $T_{ref} + \Delta T$ is the heatsink temperature, $T_{ref}$ is the reference heatsink temperature, and $\Delta T$ is the range in temperature over which the $\eta_d$ decrease can be approximated by an exponential function. For QCLs the $T_1$ parameter was initially neglected since the $\eta_d$ value is not affected by thermal backfilling, thus it hardly varies with temperature up to ∼300 K. However, the $T_1$ value, just as for interband-transition lasers [38], is the very signature of carrier leakage above RT, whose suppression is key to high CW power [3, 6, 38] and high CW wall-plug efficiency [3, 6]. For a given $\Delta T$ value,

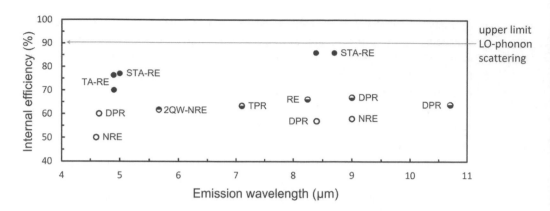

**Fig. 2.2** Internal efficiency as a function of emitting wavelength for various mid-IR QCL types. The data points correspond to references in [4] and results from [31] and [33]. Filled and empty circles correspond, respectively, to QCLs with both carrier-leakage suppression and miniband-like extraction, and to conventional QCLs. The upper-half-filled circle indicates a QCL with only carrier-leakage suppression [15]. Lower-half-filled circles indicate QCLs with only miniband-like extraction [32–35].

$T_1$ is determined by the terms in Eq. (2.5a) that are functions of temperature: $\eta_p$ and $\alpha_w$. The first term refers to carrier leakage. Since both LO-phonon- and IFR-triggered scattering from injector states and the $ul$ level to the $ul+1$ level are strong function of temperature, the $\eta_p$ term decreases with increasing temperature; thus, accounting for a large part of the $T_1$ value. However, as pointed out in [10], carrier leakage cannot alone account for the whole $T_1$ value; leaving an increase with temperature in $\alpha_w$ value as the most likely other factor determining the $T_1$ value, especially for long-wavelength (i.e., $\lambda \geq 8\,\mu m$) QCLs. For instance, Wittman [39] has found that the $\alpha_w$ of 8.4 $\mu m$-emitting conventional QCLs [7] increases with temperature. That is most likely a consequence of the fact that the part of $\alpha_w$ which is due to (nonresonant) intersubband (ISB) absorption, $\alpha_{ISB}$, increases with temperature [40]. Figure 2.3 shows that, for that QCL, carrier leakage and the increase of $\alpha_w$ with temperature have basically equal contributions to the $\eta_{sl}$ decrease with temperature; that is, to the $T_1$ value. That QCL, being of the conventional type, had significant carrier leakage, as attested by low $T_0$ and $T_1$ values: 160 K and 263 K [3] for moderately doped 8.4 $\mu m$-emitting QCLs. In contrast, when carrier leakage was suppressed in 8.4 $\mu m$-emitting moderately doped STA-RE QCLs [32] much higher $T_0$ and $T_1$ values, 219 K and 665 K, were obtained.

Carrier-leakage suppression via CB engineering of 3.9–5.6 $\mu m$-emitting QCLs [6, 15] has led to $T_1$ values in the 285–800 K range [4, 15, 21, 25, 29–32, 41, 42]. The high-end values (i.e., 650–800 K) were due either to a small value for the $\alpha_w/\alpha_m$ ratio [15, 42] or to low injector-region doping levels [3, 4, 25]. 4.9–5.0 $\mu m$-emitting STA- and TA-RE QCLs had their $T_1$ values drop from $\sim$650 K for low-doped devices to $\sim$400 K for moderately doped devices; thus confirming for 4.5–5.0 $\mu m$-emitting QCLs as well that the increase with temperature and/or doping of the ISB part of $\alpha_w$ [4] is partly responsible for the $T_1$ value.

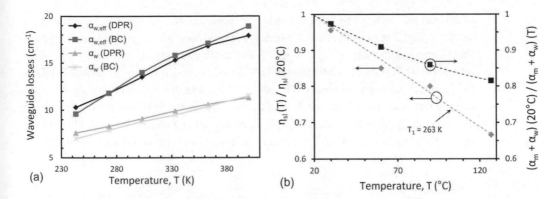

**Fig. 2.3** Temperature dependence of: (a) the $\alpha_w$ and $\alpha_w + \alpha_{bf}$ terms [39] for DPR and BC schemes; and (b) the relative slope efficiency and inverse nonresonant losses for the 8.4 µm DPR-scheme QCLs in [7]. Reproduced from [3].

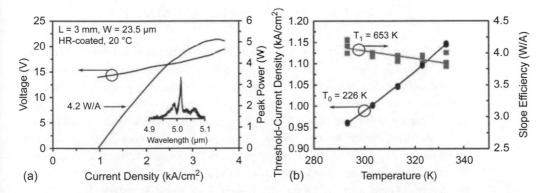

**Fig. 2.4** STA-RE QCL: (a) light– and voltage–current characteristics, and spectrum; (b) temperature dependence of the threshold-current density $J_{th}$ and the slope efficiency, as characterized by the $T_0$ and $T_1$ coefficients, respectively. Reprinted with permission from [4] © The Optical Society.

High $T_1$ values are relevant for realizing QCLs of high CW power and high CW wall-plug efficiency, as they are key to: (a) minimizing the $\eta_{sl}$ decrease from pulsed to CW operation; and (b) minimizing the core-region temperature rise [3]. This is reflected, for instance, by the fact that carrier-leakage suppression has led to virtual doubling of the single-facet RT CW power and wall-plug efficiency of QCLs emitting in the 4.5–5.0 µm wavelength range (see Section 2.5).

Figure 2.4(a) shows typical electro-optical characteristics of state-of-the-art 5.0 µm-emitting QCLs. The low $J_{th}$ value (0.96 kA/cm²) reflects a high $\eta_{inj}$ value (84.5%) brought on by strong carrier-leakage suppression [9]. The high single-facet $\eta_{sl}$ value (4.2 W/A) reflects both carrier-leakage suppression and a high $\eta_{tr}$ value (~91%) brought on by a dramatic decrease in $\tau_{3g}$ via IFR scattering in STA-type QCLs [9]. Only gas source molecular-beam-epitaxy (GSMBE)-grown TA-RE QCLs of same device geometry [30] have shown higher $\eta_{sl}$ values (5.3 W/A), mostly due to a lower $\alpha_w$ value. The high $T_0$ and $T_1$ values, 226 K and 653 K, respectively, as shown in Fig. 2.4(b), reflect carrier-leakage suppression and a relatively low injector sheet-doping density ($0.7 \times 10^{11}$ cm$^{-2}$). For same-geometry [i.e., 3 mm-long, high-reflectivity (HR)-coated)] and moderately doped ($n_s \sim 10^{11}$ cm$^{-2}$) STA-RE QCLs the $T_0$ and $T_1$ values were found to be lower: 216 K and 400 K [4]; reflecting the influence of increased backfilling on $T_0$ and the increase in the ISB-absorption part of the $\alpha_w$ value on $T_1$. Due to carrier-leakage suppression the $T_1$ value is much higher than values for moderately doped, 4.5–5.0 μm-emitting conventional QCLs ($\sim$140 K) [3].

### 2.2.3    Wall-Plug Efficiency

**Definitions, Experimental Data, and Upper Limits**

The electrical-to-optical power-conversion efficiency, so-called wall-plug efficiency $\eta_{wp}$ is defined in CW operation as the ratio of CW optical power $P_{opt}$ to CW electrical input power $IV$:

$$\eta_{wp,cw} = \frac{P_{opt}}{IV} = \frac{(h\nu/e)\eta_{d,cw}(I - I_{th,cw})}{IV}, \qquad (2.7)$$

where $\eta_{d,cw}$ and $I_{th,cw}$ are the CW external differential efficiency and threshold current, respectively. The CW terms are strong functions of the core-region temperature rise, defined as the difference between the lattice temperature, at a given drive level, and the heatsink temperature [3] (see Section 2.5.2). The (low-duty-cycle) pulsed wall-plug efficiency is the wall-plug efficiency per pulse; that is, when peak-pulsed optical power and drive current are considered. Although not of direct practical use, the maximum pulsed $\eta_{wp}$ value is useful since it provides the upper limit for the $\eta_{wp,cw}$ value. Therefore its study is relevant to maximizing the $\eta_{wp,cw}$ value and, in turn, minimizing the dissipated heat, a key requirement for the use of high-CW-power QCLs in a vast array of applications. An approximate expression for the pulsed maximum wall-plug efficiency $\eta_{wp,max}$ [3, 10], slightly modified by considering a non-unity $\eta_{inj}^{tun}$ value, is given by:

$$\eta_{wp,max} \simeq \eta_s \eta_i \frac{\alpha_{m,opt}}{\alpha_{m,opt} + \alpha_w} \frac{N_p h\nu}{eV_{wpm}} \left(1 - \frac{J_{th}}{J_{wpm}}\right), \qquad (2.8)$$

where $\eta_s$ reflects the deviation from a straight line of the pulsed L–I curve at the $\eta_{wp,max}$ point, which is likely due to carrier leakage high above threshold [3]; $\eta_i$ is the internal differential efficiency as defined in Eq. (2.5b); $\alpha_{m,opt}$ is the optimum mirror loss; $J_{wpm}$ is the current density at the $\eta_{wp,max}$ point; $h\nu$ is the photon energy and $V_{wpm}$ is the voltage at $J_{wpm}$. For state-of-the-art devices $J_{wpm}$ is generally found to be about three times $J_{th}$ at 300 K, and the $\alpha_{m,opt}$ value was experimentally found, for 4.6 μm-emitting QCLs [13], to be $\sim$2.2 cm$^{-1}$. From (2.8) it is obvious that the primary means for maximizing $\eta_{wp,max}$ are: (a) maximize the $\eta_i$ value, (b) minimize the $\alpha_w$ value, and (c) minimize the $V_{wpm}$ value. As pointed out above, $\eta_i$ can be maximized via carrier-leakage suppression, miniband-like carrier extraction and elastic-scattering enhancement of the transition efficiency. The $\alpha_w$ value can be reduced by minimizing free-carrier absorption in the cladding layers and minimizing ISB absorption in the core region [3]. The latter is primarily obtained by improving the quality of the core-region interfaces [43]. As for the $V_{wpm}$ value, it is given by:

$$V_{wpm} \simeq V_{th} + R_{diff}(I_{wpm} - I_{th}),$$ (2.9)

with $V_{th} \simeq N_p[(hv + \Delta_{inj,th})/e] + R_s I_{th}$, where $V_{th}$ is the threshold voltage; $\Delta_{inj,th}$ is the so-called voltage defect at threshold (i.e., the energy difference between the $ll$ level and the main injecting state (MIS) in the next stage); $R_{diff}$ is the differential resistance; and $R_s$ is the series resistance of the cladding layers. The $R_{diff}$ $(I_{wpm} - I_{th})$ term is significant, except for QCLs that have strong photon-induced carrier transport (PICT) [14, 44, 46] due to strong coupling (7–9 meV) between the MIS and the $ul$ level as well as to strong diagonal lasing transitions [14, 30, 46] (e.g., for [30] $z_{ul,ll} \sim 8$ Å [46]) . Then, since electron transport is primarily controlled by the cavity photon density, the $R_{diff}$ value significantly decreases and the dynamic range increases [14, 44]. Due to the proximity of the $J_{wpm}$ value to the maximum-current density $J_{max}$ value [30, 33] a good $\eta_{wp,max}$ approximation is obtained by replacing $V_{wpm}$ with $V_{max} \simeq N_p[(hv + \Delta_{inj,res})/e]$, where $\Delta_{inj,res}$ is the voltage defect at resonance, and $J_{wpm}$ with $J_{max}$ (i.e., the two changes compensate). Then, for devices with strong PICT action,

$$\eta_{wp,max} \simeq \eta_s \eta_p \eta_{tr} \frac{\alpha_{m,opt}}{\alpha_{m,opt} + \alpha_w} \left( \frac{1}{1 + \Delta_{inj,res}/(hv)} \right) \left( 1 - \frac{J_{th}}{J_{max}} \right).$$ (2.10)

This looks similar to the equation for $\eta_{wp}$ in [34], when the optimum $\alpha_m$ value for maximizing $\eta_{wp}$ is used, with the significant difference that carrier leakage is taken into account via the $\eta_s$ and $\eta_p$ terms. A comparison to experiment can be done for 4.9 μm-emitting TA-RE QCLs [30]. Taking the reported experimental data, a calculated $\Delta_{inj,res}$ value of 120 meV [46], $\eta_p$ and $\eta_{tr}$ values of 0.75 and 0.94, respectively, calculated using graded-interface modeling [46], and a measured $\eta_s$ value of 0.92 [3] one obtains an $\eta_{wp,max}$ value of 27.5%; that is, very close to the equivalent both-facets value (i.e., 27.6%). This confirms that the equation is a very good approximation for the $\eta_{wp,max}$ value of mid-IR QCLs with strong PICT action, and clarifies why prior fits to experiment [5], using the overly simplified $\eta_{wp,mnax}$ expression derived in [34], required a scaling factor of 0.7.

In addition to maximizing the $1 - J_{th}/J_{max}$ term via strong coupling and PICT action [14]; besides carrier-leakage suppression, $\eta_{tr}$ maximization and $\alpha_w$ minimization, the other condition for high $\eta_{wp,max}$ value is to minimize the $\Delta_{inj,res}$ term. Note that maximizing the $\eta_i$ value can be done independently of minimizing the $\alpha_w$ and $\Delta_{inj,res}$ values [32]. In particular, the pocket-injector design [14, 45], since it involves injection from an excited injector state, allows for both low $V_{th}$ values as well as low thermal backfilling, as needed for efficient CW operation [3]. Then it comes as no surprise that the record for CW wall-plug efficiency at RT (i.e., 21%) has been obtained from a device of pocket-injector design (i.e., the MIS at threshold is the fourth-excited injector state, while close to $J_{max}$ it is the first-excited state) [46], strong PICT action, low $\alpha_w$ value, moderate carrier-leakage suppression, and miniband-type extraction; i.e., the 4.9 μm-emitting TA-RE QCL [30, 46]. The experimental $J_{th}$ value and the V–I curve for that device were virtually reproduced [46] (Fig. 2.5(a)) by using the recently

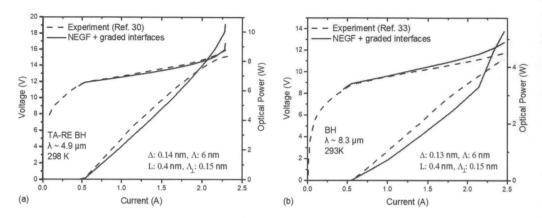

**Fig. 2.5** Experimental and calculated voltage and light vs. drive current curves for: (a) 4.9 μm-emitting TA-RE QCLs [30]; and (b) 8.3 μm-emitting [33] QCLs. Calculations performed using the generalized IFR formalism [47]. The interface width is taken to be 0.4 nm. Photon-induced carrier transport is evidenced by a differential resistance ~45% that of conventional QCLs of the same pumped area (i.e., 8 μm × 5 mm) and a $J_{max}$ value 30–50% higher than for conventional QCLs of the same injector-doping level.

published generalized IFR formalism [47] that employs nonequilibrium Green's functions (NEGF) analysis [48] and considers *graded interfaces*. The root-mean-square height, Δ (see Section 2.3.1) value is found to be 0.14 nm, which is an average value since Δ has been experimentally found [49] to increase with the barrier height within the active region. The in-plane correlation length, Λ, value is 6 nm, while interface grading is characterized by an interface width, $\mathcal{L}$, value of 0.4 nm, and there is an axial correlation length, $\Lambda_{\perp} = 0.15$ nm. For ~8.3 μm-emitting QCLs [33] of strong PICT action and record-high performance [33], the agreement with the experimental V–I curve is also good (Fig. 2.5(b)), but for smaller Δ value (0.13 nm), which is likely quite accurate since all barriers have the same height, while the Λ value is still 6 nm. Note that for both λ ~ 4.9 μm and ~ 8.3 μm QCLs the newly found Δ and Λ values are higher and significantly lower, respectively, than those used in studying, while assuming *abrupt interfaces*, high-quality 4.6 μm-emitting (i.e., Δ = 0.12 nm and Λ = 8.5 nm) [8] and 8–9 μm-emitting (i.e., Δ = 0.10 nm and Λ = 9 nm) [48] QCLs. This finding is in good agreement with the experimental Δ and Λ values extracted from atom-probe tomography (APT) analyses of InGaAs/AlInAs QCL structures [49].

For λ = 4.9 μm QCLs the $R_{diff}$ and $J_{max}$ values: ~1.6 Ω and 5.6 kA/cm², respectively, confirm that there is strong PICT action, since for conventional buried-heterostructure (BH) devices of same pumped area $R_{diff}$ ~ 3.5 Ω, and of same $n_s$ value ($0.9 \times 10^{11}$ cm⁻²) $J_{max}$ is ~3.8 kA/cm². That is, implementing PICT action has lowered $R_{diff}$ by a factor of ~2.2 and increased $J_{max}$ by ~50%. The calculated wavelength is the same as the experimental one (i.e., 4.9 μm) while without interface grading it is ~4.5 μm. For λ = 8.3 μm QCLs the $R_{diff}$ and $J_{max}$ values: ~1.6 Ω and ~5.8 kA/cm²; also confirm strong PICT action, in that $R_{diff}$ is ~2.2 times lower and $J_{max}$ is ~30% higher compared to conventional, same-area and doping ~8 μm-emitting BH devices. For diagonal-transition, weakly coupled 8.5 μm-emitting QCLs

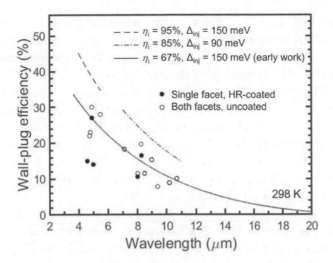

**Fig. 2.6** Fundamental limits for the wall-plug efficiency of mid-IR QCLs as a function of emission wavelength. The solid curve is for a 'voltage defect' at resonance $\Delta_{inj,res} = 150$ meV and a 70 ps dephasing time [34], and using a derived [32] internal efficiency $\eta_i$ value of 67%. The dashed curve corresponds to both abrupt- [9, 51] and graded- [46] interface TA- or STA-RE QCLs, yielding an upper-limit $\eta_i$ value of ~95% when $\Delta_{inj,res} = 150$ meV. Note that the *same curve* is obtained for the case $\eta_i \sim 90\%$ and $\Delta_{inj,res} = 120$ meV. The dot-dash curve is for both abrupt- [14, 48] and graded- [46] interface QCLs, an upper-limit $\eta_i$ value of 85% and $\Delta_{inj,res} = 90$ meV. The experimental data points are from Fig. 12 of [4], and from [33, 52–55].

[48], decreases in above-threshold voltages and an increase in dynamic range have been shown, by using a rigorous density-matrix (DM) model [50], to be caused by PICT action [50]. However, due to weak coupling, the diagonal transition is limited to the active region, which, in turn, results in moderate PICT action; i.e., $R_{diff} \sim 2.5\,\Omega$ vs. ~3.5 $\Omega$ for conventional BH QCLs of same pumped area.

We show in Fig. 2.6 experimental $\eta_{wp,max}$ values published at 298 K or adjusted for 298 K operation by using their respective $T_0$ and $T_1$ values [3], for both-facets, uncoated and single-facet, HR-coated devices; and two curves for $\Delta_{inj,res} = 150$ meV in Eq. (2.10): (a) early upper limits [34] using the highest $\eta_i$ value at the time (~67%) [32] and assuming a linear L–I curve (i.e., $\eta_s = 1$); and (b) upper limits, over the 4.0–5.5 µm wavelength range, when considering elastic scattering (i.e., the average value of 95% for $\eta_{tr}$ obtained for TA- and STA-type devices emitting in the 4–5 µm range) [9, 46, 51]. The 95%-upper-limit $\eta_i$ curve is drawn for the fundamental-limit case of $\eta_s = 1$ and $\eta_{inj} = 100\%$ (i.e., $\eta_i = \eta_{tr}$) and it results in projected $\eta_{wp,max}$ values $\geq 40\%$ for $\lambda \leq 4.75$ µm. However, the highest $\eta_{inj}$ value reported to date, over the 4.0–5.5 µm range, is only ~84.5% [9] for a STA-RE QCL of 77% $\eta_i$ value. Nonetheless, IFR-engineered 4.6–4.7 µm-emitting STA-RE QCLs, designed with pocket injection and strong carrier-leakage suppression (i.e., $\eta_{inj} \sim 95\%$) [9, 51], have provided $\eta_i$ values of 86% which lead to projected single-facet $\eta_{wp,max}$ values of 36.5% [51]. The 36.5% value is 35% higher than the highest single-facet $\eta_{wp}$ value reported to date [30] (i.e., 27% ). Most of the increase is due to the much higher $\eta_i$ value as a result of inherently stronger carrier-leakage suppression in STA-RE devices.

Further increasing the $\eta_{wp,max}$ value requires additional carrier-leakage suppression via IFR-scattering engineering and/or lowering the $\Delta_{inj,res}$ value for devices of the same high degree of carrier-leakage suppression as the IFR-engineered STA-RE device (i.e., $\eta_{inj} \sim 95\%$). Record-high, pulsed $\eta_{wp,max}$ values published at $\lambda = 4.9\,\mu m$ by Bai et al. [30] and at $\lambda = 7.1\,\mu m$ by Maulini et al. [22] were obtained for $\Delta_{inj,res}$ = 120 meV [22, 46]. Considering that the IFR-engineered, 4.7 μm-emitting STA-RE QCL has $\Delta_{inj,res}$ = 149 meV [51], lowering $\Delta_{inj,res}$ to 120 meV would raise the both-facets $\eta_{wp,max}$ value from 37.5% to 40.3%; i.e., practically the same value as that on the 95%-$\eta_i$ curve (41%). That is, even if $\eta_i \sim 90\%$, $\eta_{wp,max}$ values > 40% are attainable, over the 4.5 μm $\leq \lambda \leq$ 4.75 μm range, for PICT-action QCLs of $\Delta_{inj,res}$ = 120 meV and high degree of leakage suppression ($\eta_{inj} \sim 95\%$).

A clear distinction is made between results from single-facet, HR-coated devices and both-facets, uncoated devices. The best results from single-facet, HR-coated QCLs (i.e., 27% at $\lambda = 4.9\,\mu m$ [30] and ~16.5% at $\lambda \sim 8.25\,\mu m$ [33], at 298 K), which are also associated with record-high CW $\eta_{wp,max}$ values, have been reported from PICT-action QCLs of low $\alpha_w$ values due to high-quality interfaces typical of GSMBE, growth. For a given structure, both-facets uncoated $\eta_{wp,max}$ values are higher than single-facet values simply because of higher mirror-loss $\alpha_m$ values, especially for short-cavity (e.g., 3 mm-long) devices. However, high $\alpha_m$ values lead to high $J_{th}$ values (~2 kA/cm$^2$) in the case of structures without carrier-leakage suppression [3, 52, 53, 56], which, in turn, impair the CW performance since the core-temperature rise $\Delta T_{act}$ is directly related to $J_{th}$ [3]. Then, RT CW $\eta_{wp,max}$ values are significantly lower than in pulsed operation (e.g., 6.7% vs. 23% for 4.8 μm-emitting [53], 10% vs. 18.9% for 7.1 μm-emitting [22] and 10% vs. 16% for 9.1 μm-emitting QCLs [56]). In fact, the highest both-facets uncoated $\eta_{wp,max}$ values (at 298 K) – ~30% at $\lambda = 4.9\,\mu m$ [54], ~28.5% at $\lambda \sim 5.6\,\mu m$ [15] and ~19.8% at $\lambda \sim 8.3\,\mu m$ [33] – were obtained from QCLs unable to operate CW. That is, optimization for high pulsed $\eta_{wp,max}$ operation does not necessarily mean optimization for high CW $\eta_{wp,max}$ operation.

As for fundamental $\eta_{wp,max}$ limits for $\lambda \geq 7\,\mu m$, that is when considering the effect of elastic scattering on the lifetimes of PICT-action QCLs, two factors are taken into account. First, over the 7–10.5 μm wavelength range, Wolf [14] calculates on average optimized $\eta_{tr}$ values of 83%. However, at $\lambda = 8.25\,\mu m$, while using MBE-growth IFR parameters, calculated $\eta_{tr}$ values for STA-type devices [57] reach ~84%, and ~88% for graded-interface, PICT-action devices [33, 46]. Similarly, at $\lambda = 10\,\mu m$, graded-interface STA-type devices reach $\eta_{tr} \sim 85\%$ [46]. Thus, on average 85% is a reasonable upper limit. Second, for 7–11 μm-emitting QCLs a $\Delta_{inj,res}$ value of ~90 meV was found [14, 33] to be optimal for high-performance operation. Therefore, over the 7–11 μm range, a fundamental-limit $\eta_{wp,max}$ curve corresponding to $\eta_i$ = 85% and $\Delta_{inj,res}$ = 90 meV is plotted. Then, projected both-facets $\eta_{wp,max}$ values vary from 29.2% at $\lambda = 7\,\mu m$ to 14.7% at $\lambda = 11\,\mu m$. The only strong PICT-action QCL reported in this range (i.e., at ~8.3 μm [33]) has a relatively low $\eta_i$ value (66%) due to significant carrier leakage (i.e., $\eta_p$ = 77%) [46] mostly due to a low $E_{ul+1,ul}$

value: $\sim$62 meV. If an STA-type AR design were to be implemented for total carrier-leakage suppression ($\eta_i \sim$ 85%), the $\eta_{wp,max}$ fundamental limit becomes $\sim$25%, a value that falls right on the fundamental-limit $\eta_{wp,max}$ curve, further confirming the accuracy of Eq. (2.10). Note that Wolf has projected even higher $\eta_{wp,max}$ values (e.g., 36.5% for $\lambda = 7.8\,\mu m$) [14], however, for short-cavity (1–2 mm) devices, that is, for devices suitable only for low-power consumption operation [27].

**Temperature Dependence**

For a reference heatsink temperature of 300 K, $\eta_{wp,max}$ can be written as a function of increasing heatsink temperature, $T_h$, as such [3]:

$$\eta_{wp,max}(T_h) \approx \eta_d(300\,K) \exp\left(-\frac{T_h - 300\,K}{T_1}\right)$$

$$\times \left[1 - \frac{J_{th}(300\,K)}{J_{wpm}} \exp\left(\frac{T_h - 300\,K}{T_0}\right)\right] \frac{h\nu}{eV_{wpm}}, \qquad (2.11)$$

since $\eta_s$, $J_{wpm}$, and $V_{wpm}$ values are found to be virtually temperature insensitive above 300 K [3, 30]. Then, the temperature dependence of $\eta_{wp,max}$ is affected only by the $T_1$ and $T_0$ values and the $J_{th}/J_{wpm}$ ratio. One can define a characteristic temperature coefficient $T_2$ for $\eta_{wp,max}$ as follows:

$$\eta_{wp,max}(T_h) = \eta_{wp,max}(300\,K) \exp\left(-\frac{T_h - 300\,K}{T_2}\right). \qquad (2.12)$$

Taking the $T_0$ and $T_1$ values for conventional 3 mm-long, HR-coated QCLs [3] to be 143 K (over the 300–360 K temperature range) and a typical value of 2.5 for the $J_{th}$ (300 K)/$J_{wpm}$ ratio [3] one obtains $T_2 = 71$ K. In sharp contrast, for TA-RE-type QCLs, given $T_0 = 244$ K, $T_1 = 343$ K and 3.2 for the $J_{th}$ (300 K)/$J_{wpm}$ (300 K) ratio [30], the calculated $T_2$ value is 193 K, in very good agreement with the 195 K value obtained from the experimental data [30], thus confirming the accuracy of the approximation in Eq.(2.11). Then, for the TA-RE-type QCL the $\eta_{wp,max}$ value varies 2.2 times slower than in conventional devices, reflecting the beneficial effect that high $T_1$ and $T_0$ values bring about as far as low temperature sensitivity of the $\eta_{wp,max}$ value, a key factor for obtaining high CW power and wall-plug efficiency values.

## 2.3    Elastic Scattering

For mid-IR QCLs the elastic-scattering mechanisms that primarily affect device performance are IFR scattering [7–9, 17, 47, 57–61] and alloy-disorder (AD) scattering [9, 58, 61]. The effect of dopants/ionized impurities on scattering is negligible since the doped regions, unlike in THz QCLs, are relatively far removed from the ARs. The impact of IFR and AD scattering on device characteristics is discussed primarily for diagonal-transition QCLs since such devices lead to the maximization of the wall-plug efficiency for both high-CW-power (>1.5 W) QCLs [3, 4, 9] and for low-power-dissipation, broad-bandwidth QCLs [14].

## 2.3.1     Interface-Roughness Scattering

As described in Chapter 1, IFR scattering has been defined for *abrupt interfaces* by two IFR parameters: $\Delta$, the root-mean-square height, and $\Lambda$, the in-plane correlation length, that characterize a standard Gaussian autocorrelation of the roughness [7, 59]. However, since the actual well/barrier interface nature was unknown, the Gaussian autocorrelation function was only an assumption useful for extracting $\Delta$ and $\Lambda$ parameters that fit experimental data. An IFR study [60] did show that, for mid-IR abrupt-interface QCLs, results for Gaussian and exponential autocorrelation functions are comparable, given similar matrix elements. In turn, Gaussian autocorrelations that fit experiment were employed for abrupt-interface structures. For instance, for MBE-grown $GaAs/Al_{0.33}Ga_{0.67}As$ QCLs it was found [62] that $\Delta = 0.15$ nm and $\Lambda = 6$ nm. In contrast, for MBE-grown InGaAs/AlInAs QCLs the $\Delta$ and $\Lambda$ values were found to be quite different: 0.10 and 9 nm at $\lambda = 8$–9 μm [48], and 0.12 and $\sim$9 nm at $\lambda \sim 4.6$ μm [8], respectively. For GSMBE-grown QCLs the $\Delta$ and $\Lambda$ values were found [9] to be $\sim$0.12 and $\sim$9 nm at $\lambda \sim 4.9$ μm, while for metal-organic chemical-vapor-deposition (MOCVD)-grown QCLs the values were found to be: 0.10 and $\sim$12 nm at $\lambda \sim 5$ μm [9], and 0.11–0.12 and 10.2 nm at $\lambda \sim 8$ μm [11], respectively. However, for graded-interface QCLs the $\Delta$ and $\Lambda$ values have been recently found [46] to be higher and significantly lower, respectively, than $\Delta$ and $\Lambda$ values of abrupt-interface QCLs (see Table 2.3). These findings are consistent with experimentally APT-obtained values [49]. The IFR-parameter values are useful for designing QCL-performance enhancements via IFR-scattering engineering [9, 11, 46, 63, 64]. Since the graded-interface IFR formalism [47] with Gaussian autocorrelation has been successfully used both for matching experimental data (Fig. 2.5) as well as for obtaining good fits to APT experimental measurements [49], the Gaussian autocorrelation function is clearly a sound assumption for the $\Delta$ and $\Lambda$ values of InP-based mid-IR QCLs.

### Linewidth Broadening

The intersubband-absorption electroluminescence (EL) FWHM linewidth, $2\gamma$ in Eq. (2.1b), is dominated by intrasubband IFR scattering for wavelengths up to at least 11 μm [62]. In case the lasing transition is mostly vertical and occurs from the *ul* level to only one *ll* level, $2\gamma$ is well approximated, for abrupt-interface devices, by [7]:

$$2\gamma_{ul,ll} \approx \frac{\pi}{\hbar^2} \Delta^2 \Lambda^2 \sum_i m_{ci} \delta U_i^2 \left( \phi_{ul}^2(z_i) - \phi_{ll}^2(z_i) \right)^2, \qquad (2.13)$$

where $m_{ci}$ is the effective mass at the $i$th interface in the AR, $\delta U_i$ is the CB offset at the $i$th interface, $\varphi_{ul}(z_i)$ and $\varphi_{ll}(z_i)$ are the wavefunction amplitudes of the *ul* and *ll* levels at the $i$th interface, and $\Delta$ and $\Lambda$ are as defined above. If there is resonant extraction from the *ll* level, like in the case of STA- and TA-RE QCLs, the extractor state involved (e.g., state $3'$ when the *ll* level is state 3) [4, 11] has to be considered as a *ll* level as well, such that the overall EL linewidth is obtained by summing the Lorentzian-shaped

EL spectra of both transitions weighted by their respective oscillator strength (i.e., by their respective dipole-matrix-element squared).

However, a study of MBE-grown InGaAs/InAlAs 4.6 μm-emitting, diagonal-transition QCLs [8] has revealed that the device characteristics are significantly influenced by changes in growth temperature, in turn corresponding to $\Lambda$ varying over the 5.5–18.5 nm range. The optimal calculated $\Lambda$ value was found to be 10 nm, close to the optimal value deduced from experiment: 8.5 nm, which, in turn, is close to the value used to fit experimental data (9 nm) for ~8.5 μm-emitting MBE-grown QCLs [48, 58]. However, the experimental EL linewidth was found to be weak function of $\Lambda$, in disagreement with theory. That, in turn, was attributed to a limitation of the model used [65], which, however, does not prevent it from accurately predicting lifetimes. That is, for diagonal-transition devices the $2\gamma$ expression in Eq. (2.12) is not accurate.

A study of vertical-transition, 4.6 μm-emitting QCLs [43] has found that MBE-grown devices have lower RT EL linewidths than MOCVD-grown devices (26 meV vs. 33 meV), which was primarily associated with significantly lower $\alpha_w$ values (~0.5 cm$^{-1}$ vs. ~1.2 cm$^{-1}$) [3], most likely due to a lower $\alpha_{ISB}$ value [40, 43]. The results agree with the general tendency of MOCVD-grown devices of having somewhat higher $\Delta$ and $\Lambda$ values than MBE-grown devices, although it does not mean that MOCVD-growth conditions cannot be optimized to provide similar high-quality interfaces as those obtained by MBE. Finally, benefits in device performance were observed [66] for lattice-matched, 9 μm-emitting QCLs grown by GSMBE on (411)A-oriented InP substrates when compared to QCLs grown on conventional-orientation (i.e., (100)) substrates. The benefits were attributed to the relatively short terraces (1.8 nm) of (411)A-oriented InP substrates compared to extracted $\Lambda$ values (i.e., 9 nm) for 9 μm-emitting MBE-grown QCLs on (100)-oriented InP substrates. Thus, for lattice-matched structures changes in substrate orientation for achieving narrow linewidth are likely to bring about the dual benefit of lowering the $J_{th}$ value and increasing the slope efficiency, leading to higher wall-plug efficiency.

### Impact on the Lifetimes for Downward Transitions

The *ul-* and *ll*-level lifetimes include both inelastic (i.e., LO-phonon) and elastic scattering. Part of the elastic-scattering rate is IFR scattering to lower energy states [58, 59], treated in general in Chapter 1. The IFR scattering rate, between a state $m$ and a lower-energy state $n$, in QCL structures of varying barrier and well compositions, is given, for both abrupt-[4] and graded-[47] interfaces, by

$$\frac{1}{\tau_{mn}^{IFR}} = \frac{\pi}{\hbar^3} \Delta^2 \Lambda^2 F \sum_i m_{ci} \delta U_i^2 \varphi_m^2(z_i) \varphi_n^2(z_i) \exp\left(-\frac{\Lambda^2 m_{ci} E_{mn}}{2\hbar^2}\right). \quad (2.14)$$

The term $F$ is the scattering-rate reduction factor due to graded interfaces [47], given by: $\exp\left[\mathcal{L}^2/(16\ln(2)\Delta_\perp^2)\right]\left[erf\left[-\mathcal{L}/(4\sqrt{\ln(2)}\Delta_\perp)\right]+1\right]$. For example, for the QCL structures whose characteristics are shown in Fig. 2.5 (i.e., for $\mathcal{L} = 0.4$ nm and $\Delta_\perp = 0.15$ nm), $F$ has a value of 0.49 [46]. For abrupt interfaces, $F = 1$. For transitions from the *ul* level, $m$ is the *ul*-level state number and $n$ is the *ll*-level state number or the state number of any of the rest of low-energy AR and extractor states. For transitions from the *ll* level(s), *m* is the *ll*-level state number and *n* is the state number of any of the lower-energy AR and extractor states. For example, for 5.0 μm-emitting four-QW active-region, STA-RE QCL [4] $m = 4$ and $n = 3, 3', 2, 2', 2'', 1$ and $1'$ for transitions from the *ul* level, state 4; while $m = 3$ and $3'$ and $n = 2, 2', 1$ and $1'$ for transitions from the *ll* levels, states 3 and $3'$, where the primed states are extractor states (see Fig. 2.9(a)). $E_{mn}$ represents the energy difference between states. The same parameters as in Eq. (2.13) are used. Of course, the global *ul*-level and *ll*-level IFR scattering rates are the sum of the rates for all transitions to their respective *n* states, as derived by using Eq. (2.14). However, note that for the case of two *ll* levels (e.g., 3 and $3'$), due to the presence of IFR scattering, carrier extraction involves both coherent and incoherent tunneling, thus the carriers cannot be considered to be roughly equally shared between the states. Then, for calculating the *ll*-level IFR scattering rate only transitions from state 3 should be considered.

Due to the strong dependence on the transition energy, the IFR scattering rate has a strong impact only on low-energy transitions; that is, it controls the *ll*-level scattering rate for 4.0–5.0 μm-emitting QCLs [4, 9, 46, 51], while for 8–9 μm-emitting QCLs it is comparable to the LO-phonon scattering rate [11, 46, 55, 57]. Thus, the IFR scattering rate is a key factor as far as the depopulation of the *ll* level. As for the *ul*-level scattering rate, for InGaAs/InAlAs-based, abrupt-interface QCLs the IFR-scattering rate has a negligible impact both for 4.0–5.0 μm-emitting QCLs [4, 9, 51] and for MOCVD-grown ~8 μm-emitting QCLs [11, 55]. In turn, engineering the IFR (downward) scattering rate for improving the device performance (e.g., increasing the $\eta_{tr}$ value) involves mostly changing the *ll*-level scattering rate [4]. For instance, state-of-the-art high-power QCLs have higher CB offsets on the downstream side of the AR than on the upstream side of the AR [4, 9, 11, 30]. Then, since the IFR scattering rate is a strong function of CB offsets, the *ll*-level lifetime is significantly shortened and, in turn, very high $\eta_{tr}$ values can be achieved: 91–96% for 4.0–5.0 μm-emitting QCLs [4, 9, 51] and 85–87% for ~8.0 μm-emitting QCLs [11, 57]. However, for abrupt-interface, MBE-grown ~8 μm-emitting QCLs, as Λ decreases, the IFR and AD components of the *ul*-level lifetime become comparable [11, 60]. For graded-interface QCLs, as the Λ value further decreases to values of ~6 nm [46, 49], the IFR component overcomes the AD component [46]. Then, for ~8.0 μm-emitting QCLs IFR and LO-phonon scattering control the *ul*-level lifetime [46]. However, for 4.0–5.0 μm-emitting, graded-interface QCLs, since the lasing-transition energy increases, the IFR component becomes comparable to the AD component [46]; in turn, LO-phonon, AD and IFR scattering determine the *ul*-level lifetime (see Table 2.3).

### Carrier IFR Scattering to Upper Energy States

We described above the effect of IFR scattering on carrier relaxation (i.e., transitions from higher the lower energy states). However, as originally demonstrated by Flores *et al.* [17], IFR-scattering-assisted carrier excitation from low- to high-energy AR states does occur as well. That constitutes IFR-assisted carrier leakage, mentioned above in Section 2.2.1 and discussed below in Section 2.4.3, which, in the case of IFR-assisted scattering from the *ul* level to the next higher AR state, level *ul*+1, is characterized by the following scattering rate [9, 17]:

$$\frac{1}{\tau_{ul,ul+1}^{IFR}} = \frac{1}{\tau_{ul+1,ul}^{IFR}} I_{ul+1,ul} \exp\left(-\frac{E_{ul+1,ul}}{kT_{e,ul}}\right), \tag{2.15}$$

where $1/\tau_{ul+1,ul}^{IFR}$ has the same form as Eq. (2.14); $I_{ul+1,ul}$ is a weak function of electron temperature and $E_{ul+1,ul}$ values [17], but a strong function of $\Lambda$ [9, 17], while $E_{ul+1,ul}$ and $T_{e,ul}$ are as defined for Eq. (2.3a). This type of scattering exemplifies how the kinetic energy of hot electrons in the *ul* level and injector states is converted, via IFR scattering, to an increase in potential energy, eventually resulting in a shunt-type leakage current through high-energy AR states. In fact, IFR-scattering-assisted carrier leakage dominates carrier leakage in InGaAs/InAlAs QCLs [9, 46].

## 2.3.2    Alloy-Disorder Scattering

Just like IFR scattering, AD scattering is discussed in general in Chapter 1. Unlike IFR scattering, AD scattering is not a function of the energy-dependent scattering vector because of the short-range nature of the scatterers. If considering abrupt interfaces, AD scattering either strongly impacts *ul*-level lifetimes, since its rate dominates the elastic part of the *ul*-level scattering rate [4, 9, 11, 58], or it plays a minor role in determining the *ul*-level scattering rate for MBE-grown, diagonal-transition 8.0–9.0 μm-emitting QCLs [11, 60]. By contrast, when considering graded interfaces AD scattering has an impact only on the *ul*-level lifetimes of 4.0–5.0 μm-emitting QCLs [46] (see Table 2.3). The AD contribution to the intersubband-scattering rate between two selected states *m* and *n*, in a given well or barrier layer, is expressed as follows [9, 61]:

$$\frac{1}{\tau_{mn,w(b)}^{AD}} = \frac{1}{8} \frac{m_{c,w(b)} a_{w(b)}^3 \left(V_{w(b)}^{alloy}\right)^2 x(1-x)}{\hbar^3} \int_{alloy} \varphi_m^2(z)\varphi_n^2(z)dz, \tag{2.16a}$$

where $m_{c,w(b)}$ is the effective mass of the well/barrier; $a_{w(b)}$ is the lattice constant of the well/barrier; $V_{w(b)}^{alloy}$ is the difference between the CB minima of the well/barrier binary-alloy components; *x* is the alloy fraction in the $In_{1-x}Ga_xAs$ or $In_{1-x}Al_xAs$ alloys; and $\varphi_m(z)$ and $\varphi_n(z)$ are the wavefunction amplitudes of the states within a given alloy. In Chapter 1, by using a different derivation than in [65], the AD scattering rate is found to give values only 10% different than those derived using Eq. (2.16a). For state-of-the-art QCL structures the barriers and wells are of varying composition. Then, the overall AD scattering rate between two selected states, *m* and *n*, is given by

$$\frac{1}{\tau_{mn}^{AD}} = \frac{1}{8\hbar^3} \sum_i m_{ci} a_i^3 \left(V_i^{alloy}\right)^2 x(1-x) \int_{alloy\_i} \varphi_m^2(z)\varphi_n^2(z)dz, \tag{2.16b}$$

where *i* is the layer number in the sequence of well and barrier layers across which wavefunctions $\varphi_m(z)$ and $\varphi_n(z)$ extend. For example, just as for the IFR scattering rate in (2.14), for transitions from the $ul$ level, $m$ is the $ul$-level state number, and $n$ is the $ll$-level state number or the state number of any of the rest of low-energy AR and extractor states. Similarly, the global $ul$-level AD scattering rate is the sum of the rates for all $n$ values, as derived by using Eq. (2.16b). As expected, the AD scattering rate is found to hardly be affected by well/barrier interface grading [46, 47].

### 2.3.3  Enhancing the Laser-Transition Differential Efficiency

Considering only inelastic scattering, the $\eta_{tr}$ values are typically in the 81–86% for 4.5–5.0 μm-emitting QCLs and in the 71–78% range for 8–10 μm-emitting QCLs. As for elastic scattering, as seen in Chapter 1, when $\Lambda \geq 7$ nm and $\lambda = 4.8$ μm the IFR scattering rate for conventional, vertical-transition QCLs is ~15 times higher for the $ll$ level than for the $ul$ level. However, that translates to only an ~3% increase in the $\eta_{tr}$ value. Similarly, for conventional ~8 μm-emitting QCLs the effect of elastic scattering translates to only ~4.5% increase in the $\eta_{tr}$ value [11]. Therefore, in order to maximize the $\eta_{tr}$ value, the AR structure needs to be CB- and/or elastic-scattering-engineered in order to minimize the $\tau_{ll,g}/\tau_{ul,g}$ ratio.

Looking at the IFR scattering-rate expression (Eq. (2.14)), one can see that, for a given set of $\Delta$ and $\Lambda$ values, by decreasing the CB offset, $\delta U_i$, values on the upstream side of the AR, where most of the transitions from the $ul$ level occur, concomitant with increasing the $\delta U_i$ values on the downstream side of the AR (i.e., an STA-type device), where most of the transitions from the $ll$ level(s) occur, leads to $\tau_{ul,g}^{IFR} >> \tau_{ll,g}^{IFR}$. However, this is only one part of the $\tau_{ll,g}/\tau_{ul,g}$ minimization process. For example, for 4–5 μm-emitting and MOCVD-grown, 8 μm-emitting STA QCLs: (a) AD scattering strongly affects the elastic part of the $\tau_{ul,g}$ value; and (b) both IFR and AD scattering rates are proportional with the product of the moduli of the wavefunctions involved in the transitions, $\varphi_m(z)\varphi_n(z)$. The former leads to AD scattering dominating the elastic part of $\tau_{ul,g}$, since, unlike IFR scattering, AD scattering is hardly affected by lowering the $\delta U_i$ values. The latter implies that the $\varphi_m(z)\varphi_n(z)$ product should be minimized for transitions from the $ul$ level; that is, those transitions should be made diagonal. That helps with maximizing the overall $\tau_{ul,g}$ since the inelastic part of $\tau_{ul,g}$ also increases as the transitions become diagonal. In short, diagonal transitions both lengthen the inelastic $ul$-level lifetime and minimize the effect of AD scattering on the $ul$-level lifetime, thus leading to the longest possible $\tau_{ul,g}$ values. The same conclusion holds for MBE-grown, 8 μm-emitting STA QCLs [11, 57] since strong diagonal transitions minimize the effect of both AD and IFR scattering on the $ul$-level lifetime.

As for minimizing the $\tau_{ll,g}$ value, besides increasing the $\delta U_i$ values on the downstream side of the AR, the lower part of the AR should involve a multitude of transitions for $ll$-level depopulation; that is, it should become single-miniband-like just as in bound-to-continuum (BTC) QCLs [5].

Then, the optimal AR structure necessary for minimizing the $\tau_{ll}/\tau_{ul}$ ratio is as schematically depicted in Fig. 2.7: short barriers on the upstream side of the AR, coupled with a strong diagonal transition, for $\tau_{ul,g}$ maximization; and tall barriers on

**Table 2.1** Key parameters showing the effect of IFR and AD scattering on abrupt-interface, TA-RE QCL performance. Reproduced from [9], with the permission of AIP Publishing.

|  | LO-phonon | IFR | AD | Sum |
|---|---|---|---|---|
| $\tau_{ul,g}$ (ps) | 1.96 | 12.9 | 2.07 | 0.934 |
| $\tau_{ll,g}$ (ps) | 0.30 | 0.085 | 3.75 | 0.065 |
| $\tau_{ul-ll,g}$ (ps) | 2.99 | 16.9 | 3.21 | 1.42 |
| $\eta_{tr}$ | 85.6% | – | – | 93.2% |

**Table 2.2** Key parameters showing the effect of IFR and AD scattering on abrupt-interface, STA-RE QCL performance.

|  | LO-phonon | IFR | AD | Sum |
|---|---|---|---|---|
| $\tau_{ul,g}$ (ps) | 1.03 | 6.83 | 2.11 | 0.63 |
| $\tau_{ll,g}$ (ps) | 0.195 | 0.05 | 1.28 | 0.04 |
| $\tau_{ul-ll,g}$ (ps) | 2.62 | 10.5 | 4.75 | 1.454 |
| $\eta_{tr}$ | 82.9% | – | – | 94% |

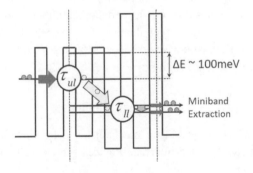

**Fig. 2.7** Schematic diagram of QCL design for maximizing both the differential laser-transition efficiency and the internal efficiency. The step-tapered active-region (STA) design allows for both large separation between the *ul* level and the energy level above it, leading to strong carrier-leakage suppression [6, 9], and minimizes the *ll*-level lifetime $\tau_{ll}$ [4] for devices with miniband-like carrier extraction. The diagonal lasing transition maximizes the *ul*-level lifetime $\tau_{ul}$.

the downstream side of the AR, coupled with miniband-like depopulation/extraction, for $\tau_{ll,g}$ minimization.

Tables 2.1 and 2.2 show the various lifetimes for abrupt-interface, GSMBE-grown 4.9 μm-emitting TA-RE devices [9, 30] and 5 μm-emitting STA-RE devices [4, 9], respectively, which confirm that the schematic AR diagram shown in Fig. 2.7 does lead to relatively high (9–13.3%) elastic-scattering enhancement of the $\eta_{tr}$ value. For graded-interface, 4.9 μm-emitting TA-RE QCLs the $\eta_{tr}$-value enhancement was found [46] to be similar: 9% (i.e., from 86% to 94%).

As clearly seen, for both cases the $\tau_{ul,g}$ value is determined by LO-phonon and AD scattering. The effect is particularly evident for the stronger diagonal-transition device (i.e., the TA-RE QCL with $z_{ul,ll} \sim 10$ Å and PICT action) in which case the LO-phonon and AD $\tau_{ul,g}$ values are basically the same. The same result (i.e., similar LO-phonon and AD $\tau_{ul,g}$ values) is obtained for other strong-diagonal-transition QCLs: $\sim$4.7 µm- and $\sim$ 4.1 µm-emitting STA-RE QCLs [51]. In contrast, the $\tau_{ll,g}$ value is dominated by IFR scattering.

Similar analyses have been performed on $\sim$8 µm-emitting, diagonal-transition conventional and STA-RE QCLs [11, 57]. Just as for diagonal-transition 4.0–5.0 µm-emitting and vertical-transition 8–9 µm-emitting QCLs [58], for MOCVD-grown devices the ul-level lifetime is determined by LO-phonon and AD scattering. Similarly, for strong STA-type, 8.2 µm-emitting devices [57] the ll-level lifetime is determined by IFR scattering, irrespective of the core-region interface quality. The net effect for those QCLs are large $\eta_{tr}$ enhancements: $\sim$21% (i.e., 72.2% to 87.3%) for MOCVD-grown core material, and $\sim$17% (i.e., 72.2% to 84.2%) for MBE-grown core material.

Other approaches have been suggested for significantly enhancing $\eta_{tr}$ by minimizing the $\tau_{ll,g}/\tau_{ul,g}$ ratio: (a) introducing a thin barrier in the rightmost QW of the AR of $\sim$7 µm-emitting devices [64]; and (b) employing strong diagonal-transition, MBE-grown devices of low $\tau_{ll,g}$ values due to ll-level depopulation via a multitude of transitions to a single miniband (i.e., BTC-type devices) [14]. For the latter, high $\eta_{tr}$ values have been projected for abrupt-interface devices: 91% at $\lambda = 4.4$ and 4.6 µm, as a result of an $\sim$0.1 value for the $\tau_{ll,g}/\tau_{ul,g}$ ratio.

### 2.3.4    Overall Effect on the Upper- and Lower-Laser Level Lifetimes

Table 2.3 summarizes which elastic scattering mechanisms determine the ul- and ll-level lifetimes, for abrupt- and graded-interface devices [46], over two different wavelength regions.

IFR scattering plays a role in determining the ul-level lifetimes for low-energy lasing transitions ($\lambda \geq 8$ µm). Thus, it should play a major role on the ul-level lifetimes of InAs/AlSb QCLs emitting in the 14–25 µm range (see Chapter 3). For graded-interface, high-energy lasing transitions (i.e., 4 µm $\leq \lambda \leq 5$ µm) both AD and IFR scattering play a role in determining the ul-level lifetimes [46]. In contrast, for graded-interface 8–9 µm-emitting QCLs the ul-level lifetimes are basically controlled

**Table 2.3** Carrier-scattering mechanisms determining lifetimes of abrupt- and graded-interface** (i.e., $\Lambda = 6$ nm) [46] InGaAs/InAlAs, diagonal-transition QCLs for different sets of IFR-parameters values.

| Wavelength | $\Delta$ | $\Lambda$ | ul-level lifetime | ll-level lifetime |
|---|---|---|---|---|
| 4.0–5.0 µm | 0.10–0.12 nm | 9–12 nm | LO + AD | IFR |
| 4.0–5.0 µm** | 0.14 nm | 6 nm | LO + AD + IFR | IFR |
| 8.0–9.0 µm | 0.10 nm | 9 nm | LO + AD + IFR | LO + IFR |
| 8.0–9.0 µm** | 0.13 nm | 6 nm | LO + IFR | LO + IFR |

by LO-phonon and IFR scattering [46]. As for *ll*-level lifetimes, the consideration of graded interfaces [46] does not change the types of carrier-scattering mechanisms which control their values.

Taking into account that carrier leakage is dominated by IFR-triggered carrier leakage [9, 46], which, in turn, is mostly suppressed in STA-RE-type QCLs [9, 11, 46, 57], IFR-scattering engineered, STA-RE devices with strong PICT action clearly appear to be the solution for reaching CW power and wall-plug efficiency values approaching upper limits (see Fig. 2.13).

## 2.4 Carrier Leakage

### 2.4.1 Background

Initial calculations of carrier leakage were performed for short-barrier, GaAs-based QCLs by considering LO-phonon-assisted excitation of hot electrons from injector to continuum states and electron-electron scattering between injector and continuum states [67, 68]. A further refinement was considering leakage of electrons scattered from $\Gamma$ continuum states to X-valley states [68]. Later, as shown in Chapter 6, the consideration of nonequilibrium phonons [69] resulted in much better agreement with experiment at heatsink temperatures below 200 K. In parallel, it was shown [70] for medium-tall-barrier, GaAs-based QCLs that carrier leakage was dominated by LO-phonon scattering from the *ul* level to continuum states. However, for tall-barrier QCLs, such as InP-based ones, LO-phonon-assisted carrier leakage was found to primarily occur from the *ul* level through AR energy states above it [10], followed by relaxation to low-energy states (see Fig. 2.1(b)), that is, the leakage current is mainly an intra-AR shunt current.

In 2013 Flores et al. [17] showed that at 80 K, in tall-barrier QCLs, IFR-assisted hot-carrier leakage occurs from injector states through high-energy AR states. Subsequently, it was shown [9, 11] for tall-barrier devices that carrier leakage through high-energy AR states is triggered by both LO-phonon and IFR scattering from both the *ul* level and injector states, and that it allows bridging the gap between theoretical and experimental $\eta_i$ values. Furthermore, that clarified that the relatively small leakage currents to the continuum found [71] at high drive levels and heatsink temperatures via a sophisticated analysis of V–I curves were primarily a consequence of IFR-assisted carrier excitation from the injector ground state to high-energy AR states.

### 2.4.2 LO-Phonon-Triggered Carrier Leakage

By considering both inelastic and elastic scattering, the equation for the LO-phonon-triggered, leakage-current density from a state $m$ through a higher-energy state $n$ [9, 16], where $m$ stands for the *ul* level or an injector-region state, and when $T_{em}/T_{\ell} \leq 1.2$ [16], is given by [9]:

$$J_{\text{leak},mn}^{\text{LO}} = \frac{en_m}{\tau_{nm}^{\text{LO}}} \frac{\tau_{n,\text{tot}}^{\text{LO,IFR,AD}}}{\tau_{n,\text{leak}}^{\text{LO,IFR,AD}}} \exp\left(-\frac{E_{nm}}{kT_{em}}\right), \qquad (2.17)$$

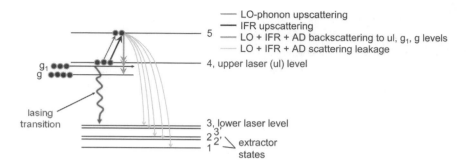

**Fig. 2.8** Schematic representation of carriers excited from the *ul* level, state 4, to the *ul* + 1 level, state 5; backscattering from state 5 to the *ul*, $g_1$ and *g* levels; and leakage from state 5 to low-energy AR and extractor states. Here, level $g_1$, the first excited injector state, is the injecting state into the *ul* level [9]. Reproduced from [9], with the permission of AIP Publishing.

where $n_m$ is the electron sheet density in state *m*, $\tau_{nm}^{LO}$ is the lifetime for LO (downward) scattering; $\tau_{n,tot}^{LO,IFR,AD}$ is the lifetime corresponding to the net electron scattering from state *n* to all energy states below it by LO-phonon, IFR and AD scattering [9]; $\tau_{n,leak}^{LO,IFR,AD}$ is the lifetime corresponding to LO-phonon, IFR and AD scattering from state *n* to all low-energy AR and extractor states [9]; $E_{nm}$ is the energy difference between states *n* and *m*; and $T_{em}$ is the electron temperature in state *m*.

Since the LO- and IFR-triggered leakage currents are interconnected [9], the sequence of events is as follows: (a) electrons from the *ul* level and key injector states (e.g., the injector ground and first excited states, *g* and $g_1$) are excited to the *ul*+1 level via LO-phonon and IFR scattering; (b) because of short $\tau_{nm}$ lifetimes (i.e., involving LO-phonon, IFR and AD scattering) a large part of the electrons return to the *ul* level and injector states; (c) the rest of electrons in the *ul*+1 level relax via LO-phonon, IFR and AD scattering to low-energy AR states and extractor states; that is, an intra-AR shunt leakage current is established. As an example, Fig. 2.8 shows schematically the scattering mechanisms involved when considering electron excitation, in a four-QW AR, only from the *ul* level (state 4). Electrons in the state 4 are excited to state 5 via LO-phonon and IFR scattering. Then, because of a relatively short (0.1–0.2 ps) backscattering lifetime, a large part of the electrons (80–90%) returns to states 4, $g_1$ and *g*. The rest of state-5 electrons (10–20%) relax to low-energy AR states (states 3, 2, and 1) and low-energy extractor states (states 3' and 2'), and it is expressed by the $\tau_{n,tot}^{LO,IFR,AD}/\tau_{n,leak}^{LO,IFR,AD}$ ratio in Eq. (2.17).

For carriers excited from state 4 or from the injector states *g* and $g_1$ (in Fig. 2.8 the main injecting state is state $g_1$) the lifetime characterizing the net scattering from state 5, to all energy states below it, is:

$$\tau_{5,tot}^{LO,IFR,AD} = \left( \sum_{i=ul,g,g1,3,3',2,2',1} 1/\tau_{5i}^{LO,IFR,AD} \right)^{-1}$$

where the scattering time from state 5 to a given state *i* is given by: $\tau_{5i}^{LO,IFR,AD} = (1/\tau_{5i}^{LO} + 1/\tau_{5i}^{IFR} + 1/\tau_{5i}^{AD})^{-1}$. The lifetime corresponding to leakage from state 5 to the low-energy states is given by: $\tau_{5,leak}^{LO,IFR,AD} = (\sum_{i=3,3',2,2',1} 1/\tau_{5i}^{LO,IFR,AD})^{-1}$ with the same definition for the scattering time from state 5 to a state $i$. Then, the total LO-triggered leakage-current density is given by: $J_{leak,tot,5}^{LO} = \sum_{i=4,g,gl} J_{leak,i5}^{LO}$. For example, the values of $J_{leak,tot,5}^{LO}/J_{th}$ and its components for 5 μm-emitting STA-RE QCLs [4] are shown in the central bar of Fig. 2.10.

## 2.4.3 IFR-Triggered Carrier Leakage

The IFR-triggered, leakage-current density from a state $m$ through a higher-energy state $n$ is [9]

$$J_{leak.mn}^{IFR} = \frac{en_m}{\tau_{nm}^{IFR}} \frac{\tau_{n,tot}^{LO,IFR,AD}}{\tau_{n,leak}^{LO,IFR,AD}} I_{nm} \left(\frac{E_{nm}}{kT_{em}}\right) \exp\left(-\frac{E_{nm}}{kT_{em}}\right), \qquad (2.18a)$$

where $\tau_{nm}^{IFR}$ is the lifetime for IFR (downward) scattering (Eq.(2.14)), and the $I_{nm}$ term is given by

$$I_{nm}(x_0) = \int_0^\infty \exp\left\{\alpha_{nm}\sqrt{(x/x_0)^2 + (x/x_0)} - x[1 + (\alpha_{nm}/x_0)]\right\} dx, \quad x_0 = E_{nm}/kT_{em},$$

$$(2.18b)$$

where $\alpha_{nm} = \bar{m}_c \Lambda^2 E_{nm}/\hbar^2$ with $\bar{m}_c$ being the effective mass averaged by the relative intensity of the $m$-state wavefunction over all wells where it dwells, and $\Lambda$ is the IFR in-plane correlation length. Unlike in [17], it is necessary to average the effective mass over all relevant wells since in TA/STA-RE devices the wells in and around the AR have different compositions [3]. The rest of the terms in Eq. (2.18a) are defined as for Eq. (2.17). Then, for the diagram shown in Fig. 2.8 the total IFR-triggered carrier leakage is given by $J_{leak,tot,5}^{IFR} = \sum_{i=4,g,gl} J_{leak,i5}^{IFR}$. The sum of LO- and IFR-triggered carrier leakages is the net carrier leakage $J_{leak,tot,5}^{net}$. For example, the $J_{leak,tot,5}^{net}/J_{th}$ value for 5 μm-emitting STA-RE QCLs [4] is the leftmost bar in Fig. 2.10.

## 2.4.4 Carrier-Leakage Estimates

### State-of-the-Art QCLs Emitting in the 4.5–5.0 μm Range

Two types of high-performance QCLs: the STA-RE [4] and TA-RE [6, 30] devices, were studied employing the comprehensive carrier-leakage formalism developed in [9]. The band diagram and relevant wavefunctions for the STA-RE QCL are shown in Fig. 2.9(a). The barrier heights in the AR increase stepwise: $x = 0.56$, 0.63, and 0.93 in $Al_xIn_{1-x}As$, and the wells depths increase stepwise: $x = 0.57$, 0.60, 0.70, and 0.70 in $In_xGa_{1-x}As$. Due to asymmetry and Stark-shift reduction [6] the energy difference between the upper level and the next higher level, $E_{54}$, increases to 98 meV from ~45 meV in conventional 4.5–5.0 μm-emitting QCLs [6]. A look at the top part of the AR (Fig. 2.9(b)) shows the states that are involved in triggering carrier leakage through the $ul+1$ level, and arrows indicating the main IFR-leakage paths. The device

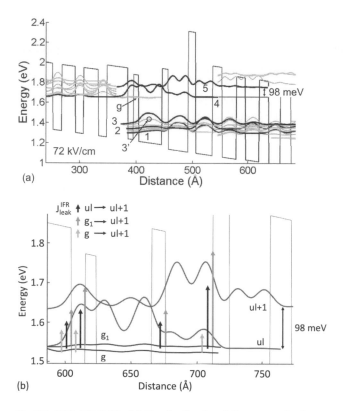

(a)

(b)

**Fig. 2.9** STA-RE QCL. (a) Conduction-band diagram and relevant wavefunctions. Reprinted with permission from [4] © The Optical Society. (b) Upper active-region band diagram, key-state wavefunctions, and arrows at interfaces indicating the main components of the IFR-triggered leakage $J_{\text{leak}}^{\text{IFR}}$ [9]. Reproduced from [9], with the permission of AIP Publishing.

has conventional injection; that is, from the injector ground state, level $g$, while level $g_1$ is a parasitic state in the AR, in that it plays a role only as far as carrier leakage.

The LO- and IFR-triggered carrier-leakage currents from the $ul$ level and the $g$ and $g_1$ levels are calculated using Eqs. (2.17) and (2.18a), and the resulting values, normalized to the $J_{\text{th}}$ value, are shown in Fig. 2.10. The electron temperature in the injector miniband $T_{\text{e,inj}}$ is calculated by using Eq. (2.3b), the fact that $\alpha_{\text{e-l}} \propto F/n_s$ [72], where $F$ is the field strength, and the measured [19] $\alpha_{\text{e-l}}$ value (i.e., 34.8 K cm$^2$/kA) for 4.8 μm-emitting QCLs with $n_s = 0.9 \times 10^{11}$ cm$^{-2}$ and threshold field $F_{\text{th}} = 78$ kV/cm. Then, for a given set of $J_{\text{th}}$, $n_s$, and $F_{\text{th}}$ values, one can extrapolate to find the $T_{\text{e,inj}}$ value at threshold: $T_{\text{e,th}}$. For the STA QCL, given $J_{\text{th}}$, $n_s$, and $F_{\text{th}}$ values [4], $T_{\text{e,th}}$ is 329 K. Then, one can use Boltzmann statistics to estimate the sheet-carrier densities in the injector states involved in leakage. The sheet-carrier density in the $ul$ level, $n_{\text{ul}}$, is obtained from the following expression:

$$J_{\text{th}} = \frac{en_{\text{ul}}}{\eta_{\text{inj}} \tau_{\text{ul,g}}},$$  (2.19)

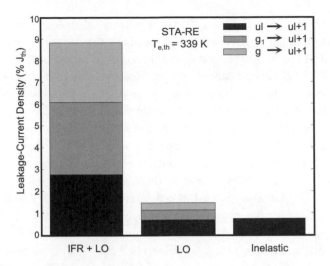

**Fig. 2.10** Bar graph of components of the relative leakage-current density for ~5.0 μm-emitting STA-RE QCL. LO and Inelastic stand for LO-phonon-triggered leakage in the presence and the absence of elastic scattering, respectively. Reproduced from [9], with the permission of AIP Publishing.

where $\tau_{ul,g}$ is the global *ul* lifetime [6] and $\eta_{inj}$ is total injection efficiency, as defined for Eq. (2.1a). This is obvious from the first rate equation in (2.1e) when considering being in steady state, the photon flux $S = 0$, and at threshold. That is, carrier leakage plays a significant role as far as the $J_{th}$ value, since it amounts at RT to values as high as 28% of $J_{th}$ [11] (i.e., $\eta_{inj} \sim 71\%$) for conventional QCLs.

As seen from Fig. 2.10, the total relative leakage is 8.9% of which ~37% is from the parasitic state $g_1$. For comparison, also shown are the LO-triggered leakage and the 'classical' case of leakage [10] (i.e., LO-phonon leakage only from the *ul* level). The first comparison reveals that most of the leakage is IFR triggered (i.e., ~83%). Since IFR leakage is dominant it is worth seeing where it happens within the AR. As shown in Fig. 2.9(b), scatterings between *ul* and *ul* + 1, and between the $g_1$ and *ul* + 1 levels are strongest at the upstream interface of the third AR barrier. That happens, as per Eqs. (2.14) and (2.15), due to both high overlap of the *ul* and $g_1$ wavefunctions with the *ul* + 1 wavefunction at that interface, and the fact that interface corresponds to the highest CB offset in the AR [4]. The second comparison reveals that the total leakage is 12 times higher than the conventionally calculated leakage (i.e., 0.74%). That is, carrier leakage had been significantly underestimated [10]. Furthermore, the ~13% 'gap' found [9], when using conventional leakage, between theoretical and experimental $\eta_i$ values is basically bridged (i.e., recalculated $\eta_i$ values, with error analysis, are 78–80.5% vs. the 77% experimental value). Similar gap bridging was found for the TA-RE QCL [9, 46]. Thus, all $\eta_i$ components have been identified.

### Conventional and STA-RE QCLs Emitting at ~8 μm

The same carrier-leakage formalism was applied to MOCVD-grown, ~8.0 μm-emitting conventional and STA-type QCLs [11, 57]. In contrast to the approach used for 4.5–5.0 μm emitting QCLs [9] leakage from all injector states penetrating the AR and through two AR energy states above the $ul$ level was considered. Figures 2.11(a) and (b) show the band diagram and relative leakages for the conventional QCL [36]. That lattice-matched structure has injection into the $ul$ level from the injector ground state $g$, while states $g_1$ through $g_4$ are parasitic injector states penetrating into the AR, in that they play a significant role in the overall carrier leakage. Also shown are the extractor states 3' and 2' which are involved in resonant-tunneling extraction. The energy difference $E_{54}$ is 57.5 meV a typical value for conventional 7–9 μm-emitting QCLs [3]. For determining the $T_{e,inj}$ value, 30 K cm$^2$/kA is taken for $\alpha_{e-1}$ at threshold, when $F_{th} = 48$ kV/cm and $n_s = 0.8 \times 10^{11}$ cm$^{-2}$, as derived from a NEGF analysis [48] of ~8.5 μm-emitting QCLs that agrees well with the experiment. For the measured $J_{th}$ and $n_s$ values [73] and the extracted $F_{th}$ value [11], $T_{e,th} = 319$ K. Then, by using the measured $J_{th}$, $2\gamma$ (i.e., 16.5 meV) [73] and the total waveguide loss, $\alpha_{tot}$, value (14.3 cm$^{-1}$) [36], which is the sum of $\alpha_w$ and $\alpha_{bf}$, $\Delta$ and $\Lambda$ parameter values of 0.11 nm and 10 nm, respectively, were extracted via iterations involving a parameter-sweep strategy similar to that performed in [4]. The LO- and IFR-triggered leakage-current densities from the $ul$ level and the $g$, $g_1$, $g_2$, $g_3$, and $g_4$ injector states through states 5 and 6 (i.e., through the $ul$ + 1 and $ul$ + 2 levels) were calculated just as in [9]. The results, normalized to $J_{th}$, are shown in Fig. 2.11(b).

The total relative leakage is 27.7% of which only 1.1% is through the $ul$ + 2 level. Of the current leaking through the $ul$ + 1 level, only 42% is due to carriers excited from the $ul$ level and from the injecting state, state $g$, with the rest being due to excitation from the parasitic states $g_1$ through $g_4$. This shows that, as part of device optimization, it is critical to minimize injector-state penetration into the AR. The total leakage current is ~6 times larger than when considering only inelastic scattering from the $ul$ level [10], similar to what was found for 4.5–5.0 μm emitting QCLs [9]. By using Eqs. (2.1a) and (2.1c), the $\eta_{inj}$ value is 71%. Then, considering a calculated $\eta_{tr}$ value of 77.5% the $\eta_i$ value is ~55% typical of conventional mid-IR QCLs with significant carrier leakage (i.e., ~50% at λ = 4.6 μm [13], ~57% at λ = 8.4 μm [7, 32] and ~58% at λ = 9.1 μm [32, 56]).

The STA QCL has injection into the $ul$ level from the first-excited state in the injector, state $g_1$, thus being of the pocket-injector type [45]; that is, excited-state injection of carriers residing in both states $g$ and $g_1$ [14, 45]. The AR consists of barrier heights increasing stepwise [11] from $x = 0.50$ and 0.51 to 0.75 in Al$_x$In$_{1-x}$As, and of well depths increasing stepwise [11] from $x = 0.55$ and 0.55 to 0.60 and 0.60 in In$_x$Ga$_{1-x}$As (Fig. 2.12(a)). This results in increasing the $E_{54}$ value to 80 meV, which, in turn, leads to substantial carrier-leakage suppression. For the measured $J_{th}$ and $n_s$ values and an extracted $F_{th}$ value [11], $T_{e,th} = 329.5$ K. Then, by using the measured $J_{th}$, $2\gamma$ (i.e., 24.7 meV), and $\eta_{sl}$ values [55], $\Delta$ and $\Lambda$ parameter values of 0.12 nm and 10 nm, respectively, were extracted iteratively [11]. As seen from Fig. 2.12(b), the total relative leakage is 12.4%; that is, 45% of the value for the conventional QCL. This is primarily due to the significantly higher $E_{54}$ value for the STA vs. the conventional QCL. The leakage is dominated by IFR leakage. Of

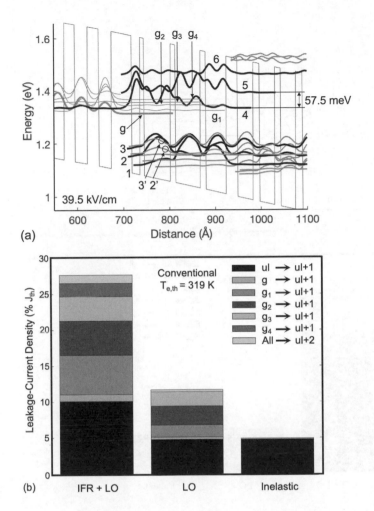

**Fig. 2.11** ~8 μm-emitting conventional QCL [36]: (a) conduction-band diagram and relevant wavefunctions; and (b) bar graph of the components of the relative leakage-current density. LO and Inelastic stand for LO-phonon-triggered leakage in the presence of elastic scattering, and in the absence of elastic scattering (from only the *ul* level) [10], respectively. Reproduced with permission from [11].

the current leaking through the *ul* + 1 level 73% is due to carriers excited from the *ul* and *g* levels, and the main injecting state, state $g_1$, thus the relative leakage from the parasitic states is significantly lower than for the conventional QCL (i.e., 27% vs. 58%). By using Eqs. (2.1a) and (2.1c), the $\eta_{inj}$ value is found to be 86.7%, which represents the highest injection efficiency value estimated to date for QCLs. Then, considering a calculated $\eta_{tr}$ value of 85%, the $\eta_i$ value is ~74% [11], that is, ~35% higher than for the conventional QCL. Subsequently, an STA-type design further optimized for carrier-leakage suppression (i.e., $\eta_{inj}$ ~ 91%) [57], and considering the IFR parameters extracted from the conventional MOCVD-grown QCL [11], resulted in an even higher $\eta_i$ value: 79.3% [57].

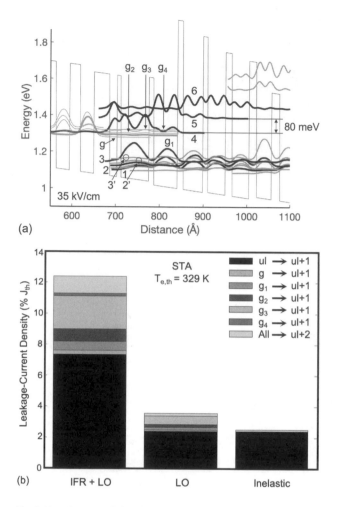

**Fig. 2.12** ~8 μm-emitting STA-type QCL [11, 55]: (a) conduction-band diagram and relevant wavefunctions; (b) bar graph of the components of the relative leakage-current density. LO and Inelastic stand for LO-phonon-triggered leakage in the presence and in the absence of elastic scattering (from only the *ul* level) [10], respectively. Reproduced with permission from [11].

## 2.4.5     Conclusions

For two types of 4.5–5.0 μm-emitting, state-of-the-art QCLs, incorporating elastic scattering into the carrier-leakage process, and considering carrier leakage from both the *ul* level and injector states has resulted in calculated $\eta_i$ values that agree very well with experimental $\eta_i$ values obtained from variable mirror-loss studies. Thus, besides bridging the gap between experimental and theoretical $\eta_i$ values, a tool was obtained for further increasing $\eta_i$ via IFR-scattering engineering (i.e., adjusting by design, within the AR, the CB offsets and the products of probability functions at interfaces in order to maximize the device performance) [9]. Furthermore, the study demonstrated that, given IFR parameters extracted from experimental device characteristics, the consideration of both elastic and inelastic scattering (for calculating the degree of carrier leakage and the laser-transition efficiency) results in accurate values for $\eta_i$. That, in turn, has allowed calculating the $\eta_i$ values for $\sim 8.0\,\mu m$-emitting conventional and STA-type QCLs [11] as well as finding which scattering mechanisms control the *ul*- and *ll*-level lifetimes.

While the DM and NEGF models agree fairly well with experiment [48, 50], since they also include the impact of elastic scattering on carrier leakage, the rate-equations model offers analytical formulas for LO- and IFR-triggered carrier leakage [9] that are useful design tools for minimizing carrier leakage and, in turn, for maximizing the wall-plug efficiency. Employing graded-interfaces modeling [46], due both to an inherently reduced IFR scattering rate [47] (see Eq. (2.14)) as well as to lower $\Lambda$ parameter values, significantly reduces the IFR-triggered upward scattering rate, but the much higher NEGF-calculated electron temperatures [46] (i.e., for both the main injecting state and the *ul* level, $T_{e,th} \sim 490\,K$ for 8.3 $\mu m$-emitting QCLs [33] and $T_{e,th} \sim 660\,K$ for 4.9 $\mu m$-emitting QCLs [30]) lead to relative carrier-leakage, $J_{leak}/J_{th}$, values (i.e., 23–24%) comparable to those calculated using the abrupt-interface, rate-equations model [9, 57]. Note that for such high $T_{e,th}$ values, the exponential in the equation for the LO-phonon-triggered leakage-current density (i.e., Eq. (2.17)) needs to be replaced by the exponential in the LO-phonon part of Eq. (2.3a).

## 2.5    Electro-Optical Characteristics under CW Operating Conditions

### 2.5.1    Threshold-Current Density

The core region heats with respect to the heatsink temperature $T_h$ by a quantity $\Delta T_{act}$ defined as:

$$\Delta T_{act} = T_\ell - T_h = R_{th}(P_{el} - P_{opt}) = R_{th}P_{el}(1 - \eta_{wp,cw}), \qquad (2.20)$$

where $T_\ell$ is the lattice temperature, $R_{th}$ is the thermal resistance, $P_{el}$ is CW input electrical power (i.e., $P_{el} = A\,J\,V$, where A is the pumped area), $P_{opt}$ is CW output optical power, and $\eta_{wp,cw}$ is the wall-plug efficiency in CW operation (see Eq.(2.7)). Then, the CW $J_{th}$ value is given by [3]:

$$J_{th,cw}(T_\ell) = J_{th}(T_h)\exp\left(\frac{\Delta T_{act,th}}{T_0}\right); \quad \Delta T_{act,th} = \frac{J_{th,cw}V_{th}}{G_{th}}, \qquad (2.21)$$

where $\Delta T_{act,th}$ is the threshold core-region temperature rise, $V_{th}$ is the threshold voltage and $G_{th}$ is the area-specific thermal conductance (i.e., $G_{th} = 1/(R_{th}A)$) [5]. In order to minimize the $J_{th}$ increase from low-duty-cycle pulsed to CW operation, one has to maximize the $T_0$ and/or $G_{th}$ values and/or minimize the threshold electrical-power density (i.e., $J_{th,cw}V_{th}$). The same requirements apply for achieving high values for the maximum CW operating temperature [5], as necessary for low-power devices needing a large tuning range [26, 74] such as DFB lasers [26]. As pointed out above, the $T_0$ value can be maximized by suppressing carrier leakage and/or minimizing backfilling. For a low $J_{th,cw}V_{th}$ value, the $J_{th}$ value can be minimized by using low doping in the injector, waveguide, and substrate [26] in order to reduce both the empty-cavity and the ISB components of $\alpha_w$.

As for the $G_{th}$ value it can be maximized via thermal engineering. From published data on BH 4.6–4.8 µm-emitting conventional, 40-stage QCLs mounted episide-down on diamond it has been found [3] that, as the buried-ridge width $w$ varies from 8.6 µm to 11.6 µm, $G_{th}$ is approximately inversely proportional with $\sqrt{w}$, in good agreement with theory [5, 39]. This is expected, since a significant portion of heat removal occurs laterally, away from the buried-core region. For example, for episide-down mounted BH devices of $w = 5$ µm a high $G_{th}$ value of $1840\,WK^{-1}cm^{-2}$ has been extracted [5]. Nevertheless, the highest CW operating temperature (i.e., 150 °C) has been achieved [26] from 12 µm-wide BH low-doped devices mounted episide down on AlN submounts.

For episide-up BH devices the $G_{th}$ values are, as expected, higher than for episide-down mounted devices, but they increase faster with decreasing $w$ value (i.e., $G_{th} \propto 1/w$) [5]. That has allowed Kato et al. [74] to reach the highest CW operating temperature for episide-up mounted BH devices (i.e., 105 °C) while using 5 µm-wide buried ridges which provided a $G_{th}$ value of $920\,WK^{-1}cm^2$ in good agreement with theory [5]. This is relevant for low power-dissipation devices since episide-up mounting provides higher device manufacturing yields than episide-down mounting.

## 2.5.2     Maximum Wall-Plug Efficiency

The maximum, single-facet CW wall-plug efficiency, at a heatsink temperature $T_h$ of a value close to 300 K, is expressed as such [3, 6]:

$$\eta_{wp,max,CW} \approx \eta_s\eta_d\,(T_h)\exp\left(-\frac{\Delta T_{act,wpm}}{T_1}\right)\left[1 - \frac{J_{th}(T_h)}{J_{wpm}(T_h)}\exp\left(\frac{\Delta T_{act,wpm}}{T_0}\right)\right]\frac{h\upsilon}{eV_{wpm}},$$
(2.22a)

where $\eta_d(T_h)$ is the (pulsed) external differential efficiency with an optimal mirror loss $\alpha_{m,opt}$, and $\Delta T_{act,wpm}$ is the core temperature rise at the $\eta_{wp,max}$ point:

$$\Delta T_{act,wpm} = T_{l,wpm} - T_h = R_{th}P_{el,wpm}\left(1 - \eta_{wp,max}\right) = \frac{J_{wpm}V_{wpm}}{G_{th}}\left(1 - \eta_{wp,max}\right),$$
(2.22b)

where $T_{\ell,wpm}$ is the lattice temperature at the CW $\eta_{wp,max}$ point, and $P_{el,wpm}$ is the electrical CW power dissipated at the $\eta_{wp,max}$ point (i.e., $P_{el,wpm} = A\,J_{wp,max}V_{wpm}$, where A is the pumped area). As evident from Eqs. (2.22a) and (2.22b), both the CW $\eta_{wp,max}$ and $\Delta T_{act,wpm}$ values are strong functions of the $T_1$ and $G_{th}$ values, and to a lesser extent functions of the $T_0$ value. Therefore, in order to maximize the CW $\eta_{wp,max}$ value and minimize the CW $\Delta T_{act,wpm}$ carrier leakage has to be suppressed (to obtain high $T_0$ and $T_1$ values), the operating voltage needs to be minimized (e.g., by using devices with strong PICT action) and $G_{th}$ needs to be maximized. Eq. (2.22a) can be simplified by using the slope-efficiency expression, $\eta_{sl} = \eta_d(h\upsilon/e)$:

$$\eta_{wp,max,CW} \approx \eta_s\frac{\eta_{sl}(T_h)}{V_{wpm}}\exp\left(-\frac{\Delta T_{act,wpm}}{T_1}\right)\left[1 - \frac{J_{th}(T_h)}{J_{wpm}(T_h)}\exp\left(\frac{\Delta T_{act,wpm}}{T_0}\right)\right].$$
(2.23)

Examples of devices with high single-facet, CW $\eta_{wp,max}$ values at RT are: the TA-RE 4.9 μm-emitting QCL [30] (i.e., 21%) and the nonresonant-extraction (NRE) 4.6 μm-emitting QCL [43] (i.e., 12.8%), having 40 stages and the same mirror- and waveguide-loss values. For the TA-RE device the $T_0$ and $T_1$ values were high (i.e., 244 K and 343 K, respectively) while the $R_{th}$ value (~3.4 K/W, as extracted from CW-curve fitting [3]) was about twice higher than that extracted [3] from conventional 5 mm-long devices of similar buried-ridge width and diamond submount (i.e., ~1.6 K/W). That was most likely due to the insertion of seven tall barriers (AlAs) per stage, in the injector region. More specifically, the $G_{th}$ value was about half that for conventional, 8 μm-wide-ridge BH devices (i.e., ~735 W/cm$^2$ K vs. 1450 W/cm$^2$ K [3]). The virtual halving of the $G_{th}$ value is mostly due to significantly more interfaces per period [19] (i.e., 34 vs. 22 interfaces) with the rest likely due to the multitude of highly strained layers [75]. The relatively high $R_{th}$ value gave $\Delta T_{act,wpm} \sim 54$ K, a rather high value for state-of-the-art BH devices. Using Eq. (2.23) with $T_h = 298$ K and the experimental (pulsed) slope-efficiency value, the calculated CW $\eta_{wp,max}$ value is 20.7%, i.e., quite close to the experimental value of 21%; proving that the equation is a good approximation. For NRE-type, 3 mm-long HR-coated QCLs the $T_0$ and $T_1$ values were smaller (i.e., 168 K and 295 K) [3] due to carrier leakage but the $R_{th}$ value (~2.4 K/W) was lower, as expected for 9.5 μm × 3 mm buried ridges of conventional 4.6 μm-emitting QCL material [3]. Then, given a smaller $J_{wpm}$ value (i.e., 2.3 vs. 3.7 kA/cm$^2$), the calculated $\Delta T_{act,wpm}$ value is significantly smaller (20.3 K) than that for the TA-RE QCL. Using Eq. (2.23) the calculated $\eta_{wp,max}$ value is 12.9%, again quite close to the experimentally obtained value of 12.8%, thus further proving the accuracy of the approximation.

As far as fundamental limits in RT CW $\eta_{wp,max}$ values for 4.5–5.0 μm-emitting QCLs one needs to consider devices having strong PICT action, high $T_1$ values (e.g., 750 K as obtained from deep-well, TA-type devices [42]), low $R_{th}$ values (i.e., the 1.6 K/W value mentioned above), 95% $\eta_{tr}$ value (see Fig. 2.6), negligible carrier leakage and unity tunneling-injection efficiency (i.e., $\eta_i \cong \eta_{tr}$). For example, as shown in Fig. 2.13(b), using as reference the $\eta_{wp}$ vs. current CW curve of the TA-RE 4.9 μm-emitting QCL [3], a 38% $\eta_{wp,max}$ value is projected, that is almost twice that reported in [30]. Lowering the emission wavelength from 4.9 μm to 4.6 μm raises the $\eta_{wp,max}$ value by ~7% [3], which means that at $\lambda = 4.6$ μm the CW $\eta_{wp,max}$ fundamental limit becomes ~41%. QCLs capable of attaining $\eta_{wp,max}$ values close to 40% would drastically decrease the dissipated heat, thus benefitting a wide range of applications, like defense-related ones, since thermal-load management drives the packaged laser system's size, weight and overall power consumption.

For ~8 μm-emitting QCLs similar conditions apply for obtaining CW $\eta_{wp,max}$ fundamental limits, except that the optimal stage number is 45 rather than 40, in order to minimize the empty-cavity loss [4, 33] and the projected pulsed $\eta_{tr}$ upper-limit value is 85% rather than 95% (Fig. 2.6). Then, as clear from Fig. 2.6, for $\lambda = 8$ μm the pulsed, both-facets $\eta_{wp,max}$ fundamental limit is 25%. Considering long-cavity, optimized facet-coated devices [33], the single-facet $\eta_{wp,max}$ upper limit decreases to

23%; that is, 40% higher than the best value reported to date [33] adjusted for 298 K heatsink temperature [3] (i.e., 16.5%). In turn, considering that for fundamental-limit devices the CW $\eta_{\text{wp,max}}$ value is only $\sim$5% lower than the pulsed one (Fig. 2.13(b)) and that $G_{\text{th}}$ slightly decreases ($\sim$5%) when $N_p$ increases from 40 to 45 [31], CW values of $\sim$21% are projected at 298 K, that is, almost twice the highest single-facet CW $\eta_{\text{wp,max}}$ value reported to date [33].

### 2.5.3    Output Optical Power

The CW optical power can be written as such:

$$P_{\text{opt}} = A \frac{h\nu}{e} \eta_{\text{d,CW}}\,(T_\ell)\,[J - J_{\text{th.CW}}\,(T_\ell)]$$

$$= A \frac{h\nu}{e} \eta_{\text{d}}\,(T_{\text{h}}) \exp\left(-\frac{\Delta T_{\text{act}}}{T_1}\right)\left[J - J_{\text{th}}\,(T_{\text{h}}) \exp\left(\frac{\Delta T_{\text{act}}}{T_0}\right)\right], \qquad (2.24\text{a})$$

taking $\eta_s(J)$, the droop in the pulsed L–I curve [3, 10] only at $T_{\text{h}}$, and with $\Delta T_{\text{act}}$ being given by:

$$\Delta T_{\text{act}} = T_\ell - T_{\text{h}} = R_{\text{th}}P_{\text{opt}}\left[(1/\eta_{\text{wp,cw}}) - 1\right], \qquad (2.24\text{b})$$

where $\eta_{\text{wp,cw}}$ is the wall-plug efficiency in CW operation as defined in Eq. (2.7).

Then, the keys for high-CW-power operation are high values for $T_1$ and low values for $R_{\text{th}}$, just as for interband-transition devices [38]. The CW L–I curve from [30] was matched, by using $T_0 = 244\,\text{K}$, $T_1 = 343\,\text{K}$, $\lambda = 4.9\,\mu\text{m}$ and curves fitting the experimental pulsed L–I and V–I curves. Best fit (i.e., 5 W CW output power at 1.73 A drive current) occurred for $R_{\text{th}} = 3.4\,\text{K/W}$ (Fig. 2.13(a)). Several cases are considered for comparison. If $R_{\text{th}} = 8\,\text{K/W}$, as may be the case for episide-up mounting, the CW power at $I = 1.73\,\text{A}$ drops from 5 W to 3.4 W, and the maximum $\eta_{\text{wp,cw}}$ value drops from 21% to 14.5%. If the $T_0$ and $T_1$ values are lowered to 143 K, typical of conventional QCLs [3], while keeping the same $R_{\text{th}}$ value, the CW power at I = 1.73 A drops from 5 W to 3.6 W, and the maximum $\eta_{\text{wp,cw}}$ value drops from 21% to 15%. Thus, low $T_1$ values have a similar effect as high $R_{\text{th}}$ values. Raising the $T_0$ value from 143 K to 244 K, for the case $T_1 = 143\,\text{K}$ and $R_{\text{th}} = 3.4\,\text{K/W}$ makes little difference as far as the $P_{\text{opt}}$ and $\eta_{\text{wp,max}}$ values (not shown) since, just as in interband-transition devices [38], at high CW drive levels above threshold ($> 2 \times I_{\text{th}}$) the $J_{\text{th}}$ variation with temperature becomes irrelevant. Next we considered a device with $T_0 = 244\,\text{K}$, $T_1 = 750\,\text{K}$ and $R_{\text{th}} = 1.6\,\text{K/W}$, as may be the case for 4.9 $\mu$m-emitting STA-RE devices; that is, devices of virtually complete carrier-leakage suppression [9] at no price in $R_{\text{th}}$ value, since there is no need for AlAs barrier inserts throughout the injector region. Then, the CW power at $I = 1.73\,\text{A}$ increases from 5 W to 6 W, and the $\eta_{\text{wp,max}}$ value increases from 21% to 25.2%.

Finally, the $\eta_i$ value experimentally obtained in [30] (i.e., 70%) is short of the theoretical upper limit value of 95% for 4.5–5.0 $\mu$m-emitting QCLs (Fig. 2.6), deduced from the analysis of STA-RE QCLs with $E_{\text{ul}+1,\text{ul}} = 120\,\text{meV}$ [9, 51]. We plot the CW L–I curve for the case: $T_0 = 244\,\text{K}$, $T_1 = 750\,\text{K}$, $R_{\text{th}} = 1.6\,\text{K/W}$ and $\eta_i = 95\%$;

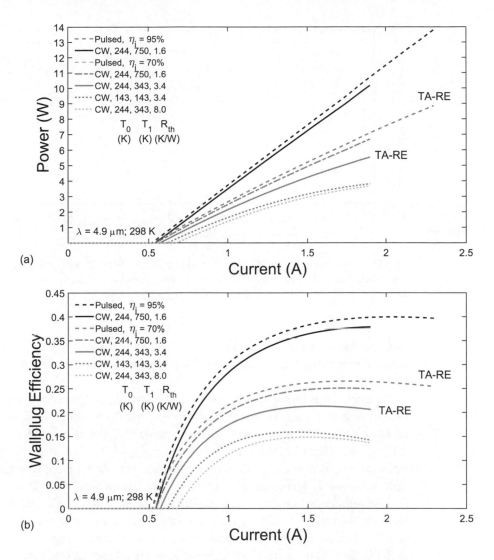

**Fig. 2.13** Room-temperature characteristics of 4.9 μm-emitting TA-RE QCL [30] and for QCLs of different $T_0$, $T_1$, and $R_{th}$ values, and for two $\eta_i$ values: 70% [30] and 95% (Fig. 2.6): (a) CW and pulsed power; (b) CW and pulsed wall-plug efficiency. Adapted from [3].

as the limiting case for 4.9 μm-emitting QCLs. Then, the CW power at $I = 1.73$ A increases from 5 W to 9 W, and the maximum projected CW power is 10.2 W; that is, twice the highest value CW obtained at 298 K heatsink temperature [30].

Having the CW L–I and $\eta_{wp}$ vs. I curves, one can estimate $\Delta T_{act}$ values. Taking as reference the 4.25 W CW power at the CW $\eta_{wp,max}$ point in Fig. 3 of [30], one can compare the $\Delta T_{act}$ values at the same CW power level for two relevant cases. While for the TA-RE QCL, at the $\eta_{wp,max}$ point, $\Delta T_{act} \sim 54$ K, it drops to ∼15 K for the case $T_0 = 244$ K, $T_1 = 750$ K, $R_{th} = 1.6$ K/W with $\eta_i = 95\%$. Notably, ∼15 K was the $\Delta T_{act}$ for the low-CW-power (0.2–0.3 W) 4.6 μm-emitting conventional QCLs

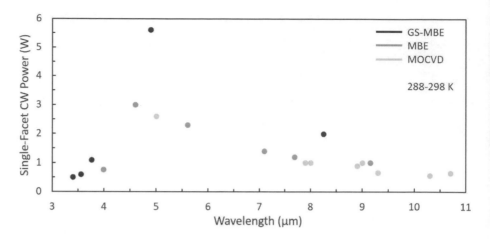

**Fig. 2.14** Single-facet-emitted, RT (288–298 K) maximum CW powers vs. emission wavelength, for QCLs grown by different crystal growth methods. The data are taken from [3, 4, 33, 35, 79–83]. © 2019 IEEE. Reprinted, with permission, from [84].

that have demonstrated long-term reliability [76]. Therefore, optimized QCLs hold the potential to reliably deliver multi-watt CW powers over long time periods (> 10,000 hrs) at RT.

Recently, it was proposed [77] that instead of scaling the power by increasing the number of stages, $N_p$, (i.e., vertically) to scale it laterally by using $N_p = 10 - 15$ for broad-area (i.e., 20–30 μm-wide) ridge-waveguide QCLs. The basic idea is that the $\Delta T_{act,th}$ value (Eq. (2.21)) decreases with decreasing $N_p$ [77, 78] since the threshold electrical-power density $J_{th,cw} V_{th}$ decreases with decreasing $N_p$. From 10-stage 30 μm-wide ridge-guide, 4.6 μm-emitting QCLs CW operation was obtained to > 100 °C [77], something that only conventional BH-type devices have achieved [26, 30, 74]. However, the maximum CW $P_{opt}$ and $\eta_{wp}$ values were only 0.9 W and 5% [78] since both $P_{opt}$ and $\eta_{wp}$ are strong functions of $N_p$. The same scaling approach was applied to BH QCLs [79] by using 15-stage, broad-area BH ridges, with the best result being 2.34 W single-facet CW power from 21 μm-wide BH ridges at $\lambda = 5.7$ μm. Again, $P_{opt}$ was limited mainly because it is dependent on $N_p$.

The maximum single-facet CW powers achieved at or near RT (i.e., 288–298 K heatsink temperature range) over the 3–11 μm wavelength range are shown in Fig. 2.14. The crystal-growth methods used are indicated. The highest CW power (i.e., 5.6 W) was obtained [80] at $\lambda = 4.9$ μm by using a GSMBE-grown TA-RE QCL with PICT action. The ~10% improvement over the best prior value [30] was claimed to reflect effective heat removal from devices with regrowth planarization. The highest single-facet values for each growth method [4, 43, 80] occur in the 4.5–5.0 μm wavelength range because of low $\alpha_w$ values and/or relatively high thermal-conductance values [3] (those devices were mounted with indium on diamond submounts). In the 7–11 μm range the highest value (2 W) is also from a GSMBE-grown, PICT-action device, in spite of relatively high carrier leakage [46, 57]. MBE-grown QCLs of NRE design (i.e., except QCLs emitting at $\lambda = 7.1$ μm [22] and 7.7 μm [81]) have yielded lower maximum $P_{opt}$ values, in spite of similarly low $\alpha_w$ values as GSMBE-grown devices, due both to lower $\eta_i$ values [3, 13, 56] and to lack of PICT action, as expected for vertical-transition devices. MOCVD-grown QCLs [4, 82] currently have somewhat higher $\alpha_w$ values than GSMBE- or MBE-grown QCLs, which for STA-RE QCLs are in large part compensated for by both very high $\eta_i$ values (77% at $\lambda = 5$ μm and 74% at $\lambda = 8$ μm) [4, 11] and very high $T_1$ values (653 K at $\lambda = 5$ μm and 478 K at $\lambda = 8$ μm) [4, 11].

Overall, just as seen above in Fig. 2.13(a) for 4.9 μm-emitting devices, the $P_{opt}$ values can be further significantly increased by combining carrier-leakage suppression with strong PICT action in high interface-quality core regions, and by maximizing the packaged-device thermal conductance.

## 2.6 Coherent-Power Scaling for Monolithic Structures

### 2.6.1 Introduction

For mid-IR QCLs the approaches previously employed for scaling the coherent power in near-IR diode lasers (i.e., phase-locked laser arrays [85], tapered amplifiers and lasers [86], and high-index-contrast (HC) photonic-crystal (PC) lasers [85, 87]) have been implemented with various degrees of success. Approaches solely employed for QCLs have been PC distributed-feedback (PC-DFB) lasers [88] tree-type phase-locked array lasers [89], and broad-area (BA) QCLs with optical loss provided at sidewalls [90].

Phase-locked laser arrays have two basic types of interelement coupling [85]: series coupling (i.e., nearest-neighbor coupling) and parallel coupling, also called global coupling. Series coupling is fundamentally unstable due to three factors [85]: (a) low intermodal discrimination; (b) multimode operation above threshold due to gain spatial hole burning (GSHB); and (c) temporal instabilities; and characterizes evanescent-wave coupled and diffraction-coupled (e.g., Talbot-filter or -cavity) arrays. Indeed, evanescent-wave coupled QCL arrays have operated in either wide single-lobe beams (e.g., 4.2 × diffraction-limited (D.L.) beam already at threshold [91]) or in near-D.L., two-lobe beam patterns to low peak powers (0.23 W) [92]. Similarly, since coupling using only the Talbot effect is inherently of the nearest-neighbor type [93] the best reported result for Talbot-filter QCLs [94] is 1.2 W peak-pulsed power with a 2 × D.L. beamwidth and evidence of other array modes lasing [93]. Actually, the Talbot effect can be used to achieve global coupling but only in conjunction with other array-mode discrimination mechanisms such as resonant leaky-wave coupling [95] or a diffractive mode-selecting mirror in an external cavity [96].

Global coupling does ensure strong intermodal discrimination and temporal stability [85]. However, in semiconductor lasers it works well only for HC-type structures that are not subject to piston errors [89] and/or radiation losses. PC-DFBs, being based on etched diffraction gratings/features, are low-index-contrast ($\Delta n \sim 0.007$) structures [97] which operated for edge emitters, at $\lambda = 4.6$ μm, in a single-lobe near-D.L. beam to only 10% above threshold and 0.5 W/facet peak power [98], and for surface emitters in a single-lobe D.L. beam, at $\lambda = 8.5\,\mu m$, only near threshold [99] with the beam broadening above threshold due to index-profile changes caused by GSHB and/or thermal lensing. HC Y-junction-combined tree arrays have operated either pulsed in a pure in-phase array mode, but, due to modal competition among branches of different length, at no increase in brightness [89] or CW in multimode, unstable beams from three-junction devices [100] (one-junction devices yielded low pulsed (1 W) and CW (0.32 W) D.L. powers [100]). Only with the implementation of multimode-interferometer (MMI) combining [33] did tree arrays operate to multi-watt range, near-D.L. peak-pulsed and CW powers (Section 2.6.3), albeit in heavily multi-lobed beam patterns. The only other HC globally coupled QCLs that have operated to multiwatt-range powers in stable, near-D.L. beams have been HC-PC edge-emitting (Section 2.6.2) and HC-PC surface-emitting lasers (Section 2.6.4).

Large emitting-aperture ($\sim$100 $\mu m$) tapered amplifiers [101] and lasers [102], while demonstrating near-D.L. beam operation to peak-pulsed powers as high as $\sim$4.3 W [102], widen to BA-ridge guides; thus are inherently vulnerable to thermal lensing in quasi-CW or CW operation. Similarly, while for an $\sim$60 $\mu m$-wide BA-ridge laser the lateral beam pattern was reduced to an $\sim$2.3 $\times$ D.L-wide single lobe by using sidewall losses via partial exposure to metal [90], maintaining high beam quality under quasi-CW or CW operation is unlikely. Even the use of 10–15 stages BA-type QCLs for which significant reduction in the $J_{th,cw}V_{th}$ value is obtained [78, 79] is unlikely to help, because of significant penalties in CW wall-plug efficiency and power. That is, above threshold the generated heat will most likely cause lateral thermal gradients which, in turn, will lead to multi-spatial-mode operation and/or beam instabilities.

## 2.6.2    High-Index Contrast, Photonic-Crystal Edge-Emitting Lasers

### Resonant Leaky-Wave Coupled Phase-Locked Arrays of Antiguides: ROW Array

Resonant leaky-wave coupling of antiguides [85] has been used for phase-locking near-IR-emitting lasers to high peak-pulsed (10 W) [103] and CW (1.6 W) [104] near-D.L. powers. Since at and near resonance the structures are in fact second-order *lateral* DFB structures [105, 106] they represent HC-PC structures with gain preferentially placed in the low-index PC regions; that is, unlike in conventional PC lasers [87]. Such devices allow global coupling between array elements in an in-phase mode of uniform intensity profile [85, 107, 108]. Thus, stable diffraction-limited-beam operation to high drive levels above threshold is assured. Furthermore, in quasi-CW and CW operation high-index contrast ($\Delta n \simeq 0.10$) prevents multimoding due to thermal lensing [104]. However, for resonant transmission between elements, the Bragg condition needs to be exactly satisfied [105] and the range in interelement width $s$, $\Delta s$, over which the in-phase mode is favored to lase is proportional to the emission wavelength [109]. That is, while for near-IR arrays the fabrication tolerance $\Delta s$ is rather small ($\sim$0.1 $\mu m$), by applying the concept to mid-IR arrays the fabrication tolerance increases by as much as an order of magnitude [109].

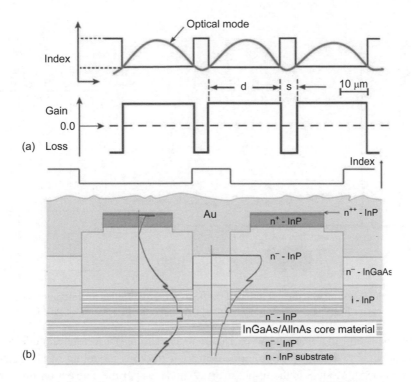

**Fig. 2.15** (a) Schematic representation of resonant leaky-wave coupled array of antiguides. The wavy line is the amplitude of the in-phase array mode at resonance. The built-in index step is 0.06 to 0.12; and the interelement spacing $s$ corresponds to one half the projected wavelength. (b) Schematic cross-section of an antiguided mid-IR QCL phase-locked array. Reproduced from [109], with the permission of AIP Publishing.

Resonant leaky-wave coupling occurs when the interelement region width $s$ is equal to an integer multiple of half the *projected* wavelength [85], $\lambda_1$, in the lateral direction (i.e., the resonance condition is: $s = m\lambda_1/2$). Figure 2.15(a) schematically shows the in-phase-mode resonance condition for coupled fundamental array-element modes when $m = 1$. The interelement regions become Fabry–Pérot resonators in the resonant condition (i.e., 100% transmission); thus, the array mode becomes the band-edge mode of a photonic-crystal laser. Since that is the opposite of ARROW-type devices, the devices were named ROW arrays [110]. The ROW-array structure for QCLs is schematically shown in Fig. 2.15(b). An InP interstack layer is inserted in the QCL core region to serve both as an etch-stop layer as well as a means for lateral heat removal [111]. The high-index, lossy interelement regions are formed by regrowing InGaAs layers and plating Au in etched trenches. As a result, QCL ROW-type arrays [109, 112, 113] of high index contrast ($\Delta n = 0.06$–$0.12$) can be formed; i.e., at least an order of magnitude higher index contrast than for PC-DFB lasers.

Figure 2.16 shows $J_{th}$ vs. $s$ curves and field-intensity profiles for relevant array modes of a five-element, 8.4 μm-emitting array with curved regrown layers in the interelement regions. The built-in index step was 0.10, the interelement loss was

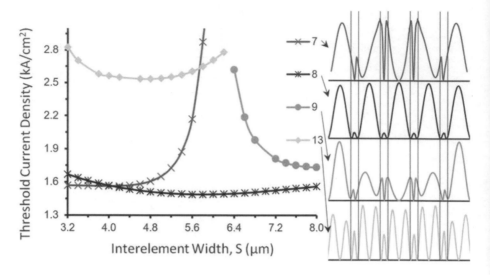

**Fig. 2.16** $J_{th}$ vs. interelement width and the field-intensity profiles for relevant modes of a five-element ROW array. Reproduced from [109], with the permission of AIP Publishing.

99 cm$^{-1}$ and the elements were 15 μm wide. The in-phase mode, mode 8, is resonant for $s \approx 6$ μm, in good agreement with rigorous modeling of antiguided laser arrays [85]. Mode 8 is the in-phase array mode, as per the convention for numbering array modes in antiguided arrays (i.e., the number of field-intensity minima) [107]. The competing modes are the adjacent array modes 7 and 9, and mode 13, an array mode composed of coupled first-order element modes. The resonant mode has 2.5% field intensity in the interelement regions; thus, it is negligibly affected by interelement loss. Modes 7 and 9, being nonresonant, have significant interelement field (22.5% and 15.2%) [107, 108]; thus, are suppressed via strong absorption to metals (Ti and Au) in the interelement regions, while mode 13 is suppressed via absorption to metals at the elements' edges. The $J_{th}$ values for modes 7 and 9 are ≥ 50% higher than the in-phase-mode $J_{th}$ value over a 1.0 μm-wide range in $s$ variation. We consider ≥50% higher $J_{th}$ value as adequate intermodal discrimination, since, as seen from Fig. 2.16, the resonant mode has a virtually uniform near-field intensity profile which prevents multimoding, due to GSHB, at high drives above threshold. The 1.0 μm-wide design window is about an order of magnitude higher than for HC-PC near-IR-emitting ROW arrays [85] thus allowing for much easier fabrication.

Figure 2.17 shows the L–I curve and lateral far-field patterns of the five-element array in pulsed operation. The threshold is 10.4 A, which corresponds to a $J_{th}$ value of ~2.2 kA/cm² , when considering current spreading to the interelement regions and to ~10 μm beyond the array edges. The emitted far-field pattern indicates in-phase-mode array operation to 1.8 × threshold and 5.5 W front-facet emitted power. The FWHM lobewidth is 6.8°, corresponding to ~1.6 × D.L. for a 105 μm-wide aperture, uniform-intensity five-element array [109] at 8.36 μm wavelength. 82% of the light is emitted in the main lobe; thus, 4.5 W is directly useful near-D.L. power, which represents the highest near-D.L., single-facet power obtained to date in an on-axis, single

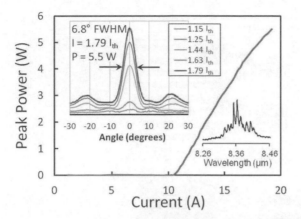

**Fig. 2.17** An 8.4 μm-emitting ROW array. L–I curve and far-field patterns of a five-element array as a function of drive above threshold, $I_{th}$. The diffraction limit is 4.11°. Inset: spectrum. Reproduced from [109], with the permission of AIP Publishing.

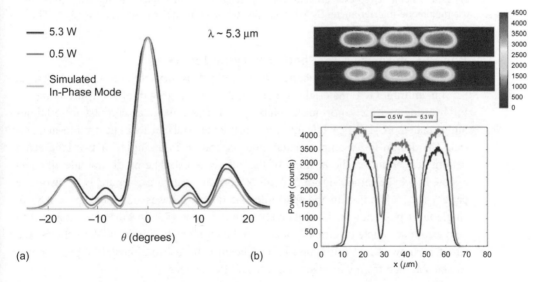

**Fig. 2.18** A 5.3 μm-emitting ROW array. (a) Measured and simulated lateral far-field patterns at two power levels. © 2019 IEEE. Reprinted, with permission, from [84]. (b) Near-field intensity patterns and line scan images of the patterns at two output power levels.

lobe from any type of large-aperture QCL. HC-PC lasers employing resonant leaky-wave coupling have also been demonstrated at $\lambda = 4.7$ μm [112]: 5.1 W near-D.L peak power with ~3.2 W in the main lobe; and at $\lambda = 5.3$ μm [113] with pure D.L.-beam operation to 5.3 W peak power (Fig. 2.18(a)) due to a nearly uniform near-field intensity profile (Fig. 2.18(b)). CW single-mode powers of 7–8 W from three- to five-element arrays have been projected [114], which would require micro-impingement coolers [33] to dissipate the generated heat involved.

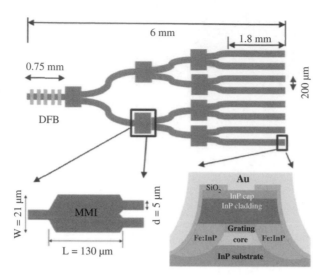

**Fig. 2.19** Schematic representation of an eight-element, buried-ridge QCL tree array combined via multimode-interferometer (MMI) sections. Reprinted with permission from [33] © The Optical Society.

## Buried-Heterostructure Photonic-Crystal Lasers

HC ($\Delta n/n \sim 10\%$) PC structures were realized by dry etching square pillars in 8.5 μm-emitting QCL material and burying them via regrowth of semi-insulating Fe-InP by using hydride vapor-phase epitaxy [115]. Thus, the index contrast is $\sim$50 times higher than in PC-DFBs, while the regrown-InP material can act as heat-removal material. The design is for single-spatial-mode operation in a single- or two-lobe beam pattern depending on the position of the cleaved facet. The mode was identified as a slow Bloch PC mode. From 550 μm × 550 μm structures, a pure D.L., two-lobe pattern (i.e., operation in the first-order lateral mode) was obtained up to 0.88 W single-facet peak power. The relatively high $J_{th}$ value (7.8 kA/cm$^2$) was assumed to be due to incomplete electrical isolation in the regrown-InP regions. Nevertheless the structure showed that diffraction-limited beam patterns, to almost 1 W peak power, can be obtained from very wide aperture HC-PC devices.

## 2.6.3    Multimode-Interferometer (MMI) – Combined Tree Arrays

Figure 2.19 schematically shows the top view of a MMI-combined eight-element, 8 μm-emitting tree array [33]. The single-element end has a DFB grating that selects a single frequency mode, as required for effective use of the MMI sections. Except for the MMI sections, the buried ridges are $\sim$5 μm wide, to ensure single-spatial-mode lasing and efficient heat removal. The MMI sections ensure that the single spatial mode from one array branch is coupled in-phase to the next two branches. Thus, the structure emits with all elements in phase with each other; that is, in the in-phase array mode. The eight emitting elements were separated 200 μm apart in order to ensure, at watt-range CW powers, a core-temperature rise of only ∼70 °C, while using micro-impingement cooling for heat dissipation. An in-phase near-D.L. beam pattern was obtained to 7 W CW power [33]; that is, the highest near-D.L. CW power reported to date from phase-locked QCL arrays. However, since the fill factor was quite low (i.e., 2.5%) the far-field beam pattern consisted of 35 lobes. As a result only ∼5% of the power (i.e., ∼0.35 W) was emitted in the main lobe. To date no monolithic-structure design, for this array type, that could garner under CW operation most of the sidelobe energy into the main lobe has been reported. Furthermore, light garnering into a single lobe cannot be efficiently done even by using external optics (i.e., via phase spatial filtering [116]) since that would require fill factors of at least 25%.

## 2.6.4  High-Index Contrast, Photonic-Crystal Surface-Emitting Lasers

### DFB/DBR Surface-Emitting (SE) ROW Arrays

By inserting in the elements of a ROW array second-order metal/semiconductor DFB/DBR gratings (Fig. 2.20(a)), designed for surface emission into a single-lobe beam pattern [117] (see Section 2.7), one obtains a 2-D HC-PC SE structure [118] that emits in a single-lobe diffraction-limited beam to high drive levels. Thus, unlike TE-polarized diode lasers [118, 119] there is no need for a central grating π phase shift to obtain a single-lobe beam in the longitudinal direction. Although spatial modes in all directions (i.e., transverse, lateral and longitudinal) need to be considered, 3-D modeling need not be employed since the periodicity in the lateral direction is much larger than that in the longitudinal direction; allowing to break the 3-D problem into

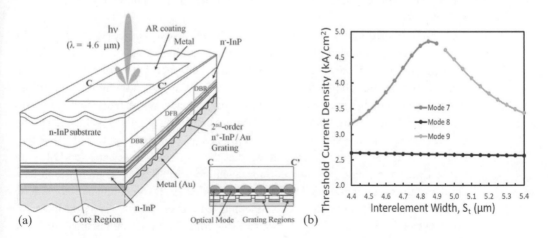

**Fig. 2.20**  (a) 3-D schematic view of 2-D grating-coupled surface-emitting (GCSE), 4.6 μm-emitting photonic-crystal laser composed of a resonant antiguided (ROW) array with second-order DFB/DBR InP/Au gratings in the array elements. Inset: cross-sectional view. (b) Intermodal discrimination for a 2-D HC-PC GCSE, five-element ROW array. Mode 8 is the in-phase array mode, which is resonant at an interelement width, $S_t$, of ∼4.85 μm.

two relatively easy 2-D problems [118]. The first is a 2-D in-plane problem for lateral and transverse modes; i.e., the model used to analyze ROW arrays [85, 109]. The second is a 2-D problem for modes in the plane normal to the core region, that solves for the longitudinal modes by using coupled-mode theory and the transfer-matrix method, as used for studying single-element GCSE DFB/DBR QCLs [117]. The two problems are coupled to each other via grating-related parameters, such as the grating coupling coefficient $\kappa$, which are array-mode dependent. Then the $J_{th}$ value for a given array mode $m$, $J_{th,m}$, can be calculated [118]. Figure 2.20 (b) shows the $J_{th}$ values for the in-phase array mode, mode 8, and the adjacent array modes, modes 7 and 9, for a 4.6 $\mu$m-emitting, five-element array with 8.65 mm-long grating, of which 4 mm is DFB grating, as a function of interelement spacing, $S_t$, around the in-phase mode resonance ($S_t = 4.85$ $\mu$m). The "window" of in-phase-mode operation, defined as the range in $S_t$ over which the $J_{th}$ values for competing modes are $\geq 50\%$ the $J_{th,8}$ value, is $\sim$0.6 $\mu$m wide, allowing for a relatively large fabrication tolerance. The array elements are 12.5 $\mu$m wide, with InP spacer layers for lateral heat removal to $\sim$5 $\mu$m-wide interelement regions which, in turn, periodically remove the generated heat to the heatsink. For seven-element devices, single-mode, single-lobe 15 W CW surface-emitted power has been projected.

### HC-PC Surface-Emitting Lasers with Gain in the High-Index PC Regions

Colombelli et al. [120] have demonstrated HC-PC SE lasers by periodically etching circular holes in 8 $\mu$m-emitting QCL material. The emitted far-field patterns were antisymmetric (i.e., four-lobed) as expected for 2-D PC SE lasers [87]. Single-lobe surface emission was obtained by introducing a $\pi$ phase shift in the crystal center for devices with elliptical holes [121, 122] or by using triangular-shaped features [87, 99].

A 3-D coupled-mode analysis [123] of HC-PC square-lattice structures (0.55x 0.55 mm$^2$ area) composed of circular nanopillars, similar to the 8.5 $\mu$m-emitting BH PC structure in [115], revealed that over a narrow fill-factor (FF) range (0.25–0.30), single-mode, single-lobe surface emission is favored to lase, although the mode loss is dominated by edge emission. 8.5 $\mu$m-emitting devices were fabricated similar to those in [115] but of larger area (1.5 $\times$ 1.5 mm$^2$) and with a 0.45 FF value [124]. The device is schematically shown in Fig. 2.21(a). A surface-emitted power of 0.6 W was obtained at 289 K. Figure 2.21(b) shows that the emission is, as expected [123], an $\sim$2 $\times$ D.L.-wide single-lobe beam. Subsequently, by employing asymmetric pillars, similar to the triangular-shaped teeth of SE PC-DFBs [99] which enhance single-lobe mode emission [87], and uniformizing the injected-current profile via a grid-like window contact [125] (Fig. 2.21(c)), 5 W surface-emitted power was obtained at 289 K (Fig. 2.21(d)). However, the device operated multimode ($\sim$4 $\times$ D.L.-wide main lobe) most probably as a result of a large ($\sim$ 10) in-plane $|\kappa|$L product, which likely led to a highly peaked 2-D mode intensity profile and subsequent GSHB (typically acceptable $|\kappa|$L values lie in the 2.5–3.5 range [118]). Future designs will focus on lowering the in-plane $|\kappa|$L product.

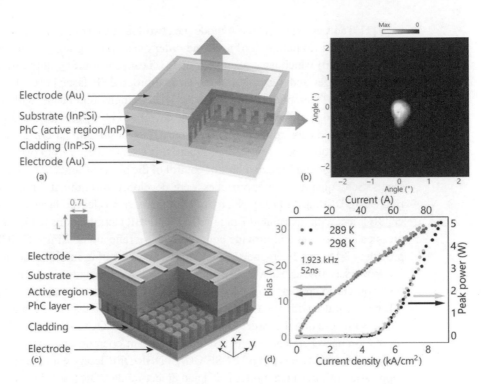

**Fig. 2.21** High-index contrast 2-D, surface-emitting (SE) photonic-crystal QCLs: (a) and (b) show the schematic diagram and SE beam pattern, respectively [124]; (c) and (d) show the schematic diagram of the structure with asymmetric pillars and grid-like window contact, and its L–I–V characteristics, respectively. Reprinted with permission from [125] ©The Optical Society.

## 2.7  Grating-Coupled Surface-Emitting Lasers

### 2.7.1  Introduction

GCSE lasers are attractive sources due to low-cost packaging, since cleaved emitting facets are not needed, and since there is no emitting-facet heating and subsequent degradation. Near-IR GCSE lasers have been studied since the early 1970s [118]. The longitudinal mode favored to lase is an antisymmetric (A) one (i.e., a mode of two-lobe far-field beam pattern) due to its inherent low radiation loss and subsequent low threshold gain [126]. Many approaches were proposed and demonstrated for obtaining single-lobe beam operation [117] with the most successful ones being those involving no penalty in efficiency: chirped grating corresponding to a distributed $\pi$ phase shift [127] and grating with a central $\pi$ phase shift [119, 128]. That is, the $\pi$ phase shift causes the lasers operating internally in an A mode to radiate into a single-lobe, diffraction-limited beam.

With the advent of QCLs in the mid-1990s, the Noll and Macomber GCSE model [126] was modified for TM-polarized light [129]. Various DFB gratings have been used [117] and, as expected, the A mode (i.e., two-lobe beam) was favored to lase, with two exceptions: (a) pumping of a second-order outcoupling grating with a first-order DFB laser [130] which gave a single-lobe D.L beam for short ($\leq 200\,\mu$m) apertures; and (b) an edge- and surface-emitting second-order DFB laser [131] that provided 0.1 W CW surface-emitted power in a near-D.L. single-lobe beam, with significant power being edge emitted. However, other devices in [131] provided two-lobe beam emission due to uncontrolled facet reflections. Thus, besides not offering a solution for stable, single-lobe operation, second-order DFB lasers suffered from two issues: relatively large edge emission and lack of control of the phase of reflections at the cleaved facets. Subsequently, two approaches were developed to eliminate those issues: (a) ring GCSE DFB lasers [132] which by definition have no cleaved facets; and (b) linear GCSE lasers with distributed Bragg reflectors (DBR) terminations [117, 133] to suppress edge emission and provide controlled-phase reflections. Ring SE QCLs, while providing 2-D, low-divergence beam patterns and operating in a symmetric mode to CW powers as high as 0.2 W [134], inherently display beam patterns composed of multiple concentric circular rings. By introducing two distributed $\pi$ phase shifts [135] a 2-D beam pattern with a central lobe was obtained, although that lobe contained $\sim$10% of the surface-emitted power. Thus, ring GCSE DFB QCLs are primarily suitable for short-range chemical sensing. Single-lobe beam operation has been achieved by design only from two types of GCSE QCLs: DBR/DFB lasers with plasmon-enabled suppression of unwanted modes [117] and defect-mode DBR lasers [133].

## 2.7.2      Single-Lobe, Single-Mode Surface Emitters with DBR Reflectors

### GCSE Defect-Mode DBR-Reflector Lasers

Feedback was provided by first-order DBR grating reflectors. The reflectors were separated by a $(60 + 3/4)\,\lambda/n_{\text{eff}}$-long spacer, where $\lambda$ is the vacuum wavelength and $n_{\text{eff}}$ is the effective guided-mode index. Thus a defect mode was created at a frequency in the middle of the stopgap [133]. A short second-order Bragg grating was formed in the center of the spacer for the purpose of outcoupling part of the laser field. The defect mode was favored to lase over the band-edge mode since the latter overlapped more with the lossy, unpumped DBR regions than the former. The net effect was that the defect-mode field was outcoupled, through an aperture in the top metal contact, as a single-lobe D.L. beam. The concept was demonstrated for each element of a monolithic ten-element array of single-mode devices, covering the 8–10 $\mu$m wavelength range. Peak powers of at most 2 mW were measured, as a result of leakage currents and the fact that most light was diffracted into the substrate. Such devices are intended as monolithically integrated sources for trace-gas sensing of multiple species with high sampling rates, over a broad wavelength range, with high accuracy and precision. However, just as in the case of edge-emitting DFB or ring SE DFB laser arrays, for the separated beams to sample the interaction volume an external beam-combining scheme is needed, which leads to a large footprint and mechanical instabilities. A solution to this problem was demonstrated by Süess et al. [136] by employing a monolithic array of GCSE defect-mode DBR lasers whose beams diffracted into the substrate and were coupled into one ridge guide. The net effect was that lasing light from seven QCLs emitting at different wavelengths was obtained through a single facet, thus eliminating the need for external beam combining.

### GCSE DFB/DBR Lasers with Plasmon-Enabled Longitudinal-Mode Suppression

A schematic representation of GCSE DFB/DBR mid-IR QCLs with metal/ semiconductor gratings is shown in Fig. 2.22. Second-order DBR gratings placed on either side of a same-period second-order DFB grating provide controlled-phase reflections as well as suppressed edge emission, just as for high-power near-IR GCSE lasers [118, 119, 127]. Furthermore, the guided-field-intensity peak-to-valley ratio in the DFB region, $R_o$, [119] can assume low values (i.e., 2-3), thus ensuring single-longitudinal-mode operation to high output powers. The metal/semiconductor grating provides both feedback and high light-outcoupling efficiency [117, 119], and ensures light extraction through the substrate, as needed for high-power CW operation. For QCLs, unlike diode lasers, one has to take into account backfilling and carrier leakage. Then, $J_{th}$ is expressed by:

$$J_{th,q} = \frac{2\alpha + (\alpha_w / \Gamma_{lg}) + \alpha_{bf}}{\eta_{inj}\Gamma g} = \frac{g_{th,q}}{\Gamma g}, \tag{2.25}$$

where $g_{th,q}$ is the *local* threshold gain (i.e., in the DFB region) [118, 119], $2\alpha$ is the grating-related (intensity) loss coefficient [118, 119], $\Gamma_{lg}$ is percentage of field intensity in DFB region, and the other terms are defined as for Eq. (2.1a). In contrast to Eq. (2.1a), $\alpha_m$ is replaced by $2\alpha$, which includes the surface-radiation loss $\alpha_{surf}$, grating-absorption losses, background absorption loss in the DBR regions and end-facets losses [119]. Similarly $\alpha_w$ is replaced by $\alpha_w/\Gamma_{lg}$ since the waveguide losses under pumping conditions are seen only in the DFB region. The $2\alpha$ value is obtained by using the transfer-matrix method for DFB/DBR structures [117]. As for the external differential efficiency $\eta_d$, it is given by [117]:

$$\eta_d = \eta_i \frac{\alpha_{surf}}{2\alpha\Gamma_{lg} + \alpha_w} N_p = \eta_i \eta_{rad} N_p, \tag{2.26}$$

where $\eta_i$ is as defined in Eq. (2.5b), and $\eta_{rad}$ is the grating outcoupling efficiency.

### *Symmetric-Mode GCSE Lasers*

Since the generated light is TM polarized, it is found that the guided optical mode can resonantly couple to the antisymmetric surface-plasmon modes of second-order DFB metal/semiconductor gratings which, in turn, results in strong A-mode absorption [117]. Then, lasing in the symmetric (S) mode, that is, emission into a single-lobe beam pattern, is strongly favored over certain ranges in grating duty cycle, $\sigma$, defined as the percentage of metal in one grating period. For example, a finite-length 4.6 μm-emitting, BH device of 8 μm-wide buried ridge and second-order Ag/InP grating was studied [117] (Fig. 2.22). Specifically, for a 7 mm-long device of 3.1 mm-long DFB region, and assuming $\eta_{inj} \sim 1$ (i.e., like for optimized STA-RE QCLs that have $\eta_{inj} \geq 90\%$ [9, 51]) the results are shown in Fig. 2.23. Figure 2.23(a) shows the $g_{th,q}$

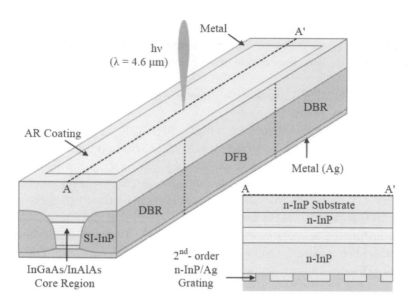

**Fig. 2.22** Schematic representation of surface-emitting, DFB/DBR BH QCL for operation at 4.6 μm wavelength. Reproduced from [117], with the permission of AIP Publishing.

values for A and S modes as a function of detuning from the reference wavelength, for $\sigma = 39\%$ and a tooth height, $h$, of 0.22 μm. One S mode is strongly favored to lase over two adjacent A modes, which have $g_{th,q}$ values 23.1 cm$^{-1}$ and 23.9 cm$^{-1}$ higher than for the S mode. Figure 2.23(b) shows the $g_{th,q}$ values for the S and the two adjacent A modes as a function of $h$, given $\sigma = 39\%$. The A modes reach $g_{th,q}$ maxima at values that correspond to resonant coupling of the guided mode to the Ag/InP-grating anti-symmetric plasmon mode, at their respective oscillation wavelength. An intermodal discrimination $\Delta g_{th,q}$ of $\geq 10$ cm$^{-1}$ is achieved over a 0.03 μm-wide variation in $h$. Figure 2.23(b) shows $\Delta g_{th,q}$ as a function of $\sigma$ and $h$. Thin black lines indicate where the $\Delta g_{th,q}$ value is 10 cm$^{-1}$, thus defining a curved-stripe shaped domain. A study performed within that domain, over the 37–41% range in $\sigma$ and the 0.20–0.23 μm range in $h$, found that $\eta_{rad}$ decreases by at most 15% (i.e., from 40% to 34%). The same $\eta_{rad}$ value (i.e., 40%) is obtained for $\sigma = 38\%$ and $h = 0.2$ μm, and for $\sigma = 40\%$ and $h = 0.23$ μm. (By contrast, the $\eta_{rad}$ value for QCLs with air-metal/semiconductor gratings is at most 17.5% [129].) Thus, the chance of getting $\eta_{rad}$ values close to 40% is relatively high, and the range of acceptable $\sigma$ values is 4% wide. By using Eqs. (2.25) and (2.26), the estimated threshold and $\eta_{sl}$ values were 0.45 A and 3.4 W/A, respectively, which led to a projected peak power of 3.06 W. The surface-emitted power can be coherently scaled (laterally) by using such gratings incorporated into the elements of ROW arrays (Section 2.6.4).

The concept was reduced to practice [137] by using shorter devices (i.e., 5.1 mm-long with 2.55 mm-long DFB regions) of $\sigma = 38.7\%$ that fell within a 5%-wide $\sigma$ window for S-mode operation: 36–41%. As a result, ~0.4 W power in a single-lobed beam pattern was obtained from a 4.75 μm-emitting QCL. The measured far-field

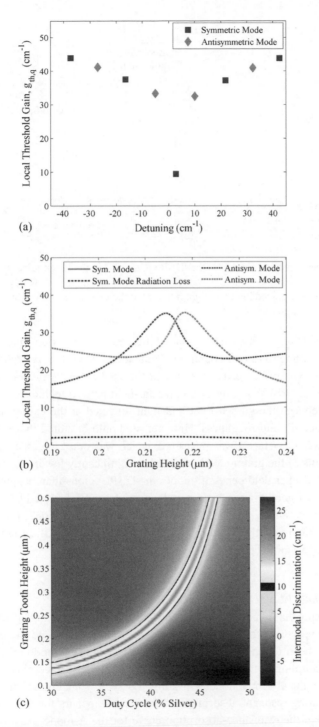

**Fig. 2.23** (a) Local threshold gain $g_{th,q}$ for S and A modes, at 39% duty cycle $\sigma$ and 0.22 $\mu$m grating tooth height $h$ as a function of detuning from the 4.6 $\mu$m reference wavelength; (b) $g_{th,q}$ for S and A modes vs. $h$ at $\sigma = 39\%$; (c) the intermodal discrimination $\Delta g_{th,q}$ vs. $h$ and $\sigma$. $\Delta g_{th,q}$ reaches maxima between the two thin curved black curves ($\Delta g_{th,q} \geq 10\,\text{cm}^{-1}$). Reproduced from [117], with the permission of AIP Publishing.

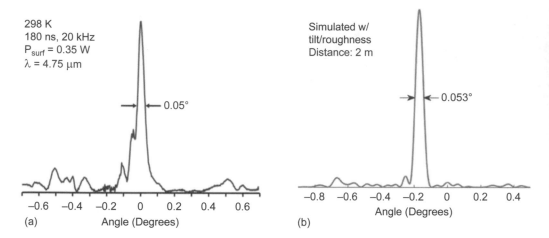

298 K
180 ns, 20 kHz
$P_{surf}$ = 0.35 W
$\lambda$ = 4.75 µm

0.05°

Simulated w/
tilt/roughness
Distance: 2 m

0.053°

(a)

-0.6   -0.4   -0.2   0   0.2   0.4   0.6
Angle (Degrees)

(b)

-0.8   -0.6   -0.4   -0.2   0   0.2   0.4
Angle (Degrees)

**Fig. 2.24** Far-field beam pattern of GCSE DFB/DBR QCL [137] in the longitudinal direction:
(a) measured at 0.35 W surface-emitted power; (b) simulated, including the effects of
interference between the guided light and light reflected at the exit surface, both due to
exit-surface tilt and roughness.

pattern at 0.35 W (Fig. 2.24(a)) has a main lobe that is diffraction-limited [137],
thus confirming symmetric-mode operation. Simulations show that reflections from
a rough-substrate exit surface cause features in the far-field pattern away from the
main lobe. Figure 2.24(b) shows the simulated far-field pattern including the effects
of interference between the guided light and light reflected at the exit surface, due
both to exit-surface tilt and roughness. However, it should be noted that substantial
edge-emitted power (1 W/facet) was measured, which was due to poor metal coverage of the sidewalls of the grating teeth, that led to significantly lower $\eta_{rad}$ value and
allowed for much deeper field penetration into the DBR regions than designed. Subsequently, devices of good grating-teeth metal coverage were fabricated [138], which
resulted in ~150 mW single-lobe, near-D.L. power from narrower ridges (i.e., ~7 µm
vs ~21 µm) and a significantly higher ratio (i.e., 5:1) between the surface-emitted and
the both-facets edge-emitted powers.

### Antisymmetric-Mode GCSE Lasers with Central $\pi$ Phase Shift

While S-mode lasing via A-mode suppression was proven to work to high peak powers
[137], it requires very tight tolerances in the $\sigma$ value [117] which lead to fabrication
challenges. It has been found [139] that at higher $\sigma$ values the S mode is resonantly
absorbed due to coupling of the guided light to the symmetric surface-plasmon mode
of a second-order metal-semiconductor grating. As seen from Fig. 2.25(a), for an
infinite-length grating resonant absorption of the S mode occurs for $\sigma \sim$ 50% and
it is less sharp than the resonance-absorption curve for the A-mode. As a result, A-
mode operation is favored over a wider range in $\sigma$ values than S-mode operation. For
instance, for a device of 5 mm-long DFB region (as needed for high output power),

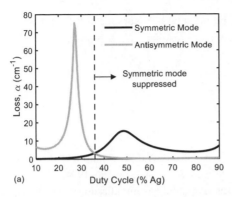 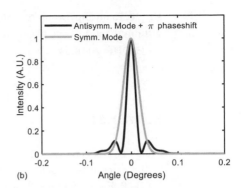

**Fig. 2.25** (a) Symmetric (S) and antisymmetric (A) modes losses as a function of grating duty cycle, for an infinite-length, metal/semiconductor grating of a GCSE laser; (b) far-field beam patterns of devices with 5 mm-long DFB regions for the S mode, and for the A mode from a grating with a central $\pi$ phase shift © 2021 IEEE. Reprinted, with permission, from [139].

in order to maintain $\Delta g_{th,q} \geq 10 \, \text{cm}^{-1}$ for the A mode, the window of desired $\sigma$ values is 7.5% wide, while for the S mode that window in $\sigma$ is zero [139]. However, the wider $\sigma$ window comes at a price: the power of the A-mode operating devices scales less efficiently with device length than for S-mode operating devices, as expected from theory [119, 126, 129]. To radiate primarily into a single-lobe beam pattern, A-mode operating devices require a central $\pi$ phase shift [119, 128]. Then, as shown in Fig. 2.25(b), the far-field beam pattern has a narrow central lobe and minor side lobes.

## 2.8    Conclusions

For state-of-the-art, mid-IR QCLs carrier leakage is primarily a shunt-type current within the active regions (ARs), as a result of electron thermal excitation via LO-phonon and IFR scattering, from the upper laser level and injector states to high-energy states within the AR, followed by relaxation to lower AR states. Carrier-leakage suppression via CB and IFR-scattering engineering is one of the keys to achieving high CW output power and wall-plug efficiency values, and ultimately long-term reliable operation at watt-range CW power levels.

Besides carrier-leakage suppression the other keys to maximizing the CW power and wall-plug efficiency are: (a) maximize the transition efficiency in diagonal-transition devices via IFR-scattering engineering (e.g., via step-tapering the barrier heights across the AR) and using miniband-like carrier extraction; (b) significantly improve the interface quality by optimizing the crystal-growth conditions; (c) minimize above-threshold voltages by designing QCLs with strong photon-induced carrier transport; and (d) maximize the packaged-device thermal conductance

For effective coherent-power scaling in 1-D and 2-D structures, to multiwatt output-power levels, global coupling is required in high-index-contrast, photonic-crystal structures. High single-mode, single-lobe powers from grating-coupled surface-emitting structures can be achieved by suppressing lasing of unwanted longitudinal modes via plasmon-enabled resonant absorption to metal/semiconductor DFB/DBR gratings.

# References

[1]  J. Faist, F. Capasso, D. L. Sivco, C. Sirtori, A. L. Hutchinson, and A. Y. Cho, "Quantum cascade laser," *Science*, vol. 264, no. 5158, pp. 553–556, 1994.

[2]  R. F. Kazarinov and R. A. Suris "Possibility of amplification of electromagnetic waves in a semiconductor with a superlattice," *Sov. Phys.-Semicond.*, vol. 5, no. 4, pp. 707–709, 1971.

[3]  D. Botez, C.-C. Chang, and L. J. Mawst, "Temperature sensitivity of the electro-optical characteristics for mid-infrared ($\lambda = 3$–16 μm)-emitting quantum cascade lasers," *J. Phys. D: Appl. Phys.*, vol. 49, no. 4, 043001, 2016.

[4]  D. Botez, J. D. Kirch, C. Boyle, K. M. Oresick, C. Sigler, H. Kim, B. B. Knipfer, J. H. Ryu, D. Lindberg III, T. Earles, L. J. Mawst, and Y. V. Flores, "High-efficiency, high-power mid-infrared quantum cascade lasers," *Opt. Mater. Express*, vol. 8, no. 5, pp. 1378–1398, 2018; Erratum: vol. 11, no. 7, p. 1970, 2021.

[5]  Jerome Faist, *Quantum Cascade Lasers*, Oxford, UK: Oxford University Press, 2013.

[6]  D. Botez, J. C. Shin, J. D. Kirch, C.-C. Chang, L. J. Mawst, and T. Earles, "Multidimensional conduction band engineering for maximizing the continuous-wave (CW) wallplug efficiencies of mid-infrared quantum cascade lasers," *IEEE J. Sel. Top. Quantum Electron.*, vol. 19, no. 4, 1200312, 2013; Correction: *IEEE J. Sel. Top. Quantum Electron.*, vol. 19, no. 4, 9700101, 2013.

[7]  A. Wittmann, Y. Bonetti, J. Faist, E. Gini, and M. Giovannini, "Intersubband linewidths in quantum cascade laser designs," *Appl. Phys. Lett.*, vol. 93, no. 14, 141103, 2008.

[8]  A. Bismuto, R. Terazzi, M. Beck, and J. Faist, "Influence of the growth temperature on the performances of strain-balanced quantum cascade lasers," *Appl. Phys. Lett.*, vol. 98, no. 9, 091105, 2011.

[9]  C. Boyle, C., K. M. Oresick, J. D. Kirch, Y. V. Flores, L. J. Mawst, and D. Botez, "Carrier leakage via interface-roughness scattering bridges gap between theoretical and experimental internal efficiencies of quantum cascade lasers," *Appl. Phys. Lett.*, vol. 117, no. 5, 051101, 2020; Erratum: *Appl. Phys. Lett.*, vol. 117, no. 10, 109901, 2020.

[10] D. Botez, S. Kumar, J. C. Shin, L.J. Mawst, I. Vurgaftman, and J. R. Meyer, "Temperature dependence of the key electro-optical characteristics for midinfrared emitting quantum cascade lasers," *Appl. Phys. Lett.*, vol. 97, no 7, 071101, 2010; Erratum: *Appl. Phys. Lett.*, vol. 97, no. 19, 199901, 2010.

[11] K. M. Oresick, J. D. Kirch, L. J. Mawst, and D. Botez, "Highly efficient long-wavelength infrared, step-tapered quantum cascade lasers," *Proc. SPIE*, vol. 11705, 1170515, 2021.

[12] P. M. Smowton and P. Blood, "The differential efficiency of quantum-well lasers," *IEEE J. Sel. Top. Quantum Electron.*, vol. 3, no. 2, pp. 491–498, 1997.

[13] R. Maulini, A. Lyakh, A. Tsekoun, R. Go, C. Pflügl, L. Diehl, F. Capasso, and C. K. N. Patel, "High power thermoelectrically cooled and uncooled quantum cascade lasers with optimized reflectivity facet coatings," *Appl. Phys. Lett.*, vol. 95, no. 15, 151112, 2009.

[14] J. M. Wolf, "Quantum cascade laser: from 3 to 26 μm," Ph.D. dissertation, ETH Zurich, 2017.

[15] A. Lyakh, M. Suttinger, R. Go, P. Figueiredo, and A. Todi, "5.6 μm quantum cascade lasers based on a two-material active region composition with a room temperature wall-plug efficiency exceeding 28%," *Appl. Phys. Lett.*, vol. 109, no 12, 121109, 2016.

[16] D. Botez, "Comment on 'Highly temperature insensitive quantum cascade lasers' [Appl. Phys. Lett. 97, 251104, (2010)]," *Appl. Phys. Lett.*, vol. 98, no. 21, 216101, 2011.

[17] Y.V. Flores, S. S, Kurlov, M. Elagin, M P. Semtsiv, and W.T. Masselink, "Leakage current in quantum-cascade lasers through interface roughness scattering," *Appl. Phys. Lett.*, vol. 103, no. 16, 161102, 2013.

[18] P. Harrison, D. Indjin, and R.W. Kelsall, "Electron temperature and mechanisms of hot carrier generation in quantum cascade lasers," *J. Appl. Phys.*, vol. 92, no 11, pp. 6921–6923, 2002.

[19] M. S. Vitiello, T. Gresch, A. Lops, V. Spagnolo, G. Scamarcio, N. Hoyler, M. Giovannini, and J. Faist, "Influence of InAs, AlAs δ layers on the optical, electronic, and thermal characteristics of strain-compensated GaInAs/AlInAs quantum-cascade lasers," *Appl. Phys. Lett.*, vol. 91, no 16, 161111, 2007.

[20] J. C. Shin, M. D'Souza, Z. Liu, J. Kirch, L. J. Mawst, D. Botez, I. Vurgaftman, and J. R. Meyer, "Highly temperature insensitive, deep-well 4.8 μm emitting quantum cascade semiconductor lasers," *Appl. Phys. Lett.*, vol. 94, no. 20, 201103, 2009.

[21] Y. V. Flores, M. Elagin, S. S. Kurlov, G. Monastyrskyi, A. Aleksandrova, J. Kischkat, and W. T. Masselink, "Thermally activated leakage current in high-performance short-wavelength quantum cascade lasers," *J. Appl. Phys.*, vol. 113, no. 13, 134506, 2013.

[22] R. Maulini, A. Lyakh, A. Tsekoun, and C. K.N. Patel, "λ ~ 7.1 μm quantum cascade lasers with 19% wall-plug efficiency at room temperature," *Opt. Express*, vol. 19, no 18, pp. 17203–17211, 2011.

[23] M. Beck, D. Hofstetter, T. Aellen, J. Faist, U. Oesterle, M. Ilegems, E. Gini, and H. Melchior, "Continuous-wave operation of a mid-infrared semiconductor laser at room temperature," *Science*, vol. 295, no. 5553, pp. 301–305, 2002.

[24] Jae Cheol Shin, "Tapered active-region quantum cascade laser," Chap. 6, pp. 92–103, Ph.D thesis, University of Wisconsin-Madison, Madison WI, 2010. http://digital.library.wisc.edu/1793/52493

[25] Y. Bai, N. Bandyopadhyay, S. Tsao, E. Selcuk, S. Slivken, and M. Razeghi, "Highly temperature insensitive quantum cascade lasers," *Appl. Phys. Lett.*, vol. 97, no 25, 251104, 2010.

[26] A. Wittmann, Y. Bonetti, M. Fischer, J. Faist, S. Blaser, and E. Gini, "Distributed-feedback quantum-cascade lasers at 9 μm operating in continuous wave up to 423 K," *IEEE Photon. Technol. Lett.*, vol. 21, no. 12, pp. 814–816, 2009.

[27] A. Bismuto, S. Blaser, R. Terazzi, T. Gresch, and A. Muller, "High performance, low dissipation quantum cascade lasers across the mid-IR range," *Opt. Express*, vol. 23, no. 5, pp. 5477–5484, 2015.

[28] K. Fujita, T. Edamura, S. Furuta, and M. Yamanishi, "High-performance, homogeneous broad-gain quantum cascade lasers based on dual-upper-state design," *Appl. Phys. Lett.*, vol, 96, no. 24, 241107, 2010.

[29] D. Botez, J. C. Shin, S. Kumar, J. Kirch, C.-C. Chang, L. J. Mawst, I. Vurgaftman, J. R. Meyer, A. Bismuto, B. Hinkov, and J. Faist, "The temperature dependence of key

electro-optical characteristics for midinfrared emitting quantum cascade lasers," *Proc. SPIE*, vol. 7953, 79530N, 2011.

[30]  Y. Bai, N. Bandyopadhyay, S. Tsao, S. Slivken, and M. Razeghi, "Room temperature quantum cascade lasers with 27% wall plug efficiency," *Appl. Phys. Lett.*, vol. 98, no. 18, 181102, 2011.

[31]  F. Wang, S. Slivken, D. Wu, and M. Razeghi, "Room temperature quantum cascade lasers with 22% wall plug efficiency in continuous-wave operation," *Opt. Express*, vol. 28, no. 12, pp. 17532–17538, 2020.

[32]  J. D. Kirch, C.-C. Chang, C. Boyle, L. J. Mawst, D. Lindberg, T. Earles, and D. Botez, "86% internal differential efficiency from 8 to 9 μm-emitting, step-taper active-region quantum cascade lasers," *Opt. Express*, vol. 24, no. 21, pp. 24483–24494, 2016.

[33]  W. Zhou, Q.-Y. Lu, D.-H. Wu, S. Slivken, and M Razeghi, "High-power, continuous-wave, phase-locked quantum cascade laser arrays emitting at 8 μm," *Opt. Express*, vol. 27, no. 11, pp. 15776–15785, 2019.

[34]  J. Faist, "Wallplug efficiency of quantum cascade lasers: Critical parameters and fundamental limits," *Appl. Phys. Lett.*, vol. 90, no. 25, 253512, 2007.

[35]  F. Xie, C. Caneau, H. P. Leblanc, D.P. Caffey, L.C. Hughes, T. Day, and C. Zah, "Watt-level room temperature continuous-wave operation of quantum cascade lasers with λ > 10 μm," *IEEE J. Sel. Top. Quantum Electron.*, vol. 19, no. 4, 1200407, 2013.

[36]  Z. Liu, D. Wasserman, S.S. Howard, A.J. Hoffman, C. F. Gmachl, X. Wang, T. Tanbun-Ek, L. Cheng, and F.-S. Choa, "Room-temperature continuous-wave quantum cascade lasers grown by MOCVD without lateral regrowth," *IEEE Photon. Technol. Lett.*, vol. 18, no. 12, pp. 1347–1349, 2006.

[37]  A. Tredicucci, F. Capasso, C. Gmachl, D. L. Sivco, A. L. Hutchinson, and A. Y. Cho, "High performance interminiband quantum cascade lasers with graded superlattices," *Appl. Phys. Lett.*, vol. 73, no. 15, pp. 2101–2103, 1998.

[38]  D. Botez, "Design considerations and analytical approximations for high continuous-wave power, broad-waveguide diode lasers," *Appl. Phys. Lett.*, vol. 74, no. 21, pp. 3102–3104, 1999.

[39]  Andreas Wittmann 2009, "High-performance quantum cascade laser sources for spectroscopic applications," Dissertation ETH Nr. 18363, pp. 112–113, Swiss Federal Institute, Zurich, 2009.

[40]  A. Wittmann, T. Gresch, E. Gini, L. Hvozdara, N. Hoyler, M. Giovannini, and J. Faist, "High-performance bound-to-continuum quantum-cascade lasers for broad-gain applications," *IEEE J. Quantum Electron.*, vol. 44, no. 1, 36–40, 2008.

[41]  J. C. Shin, L. J. Mawst, D. Botez, I. Vurgaftman, and J. R. Meyer, "Ultra-low temperature sensitive deep-well quantum cascade lasers (λ = 4.8 μm) via uptapering conduction band edge of injector region," *Electron. Lett.*, vol. 45, no. 14, pp. 741–743, 2009.

[42]  J. D. Kirch, J. C. Shin, C.-C. Chang, L.J. Mawst, D. Botez, and T. Earles, "Tapered active-region quantum cascade lasers (λ = 4.8 μm) for virtual suppression of carrier-leakage currents," *Electron. Lett.*, vol. 48, no. 4, pp. 234–235, 2012.

[43]  A. Lyakh, R. Maulini, A. Tsekoun, R. Go, C. Pflugl, L. Diehl, Q. J. Wang, F. Capasso, and C. K. N. Patel, "3 W continuous wave room temperature single-facet emission from quantum cascade lasers based on nonresonant extraction design approach," *Appl. Phys. Lett.*, vol. 95, no. 14, 141113, 2009.

[44]  H. Choi, L. Diehl, Z.-K. Wu, M. Giovannini, J. Faist, F. Capasso, and T. B Norris, "Time-resolved investigations of electronic transport dynamics in quantum cascade lasers based

on diagonal lasing transition," *IEEE J. Quantum Electron.*, vol. 45, no. 16, no. 4, pp. 307–321, 2009

[45] A. Bismuto, R. Terazzi, B. Hinkov, M. Beck, and J. Faist, "Fully automatized quantum cascade laser design by genetic optimization," *Appl. Phys. Lett.*, vol. 101, no 2, 021103, 2012.

[46] S. Suri, H. Gao, T. Grange, B. Knipfer, J. Kirch, L. Mawst, R. Marsland, and D. Botez, "Effect of graded interfaces on the performance of record-high-efficiency 4.9 µm- and 8.3 micron-emitting QCLs," International Quantum Cascade Lasers School & Workshop, Ascona, Switzerland, August 23–28, 2022, available at: https://botezgroup.wiscweb.wisc.edu/wp-content/uploads/sites/1905/2022/10/Suri-et-al.-IQCLSW-2022-UW-Madison-Intraband-poster-2.pdf.

[47] T. Grange, S. Mukherjee, G. Capellini, M. Montanar, L. Persichetti, L. Di Gaspare, S. Birner , A. Attiaoui, O. Moutanabbir, M. Virgilio, and M. De Seta, "Atomic-scale insights into semiconductor heterostructures: from experimental three-dimensional analysis of the interface to a generalized theory of interfacial roughness scattering," *Phys. Rev. Applied*, vol. 13, no 4, 044062, 2020.

[48] M. Lindskog, J. M. Wolf, V. Trinite, V. Liverini, J. Faist, G. Maisons, M. Carras, R. Aidam, R. Ostendorf, and A. Wacker, "Comparative analysis of quantum cascade laser modeling based on density matrices and non-equilibrium Green's functions," *Appl. Phys. Lett.*, vol. 105, no. 10, 103106, 2014.

[49] B. Knipfer, S. Xu, J. D. Kirch, D. Botez, and L. J. Mawst, "Analysis of interface roughness in strained InGaAs/AlInAs quantum cascade laser structures ($\lambda \sim 4.6$ µm) by atom probe tomography," *J. Cryst. Growth*, vol. 583, 126531, 2022.

[50] S. Soleimanikahnoj, M. L. King, and I. Knezevic, "Density-matrix model for photon-driven transport in quantum cascade lasers", *Phys. Rev. Applied*, vol. 15, no. 3, 034045, 2021.

[51] S. Suri, B. B. Knipfer, L. Mawst, and D. Botez, unpublished work on pocket-injector 4.1 µm- and 4.7 µm-emitting STA-type QCL designs of strong diagonal lasing transition.

[52] Q. Yang, R. Losch, W. Bronner, S. Hugger, F. Fuchs, R. Aidam, and J. Wagner, "High-peak-power strain-compensated GaInAs/AlInAs quantum cascade lasers ($\lambda \sim 4.6$ µm) based on a slightly diagonal active region design," *Appl. Phys. Lett.*, vol. 93, no. 25, 251110, 2008.

[53] Y. Yao, X. Wang, J.-Y. Fan, and C. F. Gmachl, "High performance 'continuum-to-continuum' quantum cascade lasers with a broad gain bandwidth of over 400 cm$^{-1}$," *Appl. Phys. Lett.*, vol. 97, no. 8, 081115, 2010.

[54] F. Wang, S. Slivken, D. H. Wu, and M. Razeghi, "Room temperature quantum cascade laser with ~31% wall-plug efficiency," *AIP Advances*, vol. 10, no. 7, 075012, 2020.

[55] K. M. Oresick, J. D. Kirch, L. J. Mawst, and D. Botez, "Highly efficient ~8 µm-emitting, step-taper active-region quantum cascade lasers," *AIP Advances*, vol. 11, no. 2, 025004, 2021.

[56] A. Lyakh, R. Maulini, A. Tsekoun, R. Go, and C. K. N. Patel, "Multiwatt long wave-length quantum cascade lasers based on high strain composition with 70% injection efficiency," *Opt. Express*, vol. 20, no. 22, pp. 24272–24279, 2012.

[57] Kevin M. Oresick, "Highly efficient long-wavelength infrared, step-taper active-region quantum cascade lasers," Ph.D. Thesis, University of Wisconsin-Madison, 2021.

[58] A. Vasanelli, A. Leuliet, C. Sirtori, A. Wade, G. Fedorov, D. Smirnov, G. Bastard, B. Vinter, M. Giovannini, and J. Faist, "Role of elastic scattering mechanisms in

GaInAs/AlInAs quantum cascade lasers," *Appl. Phys. Lett.*, vol. 89, no 17, 172120, 2006.

[59] Y. T. Chiu, Y. Dikmelik, P. Q. Liu, N. L. Aung, J. B. Khurgin, and C. F. Gmachl, "Importance of interface roughness induced intersubband scattering in mid-infrared quantum cascade lasers," *Appl. Phys. Lett.*, vol. 101, no. 17, 171117, 2012.

[60] M. Franckie, D. O. Winge, J. Wolf, V. Liverini, E. Dupont, V. Trinite, J. Faist, and A. Wacker, "Impact of interface roughness distributions on the operation of quantum cascade lasers," *Opt. Express,* vol. 23, no. 4, pp. 5201–5212, 2015.

[61] R. Terazzi, "Transport in quantum cascade lasers," Ph.D. dissertation, ETH Zurich, pp. 103–106, 2011.

[62] A. Leuliet, A. Vasanelli, A. Wade, G. Fedorov, D. Smirnov, G. Bastard, and C. Sirtori, "Electron scattering spectroscopy by a high magnetic field in quantum cascade lasers," *Phys. Rev. B*, vol. 73, no. 8, 085311, 2006.

[63] M. P. Semtsiv, Y. Flores, M. Chashnikova, G. Monastyrskyi, and W. T Masselink, "Low-threshold intersubband laser based on interface-scattering-rate engineering," *Appl. Phys. Lett.*, vol. 100, no. 16, 163502, 2012.

[64] Y. T. Chiu, Y. Dikmelik, Q. Zhang, J. B. Khurgin, and C. F. Gmachl, "Engineering the intersubband lifetime with interface roughness in quantum cascade lasers," in *Conference on Lasers and Electro-Optics 2012*, OSA Technical Digest Series (Optical Society of America, 2012), paper CTh3N.1.

[65] T. Unuma, M. Yoshita, T. Noda, H. Sakaki, and H. Akiyama, "Intersubband absorption linewidth in GaAs quantum wells due to scattering by interface roughness, phonons, alloy disorder, and impurities," *J. Appl. Phys.*, vol. 93, no. 3, pp. 1586–1597, 2003.

[66] M. P. Semtsiv, S. S. Kurlov, D. Alcer, Y. Matsuoka, J.-F. Kischkat, O. Bierwagen, and W. T. Masselink, "Reduced interface roughness scattering in InGaAs/InAlAs quantum cascade lasers grown on (411)A InP substrates," *Appl. Phys. Lett.*, vol. 113, no. 12, 121110, 2018.

[67] D. Indjin, P. Harrison, R. W. Kelsall, and Z. Ikonić, "Influence of leakage current on temperature performance of GaAs/AlGaAs quantum cascade lasers," *Appl. Phys. Lett.*, vol. 81, no. 3, pp. 400–402, 2002.

[68] X. Gao, D. Botez, and I. Knezevic, "X-valley leakage in GaAs/AlGaAs quantum cascade lasers," *Appl. Phys. Lett.*, vol. 89, no. 19, 191119, 2006.

[69] Y. B. Shi and I. Knezevic, "Nonequilibrium phonon effects in midinfrared quantum cascade lasers", *J. Appl. Phys.*, vol. 116, no. 12, 123105, 2014

[70] S. R. Jin, C. N. Ahmad, S. J. Sweeney, A. R. Adams, B. N. Murdin, H. Page, X. Marcadet, C. Sirtori, and S. Tomic, "Spectroscopy of GaAs/AlGaAs quantum-cascade lasers using hydrostatic pressure," *Appl. Phys. Lett.*, vol. 89, no. 22, 221105, 2006.

[71] C. Pflügl, L. Diehl, A. Lyakh, Q. J. Wang, R. Maulini, A. Tsekoun, C. K. Patel, X. Wang, and F. Capasso, "Activation energy study of electron transport in high-performance short wavelengths quantum cascade lasers," *Opt. Express*, vol. 18, no. 2, pp. 746–753, 2010.

[72] J. Mc Tavish, D. Indjin, and P. Harrison, "Aspects of the internal physics of InGaAs/InAlAs quantum cascade lasers," *J. Appl. Phys.*, vol. 99, no. 11, 114505, 2006.

[73] S. S. Howard, Z. Liu, D. Wasserman, A. J. Hoffman, T. S. Ko, and C. F. Gmachl, "High-performance quantum cascade lasers: optimized design through waveguide and thermal modeling," *IEEE J. Sel. Top. Quantum Electron.*, vol. 13, no. 5, pp. 1054–1063, 2007.

[74] T. Kato, H. Mori, H. Yoshinaga, and S. Souma, "High-temperature operation of a quantum cascade laser with definite parity of the wave functions," *IEEE Photon. Technol. Lett.*, vol. 33, no. 10, pp. 507–510, 2021.

[75] H. K Lee and J. S. Yu, "Thermal analysis of short wavelength InGaAs/InAlAs quantum cascade lasers," *Solid-State Electron.*, vol. 54, no. 8, pp. 769–776, 2010.

[76] F. Xie H.-K. Nguyen, H. Leblanc, L. Hughes, J. Wang, J. Wen, D. J. Miller, and K. Lascola, "Long term reliability study and life time model of quantum cascade lasers," *Appl. Phys. Lett.*, vol. 109, no. 12, 121111, 2016.

[77] M. P. Semtsiv and W. T. Masselink, "Above room temperature continuous wave operation of a broad-area quantum-cascade laser," *Appl. Phys. Lett.*, vol. 109, no. 20, 203502, 2016.

[78] A. Aleksandrova, Y. V. Flores, S, S. Kurlov, M. P. Semtsiv, and W. T. Masselink, "Impact of cascade number on the thermal properties of broad-area quantum cascade lasers," *Phys. Status Solidi A*, vol. 215, no. 8, 1700441, 2017.

[79] M. Suttinger, R. Go, P, Figueiredo, A. Todi, H. Shu, J, Leshin, and A. Lyakh, "Power scaling and experimentally fitted model for broad area quantum cascade lasers incontinuous wave operation," *Opt. Eng.*, vol 57, no. 1, 011011, 2018.

[80] F. Wang, S. Slivken, D. H. Wu, Q. Y. Lu, and M. Razeghi, "Continuous wave quantum cascade lasers with 5.6 W output power at room temperature and 41% wall-plug efficiency in cryogenic operation," *AIP Advances*, vol. 10, no. 5, 055120, 2020.

[81] H. Wang, J. Zhang, F. Cheng, N. Zhuo, S. Zhai, J. Liu, L. Wang, S. Liu, F. Liu, and Z. Wang, "Watt-level, high wall plug efficiency, continuous-wave room temperature quantum cascade laser emitting at 7.7 μm," *Opt. Express*, vol. 28, no. 26, pp. 40155–40163, 2020.

[82] B. Schwarz, C. A. Wang, L. Missaggia, T. S. Mansuripur, P. Chevalier, M. K. Connors, D. McNulty, J. Cederberg, G. Strasser, and F. Capasso, "Watt-level continuous-wave emission from a bifunctional quantum cascade laser/detector," *ACS Photonics*, vol. 4 no. 5, pp. 1225–1231, 2017.

[83] A. Lyakh, R, Maulini, A. Tsekoun, R. Go, S. Von der Porten, C. Pflügl, L. Diehl, F. Capasso, and C. K. N. Patel, "High-performance continuous-wave room temperature 4.0-μm quantum cascade lasers with single-facet optical emission exceeding 2 W," *Proc. Natl. Acad. Sci. U.S.A.*, vol. 107, no. 44, pp. 18799–18802, 2010.

[84] D. Botez, C. Boyle, J. Kirch, K. Oresick, C. Sigler, L. Mawst, D. Lindberg, and T. Earles, "High-power mid-infrared quantum cascade semiconductor lasers," *2019 IEEE Photonics Conference (IPC)*, 2019, pp. 1–2, doi: 10.1109/IPCon.2019.8908354.

[85] D. Botez, "Monolithic phase-locked semiconductor laser arrays," in *Diode Laser Arrays*, D. Botez and D. R. Scifres, Eds., Cambridge: Cambridge University Press, 1994, pp. 1–71.

[86] J. N. Walpole, "Semiconductor amplifiers and lasers with tapered gain regions," *Opt. Quantum Electron.*, vol. 28, no. 6, pp. 623–645, 1996.

[87] E. Miyai, K. Sakai, T. Okano, W. Kunishi, D. Ohnishi, and S. Noda, "Lasers producing tailored beams," *Nature*, vol. 441, p. 946, 2006.

[88] I. Vurgaftman and J. R. Meyer, "Photonic-crystal distributed-feedback quantum cascade lasers," *IEEE J. Quantum Electron.*, vol. 38, no. 6, pp. 592–602, 2002.

[89] L. K. Hoffmann, M. Klinkmuller, E. Mujagic, M. P. Semtsiv, W. Schrenk, W. T. Masselink, and G. Strasser, "Tree array quantum cascade laser," *Opt. Express*, vol. 17, no. 2, pp. 649–657, 2009.

[90]   R. Kaspi, S. Luong, T. Bate, C. Lu, T. Newell, and C. Yang, "Distributed loss method to suppress high order modes in broad area quantum cascade lasers," *Appl. Phys. Lett.*, vol. 111, no. 20, 201109, 2017.

[91]   F. Yan, J. Zhang, Z. Jia, N. Zhuo, S. Zhai, S. Liu, F. Liu, and Z. Wang, "High-power phase-locked quantum cascade laser array emitting at $\lambda \sim 4.6$ µm," *AIP Advances*, vol. 6, no. 3, 035022, 2016.

[92]   G. M. de Naurois, M. Carras, B. Simozrag, O. Patard, F. Alexandre, and X. Marcadet, "Coherent quantum cascade laser micro-stripe arrays," *AIP Advances*, vol. 1, no. 3, 032165, 2011.

[93]   D. Botez, "Comment on "Phase-locked array of quantum cascade lasers with an intra-cavity spatial filter" [Appl. Phys. Lett. 111, 061108 (2017)]," *Appl. Phys. Lett.*, vol. 111, no. 25, 256101, 2017.

[94]   Z. Jia, L. Wang, J. Zhang, Y. Zhao, C. Liu, S. Zhai, N. Zhuo, J. Liu, L. Wang, S. Liu, F. Liu, and Z. Wang, " Phase-locked array of quantum cascade lasers with an intracavity spatial filter," *Appl. Phys. Lett.*, vol. 111, no. 6, 061108, 2017.

[95]   L. J. Mawst, D. Botez, T. J. Roth, W. W. Simmons, G. Peterson, M. Jansen, J. Z. Wilcox, and J. J. Yang, " Phase-locked array of antiguided lasers with monolithic spatial filter," *Electron. Lett.*, vol. 25, no. 5, pp. 365–366, 1989.

[96]   J. R. Leger and G. Mowry, "External diode-laser-array cavity with mode-selecting mirror," *Appl. Phys. Lett.*, vol. 63, no. 21, pp. 2884–2886, 1993.

[97]   Y. Bai, S. R. Darvish, S. Slivken, P. Sung, J. Nguyen, A. Evans, W. Zhang, and M. Razeghi, "Electrically pumped photonic crystal distributed feedback quantum cascade lasers," *Appl. Phys Lett.*, vol. 91, no. 14, 141123, 2007.

[98]   Y. Bai, B. Gokden, S. R. Darvish, S. Slivken, and M. Razeghi, "Photonic crystal distrib-uted feedback quantum cascade lasers with 12 W output power," *Appl. Phys Lett.*, vol. 95, no. 3, 031105, 2009.

[99]   Y. Liang, Z. Wang, J. Wolf, E. Gini, M. Beck, B. Meng, J. Faist, and G. Scalari, "Room temperature surface emission on large-area photonic crystal quantum cascade lasers," *Appl. Phys Lett.*, vol. 114, no. 3, 031102, 2019.

[100]  A. Lyakh, R. Maulini, A. Tsekoun, R. Go, and C. K. N. Patel, "Continuous wave opera-tion of buried heterostructure 4.6µm quantum cascade laser Y-junctions and tree arrays," *Opt. Express*, vol. 22, no. 1, pp. 1203–1208, 2014.

[101]  P. Rauter, S. Menzel, A. Goyal, B. Gokden, C. A. Wang, A. Sanchez, G. W. Turner, and F. Capasso, "Master-oscillator power-amplifier quantum cascade laser array," *Appl. Phys Lett.*, vol. 101, no. 26, 261117, 2012.

[102]  R. Blanchard, T. S. Mansuripur, B. Gokden, N. Yu, M. Kats, P. Genevet, K. Fujita, T. Edamura, M. Yamanishi, and F. Capasso, "High-power low-divergence tapered quantum cascade lasers with plasmonic collimators," *Appl. Phys Lett.*, vol. 102, no. 19, 191114, 2013.

[103]  H. Yang, L. J. Mawst, M. Nesnidal, J. Lopez, A. Bhattacharya, and D. Botez, "10 W near-diffraction-limited pulsed power from 0.98 µm-emitting, Al-free phase-locked antiguided arrays," *Electron Lett.*, vol. 33, no. 22, pp. 136–138, 1997.

[104]  H. Yang, L. J. Mawst, and D. Botez, "1.6 W continuous-wave coherent power from large-index-step ($\Delta$n $\approx$ 0.1) near-resonant, antiguided diode laser arrays," *Appl. Phys. Lett.*, vol. 76, no. 10, pp. 1219–1221, 2000.

[105] C. A. Zmudzinski, D. Botez, and L. J. Mawst, "Simple description of laterally resonant, distributed-feedback-like modes of arrays of antiguides," *Appl. Phys. Lett.*, vol. 60, no. 9, pp. 1049–1051, 1992.

[106] A. P. Napartovich and D. Botez, "Analytical theory of phase-locked arrays of antiguided diode lasers," *Proc. SPIE*, vol. 2994, pp. 600–610, 1997.

[107] D. Botez, L. J. Mawst, G. Peterson, and T. J. Roth, "Phase-locked arrays of antiguides: modal content and discrimination," *IEEE J. Quantum Electron.*, vol. 26, no. 3, pp. 482–495, 1990.

[108] C. Zmudzinski, D. Botez, L. J. Mawst, A. Bhattacharya, M. Nesnidal, and R. F. Nabiev, "Three-core ARROW-type diode laser: novel high-power, single-mode device, and effective master oscillator for flared antiguided MOPA's," *IEEE J. Sel. Top. Quantum Electron.*, vol. 1, no. 2, pp. 129–137, 1995.

[109] J. D. Kirch, C.-C. Chang, C. Boyle, L. J. Mawst, D. Lindberg, T. Earles, and D. Botez, "5.5 W near-diffraction-limited power from resonant leaky-wave coupled phase-locked arrays of quantum cascade lasers," *Appl. Phys. Lett.*, vol. 106, no. 6, 061113, 2015.

[110] D. Botez, L. J. Mawst, G. Peterson, and T. J. Roth, "Resonant optical transmission and coupling in phase-locked diode-laser arrays of antiguides: The resonant-optical-waveguide array," *Appl. Phys. Lett.*, vol. 54, no. 22, pp. 2183–2185, 1989.

[111] A. Bismuto, T. Gresch, A. Bachle, and J. Faist, "Large cavity quantum cascade lasers with InP interstacks," *Appl. Phys. Lett.*, vol. 93, no. 23, 231104, 2008.

[112] C. Sigler, C. A. Boyle, J. D. Kirch, D. Lindberg III, T. Earles, D. Botez, and L. J. Mawst, "4.7 μm-emitting near-resonant leaky-wave-coupled quantum cascade laser phase-locked arrays," *IEEE J. Sel. Top. Quantum Electron.*, vol. 23, no. 6, 1200706, 2017.

[113] C. Sigler, R. Gibson, J. H. Ryu, C. A. Boyle, J. D. Kirch, D. Lindberg III, T. Earles, D. Botez, L. J. Mawst, and R. Bedford, "5.3 μm-emitting diffraction-limited leaky-wave-coupled quantum cascade laser phase-locked array," *IEEE J. Sel. Top. Quantum Electron.*, vol. 25, no. 6, 1200509, 2019.

[114] C. A. Sigler, "Coherent power Scaling of mid-infrared quantum cascade lasers," Ph.D. Thesis, University of Wisconsin-Madison, 2018.

[115] R. Peretti, V. Liverini, M. J. Suess, Y. Liang, P.-B. Vigneron, J. M. Wolf, C. Bonzon, A. Bismuto, W. Metaferia, M. Balaji, S. Lourdudoss, E. Gini, M. Beck, and J. Faist, "Room temperature operation of a deep etched buried heterostructure photonic crystal quantum cascade laser," *Laser Photonics Rev.*, vol. 10, no. 5, pp. 843–848, 2016.

[116] G. J. Swanson, J. R. Leger, and M. Holz, "Aperture filling of phase-locked laser arrays," *Opt. Lett.*, vol. 12, no. 4, pp. 245–247, 1987.

[117] C. Sigler, J. D. Kirch, T. Earles, L. J. Mawst, Z. Yu, and D. Botez, "Design for high-power, single-lobe, grating-surface-emitting quantum cascade lasers enabled by plasmon-enhanced absorption of antisymmetric modes," *Appl. Phys. Lett.*, vol. 104, no. 13, 131108 (2014).

[118] S. Li and D. Botez, "Analysis of 2-D surface-emitting ROW-DFB semiconductor lasers for high-power, single-mode operation," *IEEE J. Quantum Electron.*, vol. 43, no. 8, pp. 655–668, 2007.

[119] S. Li, G. Witjaksono, S. Macomber, and D. Botez, "Analysis of surface-emitting second-order distributed feedback lasers with central grating phaseshift," *IEEE J. Sel. Top. Quantum Electron.*, vol. 9, no. 5, pp. 1153–1165, 2003.

[120] R. Colombelli, K. Srinivasan, M. Troccoli, O. Painter, C. F. Gmachl, D. M. Tennant, A. M. Sergent, D. L. Sivco, A. Y. Cho, and F. Capasso, "Quantum cascade surface-emitting photonic crystal laser," *Science*, vol. 302, no. 5649, pp. 1374–1377, 2003.

[121] E. Miyai and S. Noda, "Phase-shift effect on a two-dimensional surface-emitting photonic-crystal laser," *Appl. Phys. Lett.*, vol. 86, no. 11, 111113, 2005.

[122] G. Xu Y. Chassagneux, R. Colombelli, G. Beaudoin, and I. Sagnes, "Polarized single-lobed surface emission in mid-infrared, photonic-crystal, quantum-cascade lasers," *Opt. Lett.*, vol. 35, no. 6, pp. 859–861, (2010).

[123] Z. Wang, Y. Liang, X. Yin, C. Peng, W. Hu, and J. Faist, "Analytical coupled-wave model for photonic crystal surface-emitting quantum cascade lasers," *Opt. Express*, vol. 25, no. 10, pp. 11997–12007, 2017.

[124] Z. Wang, Y. Liang, B. Meng, Y. Sun, G. Omanakuttan, E. Gini, M. Beck, I. Sergachev, S. Lourdudoss, J. Faist, and G. Scalari, "Over 2W room temperature lasing on a large area photonic crystal quantum cascade laser," in *Conference on Lasers and Electro-Optics 2019*, OSA Technical Digest Series (Optical Society of America, 2019), paper SW4N.4.

[125] Z. Wang, Y. Liang, B. Meng, Y. Sun, G. Omanakuttan, E. Gini, M. Beck, I. Sergachev, S. Lourdudoss, J. Faist, and G. Scalari, "Large area photonic crystal quantum cascade laser with 5 W surface-emitting power," *Opt. Express*, vol. 27, no. 16, pp. 22708–22726, 2019.

[126] R. J. Noll and S. H. Macomber, "Analysis of grating surface emitting lasers," *IEEE J. Quantum Electron.*, vol. 26, no. 3, pp. 456–466, 1990.

[127] S. H. Macomber, J. S. Mott, B. D. Schwartz, R. S. Setzko, J. J. Power, P. A. Lee, D. P. Kwo, R. M. Dixon, and J. E. Logue, "Curved-grating surface-emitting DFB lasers and arrays," *Proc. SPIE*, vol. 3001, pp. 42–54, 1997.

[128] G. Witjaksono, S. Li, J. J. Lee, D. Botez, and W. K. Chan, "Single-lobe, surface-normal beam surface emission from second-order distributed feedback lasers with half-wave grating phase shift," *Appl. Phys. Lett.*, vol. 83, no. 26, pp. 5365–5367, 2003.

[129] N. Finger, W. Schrenk, and E. Gornik, "Analysis of TM-polarized DFB laser structures with metal surface gratings," *IEEE J. Quantum Electron.*, vol. 36, no. 7, pp. 780–786, 2000.

[130] G. Maisons, M. Carras, M. Garcia, B. Simozrag, and X. Marcadet, "Directional single mode quantum cascade laser emission using second-order metal grating coupler," *Appl. Phys. Lett.*, vol. 98, no. 2, 021101, 2011.

[131] D.-Y. Yao, J.-C. Zhang, F.-Q. Liu, N. Zhuo, F.-L. Yan, L.-J. Wang, J.-Q. Liu, and Z.-G. Wang, "Surface emitting quantum cascade lasers operating in continuous-wave mode above 70 °C at $\lambda \sim 4.6$ μm," *Appl. Phys. Lett.*, vol. 103, no. 4, 041121, 2013.

[132] E. Mujagić, M. Nobile, H. Detz, W. Schrenk, J. Chen, C. Gmachl, and G. Strasser, "Ring cavity induced threshold reduction in single-mode surface emitting quantum cascade lasers," *Appl. Phys. Lett.*, vol. 96, no. 3, 031111, 2010.

[133] P. Jouy, C. Bonzon, J. Wolf, E. Gini, M. Beck, and J. Faist, "Surface emitting multi-wavelength array of single frequency quantum cascade lasers," *Appl. Phys. Lett.*, vol. 106, no. 7, 071104, 2015.

[134] D. H. Wu and M. Razeghi, "High power, low divergent, substrate emitting quantum cascade ring laser in continuous wave operation," *APL Materials*, vol. 5, no. 3, 035505. 2017.

[135]  R. Szedlak, M. Holzbauer, D. MacFarland, T. Zederbauer, H. Detz, A. M. Andrews, C. Schwarzer, W. Schrenk, and G. Strasser, "The influence of whispering gallery modes on the far field of ring lasers," *Sci. Rep.*, vol. 5, no. 1, pp. 1–8, 2015.

[136]  M. J. Süess, P. Jouy, C. Bonzon, J. M. Wolf, E. Gini, M. Beck, and J. Faist, "Single-mode quantum cascade laser array emitting from a single facet," *IEEE Photon. Technol. Lett.*, vol. 28, no. 11, pp. 1197–1200, 2016.

[137]  C. Boyle, C. Sigler, J.D. Kirch, D.F. Lindberg III, T. Earles, D. Botez, and L.J. Mawst, "High-power, surface-emitting quantum cascade laser operating in a symmetric grating mode," *Appl. Phys. Lett.*, vol. 108, no. 12, 121107, 2016.

[138]  J. H. Ryu, C. Sigler, C. Boyle, J. D. Kirch, D. Lindberg, T. Earles, D. Botez, and L. J. Mawst, "Surface-emitting quantum cascade lasers with 2nd-order metal/semiconductor gratings for high continuous-wave performance," *Proc. SPIE*, vol. 11301, 113011P, 2020.

[139]  J. H. Ryu, C. Sigler, J. D. Kirch, T. Earles, D. Botez, and L. J. Mawst, "Fabrication-tolerant design for high-power, single-lobe, surface-emitting quantum cascade lasers." *IEEE Photon. Technol. Lett.*, vol. 33, no. 16, pp. 820–823, 2021.

# 3 Long-Wavelength Mid-Infrared Quantum Cascade Lasers

Alexei Baranov, Michael Bahriz, and Roland Teissier

University of Montpellier

## 3.1 Introduction

Quantum cascade lasers (QCLs) exploit optical transitions between electron states in the conduction band of coupled quantum wells [1]. The energy of intersubband transitions can be varied over a very large range by adjusting the width of the active quantum wells, contrary to interband diode lasers, where the transition energy cannot be smaller than the bandgap of the well material. Lattice vibrations in III-V alloys constituting a QCL play an important role in the device operation. Emission of a longitudinal optical (LO) phonon is the dominant non-radiative relaxation mechanism in mid-infrared QCLs, where the distance between the transition states exceeds the LO-phonon energy $E_{LO}$ [2]. Such relaxation processes as interface-roughness (IFR) scattering and/or alloy-disorder scattering can compete with the LO-phonon scattering under certain circumstances, which will be considered later in this chapter. In far-infrared THz QCLs exploiting transitions with energies smaller than $E_{LO}$, direct emission of LO-phonons is not allowed at low temperatures and device operation is governed by other relaxation mechanisms, such as thermally assisted LO-phonon scattering and electron–electron interaction [2]. Obviously, no lasing can be obtained at energies in the vicinity of $E_{LO}$. The light generated in a QCL can be coupled with other types of lattice vibrations in the materials used not only in the active quantum wells but in all layers inside the optical waveguide as well. The complex phonon structure in III-V compounds and alloys results in formation of a large Reststrahlen band separating mid- and far-infrared, consisting of zones of strong absorption prohibiting lasing. From the short wavelength side, the Reststrahlen band of III-V materials used in QCLs starts at about 28 μm for AlAs (see Table 3.1). Light absorption caused by multiphonon modes is weaker but it can nevertheless significantly trouble the laser operation at wavelengths nearly two times shorter. Different absorption bands due to two-phonon absorption in AlSb are indicated in the transmission spectrum of AlSb shown in Fig. 3.1. This spectrum gives a general idea about the relative spectral position of different absorption features and their strength because the multiphonon absorption spectra of III-V materials are quite similar if plotted as a function of energy normalized to the frequency of the fundamental phonon absorption [3]. Wavelengths corresponding to the energies of phonons and their combinations for III-V materials used in QCLs are listed in Table 3.1. It should be noted that in alloys the phonon structure is more complicated than in binary compounds. In particular, GaInAs and InAlAs

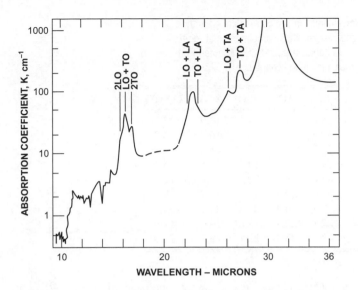

**Fig. 3.1** Absorption coefficient of AlSb vs wavelength at 300 K. Reprinted with permission from [6]. Copyright (1962) by the American Physical Society.

alloys exhibit two-mode phonon behavior [4, 5], which is reflected in Table 3.1. In this chapter we review QCLs emitting between 15 and 28 µm, in the spectral region corresponding to the multiphonon absorption. Historically, 15 µm was a frontier wavelength above which the QCL performances dramatically degraded. Three major material systems are currently used to fabricate QCLs – InGaAs/AlInAs alloys grown on InP, the GaAs/AlGaAs family, and the InAs /AlSb compounds employing InAs as a substrate. The correct choice of the QCL materials can allow nearly full covering of the mid-infrared wavelength range containing regions of strong multiphonon absorption below the Reststrahlen band. We will consider specific properties of the long-wavelength mid-infrared QCLs based on different materials, as well as more general issues related to the QCL design in this long-wavelength frontier of the mid-infrared. Almost all the devices have been grown using molecular-beam epitaxy (MBE); other growth techniques will be mentioned if necessary. Discussing the results, presented in this chapter in the chronological order inside each section, we mostly give the point of view of the authors of the publications, which allows the reader to follow the advance in the QCL philosophy. Some comments on the design of the considered devices based on the modern understanding of the QCL operation are given in the Section 3.6.

## 3.2    InGaAs/AlInAs QCLs

The issues associated with multiphonon absorption were directly observed for the first time in the InP-based QCLs emitting near 16 µm [7]. These lasers were based on the bound-to-continuum scheme [8] with slightly diagonal transitions. Optical confinement in the dielectric waveguide was provided by the low doped InP substrate and a

**Table 3.1** Wavelengths in μm corresponding to the energies of phonons and their combinations at room temperature in materials used in QCLs. Adapted from different sources cited in the chapter.

|                      | LO   | TO   | 2LO  | 2TO  | LO+TO |
|----------------------|------|------|------|------|-------|
| AlAs                 | 25.0 | 27.8 | 12.5 | 13.9 | 13.2  |
| AlSb                 | 31.6 | 33.7 | 15.8 | 16.8 | 16.3  |
| GaAs                 | 34.2 | 37.3 | 17.1 | 18.6 | 17.8  |
| InP                  | 29.0 | 32.9 | 14.5 | 16.5 | 15.4  |
| InAs                 | 41.6 | 46.0 | 20.8 | 23.0 | 21.9  |
| GaInAs$_{GaAs}$      | 37.0 | 39.4 | 18.5 | 19.7 | 19.1  |
| GaInAs$_{InAs}$      | 42.7 | 44.2 | 21.4 | 22.1 | 21.7  |
| AlInAs$_{AlAs}$      | 27.1 | 28.9 | 13.6 | 14.4 | 14.0  |
| AlInAs$_{InAs}$      | 42.2 | 45.4 | 21.1 | 22.7 | 21.9  |

600-nm-thick InGaAs contact layer, heavily doped with Si ($n = 1 \times 10^{18}$ cm$^{-3}$), grown on the top of the structure. The pulsed threshold-current density of 75-μm-wide and 1.6-mm-long Fabry–Pérot lasers was $J_{th} \approx 9$ kA/cm$^2$ at room temperature (RT) and increased up to 10 kA/cm$^2$ at 60 °C, the maximum temperature of the measurements. Between $-40$ and 60 °C, $J_{th}$ increased exponentially with a characteristic temperature of this dependence $T_0 = 234$ K and peak optical power dropped from 400 to 150 mW.

The laser emission spectra, shown in Fig. 3.2(a), consisted of two groups of Fabry–Pérot modes situated around 644 cm$^{-1}$ ($\lambda \approx 15.5$ μm) and 607 cm$^{-1}$ ($\lambda \approx 16.5$ μm, see arrow). The dotted line in this figure represents the two-phonon absorption in InP on a linear arbitrary scale from [3]. A correlation between the minima of the absorption and the lasing frequencies is clearly seen. Distributed feedback lasers (DFB) with gratings etched in the top of the ridges [9] were then fabricated to force laser emission at $\lambda \approx 16.5$ μm and $\lambda \approx 15.5$ μm. The first device exhibited single frequency emission at 16.6 μm at low injection (Fig. 3.2(b)) but the peak at 15.5 μm reappeared with increasing current. Emission of the second device also became multimode at high currents but was always positioned near 15.5 μm (Fig. 3.2(c)). The failure to obtain laser action in the region between the observed two group of modes, which corresponded presumably to the middle of the QCL gain curve, confirmed the presence of a specific loss at these wavelengths highly likely due to the two-phonon absorption in the InP part of the laser waveguide. These results were reproduced in 2017 in the work of A. Szerling et al. [10]. The same QCL structure was grown by gas source molecular-beam epitaxy; the device construction being slightly different. The lasers exhibited higher peak optical power, reaching 720 mW at 303 K, but also higher threshold-current densities, 11.9 kA/cm$^2$ at 298 K for the best devices, and $T_0 = 272$ K. Two distinct mode groups at 15.6 and 16.5 μm could also be visible in their emission spectra whereas no emission was observed in between.

An original *indirect pump* scheme was employed in an attempt to improve performances of QCLs operating near 15 μm [11]. The diagonal bound-to-continuum design incorporated an intermediate state 4 coupled with the last injector level and

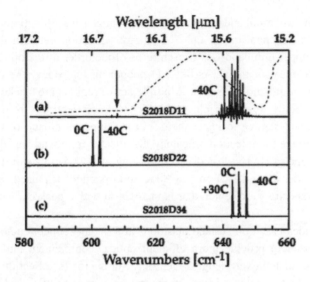

**Fig. 3.2** (a) Emission spectrum of a laser with a Fabry–Pérot resonator; (b) with a grating designed to force lasing at $\lambda \approx 16.5\,\mu m$; (c) with a grating designed for $\lambda \approx 15.5\,\mu m$. Reprinted from [7], with the permission of AIP Publishing.

located above the upper level of the laser transition 3. Electrons injected into this intermediate state rapidly relaxed to state 3 because the distance $E_{43}$ was chosen to be close to the LO-phonon energy that was considered to be $E_{LO} = 30\,meV$ in this work. This $E_{LO}$ corresponds to the energy of the InAs-like phonon in InGaAs lattice matched with the InP substrate (Table 3.1). Compared with [7], in these QCLs $J_{th}$ was reduced to 3.5 kA/cm$^2$ and the slope efficiency was increased from ~40 mW/A to 346 mW/A at RT resulting in a maximum output power of 216 mW in a device with an HR-coated rear facet. The lasers demonstrated remarkable properties in the temperature dependence of the threshold current: the $T_0$ value increased from 150 K in the 77–200 K range up to 450 K between 320 and 380 K. The authors explain this high performance of the IDP QCLs by lower electron populations in injectors at high currents and temperatures compared with the traditional direct pump designs. It is worth noting the considerable shift of the emission line from ~15.5 μm to ~ 14.0 μm when the temperature increased from 290 to 320 K. Such shift provoked by a relatively small increase in temperature could possibly indicate a change in the transition scheme. It should also be noted that a region of strong multiphonon absorption in the employed InP-based waveguide is situated between these wavelengths [3], which could affect the QCL operation.

In 1997 the spectral range accessible with QCLs, initially emitting at 4.26 μm [1], has been extended to 8 μm [12]. A giant step towards the long wavelength mid-infrared has been made in 1999 by demonstration of QCLs emitting at 17 μm [13]. The choice of the photon energy was made considering weak two-phonon absorption in the InP/AlInAs waveguide in this spectral region [3]. The QCL was based on the "chirped" superlattice design. Laser action takes place at the edge of the minigap between two minibands, where well delocalized wavefunctions provide a large dipole matrix element of 4.0 nm. The energy separation between the individual states of the upper miniband was chosen to be of the order of the optical phonon energy $E_{LO} = 34$ meV of the GaAs-like LO phonon in $In_{0.53}Ga_{0.47}As$ (Table 3.1) to ensure fast intraminiband relaxation. Multiple active regions are stacked together, connected by specially designed injectors which extract electrons from the lower miniband of one superlattice to inject them into the upper miniband of the following one. To maximize the injection efficiency the injector miniband was designed to have a funnel shape [14]. The tested 0.72-mm-long and 22.5-μm-wide laser had a threshold-current density ~14 kA/cm$^2$ at 80 K and operated in pulsed mode up to 150 K. Emission spectra were centered around 16.7 μm with a peak optical power of 12 mW at 5 K.

Laser waveguides based on surface plasmons at a metal–semiconductor interface have been proposed to provide more efficient optical confinement and to reduce the total thickness of long wavelengths QCL structures [15] (see Section 3.7). This approach was applied to devices based on the design similar to that used previously in the QCL emitting at 17 μm [13]. The total thickness of the QCL structure with the surface plasmon waveguide was reduced from almost 9 μm to less than 4 μm, which was accompanied by an increase in optical confinement factor $\Gamma$ from 0.41 to 0.81 compared with the devices based on the conventional dielectric waveguide. A metal layer was deposited on the top of the laser ridges to form the waveguide. From the bottom side the optical confinement was provided by the InP substrate. In the lasers with a gold surface-plasmon layer the threshold-current density was 6 kA/cm$^2$, approximately two times lower than in [13]. The peak optical power was significantly increased up to 38 mW at 5 K and the maximum operation temperature $T_{max}$ reached 240 K. The laser line was centered at 16.7 μm at low injection and near 17.1 μm at high currents. The authors then applied to these lasers the concept of a distributed feedback (DFB). A two-metal grating deposited on top of the laser ridges produced a strong index contrast due to the large spatial modulation of the skin depth, thus providing selection of a single-frequency mode with a high side mode suppression ratio [16]. A first-order DFB grating of 10-nm-thick Ti with a period $\Lambda = 2.05$ μm was first formed by optical contact lithography. The subsequent evaporation of 300 nm of gold resulted in an alternate sequence of Ti/Au and pure Au stripes along the surface plasmon propagation. Single-frequency emission at 16.18 μm with a side mode suppression ratio of 30 dB has been obtained from these lasers. DFB QCLs were also fabricated from another wafer with 40 periods of same core-region design, instead of 35. A grating period $\Lambda = 2.0$ μm was used to fabricate these lasers, which should result in a shift of the emission wavelength to about 15.8 μm provided the effective refractive index in the devices did not change. However, the emission wavelength of the new lasers was also close to 16.18 μm at 5 K, increasing to 16.25 μm at 120 K. The authors did not discuss this issue, but it is worth noting that $\lambda = 15.8$ μm corresponds to the middle of the multiphonon absorption feature in InP in this region.

A combination of the chirped-superlattice design and the surface-plasmon wave-guide was used to demonstrate QCLs emitting at 19 μm [17]. The devices emitted peak optical powers of 14 mW at 20 K with a threshold-current density of 4.5 kA/cm$^2$, even smaller than for previous surface-plasmon QCLs operating at 17 μm [16], but had a lower $T_{max}$ (145 K). An important milestone was achieved in 2001 by demonstration of the continuous-wave (CW) operation of QCLs emitting at 19 μm [18]. These devices also exploited the chirped-superlattice design and the gold surface-plasmon layer on the top of the laser ridges. The doping level of the structure was decreased by approximately two times (to $n_s \sim 1.5 \times 10^{11}$ cm$^{-2}$ per period) with respect to [17], which was expected to lead to a reduction in optical losses, weaker thermal backfill-ing, and improved injection efficiency. In pulsed mode, the threshold-current density at liquid helium temperature was reduced to 2 kA/cm$^2$, and at the same time $T_{max}$ increased from 145 to 170 K. The lasers operated in the CW regime up to T = 10 K with $J_{th} = 3.2$ kA/cm$^2$ and the highest output power of about 1 mW. The CW lasing occurred in a single longitudinal mode with a side mode suppression of at least 20 dB.

The first QCLs emitting above 20 μm were reported in 2001 [19] (these results have also been presented in [20]). The surface-plasmon devices grown on InP operated at λ ≈ 21.5 μm and λ ≈ 24 μm. The active regions of the lasers were based on the chirped-superlattice design but the superlattice was formed by only four quantum wells in this case instead of five wells in all previous long-wavelength QCLs using this transition scheme. The core region of the 21.5 and 24 μm QCLs contained 40 and 65 periods, respectively. The optical waveguide was formed from the bottom side by the InP substrate followed by a 1.5-μm-thick low doped n-InGaAs layer and a surface-plasmon carrying gold layer on the top of the devices. A high confinement factor of $\Gamma = 0.76$ and a waveguide loss of $\alpha_w = 42$ cm$^{-1}$ were calculated for the first structure whereas the values for the 24 μm lasers were $\Gamma = 0.88$ and $\alpha_w = 41$ cm$^{-1}$. At liquid-helium temperatures the threshold-current densities were $J_{th} = 4.1$ and 4.6 kA/cm$^2$, respectively, in these lasers, both operating up to 140 K in pulsed mode. The specific choice of the wavelengths in this work was explained by an attempt to match the minima in the In$_{0.53}$Ga$_{0.47}$As and InP two-phonon absorption spectrum. Indeed, the multiphonon absorption in the In$_{0.53}$Ga$_{0.47}$As alloy is weak at these wavelengths [21] and a local minimum of two-phonon absorption in InP is located between 23 and 24 μm [3]. On the other hand, the emission wavelength of the λ ≈ 21.5 μm lasers is located close to the region of two-phonon absorption of InP, but no performance degradation was noticed in these lasers compared to the 24 μm devices. This find-ing can be explained by a relatively weak effect of the substrate absorption in this waveguide. Obviously, the use of the plasmon confinement on the bottom side of the waveguide should allow suppressing optical losses due to the absorption in InP. The first step towards a double-sided surface-plasmon waveguide was also made in this work by replacing the low-doped InGaAs layer below the core region of the 21.5 μm laser by a 0.75-μm-thick heavily doped n$^{++}$-InGaAs layer. The n$^{++}$-layer acted as a "metal," thus forming surface-plasmon optical confinement from the bottom side of the waveguide. The confinement factor of this double-sided surface-plasmon wave-guide was calculated to be extremely high, $\Gamma \sim 0.98$. To keep calculated losses at the same level as in the original device, the number of periods in the core region was doubled. The new lasers exhibited a higher threshold-current density of 7.1 kA/cm$^2$, which was explained by the free-carrier absorption in the heavily doped n$^{++}$-layer. This first working double-plasmon laser validated the concept of such a waveguide configuration that was expected to have better performances at longer wavelengths (see Section 3.7).

At long wavelengths the problems associated with high losses in a plasmon layer made of a doped semiconductor can be resolved by replacing it with a metal. QCLs based on double-metal waveguides emitting at 19, 21, and 24 μm have been reported in [22]. Again, the emission wavelengths have been chosen to lay at pronounced minima of the InGaAs two-phonon absorption spectrum [21]. The active regions and injectors, which were identical to ones employed in [18] and [19], were embedded between 40-nm-thick InGaAs layers doped to $n = 1 \times 10^{17}$ cm$^{-3}$ and sandwiched between n$^{++}$-InGaAs layers (10 nm thick on top and 20 nm thick on bottom) for contact purposes. Seventy-five periods have been grown for the $\lambda = 19$ μm structure, and 80 periods each for the $\lambda = 21$ μm and $\lambda = 24$ μm ones. The total thickness of the structures was between 5.9 and 6.24 μm. To fabricate lasers with the DM waveguide the epitaxial structure was transferred onto another substrate using metal bonding.

After the removal of the initial substrate the device processing was performed on this side of the sample in order to produce ridge lasers with a gold surface-plasmon layer on top. The DM QCLs emitting at 21 μm were compared with conventional surface-plasmon (SP) devices fabricated from the same wafer. The optical confinement factor and waveguide losses were calculated to be $\Gamma = 0.98$ and 0.92, $\alpha_w = 38$ and 40 cm$^{-1}$ for the DM and SP lasers, respectively. The thresholds and their temperature dependence were similar in the two cases (Fig. 3.3), in agreement with the similar waveguide characteristics, but the DM devices exhibited a little bit lower $T_{max}$ (170 K instead of 180 K), which was explained by worse thermal sinking. The threshold-current density at liquid helium temperature was about 2.2 kA/cm$^2$ for both types of the lasers, and 2.5–2.7 kA/cm$^2$ at 80 K, slightly higher for the DM devices. The maximum output optical power, estimated to be a few mW, was similar. The obtained results demonstrated that, for wavelengths around 20 μm, the double-metal waveguide is performing well and is comparable to the single-plasmon one.

Molecular-beam epitaxy is predominantly used for QCL growth due to better control of the layer thickness and interface formation compared with metal-organic vapor-phase epitaxy (MOVPE). The design of long-wavelength QCLs requires the use of very thin barrier layers, often just a few ångströms thick, which further strengthen the requirement for high growth quality. However, MOVPE has been successfully employed to fabricate QCLs emitting at 19 μm [23]. The active region design was based on the layer sequence described in [17], but the injector doping was reduced by a factor of two, as it was done in [18] in order to improve the temperature performance of the devices. Two wafers were grown: a "fast-grown" sample was grown at a rate of 6 Å/s and a "slow-grown" sample was grown at a rate of 1 Å/s. The wafers were processed into ridge lasers with a double-metal waveguide, following the

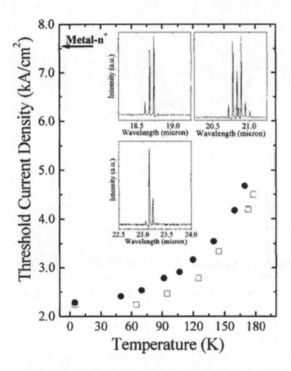

**Fig. 3.3** Temperature dependence of the threshold-current density of a QCL emitting at 21 μm with double-metal waveguide (full circles) and single-sided plasmon waveguide (open squares). Devices are 36 μm wide and 800 μm long. The arrow indicates the low-temperature threshold-current density of an identical device with a double-sided surface-plasmon waveguide [19]; i.e., where a very thick $n^{++}$-type bottom layer acts as a "metal." The insets show emission spectra of the studied devices. Reprinted from [22], with the permission of AIP Publishing.

procedure detailed in [22]. The threshold currents for both structures were comparable, $J_{th} = 1.85$ and $1.75$ kA/cm$^2$ at 80 K for the slow-grown and fast-grown devices, respectively, which are 10% smaller than in [18]. The emission wavelength of the fast-grown laser was shifted to ~18 μm in comparison with the slow-grown device and the MBE-grown QCLs reported in [17] and [18]. This shift could be attributed to larger interface grading at high growth rates. The temperature performance of the devices was comparable to that of the similar MBE-grown structures processed as SP [18] or DM waveguide [22], with $T_{max} = 170$ K for the slow-grown devices and 160 K for the fast-grown lasers. A few years later, a QCL emitting at 23 μm was demonstrated [24]. It is the longest wavelength of lasers grown by MOVPE. This growth technique allowed a waveguide with thick InP cladding layers to be used. The active regions were based on a bound-to-continuum design. The lasers operated only at low temperatures up to 100 K with a $J_{th}$ of about 3.5 kA/cm$^2$. The low $T_{max}$ was attributed to the high losses due to the tail of absorption by the AlAs-like phonon in InAlAs. The obtained results demonstrated that MOVPE can be successfully used for growth of long-wavelength QCLs.

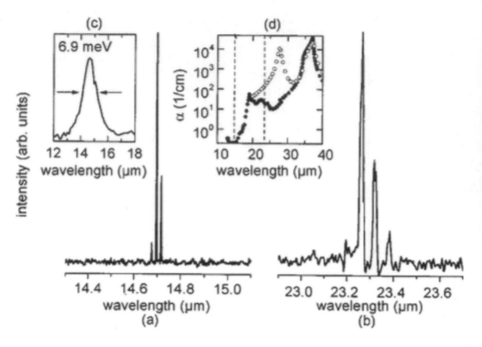

**Fig. 3.4** Laser spectra at a temperature of 5 K (a) of sample A at a current density of 5.1 kA/cm$^2$ and (b) of sample B at a current density of 11.9 kA/cm$^2$; (c) spontaneous emission spectrum of the sample A; and (d) intrinsic absorption coefficient of undoped GaAs (full symbols) and Al$_{0.14}$Ga$_{0.86}$As (open symbols) after Palik [26]. Emission wavelengths of the two samples are marked with dashed lines. Reprinted from [25], with the permission of AIP Publishing.

## 3.3    GaAs/AlGaAs QCLs

Due to the heavy electron effective mass resulting in a lower intersubband gain and the smaller practical conduction-band offset compared with other materials, the GaAs/AlGaAs material system is less employed for mid-infrared QCLs despite its technological maturity. The only long-wavelength, mid-infrared GaAs-based QCLs have been reported in [25]. These lasers based on the chirped superlattice active regions were designed to emit at 15 (sample A) and 23 μm (sample B). The sample A was intended to demonstrate the potential of the band structure scheme in the spectral region of weak free-carrier and multiphonon absorption in GaAs (Fig. 3.4), and where a conventional dielectric waveguide could be still employed without excessively long epitaxial growth. The choice of $\lambda = 23$ μm avoided the multiphonon absorption in GaAs at 19 μm as well as the AlAs-like phonon absorption in AlGaAs at 28 μm. Both lasers were grown by MBE on n-GaAs substrates ($n = 2 \times 10^{18}$ cm$^{-3}$). The Al content of the barriers of $x = 0.45$ for the sample A yielded the maximum $\Gamma$-$\Gamma$ band offset of 390 meV without introducing indirect-barriers states [27]. In sample B, for which carrier leakage into the continuum played a minor role, the Al concentration was chosen to be $x = 0.35$ in order to avoid ultrathin barriers. The plasmon enhanced dielectric waveguide of the sample A consisted of the core region being embedded in two low-doped 3.5-µm-thick GaAs spacers and sandwiched between n$^+$-doped GaAs cladding layers ($n = 2 \times 10^{18}$cm$^{-3}$). Calculated values for the optical confinement factor and waveguide losses were $\Gamma = 0.58$ and $\alpha_w = 26$ cm$^{-1}$, respectively. The surface-plasmon waveguide for sample B also employed the bottom optical confinement formed by the plasma effect of the n$^+$-doping of the GaAs substrate, separated from the core region by a 5-µm-thick low-doped spacer layer. The calculated characteristics of this waveguide were $\Gamma = 0.8$ and $\alpha_w = 86$ cm$^{-1}$. The A-lasers emitted at 14.7 µm with a threshold-current density of 2.2 kA/cm$^2$ at cryogenic temperatures and operated up to $T_{max} = 220$ K in pulsed mode. The sample B emitting around 23.3 µm was characterized by a much higher threshold-current density of 10.2 kA/cm$^2$ and the temperature range of operation was limited to 100 K. The higher threshold of the sample B was explained by the shorter carrier lifetimes at the upper laser-transition level and by the increased waveguide loss from free carriers and multiphonon absorption. Performances of the B-type lasers were poorer compared with their InP-based counterparts emitting at a similar wavelength with $J_{th} = 4.6$ kA/cm$^2$ and $T_{max} = 140$ K [19]. For the sample A the difference vs. the contemporaneous InP-based QCLs [7] was smaller. It is worth noting that the better performances of the InP-based devices could partially be attributed to greater efforts invested in this material system.

## 3.4     InAs/AlSb QCLs

The InAs/AlSb material system is very attractive for use in QCLs due to the small electron effective mass for InAs. As the effective mass increases with energy because of the nonparabolicity of the conduction band, this advantage is not significant in short-wavelength devices but it becomes decisive at long wavelengths, when all transition levels are close to the bottom of the conduction band and the effect of nonparabolicity is weak [28]. Bands of multiphonon absorption in these materials are shifted towards long wavelengths compared with the GaAs/AlGaAs and InGaAs/InAlAs systems, which gives another degree of freedom for the spectral coverage. No two-phonon absorption has been observed in InAs below 20 µm [21], whereas AlSb is free of two-phonon absorption between 17 and 22 µm [3].

The first InAs/AlSb long-wavelength QCLs have been reported in 2013 [29]. The lasers employed a double-phonon extraction scheme with a narrow quantum well next to the injection barrier. This slightly diagonal design was intended to improve the upper-level lifetime experiencing a fast decrease when the transition energy is approaching the LO phonon energy with increasing the emission wavelength. The QCL active region consisted of 80 cascades doped with Si to $n_s = 1 \times 10^{11}$ cm$^{-2}$ per period without considering residual n-type doping of InAs layers. The QCL wafer has been grown by MBE on an n-InAs (100) substrate in a Riber Compact 21 solid-source machine. A 2-µm-thick AlSb$_{0.84}$As$_{0.16}$ stop-etch layer, required for further laser processing, has been grown before the core region of the laser. The DM waveguide was formed by transfer of the laser core region onto a host InAs substrate using wafer bonding with indium. After wafer bonding and substrate removal the samples were

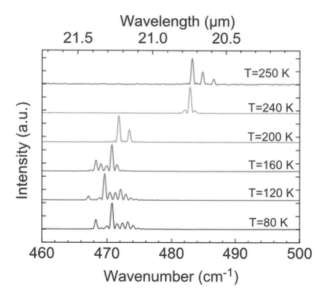

**Fig. 3.5** Pulsed emission spectra of an InAs-based DM QCL emitting above 21 μm. Reprinted from [30], with the permission of AIP Publishing.

processed into 25-μm-wide deep-mesa ridge lasers by contact optical lithography and wet etching. Hard-baked, 4-μm-thick photoresist was used for electrical insulation. The fabricated lasers emitted at 19.5–20.0 μm at 80 K with a threshold-current density of 1.6 kA/cm² at 80 K, thus outperforming the best reported InP-based operating in the same spectral region with ~2 kA/cm² at 19 μm [18] and 2.5–2.7 kA/cm² at 21 μm [22]. The threshold-current density of the fabricated lasers increased up to 2.6 kA/cm² at their maximum operating temperature of 170 K, similar to that of the InP-based QCLs mentioned above. These first results confirmed the advantage of the InAs/AlSb material system for use in long-wavelength QCLs.

A similar active region with doping increased by 50% to $n_s = 1.5 \times 10^{11}$ cm⁻² per period was used in [30] in attempt to enlarge the drive-current dynamic range and thus to increase the QCL operating temperature. In this study the DM waveguide of the lasers was formed using Au-Au thermo-compressive bonding. The threshold-current density for these devices was slightly lower compared with the original design, 1.3 instead of 1.6 kA/cm² at 80 K, whereas the maximum available current density $J_{max}$ increased from 3 to 4 kA/cm² because of the higher doping level. It should be noted that the increased doping had no negative impact on the laser threshold in these lasers. The modifications of the active region resulted in a higher operating temperature $T_{max} = 250$ K. The laser-emission wavelength shifted from 21.3 μm at 80 K to 20.7 μm at 250 K (Fig. 3.5). The longer wavelength relative to the initial design was explained by an 8% higher InAs growth rate. The evolution of the emission wavelength with temperature exhibited a jump around $\lambda = 21$ μm between 200 and 240 K (Fig. 3.5). This spectral band corresponds to the doubled energy of the LO-phonon in InAs, which can be a reason of the unwillingness of the QCL to operate in this region of increased absorption. Another active region, with a vertical transition maximizing the oscillator strength has been studied in this work. The upper-level lifetime limited by LO-phonon scattering was calculated to be 0.76 ps for this design and 1.2 ps for the diagonal-transition structure at 80 K. Calculations taking into account IFR scattering showed its strong effect on the upper-level lifetime, which decreases in this case down to 0.23 and 0.46 ps in the vertical- and diagonal-transition structures, respectively. The QCL active region consisted of 70 cascades doped with Si to $n_s = 3 \times 10^{11}$ cm$^{-2}$ per period. The DM devices fabricated from this wafer emitted around 19 μm with a low $J_{th} = 0.6$ kA/cm$^2$ at 80 K in pulsed mode and $T_{max} = 291$ K, which was the first demonstration of RT QCL operation at wavelengths above 16 μm. The threshold-current density of these lasers at 80 K was three times lower compared with the best InP-based QCLs in this spectral range [18, 23]. It was also two times lower than the $J_{th}$ value of the QCLs fabricated from the first wafer of this study, based on a diagonal design, which was in contradiction with an analysis performed in this publication [30]. The authors explained the obtained result by possibly lower internal losses in the lasers based on the vertical design. This should be reconsidered today, considering the contribution of the thermal backfilling into the lower level and the possible leakage currents, since the injectors are very different in the two designs (see also Section 3.6). Far-field patterns of the emitted laser beam were also measured in this work. It is well known for THz QCLs that the emission from the devices with DM waveguides is very divergent. The origin of this problem stems from the large difference between the thickness $L$ of the active region and the emission wavelength λ. In the THz domain, the ratio $L/\lambda$ is typically 0.1 or less. This very low value results in elevated facet reflectivity of 0.7–0.8 and strong diffraction effects. Near 20 μm, $L/\lambda$ is about 0.36, close to its value for a plane wave and better beam quality was expected for DM devices operating in this range compared with that from THz QCLs. This conclusion was confirmed by the far-field measurements. Output beams of several tested lasers demonstrated a 40° × 60° divergence with a single lobe far-field beam pattern. This result shows that the power extraction from DM QCLs operating near 20 μm can be as efficient as that for mid-IR lasers with dielectric waveguides.

InAs/AlSb QCLs operating at 17.8 μm up to $T_{max} = 333$ K have been reported in [31]. The vertical design from [30] was modified to further enlarge the current dynamical range. For that purpose, the injection-barrier width was reduced from 21 to 18 Å. The fabricated DM devices with Fabry–Pérot resonators exhibited $J_{th} = 1.6$ kA/cm$^2$ at 78 K increasing up to 7 kA/cm$^2$ at $T_{max}$. The higher $J_{th}$ compared with the original design from [30] was explained by increased leakage through the thinner injection barrier. DFB lasers were fabricated from the same wafer using a first-order grating with a period of 2.634 μm inscribed in the sole top metallic contact. The threshold-current density of the DFB lasers increased to 2 kA/cm$^2$ compared with the Fabry–Pérot devices but at higher temperatures the difference was smaller resulting in the same $T_{max} = 333$ K. The lasers demonstrated single-frequency emission at wavelengths increasing from 17.37 to 17.75 μm, in the range of 77–333 K.

The next step in the performance improvement of InAs-based QCLs, in the considered spectral range, was provided by the use of dielectric waveguides similar to those employed in short-wavelength QCLs in this material system [28 and refs

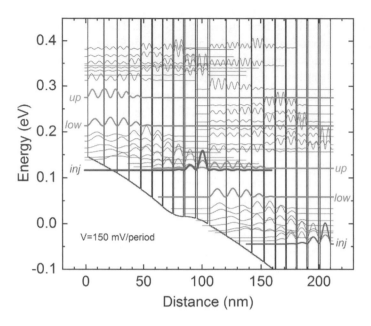

**Fig. 3.6** Electronic band structure of a portion of the laser active region. One period consists of the following layers (in Å and starting from the injection barrier): **18**/139/**1.1**/111/**1.1**/102/**2.6**/ 99/**5.6**/91/**5.6**/85/**5.6**/<u>84</u>/**8.6**/<u>84</u>/**12.6**/90/**15.6**/91, where AlSb layers are in bold and the Si-doped layers ($n = 1.6 \times 10^{17}$ cm$^{-3}$) are underlined. Reprinted with permission from [32], The Optical Society.

therein]. In such plasmon-enhanced waveguides the optical confinement is provided by n-doped InAs cladding layers, separated from the active region by low-doped spacers required to decrease free-carrier absorption. Simulations showed that propagation losses could be considerably decreased in such a waveguide compared with a DM one (see Section 3.7). The first long-wavelength, InAs-based QCLs with a dielectric waveguide (DW) have been reported in [32]. A detailed description of this waveguide and its analysis are also given in Section 3.7. The active region of these lasers had a structure similar to the vertical design used in [30] in QCLs emitting near 18 μm, except for small adjustments of the barrier widths required to shift the transition energy to the desired wavelength of 20 μm. The structure of the active region and its band diagram are shown in Fig. 3.6. Two injector layers were n-doped with silicon, leading to an average dopant concentration per period of $2.7 \times 10^{11}$ cm$^{-2}$. Considering residual n-type doping of InAs layers of the order of $1 \times 10^{16}$ cm$^{-3}$, the total electron sheet density per period was about $4 \times 10^{11}$ cm$^{-2}$. The threshold current density of the lasers emitting at 20 μm was as low as 1.3 kA/cm$^2$ at 80 K, 4.3 kA/cm$^2$ at RT, and increased up to 5.3 kA/cm$^2$ at 350 K with emission around 20 μm. These values were improved compared with the DM QCLs from [30]. The maximum operating temperature of the devices with the dielectric waveguide reached 353 K and RT operation was demonstrated for the first time for any semiconductor laser emitting at such long wavelengths. A high differential gain of the active region of 40 cm/kA at RT was found from the analysis of the threshold-current density in lasers of different length.

However, the high gain was accompanied by a large transparency current that was presumed to be due to both intersubband absorption in the active region and to leakage currents: (a) through the injection barrier to the lower laser level; (b) IFR scattering from the ground injector state through active-region states lying above the upper laser level (see Chapter 2 of this text). Both mechanisms are proportional to the number of electrons, which lead to the conclusion that these losses could be reduced by decreasing the doping level.

InAs/AlSb DW QCLs with a doping level reduced by five times compared with previous lasers were reported in [33]. The design similar to that from [32] was revised to take into account a shift of the quantum levels in the structure. doped only to $8 \times 10^{10}$ cm$^{-2}$ per period with a much weaker band bending. The laser was designed to emit at 15 μm. The fabricated QCLs exhibited significantly better performances compared with the previous devices. The pulsed threshold-current density at RT dropped to 0.73 kA/cm$^2$ for 3.6-mm-long devices and $T_{max}$ exceeded 400 K. The peak optical power from one facet was measured to be 100 mW at 80 K, 50 mW at RT and $\geq 5$ mW at 400 K (without any correction). Both the threshold current and the slope efficiency of the lasers changed slowly at low temperatures, but beyond 200 K their evolution was nearly exponential with the corresponding characteristic temperatures $T_0 = 165$ K and $T_1 = 260$ K. These values are probably limited by the thermally activated LO-phonon- and IFR-assisted carrier leakages through the energy state above the upper laser level, since its energy distance is relatively small (40 meV). The fabricated lasers were tested in the CW regime. The best devices soldered epi-side down were able to operate continuously at temperatures up to 20 °C near 15.1 μm. It should be noted that the longest reported wavelength of RT CW operation of InP-based QCLs is 12.4 μm. The voltage–current and light–current characteristics of a laser measured in the CW regime are shown in Fig. 3.7. In the CW regime the lasers demonstrated single-frequency emission, except at conditions of coexistence of several spatial modes appeared as kinks in the light current curves (Fig. 3.7, at 80 K).

As it was mentioned above, InP-based QCLs failed to emit between 15.6 and 16.5 μm presumably because of two-phonon absorption in InP in this spectral interval [7, 10]. It was interesting to check if the InAs/AlSb QCLs could fill this gap. It should be noted, however, that in this region there is a zone of two-phonon absorption in AlSb. The absorption bands corresponding to 2 LO, LO+TO and 2 TO phonon absorption at RT are located at 15.80, 16.31, and 16.81 μm, respectively [6], which could affect QCL operation in the vicinity of these wavelengths. We scaled the design of the lasers which demonstrated high performances at 15 μm to shift the gain curve to 16 μm. Devices fabricated from the new wafer exhibited similar high performances. The threshold current density of 3.6-mm-long lasers with Fabry–Pérot resonators was as low as 0.8 kA/cm$^2$ in pulsed mode at RT, and the CW regime was obtained up to 290 K. Emission spectra of the lasers were centered near 16.5 μm at RT. To obtain emission in the region of interest, we fabricated DFB lasers with a first-order surface metal grating [34, 35]. The grating was fabricated using electron-beam lithography in combination with dry and wet etching. The DFB lasers mounted epi-side down operated in the CW regime up to 240 K. The single-frequency emission was obtained

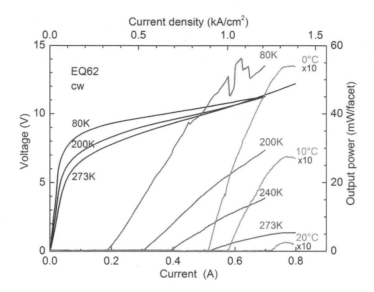

**Fig. 3.7** Voltage–current and light–current characteristics of a 3.6-mm-long and 16-μm-wide laser in the continuous wave regime at different temperatures. Output power at temperatures expressed in °C is multiplied by 10. Reprinted with permission from [33], The Optical Society.

from devices of grating period $\Lambda = 2.39\,\mu m$, and could be continuously tuned by varying the temperature and current from 15.7 to 16.0 μm without any irregularity near the wavelength of 15.8 μm corresponding to the 2 LO phonon absorption in AlSb (Fig. 3.8). The effect of the two-phonon absorption on the optical loss in the studied QCLs can be estimated using the data from [6]. The 2 LO absorption in AlSb is characterized at RT by a barely visible shoulder of about $20\,cm^{-1}$ or even less. The AlSb barriers represent 9% of the total thickness of the core region and the overlap of the optical mode with the core region is 0.56 for this structure. The phonon absorption should therefore result in an additional loss of about $1\,cm^{-1}$ at 15.8 μm. This value is smaller than the empty waveguide absorption of $4\,cm^{-1}$ for the given structure and considerably smaller than the total propagation loss, estimated to be $15\,cm^{-1}$. For this reason, the influence of the two-phonon absorption on the laser operation is quite weak. In the InP-based QCLs which fail to lase between 15.6 and 16.5 μm the additional phonon absorption is due to the 2 TO absorption being stronger ($\approx 50\,cm^{-1}$) than the 2 LO process. Besides, in the asymmetric waveguide of those lasers up to 30% of the optical mode is located in the InP substrate and therefore an additional loss due to the 2 TO phonon absorption is about $15\,cm^{-1}$, which likely forces lasing at wavelengths outside that spectral region.

    Another series of InAs-based DFB QCLs was fabricated from a wafer with a gain curve centered at 17.7 μm [36]. Fabry–Pérot lasers exhibited the same performances as the devices operating near 16 μm. The first-order DFB lasers with a grating period $\Lambda = 2.665\,\mu m$ were designed to provide emission corresponding to the maximum of the gain curve at RT. This favorable configuration allowed us to achieve

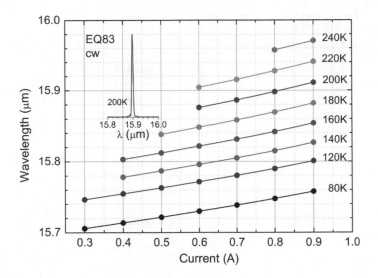

**Fig. 3.8** Emission spectra and tuning range of the InAs/AlSb DFB QCL emitting close to 16 μm measured in the CW regime (our data).

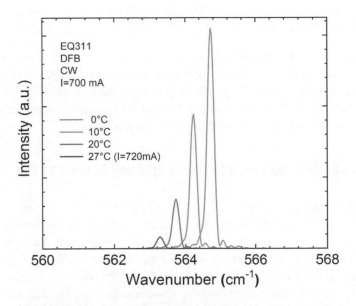

**Fig. 3.9** Emission spectra an InAs/AlSb DFB QCL demonstrating CW operation at 17.8 μm at a temperature of 300 K. Adapted from [36].

single-frequency CW operation at 300 K at a wavelength of 17.8 μm (Fig. 3.9), thus establishing a new record for the wavelength of RT CW operation for semiconductor lasers.

A design like that of the mentioned InAs/AlSb QCLs operating continuously at RT was used to fabricate QCLs emitting above 20 μm [37]. In pulsed mode these lasers with Fabry–Pérot resonators demonstrated threshold-current densities as low

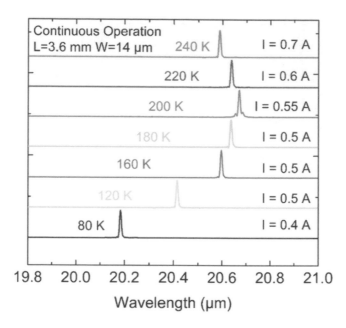

**Fig. 3.10** Emission spectra of an InAs/AlSb QCL emitting beyond 20 μm measured in the CW regime, at different temperatures (our data).

as $1.15\,\text{kA/cm}^2$ at RT and operated up to $T_{max} = 380\,\text{K}$. The emission wavelength increased from 20.4 to 20.75 μm in the temperature range 80–300 K and then decreased again to 20.6 μm at 380 K. The devices were able to work in the CW regime up to $T_{max} = 240\,\text{K}$ (Fig. 3.10). Both pulsed and CW $T_{max}$ values represent the best results to date for semiconductor lasers emitting beyond 20 μm.

## 3.5     Mid-Infrared Quantum Cascade Lasers Emitting Beyond 24 μm

Extending the spectral range of QCL operation beyond 20 μm is a challenging task because one approaches the Reststrahlen band. The fundamental phonon absorption in materials constituting QCLs, which is much stronger than the multiphonon processes, prohibits lasing in the vicinity of these resonances. Besides, the upper-level lifetime in long-wavelength mid-infrared QCLs dramatically decreases as the laser transition energy approaches the LO phonon energy, which makes it difficult to achieve population inversion. From these points of view, both well and barrier materials with small phonon energies are better suited for long-wavelength operation of QCLs in the mid-infrared. As it was already mentioned above, in the 23–24 μm range InP-based QCLs exhibited better performances than GaAs/AlGaAs devices, which operated closer to the Reststrahlen band of the used materials. However, any further wavelength increase is difficult to achieve for InP-based QCLs as well because of the fundamental AlAs-like phonon absorption in the AlInAs barrier layers.

A study of lattice-matched $In_{0.53}Ga_{0.47}As/Al_{0.48}In_{0.52}As$ QCLs with DM waveguides emitting around 24 μm was presented in [38]. The dielectric constant of the

QCL core region was computed considering the two bands of phonon energies in both alloys. As shown in Fig. 3.11(a), three resonant features corresponding to InAs-like, GaAs-like, and AlAs-like phonons, are seen on the real and imaginary parts of the refractive index in the core region. A tail of dispersive feature in the real part by the AlAs-like phonon is visible even at the target energy of 52 meV (24 μm). The computed waveguide loss $\alpha_w$, presented in Fig. 3.11(b) as a function of energy, exhibits a significant increase as the emission energy approaches the AlAs-like phonon resonance. The waveguide loss and the effective refractive index were calculated to be 53 cm$^{-1}$ and 3.17, respectively. The 60-periods QCL core region based on a diagonal bound-to-continuum transition was doped at $2.6 \times 10^{11}$ cm$^{-2}$ per period. The grown wafer was processed into 30-μm-wide Au-Au DM ridge lasers. The lasers operated in pulsed mode up to $T_{max} = 240$ K with threshold-current densities of 5.7 and 18 kA/cm$^2$ at 50 and 240 K, respectively. Emission wavelengths of these lasers exceeded 24 μm for the first time for mid-infrared QCLs operating below the Reststrahlen band. The lasers emitted at 24.1 μm at 50 K and at 23.5 μm at 240 K. Devices fabricated from another wafer with 3% thicker InGaAs wells emitted at 24.4 μm at cryogenic temperatures. From the analysis of the experimental data the gain coefficient in the studied lasers was found to be 11.5 cm/kA. With the assumption that the gain coefficient is constant in the wavelength range from 23 to 28 μm, $J_{th} = 17$ kA/cm$^2$ at a wavelength of 27.3 μm was estimated. To avoid AlAs-like phonon absorption limiting the QCL performances in this spectral region the authors proposed to employ Al-free alloys such as In$_{0.53}$Ga$_{0.47}$As and GaAs$_{0.51}$Sb$_{0.49}$, which would be suited materials for longer wavelengths until absorption of the GaAs-like phonon mode significantly increases.

Such QCLs designed to emit at 28 μm have been fabricated and studied in [39]. The 60-period In$_{0.53}$Ga$_{0.47}$As /GaAs$_{0.51}$Sb$_{0.49}$ core region with a total thickness of 4.2 μm was based on the bound-to-continuum scheme with diagonal transitions.

The wafer was processed into 30-μm-wide Au-Au DM ridge lasers. At the temperature of 10 K, laser emission centered at 28.3 μm was observed with a threshold-current density of 5.5 kA/cm$^2$ (Fig. 3.12). The lasers thus demonstrated the longest emission wavelength for mid-infrared QCLs. The maximum operating temperature of the studied lasers was 175 K, limited by the relatively small available current of 9 kA/cm$^2$. Al-free QCLs emitting at 26.5 and 27.2 μm with similar performances were also reported in this work, as well as traditional In$_{0.53}$Ga$_{0.47}$As/Al$_{0.48}$In$_{0.52}$As lasers operating at 26.3 μm with a threshold-current density of 10 kA/cm$^2$ at cryogenic temperatures. Using an assumed gain coefficient of 11.5 cm/kA and taking the same $J_{max}$ as for the lasers emitting at 28 μm, the longest operation wavelength for the Al-free QCLs was estimated to be 32 μm, limited by a strong absorption tail of the GaAs-like phonons in the materials of the core region. The authors considered alternative materials to further increase the emission wavelength of QCLs. InSb would be attractive for this purpose because of the small phonon energies (22.3 and 23.8 meV for TO and LO phonons, respectively) and also having the smallest effective mass among the III-V semiconductors suited to obtain high intersubband gain. As shown in Fig. 3.13, for similar values of the gain coefficient $g = 11.5$ cm/kA

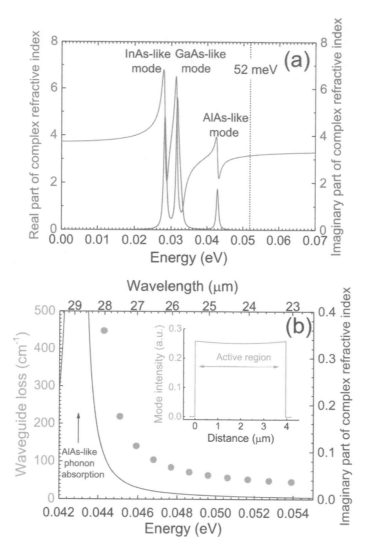

**Fig. 3.11** (a) Computed complex refractive index of the active region. The dielectric constant is linearly interpolated by the relative amount of volume of each alloy semiconductor and the damping coefficient (4 meV) is independent on the energy. The reported TO and LO phonon energies are used; (b) Computed waveguide loss as a function of wavelength. The core-region thickness was chosen to be 4 μm. The inset depicts the mode intensity of the one-dimensional Au-Au DM waveguide. The computed mode confinement factor was 1 in this wavelength range. Reprinted from [38], with the permission of AIP Publishing.

and $J_{max} = 9 \, \text{kA/cm}^2$, the authors predicted the longest operating wavelength around 41 μm ($h\nu = 30 \, \text{meV}$). Other possible candidates are InP quantum wells with lattice-matched $AlAs_{0.56}Sb_{0.44}$ barriers, which are attractive for covering photon energies smaller than the phonon ones. As illustrated in Fig. 3.13, these materials can be more transparent than GaAs/AlGaAs thanks to the larger optical phonon energies of InP (TO = 38 meV, LO = 43 meV) and AlAsSb (TO = 42 meV, LO = 46 meV).

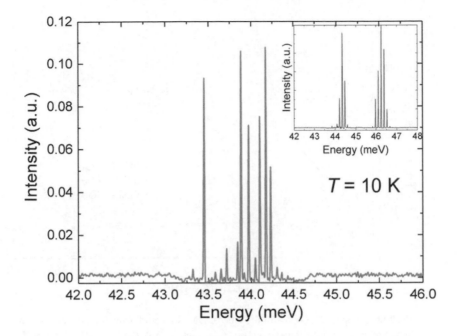

**Fig. 3.12** Laser emission spectra measured below (1.85 A) and above (2.75 A, the inset) the negative differential resistance region, at a wavelength close to 28 μm. Reprinted with permission from [39]. Copyright (2016) American Chemical Society.

The InAs/AlSb system should also be considered as a candidate for QCLs emitting beyond 24 μm and the high performance of the InAs-based QCLs emitting at 20 μm was a good motivation for further increase in the emission wavelength. It is necessary, however, to design the new lasers avoiding spectral bands corresponding to two-phonon absorption in InAs, which are clearly visible in the transmission spectrum of undoped InAs shown in Fig. 3.14. The assignment of the phonon combinations responsible for the observed transmission minima was made following the approach used for AlSb in [6]. The absorption strength of the indicated two-phonon processes is 50–60 cm$^{-1}$ [40] except for a weaker 2LO absorption. It is also necessary to mention the presence of LO+LA and TO+LA absorption in AlSb between 22 and 24 μm [6]. For all these reasons, the wavelength of 25 μm, being less influenced by multiphonon absorption, was chosen as the next target wavelength for InAs/AlSb QCLs [39].

The QCL active region designed to emit at 25 μm was also inspired by the design from [32] and the fruitful approach of reducing the doping level, which was successfully employed in [33]. The upper and lower levels of the laser transition are delocalized over six quantum wells with vertical transitions to maximize the oscillator strength; the injector miniband width was designed to be larger than 120 meV in order to minimize thermal population of the lower laser level. The employed dielectric waveguide was characterized by the overlap of the fundamental mode with the core region of $\Gamma = 57\%$ and internal optical losses $\alpha_w = 8.5$ cm$^{-1}$. The higher

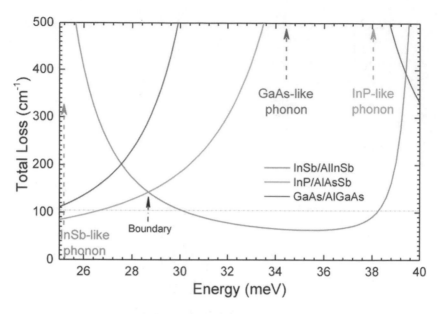

**Fig. 3.13** Computed total loss of the Au-Au DM waveguides for InP/AlAsSb and InSb/AlInSb active regions with the thickness ratio ($L_{btotal}/(L_{wtotal} + L_{btotal}) = 15\%$), where $L_{wtotal}$ and $L_{btotal}$ are the total thickness of wells and barriers, respectively. The core-region thickness is 4.0 μm. The phonon damping constant is assumed to be 8 cm$^{-1}$, being independent of the materials. Lasing would be possible when the loss is below 100 cm$^{-1}$. Reprinted with permission from [39]. Copyright (2016) American Chemical Society.

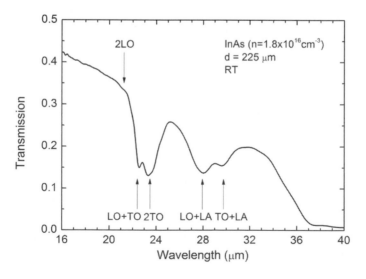

**Fig. 3.14** Transmission spectrum of undoped InAs (our data).

loss compared with previously reported InAs-based QCLs emitting at shorter wavelengths was due to stronger free-carrier absorption in the metal contact. Simulations made for the same core region with a metal–metal waveguide showed the optical

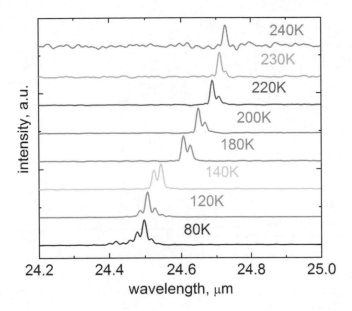

**Fig. 3.15** Pulsed emission spectra of an InAs/AlSb QCL operating above 24 μm. Reprinted with permission from [41]. Copyright (2016) The Institution of Engineering and Technology.

confinement close to unity with $\Gamma = 99\%$ and the optical loss $\alpha_w = 15\,\text{cm}^{-1}$. The $\Gamma/\alpha_w$ ratio is nearly the same in both cases, thus the dielectric waveguide was chosen because of simpler and more reliable device processing. The threshold-current density of the fabricated devices with Fabry–Pérot resonators, in pulsed mode, was $1\,\text{kA/cm}^2$ at 80 K and $1.6\,\text{kA/cm}^2$ at 240 K, the maximum operating temperature of the lasers. The laser emission wavelength varied from 24.4 to 24.7 μm in this temperature range (Fig. 3.15). The multiphonon absorption in this spectral region (Figs. 3.1 and 3.14) and the increased internal losses in the QCLs are certainly responsible for the lower performances compared with operation at $\lambda = 20$ μm. The $J_{\text{th}}$ value was nevertheless much lower compared with those for InP-based QCLs emitting around 24 μm (i.e., $5.7\,\text{kA/cm}^2$ at 50 K in [36]).

Another attempt was made to shift the emission of the InAs-based QCLs further towards longer wavelengths. A QCL structure with a dielectric waveguide was designed to emit at 27 μm. The doping level of $1 \times 10^{11}\,\text{cm}^{-2}$ per period was doubled compared with the previous devices and the thickness of the injection barrier was decreased from 18 to 15 Å in order to increase the maximum available current and thus the operating temperature. The lasers operated in pulsed mode up to 230 K with a threshold-current density of $2.4\,\text{kA/cm}^2$ at 80 K and $4.2\,\text{kA/cm}^2$ at 220 K (Fig. 3.16). The high leakage current observed around a voltage of 5–8 V in these devices is probably due to the increased direct injection to the bottom transition level because of the thinner injection barrier. The emission wavelength was significantly shifted from the targeted value of 27 μm to 25.3 μm (Fig. 3.16), to the region of weaker two phonon absorption in InAs (Fig. 3.14). It should also be mentioned LO+LA and TO+TA absorption in AlSb near 26.3 and 27.5 μm, respectively (Fig. 3.1). Considering also

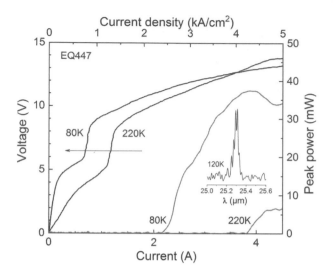

**Fig. 3.16** Voltage–current and light–current characteristics of an InAs/AlSb QCL emitting above 25 μm (our data).

the fundamental phonon absorption in AlSb we can conclude that the 27–32 μm spectral range region is prohibited for InAs/AlSb QCLs (Fig. 3.1). On the other hand, the 33–36 μm region seems to be within reach for such lasers because it is still located away from both LO and TO phonons in InAs (see Table 3.1).

## 3.6        Design of the Active Region of Long-Wavelength Mid-Infrared QCLs

The design of QCL relies on simple principles that can be applied to wavelengths spanning from 2.6 to 200 μm. Still, each spectral range has its own particularities. We will discuss in the following the specificities and challenges associated with the realization of long-wavelength QCLs at wavelengths above 15 μm. The first one consists in the rapid decrease of the upper-state lifetime with increasing wavelength. The second one is the presence of resonant losses in the core region and problems related to the achievement of efficient carrier injection and extraction when the subbands get closer. Finally, we will discuss the issues associated with the waveguide: higher free-carrier absorption, thicker core and cladding layers, and the possibility of using surface plasmons.

An efficient QCL design should provide population inversion between the upper and lower subbands, which generates an optical gain large enough to overcome the global loss of the guided optical mode. This threshold gain value is noted $g_{th}$. In the QCL active region, the peak material gain associated to the transition from an upper to a lower subband is commonly given by

$$G = \frac{e^2\, \hbar f_{ul}}{\varepsilon_0\, n\, c\, m_0\, 2\gamma\, L_p} \left(n_{up} - n_{low}\right), \tag{3.1}$$

where $n$ is the refractive index, $f_{ul}$ the oscillator strength of the transition, $2\gamma$ the full width of the electroluminescence spectrum at half maximum, $L_p$ the length of one period of the active region, and $n_{up}$ (resp. $n_{low}$) the population in the upper (resp. lower) subband. The population inversion, in a simplified three level picture, reads

$$n_{up} - n_{low} = \frac{\eta_i}{e} \tau_{up} \left( 1 - \frac{\tau_{low}}{\tau_{ul}} \right) \cdot J - n_{low}^{therm}, \qquad (3.2)$$

where $\eta_i$ is the injection efficiency, $\tau_{up}$ (resp. $\tau_{low}$) the electron lifetime in the upper (resp. lower) subband, $\tau_{ul}$ the scattering time from upper to lower subband, and $n_{low}^{therm}$ the quasi-equilibrium thermal population of the lower level not coming from sequential tunneling from the upper level (thermal backfilling). Note that $\eta_i$ takes into account carrier leakage from both the injector ground state to the lower subband as well as from the injector ground state and the upper subband through subbands lying above the upper subband (see Chapter 2 of this text).

The upper-state lifetime is set by LO-phonon and IFR scattering to the lower subbands. While IFR scattering depends on the heterostructure design and on the growth quality, LO-phonon scattering is a fundamental mechanism that cannot be avoided. The latter mechanism becomes more and more efficient at longer wavelengths when the intersubband energy decreases and gets closer to the LO-phonon energy. This is shown in Fig. 3.17(a), where the LO-phonon scattering time is calculated as a function of the intersubband transition wavelength in a single QW, for different material systems. Above 15 μm, $\tau_{up}$ is lower than 0.3 ps for GaAs, 0.4 ps for InGaAs on InP, and 0.5 ps for InAs. These values are further reduced by IFR scattering, as shown in [43]. Consequently, a larger current density is required to achieve a given electron population inversion between the upper- and lower-laser states. This effect is compensated by an increase of the oscillator strength (Fig. 3.17(b)), however, and also by the reduction of the intersubband transition linewidth (Fig. 3.17(c)) at longer wavelength. Altogether, the differential gain $g = dG/dJ$ is larger for QCL emitting above 15 μm, as compared to the standard mid-IR QCLs emitting below 10 μm (Fig. 3.17(d)).

This picture also depends on the employed materials. It has been shown [42] that the experimental data on the intersubband peak gain followed the expected dependence on the electron effective mass of the QW material. This is illustrated in Fig. 3.17, where we separated the different contributions to the intersubband gain: upper state lifetime, oscillator strength, and EL linewidth for three material systems GaAs/AlGaAs, InGaAs/AlInAs, and InAs/AlSb. In this calculation, we consider the transition between the first and second subbands in a single QW. This is of course an oversimplified situation, but we will see in the following that it correctly describes the trends observed in the active regions reported in the literature. The upper-state lifetime considered here is the LO-phonon scattering time. In real structures it is further reduced by interface-roughness [43] or alloy-disorder [44] scattering. The oscillator strength is inversely proportional to the effective mass and increases when wider QWs are employed for longer wavelength. The transition linewidth decreases with emission wavelength because it depends predominantly on interface roughness, and its influence becomes weaker in wider active quantum wells of long-wavelength QCLs

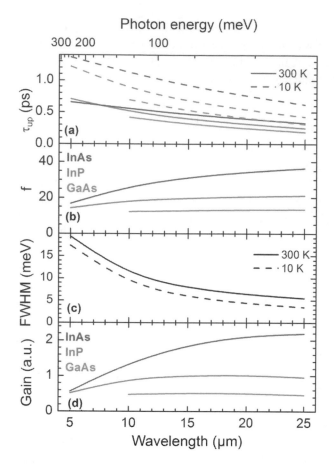

**Fig. 3.17** Estimated intersubband gain and the related parameters for the two first subbands of a single QW as a function of emission wavelength for the three main material systems. The upper state lifetime ($\tau_{up}$) is calculated at 10 K (dashed lines) and 300 K (solid lines) considering only LO-phonon scattering. The full width at half-maximum (FWHM) of the EL linewidth is estimated from empirical values, and supposed not to depend on the material system (our data).

[45]. Typical experimental values of the FWHM at room temperature are 5–10% of the transition energy [47, 48]. The thermally activated intrasubband phonon scattering contributes to a broadening of a few meV at room temperature [45]. This mechanism is almost independent on the transition energy. It becomes therefore a major contribution to the linewidth in long-wavelength QCLs at high temperature. It must be noted that the linewidth is almost independent on the material system, as shown in [42]. This analysis clearly shows that the use of a material with lower effective mass, such as InAs, is beneficial for long-wavelength QCLs thanks to both larger $f_{ul}$ and $\tau_{up}$.

This discussion highlights a very important point for the design of a long-wavelength (LW) QCL active region: in order to compensate for the small upper-state lifetime, a large oscillator strength must be obtained. This is why the first LW QCL [13] employed the concept of a superlattice active region [12], which maximized the oscillator strength for a given intersubband transition energy. With a chirped superlattice active region, the typical oscillator strength is about 50–55 in InP-based QCLs [13, 16, 17, 19, 22, 23] and 30–35 in GaAs-based QCLs [25]. It is in the range 70–75 for similar designs in the InAs/AlSb system using four QW and a three-level resonant-phonon extraction scheme [30–33]. The corresponding differential gain is estimated to be of the order of 20, 30, and 40 cm/kA for GaAs-, InP-, and InAs-based QCLs respectively. This difference between different material systems is not as large as what is shown in Fig. 3.17(d), because of the additional scattering mechanisms, in particular, interface roughness, which is larger in high conduction-band offset systems, and also the larger $L_p$ resulting from the use of wider QWs when the effective mass is smaller.

If high gain values can be achieved at long wavelengths, an important point is that they are obtained for very low population inversion values. For example, a value of $g = 40$ cm/kA was measured at room temperature at $\lambda = 19\,\mu m$ in an InAs/AlSb QCL [32]. With a typical lifetime $\tau_{up}$ of 0.25 ps (0.32 ps for LO-phonon scattering and 1.2 ps for IFR scattering), it means that a gain of 20 cm$^{-1}$ is obtained with a population inversion of only $0.008 \times 10^{11}$ cm$^{-3}$, to be compared to the number of electrons per period of about $3 \times 10^{11}$ cm$^{-3}$. Conversely, a similar population of $0.008 \times 10^{11}$ cm$^{-3}$ in the lower subband will lead to an optical absorption loss of 20 cm$^{-1}$ in the active region. For this reason, $n_{low}^{therm}$ in Eqn. (3.2) may become a dominant term that strongly affects the efficiency of the active region. This mechanism usually referred to as thermal backfilling limits the operation temperature of long-wavelength QCLs. This resonant intersubband absorption in the active region was interpreted as free-carrier losses in the earlier papers of 1998–2002. According to Eqs. (3.1) and (3.2), such resonant losses rapidly increase with the wavelength since they are proportional to $f_{ul}/2\gamma$. They also scale with the doping level, given by the electron sheet density per period $n_s$, and depend exponentially on the injector miniband width ($\Delta$). In the approximation of the Boltzmann electron distribution, the thermal population in the lower subband is given by [46]

$$n_{low}^{therm} = n_s \left( -\frac{\Delta}{2kT} \right) \frac{\sinh\left[ \Delta / \left( 2N_{inj}kT \right) \right]}{\sinh\left[ \left( N_{inj} + 1 \right) \Delta / \left( 2N_{inj}kT \right) \right]}, \tag{3.3}$$

where $N_{inj}$ is the number of subbands in the injector. Quantitatively, for $n_s = 3 \times 10^{11}$ cm$^{-3}$ and $\Delta = 100$ meV we obtain parasitic losses at room temperature as high as 50 cm$^{-1}$ for the above example. This is one of the reasons why the maximum operating temperature $T_{max}$ of most of the long-wavelength QCLs is low. The InP [13] or GaAs [25] SL designs typically employed a value of $\Delta = 70$ meV and $n_s = 2 - 3 \times 10^{11}$ cm$^{-3}$. A solution demonstrated in [33], which showed low $J_{th}$ and $T_{max} = 400$ K, is to use a large $\Delta$ ($>100$ meV) and a significant reduction of $n_s$ ($< 1 \times 10^{11}$ cm$^{-3}$). The reduction of $n_s$ cannot be pushed too far, since it limits the current dynamic range and it leads to an increase of the electron temperature for a given current density. These considerations impose a careful design of the injector region in order to ensure a fast extraction from the lower subband below the lower laser level, e.g., by using successive resonant phonon scattering [33] or resonant phonon scattering to a miniband [50] in an injector with large $\Delta$. Also, one has to avoid parasitic intersubband absorptions which are likely to arise in the injector miniband when $\Delta$ becomes larger than the photon energy.

Finally, one essential parasitic mechanism is the current injection from the last injector states directly to the lower miniband ($J_{leak}$). It increases with the wavelength as the upper and lower subbands get closer. It is experimentally seen from the shoulder at low voltage on the low temperature I–V curves: in [19], we have $J_{leak} \approx 1\,kA/cm^2$ for $\lambda = 21\,\mu m$ and $J_{leak} \approx 2\,kA/cm^2$ for $\lambda = 24\,\mu m$; in [25] $J_{leak}$ is negligible for $\lambda = 15\,\mu m$ but very large ($5\,kA/cm^2$) for $\lambda = 23\,\mu m$. This leakage current increases rapidly at higher temperatures, as seen on the I–V curves in [32]. At room temperature it is responsible for the large conductance at low bias [10]. It can be strongly reduced, with a demonstrated value of $J_{leak} < 0.2\,kA/cm^2$ at 400 K, by using low $n_s$ and an optimized injection region [33]. The leakage or scattering to the subband lying above the upper state is another possible parasitic transport channel, as shown in short wavelength mid-IR QCLs [49, 50]. In long-wavelength QCLs, the distance between the upper laser level ($ul$) and the first above-lying level ($ul$+1) cannot be freely increased like in short wavelength lasers. First, it is difficult to obtain a large level separation when using large QWs; second, the optical transition from $ul$ to $ul$+1 would result in large loss if the level separation gets close to the laser photon energy. In practical designs, this value is set between the phonon energy and the photon energy, with typical values of 30 to 50 meV. However, the electrons in the $ul$+1 subband are partly scattered back to the upper state $ul$ by LO-phonon emission, because the transition energy is close to the phonon energy. This reduces the impact of the proximity of the $ul$+1 level. In addition, IFR scattering also contributes to the fast $ul$+1 $\rightarrow$ $ul$ return [50], while IFR-assisted relaxation from the $ul$+1 level to the lower laser level is less efficient due to the their small probability-functions products at the interfaces in the employed vertical-transition design. Using a rate-equations transport model, we calculated that the LO-phonon-assisted leakage current through the $ul$+1 level is only 5–10% in the InAs/AlSb long-wavelength QCLs at room temperature. However, as shown in [50], IFR-scattering-assisted leakage current through the $ul$+1 level can be several times larger than LO-phonon-assisted leakage; thus, the total leakage current through the $ul$+1 level may well be larger than 5–10% of the threshold current.

The parasitic-injection current leakage can also be limited by using a diagonal transition with an upper state bound close to the injection barrier. It has the benefit of a better tunneling injection efficiency with an expected reduction of $J_{leak}$. Such a design also allows increasing $\tau_{up}$ and the population inversion. The associated reduction of $f_{ul}$ is beneficial to reduce the resonant loss related to the thermal backfilling and does not impact much the differential gain since it is compensated by the longer $\tau_{up}$. This solution was used in a InGaAs/InAlAs QCL emitting at 16 μm with a bound-to-continuum scheme [7]. The $T_{max}$ was improved up to 350 K, but the threshold-current density was still large ($9\,kA/cm^2$ at room temperature) probably because of a too large doping level. The continuum of lower states was advantageously replaced by a unique delocalized subband at similar wavelength [11, 51], which resulted in concentration of the oscillator strength in a single transition thus providing an increase in the gain. In addition, the electron density was reduced in the latter work to about $n_s = 1 \times 10^{11}$ cm$^{-3}$. These improvements have led to very high $T_{max} > 390$ K and low $J_{th}$ at room temperature of 2 kA/cm$^2$ at $\lambda = 14$ μm [51] and 3.5 kA/cm$^2$ at $\lambda = 15$ μm [11].

In the InAs/AlSb system, a comparison of vertical and diagonal transitions at $\lambda \approx 20$ μm has shown that, despite better theoretical predictions, the diagonal design exhibited lower performances [30]. The increased interface roughness scattering in the diagonal structure was considered by the authors and could not explain the experimental difference. Here, we think that this particular result could be due to the much lower injector width in the diagonal design ($\Delta \approx 60$ meV) as compared to the vertical design ($\Delta \approx 110$ meV) or to a broader transition in the diagonal case. Besides, in the analysis given in this publication the IFR-related carrier-leakage processes involving the $ul+1$ level [50] were not considered. They can, however, be stronger in the diagonal design compared with the previously discussed vertical structures. We still believe that a slightly diagonal design would indeed be beneficial if the other aspects of the design are also optimized. It should finally be noted that the transition energy in diagonal designs is much more sensitive to the exact operating electric field in long-wavelength QCLs as compared to standard mid-IR QCLs because of the much wider QW in the active region.

In summary, the ideal QCL active region for long-wavelength operation must satisfy several equally important conditions. The doping level must be low ($n_s < 1 \times 10^{11}$ cm$^{-3}$); the injector miniband must be large ($\Delta > 100$ meV); the laser transition must be slightly diagonal but preferentially between two single subbands; the injector structure must prevent intra-miniband absorption and parasitic injection to the lower subbands. Also, the use of materials with lower electron effective mass is beneficial to obtain larger intersubband gain.

As an illustration, we present in the following an analysis of the main characteristics of state-of-the-art InAs/AlSb QCLs emitting around 17 μm. We used a transport model based on the rate equations method, considering the scattering mechanisms between all subbands of one period of the core region and the tunneling between subbands of adjacent periods. The total gain (including resonant and non-resonant intersubband absorption) was derived from the populations of the subbands. The calculated gain and voltage are plotted in Fig. 3.18 as a function of the current density for different laser temperatures. It was confronted to experimental data. A good agreement was obtained for the I–V curves. Most interestingly, we compared the calculated gain with the experimental points defined by $g_{th}$ and $J_{th}$ obtained for different cavity lengths and different temperatures. An excellent agreement is obtained, using the nominal waveguide parameters $\alpha_w = 3$ cm$^{-1}$ and $\Gamma = 0.63$ (cf. next section) and assuming that the intersubband linewidth increased from 6.3% to 12% of the transition energy when the temperature varied from 80 K to 400 K. This confirmed that very large differential gain of 50 cm/kA at low T and 20 cm/kA at RT has been achieved, together with the control of low leakage currents ($<0.2$ kA/cm$^2$) and low resonant loss, as shown by the low values of the transparency current that were obtained even at high temperature.

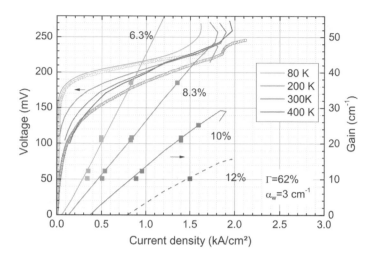

**Fig. 3.18** Calculated gain and V–I curves for an InAs/AlSb QCL at $\lambda \approx 17\,\mu m$ at different temperatures (solid lines). The experimental V–I curves are shown for 80 K and 300 K (circles). The squares represent the experimental points $(J_{th}, g_{th})$, where $J_{th}$ is the measured threshold and $g_{th}$ is equal to the total loss in the waveguide, for different cavity lengths and temperatures. The gain linewidth is indicated for each temperature as a percentage of the ISB transition energy (our data).

## 3.7    Waveguide Issues

The waveguide is, after the core region, the second key element for a QCL. It is characterized by the propagation loss $\alpha_w$ and the overlap factor with the core region $\Gamma$ for the fundamental confined mode. The optical cavity formed by the cleaved facets also introduces mirror losses $\alpha_m$. These optical parameters allow defining a threshold gain $g_{th} = (\alpha_w + \alpha_m)/\Gamma$ as a figure of merit for a given waveguide structure.

In the mid-IR the waveguide is ideally composed of low-loss dielectric cladding layers on both sides of the core region (CR) often separated by low doped spacer layers. When the wavelength is increased, the thickness of all these layers must scale with the wavelength. This eventually becomes difficult to handle with the usual MBE growth techniques. Another issue with long-wavelength waveguides is the emergence of large optical losses due to intrinsic material properties such as multiphonon absorption, as already thoroughly discussed above, free-carrier absorption in the doped or metallic layers.

For mid-IR InP-based QCLs the best solution for the waveguide consists in using low doped InP cladding layers which have a low refractive index of about 3.1. The substrate is generally used as a bottom cladding, whereas the upper cladding can be made of AlInAs or InP, depending on the available growth techniques. The first long-wavelength QCL, emitting at 17 μm, employed an InP/CR/AlInAs waveguide [13] with a 3.7 μm-thick top cladding n-doped to $1\ 2 \times 10^{17}\ cm^{-3}$. The authors estimated the loss due to free carriers to $\alpha_w = 140\ cm^{-1}$ and $\Gamma = 0.48$, which gave a very large value $g_{th} = 330 \, \text{cm}^{-1}$. Air has been used for the top optical confinement in the work of [7], with the core region embedded between two low-doped InGaAs waveguiding layers. The loss was then reduced to $\alpha_w = 30 \, \text{cm}^{-1}$ at the wavelength of 16 μm. However, such approach required to implement lateral current injection on the top of the ridges, which was a major drawback for injection uniformity and thermal management. A variant of such a waveguide has been reported in [10], with the use of $SiO_2$ as a cladding on the top of the ridges. The loss could be significantly reduced in an InP/CR/InAlAs waveguide by lowering the doping level of the waveguide and cladding layers to only $5 \times 10^{16} \, \text{cm}^{-3}$ [51]. It allowed reducing the threshold-current density to $2 \, \text{kA/cm}^2$ at room temperature with a $T_{max}$ of 390 K in pulsed mode in a QCL emitting at 14 μm. This approach has not been extended to longer wavelength, probably because of the difficulty to handle the growth of much thicker InAlAs layers. Also, InAlAs is not very attractive because it has very poor thermal conductivity. A thick InP top cladding layer grown by MOVPE was successfully used for the waveguide of QCL emitting at 15 μm [11]. Thanks to the low loss of the empty waveguide, estimated to be only $3 \, \text{cm}^{-1}$, the lasers operated up to 390 K in pulsed mode and exhibited a threshold-current density of $3.5 \, \text{kA/cm}^2$ at room temperature. The same waveguide structure, also grown by MOVPE, has been used in QCLs emitting at 23 μm that operated at cryogenic temperatures with a reduced threshold-current density ($3.5 \, \text{kA/cm}^2$ at 80 K) [24], as compared to previous results using surface-plasmon waveguides ($5 \, \text{kA/cm}^2$ at 80 K) [19].

The use of surface plasmons at the interface of a semiconductor and a metal has been proposed by the Bell Labs group as an efficient optical confinement solution for long-wavelength QCLs [16]. The major advantage of this solution is the strong reduction of the total thickness of semiconductor to be grown that permits its use up to very long wavelength, such as in THz QCLs. In the QCLs with single-sided surface-plasmon (SP) waveguides, the modes are guided by the metal of the top contact. On the bottom side, additional dielectric confinement of the mode is achieved by the refractive index contrast between the CR and the InP substrate. The mode overlap with CR is very high with Γ ranging from 0.8 to 0.95. The propagation losses are partly due to the metal but also to significant absorption in the substrate and in the doped contact layers. The total loss was as high as $60 \, \text{cm}^{-1}$ at wavelengths close to 20 μm [17], but later it could be reduced to $40 \, \text{cm}^{-1}$ [19, 22] or even $30 \, \text{cm}^{-1}$ [18] with doping optimizations at similar wavelengths.

An alternative solution is the double metal (DM) surface-plasmon waveguide [22]. In this configuration that allows suppressing losses in the substrate the core region is embedded between two metal layers. The device fabrication is more demanding, since it requires a metallic bonding of the epitaxial wafer on a host substrate and removal of the initial substrate. Nevertheless, similar performances were obtained when comparing SP and DM lasers [22] at a wavelength of 21 μm. It has been used up to $\lambda = 28$ μm for InP-based QCLs [39] and also for the first InAs-based QCLs operating close to 20 μm [29, 30].

Performance of SP and DM waveguides are ultimately limited by the absorption in the metal layers. The corresponding $g_{th}$ can be calculated for empty waveguides

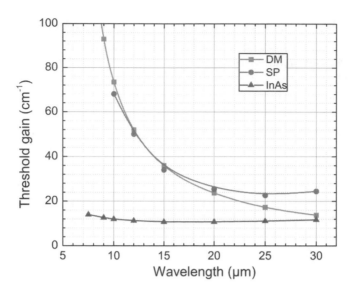

**Fig. 3.19** Calculated $g_{th}$ for 3-mm-long SP and DM empty waveguides, i.e., considering only the propagation losses in the metal. The metal layers are 20 nm Ti/200 nm Au and the thickness of the CR is $\lambda/n$, where $n \approx 3.3$ is the refractive index of the AZ. The threshold gain $g_{th}$ is also calculated for a 3-mm-long plasmon-enhanced InAs waveguide consisting of a CR of the same thickness, enclosed between low doped $\lambda/2n$-thick InAs spacers and InAs cladding layers n-doped to a level such as the refractive index difference with the CR is 0.4. The CR losses and phonon absorption are not taken into account (our data).

(Fig. 3.19). These values do not depend on the materials used in the core region. In InP- or GaAs-based lasers, unavoidable additional losses come from the doped contact layers. This is not the case for InAs QCLs, where heavy doping of contact layers is not necessary thanks to the absence of a Schottky barrier at the semiconductor/metal interface. Although the data in Fig. 3.19 correspond to the best achievable performances for a metal waveguide, such values of $g_{th}$ – significantly larger than 20 cm$^{-1}$ up to 25 μm – are the major obstacle for the achievement of low-threshold QCLs operating at high temperature.

For GaAs- or InAs-based QCLs, n-doped semiconductors can be used as cladding layers, to form the so-called plasmon-enhanced waveguide [52]. Their refractive index depending on the doping level via the change in the plasma frequency is well described by the Drude model. However, this effect is intrinsically associated with strong free-carrier absorption in the doped material. For that reason, low-doped layers are usually inserted between the core region and the doped cladding layers in order to reduce the propagation loss of the confined mode. Plasmon-enhanced waveguides were used in GaAs-based QCLs up to a wavelength of 15 μm [25]. This is also the well-established best solution for InAs/AlSb QCLs in the mid-IR [28, 52]. We also demonstrated that it could be used for long-wavelength QCL with advantageous performances as compared to DM waveguides [32, 33]. Its drawback is still the challenging growth of very thick structures with a total thickness of the order of the wavelength. However, the waveguide material is a binary compound which does not require any composition

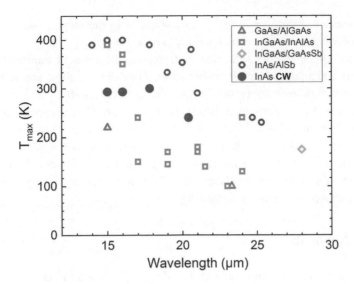

**Fig. 3.20** Maximum operating temperature of mid-infrared QCLs emitting above 15 μm as function of wavelength.

control to be lattice matched with the substrate. Hence, thick InAs or GaAs layers are much easier to grow as compared to InGaAs or InAlAs on InP substrate.

A good tradeoff can be obtained between low loss/low overlap with thick spacers and high loss/high overlap with thin spacers, giving typical value of $\alpha_w = 3 \, \text{cm}^{-1}$ and $\Gamma = 0.6$ (see Fig. 3.18). This tradeoff is almost independent on the wavelength when considering normalized thicknesses as shown in Fig. 3.19. Hence, in the range 15–25 μm, the InAs plasmon-enhanced waveguide has very attractive properties as compared to SP or DM waveguides.

## 3.8    Conclusion

We could see from the above review and discussions that long-wavelength mid-infrared QCLs have many particularities as compared to more conventional QCLs emitting below 10 μm. It can be summarized by the following conclusion: they offer high gain, but have to face high optical and carrier losses of different origins. The most important source of optical losses is the phonon absorption in the spectral region approaching the Reststrahlen band. This intrinsic limitation sets forbidden or favorable spectral areas depending on the employed materials. The main experimental results were obtained using InP- or InAs-based material systems. Figure 3.20 represents the state of the art of long-wavelength mid-infrared quantum cascade lasers. It displays the maximum operating temperature of mid-infrared QCLs emitting above 15 μm.

InP-based QCLs have demonstrated room-temperature pulsed operation up to a wavelength of 16 μm and the longest emission wavelength of 28 μm at low temperatures. They did not allow to cover some spectral bands around 16 and 20 μm. Besides, strong free-carrier absorption in the contact layers or in the metal is still a limiting factor for reducing threshold-current densities above 15 µm.

InAs-based QCLs exhibited the best performances in terms of threshold-current densities and maximum operating temperatures. They did not show any restrictions due to phonon absorptions up to 21 µm, the current spectral limit for room-temperature operation. Forbidden bands in InAs/AlSb QCLs appear at longer wavelengths, between 22 and 24 µm and in the 27–32 µm range. Thanks to the low loss plasmon-enhanced waveguide and high gain core region, continuous wave operation was achieved at room temperature up to 18 µm and up to 20 µm at temperatures accessible using a Peltier cooler. This opens the way to new applications based on molecular spectroscopy in this spectral range.

## Acknowledgements

Some previously unpublished results on InAs/AlSb QCLs were obtained with technical assistance of Zeineb Loghmari and Nguyen-Van Hoang. We acknowledge our colleagues for their help.

## References

[1]  J. Faist, F. Capasso, D. L. Sivco, C. Sirtori, A. L. Hutchinson, and A. Y. Cho, "Quantum cascade laser," *Science*, vol. 264, pp. 553–556, 1994.

[2]  R. Ferreira and G. Bastard, "Evaluation of some scattering times for electrons in unbiased and biased single- and multiple-quantum-well structures," *Physical Review B*, vol. 40, no. 2, pp. 1074–1086, 1989.

[3]  E. Koteles and W. Datars, "Two-phonon absorption in InP and GaP," *Solid State Communications*, vol. 19, no. 3, pp. 221–225, 1976.

[4]  J. Groenen, R. Carles, G. Landa, C. Guerret-Piécourt, C. Fontaine, and M. Gendry, "Optical-phonon behavior in $Ga_{1-x}In_xAs$: The role of microscopic strains and ionic plasmon coupling," *Physical Review B*, vol. 58, no. 16, pp. 10452–10462, 1998.

[5]  A. Milekhin, A. Kalagin, A. Vasilenko, A. Toropov, N. Surovtsev, and D. Zahn, "Vibrational spectroscopy of InAlAs epitaxial layers," *Journal of Applied Physics*, vol. 104, no. 7, p. 073516, 2008.

[6]  W. Turner and W. Reese, "Infrared lattice bands in AlSb," *Physical Review*, vol. 127, no. 1, pp. 126–131, 1962.

[7]  M. Rochat, D. Hofstetter, M. Beck, and J. Faist, "Long-wavelength ($\lambda \approx 16\,\mu m$), room-temperature, single-frequency quantum-cascade lasers based on a bound-to-continuum transition," *Applied Physics Letters*, vol. 79, no. 26, pp. 4271–4273, 2001.

[8]  J. Faist, M. Beck, T. Aellen, and E. Gini, "Quantum-cascade lasers based on a bound-to-continuum transition," *Applied Physics Letters*, vol. 78, no. 2, pp. 147–149, 2001.

[9]  D. Hofstetter, J. Faist, M. Beck, A. Müller, and U. Oesterle, "Demonstration of high-performance 10.16 µm quantum cascade distributed feedback lasers fabricated without epitaxial regrowth," *Applied Physics Letters*, vol. 75, no. 5, pp. 665–667, 1999.

[10]  A. Szerling, S. Slivken, and M. Razeghi, "High peak power 16 µm InP-related quantum cascade laser," *Opto-Electronics Review*, vol. 25, no. 3, pp. 205–208, 2017.

[11] K. Fujita, M. Yamanishi, T. Edamura, A. Sugiyama, and S. Furuta, "Extremely high $T_0$-values ($\sim$450 K) of long-wavelength ($\sim$15 μm), low-threshold-current-density quantum-cascade lasers based on the indirect pump scheme," *Applied Physics Letters*, vol. 97, no. 20, p. 201109, 2010.

[12] G. Scamarcio, "High-power infrared (8-μm wavelength) superlattice lasers," *Science*, vol. 276, no. 5313, pp. 773–776, 1997.

[13] A. Tredicucci, C. Gmachl, F. Capasso, D. Sivco, A. Hutchinson, and A. Cho, "Long wavelength superlattice quantum cascade lasers at $\lambda \simeq 17$ μm," *Applied Physics Letters*, vol. 74, no. 5, pp. 638–640, 1999.

[14] J. Faist, F. Capasso, C. Sirtori, D. Sivco, J. Baillargeon, A. Hutchinson, S. Chu, and A. Cho, "High power mid-infrared ($\lambda \sim 5$ μm) quantum cascade lasers operating above room temperature," *Applied Physics Letters*, vol. 68, no. 26, pp. 3680–3682, 1996.

[15] C. Sirtori, C. Gmachl, F. Capasso, J. Faist, D. Sivco, A. Hutchinson, and A. Cho, "Long-wavelength ($\lambda \approx 8$–11.5 μm) semiconductor lasers with waveguides based on surface plasmons," *Optics Letters*, vol. 23, no. 17, p. 1366, 1998.

[16] A. Tredicucci, C. Gmachl, F. Capasso, A. Hutchinson, D. Sivco, and A. Cho, "Single-mode surface-plasmon laser," *Applied Physics Letters*, vol. 76, no. 16, pp. 2164–2166, 2000.

[17] A. Tredicucci, C. Gmachl, M. Wanke, F. Capasso, A. Hutchinson, D. Sivco, S. Chu, and A. Cho, "Surface plasmon quantum cascade lasers at $\lambda \sim 19$ μm," *Applied Physics Letters*, vol. 77, no. 15, pp. 2286–2288, 2000.

[18] R. Colombelli, A. Tredicucci, C. Gmachl, F. Capasso, D. Sivco, A. Sergent, A. Hutchinson, and A. Cho, "Continuous wave operation of $\sim$19 μm surface-plasmon quantum cascade lasers," *Electronics Letters*, vol. 37, no. 16, p. 1023, 2001.

[19] R. Colombelli, F. Capasso, C. Gmachl, A. Hutchinson, D. Sivco, A. Tredicucci, M. Wanke, A. Sergent, and A. Cho, "Far-infrared surface-plasmon quantum-cascade lasers at 21.5 μm and 24 μm wavelengths," *Applied Physics Letters*, vol. 78, no. 18, pp. 2620–2622, 2001.

[20] R. Colombelli, F. Capasso, A. Straub, C. Gmachl, M. Blakey, A. Sergent, D. Sivco, A. Cho, K. West, and L. Pfeiffer, "FIR quantum cascade lasers at $\lambda > 20$μm and THz emitters at $\lambda = 80$ μm," *Physica E: Low-dimensional Systems and Nanostructures*, vol. 13, no. 2–4, pp. 848–853, 2002.

[21] E. Koteles and W. Datars, "Two-phonon absorption in InSb, InAs, and GaAs," *Canadian Journal of Physics*, vol. 54, no. 16, pp. 1676–1682, 1976.

[22] K. Unterrainer, R. Colombelli, C. Gmachl, F. Capasso, H. Hwang, A. Sergent, D. Sivco, and A. Cho, "Quantum cascade lasers with double metal-semiconductor waveguide resonators," *Applied Physics Letters*, vol. 80, no. 17, pp. 3060–3062, 2002.

[23] J. Fan, M. Belkin, M. Troccoli, S. Corzine, D. Bour, G. Höfler, and F. Capasso, "Double-metal waveguide $\lambda \simeq 19$μm quantum cascade lasers grown by metal organic vapour phase epitaxy," *Electronics Letters*, vol. 43, no. 23, p. 1284, 2007.

[24] F. Castellano, A. Bismuto, M. Amanti, R. Terazzi, M. Beck, S. Blaser, A. Bächle, and J. Faist, "Loss mechanisms of quantum cascade lasers operating close to optical phonon frequencies," *Journal of Applied Physics*, vol. 109, no. 10, p. 102407, 2011.

[25] J. Ulrich, J. Kreuter, W. Schrenk, G. Strasser, and K. Unterrainer, "Long wavelength (15 and 23 μm) GaAs/AlGaAs quantum cascade lasers," *Applied Physics Letters*, vol. 80, no. 20, pp. 3691–3693, 2002.

[26] E. D. Palik, *Handbook of Optical Constants of Solids*, Academic, San Diego, 1985.

[27]  H. Page, C. Becker, A. Robertson, G. Glastre, V. Ortiz, and C. Sirtori, "300 K operation of a GaAs-based quantum-cascade laser at $\lambda \approx 9\,\mu m$," *Applied Physics Letters*, vol. 78, no. 22, pp. 3529–3531, 2001.

[28]  A. N. Baranov and R. Teissier, "Quantum Cascade Lasers in the InAs/AlSb Material System," *IEEE Journal of Selected Topics in Quantum Electronics*, vol. 21, no. 6, pp. 85–96, 2015.

[29]  M. Bahriz, G. Lollia, A. Baranov, P. Laffaille, and R. Teissier, "InAs/AlSb quantum cascade lasers operating near 20 µm," *Electronics Letters*, vol. 49, no. 19, pp. 1238–1240, 2013.

[30]  D. Chastanet, G. Lollia, A. Bousseksou, M. Bahriz, P. Laffaille, A. N. Baranov, F. Julien, R. Colombelli, and R. Teissier, "Long-infrared InAs-based quantum cascade lasers operating at 291 K ($\lambda = 19\,\mu m$) with metal-metal resonators," *Applied Physics Letters*, vol. 104, no. 2, p. 021106, 2014.

[31]  D. Chastanet, A. Bousseksou, G. Lollia, M. Bahriz, F. Julien, A. N. Baranov, R. Teissier, and R. Colombelli, "High temperature, single mode, long infrared ($\lambda = 17.8\,\mu m$) InAs-based quantum cascade lasers," *Applied Physics Letters*, vol. 105, no. 11, p. 111118, 2014.

[32]  M. Bahriz, G. Lollia, A. N. Baranov, and R. Teissier, "High temperature operation of far infrared ($\lambda \approx 20\,\mu m$) InAs/AlSb quantum cascade lasers with dielectric waveguide," *Optics Express*, vol. 23, no. 2, p. 1523, 2015.

[33]  A. N. Baranov, M. Bahriz, and R. Teissier, "Room temperature continuous wave operation of InAs-based quantum cascade lasers at 15 µm," *Optics Express*, vol. 24, no. 16, p. 18799, 2016.

[34]  J. Faist, C. Gmachl, F. Capasso, C. Sirtori, D. Sivco, J. Baillargeon, and A. Cho, "Distributed feedback quantum cascade lasers," *Applied Physics Letters*, vol. 70, no. 20, pp. 2670–2672, 1997.

[35]  O. Cathabard, R. Teissier, J. Devenson, and A. N. Baranov, "InAs-based distributed feedback quantum cascade lasers," *Electron. Lett.*, vol. 45, pp. 1028–1030, 2009.

[36]  H. N. Van, Z. Loghmari, H. Philip, A. N. Baranov, and R. Teissier, "Long wavelength ($\lambda > 17\,\mu m$) distributed feedback quantum cascade lasers operating in a continuous wave at room temperature," *Photonics*, vol. 6, no. 1, 31, 2019.

[37]  Z. Loghmari, M. Bahriz, A. Meguekam, H. Nguyen Van, R. Teissier, and A. N. Baranov, "Continuous wave operation of InAs-based quantum cascade lasers at 20 µm," *Applied Physics Letters*, vol. 115, no. 15, p. 151101, 2019.

[38]  K. Ohtani, M. Beck, and J. Faist, "Double metal waveguide InGaAs/AlInAs quantum cascade lasers emitting at 24 µm," *Applied Physics Letters*, vol. 105, no. 12, p. 121115, 2014.

[39]  K. Ohtani, M. Beck, M. Süess, J. Faist, A. Andrews, T. Zederbauer, H. Detz, W. Schrenk, and G. Strasser, "Far-Infrared quantum cascade lasers operating in the AlAs phonon Reststrahlen band," *ACS Photonics*, vol. 3, no. 12, pp. 2280–2284, 2016.

[40]  O. Lorimor and W. Spitzer, "Infrared refractive index and absorption of InAs and CdTe," *Journal of Applied Physics*, vol. 36, no. 6, pp. 1841–1844, 1965.

[41]  M. Bahriz, R. Teissier, A. N. Baranov, Z. Loghmari, and A. Meguekam-Sado, "InAs-based quantum cascade lasers emitting close to 25 µm," *Electron. Lett.*, vol. 55, pp. 144–146, 2019.

[42]  E. Benveniste, A. Vasanelli, A. Delteil, J. Devenson, R. Teissier, A. N. Baranov, A. M. Andrews, G. Strasser, I. Sagnes, and C. Sirtori, "Influence of the material parameters on quantum cascade devices," *Appl. Phys. Lett.*, vol. 93, pp. 131108-1–131108-3, 2008.

[43] Y. Chiu, Y. Dikmelik, P. Liu, N. Aung, J. Khurgin, and C. Gmachl, "Importance of interface roughness induced intersubband scattering in mid-infrared quantum cascade lasers," *Applied Physics Letters*, vol. 101, no. 17, p. 171117, 2012.

[44] A. Vasanelli, A. Leuliet, C. Sirtori, A. Wade, G. Fedorov, D. Smirnov, G. Bastard, B. Vinter, M. Giovannini, and J. Faist, "Role of elastic scattering mechanisms in GaInAs/AlInAs quantum cascade lasers," *Appl. Phys. Lett.*, vol. 89, no. 17, p. 172120, 2006.

[45] T. Unuma, M. Yoshita, T. Noda, H. Sakaki, and H. Akiyama, "Intersubband absorption linewidth in GaAs quantum wells due to scattering by interface roughness, phonons, alloy disorder, and impurities," *Journal of Applied Physics*, vol. 93, p. 1586, 2003.

[46] D. Barate, R. Teissier, Y. Wang, and A. N. Baranov, "Short wavelength intersubband emission from InAs/AlSb quantum cascade structures," *Applied Physics Letters*, vol. 87, no. 5, p. 051103, 2005.

[47] A. Wittmann, Y. Bonetti, J. Faist, E. Gini, and M. Giovannini, "Intersubband linewidths in quantum cascade laser designs," *Applied Physics Letters*, vol. 93, no. 14, p. 141103, 2008.

[48] R. Maulini, A. Lyakh, A. Tsekoun, and C. K. N. Patel, "$\lambda \sim 7.1\,\mu m$ quantum cascade lasers with 19% wall-plug efficiency at room temperature," Optics Express 19(18), 17203–17211, 2011.

[49] D. Botez, S. Kumar, J. Shin, L. Mawst, I. Vurgaftman, and J. Meyer, "Temperature dependence of the key electro-optical characteristics for midinfrared emitting quantum cascade lasers," *Applied Physics Letters*, vol. 97, no. 7, p. 071101, 2010.

[50] C. Boyle, K. M. Oresick, J. D. Kirch, Y. V. Flores, L. J. Mawst, and D. Botez, "Carrier leakage via interface-roughness scattering bridges gap between theoretical and experimental internal efficiencies of quantum cascade lasers," *Applied Physics Letters*, vol. 117, no. 5, p. 051101, 2020.

[51] X. Huang, W. Charles, and C. Gmachl, "Temperature-insensitive long-wavelength ($\lambda \approx 14\,\mu m$) quantum cascade lasers with low threshold," *Optics Express*, vol. 19, no. 9, p. 8297, 2011.

[52] C. Sirtori, P. Kruck, S. Barbieri, H. Page, J. Nagle, M. Beck, J. Faist, and U. Oesterle, "Low-loss Al-free waveguides for unipolar semiconductor lasers," *Applied Physics Letters*, vol. 75, no. 25, pp. 3911–3913, 1999.

# 4 Overview of the State-of-the-Art Terahertz QCL Designs

Qi Jie Wang and Yongquan Zeng

Nanyang Technological University, Singapore

## 4.1 Introduction

Lying between the microwave and mid-infrared (mid-IR) frequencies, the electromagnetic spectral band 0.3–10 THz has been referred to as the "terahertz (THz) gap" since the technology of efficient wave generation, manipulation, and detection is the least developed in this spectral range compared to that in the neighboring spectral regions. However, driven by the great potential for diverse applications ranging from non-destructive imaging [1–3], to spectroscopic sensing [4–6], to ultra-high bit rate wireless communication [7, 8], continuous efforts have been devoted to this field in the past few decades, especially to the development of a compact, coherent, and high-power radiation source. Due to significant thermal broadening of the gain peak in the interband diode lasers and the general lack of suitable materials with sufficiently small bandgaps and sufficiently long carrier lifetimes, the development of interband THz lasers, analogous to the visible and the near-infrared diode lasers, was not successful. Compact semiconductor sources of THz radiation based on electronic technologies and optical down-conversion have been developed but they face limitations. For example, solid-state electronic devices such as transistors, Gunn oscillators, or Schottky diode multipliers [9] are intrinsically inefficient in generating THz radiation in the middle- and high-frequency range of the THz band (above about 2 THz) due to the limitation of the carrier (electron or hole) transit time and the resistance–capacitance effects. Photonic approaches of down-conversion of near-IR waves utilizing nonlinear optical effects or photoconductive effects are inherently limited either by low efficiency, limited tunability, or large system size [10, 11]. Optically pumped molecular gas lasers [12–14] and free-electron lasers [15] are complex, expensive, and bulky. Although semiconductor p-Ge lasers are capable of generating high peak powers (up to 40 W) in the 1.5–4.2 THz spectral range [16], they require very low temperature (<40 K) and a magnetic field of over 1 T to operate.

Since the first intersubband laser proposal in 1971, intersubband transitions in a semiconductor superlattice [17–20] were considered promising to build a compact and powerful THz semiconductor laser [21–24]. The first quantum cascade laser (QCL) was demonstrated in the mid-IR frequency range in the GaInAs/AlInAs/InP material system [25] and rapid progress was made in improving its performance [26]. The first THz QCL was demonstrated eight years later based on the GaAs/AlGaAs material system [27]. Compared to the mid-IR QCL, the development of a THz counterpart mainly confronts three challenges. First, long THz wavelengths presented challenges in confining the laser mode in relatively thin epitaxially grown active regions. Second, the THz wave experiences high optical loss even in lightly doped semiconductor layers as the free carrier loss scales as $\lambda^2$. Therefore, novel waveguides had to be developed to minimize the modal overlap with the doped cladding regions. Last, small THz photon energy presented challenges in selective injection and extraction of carriers into and out of the closely spaced upper and lower laser levels. Hence, the development of THz QCLs required the invention of new active region and waveguide designs, compared to those of mid-IR QCLs.

In this chapter, we discuss the current state of the art of the THz QCL designs with focus on the design of the waveguides and active regions. Since the state-of-the-art THz QCLs are all based on GaAs/AlGaAs heterostructures, our description focuses on these devices. We note that THz QCLs have been demonstrated in a number of materials systems, including InGaAs/AlInAs/InP and InGaAs/GaAsSb/InP. However, their active region and waveguide designs are conceptually similar to that of GaAs/AlGaAs devices.

## 4.2 THz QCL Waveguides

For QCLs with typical ridge waveguides, the threshold gain $g_{th}$ required for lasing is given as $g_{th} = (\alpha_w + \alpha_m)/\Gamma$, where $\Gamma$ is the confinement factor which describes the overlap of the mode with the active region, $\alpha_w$ is the waveguide loss due to light scattering and absorption, and $\alpha_m$ is the mirror loss which accounts for the optical out-coupling losses, resulting, e.g., from the finite facet reflectivity. Mid-IR QCLs use dielectric waveguides similar to that used in interband semiconductor lasers. The mode confinement there is achieved by refractive index contrast between the materials in the active region and in the cladding layers. However, the scaling of such waveguide architecture directly to the THz frequency range is not feasible since it would require very thick cladding layers (scalable to the wavelength of light, e.g., ~100 μm) exceeding the practical limits of epitaxial growth. In addition, doping of the cladding layers, which is required for electrical conductivity, leads to high waveguide losses at THz frequencies because of the strong free carrier absorption at long wavelengths. To address these challenges, two particular types of waveguides have been developed for THz QCLs: the semi-insulating surface-plasmon (SISP) waveguide and the double-metal (DM) waveguide. These waveguides avoid thick cladding layers, enable high overlap of the laser mode with the active region, and minimize waveguide losses.

### 4.2.1 SISP Waveguide

The SISP waveguide is shown in Fig. 4.1(a). It employs a thin (0.2–0.8 μm) highly doped semiconductor layer (typically GaAs with n-doping of approximately

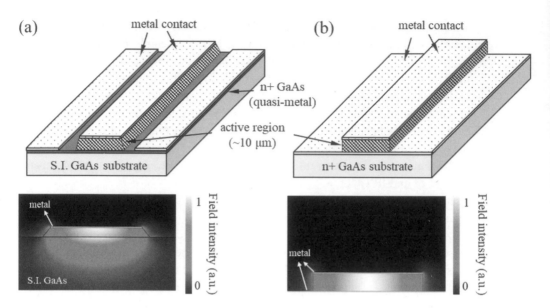

**Fig. 4.1** Schematic diagrams of (a) the semi-insulating surface-plasmon (SISP) waveguide and (b) the double-metal (DM) waveguide used in THz QCLs. Typical modulus of electric field intensity of $TM_{00}$ waveguide modes is plotted.

$10^{18}$ cm$^{-3}$) between the QCL active region and semi-insulating substrate (typically also GaAs). The doping of the highly doped semiconductor layer is chosen to produce a negative dielectric constant in the THz frequency range. Such a "quasi-metal" layer supports a transverse-magnetic (TM) surface plasmon mode on the layer interfaces with the QCL active region and the substrate. With the existence of the top metal contact layer, a compound waveguide mode arising from the coupling of the two surface plasmon modes is formed as shown in Fig. 4.1(a). A prominent portion of this mode is confined in the QCL active region (which is typically 10–12 μm in thickness). Even though the mode extends substantially into the substrate, the overlap with the highly-doped layer is small due to relatively small layer thickness, which minimizes the modal loss. The highly doped layer also works as a bottom contact layer that provides lateral current extraction to the metal contact as shown in Fig. 4.1(a).

The advantage of SISP waveguide lies in the high figure of merit $\alpha_w/\Gamma$ in the 2–5 THz spectral range and a relatively low facet reflectivity of $R \sim 0.32$ (compared to that in DM waveguides discussed further below), which is close to the Fresnel value given by the refractive index mismatch. Low facet reflectivity implies relatively high outcoupling efficiency of THz radiation from the laser into free space. The SISP waveguide provides a good balance between a fairly low threshold gain and efficient light outcoupling to free space with a good far field emission profile. However, when it comes to longer operation wavelengths (frequencies below approximately 2–3 THz) and/or reduced waveguide widths below approximately 100–150 μm, this waveguide scheme becomes less useful as the fundamental waveguide mode is squeezed into the substrate, decreasing the modal overlap with the active region.

## 4.2.2    DM Waveguide

The SISP waveguides provide relatively low confinement of the optical mode within the QCL active region ($\Gamma \approx 0.2$–$0.3$ typically). The alternative DM waveguides have been developed to provide nearly 100% confinement of the laser mode in the active region. The schematic diagrams of the DM waveguide and the laser mode profile in the waveguide are shown in Fig. 4.1(b). The DM waveguides substitute the highly doped semiconductor layer in the SISP waveguides with a metal layer. This is accomplished by an episide-down thermocompression metal–metal bonding of a metal-coated THz QCL wafer with a metal-coated receptor wafer, followed by the QCL wafer substrate removal using wet chemical etching [28]. The exposed QCL active region is then patterned by conventional photolithography and dry etching processes to form a ridge waveguide structure with a metal contact layer on top. The optical mode in the DM waveguides is completely confined to the subwavelength active region layer with $\Gamma \approx 1$ independent of the laser frequency. As the doped contact layers in DM QCLs can be made very thin, light absorption in the metal and the active region dominates the waveguide losses.

Despite significant differences in the modal confinement, the figures of merit $\alpha_w / \Gamma$ for the SISP and DM waveguides for the laser ridge widths of over 150 µm are comparable. However, unlike SISP waveguides, DM waveguides maintain their high figure of merit even for deeply subwavelength ridge widths. The subwavelength confinement of optical modes in DM waveguide result in a large impedance mismatch with the free space propagating modes, leading to an enhanced facet reflectivity of $R \sim 0.9$ (for a 10 µm × 100 µm waveguide facet and $\lambda = 100$ µm) and hence a smaller mirror loss. As a result, the THz QCLs employing DM waveguides have smaller threshold current densities and achieve higher operating temperatures, compared to SISP devices. Gold and copper are typically used metal layers along with necessary sticking and diffusion blocking layers, typically made of 10–20 nm Ti and/or Ta materials. DM waveguides made of a Ta/Cu metal layer combination have slightly lower losses compared to DM waveguides made of Ti/Au metal layers [29–31]

DM waveguides can provide strong mode confinement not only in the vertical but also in the lateral direction. Microstrip DM THz QCLs [32, 33] with ridge widths of only 20–30 µm have been demonstrated in order to lower the threshold current and achieve a more efficient thermal dissipation, compared to THz QCLs with ridge width comparable to the light wavelength. As a result, DM microstrip QCLs achieve the highest operating temperature in the continuous wave (CW) regime [34]. The so-called photonic wire laser [35] has also been implemented using DM waveguides with various interesting functionalities, including beam collimation [36], broadband tunability [37], and unidirectional emission [38]. The mirror loss in typical DM THz QCLs ($\alpha_m = 1$–$2$ cm$^{-1}$) is much smaller than the waveguide loss ($\alpha_w = 10$–$20$ cm$^{-1}$)

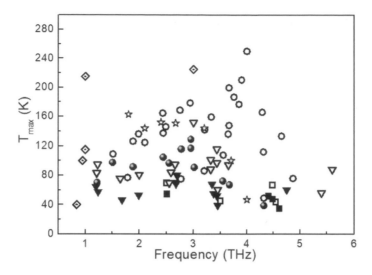

**Fig. 4.2** Schematic of the maximum operating temperatures of THz QCLs reported to date based on various active region designs in GaAs/AlGaAs material systems. ■, chirped superlattice designs in continuous wave (CW) mode; □, chirped superlattice designs in pulsed mode; ▼, bound-to-continuum designs in CW mode; ▽, bound-to-continuum designs in pulsed mode; ◕, resonant-phonon designs in CW mode; ○, resonant-phonon/direct-phonon designs in pulsed mode; ☆, scattering-assisted injection designs in pulsed mode; and ◈, THz QCLs operating with the assistance of a strong magnetic field (10–15 T). The data is mainly extracted from [40] and is updated to include the latest results.

[39] and results in low out-coupling efficiency $\sim \alpha_m/(\alpha_m+\alpha_w)$. As a result, the power output of DM THz QCLs is typically more than a factor of ten smaller than that of devices with SISP waveguides. This problem has been addressed, however, using various surface-outcoupling techniques as discussed, e.g., in Chapters 9 and 10.

## 4.3    Active Region Design of THz QCLs

As the photon energy of THz wave is smaller than the longitudinal-optical (LO) phonon energy ($\sim$36 meV for GaAs) and also closer to the value of thermal energy even at low temperatures, a straightforward scaling of mid-IR QCL active region designs down to THz range ($\sim$4–20 meV) is not effective and extra constraints need to be considered. Special attention has been paid to selective injection of carriers into the upper laser state and efficient depopulation of the lower laser state, which are positioned very close to each other in energy space. In this section, we will review several mainstream active region designs that produced high-performance THz QCL operation. Figure 4.2 summarizes the temperature performance of various THz QCLs reported to date as a function of their operating frequencies. Figures 4.3 and 4.4 show the band structures of the specific designs.

### 4.3.1 Chirped Superlattice Design

The first THz QCL was historically a chirped superlattice active region design shown in Fig. 4.3(a) [27]. In this design, the upper and lower minibands (shaded areas in Fig. 4.3(a)) extend over several coupled quantum wells when the appropriate electric field is applied. The carriers are injected into the tightly coupled states of the upper miniband via resonant tunneling from the injector ground state $g'$ of the previous stage. The radiative transition takes place vertically across the minigap with photon energy $hv \approx 10$–$20$ meV. This active region design exhibits a large dipole matrix element for the laser transition due to the large extent of the upper and lower minibands and their high overlap in real space. The electrons in the lower miniband relax into the ground state $g$ via several fast intra-miniband scattering processes such as LO-phonon scattering, ionized impurity scattering, electron–electron scattering, and interface roughness scattering. As a result, the lower laser state 1 is quickly depleted, the electrons are accumulated in the ground state g and then selectively injected into state 2 of the next miniband. Despite its success in the first THz QCL demonstrations at a frequency of ∼4.5 THz [41, 42], the chirped superlattice laser devices have limited performance, namely the poor slope efficiency ($dP/dI$) and the low maximum operating temperatures [43], owing to parasitic THz absorption within minibands, thermal backfilling of the lower miniband, and relatively slow relaxation times of the electrons in the lower laser state.

### 4.3.2 Bound-to-Continuum Design

The bound-to-continuum (BTC) design shown in Fig. 4.3(b) is similar to the chirped superlattice design except that the upper radiative state is replaced by a "bound" state lying in the minigap, created by a thin well adjacent to the injection barrier [43, 46]. The electrons are injected into the upper radiative state whose wavefunction has a maximum close to the injection barrier and rolls off in the active region. The lower miniband is depopulated through the miniband transport and resonant tunneling as in the chirped superlattice design. Due to the diagonal nature of the laser transition, the design has significantly smaller dipole matrix element compared to the chirped superlattice designs [43]; however, this is partially compensated by the higher injection efficiency due to the strong coupling of the injector state and the upper laser state as well as longer upper laser state lifetime as non-radiative scattering rates are typically reduced for diagonal transitions. The barrier thickness in this design is an important parameter for the laser performance, which can be optimized to achieve a high injection efficiency into the upper laser state while reducing the parasitic current [47]. Overall, the population inversion is enhanced comparing to the chirped superlattice designs and thus the BTC designs yield better power and temperature performance [48, 49]. The higher injection selectivity enables this active region design to have larger frequency coverage, including very low THz frequencies (see Fig. 4.2). However, the performance of BTC designs degrades rapidly with temperature as thermal backfilling of the lower miniband becomes prominent at high temperature, greatly reducing the population inversion. The maximum operation temperature of BTC THz QCLs has been limited to ∼100 K for a number of years.

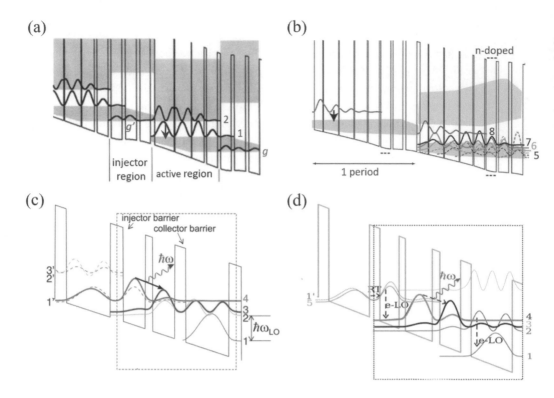

**Fig. 4.3** Conduction band diagrams for THz quantum cascade active region designs. (a) Chirped superlattice design, reprinted from [44]. (b) Bound-to-continuum design, reprinted from [43]. (c) Resonant-phonon design, reprinted from [30]. (d) Scattering-assisted injection design, reprinted from [45].

To further enhance the high-temperature performance, refined BTC designs incorporating optical phonon scattering to enable fast carrier depopulation of the lower miniband have been proposed [50, 51]. The "hybrid" active region design inserts additional extractor/injector quantum wells at the end of original BTC active region. When a proper electric field is applied, the extraction miniband is resonantly aligned with the lower laser transition states, with its energy ~32–36 meV (close to the LO-phonon energy in GaAs) above the ground/injector state. Through this channel, the carriers in the lower radiative transition miniband can be selectively extracted into the extraction miniband and then relaxed down to the ground/injector state through a fast electron–LO-phonon scattering process. This approach significantly reduces the lifetime of lower laser state and suppresses the thermal backfilling. With this design, the operating temperature of BTC designs can be lifted to 152 K [51] and high pulsed/CW power performance can also be achieved [52, 53].

### 4.3.3    Resonant-Phonon Design

The resonant-phonon (RP) design (see Fig. 4.3(c)) with relatively few quantum wells (typical three quantum wells) and a simple bandstructure currently delivers devices with the highest performances in the higher frequency range (>2.5 THz). In this design, electrons are injected into upper laser state 4 from the ground state 1' of the previous stage via resonant tunneling. Radiative transition occurs from energy states 4 to 3 and the depopulation of state 3 occurs though a combination of resonant tunneling between energy states 3 and 2 and fast resonant LO-phonon scattering of electrons from energy state 2 to ground state 1 [39]. The laser transition between states 4 and 3 can be vertical or diagonal. Diagonal transitions have been shown to provide improved temperature performance due to a combination of improved injection efficiency and reduced non-radiative scattering from state 4 to state 3 [54, 55]. In addition, the spatial separation between the upper and lower laser states also contributes to a better electron extraction selectivity of the lower laser state. The amount anticrossing between states 1' and 4 and states 3 and 2 is carefully adjusted by controlling the thickness of injection and extraction barriers (see Fig. 4.3(c)) to achieve high operating temperatures [55].

A systematic optimization of the oscillator strength between upper and lower laser states and the barrier thicknesses for electron tunneling coupling strengths was done for the three-well RP design shown in Fig. 4.3(c) and a maximum operation temperature of 199.5 K had been achieved [54], which was a record point for several years.

Recently, THz QCL active region design involving higher aluminum fraction in the barriers and fewer energy levels based on a two-well direct-phonon depopulation scheme have been proposed as shown in Fig. 4.4 [56, 57]. The increase of the aluminum fraction in the AlGaAs barriers in the GaAs/AlGaAs RP designs from 15% to 25–30% have been shown to reduce carrier leakage into the continuum [57–60]. Optimization of the wells and barriers thicknesses in the two-well direct-phonon depopulation designs using transport simulations and the use of more accurate materials parameters in the modeling has led to the increase of the $T_{max}$ to 210.5 K [31] and 250 K [61] at $\sim$4 THz. In the two-well design (as shown in Fig. 4.4), the lower laser state $l_n$ is directly depleted via LO-phonon scattering to the ground state $i_n$ in the same well, where the electrons in the ground state $i_n$ are then injected into the upper laser state $u_{n+1}$ of the next module through resonant tunneling. Compared to the prevailing three-well RP design utilizing resonant tunneling and LO-phonon scattering for depopulation, this direct-phonon depopulation of the lower laser state is less impacted by the band misalignment and insensitive to dephasing caused by interface roughness and impurities. In addition, in [61], a taller barrier based on $Al_{0.3}Ga_{0.7}As$ was used to increase the energy spacing between upper laser state $u_n$ and higher-energy bound states $p_{1,n}, p_{2,n}$ while maintaining a reasonable oscillator strength between the upper and lower laser states. As a result, only three energy states are involved in the laser operation with suppressed carrier leakage to the bound states as well as the continuum. It should also be noted that the high-quality growth of the active region with high Al composition (30%) in the barrier is another contributing factor for the achievement of 250 K operation. The layer thickness variation was well-controlled to within $\pm0.05\%$.

### 4.3.4    Scattering-Assisted Injection Design

In the active region designs discussed above, electrons are injected into the upper laser states by resonant tunneling. Since the energy separation between the upper and lower

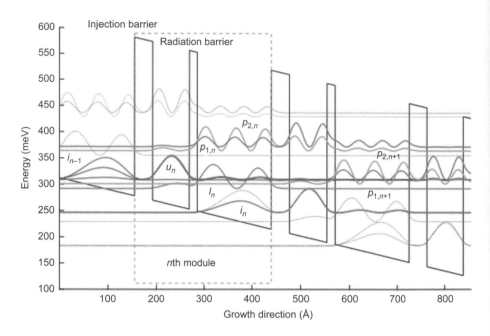

**Fig. 4.4** Conduction band diagrams for THz quantum cascade active region based on the two-quantum-well RP design based on direct phonon depopulation, reprinted from [61]. Lower laser state $l_n$ is directly depopulated to ground state $i_n$ via LO-phonon scattering. An $Al_{0.3}Ga_{0.7}As$ barrier is used and the energy gap $\Delta E_{u_n,p_n}$ is optimized to be over 50 meV. The QCL processed with an Au-Au waveguide operates up to 250 K.

laser states in THz QCLs is relatively small ($\sim$4–20 meV), it is difficult to selectively inject electrons only into the upper laser state particularly for low-frequency devices. To enable selective injection into the upper laser state, the injector barrier in resonant-tunneling injection QCLs is relatively thick so that off-resonant tunneling into states other than the upper laser state is suppressed. Typically, the width of the injector barrier is chosen such that the anti-crossing between the injector state and the upper laser state $2\hbar\Omega_{iu}$ (where $\Omega_{iu}$ is the Rabi frequency of the coupled states) is smaller than $\hbar\Delta\nu_{scatt}$, i.e., the FWHM linewidth of the laser transition, determined by dephasing rates due to scattering [62]. The disadvantage of a small $\Omega_{iu}$ and consequently a weak coupling between the injector and the upper laser state is the long electron tunneling time into the upper laser state. The scattering-assisted (SA) injection (or indirect pumping) scheme shown in Fig. 4.3(d) avoids injection electrons by tunneling. Instead, electrons are injected into the upper laser state by resonantly emitting phonons from the injector level [45, 63–69]. In this case, off-resonant tunneling from the injector level to the lower laser state is strongly suppressed and the injector barriers can be kept thinner than those in the resonant-injection scheme without the risk of increasing parasitic current injection to the lower laser state and without broadening of the laser gain spectrum. This approach is particularly useful for THz QCLs operating at low THz frequencies ($<$2 THz) and indeed THz QCLs based on the SA injection scheme have the best temperature performance compared to competing designs at frequencies below 2–2.5 THz.

In 2011, the first SA injection scheme based on a four-well active region design was reported to work at a frequency of 1.8 THz [63]. The THz QCL operated up to a maximum temperature $T_{max} \sim 163$ K using a GaAs/AlGaAs material system. Thereafter, a wide variety of active region designs with this scheme were intensively investigated to enhance the temperature performance. A five-well SA injection with a high injection coupling strength and RP depopulation in an InGaAs/InAlAs/InP material system was reported at 3.7 THz and provided lasing up to 100 K [66]. A four-well SA injection scheme named phonon-photon-phonon (3P) with a low injection coupling strength and a direct phonon scattering depletion without the mediation of resonant tunneling was demonstrated in a GaAs/AlGaAs material system at an operation frequency of 3.2 THz and $T_{max} = 138$ K [67]. The low injection coupling strength of this design is intended to minimize early negative differential resistance prior to the lasing threshold. Utilizing the same 3P four-well SA injection scheme but unusually low injection coupling strength, a 2.4 THz QCL working up to 152.5 K was reported [68]. Subsequently, in order to render the coherent electron injection and enhance lasing dynamic range, a 3P five-well SA injection design with strong injection coupling was proposed, where the extra well was included in the upper phonon stream to minimize the rate of wrong injection by scattering to the lower laser state [69]. The laser was demonstrated to work at 2.67 THz and $T_{max} = 151$ K. Most recently, a 2.1 THz QCL based on a four-well SA injection scheme with a high injection coupling strength and RP depopulation was also realized up to a maximum operating temperature of 144 K, showing a comparatively lower threshold current density than other reported THz QCLs with the same injection scheme in the GaAs/AlGaAs material system [45]. Using a similar active region structure with a slightly higher Al composition of the AlGaAs barriers (from 15% to 17%) to provide a suppression of the current leakage to continuum band, the threshold current density has been further reduced by 26% with a broader lasing spectra coverage of 2.3–2.6 THz while the same maximum operation temperature is maintained [70].

## 4.4    Conclusion

With the unrelenting development of waveguide and active region designs mainly based on the GaAs/AlGaAs materials system, THz QCLs have demonstrated a remarkable progress in performance. To date, THz QCLs have covered the spectral range of 1.2–56 THz without the use of magnetic fields. With regard to temperature performance, the $T_{max}$ has already been lifted to aforementioned 250 K, allowing cryogenic-free THz coherent radiation with sufficient power for portable real applications [31, 61].

A single THz QC ridge laser could deliver over 2.4 W peak power at 10 K heatsink temperature [71] and a coupled microcavity THz QCL achieved $\sim 2$ W output at 58 K heatsink temperature [72] in pulsed mode operation. Meanwhile, 230 mW in CW mode at 15 K heatsink temperature was also reported [53]. Some of the specific designs of high-power THz QCLs are discussed in Chapters 9 and 10 of this text.

Significant efforts are now devoted to developing room temperature THz QCLs. Referring to the most recent RP THz QCL design based on two-well direct-phonon depopulation scheme [61], potential strategies to further improve the temperature performance of THz QCLs may rely on the oscillator strength optimization, doping density engineering, and on reducing waveguide optical losses. The detrimental factors for population inversion, such as temperature broadening of re-absorption transitions, thermal backfilling, and electron–electron scattering, should also be systematically considered, especially at elevated temperatures. In addition, the RP design has a limit of maximum population inversion of 50% [67], intrinsically determined by resonant-tunneling injection processes. Theoretically, the SA injection design could achieve a higher population inversion, and thus may lead to a higher gain coefficient. With the maturity of high-barrier material growth and better understanding of various carrier leakage mechanisms, the SA injection design is also highly promising for high-temperature THz QCL operation.

Our discussion here focused on the THz QCL active region designs based on GaAs/AlGaAs heterostructures, whereas a number of studies have also been conducted on other material systems like InGaAs/GaAsSb/InP [73], InGaAs/InAlAs/InP [74], InGaAs/AlInGaAs/InP [75], and GaN/AlGaN [76]. InGaAs has a lower effective electron mass $m_e^*$ and can be used as a quantum well material with larger optical gain coefficient to achieve better performance [77]. The high barrier of these material systems would also provide design flexibility to suppress the carrier leakage into high-energy bound states and the continuum. GaN/AlGaN seems to be a promising candidate of THz QCLs for high-temperature operation as well as wide frequency coverage (i.e., 1–15 THz) thanks to its much larger LO-phonon energy (90 meV) as compared to the traditional GaAs/AlGaAs heterostructures. The critical issue of this material system however is its low material growth quality. Nevertheless, performance breakthroughs, not only in high operation temperatures but also in wallplug efficiency and operation frequency, are still expected with continuous progress in material growth technologies.

Last, but not least, difference-frequency generation (DFG) in high-power dual-wavelength mid-IR QCLs [78–83] is another method for achieving THz wave generation in QCLs. In this approach, the THz wave ($\omega_{THz} = \omega_1 - \omega_2$) is generated based on a nonlinear optical process in which two mid-IR pump beams at frequencies $\omega_1$ and $\omega_2$ interact in a QCL active region with second-order nonlinear susceptibility $\chi^{(2)}$. Several appealing aspects of the THz DFG-QCLs have attracted research interests, including room temperature operation, monolithic packaging, and wide THz frequency tunability obtained by modest tuning of the mid-IR modes. Intrinsically limited by the requirement of phase matching and large second-order nonlinear susceptibility $\chi^{(2)}$, the main challenges of THz DFG-QCLs are low conversion efficiency and small wallplug efficiency. The state of the art and the prospect for THz DFG-QCLs are discussed in Chapter 11.

# References

[1] W. L. Chan et al., "Imaging with terahertz radiation," *Reports Prog. Phys.*, vol. 70, no. 8, pp. 1325–1379, 2007.

[2] S. M. Kim et al., "Biomedical terahertz imaging with a quantum cascade laser," *Appl. Phys. Lett.*, vol. 88, no. 15, pp. 81–83, 2006.

[3] A. W. M. Lee et al., "Real-time terahertz imaging over a standoff distance (> 25 meters)," *Appl. Phys. Lett.*, vol. 89, no. 14, p. 141125, 2006.

[4] D. M. Mittleman et al., "Gas sensing using terahertz time-domain spectroscopy," *Appl. Phys. B Lasers Opt.*, vol. 67, pp. 379–390, 1998.

[5] J. F. Federici et al., "THz imaging and sensing for security applications—explosives, weapons and drugs," *Semicond. Sci. Technol.*, vol. 20, no. 7, pp. S266–S280, 2005.

[6] Y. Ren et al., "High-resolution heterodyne spectroscopy using a tunable quantum cascade laser around 3.5 THz," *Appl. Phys. Lett.*, vol. 98, no. 23, pp. 3–6, 2011.

[7] S. Koenig et al., "Wireless sub-THz communication system with high data rate," *Nat. Photonics*, vol. 7, no. 12, pp. 977–981, 2013.

[8] T. Kleine-Ostmann and T. Nagatsuma, "A review on terahertz communications research," *J. Infrared, Millimeter, Terahertz Waves*, vol. 32, no. 2, pp. 143–171, 2011.

[9] A. Maestrini et al., "A 1.7-1.9 THz local oscillator source," *IEEE Microw. Wirel. Components Lett.*, vol. 14, no. 6, pp. 253–255, 2004.

[10] C. Staus et al., "Continuously phase-matched terahertz difference frequency generation in an embedded-waveguide structure supporting only fundamental modes," *Opt. Express*, vol. 16, no. 17, pp. 13296–13303, 2008.

[11] K. Saito et al., "THz-wave generation via difference frequency mixing in strained silicon based waveguide utilizing its second order susceptibility $\chi^{(2)}$," *Opt. Express*, vol. 22, no. 14, p. 16660, 2014.

[12] M. Inguscio et al., "A review of frequency measurements of optically pumped lasers from 0.1 to 8 THz," *J. Appl. Phys.*, vol. 60, no. 12, pp. R161–R192, 1986.

[13] A. Pagies et al., "Low-threshold terahertz molecular laser optically pumped by a quantum cascade laser," *APL Photonics*, vol. 1, no. 3, 2016.

[14] P. Chevalier et al., "Widely tunable compact terahertz gas lasers," *Science*, vol. 366, no. 6467, pp. 856–860, 2019.

[15] G. P. Williams, "FAR-IR/THz radiation from the Jefferson Laboratory, energy recovered linac, free electron laser," *Rev. Sci. Instrum.*, vol. 73, no. 3 II, p. 1461, 2002.

[16] A. V. Muravjov et al., "Injection-seeded internal-reflection-mode p-Ge laser exceeds 10W peak terahertz power," *J. Appl. Phys.*, vol. 103, no. 8, p. 083112, 2008.

[17] R. Kazarinov, "Possibility of amplification of electromagnetic waves in a semiconductor with superlattice," *Sov. Phys.-Semicond.*, vol. 5, no. 4, pp. 707–709, 1971.

[18] F. Capasso et al., "Resonant tunneling through double barriers, perpendicular quantum transport phenomena in superlattices, and their device applications," in *Electronic Structure of Semiconductor Heterojunctions*, Springer, 1988, pp. 99–115.

[19] P. F. Yuh and K. L. Wang, "Novel infrared band-aligned superlattice laser," *Appl. Phys. Lett.*, vol. 51, no. 18, pp. 1404–1406, 1987.

[20] H. C. Liu, "A novel superlattice infrared source," *J. Appl. Phys.*, vol. 63, no. 8, pp. 2856–2858, 1988.

[21] S. I. Borenstain and J. Katz, "Evaluation of the feasibility of a far-infrared laser based on intersubband transitions in GaAs quantum wells," *Appl. Phys. Lett.*, vol. 55, no. 7, pp. 654–656, 1989.

[22] A. Kastalsky et al., "Possibility of infrared laser in a resonant tunneling structure," *Appl. Phys. Lett.*, vol. 59, no. 21, pp. 2636–2638, 1991.

[23] A. Kastalsky et al., "Feasibility of far-infrared lasers using multiple semiconductor quantum wells," *Appl. Phys. Lett.*, vol. 59, no. 23, pp. 2923–2925, 1991.

[24] M. Helm et al., "Observation of grating-induced intersubband emission from GaAs/AlGaAs superlattices," *Appl. Phys. Lett.*, vol. 53, no. 18, pp. 1714–1716, 1988.

[25] J. Faist et al., "Quantum cascade laser," *Science*, vol. 264, no. 5158, pp. 553–556, 1994.

[26] C. Gmachl et al., "Recent progress in quantum cascade lasers and applications," *Rep. Prog. Phys.*, vol. 64, pp. 1533–1601, 2001.

[27] R. Kohler et al., "Terahertz semiconductor-heterostructure laser," *Nature*, vol. 417, May, pp. 156–159, 2002.

[28] B. S. Williams et al., "Terahertz quantum-cascade laser at $\lambda \approx 100$ μm using metal waveguide for mode confinement," *Appl. Phys. Lett.*, vol. 83, no. 11, pp. 2124–2126, 2003.

[29] M. A. Belkin et al., "Terahertz quantum cascade lasers with copper metal-metal waveguides operating up to 178 K," *Opt. Express*, vol. 16, no. 5, p. 3242, 2008.

[30] S. Kumar et al., "186 K operation of terahertz quantum-cascade lasers based on a diagonal design," *Appl. Phys. Lett.*, vol. 94, no. 13, pp. 13–16, 2009.

[31] L. Bosco et al., "Thermoelectrically cooled THz quantum cascade laser operating up to 210 K," *Appl. Phys. Lett.*, vol. 115, no. 1, p. 010601, 2019.

[32] S. Dhillon et al., "Ultralow threshold current terahertz quantum cascade lasers based on double-metal buried strip waveguides," *Appl. Phys. Lett.*, vol. 87, no. 7, p. 071107, 2005.

[33] B. Williams et al., "Operation of terahertz quantum-cascade lasers at 164 K in pulsed mode and at 117 K in continuous-wave mode," *Opt. Express*, vol. 13, no. 9, pp. 3331–3339, 2005.

[34] M. Wienold et al., "High-temperature, continuous-wave operation of terahertz quantum-cascade lasers with metal-metal waveguides and third-order distributed feedback," *Opt. Express*, vol. 22, no. 3, p. 3334, 2014.

[35] E. E. Orlova and J. L. Reno, "Antenna model for wire lasers," *Phys. Rev. Lett.*, vol. 96, May, pp. 1–4, 2006.

[36] M. I. Amanti et al., "Low divergence terahertz photonic-wire laser," *Opt. Express*, vol. 18, no. 6, pp. 6390–6395, 2010.

[37] Q. Qin et al., "Tuning a terahertz wire laser," *Nat. Photonics*, vol. 3, no. 12, pp. 732–737, 2009.

[38] A. Khalatpour et al., "Unidirectional photonic wire laser," *Nat. Photonics*, August, pp. 1–6, 2017.

[39] B. S. Williams, "Terahertz quantum-cascade lasers," *Nat. Photonics*, vol. 1, no. 9, pp. 517–525, 2007.

[40] M. A. Belkin and F. Capasso, "New frontiers in quantum cascade lasers: high performance room temperature terahertz sources," *Phys. Scr.*, vol. 90, no. 11, p. 118002, 2015.

[41] M. Rochat et al., "Low-threshold terahertz quantum-cascade lasers," *Appl. Phys. Lett.*, vol. 81, no. 8, pp. 1381–1383, 2002.

[42] L. Ajili et al., "Continuous-wave operation of far-infrared quantum cascade lasers," *Electron. Lett.*, vol. 38, no. 25, pp. 1675–1676, 2002.

[43] G. Scalari et al., "Far-infrared ($\lambda \simeq 87$ μm) bound-to-continuum quantum-cascade lasers operating up to 90 K," *Appl. Phys. Lett.*, vol. 82, no. 19, pp. 3165–3167, 2003.

[44] R. Köhler et al., "High-intensity interminiband terahertz emission from chirped superlattices," *Appl. Phys. Lett.*, vol. 80, no. 11, pp. 1867–1869, 2002.

[45] S. Khanal et al., "2.1 THz quantum-cascade laser operating up to 144 K based on a scattering-assisted injection design," *Opt. Express*, vol. 23, no. 15, p. 19689, 2015.

[46] J. Faist et al., "Quantum-cascade lasers based on a bound-to-continuum transition," *Appl. Phys. Lett.*, vol. 78, no. 2, pp. 147–149, 2001.

[47] J. Alton et al., "Optimum resonant tunnelling injection and influence of doping density on the performance of THz bound-to-continuum cascade lasers," in *Proc. SPIE*, vol. 5727, no. 0, p. 65, 2005.

[48] L. Ajili et al., "High power quantum cascade lasers operating at $\lambda \simeq 87$ and $130\,\mu m$," *Appl. Phys. Lett.*, vol. 85, no. 18, pp. 3986–3988, 2004.

[49] S. Barbieri et al., "2.9 THz quantum cascade lasers operating up to 70 K in continuous wave," *Appl. Phys. Lett.*, vol. 85, no. 10, pp. 1674–1676, 2004.

[50] G. Scalari et al., "Terahertz bound-to-continuum quantum-cascade lasers based on optical-phonon scattering extraction," *Appl. Phys. Lett.*, vol. 86, no. 18, pp. 1–3, 2005.

[51] M. I. Amanti et al., "Bound-to-continuum terahertz quantum cascade laser with a single-quantum-well phonon extraction/injection stage," *New J. Phys.*, vol. 11, no. 12, p. 125022, 2009.

[52] L. Li et al., "Terahertz quantum cascade lasers with > 1 W output powers," *Electron. Lett.*, vol. 50, no. 4, pp. 309–311, 2014.

[53] X. Wang et al., "High-power terahertz quantum cascade lasers with ~0.23 W in continuous wave mode," *AIP Adv.*, vol. 6, no. 7, pp. 2–7, 2016.

[54] S. Fathololoumi et al., "Terahertz quantum cascade lasers operating up to ~200 K with optimized oscillator strength and improved injection tunneling," *Opt. Express*, vol. 20, no. 4, p. 3866, 2012.

[55] E. Dupont et al., "Simplified density-matrix model applied to three-well terahertz quantum cascade lasers," *Phys. Rev. B*, vol. 81, no. 20, p. 205311, 2010.

[56] A. Albo et al., "Two-well terahertz quantum cascade lasers with suppressed carrier leakage," *Appl. Phys. Lett.*, vol. 111, no. 11, pp. 1–6, 2017.

[57] M. Franckié et al., "Two-well quantum cascade laser optimization by non-equilibrium Green's function modelling," *Appl. Phys. Lett.*, vol. 112, no. 2, 2018.

[58] S. Kumar et al., "Two-well terahertz quantum-cascade laser with direct intrawell-phonon depopulation," *Appl. Phys. Lett.*, vol. 95, no. 14, pp. 32–34, 2009.

[59] A. Albo and Q. Hu, "Carrier leakage into the continuum in diagonal GaAs/Al$_{0.15}$GaAs terahertz quantum cascade lasers," *Appl. Phys. Lett.*, vol. 107, no. 24, p. 241101, 2015.

[60] A. Albo et al., "Room temperature negative differential resistance in terahertz quantum cascade laser structures," *Appl. Phys. Lett.*, vol. 109, no. 8, pp. 30–35, 2016.

[61] A. Khalatpour et al., "High-power portable terahertz laser systems," *Nat. Photonics*, vol. 15, no. 1, pp. 16–20, 2021.

[62] S. Kumar and Q. Hu, "Coherence of resonant-tunneling transport in terahertz quantum-cascade lasers," *Phys. Rev. B – Condens. Matter Mater. Phys.*, vol. 80, no. 24, pp. 1–14, 2009.

[63] S. Kumar et al., "A 1.8-THz quantum cascade laser operating significantly above the temperature of $\hbar\omega / k_B$," *Nat. Phys.*, vol. 7, no. 2, pp. 166–171, 2011.

[64] T. Kubis et al., "Design concepts of terahertz quantum cascade lasers: proposal for terahertz laser efficiency improvements," *Appl. Phys. Lett.*, vol. 97, no. 26, pp. 1–4, 2010.

[65] M. Yamanishi et al., "Indirect pump scheme for quantum cascade lasers: dynamics of electron-transport and very high $T_0$-values," *Opt. Express*, vol. 16, no. 25, pp. 20748–20758, 2008.

[66] K. Fujita *et al.*, "Indirectly pumped 3.7 THz InGaAs/InAlAs quantum-cascade lasers grown by metal-organic vapor-phase epitaxy," *Opt. Express*, vol. 20, no. 18, pp. 20647–58, 2012.

[67] E. Dupont *et al.*, "A phonon scattering assisted injection and extraction based terahertz quantum cascade laser," *J. Appl. Phys.*, vol. 111, no. 7, p. 073111, 2012.

[68] S. G. Razavipour *et al.*, "An indirectly pumped terahertz quantum cascade laser with low injection coupling strength operating above 150 K," *J. Appl. Phys.*, vol. 113, no. 20, p. 203107, 2013.

[69] S. G. Razavipour *et al.*, "A high carrier injection terahertz quantum cascade laser based on indirectly pumped scheme," *Appl. Phys. Lett.*, vol. 104, no. 4, p. 041111, 2014.

[70] B. Wen *et al.*, "Dual-lasing channel quantum cascade laser based on scattering-assisted injection design," *Opt. Express*, vol. 26, no. 7, p. 9194, 2018.

[71] L. H. Li *et al.*, "Multi-watt high-power THz frequency quantum cascade lasers," *Electron. Lett.*, vol. 53, no. 12, pp. 799–800, 2017.

[72] Y. Jin *et al.*, "Phase-locked terahertz plasmonic laser array with 2 W output power in a single spectral mode," *Optica*, vol. 7, no. 6, p. 708, 2020.

[73] C. Deutsch *et al.*, "Terahertz quantum cascade lasers based on type II InGaAs/GaAsSb/InP," *Appl. Phys. Lett.*, vol. 97, no. 26, pp. 73–76, 2010.

[74] M. Fischer *et al.*, "Scattering processes in terahertz InGaAs/InAlAs quantum cascade lasers," *Appl. Phys. Lett.*, vol. 97, no. 22, p. 221114, 2010.

[75] K. Ohtani *et al.*, "Terahertz quantum cascade lasers based on quaternary AlInGaAs barriers," *Appl. Phys. Lett.*, vol. 103, no. 4, pp. 4–8, 2013.

[76] H. Hirayama and W. Terashima, "Recent progress toward realizing GaN-based THz quantum cascade laser," in *Quantum Sensing and Nanophotonic Devices XI, 2013*, January, p. 89930G, 2014.

[77] E. Benveniste *et al.*, "Influence of the material parameters on quantum cascade devices," *Appl. Phys. Lett.*, vol. 93, no. 13, p. 131108, 2008.

[78] M. Razeghi *et al.*, "Quantum cascade lasers: from tool to product," *Opt. Express*, vol. 23, no. 7, pp. 8462–8475, 2015.

[79] Q. Y. Lu *et al.*, "Continuous operation of a monolithic semiconductor terahertz source at room temperature," *Appl. Phys. Lett.*, vol. 104, no. 22, pp. 1–6, 2014.

[80] M. A. Belkin *et al.*, "Terahertz quantum-cascade-laser source based on intracavity difference-frequency generation," *Nat. Photonics*, vol. 1, no. 5, pp. 288–292, 2007.

[81] K. Vijayraghavan *et al.*, "Broadly tunable terahertz generation in mid-infrared quantum cascade lasers," *Nat. Commun.*, vol. 4, May, p. 2021, 2013.

[82] K. Vijayraghavan *et al.*, "Terahertz sources based on Čerenkov difference-frequency generation in quantum cascade lasers," *Appl. Phys. Lett.*, vol. 100, no. 25, p. 251104, 2012.

[83] C. Pflügl *et al.*, "Surface-emitting terahertz quantum cascade laser source based on intracavity difference-frequency generation," *Appl. Phys. Lett.*, vol. 93, no. 16, pp. 1–3, 2008.

# 5 Simulating Quantum Cascade Lasers: The Challenge to Quantum Theory

Andreas Wacker

Lund University, Sweden

Quantum cascade lasers (QCLs), as first realized in 1994 [1], are technologically important devices for the generation of radiation in the mid- and long-wavelength infrared (IR) region. In addition, they operate also in the THz region [2] albeit the requirement of cooling still limits the range of applications here. Further details on their realization, different operations schemes, and applications are given in the other chapters of this book.

The basic operation of QCLs relies on three key issues:

1. Laser operation by stimulated emission between quantized levels in a semiconductor heterostructure. This allows to define the transition energy by designing the semiconductor layers, actually covering two decades of the electromagnetic spectrum.
2. Electrical pumping by a specifically designed current path through the device. Here a combination of resonant tunneling and scattering processes fills the upper laser level and empties the lower laser level, thereby achieving population inversion.
3. The periodic repetition of basic modules in order to fill the waveguide with active material (i.e., the cascading process).

These issues require a careful design of the underlying semiconductor heterostructure and set high demands on growth perfection. As for any laser, the design and fabrication of the optical waveguide is of equal importance.

In order to design the layer structure, detailed modeling on different levels is required to enable and further optimize QCLs. In this chapter general problems of modeling is addressed and a review of different existing approaches is given. Here the calculation of electronic states in the QCL is addressed in Section 5.1. Section 5.2 refers to the description of electric transport and Section 5.3 to the gain spectrum. Finally, Section 5.4 addresses the problem of domain formation in QCL structures.

There exists an excellent textbook on the general physics behind QCLs [3] and a thorough review article on transport models [4]. Technical details for the practical modeling can be found in [5] and aspects of different QCL designs together with some general modeling issues were reviewed in [6]. This chapter is more focused on the underlying problems and ideas behind the different approaches used, while referring to the literature for details.

## 5.1        Quantum Levels in Biased Semiconductor Heterostructures

The physics in the conduction band of semiconductors is naturally described by envelope functions $\Psi(\mathbf{r})$, which disregard the lattice-periodic part of the Bloch functions. In its simplest form of effective mass approximation this provides the stationary Schrödinger-like equation [7]

$$\left( \frac{\hbar^2}{2} \nabla \frac{1}{m_c(z)} \nabla + E_c(z) - e\phi(z) \right) \Psi(\mathbf{r}) = E\Psi(\mathbf{r}), \tag{5.1}$$

where $m_c(z)$ and $E_c(z)$ are the effective mass and energy at the $\Gamma$-point of the conduction band, respectively. Their $z$-dependence reflects layers of different semiconductor materials stacked in the $z$-direction. $\phi(z)$ is the electrostatic potential corresponding to the bias drop $Fz$ along the QCL, where ionized dopants and the electron distribution can be easily included on a mean-field level. Note that the physical electric field $F$ (like the current density $J$, the dipole moment $d$, and the polarization $P$) is here defined to point in the $-\mathbf{e}_z$-direction. Thus positive values of $F/J$ relate to forces on/motion of electrons in the $\mathbf{e}_z$-direction, respectively.

The translational invariance of Eq. (5.1) in the $(x, y)$direction, suggests the ansatz

$$\Psi(\mathbf{r}) = \varphi_n(z) \frac{1}{\sqrt{A}} e^{i(k_x x + k_y y)}, \tag{5.2}$$

resulting in eigenvalues $E = E_n + \hbar^2 k^2 / 2m_n$. Here $A$ is the cross-section of the QCL and we introduced the two-dimensional vector $\mathbf{k} = k_x \mathbf{e}_x + k_y \mathbf{e}_y$.

For the simulation of QCLs, we require levels in an extended heterostructure under an applied electric field $F$ with wavefunctions $\varphi_n(z)$ and energies $E_n$, which are used as a basis. In order to exploit the periodicity of the QCL structure (with period $d$ reflecting the length of a single module), they are commonly chosen such that the levels for the neighboring module to the left satisfy $\varphi_{n'}(z) = \varphi_n(z + d)$ and $E_{n'} = E_n + eFd$, as illustrated in Fig. 5.1(a) and (d) by full and dashed lines (*periodicity condition*). It appears natural to chose these levels as eigenstates in Eq. (5.1) using the heterostructure potential for the extended QCL. However, the wavefunctions of such states would leak out for $z \rightarrow \infty$ and form a continuum in energies. This is a general problem in periodic structures, such as superlattices, where it has been thoroughly discussed [8, 9]. Localized states within an extended heterostructure under an applied electric field can actually only be defined as resonant states with a finite lifetime by tunneling to the continuum in this context. Determining such states is, however, quite cumbersome and I am not aware of such approaches for QCLs. Furthermore, the tunneling lifetime to the continuum is in well-defined structures much larger than other scattering lifetimes, and thus negligible. Thus, sets of periodically repeated localized states are physically reasonable and have been observed in superlattices [10, 11]. In order to construct such levels, satisfying the periodicity condition, three different approaches are typically used.

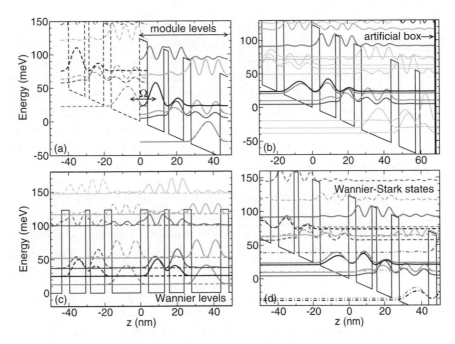

**Fig. 5.1** Calculation of electronic levels for the QCL structure of [13]. (a) Levels calculated for a single module extending from $z = 0$ to $z = 43.91$ nm including the terminating barrier ending at $z = 48.21$ nm. (b) States calculated with an artificial box ending at $L = 68$ nm. (c) Wannier levels of an unbiased structure. (d) Wannier–Stark states obtained by diagonalizing the basis of Wannier levels in the presence of an electric potential $\phi(z) = Fz$. The bias is 53 mV per module in panels (a, b, d). The thicker full lines are calculated levels attributed to the central module, while dashed and dash-dotted levels are obtained by shifting these levels by one period. In all cases the absolute square of the wavefunctions is shown, where the offset matches the level energy. The data of (a) and (b) was produced using the Aestimo package [14].

The first option is to assume that the thickest barrier separates states from different modules. Extending the thickness of this barrier to infinity (and neglecting the potential drop) or using $\varphi(z) = \infty$ at the end of the barrier (as done here), we have the standard problem of a multiple quantum well as shown in Fig. 5.1(a). With increasing energy, we observe essentially the ground state of the wide well (injector level), the ground state of the middle well (lower laser level), the first excited state of the wide well (extractor level), the ground state of the first well (upper laser level), the second excited state of the wide well, and an excited state of the first two wells. These *module levels* can then be shifted to neighboring periods using the periodicity condition as shown by the dashed lines. As the levels in neighboring periods are eigenstates of different Hamilton operators they are not orthogonal to each other. Using the Bardeen tunneling Hamiltonian [12] one can construct the tunneling matrix elements $\Omega_{ij}$ between such levels. Details can be found in [5].

A second option is to use an artificial box to confine the electrons. In its simplest form, this is a boundary condition at $z = L$ with $\varphi(L) = \infty$ as shown in Fig. 5.1(b). (Alternatively, a basis of wavefunctions with a finite extend in the $z$-direction can be used [15].) Then one identifies the states, which are centered within one module, and uses these to construct all states of the periodic structure by the periodicity condition (not shown here). Here $L$ should be large enough, so that the boundary condition does not affect the states in the module too much. On the other hand, if $L$ is too large, the module wavefunctions leak out. Thus, some experience and intuition is required for a good choice. For example, in Fig. 5.1(b), it is not a priori clear whether the state around 117 meV or at 72 meV is a better choice for the second excited state of the wide well. Furthermore, note that the wavefunctions constructed via the periodicity condition are not exactly orthogonal to the original wavefunctions. However, the deviations are often negligible [16].

As a third option, we consider first the QCL with $\phi(z) \equiv 0$, i.e., without any applied bias. Then the modules are periodically repeated and Bloch's theorem allows to determine energy bands $E_\nu(q)$ with the Bloch functions $\varphi_{\nu q}(z)$ applying the normalization $\int_0^d dz |\varphi_{\nu q}(z)|^2 = 1$. Based on these Bloch functions, we can construct *Wannier functions*

$$\Psi_\nu(z) = \frac{d}{2\pi} \int dq \, e^{i\alpha(q)} \varphi_{\nu q}(z),$$

where the phase $\alpha(q)$ has to be chosen to maximize their localization [17, 18]. Then the set of shifted Wannier functions $\Psi_{\ell\nu}(z) = \Psi_\nu(z - \ell d)$, where the integer $\ell$ is the module index, constitute a complete orthonormal set of functions. Limiting to bands with energies lower than $E_{\max}$ and a finite number of modules $\ell$ (typically 3–9), we thus obtain a finite orthonormal basis set, which is complete within the energy and spatial range chosen. An example is shown in Fig. 5.1(c). The Hamilton operator is then given by its matrix elements

$$\langle \Psi_{\ell'\nu'} | \hat{H} | \Psi_{\ell\nu} \rangle = H_\nu^{\ell'-\ell} \delta_{\nu\nu'} + \langle \Psi_{\ell'\nu'} | - e\phi(z) | \Psi_{\ell\nu} \rangle$$

$$\text{with } H_\nu^h = \frac{d}{2\pi} \int_{-\pi/d}^{\pi/d} dq \, e^{iqhd} E_\nu(q),$$

which can be easily extracted from the band structure $E_\nu(q)$ and the electrostatic potential. Diagonalizing this Hamilton operator provides an exact sequence of states satisfying the periodicity condition, as long as the states are within the periods covered by the finite number of modules used. These are the *Wannier–Stark* (WS) states as originally studied for single bands [19, 20]. An example is shown in Fig. 5.1(d). Comparing with the box states, we find a good agreement for the particular energies and wavefunctions chosen in Fig. 5.1(b). While computationally slightly more demanding, this scheme has the strong advantage that no judgment is needed to choose the external box and to select the appropriate states. Furthermore, the orthonormality of all states is guaranteed.

There is a fundamental difference between the module levels and the WS states (or the very similar periodically repeated box states): while the WS states diagonalize the system Hamiltonian of the pure heterostructure potential, the module levels exhibit tunneling matrix elements with levels in neighboring modules. Correspondingly, in Fig. 5.1(d) the two WS-states slightly above 20 meV correspond to the binding and anti-binding linear combination of the injector (dashed) and upper laser level (full) in Fig. 5.1(a), with approximately equal energies of 22 meV. The splitting of 2.9 meV between these WS states agrees well with twice the tunnel coupling $\Omega = 1.38$ meV (as given in [13]).

The constant effective mass in Eq. (5.1) is only a good approximation for the band structure a few 100 meV around the conduction band minimum. Thus, in particular for mid-IR-QCLs, nonparabolicity effects are of relevance. This can be done phenomenologically by using an energy-dependent effective mass [21], which, however, does not guarantee the orthogonality of eigenstates at different energies. A more microscopic approach is based on the coupling between different bands following [22] or more advanced $k \cdot p$ approaches, see also chapter 2 of [5]. Alternatively, tight-binding models can be used, which allow for the exploration of further valleys in the conduction band [23]. Such improvements are routinely used for all approaches addressed above.

## 5.2     Electric Transport in QCLs

The key task for simulating QCLs is to quantify the electron kinetics in the structure and thereby predict the occurrence of population inversion between the lasing levels. In this context it is most important to realize that the energy eigenstates, as evaluated in Section 5.1, are stationary states of the semiconductor heterostructure. Therefore these states themselves cannot provide any dynamical evolution such as current flow. Thus deviations from the ideal structure, such as lattice vibrations, lateral fluctuations in barrier thicknesses (often called interface roughness), alloy disorder for ternary semiconductor compounds, or the electric potentials of randomly implanted ionized dopants, play a crucial role by enabling transitions between these eigenstates. It is the task for theory to quantify this interplay between the ideal semiconductor heterostructure and these fluctuating potentials. More formally, we express the total Hamiltonian as

$$\hat{H} = \hat{H}_0 + \hat{H}_{\text{scatt}}, \tag{5.3}$$

where the effective Hamiltonian $\hat{H}_0$ represents the ideal semiconductor heterostructure including the mean field potential, such as the left-hand side of Eq. (5.1) for a parabolic band structure. Due to the translational invariance of $\hat{H}_0$ in the lateral $(x, y)$-direction, $\hat{H}_0$ is diagonal in $\mathbf{k}$. In contrast, $\hat{H}_{\text{scatt}}$, describing all deviations from the ideal structure, provides scattering transitions between states of different $\mathbf{k}$. In practice, one derives a quantum kinetic equation for the key variables within the conceptual approaches addressed below, and solves it in the stationary state, neglecting transient effects.

As it is quite cumbersome to simulate 30–200 modules, one typically applies periodic boundary conditions. This means that the levels with dashed (or dashed-dotted) lines in Fig. 5.1 have the same occupations as the corresponding levels with full lines. Then it is sufficient to determine the occupations in a single module in the presence of scattering within the module as well as the transitions to and from the neighboring modules. (For short modules with two or three barriers, even next-neighboring modules can be of relevance.)

## 5.2.1    Semiclassical Approaches

The most straightforward concept is to evaluate scattering rates by Fermi's golden rule as

$$\Gamma_{n\mathbf{k}\to n'\mathbf{k}'} = \frac{2\pi}{\hbar} \sum_{q_z} \left| \left\langle \Psi_{n'\ \mathbf{k}'} | \hat{H}_{scatt} | \Psi_{n'\mathbf{k}'} \right\rangle \right|^2 \delta \left( E_{n'\mathbf{k}'} - E_{n\mathbf{k}} + \Delta E_{scatt} \right), \qquad (5.4)$$

where $\Delta E_{scatt}$ is the energy transferred to a bath (usually phonons) during the scattering considered. Details are given in textbooks, e.g., [24], and in [4]. These transitions provide rate equations for the occupation probabilities $f_n(\mathbf{k})$ for the eigenstates which form a large system of differential equations. Commonly, these are solved with Monte-Carlo methods [25–29], so that this approach is often referred to as *Monte-Carlo simulations*. A strong advantage is that it allows for a microscopic treatment of electron–electron scattering [30] together with fully self-consistent distribution functions $f_n(\mathbf{k})$ in the stationary state.

A significant simplification can be obtained by considering the electron densities

$$n_n = \frac{2(\text{for spin})}{A} \sum_{\mathbf{k}} f_n(\mathbf{k}). \qquad (5.5)$$

By averaging the scattering rates between two levels occupied by some thermal distribution $f^{therm}$ (typically using an estimation for an effective electron temperature),

$$R_{n\to n'} = \frac{\sum_{\mathbf{k},\mathbf{k}'} f^{therm}(\mathbf{k}) \Gamma_{n\mathbf{k}\to n'\ \mathbf{k}'}}{\sum_{\mathbf{k}} f^{therm}(\mathbf{k})}, \qquad (5.6)$$

we obtain *rate equations*

$$\frac{d}{dt} n_n = \sum_{n'} n_{n'} R_{n'\to n} - n_n R_{n\to n'}, \qquad (5.7)$$

which allow for a self-consistent solution [31] of the transport problem.

## 5.2.2    The Relevance of Coherence

The semiclassical model fails for the case of weak to moderate tunnel couplings, where the energy splitting of the states in resonance becomes smaller than the relevant scattering rates times $\hbar$. Such a situation is shown in Fig. 5.1(b) and (d) with the doublet of states with energies slight above 20 meV centered around the tunneling barrier at $z \approx 0$. Here the semiclassical approach overestimates the current as illustrated in Fig. 5.2, which is well known for tunneling in double-dot systems and superlattices [32]. For QCLs, this has been thoroughly analyzed in [33].

A common solution to the problem is to apply the module levels from Fig. 5.1(a) and restrict the rate equations to intra-module transitions [34–36]. Inter-module transitions are treated by a density-matrix based approach [37, 38]. This results in a transition rate through a double quantum well such as in Fig. 5.2:

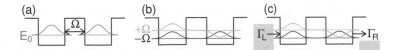

**Fig. 5.2** The problematics of rate equations for tunneling: (a) depicts the situation of two quantum wells, where the individual levels are in resonance with each other; (b) shows the corresponding energy eigenstates as binding and anti-binding states; and (c) carriers are supplied from the left by a reservoir and are extracted to the right of the double well. Within a rate equation scheme, the transitions occur between the reservoirs and the individual levels, where each of them connects both reservoirs. Thus the current just depends on the rates $\Gamma_L$ and $\Gamma_R$ but not on the tunnel coupling $\Omega$, which is nonphysical. In particular, the current is highly overestimated at resonance in the case $\Omega < \hbar\Gamma_{L/R}$.

$$R_{L\to R} = \frac{2\Omega^2\Gamma}{\hbar^2\Gamma^2 + \Delta E^2 + 2\Omega^2\Gamma\left(\Gamma_L^{-1} + \Gamma_R^{-1}\right)}, \tag{5.8}$$

as derived in the appendix. Here the detuning $\Delta E$ between the levels in the left and right well has been considered as well. $\Gamma$ is the total dephasing rate, which contains the term $(\Gamma_L + \Gamma_R)/2$ as well as contributions from intra-level scattering. For $\Omega \gg \hbar\Gamma, \Delta E$, we find that $R_{L\to R}$ only depends on the rates $\Gamma_{L/R}$, as correctly predicted by the semiclassical approaches. Also the drop in current with increasing $\Delta E$ is recovered in such models by the decreasing overlap of the eigenstates. However, the case $\hbar\Gamma > \Delta E, \Omega$, where the energy spacing of the eigenstates is smaller than $2\hbar\Gamma$, requires such a quantum treatment. A similar approach has also been incorporated into Monte-Carlo simulations [39].

## 5.2.3    Quantum Approaches

The aim of these approaches is to provide a self-consistent quantum approach without any arbitrary division of the system as in the density-matrix complemented rate equations addressed above. The key quantity is the single-particle density matrix

$$\rho_{mn}(\mathbf{k}) = \mathrm{Tr}\{\hat{\rho}\hat{a}_{n\,\mathbf{k}}^{\dagger}\hat{a}_{m\,\mathbf{k}}\}, \tag{5.9}$$

where $\hat{\rho}$ is the density operator and $\hat{a}_{n\mathbf{k}}^{\dagger}/\hat{a}_{n\mathbf{k}}$ are the creation/annihilation operators for the three-dimensional states, Eq. (5.2), respectively. All elements with different $\mathbf{k}$ are neglected, as they provide vanishing results upon integrating over the cross section of the QCL. The diagonal elements of the single-particle density matrix, $f_n(\mathbf{k}) = \rho_{nn}(\mathbf{k})$, are the occupation probabilities addressed above.

The dynamical evolution due to the Hamilton operator, Eq. (5.3), is typically written as

$$\frac{\mathrm{d}}{\mathrm{d}t}\rho_{mn}(\mathbf{k}, t) = \frac{i}{\hbar}\mathrm{Tr}\left\{\hat{\rho}(t)\left[\hat{H}_0, \hat{a}_{n\,\mathbf{k}}^{\dagger}\hat{a}_{m\mathbf{k}}\right]\right\}$$
$$+ \sum_{m'n'\mathbf{k'}} K_{mnm'n'}(\mathbf{k}, \mathbf{k'})\rho_{m'n'}(\mathbf{k'}, t), \tag{5.10}$$

where the scattering part of the Hamiltonian is subsumed in the scattering tensor $K_{mnm'n'}(\mathbf{k}, \mathbf{k}')$. Here we used the Markovian approximation neglecting all memory terms (i.e., $\rho_{m'n'}(\mathbf{k}', t')$ with $t' < t$ on the right-hand side), which simplifies the numerics significantly. In contrast to the semiclassical approach, this approach is, in principle, invariant under basis transformations, so that box, Wannier, or Wannier–Stark states can be applied as a basis. However, specific approximate evaluations of the scattering tensor impair this property. The diagonal elements $K_{nnn'n'}(\mathbf{k}, \mathbf{k}')$ can be related to the scattering rates, Eq. (5.4), but the choice of the non-diagonal elements is an intricate problem. The straightforward evaluation by second-order perturbation theory (following Wangsness and Bloch [40] and Redfield [41]) has been used by several authors [25, 42–45], which often provides good quantitative results.

Most physical quantities, such as the current density

$$J(z) = e \sum_{mn} \mathrm{Re}\left\{ \rho_{mn} \psi_n^*(z) \frac{\hbar}{m_c(z)\mathrm{i}} \frac{\partial \psi_m(z)}{\partial z} \right\}, \tag{5.11}$$

are actually given by the average density matrix

$$\rho_{mn} = \frac{2(\text{for spin})}{A} \sum_{\mathbf{k}} \rho_{mn}(\mathbf{k}), \tag{5.12}$$

which generalizes the electron density, Eq. (5.5). The dynamical evolution of $\rho_{mn}(t)$ is fully equivalent to Eq. (5.10). Here, the scattering tensor is typically established in a more phenomenological way based on the average scattering rates, Eq. (5.6), see, e.g., [46].

A severe shortcoming of the Wangsness–Bloch–Redfield approach is the fact that it can provide negative occupations [47] as specifically demonstrated in [43]. As average quantities like the current density or level densities are usually well-behaved, this conceptional problem may be ignored for many applications. Alternatively, one may use a phenomenologically modified tensor $K$, as proposed by Gordon and Majer [48] and generalized in terms of the PERLind approach in [49].

The inherent problem of Markovian density-matrix approaches is the fact that the energy of correlations (i.e., non-diagonal elements of the density matrix $\rho_{nm}$) is not defined. However, for scattering processes this energy information is crucial as the energy transfer appears in the spectral function of the bath, as, e.g., visible in Fermi's golden rule, Eq. (5.4). Thus, the secular approximation, which restricts the tensor $K_{mnm'n'}$ to elements with a defined energy transfer $\Delta E, = E_m - E_{m'} = E_n - E_{n'}$, is a popular way to generate a quantum rate equation which guarantees the positivity of $\rho_{nn}$ for all times [47]. However, this restriction neglects a lot of important physics and, in particular, the problem with tunneling addressed in Fig. 5.2 reappears [49].

Within the theory of nonequilibrium Green's functions (NEGF) [50], the density matrix is related to the lesser Green's function $G^<$ by

$$\rho_{mn}(\mathbf{k}) = \frac{1}{2\pi\mathrm{i}} G_{mn}^<(E, \mathbf{k}), \tag{5.13}$$

so that the $G^<$ carries the energy information lacking in the density matrix. Thus a fully consistent set of equations can be established with NEGF, where scattering can be treated perturbatively based on microscopic input. This concept has been applied to QCL simulations by a variety of groups [51–56]. However, the additional continuous variable $E$ increases the numerical effort significantly. Thus some approximations have to be done to achieve tractable schemes which have different shortcomings as outlined in [57]. While most NEGF implementations use periodic boundary conditions, others consider short QCL segments with injection from a contact with thermal distribution [52].

Recently, a detailed comparison with a large number of different samples has been done using the identical simulation scheme [58], which demonstrates the quantitative reliability for current and gain of this approach. Note that only the nominal device parameters, such as layer thicknesses and doping profile, enter these simulations as input parameter, while all other model approximations are identical for all samples studied, so that the simulation is of predictive nature, see also [59]. The temperature used in these simulations determines the occupation of the phonon modes and should be chosen somewhat higher than the experimental heat sink temperature even for pulsed operation, as the most relevant optical phonons are heated on a short time scale [60, 61].

Figure 5.3 shows results for the THz device of [13] as an example of this scheme. We find good agreement for the current peak around 39 mV/module. For higher bias, the current drops slightly in the simulation with the standard scheme, which is not visible in the experimental data. Such a drop of current with increasing local electrical field is not uncommon for tunneling in heterostructures if the levels get detuned with

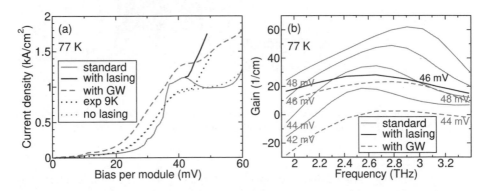

**Fig. 5.3** Results of the NEGF simulations for the device of [13]. The full/dashed lines show results without/with electron–electron scattering within a rudimentary GW approximation. A phonon temperature of 77 K has been applied. (a) Current bias characteristics together with experimental data (dotted lines, courtesy of E. Dupont) for a heatsink temperature of 9 K. For the laser with AuAu waveguide a 0.8 V drop at the Schottky contact is subtracted. Here the kink at $\approx 44$ mV indicates the onset of lasing. The non-lasing device with PdGe contacts has no Schottky contact. The dark line shows the current under lasing, where $F_{ac}$ is adjusted so that the peak gain matches the waveguide losses of $\approx 24$/cm [13]. (b) Gain spectra at different operation points. The gray lines are calculated for weak $F_{ac}$, while the black line is calculated for $F_{ac}d = 10$ mV at 46 mV per module. Experimentally, the onset of lasing is observed at 2.6 THz [13].

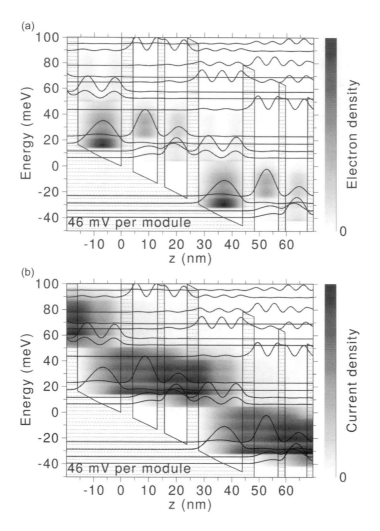

**Fig. 5.4** Spatially and energetically resolved electron (a) and current (b) density evaluated from the NEGF model for the device of [13] at a bias of 46 mV per module.

increasing field above resonance [37, 62, 63]. This situation is referred to as negative differential conductivity (NDC) or negative differential resistance (NDR) and leads to electrical instabilities as addressed in Section 5.4. Considering electron–electron scattering within the highly simplified GW approach of [64], the NDC vanishes in the bias range considered and the current-bias relation is much smoother. Comparison with experimental data indicates that the effect is too strong, so that this simplified GW approach is probably overestimating electron–electron scattering. Furthermore, neither in the experimental data nor in the simulation a current peak at the design bias of 53 mV per module is visible. Here the electrons are essentially trapped in the upper laser level, so that the tunneling resonance does play a minor role without lasing for this structure.

The energy dependence of $G^<$, see Eq. (5.13), allows to extract the spatially and energetically resolved electron and current densities as shown in Fig. 5.4, where the energy is the argument of $G^<_{nm}(E, \mathbf{k})$. These allow for a very detailed characterization of the device. Note the variations in the densities with a spacing of $\approx 10$ meV. They result from the fact that optical phonon emission dominates the energy loss with portions of 36 meV, while the energy increases by 46 meV while transversing one module. These features vanish if electron–electron scattering is included within the GW approximation, as this constitutes further energy transfer between the carriers.

### 5.2.4 Comparison Between Approaches

The validation of different approaches is a difficult task. At first, in many cases only successful simulations, where reasonable agreement with experimental data is found, are published. In this context, it is also important to realize that many small details can change the outcome. For example, the chosen model for screening can significantly modify electron–electron or electron–impurity scattering rates. In the same way, the parameters for interface roughness are usually not known and can affect the results. Thus a model can only be reasonably validated if simulations for a larger number of different devices are shown using precisely the same scheme as in [16, 58]. A second type of validation is the comparison of different approaches with each other, where the same assumptions are made. Here [65] compared Monte-Carlo simulations with the NEGF scheme for a THz structure and found good agreement as long as the individual levels are not to close, which agrees with the arguments given in Section 5.2.2. The density-matrix complemented rate equations have been compared to the NEGF scheme in [66] for an infrared QCL. In this case, the much simpler rate equation model showed good results provided second-order currents are implemented, which are inherently present in the NEGF simulations.

## 5.3 Gain and Lasing Field

The central property of any laser is the amplification of the light field as manifested by the optical gain. Its calculation is the central task of any theoretical description of QCLs. The easiest approach is to consider a two-level model, where the net rate of stimulated emission and absorption between the upper (up) and the lower (low) laser level due to an oscillating field $F_z(t) = F_{ac} \cos(\omega t)$ is given by

$$R^{opt}_{up \to low} = \frac{1}{\hbar} \left| \frac{F_{ac} d_{up,low}}{2} \right|^2 (n_{up} - n_{low}) \frac{\gamma}{(E_{up} - E_{low} - \hbar\omega)^2 + \gamma^2/4} . \tag{5.14}$$

This rate can directly be translated to the gain by relating the number of generated photons to the photon flux given by the optical intensity ($\propto F^2_{ac}$), see, e.g., [6] for details. $\gamma$ is the width of the transition, which is related to the lifetimes of the states. As the optical transitions change the balance between the occupations in the rate equations, the optical intensity has to be calculated self-consistently with the electron occupations, which was first done in [67] for the rate equation model. The key result is that the inversion $n_{up} - n_{low}$ is reduced by the lasing field until the waveguide losses compensates the remaining gain, which is called gain saturation. Further transitions between other levels can be easily added in the same manner, which provide, e.g., detrimental absorption effects.

Within a quantum kinetic approach, the oscillating electric field $F_z(t)$ drives the system resulting in a time-dependent density matrix $\rho_{mn}(t)$. The coherence between the upper and lower laser states, $\rho_{low,up}$, provides the polarization

$$P_z(t) = \frac{1}{d}\rho_{low,up}(t)d_{up,low} + cc, \qquad (5.15)$$

where $d_{up,low}$ is the electric dipole matrix element between the upper and lower laser states. This allows for the identification of the frequency-dependent dielectric susceptibility $\chi(\omega) = P_z(\omega)/F_z(\omega)\epsilon_0$, whose imaginary part is related to the gain/absorption coefficient by standard electrodynamics. Such approaches are very suitable to study the broadening and Coulomb shifts on the gain spectra [68–70]. Using higher-order perturbation theory, dispersive gain can be studied as well [71, 72].

A third possibility is to consider the time-dependent electrical conductivity evaluated from the current, Eq. (5.11), see [49, 55], which provides the gain coefficient by

$$G(\omega) = -\frac{1}{cn_r\epsilon_0}\sigma(\omega) \quad \text{with } \sigma(\omega) = \frac{\langle J_z(\omega)\rangle}{F_z(\omega)}, \qquad (5.16)$$

where $n_r$ is the refractive index. Here the ac current component $J_z(z, \omega)$ is averaged over the module. Identifying $\sigma(\omega) = -i\omega\epsilon_0\chi(\omega)$, where polarization currents are replaced by real currents; this is analogous to the approach via the susceptibility addressed above. However, for extended states, the proper identification of transitions is less clear for the polarization, so that the approach via the conductivity is better defined in general.

One of the particular strengths of the NEGF scheme is the self-consistent determination of the laser spectrum without any phenomenological broadening parameter $\gamma$. This is due to the fact that the full energy dependence of all relevant quantities is kept, as can be seen for the lesser Green's function, Eq. (5.13). In particular, dispersive features of gain, as dominating the gain in superlattices, are fully contained [73, 74].

Figure 5.3 shows the gain spectrum at different bias points calculated by the NEGF scheme which evaluates gain via the frequency-dependent conductivity, Eq. (5.16). We see that the gain overcomes the waveguide losses of 24/cm [13] at $\approx 43\,\text{mV}$ per module for frequencies around 2.6 THz. This matches the experimental lasing frequency at ignition and the bias is only slightly smaller than the experimental value of $\approx 44.5\,\text{mV}$ per module. Including electron–electron scattering, the gain is reduced and the threshold bias increases slightly.

In Fig. 5.5 we compare the gain calculated by the full NEGF treatment based on the conductivity with the simple approach used in Eq. (5.14). For the latter, the Wannier–Stark states are evaluated and the population of these states is extracted from the NEGF data. Here the width of the transitions was determined in two ways: (i) the inverse scattering lifetimes times $\hbar$ of the individual states (as extracted from the

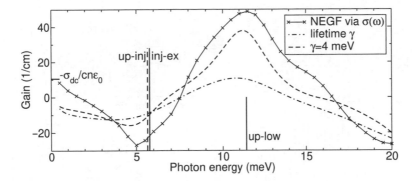

**Fig. 5.5** Comparison of gain spectrum for the operation point of Fig. 5.4 calculated by the conductivity from the NEGF model (full line with crosses) and by transition rates, Eq. (5.14). For the latter case the dashed and dash-dotted line use different values of $\gamma$. The vertical lines indicate the most important transitions providing gain (full line) or absorption (dashed). The vertical lines depict the energy difference for specific pairs of states. These states are displayed in Fig. 5.4 for $0 < E < 25$ meV in the sequence low, ex, inj, up.

imaginary part of the self energy) were added for both states involved in the transition to obtain $\gamma$. This overestimates the width due to correlation effects as discussed in [75] and provides rather wide peaks with little gain or absorption. (ii) A phenomenological width of 4 meV was used, which provides reasonable agreement with the full NEGF treatment. Finally, note that the NEGF gain spectrum approaches the value $-\sigma_{dc}/cn_r\epsilon_0$, where $\sigma_{dc}$ is extracted from the current-bias relation in Fig. 5.3(a). This physical feature is hardly achieved by the other approaches. It also demonstrates the tight relation between the electrical performance and the optical spectrum in the THz range.

Once the gain of the QCL material overcomes the waveguide losses, the lasing field ignites and becomes significant. Then the stimulated transitions strongly affect the operation of the QCL in two ways: (i) the inversion is reduced and gain diminishes, see the thick line in Fig. 5.3(b) as an example; and (ii) the carriers are commonly more easily transported though the module and the current increases. All these effects can be modeled in detail. In Fig. 5.3(a), the thick line shows the calculated current for the lasing field which saturates the gain coefficient to the waveguide absorption. The increase of current is very similar to the experimental data.

## 5.4    Operation of the Cascaded System

The cascading scheme for QCLs implies that the same bias drops over each module as shown in Fig. 5.6(a). This implies that the electron density per module matches the sheet-doping density $n_D$ per module (both in units cm$^{-2}$). If this operation point shows negative differential conductivity (NDC), such a situation is unstable as spontaneous charge fluctuations increase exponentially. Let us assume that there is a slight additional electron density $\delta n_\ell$ in module $\ell$. Then Poisson's equation implies that $F$ is smaller on the left and larger on the right side of module $\ell$. In the presence of

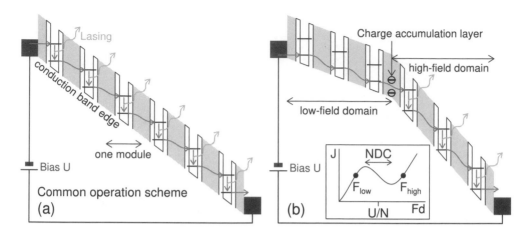

**Fig. 5.6** (a) QCL operation, where the applied bias $U$ is equally distributed over all modules. The gray regions indicate specifically designed heterostructures guiding the electron flow as indicated by a sequence of alternating vertical and downward sloping arrows. (b) If the average bias per module $U/N$ is located at an operation point with negative differential conductivity (NDC), see inset, a low-field and a high-field domain form, where both have the same current density. In between, an electron accumulation layer is required by Poisson's equation.

NDC, the current density $J$ increases for the modules to the left and decreases to the right, so that the carrier imbalance $\delta n_\ell$ increases in time due to the continuity equation. (Note the convention used here, that the physical currents $J$ and fields $F$ point in the negative direction, thus $J$ and $F$ point in the particle-flow direction.) This charge accumulation increases until the current is the same on both sides, as depicted in the inset of Fig. 5.6(b). While for continuous systems, such as Gunn diodes, these charge accumulation layers travel through the device with an average drift velocity [76], they can become stationary in highly doped superlattices where the accumulation layer is essentially located in a single quantum well [77]. In analogy to the related superlattices, such electric field domains have been observed and analyzed in several QCL structures [78–82]. Typically, this situation is avoided, as the periodic repetition of identically operating modules is one of the core design ideas of the QCL. Thus, generally, a lasing operation point with positive differential conductivity is looked for. However, laser ignition can also occur in a domain state and might provide a better power efficiency [83].

## Acknowledgements

I thank D. Winge, M. Franckie, and T. Almqvist for their contributions to the method development used here and E. Dupont for providing the experimental data in Fig. 5.3. Extended discussions with F. Capasso, E. Dupont, J. Faist, H. Grahn, Q. Hu, T. Kubis, S. Kumar, M. Pereira, C. Sirtori, G. Strasser, R. Terazzi, and K. Unterrainer over the last 15 years contributed essentially to the understanding of the concepts addressed here. This work was supported by the Swedish Science council within project 2017-04287.

## Appendix: Derivation of Eq. (5.8)

The two-level system in Fig. 5.2 is described in occupation number formalism by the Hamiltonian

$$\hat{H} = \Delta E \hat{a}_r^\dagger \hat{a}_r + \Omega \hat{a}_l^\dagger \hat{a}_r + \Omega \hat{a}_r^\dagger \hat{a}_l,$$

where the indices l/r refer to the level in the left/right well and $\Delta E$ is the detuning between the levels. The kinetics of carriers due to injection and extraction as well as dephasing by intra-level scattering processes can be treated by a Lindblad–Gorini–Kossakowski–Sudarshan master equation (see, e.g., [47]) for the reduced density operator $\hat{\rho}$ of the two-well system

$$\frac{d\hat{\rho}}{dt} = \frac{i}{\hbar}[\hat{\rho}, \hat{H}] + \sum_j \Gamma_j \left( \hat{J}_j \hat{\rho} \hat{J}_j^\dagger - \frac{1}{2}\hat{J}_j^\dagger \hat{J}_j \hat{\rho} - \frac{1}{2}\hat{\rho}\hat{J}_j^\dagger \hat{J}_j \right).$$

Here the coupling to the environment is treated by jump operators $\hat{J}_1 = \hat{a}_1^\dagger$, $\hat{J}_2 = \hat{a}_r$, $\hat{J}_3 = \hat{a}_1^\dagger \hat{a}_1$, and $\hat{J}_4 = \hat{a}_r^\dagger \hat{a}_r$ with rates $\Gamma_1 = \Gamma_L$, $\Gamma_2 = \Gamma_R$, $\Gamma_3 = \Gamma_{dephL}$, and $\Gamma_4 = \Gamma_{dephR}$, respectively. Without Coulomb interaction, we can project to the expectation values $\rho_{mn} = \text{Tr}\left\{\hat{a}_m \hat{\rho} \hat{a}_n^\dagger\right\}$ providing after some algebra (applying cyclic permutations of operators under the trace and the anti-commutation rules of $\hat{a}_i, \hat{a}_j^\dagger$) the equations of motion for the populations $\rho_{ll}/\rho_{rr}$ of the left/right level, respectively, and the coherence $\rho_{rl}$ between these levels:

$$\frac{d\rho_{ll}}{dt} = \frac{i\Omega}{\hbar}(\rho_{lr} - \rho_{rl}) + \Gamma_L(1 - \rho_{ll}),$$

$$\frac{d\rho_{rr}}{dt} = \frac{i\Omega}{\hbar}(\rho_{rl} - \rho_{lr}) - \Gamma_R\rho_{rr},$$

$$\frac{d\rho_{rl}}{dt} = \frac{i\Omega}{\hbar}(\rho_{rr} - \rho_{ll}) - \frac{i\Delta E}{\hbar}\rho_{rl} - \Gamma\rho_{rl},$$

with $\Gamma = (\Gamma_L + \Gamma_R + \Gamma_{dephL} + \Gamma_{dephR})/2$. These terms have a straightforward interpretation: $\Gamma_L(1 - \rho_{ll})$ describes electrons entering the left level (if it is empty) from the left filled reservoir. $\Gamma_R\rho_{rr}$ describes electrons leaving the right level to the right empty reservoir. $\frac{i\Omega}{\hbar}(\rho_{rl} - \rho_{lr})$ is the transition rate between the dots, which is related to coherent superposition of the dot levels. These coherencies $\rho_{rl}$ and $\rho_{lr} = \rho_{rl}^*$ are driven by differences in the dot occupations ($\rho_{rr} - \rho_{ll}$), oscillate by detuning $\Delta E$, and decay due to all scattering processes subsumed in $\Gamma$. Solving these equations in the stationary state provides

$$R_{L\rightarrow R} = \frac{i\Omega}{\hbar}(\rho_{rl} - \rho_{lr}) = \frac{2\Omega^2\Gamma}{\hbar^2\Gamma^2 + \Delta E^2 + 2\Omega^2\Gamma\left(\Gamma_L^{-1} + \Gamma_R^{-1}\right)},$$

as given in Eq. (5.8) and originally derived by Kazarinov and Suris [37] using $\Gamma = 1/\tau_\perp$ and $\Gamma_L = \Gamma_R = 1/\tau_\parallel$.

## References

[1]  J. Faist, F. Capasso, D. L. Sivco, C. Sirtori, A. L. Hutchinson, and A. Y. Cho, "Quantum cascade laser," *Science*, vol. 264, pp. 553–556, 1994.

[2]  R. Köhler, A. Tredicucci, F. Beltram, H. E. Beere, E. H. Linfield, A. G. Davies, D. A. Ritchie, R. C. Iotti, and F. Rossi, "Terahertz semiconductor-heterostructure laser," *Nature*, vol. 417, p. 156, 2002.

[3]  J. Faist, *Quantum Cascade Lasers*. Oxford: Oxford University Press, 2013.

[4]  C. Jirauschek and T. Kubis, "Modeling techniques for quantum cascade lasers," *Appl. Phys. Rev.*, vol. 1, p. 011307, 2014.

[5]  R. Terazzi, "Transport in quantum cascade lasers," Ph.D. dissertation, ETH Zürich, 2011.

[6]  A. Wacker, "Quantum cascade laser: An emerging technology," in *Nonlinear Laser Dynamics*, K. Lüdge, Ed. Berlin: Wiley-VCH, 2012.

[7]  D. J. BenDaniel and C. B. Duke, "Space-charge effects on electron tunneling," *Phys. Rev.*, vol. 152, pp. 683–692, 1966.

[8]  J. Zak, "Stark ladder in solids?" *Phys. Rev. Lett.*, vol. 20, pp. 1477–1481, 1968.

[9]  G. Nenciu, "Dynamics of band electrons in electric and magnetic fields: Rigorous justification of the effective Hamiltonians," *Rev. Mod. Phys.*, vol. 63, pp. 91–127, 1991.

[10]  E. E. Mendez, F. Agulló-Rueda, and J. M. Hong, "Stark localization in GaAs-GaAlAs superlattices under an electric field," *Phys. Rev. Lett.*, vol. 60, pp. 2426–2429, 1988.

[11]  P. Voisin, J. Bleuse, C. Bouche, S. Gaillard, C. Alibert, and A. Regreny, "Observation of the Wannier-Stark quantization in a semiconductor superlattice," *Phys. Rev. Lett.*, vol. 61, pp. 1639–1642, 1988.

[12]  J. Bardeen, "Tunnelling from a many-particle point of view," *Phys. Rev. Lett.*, vol. 6, p. 57, 1961.

[13]  S. Fathololoumi, E. Dupont, C. Chan, Z. Wasilewski, S. Laframboise, D. Ban, A. Mátyás, C. Jirauschek, Q. Hu, and H. C. Liu, "Terahertz quantum cascade lasers operating up to ∼ 200 K with optimized oscillator strength and improved injection tunneling," *Opt. Express*, vol. 20, pp. 3866–3876, 2012.

[14]  R. Steed, H. Hebal, and S. B. Lisesivdin, "Aestimo v.1.2," 2017, https://doi.org/10.5281/zenodo.1042657.

[15]  O. Jonasson, F. Karimi, and I. Knezevic, "Partially coherent electron transport in terahertz quantum cascade lasers based on a markovian master equation for the density matrix," *J. Comput. Electron.*, vol. 15, pp. 1192–1205, 2016.

[16]  B. A. Burnett, A. Pan, C. O. Chui, and B. S. Williams, "Robust density matrix simulation of terahertz quantum cascade lasers," *IEEE T. THz Sci. Techn.*, vol. 8, pp. 492–501, 2018.

[17]  A. Bruno-Alfonso and D. R. Nacbar, "Wannier functions of isolated bands in one-dimensional crystals," *Phys. Rev. B*, vol. 75, p. 115428, 2007.

[18]  M. Maczka, G. Hałdaś, and S. Pawłowski, "Study of quantum states maximal localization in nonsymmetrical semiconductor superlattice structures," in *2016 13th Selected Issues of Electrical Engineering and Electronics (WZEE)*, 2016, pp. 1–5.

[19]  G. H. Wannier, "Wave functions and effective Hamiltonian for Bloch electrons in an electric field," *Phys. Rev.*, vol. 117, pp. 432–439, 1960.

[20]  D. Emin and C. F. Hart, "Existence of Wannier-Stark localization," *Phys. Rev. B*, vol. 36, pp. 7353–7359, 1987.

[21] J. D. Cooper, A. Valavanis, Z. Ikonic, P. Harrison, and J. E. Cunningham, "Finite difference method for solving the Schrödinger equation with band nonparabolicity in mid-infrared quantum cascade lasers," *J. Appl. Phys.*, vol. 108, p. 113109, 2010.

[22] C. Sirtori, F. Capasso, J. Faist, and S. Scandolo, "Nonparabolicity and a sum rule associated with bound-to-bound and bound-to-continuum intersubband transitions in quantum wells," *Phys. Rev. B*, vol. 50, pp. 8663–8674, 1994.

[23] J. Green, T. B. Boykin, C. D. Farmer, M. Garcia, C. N. Ironside, G. Klimeck, R. Lake, and C. R. Stanley, "Quantum cascade laser gain medium modeling using a second-nearest-neighbor sp3s* tight-binding model," *Superlattices and Microstructures*, vol. 37, pp. 410–424, 2005.

[24] P. Harrison and A. Valavanis, *Quantum Wells, Wires and Dots*, 4th ed. Hoboken: John Wiley & Sons, 2016.

[25] R. C. Iotti and F. Rossi, "Nature of charge transport in quantum-cascade lasers," *Phys. Rev. Lett.*, vol. 87, p. 146603, 2001.

[26] H. Callebaut, S. Kumar, B. S. Williams, Q. Hu, and J. L. Reno, "Importance of electron-impurity scattering for electron transport in terahertz quantum-cascade lasers," *Appl. Phys. Lett.*, vol. 84, p. 645, 2004.

[27] X. Gao, D. Botez, and I. Knezevic, "X-valley leakage in GaAs-based midinfrared quantum cascade lasers: A Monte Carlo study," *J. Appl. Phys.*, vol. 101, p. 063101, 2007.

[28] C. Jirauschek and P. Lugli, "Monte-Carlo-based spectral gain analysis for terahertz quantum cascade lasers," *J. Appl. Phys.*, vol. 105, p. 123102, 2009.

[29] P. Borowik, J.-L. Thobel, and L. Adamowicz, "Monte Carlo modeling applied to studies of quantum cascade lasers," *Opt. Quant. Electron.*, vol. 49, p. 96, 2017.

[30] O. Bonno, J. Thobel, and F. Dessenne, "Modeling of electron-electron scattering in Monte Carlo simulation of quantum cascade lasers," *J. Appl. Phys.*, vol. 97, p. 043702, 2005.

[31] K. Donovan, P. Harrison, and R. W. Kelsall, "Self-consistent solutions to the intersubband rate equations in quantum cascade lasers: Analysis of a GaAs/AlGaAs device," *J. Appl. Phys.*, vol. 89, pp. 3084–3090, 2001.

[32] A. Wacker and A.-P. Jauho, "Quantum transport: The link between standard approaches in superlattices," *Phys. Rev. Lett.*, vol. 80, pp. 369–372, 1998.

[33] H. Callebaut and Q. Hu, "Importance of coherence for electron transport in terahertz quantum cascade lasers," *J. Appl. Phys.*, vol. 98, p. 104505, 2005.

[34] S. Kumar and Q. Hu, "Coherence of resonant-tunneling transport in terahertz quantum-cascade lasers," *Phys. Rev. B*, vol. 80, p. 245316, 2009.

[35] R. Terazzi and J. Faist, "A density matrix model of transport and radiation in quantum cascade lasers," *New J. Phys.*, vol. 12, p. 033045, 2010.

[36] E. Dupont, S. Fathololoumi, and H. C. Liu, "Simplified density-matrix model applied to three-well terahertz quantum cascade lasers," *Phys. Rev. B*, vol. 81, p. 205311, 2010.

[37] R. F. Kazarinov and R. A. Suris, "Possibility of the amplification of electromagnetic waves in a semiconductor with a superlattice," *Sov. Phys. Semicond.*, vol. 5, p. 707, 1971.

[38] C. Sirtori, F. Capasso, J. Faist, A. Hutchinson, D. L. Sivco, and A. Y. Cho, "Resonant tunneling in quantum cascade lasers," *IEEE J. Quantum Elect.*, vol. 34, p. 1772, 1998.

[39] C. Jirauschek, "Density matrix Monte Carlo modeling of quantum cascade lasers," *Journal of Applied Physics*, vol. 122, p. 133105, 2017.

[40] R. K. Wangsness and F. Bloch, "The dynamical theory of nuclear induction," *Phys. Rev.*, vol. 89, pp. 728–739, 1953.

[41] A. G. Redfield, "On the theory of relaxation processes," *IBM J. Res. Dev.*, vol. 1, pp. 19–31, 1957.

[42] R. C. Iotti and F. Rossi, "Microscopic theory of semiconductor-based optoelectronic devices," *Rep. Prog. Phys.*, vol. 68, p. 2533, 2005.

[43] C. Weber, A. Wacker, and A. Knorr, "Density-matrix theory of the optical dynamics and transport in quantum cascade structures: The role of coherence," *Phys. Rev. B*, vol. 79, p. 165322, 2009.

[44] O. Jonasson, S. Mei, F. Karimi, J. Kirch, D. Botez, L. Mawst, and I. Knezevic, "Quantum transport simulation of high-power 4.6-µm quantum cascade lasers," *Photonics*, vol. 3, 2016.

[45] A. Pan, B. A. Burnett, C. O. Chui, and B. S. Williams, "Density matrix modeling of quantum cascade lasers without an artificially localized basis: A generalized scattering approach," *Phys. Rev. B*, vol. 96, p. 085308, 2017.

[46] W. Freeman, "Self-consistent calculation of dephasing in quantum cascade structures within a density matrix method," *Phys. Rev. B*, vol. 93, p. 205301, 2016.

[47] H.-P. Breuer and F. Petruccione, *Open Quantum Systems*. Oxford: Oxford University Press, 2006.

[48] A. Gordon and D. Majer, "Coherent transport in semiconductor heterostructures: A phenomenological approach," *Phys. Rev. B*, vol. 80, p. 195317, 2009.

[49] G. Kiršanskas, M. Franckié, and A. Wacker, "Phenomenological position and energy resolving Lindblad approach to quantum kinetics," *Phys. Rev. B*, vol. 97, p. 035432, 2018.

[50] H. Haug and A.-P. Jauho, *Quantum Kinetics in Transport and Optics of Semiconductors*. Berlin, Heidelberg: Springer, 1998.

[51] S.-C. Lee, F. Banit, M. Woerner, and A. Wacker, "Quantum mechanical wavepacket transport in quantum cascade laser structures," *Phys. Rev. B*, vol. 73, p. 245320, 2006.

[52] T. Kubis, C. Yeh, P. Vogl, A. Benz, G. Fasching, and C. Deutsch, "Theory of nonequilibrium quantum transport and energy dissipation in terahertz quantum cascade lasers," *Phys. Rev. B*, vol. 79, p. 195323, 2009.

[53] T. Schmielau and M. F. Pereira, "Nonequilibrium many body theory for quantum transport in terahertz quantum cascade lasers," *Appl. Phys. Lett.*, vol. 95, p. 231111, 2009.

[54] G. Hałdaś, A. Kolek, and I. Tralle, "Modeling of mid-infrared quantum cascade laser by means of nonequilibrium Green's functions," *IEEE J. Quantum Elect.*, vol. 47, pp. 878–885, 2011.

[55] A. Wacker, M. Lindskog, and D. O. Winge, "Nonequilibrium Green's function model for simulation of quantum cascade laser devices under operating conditions," *IEEE J. Sel. Top. Quant.*, vol. 19, p. 1200611, 2013.

[56] T. Grange, "Contrasting influence of charged impurities on transport and gain in terahertz quantum cascade lasers," *Phys. Rev. B*, vol. 92, p. 241306, 2015.

[57] T. Kubis and P. Vogl, "Assessment of approximations in nonequilibrium Green's function theory," *Phys. Rev. B*, vol. 83, p. 195304, 2011.

[58] D. O. Winge, M. Franckié, and A. Wacker, "Simulating terahertz quantum cascade lasers: Trends from samples from different labs," *Journal of Applied Physics*, vol. 120, p. 114302, 2016.

[59] M. Franckié, L. Bosco, M. Beck, C. Bonzon, E. Mavrona, G. Scalari, A. Wacker, and J. Faist, "Two-well quantum cascade laser optimization by non-equilibrium Green's function modelling," *Appl. Phys. Lett.*, vol. 112, p. 021104, 2018.

[60] M. S. Vitiello, R. C. Iotti, F. Rossi, L. Mahler, A. Tredicucci, H. E. Beere, D. A. Ritchie, Q. Hu, and G. Scamarcio, "Non-equilibrium longitudinal and transverse optical phonons in terahertz quantum cascade lasers," *Appl. Phys. Lett.*, vol. 100, 2012.

[61] Y. B. Shi and I. Knezevic, "Nonequilibrium phonon effects in midinfrared quantum cascade lasers," *J. Appl. Phys.*, vol. 116, p. 123105, 2014.

[62] L. Esaki and L. L. Chang, "New transport phenomenon in a semiconductor 'superlattice'," *Phys. Rev. Lett.*, vol. 33, pp. 495–498, 1974.

[63] F. Capasso, K. Mohammed, and A. Y. Cho, "Sequential resonant tunneling through a multiquantum well superlattice," *Appl. Phys. Lett.*, vol. 48, p. 478, 1986.

[64] D. O. Winge, M. Franckié, C. Verdozzi, A. Wacker, and M. F. Pereira, "Simple electron-electron scattering in non-equilibrium Green's function simulations," *J. Phys.: Conf. Ser.*, vol. 696, p. 012013, 2016.

[65] A. Mátyás, T. Kubis, P. Lugli, and C. Jirauschek, "Comparison between semiclassical and full quantum transport analysis of thz quantum cascade lasers," *Physica E*, vol. 42, pp. 2628–2631, 2010.

[66] M. Lindskog, J. M. Wolf, V. Trinite, V. Liverini, J. Faist, G. Maisons, M. Carras, R. Aidam, R. Ostendorf, and A. Wacker, "Comparative analysis of quantum cascade laser modeling based on density matrices and non-equilibrium Green's functions," *Appl. Phys. Lett.*, vol. 105, p. 103106, 2014.

[67] D. Indjin, P. Harrison, R. W. Kelsall, and Z. Ikonic, "Self-consistent scattering theory of transport and output characteristics of quantum cascade lasers," *J. Appl. Phys.*, vol. 91, p. 9019, 2002.

[68] M. F. Pereira, Jr., S.-C. Lee, and A. Wacker, "Controlling many-body effects in the mid-infrared gain and THz absorption of quantum cascade laser structures," *Phys. Rev. B*, vol. 69, p. 205310, 2004.

[69] I. Waldmueller, W. W. Chow, E. W. Young, and M. C. Wanke, "Nonequilibrium many-body theory of intersubband lasers," *IEEE J. Quantum Elect.*, vol. 42, pp. 292–301, 2006.

[70] T. Liu, K. E. Lee, and Q. J. Wang, "Microscopic density matrix model for optical gain of terahertz quantum cascade lasers: Many-body, nonparabolicity, and resonant tunneling effects," *Phys. Rev. B*, vol. 86, p. 235306, 2012.

[71] H. Willenberg, G. H. Döhler, and J. Faist, "Intersubband gain in a Bloch oscillator and quantum cascade laser," *Phys. Rev. B*, vol. 67, p. 085315, 2003.

[72] R. Terazzi, T. Gresch, M. Giovannini, N. Hoyler, N. Sekine, and J. Faist, "Bloch gain in quantum cascade lasers," *Nat. Physics*, vol. 3, p. 329, 2007.

[73] A. Wacker, "Gain in quantum cascade lasers and superlattices: A quantum transport theory," *Phys. Rev. B*, vol. 66, p. 085326, 2002.

[74] A. Wacker, "Lasers: Coexistence of gain and absorption," *Nat. Phys.*, vol. 3, p. 298, 2007.

[75] F. Banit, S.-C. Lee, A. Knorr, and A. Wacker, "Self-consistent theory of the gain linewidth for quantum cascade lasers," *Appl. Phys. Lett.*, vol. 86, p. 041108, 2005.

[76] M. P. Shaw, V. V. Mitin, E. Schöll, and H. L. Grubin, *The Physics of Instabilities in Solid State Electron Devices*. New York: Plenum Press, 1992.

[77] A. Wacker, M. Moscoso, M. Kindelan, and L. L. Bonilla, "Current-voltage characteristic and stability in resonant-tunneling n-doped semiconductor superlattices," *Phys. Rev. B*, vol. 55, pp. 2466–2475, 1997.

[78] S. L. Lu, L. Schrottke, S. W. Teitsworth, R. Hey, and H. T. Grahn, "Formation of electric-field domains in GaAs – $Al_xGa_{1-x}As$ quantum cascade laser structures," *Phys. Rev. B*, vol. 73, p. 033311, 2006.

[79]  M. Wienold, L. Schrottke, M. Giehler, R. Hey, and H. T. Grahn, "Nonlinear transport in quantum-cascade lasers: The role of electric-field domain formation for the laser characteristics," *J. Appl. Phys.*, vol. 109, p. 073112, 2011.

[80]  H. Yasuda, I. Hosako, and K. Hirakawa, "High-field domains in terahertz quantum cascade laser structures based on resonant-phonon depopulation scheme," in *38th International Conference on Infrared, Millimeter, and Terahertz Waves*, 2013, pp. 1–2.

[81]  R. S. Dhar, S. G. Razavipour, E. Dupont, C. Xu, S. Laframboise, Z. Wasilewski, Q. Hu, and D. Ban, "Direct nanoscale imaging of evolving electric field domains in quantum structures," *Sci. Rep.*, vol. 4, p. 7183, 2014.

[82]  T. Almqvist, D. O. Winge, E. Dupont, and A. Wacker, "Domain formation and self-sustained oscillations in quantum cascade lasers," *Eur. Phys. J. B*, vol. 92, p. 72, 2019.

[83]  D. O. Winge, E. Dupont, and A. Wacker, "Ignition of quantum cascade lasers in a state of oscillating electric field domains," *Phys. Rev. A*, vol. 98, p. 023834, 2018.

# 6    Coupled Simulation of Quantum Electronic Transport and Thermal Transport in Mid-Infrared Quantum Cascade Lasers

Michelle L. King, Farhad Karimi, Sina Soleimanikahnoj, Suraj Suri, Song Mei, Yanbing Shi, Olafur Jonasson, and Irena Knezevic

University of Wisconsin–Madison

## 6.1    Introduction

The mid-infrared (mid-IR) part of the electromagnetic spectrum, with wavelengths in the 3–20 μm range, is of great industrial, medical, and military importance. The atmospheric low-absorption windows at 3–5 μm and 8–13 μm enable free-space applications, such as remote sensing of chemical and biological species, hard-target imaging, range finding, target illumination, and free-space communication. Many rotational-vibrational molecular transitions occur in the mid-IR range. In particular, the strong vibrational bands associated with the important C-H, N-H, and O-H bonds fall within 3–5 μm or 8–13 μm, so these wavelength ranges are also useful for trace-gas analysis in environmental or medical monitoring. These applications require sources of watt-level coherent optical power that are portable and operate reliably for many hours at room temperature and in continuous-wave mode.

Quantum cascade lasers (QCLs) are the highest-power monolithic coherent light sources in the mid-IR [1–3]. The QCL active core consists of tens of stages, each containing a carefully tailored superlattice of III-V semiconductor layers of different band gaps (Fig. 6.1). The structures are grown via molecular-beam epitaxy (MBE) or metal-organic chemical vapor deposition (MOCVD), which both enable atomic-layer precision in the growth process. The resulting material system has a conduction band with wells and barriers of precisely controlled thickness. Electronic states are well confined in the growth direction and the associated two-dimensional energy subbands have minima at specific, designed energies. An external electric field is needed to achieve the appropriate alignment of subband energy levels. Stimulated light emission (lasing) stems from electron transitions between two specific energy levels, usually referred to as the upper and lower lasing levels, when population inversion between them is achieved. Having emitted a photon, an electron ends up in the lower lasing level, from which it quickly departs by emitting an optical phonon (this step helps maintain population inversion); the electron then gets injected into the next stage, where it can emit a photon again. This cascading scheme, in which an electron can emit multiple photons, is the basis for the high optical power of QCLs. The use of

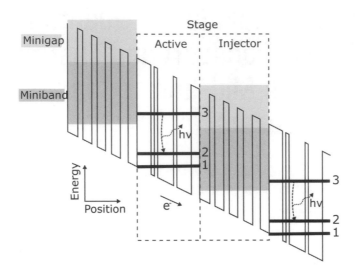

**Fig. 6.1** Band diagram of two QCL stages in an applied electric field.

different III-V materials and their alloys enables great flexibility in QCL design. At present, the InGaAs (wells)/InAlAs (barriers) material system grown on InP is the system of choice in the mid-IR.

Today, room-temperature (RT), continuous-wave (CW) operation has been demonstrated throughout the mid-IR using QCLs based on the InGaAs (wells)/InAlAs (barriers) material system on the InP substrate [1–4]. The unstrained InGaAs/InAlAs system, where both wells and barriers are lattice matched to the InP substrate, has a band offset of only 520 meV between the wells and barriers [4], which is not enough quantum confinement for room-temperature light emission at wavelengths below 5 μm. With the use of strain, band offsets up to 1 eV have been achieved in the InGaAs/InAlAs material system and shorter-wavelength devices have been developed [5–10]. While QCL structures with wavelengths in the 4–11 μm range have shown watt-level CW optical power and high wallplug efficiency WPE (the WPE is the fraction of electrical power converted to optical power), the same is not yet true at shorter wavelengths. Furthermore, long-term-reliable CW operation at power levels above 200 mW remains a critical open problem throughout the first atmospheric window [11], as devices are prone to catastrophic breakdown owing to high thermal stress [12]. Finally, the CW temperature performance needs improvement by implementing designs for carrier-leakage suppression [3, 11, 14] (see Chapter 2), with weaker temperature dependencies of the threshold-current density $J_{th} \sim \exp(T/T_0)$ and of the external differential efficiency $\eta_d \sim \exp(-T/T_1)$, but the underlying microscopic physics is not yet fully understood [11, 13, 14].

Under high-power, CW operation, the electron and phonon systems in QCLs are both very far from equilibrium and strongly coupled to one another, which makes them very challenging to accurately model. The problem of their coupled dynamics is both multiphysics (coupled electronic and thermal) and multiscale (bridging between a single stage and device level; see Fig. 6.2). During typical QCL operation, large

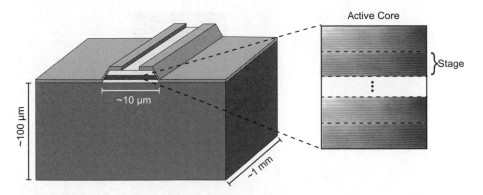

**Fig. 6.2** The multiscale nature of the QCL transport problem. While electron transport and optical-field emission occur in the active core of the device and can be electrically controlled, thermal transport involves the entire large device and is only controlled via thermal boundary conditions that can be far from the active core.

amounts of energy are pumped into the electronic system, of which a small fraction is given back through the desired optical transitions, while the bulk of it is deposited into the optical-phonon system. Longitudinal optical (LO) phonons decay into longitudinal acoustic (LA) phonons; this three-phonon process is often referred to as anharmonic decay. LA phonons have high group velocity and are the dominant carriers of heat. Figure 6.3 depicts a typical energy flow in a QCL.

Anharmonic decay is typically an order of magnitude slower than the rate at which the electron system deposits energy into the LO-phonon system. The fast relaxation of electrons into LO phonons, followed by the LO phonon slower decay into LA phonons, results in excess nonequilibrium LO phonons that can have appreciable feedback on electronic transport, population inversion, and the QCL figures of merit. As a result, different stages in the active core will have temperatures different from one another and drastically differ from the heat sink (see Fig. 6.2). The electron temperatures are higher still, differing among subbands, and affecting leakage paths and thus QCL performance [14–16]. In order to accurately describe QCL performance in the far-from-equilibrium conditions of CW operation, a multiscale electrothermal simulation is needed.

In this chapter, we overview our recent work on developing a simulation framework capable of capturing the highly nonequilibrium physics of the strongly coupled electron and phonon systems in QCLs. In mid-IR devices [17, 18] both electronic and optical phonon systems are largely semiclassical and described by coupled Boltzmann transport equations, which we solve using an efficient stochastic technique known as ensemble Monte Carlo [19]. The optical phonon system is strongly coupled to acoustic phonons, the dominant carriers of heat, whose dynamics and thermal transport throughout the whole device are described via a global heat-diffusion solver. We discuss the roles of nonequilibrium optical phonons in QCLs at the level of a single stage [20], anisotropic thermal transport of acoustic phonons in QCLs [21], outline the algorithm for multiscale electrothermal simulation, and present data for a mid-IR QCL based on this framework [22, 23].

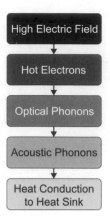

**Fig. 6.3** Flow of energy in a quantum cascade laser.

## 6.2 The Big Picture: Algorithm for Device-Level Electrothermal Simulation

The simulated structure [20, 22, 23] is an older and longer-wavelength QCL: the well-known 9 μm GaAs-based structure by Page et al. [24]. This test device has been well characterized in the literature and our group published several papers [25–27] on its electronic properties without lattice heating; these serve as a benchmark.

Figure 6.4 shows the schematic of a typical device structure (not to scale). Plasmonic waveguides (cladding layers) are typically employed in the mid-IR [28]. As the "depth" of a QCL device (dimension normal to the page) is much greater than its width or height, we can carry out a 2D device-level electrothermal simulation. What we know (i.e., can measure or directly control in experiment) are: bias across the device, electrical current, and the temperature boundary conditions. Typically, the bottom boundary of the device is connected to a heat sink while other boundaries have the convective boundary condition at the environment temperature (single-device case) or the adiabatic (i.e., zero heat flux) boundary condition (array case). Inside the active core, we cannot a priori tell much about the lattice temperature, other than qualitatively expecting an active region hotter than the rest of the device because of all the transfer of energy from the electron to the phonon system. In the same vein, there is no guarantee that the electric field will be uniform across different stages in the active core. In fact, electric-field variation between stages is a staple of superlattices [17, 29, 30]. What we *do* know is that the charge–current continuity equation must hold and that, if we approximate the current flow as 1D through the active core (vertical direction in Figs. 6.4 and 6.5) and assume nearly constant cross-sections (as in Fig. 6.5), then the current density must be constant in the steady state.

This key insight informs **the algorithm** in Fig. 6.6 for device-level electrothermal simulation. (We discuss the limits of these assumptions under 2 below.)

**1. Information needed from single-stage simulation.** A stage inside the active core has an assumed electric field $F$ and lattice temperature $T_L$, the latter coinciding with the acoustic-phonon-ensemble temperature. $T_L$ gives baseline phonon occupations and

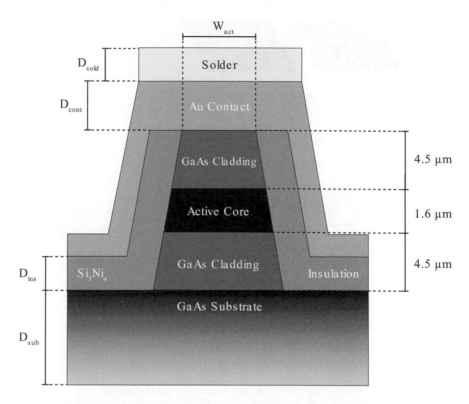

**Fig. 6.4** Schematic of a typical GaAs-based mid-IR QCL structure with a substrate (not to scale).

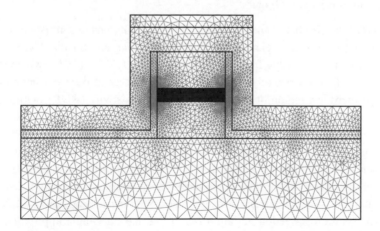

**Fig. 6.5** Finite-element mesh for solving the heat-diffusion equation in a QCL.

electron–phonon scattering rates. For each field $F$ and lattice temperature $T_L$, the output of single-stage simulation consists of electrical-current density $J$ and the heat-generation rate $Q$. $Q$ is proportional to the rate at which acoustic phonons are generated by the decaying optical phonons and is easily recorded in single-stage simulation. By

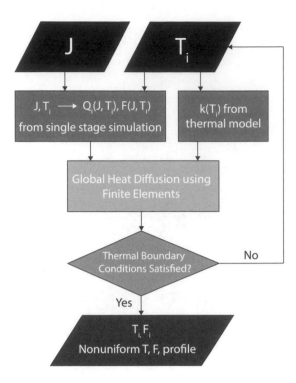

**Fig. 6.6** Flowchart of the device-level electrothermal simulation.

sweeping $F$ and $T_L$, single-stage coupled simulation yields a "table" that links $(F, T_L)$ pairs to appropriate $(J, Q)$ pairs.

**2. Current continuity.** In the steady state, the current density $J$ must be the same in every stage, so it makes sense to use $J$ as an input variable for the device-level simulation. We "flip" the table from $(T_L, F) \rightarrow (J, Q)$ to $(J, T_L) \rightarrow (Q, F)$, which can really be thought of as a series of $F$ vs. $J$ and $Q$ vs. $J$ curves at different lattice temperatures. (If there is a region of negative differential resistivity (NDR), with multiple $F$ corresponding to a given $J$, microscopically this means that $F$ is highly nonuniform in the active core, stemming from electric-field-domain formation [29, 30]. Thus, when "flipping" the table in this case, we assign the midpoint value of $F$ from the NDR to the given $J$.) The flipped table $(J, T_L) \rightarrow (Q, F)$ is key information needed from single-stage simulation for device-level simulation. The assumption of constant $J$ stems from the assumptions of 1D current flow and a nearly constant stage cross-section, as in Fig. 6.5. With cross-section variation (as in Fig. 6.4), a simple approximation is to assume 1D current flow and constant total current, so the current density averaged over the stage area would go as the inverse of the area. The algorithm above would not fundamentally change; we would assume a current $I$, compute a stage-specific $J$, and assign it to a stage as we do for $T_L$.

**3. Heat flow through the whole device.** In essence, the global simulation is the solution to the heat-diffusion equation, with each stage in the active core acting as a current-dependent heat source. A typical nonuniform mesh for the finite-element solution to the global heat-diffusion equation is depicted in Fig. 6.5; it is dense near materials interfaces. The mesh has one point per stage inside the active core, as the temperature is assumed uniform within a stage. We do not know what temperature each stage might have; all we can assume are a current density $J$ and certain thermal boundary conditions (heat-sink temperatures or convection boundary conditions), but we can calculate the temperature-dependent thermal-conductivity tensor everywhere in the device. Inside the substrate, we can assume the material's temperature-dependent bulk thermal conductivity [23]. Inside the active core, the thermal conductivity tensor is far from the bulk value of the constituent materials and is very anisotropic. It can be accurately calculated by solving the phonon Boltzmann transport equation [21]. In strained-layer devices and near facets, an atomistic picture of mechanical properties is needed to capture the damage due to thermal stress. The thermal-conductivity tensor $\kappa(T)$ at every point in the device is required for device-level simulation. This aspect will be discussed in Section 6.5.

**4. Putting it all together** (see flow chart in Fig. 6.6). Based on single-stage simulation, we have created a large table of $(J, T_L) \rightarrow (Q, F)$ maps. Then, for a given current density $J$ and a given set of thermal boundary conditions (heat sink temperatures or convection boundary conditions at exposed facets), we assume a temperature profile throughout the device (e.g., we could assume the whole device to be at the heat sink temperature). In each stage $i$, the $J$ and the guess for the stage temperature $T_i$ yield the appropriate heat generation rate in that stage, $Q_i(J, T_i)$, based on the table. The temperature-profile guess $T_i$, heat-generation rate profile $Q_i$, and the thermal model yielding the thermal conductivity tensor everywhere are used in the heat-diffusion equation, which is solved via the standard finite-elements technique to calculate an updated temperature profile. The process is iterated until the obtained temperature profile agrees well with the imposed thermal-boundary conditions. After we have established the final temperature profile, we read off the corresponding field profile $F_i(J, T_i)$ and calculate the total voltage drop across the whole device, which then gives us an I–V curve directly comparable to experiment.

## 6.3      Stage-Level Coupled Simulation

In QCLs, owing to their multiple-quantum-well structure, electrons are confined in the QCL growth direction (also known as the cross-plane direction). Electronic states are denoted as $|k_\parallel, i\rangle$, with $k_\parallel$ being the in-plane wave vector and $i$ being the subband index. Electron transport in the cross-plane direction is governed by transitions between different localized electronic wavefunctions (associated with discrete energy subbands) while electrons move freely in plane owing to the absence of in-plane confinement or an electric field. We assume here that inelastic scattering between quasibound states dominates over coherent tunneling as a mechanism of electron transport, which is a decent approximation in the mid-IR and enables us to simulate electron transport semiclassically [15, 18, 20]. In reality, tunneling is partially responsible for the filling of the upper laser level from the injector and for the extraction of carriers from lower active-region levels. Accurate theoretical description of these processes requires quantum-transport simulation [15, 18].

Lasing action is based on electrons depopulating the upper lasing level to the lower lasing level via radiative transitions, which requires that the population inversion between the upper and lower lasing levels be maintained. To do so, electrons must quickly depopulate the lower lasing level and relax their excess in-plane kinetic energy by emitting LO phonons (interface roughness scattering [3] is neglected here). This process drives the phonon energy distribution away from equilibrium. In other words, in QCLs, both electrons and phonons are far from equilibrium (often referred to as "hot") and strongly coupled to one another [20].

The coupled Boltzmann transport equations (BTEs) of the electron and LO-phonon subsystems read

$$\frac{\partial f\left(k_{\|}, i\right)}{\partial t} = \frac{\partial f\left(k_{\|}, i\right)}{\partial t}\bigg|_{e-ph}, \quad \frac{\partial N\left(\omega_q\right)}{\partial t} = \frac{\partial N\left(\omega_q\right)}{\partial t}\bigg|_{e-ph} + \frac{\partial N\left(\omega_q\right)}{\partial t}\bigg|_{ph-ph}, \quad (6.1)$$

where $f\left(k_{\|}, i\right)$ denotes the electron distribution function corresponding to $\left|k_{\|}, i\right\rangle$, and $N\left(\omega_q\right)$ is the occupation number of phonons of energy $\omega_q$ and with wave vector $q$. $\frac{\partial f\left(k_{\|}, i\right)}{\partial t}\bigg|_{e-ph}$ and $\frac{\partial N\left(\omega_q\right)}{\partial t}\bigg|_{e-ph}$ are the electron–LO-phonon collision integral for electron and phonon energy distributions, respectively. $\frac{\partial N\left(\omega_q\right)}{\partial t}\bigg|_{ph-ph}$ denotes the phonon–phonon collision integral. The dominant phonon–phonon interaction is the anharmonic decay of LO phonons into LA phonons; the direct electron-to-LA–phonon collision integral is negligible. In QCLs, the electron–electron interaction is relatively weak because the wells and barriers in the active region are either slightly doped or not doped. Here, for educational purposes, we neglect electron–electron interaction in the BTE.

LO phonons can be treated as bulk-like dispersionless phonons with energy $\hbar\omega_{LO}$. The wave vector of LO phonons ($q$) can be decomposed as $q = q_{\|} + q_z$, where $q_{\|}$ and $q_z$ are the in-plane and cross-plane wave vectors, respectively. From Fermi's golden rule, the transition rate between an initial electronic state $\left|k_{\|}, i\right\rangle$ with energy $E$ to a final electronic state $\left|k_{\|}', f\right\rangle$ with energy $E'$ is

$$S^{(\mp)}\left(k_{\|}, i; k_{\|}', f\right) = \frac{\pi e^2 \omega_{LO}}{V}\left(\frac{1}{\epsilon^\infty} - \frac{1}{\epsilon^0}\right)\left(N(\omega_{LO}) + \frac{1}{2} \mp \frac{1}{2}\right)$$

$$\times \frac{\left|\mathcal{J}_{if}(q_z)\right|^2}{q_{\|}^2 + q_z^2}\delta_{k_{\|}', k_{\|} \pm q_{\|}}\delta(E' - E \mp \hbar\omega_{LO}). \quad (6.2)$$

The $-$ and $+$ signs denote the absorption and emission processes, respectively. $\epsilon^\infty$ and $\epsilon^0$ are the high-frequency and low-frequency permittivities, respectively. Note that the Kronecker delta function forces the in-plane momentum conservation by setting $k_{\|}' = k_{\|} \pm q_{\|}$. If we describe the spatial representation of $\left|k_{\|}, i\right\rangle$ as $A^{-1/2}\phi_i(z)\exp(ik_{\|} \cdot r_{\|})$, with $A$ being the normalization area, $\mathcal{J}_{if}(q_z)$ reads

$$\mathcal{J}_{if}(q_z) = \int_{-\infty}^{\infty} dz \phi_f^*(z) \, \phi_i(z) \, e^{-iq_z z}. \tag{6.3}$$

Although the integration limits are infinite, the electronic wavefunctions are quasi-bound in the cross-plane direction, i.e., they are mainly bounded to a single stage. Therefore, for computational purposes, the integration interval can be shrunk significantly. Now, by taking an integral over all final states and a few mathematical steps, the scattering rate corresponding to the electronic state $|k_{\parallel}, i\rangle$ is

$$\Gamma^{(\mp)}(k_{\parallel}, i) = \frac{e^2 \omega_{LO}}{8\pi^2} \left( \frac{1}{\epsilon^{\infty}} - \frac{1}{\epsilon^0} \right) \int_0^{2\pi} d\varphi \int_{-\infty}^{\infty} dq_z \int_0^{\infty} dk_{\parallel}' k_{\parallel}'$$

$$\times \left( N(\omega_{LO}) + \frac{1}{2} \mp \frac{1}{2} \right) \frac{\left| \mathcal{J}_{if}(q_z) \right|^2}{\left| k_{\parallel}' - k_{\parallel} \right|^2 + q_z^2}$$

$$\times \delta \left( E' - E \mp \hbar \omega_{LO} \right), \tag{6.4}$$

where $\varphi$ denotes the angle between $k_{\parallel}'$ and $k_{\parallel}$. From $k_{\parallel}' dk_{\parallel}' = \left( m_f^*/\hbar^2 \right) dE_{k_{\parallel}'}$, one can change variables and rewrite the above equation as an integration over $\varphi$, $q_z$, and the kinetic energy of the final state $(E_{k_{\parallel}'})$.

Simulating the coupled electron-phonon dynamics in QCLs requires a model for the LO-phonon anharmonic decay. An LO phonon can anharmonically decay into multiple phonons in longitudinal-optical, transverse-optical, longitudinal-acoustic, and transverse-acoustic branches. The energies and momenta of the output phonons of the decay process are constrained by the corresponding conservation laws. Among all possible relaxation processes, the relaxation of zone-center LO phonons is generally dominated by a three-phonon process known as the Klemens channel. In this process, an LO phonon splits into two LA phonons with equal energy ($\frac{\omega_{LO}}{2}$) and equal and opposite momenta.

Using a cubic anharmonic model for the crystal potential and Fermi's golden rule, Usher and Srivastava [31] derived an analytical expression for the relaxation rate of the three-phonon anharmonic decay process

$$\Gamma = \frac{\hbar \gamma^2 \tilde{q} \omega_{LO}^3}{8\pi^2 \rho c^2 \bar{\omega}} \times \frac{q_{LA}^2}{\cos(\pi q_{LA}/2\tilde{q})} \times \frac{N_0(\omega_{LA})^2}{N_0(\omega_{LO})}, \tag{6.5}$$

where $\gamma$ is the average, mode-independent Gruneisen constant describing the effect of volume change of a crystal lattice on the vibration frequency; and $\bar{c}$, $\rho$, and $\tilde{q}$ are the average sound velocity, the mass density, and the effective radius of the Brillouin zone, respectively. Also, $\omega_{LA}$ denotes the LA-phonon energy and equals $\omega_{LO}/2$, and $N_0(\omega)$ is the equilibrium occupation number of phonons with frequency $\omega$. In the derivation of Eq. (6.5), the energy dispersion of LA phonons is assumed as $\omega_{LA} = \bar{\omega} \sin(\pi q_{LA}/2\tilde{q})$, with $\bar{\omega}$ denoting the average zone-boundary frequency of LA phonons. It is very important to not treat the LA phonons in the isotropic-continuum approximation because LA phonons decayed from an LO phonon have wave vectors on the order of half the Brillouin-zone edge.

## 6.3.1      Ensemble Monte Carlo

Ensemble Monte Carlo (EMC) is a stochastic approach to solve the BTE. The BTE treats the electrons and phonons as semiclassical particles that fly freely and scatter intermittently in real space, and these mechanics are similarly employed in an EMC simulation. The EMC simulation tracks the time evolution of a large ensemble of carriers (typically $\sim 10^5$) while the free-flight times, scattering mechanisms, and corresponding after-scattering final states of all particles are obtained via statistically appropriate stochastic methods. Throughout the simulation, the particles' states are sampled at consecutive time steps. By knowing the dynamical evolution of electrons and phonons, the dynamical evolution of the macroscopic physical properties of interest, such as the carrier average drift velocity or kinetic energy, can be readily calculated by averaging over the ensemble.

In the EMC simulation of QCLs, we keep track of the in-plane momentum ($k_\parallel$) and subband index ($i$) of electrons, and in-plane momentum ($q_\parallel$) and cross-plane momentum ($q_z$) of phonons. Electrons go through successive free flights and instantaneous random scattering events. Based on the electron-scattering rates, the free-flight time of electrons can be stochastically obtained. Then, the following electron-scattering event and the corresponding final state are determined by a set of statistically appropriate random numbers. Following this pattern, one calculates the time evolution of electrons; however, the dynamical evolution of phonons in QCLs needs to be handled rigorously. To track and sample the nonequilibrium LO-phonon distribution ($N_q$), we set up a phonon histogram defined over the discretized phonon wave vector. Subsequent to the absorption or emission of an LO phonon, the histogram should be updated accordingly, for which we need to know the momentum of that phonon. After an electron–LO-phonon scattering event, when the final state has been determined, one can only obtain the in-plane momentum of the LO phonon involved in the scattering by using the momentum-conservation law because only the in-plane momentum of electrons is well-defined. The heterostructure of QCLs confines electrons in the cross-plane direction, which, according to the uncertainty principle, results in an indefinite electron cross-plane momentum. Previously, the cross-plane momentum of phonons ($q_z$) has been determined via the momentum-conservation approximation (MCA). In MCA, the equivalent cross-plane momentum of an electron with state $|k_\parallel, i\rangle$ is defined as $k_z \equiv \sqrt{2m_i^* E_i / \hbar^2}$, with $m_i^*$ and $E_i$ being the effective mass and energy of the $i$th subband, respectively. Then, as the name indicates, the MCA assumes $q_z = k_z' - k_z$, where $k_z$ and $k_z'$ are respectively the equivalent cross-plane momenta of the initial- and final-state electrons. Consequently, for intrasubband transitions, $q_z = 0$ and for intersubband transitions, $q_z \neq 0$. In other words, the MCA forbids LO phonons emitted via an intrasubband transition to participate in any intersubband transition. Also, a phonon emitted between subbands $i$ and $f$ cannot be re-absorbed by any other transition between $i'$ and $f'$ unless $i' = i$ or $f' = f$. By posing very strict cross-plane momentum conservation, the MCA might underestimate the electron-LO interaction strength. Shi et al. [20] proposed a more accurate method for obtaining $q_z$. The $i{\rightarrow}f$ transition rate corresponding to an electron–LO-phonon scattering event is proportional to

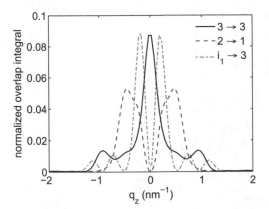

**Fig. 6.7** Normalized overlap integral from Eq. (6.3) versus cross-plane phonon wave vector ($q_z$) for intersubband ($1 \to 3$ and $2 \to 1$) and intrasubband ($3 \to 3$) transitions. Reproduced from [20], Y. B. Shi and I. Knezevic, *J. Appl. Phys.* 116, 123105 (2014), with the permission of AIP Publishing.

$\mathcal{J}_{if}(q_z)$, the overlap integral between the initial ($i$) and final ($f$) states (Fig. 6.7). In other words, the probability distribution of the cross-plane momentum of phonons absorbed/emitted in an $i \to f$ transition is proportional to $\mathcal{J}_{if}(q_z)$. Given this, in each electron–LO-phonon scattering event, we randomly select a $q_z$ value following the distribution from the overlap integral. Figure 6.7 shows the typical overlap integrals for both intersubband ($1 \to 3$ and $2 \to 1$) and intrasubband ($3 \to 3$) transitions. By knowing the momentum of each phonon ($q_\parallel, q_z$) mediating an electronic transition, we can add (remove) a phonon to the histogram if the LO phonon is emitted (absorbed). To avoid negative values in the histogram, once the phonons with a certain momentum are depleted, such phonons are not allowed to be involved in any transition.

In addition to electron–LO-phonon scattering, the three-phonon anharmonic decay of LO phonons affects $N_q$. In the EMC simulation, we model the decay process of LO phonons similar to the free-flight process of electrons. Based on the relaxation of the anharmonic-decay process ($\tau_d$), we stochastically pick a decay time, the duration an LO phonon survives after generation and before decay to LA phonons. We store the decay time of each LO phonon, as well. At the end of each time step, one needs to traverse the LO-phonon histogram and remove the phonons decayed within the time step.

In order to couple the electron- and phonon-dynamic evolution self-consistently, we recalculate the electron–LO-phonon scattering rate based on the updated phonon distribution ($N_q$). However, according to Eq. (6.4), calculating the scattering rate involves computationally intensive numerical integrals. Therefore, we do not update the scattering table in every time step.

## 6.4    Stage-Level Simulation Results

The coupled EMC simulation relies on accurately calculated quasi-bound electronic states and their associated energies in the cross-plane direction. Due to a moderate

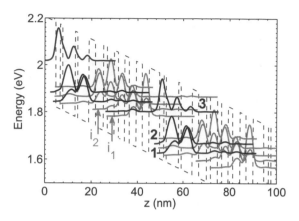

**Fig. 6.8** Wavefunction moduli squared for the key states in the QCL (3: upper laser level; 2: lower laser level; 1: ground state; $i_1$ and $i_2$: injector states). Reproduced from [20], Y. B. Shi and I. Knezevic, *J. Appl. Phys.* 116, 123105 (2014), with the permission of AIP Publishing.

doping, we use a $k \cdot p$ Schrödinger solver coupled with a Poisson solver to obtain the electronic wavefunctions and energies (Fig. 6.8). The far-from-equilibrium operation of a QCL changes the electron distribution profile in the electronic states over time. To self-consistently reflect this change in the $k \cdot p$ Poisson solver, the solver should be coupled with the EMC kernel.

To achieve lasing in QCLs, it is important to maintain population inversion ($\Delta n$); i.e., the electron population in upper lasing level ($n_3$) is greater than in the lower lasing level ($n_2$). Figure 6.9 shows the percentage of electrons at various levels as a function of electric field $F$ at 77 K with and without nonequilibrium phonons. While $n_3$ is significantly enhanced, $n_2$ is only slightly enhanced by nonequilibrium phonons. As a result, the population inversion is considerably higher than in thermal equilibrium. Also, the electron population in $i_1$ is very high with nonequilibrium phonons as compared with $i_2$ and 3 (the population of $i_1$ is three times that of 3).

So how do nonequilibrium phonons help in achieving this increase in $\Delta n$? At low temperatures, the occupation number of nonequilibrium phonons is one to two orders of magnitude higher than thermal phonons, depending on the applied field. The consequence of this is a dramatic increase in the rate of absorption of phonons by electrons [20]. This increase in absorption rate increases the intersubband transition rate. The intersubband transition rate, $i_1 \rightleftarrows 3$ is enhanced 40 times when compared to the transition rate by thermal phonons. It means the average relaxation time is very low for this transition; 8.6 ps for $i_1 \rightleftarrows 3$ with nonequilibrium phonons vs. 349 ps with thermal phonons [20]. As discussed earlier, the electron population in $i_1$ is very high with nonequilibrium phonons as compared with $i_2$ and 3. So, along with the fact that $i_1 \rightleftarrows 3$ has very small relaxation time, we can say that the current component due to the $i_1 \rightarrow 3$ transition is significantly enhanced by nonequilibrium phonons with phonon absorption. So, the nonequilibrium phonons selectively enhance the rate of interstage

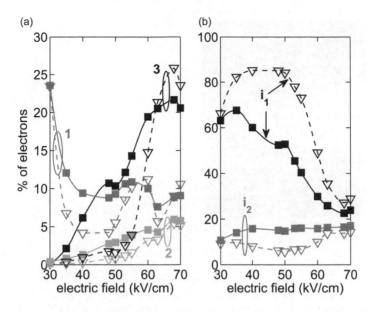

**Fig. 6.9** Population of the active region levels 3, and 2, and 1 (a) and the bottom two injector levels $i_2$ and $i_1$ (b) versus applied electric field obtained with nonequilibrium (solid curves) and thermal (dashed curves) phonons at 77 K. Reproduced from [20], Y. B. Shi and I. Knezevic, *J. Appl. Phys.* 116, 123105 (2014), with the permission of AIP Publishing.

injector-upper lasing level ($i_1 \to 3$) electron scattering by phonon absorption. This enhancement effects key QCL parameters such as $J, Q$, and $G_m$, which are discussed further ahead in the section.

Figure 6.10 shows the nonequilibrium phonon-occupation number at 77 K and at a high applied field. This distribution is based on the overlap integral for interband and intraband transitions. Considering that $i_1$ is lower than 3, $i_1 \gtrless 3$ is possible by absorption of phonons by electrons. The regions formed on either side of $q_z = 0$ (near $q_{||} = 0$) show this intersubband transitions. A third region formed at $q_z = 0$ (further away from $q_{||} = 0$) is due to the $3 \leftrightarrows 3$ intrasubband transitions.

## 6.4.1    *J–F* Curve

One key QCL characteristic is the relationship between current density $J$ and electric field $F$, shown by the *J–F* curve. Earlier, we explained how to use the EMC simulation to calculate $J$ as a function of $F$ at a given temperature $T$. For a given $F$, we calculate the electron wavefunction, energies, and corresponding effective masses and feed them into the EMC kernel. As pointed out above, in our EMC simulation, electron–LO-phonon scattering is considered as the only scattering mechanism to explore its effects within the active core. The electron-electron scattering is turned off considering the moderate doping used in n-doped regions. By applying the periodic boundary conditions at each stage boundary, we could calculate $J$ by counting

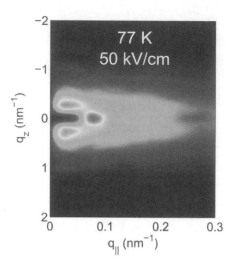

**Fig. 6.10** Occupation number of excess nonequilibrium phonons, $N_q - N_0$, versus the cross plane wave number $q_z$ and in-plane wave number $q_{||}$ at 77 K and 70 kV/cm in the 9 μm QCL. Reproduced from [20], Y. B. Shi and I. Knezevic, *J. Appl. Phys.* 116, 123105 (2014), with the permission of AIP Publishing.

the number of electrons crossing the stage boundaries in a certain amount of time. The current density is calculated as: $J = \frac{en_{net}}{A_{eff}\Delta t}$, where $n_{net} = n_{forward} - n_{backward}$ is the net electron flow, with $n_{backward}$ being the flow from the center stage to the previous stage and $n_{forward}$ being flow from the center stage to the next stage. $\Delta t$ is the simulation time step. $A_{eff}$ is the effective in-plane area of the simulated device. We could calculate $A_{eff}$ from $n_s = \frac{n_{sim}}{A_{eff}}$, where, $n_{sim}$ is the number of simulated electrons and $n_S$ is the sheet-carrier density; $n_{sim} = 50,000$ and $n_S = 3.8 \times 10^{11}$ cm$^{-2}$. It should be noted that, because the EMC simulation is a stochastic process, we could not rely on the calculation of $J$ within one time step. In fact, we need to calculate the current density over many time steps and take an average to minimize numerical noise.

We carry out the EMC simulation for different values of the electric field to obtain the $J$–$F$ curve. To understand the effects of non-equilibrium phonons, we performed the simulation for two different cases: (a) the coupled EMC simulation with nonequilibrium phonons, and (b) the uncoupled EMC simulation with thermal phonons. As seen in Fig. 6.11, the current density with nonequilibrium phonons is significantly higher than in the case with thermal phonons at 77 K. This difference vanishes as the temperature rises to 300 K. As discussed earlier, at low temperatures, the increase in current density is attributed to nonequilibrium phonons for enhanced-injection selectivity ($i_1 \rightarrow 3$). We observe that this increase holds up till 60 kV/cm for this QCL design. With increasing field, $i_1$ moves upwards with respect to 3, crossing it at 60 kV/cm. So, the $3 \rightarrow i_1$ transition is amplified by nonequilibrium phonons, and since it is negative, the overall change is reduced.

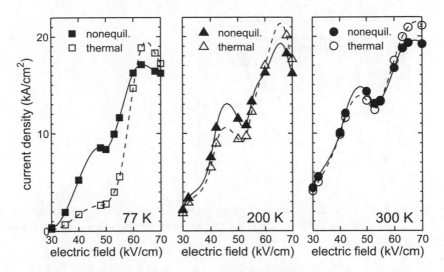

**Fig. 6.11** The current density $J$ vs. electric field $F$ curve at 77 K, 200 K, and 300 K for the simulation with thermal phonons (empty squares) and with non-equilibrium phonons (filled squares). Reproduced from [20], Y. B. Shi and I. Knezevic, *J. Appl. Phys.* 116, 123105 (2014), with the permission of AIP Publishing.

## 6.4.2    The Heat-Generation Rate

The heat-generation rate $Q$ can be calculated in a similar fashion to $J$ via the EMC simulation. In place of counting electrons, as we did for calculating $J$, we count the number of phonons emitted and absorbed to calculate $Q$. Because EMC is stochastic in nature, we calculate $Q$ over multiple time steps, and average it to obtain an accurate result. The coupled EMC simulation with nonequilibrium phonons results in lower heat generation than that with thermal phonons (Fig. 6.12). The nonequilibrium phonons enhance the rate of absorption of phonons by electrons; this, in turn, reduces the calculated $Q$.

## 6.4.3    Modal Gain

The modal gain is another important parameter for QCL analysis. It encapsulates the confinement of photons generated in the cavity. When the confinement is 100%, we get all the photons generated out of the device, which is an ideal case. But in real world conditions, we encounter losses, in this case, losses due to the cavity. There is a high chance of a generated photon to interact with the cavity and count it as a cavity loss. But we can still achieve a higher modal gain by maintaining a high population inversion, which enhances the photon-generation rate. So, the modal gain is also directly proportional to $\Delta n$.

Modal gain is calculated as follows:

$$G_m = \frac{4\pi e^2}{\varepsilon_0 \underline{n}} \frac{\langle z_{32} \rangle^2}{2\gamma_{32} L_p \lambda} \Gamma_w \Delta n, \tag{6.6}$$

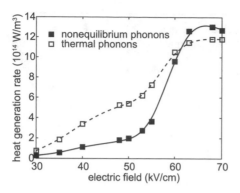

**Fig. 6.12** The heat generation rate ($Q$) vs. electric field ($F$) curve at 77 K for the simulation with thermal phonons (empty squares) and with non-equilibrium phonons (filled squares).

where $\varepsilon_0$ is the permittivity of free space; and $\Gamma_w$, $L_p$, $\underline{n}$, and $2\gamma_{32}$ are respectively the waveguide confinement factor, stage length, optical-mode refractive index, and the full width at half maximum broadening of the transition. The values of these parameters can be extracted from the experiment: $\Gamma_w = 0.31$, $L_p = 45$, $\underline{n} = 3.21$, $\gamma_{32}\,(T_L) \approx 8.68\,\text{meV} + 0.045\,\frac{\text{meV}}{\text{K}} \times T_L$. The emission wavelength $\lambda$ can be calculated from the energy difference between the upper and lower lasing levels; $\langle z_{32} \rangle$ is the dipole matrix element of the lasing transition (i.e., the transition between the upper and lower lasing levels) and can be calculated via $\langle z_{32} \rangle = \int_0^d z\phi_3^*\,(z)\,\phi_2\,(z)\,dz$, where $d$ is the length of the stage, and $\phi_2$ and $\phi_3$ are the lower- and upper-lasing states wavefunctions, respectively. $\langle z_{32} \rangle$ and $\lambda$ are both dependent on the electronic wavefunctions and energies. Therefore, they need to be calculated for each given field; e.g., for 48 kV/cm, $\langle z_{32} \rangle = 1.997\,\text{nm}$ and $\lambda = 8.964\,\mu\text{m}$. It is worth comparing our calculated values with the estimation values used in experiment: $\lambda = 9\,\mu\text{m}$ and $\langle z_{32} \rangle = 1.7\,\text{nm}$. In a similar way to the current-density calculation, the population inversion can be obtained via the EMC simulation by counting the number of electrons in the upper and lower lasing levels and averaging over time.

Figure 6.13 shows $G_m$ as a function of $F$ for the simulation with thermal phonons (filled) and with non-equilibrium phonons (open) at different temperatures. From Eq. (4.6), the population inversion ($\Delta n$) is a key quantity to calculate the modal gain. As discussed earlier, nonequilibrium phonons help in maintaining a high $\Delta n$, hence, this increases the modal gain. Like the $J$–$F$ curve, the effect of nonequilibrium phonons is particularly strong at lower temperatures, particularly below 200 K.

Now, from the modal gain, we could calculate the lasing threshold. After a photon is generated through an intersubband radiative transition, the photon can undergo two processes, either interact with the cavity and account for a loss or move out of the device, in the form of light. We define the lasing threshold as the electric field ($F_{\text{th}}$) at which the modal gain ($G_m$) equals total loss ($\alpha_{\text{tot}}$) of photons in the cavity. There are two sources of loss in the cavity, mirror ($\alpha_m = 5\,\text{cm}^{-1}$) and waveguide ($\alpha_w = 20\,\text{cm}^{-1}$), so the total loss $\alpha_{\text{tot}} = 25\,\text{cm}^{-1}$. The horizontal dotted line in Fig. 6.13 represents the total loss ($\alpha_{\text{tot}}$) in the device. The total loss line remains

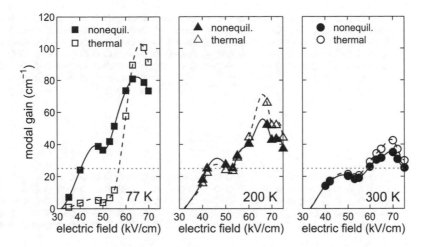

**Fig. 6.13** The modal gain ($G_m$) vs electric field ($F$) curve at 77 K for the simulation with thermal phonons (empty squares) and with non-equilibrium phonons (filled squares). Reproduced from [20], Y. B. Shi and I. Knezevic, *J. Appl. Phys.* 116, 123105 (2014), with the permission of AIP Publishing.

constant because there are no changes affecting the waveguide itself. The intersection between the total-loss line and the $G_m$ vs $F$ gives the threshold electric field $F_{th}$.

## 6.5  Thermal Conductivity in QCLs

Calculating the thermal map of the QCL requires thorough knowledge of the temperature-dependent thermal conductivity of different components of the device. The thermal conductivity stems from phonon scattering, and three-phonon scattering (dominant in the absence of boundaries, interfaces, or alloying) is critically dependent on the temperature. The temperature varies significantly across the device, typically being highest in the active-core region and lowest near the heat sink.

For the cladding layer, substrate, insulation, contact, and solder, it is appropriate to use the bulk temperature-dependent thermal conductivity of constituent materials [22]. However, the thermal conductivity of the active-core region needs to be calculated more carefully. The superlattice structure of the active-core region (Fig. 6.14) raises anisotropy in its thermal-conductivity tensor, which is verified by experimental results. However, due to the in-plane rotational symmetry, we can reduce the thermal conductivity tensor of the active region to the in-plane thermal conductivity ($\kappa_{\parallel}$) and the cross-plane thermal conductivity ($\kappa_{\perp}$), where $\kappa_{\parallel} \gg \kappa_{\perp}$ and both are much lower than the weighted average of the constituent compounds of the superlattice [32–34]. It has been theoretically shown that both effects (i.e., the overall reduction and the anisotropy of the thermal conductivity) are due to the interfaces between adjacent layers, specifically the twofold influence of interface-roughness thermal conductivity [22].

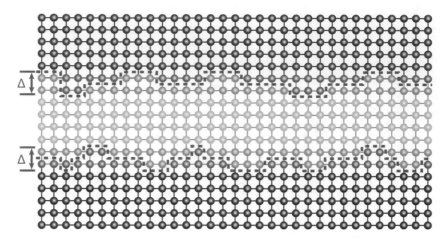

**Fig. 6.14** Even between lattice-matched crystalline materials, there exist nonuniform transition layers that behave as an effective atomic-scale interface roughness. This effective interface roughness leads to phonon momentum randomization and to interface resistance in cross-plane transport. Reproduced from [21], S. Mei and I. Knezevic, *J. Appl. Phys.* 118, 175101 (2015), with the permission of AIP Publishing.

To calculate the thermal conductivity of a superlattice (Fig. 6.14), we first solve the phonon Boltzmann-transport equation (PBTE) via the relaxation-time approximation (RTA) inside each layer. The thermal-conductivity tensor of a crystalline semiconductor at temperature T can be calculated according the following expression, and involves contributions from all phonon wavevectors and branches,

$$\kappa^{\alpha\beta}(T) = \sum_{b,q} C_b(q, T)\, \tau_b(q, T)\, v_b^{\alpha}(q)\, v_b^{\beta}(q), \tag{6.7}$$

where the summation is carried out over all phonon wavevectors ($q$) and branches ($b$), and $C_b(q, T)$ is the phonon heat capacity for mode $b$, given by

$$C_b(q, T) = \frac{\hbar^2 \omega_b(q)^2}{k_B T^2} e^{\frac{\hbar\omega_b(q)}{k_B T}} \left[ e^{\frac{\hbar\omega_b(q)}{k_B T}} - 1 \right]^{-2}, \tag{6.8}$$

where $\tau_b(q, T)$ denotes the total phonon relaxation time and $v_b^{\alpha}(q)$ is the phonon group velocity along the $\alpha$-direction. The full phonon dispersion is the key quantity to calculate both the heat capacity and the group velocity. The exact phonon-dispersion relationship can be obtained by using the adiabatic bond-charge model (ABCM) in the virtual-crystal approximation (VCA) [34–38]. The isotropic approximation of the phonon dispersion underestimates the thermal conductivity of binary compounds and overestimates the thermal conductivity of ternaries [21].

We calculate the total phonon relaxation time by evaluating the harmonic average of the phonon relaxation times due to all phonon scattering mechanisms, e.g., the three-phonon umklapp [39] and normal [40, 41] scattering processes (often referred to together simply as phonon–phonon scattering), mass-difference scattering (due to naturally occurring isotopes or alloying) [42–45] and scattering with charge carriers and ionized dopants [46, 47]. The three-particle phonon–phonon scattering is the dominant mechanism over a wide temperature range.

To calculate the thermal-conductivity tensor of superlattices, we amend the method explained above by including partially diffuse scattering from the interfaces [21]. The probability of specular reflection of a phonon from a rough interface can be described by a momentum-dependent specularity parameter:

$$p_{\text{spec}}(q) = \exp\left(-4\Delta^2 q^2 \cos(\theta)^2\right), \tag{6.9}$$

where $q = |q|$ and $\theta$ represents the angle between $q$ and the normal direction to the interface. $\Delta$ is the effective interface root mean square (rms) roughness. It is noteworthy that that $p_{\text{spec}}$ has a single fitting parameter: the effective interface rms roughness $\Delta$. The most conventional techniques in III-V QCLs are MBE or MOCVD, which both are well-controlled techniques allowing consistent atomic-level precision; therefore, using one value for $\Delta$ to describe all the interfaces is justified but will vary for different materials and heterostructures (for example, for InAlAs/InGaAs it typically is 0.10–0.12 nm, while for GaAs/AlGaAs it typically is 0.15–0.20 nm). Based on $p_{\text{spec}}$, we could derive an effective interface-scattering rate that captures the interplay between internal mechanisms and interface roughness. For the details, see [21]. We use the additional effective interface scattering rate along with the internal scattering rates to calculate the layer thermal conductivity. The effective interface scattering, by affecting the acoustic-phonon population close to the interfaces, reduces the layer thermal conductivity [48].

Like phonon reflection from interfaces, phonon transmission through interfaces is affected by the interface roughness and the degree of interface grading. Phonons must cross interfaces to transfer heat along the cross-plane direction. Because there are two different materials on the two sides of the interface, the phonons that attempt to pass the interface undergo scattering. The interface scattering results in the partial transmission through the interface and consequently the emergence of thermal boundary resistance (TBR) [32, 48]. There are two common models to calculate the TBR: the acoustic mismatch model (AMM) and the diffuse mismatch model (DMM). The AMM assumes that the interface is perfectly smooth (i.e., $\Delta = 0$) and only considers the acoustic mismatch between the materials on the two sides of the interface. This model yields the lower limit of the TBR. On the other hand, the DMM assumes that, after phonons hit the interface, they lose all memory about their momentum, and their final momentum is completely randomized. The DMM model overestimates the TBR and yields its upper limit. To find an accurate model, Mei et al. used the specularity parameter ($p_{\text{spec}}$) to interpolate between the AMM and DMM transmission coefficients and then, calculated the TBR based on the interpolated value of the transmission coefficient [21].

With properly calculated layer thermal conductivity, we obtain the $\kappa_\parallel$ of a superlattice by taking the weighted average of the thermal conductivity of constituent layers, with the thickness of the layers being the weights. $\kappa_\perp$ of a superlattice is calculated as a modified weighted harmonic average of the thermal conductivity of constituent layers

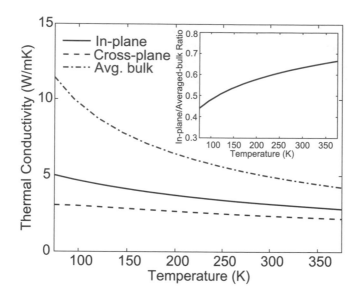

**Fig. 6.15** Thermal conductivity of a typical QCL active region as a function of temperature. A single stage consists of 16 alternating layers of $In_{0.53}Ga_{0.47}As$ and $In_{0.52}Al_{0.48}As$. The solid curve, dashed curve, and dashed-dotted curve show the calculated in-plane, cross-plane, and averaged bulk thermal conductivity, respectively. $\Delta = 1$ Å in the calculations. The inset shows the ratio between the calculated in-plane and the averaged bulk thermal conductivities. Reproduced from [21], S. Mei and I. Knezevic, *J. Appl. Phys.* 118, 175101 (2015), with the permission of AIP Publishing.

which takes TBRs between layers into account [21] (Fig. 6.15). As mentioned earlier, the interface rms roughness ($\Delta$) is the only fitting parameter in this model to calculate the thermal conductivity tensor of superlattices. By adjusting only $\Delta$, typically between 1 and 2 Å, very good agreement with experimental results can be achieved [49–53].

## 6.6    Device-Level Electrothermal Simulation

Typical QCLs operate at high electric fields. High operating electric field pumps considerable energy into the electronic system. Electrons with large kinetic energies relax energy mainly by emitting LO phonons. LO phonons have high energies but, owing to their flat dispersion, have very low group velocities, i.e., LO phonons have a negligible contribution to thermal conductivity. However, via the anharmonic decay process, an LO phonon can decay into two LA phonons. LA phonons have larger group velocity than LO phonons and, consequently, are the main carriers of heat in QCLs. The energy balance equations governing the LA phonons and LO phonons can be written as:

$$\frac{\partial \mathcal{W}_{LA}}{\partial t} = \nabla \cdot (\kappa_A \nabla T) + \mathcal{Q}_{LO \to LA},$$
$$\frac{\partial \mathcal{W}_{LO}}{\partial t} = \mathcal{Q}_{e \to LO} - \mathcal{Q}_{LO \to LA},$$

(6.10)

where $\mathcal{W}_{LA}$ and $\mathcal{W}_{LO}$ are the LA-phonon and LO-phonon energy densities, respectively. $\kappa_A$ and $T$ denote the thermal conductivity and acoustic-phonon (lattice) temperature. The term $\mathcal{Q}_{LO\rightarrow LA}$ describes the heat generation rate due to anharmonic decay of LO phonons to LA phonons, and $\mathcal{Q}_{e\rightarrow LO}$ represents the energy transfer rate from the electron subsystem to the LO-phonon subsystem.

In the steady state, $\mathcal{W}_{LA}$ and $\mathcal{W}_{LO}$ are constant. Consequently, the rate of LO-phonon generation via electron–LO-phonon scattering is equal to the anharmonic decay rate of LO phonons to LA phonons, i.e., $\mathcal{Q}_{e\rightarrow LO} = \mathcal{Q}_{LO\rightarrow LA}$. Incorporating this into the LA phonon's energy-balance equation in steady state, one can obtain the heat-diffusion equation:

$$- \nabla \cdot (\kappa_A \nabla T) = \mathcal{Q}_{e\rightarrow LO} (\equiv Q),\qquad(6.11)$$

where $\mathcal{Q}_{e\rightarrow LO}$ is, the heat-generation rate defined in the single-stage evolution in Sections 6.3 and 6.4. To solve Eq. (6.11), we need to know $\kappa_A (T)$ and $Q$. From the EMC calculation, we can calculate $Q$ by counting the number of absorbed (abs) and emitted (ems) LO phonons in a time step in steady state [5, 6]:

$$Q = \frac{n}{n_{sim}} \frac{1}{\Delta t_{sim}} \left( \sum_{ems} \hbar\omega_{LO} - \sum_{abs} \hbar\omega_{LO} \right),\qquad(6.12)$$

where $n$ is the electron density in a single stage and $n_{sim}$ is the number of simulation electrons. $\Delta t_{sim}$ denotes the simulation time step. In the next subsection, we describe how to calculate $\kappa_A (T)$ in a QCL device.

As discussed earlier, the two key quantities in the device-level simulation are the current density $J$ and the heat-generation rate $Q$. We can calculate both quantities via single-stage EMC simulation. However, bridging the EMC simulation and the device-level simulation is not trivial. On one side, the EMC simulation is carried out in a single stage of the QCL and for a fixed lattice temperature ($T$) and fixed electric field ($F$). On the other side, we know that $T$ and $F$ are not constant across all the stages in the active core and we have no a priori knowledge of how they change from one stage to another. One approach to solve this problem is to implement an isothermal EMC simulation for the entire QCL structure, instead of the single-stage EMC simulation. This approach is computationally extremely challenging and demanding because the QCL structure has many stages (typically 30–70).

Shi et al. [22] implemented the computationally affordable yet accurate solution to bridge the single-stage and device-level simulations, as described in Section 6.2. The single-stage EMC simulation outputs the current density $J (F, T)$ and the heat-generation rate $Q (F, T)$ for given temperature ($T$) and electric field ($F$) values. By carrying out the single-stage EMC for pairs of $(F, T)$ in the range of interest, we obtain a table connecting $(F, T)$ pairs to their corresponding $(J, Q)$ pairs, $[(F, T) \rightarrow (J, Q)]$. However, as discussed earlier, the steady-state current density must be the same in all stages. In fact, $J$ is a key input in the thermal simulation. Therefore, we flip the precalculated $[(F, T) \rightarrow (J, Q)]$ table to obtain the so-called device table, $[(J, T) \rightarrow (F, Q)]$. Using the device table along with the calculated temperature-dependent thermal conductivity ($\kappa_\parallel$ and $\kappa_\perp$) of the active region and the bulk thermal conductivity of other materials in the device (cladding layers, substrate, insulation, contact, and solder), one can solve the heat-diffusion equation, Eq. (6.11). For a given current density and boundary conditions, we begin with an initially assigned temperature profile and use a finite-element method to iteratively solve Eq. (6.11) for the thermal map of the whole device until convergence. Then we use the device table to calculate the electric field profile in the active-core region corresponding to the injected-current density and the calculated temperature profile.

We present the coupled EMC simulation of a 9 μm GaAs/$Al_{0.45}Ga_{0.55}As$ mid-IR QCL, which is a well-known GaAs-based mid-IR design [24]. In our simulation, we use a 1.63-μm-thick active core with 36 individual stages and 16 layers in each stage. The active core is sandwiched between 4.5-μm-thick cladding layers. The waveguide width $W_{act} = 15$ μm, the thickness of insulation layer $D_{ins} = 0.3$ μm, and the thickness of the gold contacts is $D_{cont} = 1.5$ μm. Below the active core, the cladding is in contact with the GaAs substrate which forms the base for the waveguide with a thickness of $D_{sub} = 50$ μm. This entire structure is mounted on a heat sink maintained at 77 K.

In the active core, each stage is a sequence of GaAs wells and $Al_{0.45}Ga_{0.55}As$ barriers with thicknesses of **46**/19/**11**/54/**11**/48/**28**/34/**17**/30/<u>18</u>/**28**/<u>20</u>/**30**/26/**30** in Å. The wells and barriers are represented by normal and bold font, respectively, and the underlined numbers represent moderately n-doped regions with the sheet density of $n_S = 3.8 \times 10^{11}$ cm$^{-2}$. The cladding layers which are part of the waveguide cavity keep photons confined within the core, but do not stop LA phonons from propagating through. This leads to heat dissipation from the active core.

First, the single-stage coupled simulation has to be performed at different temperatures, as in Fig. 6.16(a). We note the calculated $J$–$F$ curves show a negative-differential-conductance region, which is typical for calculations, but generally not observed in experiment. Instead, a flat $J$–$F$ dependence is typically recorded. At every temperature and field, we also record the heat-generation rate, as depicted in Fig. 6.16(b).

Second, the thermal model for the whole structure is developed. Considering that growth techniques improve over time, structures grown around the same time should have similar properties. Since the device studied here was built in 2001 [24], we assume the active core should have similar effective rms roughness Δ to other lattice-matched GaAs/AlAs SLs built around the same time [51, 54]. From our previous simulation work on fitting the SL thermal conductivities [21], we choose an effective rms roughness Δ = 5 Å in this calculation. Figure 6.17 shows the calculated in-plane and cross-plane thermal conductivities, along with the calculated bulk thermal conductivity for the GaAs substrate.

The structure we considered operated in pulsed mode at 77 K. Depending on the duty cycle, the temperature distribution in the device can differ considerably. Figure 6.18 depicts the profile across the active core alone at duty cycles of 100% (essentially continuous wave lasing, if the device achieved it) and 0.01% (as in experiment [24]). Clearly, CW operation would result in dramatic heating of the active

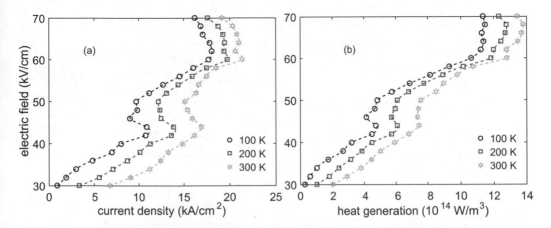

**Fig. 6.16** The field vs. current density (a) and heat generation rate vs. current density (b) characteristics for the simulated device at 100, 200, and 300 K, as obtained from single-stage simulation with nonequilibrium phonons.

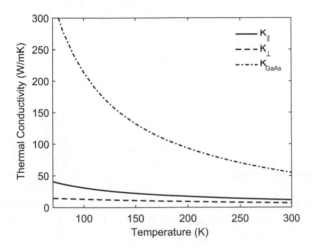

**Fig. 6.17** Calculated in-plane (solid line) and cross-plane (dashed line) thermal conductivities of the active core, along with the bulk thermal conductivity of the GaAs substrate (dash-dotted line). The effective rms roughness $\Delta$ is 5 Å.

region. Finally, Fig. 6.19 shows the $J$–$V$ curve of the entire simulated device at 77 K with a duty cycles of 0.01%, 100%, and as observed in experiment [24].

## 6.7    Conclusion

We overviewed electronic and thermal transport simulation of QCLs, as well as recent efforts in device-level electrothermal modeling of these structures, which is appropriate for transport below threshold, where the effects of the optical field are negligible. We specifically focused on mid-IR QCLs in which electronic transport is largely incoherent and can be captured by the ensemble Monte Carlo technique. The future of QCL

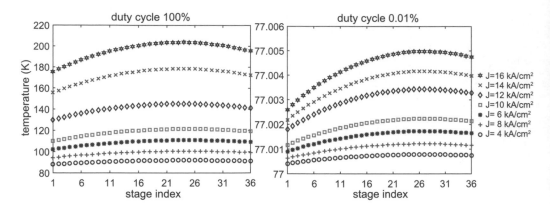

**Fig. 6.18** Temperature profile inside the active region at 100% duty cycle (left) and 0.01% duty cycle (right) for the QCL of [24].

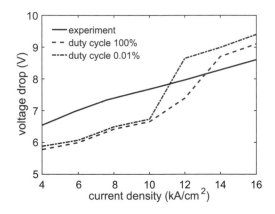

**Fig. 6.19** The current density vs. voltage drop for the simulated device in experiment (solid curve) and as calculated at 100% (dashed curve) and 0.01% (dot-dashed curve) duty cycles. The bottom of the device is placed on a heat sink held at 77 K while adiabatic boundary conditions are assumed on the rest of the boundaries.

modeling, especially for near-RT CW operation, will likely include improvements on several fronts: (1) further development of computationally efficient yet rigorous quantum-transport techniques for electronic transport, to fully account for coherent transport features that are important in short-wavelength mid-IR devices; (2) a better understanding and better numerical models for describing the role of electron–electron interaction, alloy disorder, impurities, and interface roughness on device characteristics; and (3) holistic modeling approaches in which electrons, phonons, and photons are simultaneously and self-consistently captured within a single simulation. The goal of QCL simulation should be nothing less than excellent predictive value of device operation across a range of temperatures and biasing conditions, along with unprecedented insight into the fine details of exciting nonequilibrium physics that underscores the operation of these devices.

## Acknowledgement

The authors gratefully acknowledge support by the DOE (DE-SC0008712), AFOSR (FA9550-18-1-0340), NSF (1702561), and the UW CoE Splinter Professorship. The work was performed using the compute resources of the UW-Madison Center for High Throughput Computing (CHTC).

## References

[1] Y. Yao, A. J. Hoffman, and C. F. Gmachl, "Mid-infrared quantum cascade lasers," *Nature Photonics* 6, 432 (2012).

[2] M. Razeghi, N. Bandyopadhyay, Y. Bai, Q. Lu, and S. Slivken, "Recent advances in mid infrared (3-5 μm) quantum cascade lasers," *Opt. Mater. Express* 3, 1872 (2013).

[3] D. Botez, J. D. Kirch, C. Boyle, K. M. Oresick, C. Sigler, H. Kim, B. B. Knipfer, J. H. Ryu, D. Lindberg, T. Earles, L. J. Mawst, and Y. V. Flores, "High-efficiency, high-power mid-infrared quantum cascade lasers," *Opt. Mater. Express*, 8, 1378 (2018).

[4] F. Xie, C. Caneau, H. P. LeBlanc, N. J. Visovsky, S. C. Chaparala, O. D. Deichmann, L. C. Hughes, C. Zah, D. P. Caffey, and T. Day, "Room temperature CW operation of short wavelength quantum cascade lasers made of strain balanced $Ga_xIn_{1-x}As/Al_yIn_{1-y}As$ material on InP substrates," *IEEE Journal of Selected Topics in Quantum Electronics* 17, 1445–1452 (2011).

[5] Y. Bai, S. Slivken, S. Kuboya, S. R. Darvish, and M. Razeghi, "Quantum cascade lasers that emit more light than heat," *Nat. Photonics* 4(2), 99–102 (2010).

[6] Y. Bai, S. R. Darvish, S. Slivken, W. Zhang, A. Evans, J. Nguyen, and M. Razeghi, "Room temperature continuous wave operation of quantum cascade lasers with watt-level optical power," *Appl. Phys. Lett.* 92, 101105 (2008).

[7] A. Lyakh, R. Maulini, A. Tsekoun, R. Go, S. Von der Porten, C. Pflügl, L. Diehl, F. Capasso, and C. K. N. Patel, "High-performance continuous-wave room temperature 4.0-μm quantum cascade lasers with single-facet optical emission exceeding 2 W," *PNAS* 107, 18799–18802 (2010).

[8] N. Bandyopadhyay, Y. Bai, B. Gokden, A. Myzaferi, S. Tsao, S. Slivken, and M. Razeghi, "Watt level performance of quantum cascade lasers in room temperature continuous wave operation at λ ~ 3.76 μm," *Appl. Phys. Lett.* 97, 131117 (2010).

[9] N. Bandyopadhyay, S. Slivken, Y. Bai, and M. Razeghi, "High power continuous wave, room temperature operation of λ ~ 3.4 μm and λ ~ 3.55 μm InP-based quantum cascade lasers," *Appl. Phys. Lett.* 100(21), 212104 (2012).

[10] N. Bandyopadhyay, Y. Bai, S. Tsao, S. Nida, S. Slivken, and M. Razeghi, "Room temperature continuous wave operation of λ ~ 3-3.2 μm quantum cascade lasers," *Appl. Phys. Lett.* 101, 241110 (2012).

[11] D. Botez, J. C. Shin, J. D. Kirch, C.-C. Chang, L. J. Mawst, and T. Earles, "Multidimensional conduction-band engineering for maximizing the continuous-wave (CW) wallplug efficiencies of mid-infrared quantum cascade lasers," *IEEE J. Sel. Topics Quantum Electron.* 19, 1200312 (2013).

[12] Q. Zhang, F.-Q. Liu, W. Zhang, Q. Lu, L. Wang, L. Li, and Z. Wang, "Thermal induced facet destructive feature of quantum cascade lasers," *Appl. Phys. Lett.* 96, 141117 2010.

[13]  J. D. Kirch, J. C. Shin, C.-C. Chang, L. J. Mawst, D. Botez, and T. Earles, "Tapered active-region quantum cascade lasers ($\lambda = 4.8\,\mu$m) for virtual suppression of carrier-leakage currents," *Electron. Lett.* 48, 234 (2012).

[14]  Dan Botez, Chun-Chieh Chang, and Luke J. Mawst, "Temperature sensitivity of the electro-optical characteristics for mid-infrared ($\lambda = 3$-16 $\mu$m)-emitting quantum cascade lasers," *IOP J. Phys. D: Applied Physics (Topical Review)* 49, 043001 2016.

[15]  O. Jonasson, S. Mei, F. Karimi, J. Kirch, D. Botez, L. Mawst, and I. Knezevic, "Quantum transport simulation of high-power 4.6-$\mu$m quantum cascade lasers," *Photonics* 3, 38 (2016).

[16]  M. Bugajski, P. Gutowski, P. Karbownik, A. Kolek, G. Hałdaś, K. Pierściński, D. Pierścińska, J. Kubacka-Traczyk, I. Sankowska, A. Trajnerowicz, K. Kosiel, A. Szerling, J. Grzonka, K. Kurzydłowski, T. Slight, and W. Meredith, "Mid-IR quantum cascade lasers: device technology and non-equilibrium Green's function modeling of electro-optical characteristics," *Phys. Status Solidi* 251, 1144–1157 (2014).

[17]  A. Wacker, "Semiconductor superlattices: a model system for nonlinear transport," *Physics Reports* 357, 1–111 (2002).

[18]  C. Jirauschek and T. Kubis, "Modeling techniques for quantum cascade lasers," *Appl. Phys. Rev.* 1, 011307 (2014).

[19]  C. Jacoboni and L. Reggiani, "The Monte Carlo method for the solution of charge transport in semiconductors with applications to covalent materials," *Rev. Mod. Phys.* 55, 645 (1983).

[20]  Y. B. Shi and I. Knezevic, "Nonequilibrium phonon effects in midinfrared quantum cascade lasers," *J. Appl. Phys.* 116, 123105 (2014).

[21]  S. Mei and I. Knezevic, "Thermal conductivity of III-V semiconductor superlattices," *J. Appl. Phys.* 118, 175101 (2015).

[22]  Y. B. Shi, S. Mei, O. Jonasson, and I. Knezevic, "Modeling quantum cascade lasers: Coupled electron and phonon transport far from equilibrium and across disparate spatial scales," *Fortschr. Phys.* 65, 1600084 (2017).

[23]  S. Mei, Y. B. Shi, O. Jonasson, and I. Knezevic, "Quantum cascade lasers: electrothermal simulation," in *Handbook of Optoelectronic Device Modeling and Simulation*, Ed. Joachim Piprek, Boca Raton, FL: CRC Press, 2017.

[24]  H. Page, C. Becker, A. Robertson, G. Glastre, V. Ortiz, and C. Sirtori, "300 K operation of a GaAs-based quantum-cascade laser at $\lambda \approx 9\,\mu$m," *Appl. Phys. Lett.* 78, 3529 (2001).

[25]  X. Gao, D. Botez, and I. Knezevic, "X valley leakage in GaAs/AlGaAs quantum cascade lasers," *Appl. Phys. Lett.* 89, 191119 (2006).

[26]  X. Gao, D. Botez, and I. Knezevic, "X-valley leakage in GaAs-based mid-infrared quantum cascade lasers: a Monte Carlo study," *J. Appl. Phys.* 101, 063101 (2007).

[27]  X. Gao, D. Botez, and I. Knezevic, "Phonon confinement and electron transport in GaAs-based quantum cascade structures," *J. Appl. Phys.* 103, 073101 (2008).

[28]  S. L. Chuang, *Physics of Photonic Devices*, 2nd ed., New York: Wiley, 2009.

[29]  R. S. Dhar, S. G. Razavipour, E. Dupont, C. Xu, S. Laframboise, Z. Wasilewski, Q. Hu, and D. Ban, "Direct nanoscale imaging of evolving electric field domains in quantum structures," *Scientific Reports* 4, 7183 (2014).

[30]  M. Wienold, L. Schrottke, M. Giehler, R. Hey, and H. T. Grahn, "Nonlinear transport in quantum-cascade lasers: the role of electric-field domain formation for the laser characteristics," *J. Appl. Phys.* 109, 073112 (2011).

[31] S. Usher and G. P. Srivastava, "Theoretical study of the anharmonic decay of nonequilibrium LO phonons in semiconductor structures," *Physical Review B* 50(19), 14179–14186 (1994).

[32] D. G. Cahill, P. V. Braun, G. Chen, D. R. Clarke, S. Fan, K. E. Goodson, P. Keblinski, W. P. King, G. D. Mahan, A. Majumdar, H. J. Maris, S. R. Phillpot, E. Pop, and L. Shi, "Nanoscale thermal transport. II. 2003–2012," *Appl. Phys. Rev.* 1, 011305 (2014).

[33] G. Chen, "Size and interface effects on thermal conductivity of superlattices and periodic thin-film structures," *J. Heat Transfer* 119, 220 (1997).

[34] G. Chen, "Thermal conductivity and ballistic-phonon transport in the cross-plane direction of superlattices," *Phys. Rev. B* 57, 14958 (1998).

[35] W. Weber, "New bond-charge model for the lattice dynamics of diamond-type semiconductors," *Phys. Rev. Lett.* 33(6), 371–374 (1974).

[36] K. C. Rustagi and W. Weber, "Adiabatic bond charge model for the phonons in {A3B5} semiconductors," *Solid State Commun.* 18, 673–675 (1976).

[37] H. M. Tütüncü and G. P. Srivastava, "Phonons in zinc-blende and wurtzite phases of GaN, AlN, and BN with the adiabatic bond-charge model," *Phys. Rev. B* 62(8), 5028–5035 (2000).

[38] B. Abeles, "Lattice thermal conductivity of disordered semiconductor alloys at high temperatures," *Phys. Rev.*, 131(5), 1906–1911 (1963).

[39] S. Adachi, "GaAs, AlAs, and $Al_xGa_{1-x}As$: material parameters for use in research and device applications," *J. Appl. Phys.* 58, R1–R29 (1985).

[40] G. A. Slack and S. Galginaitis, "Thermal conductivity and phonon scattering by magnetic impurities in CdTe," *Phys. Rev.* 133(1A), A253–A268 (1964).

[41] D. T. Morelli, J. P. Heremans, and G. A. Slack, "Estimation of the isotope effect on the lattice thermal conductivity of group IV and group III-V semiconductors," *Phys. Rev. B* 66(19), 195304 (2002).

[42] M. Asen-Palmer, K. Bartkowski, E. Gmelin, M. Cardona, A. P. Zhernov, A. V. Inyushkin, A. Taldenkov, V. I. Ozhogin, K. M. Itoh, and E. E. Haller, "Thermal conductivity of germanium crystals with different isotopic compositions," *Phys. Rev. B* 56(15), 9431–9447 (1997).

[43] S.-I. Tamura, "Isotope scattering of dispersive phonons in Ge," *Phys. Rev. B* 27(2), 858–866 (1983).

[44] H. J. Maris, "Phonon propagation with isotope scattering and spontaneous anharmonic decay," *Phys. Rev. B*, 41(14), 9736–9743 (1990).

[45] M. G. Holland, "Phonon scattering in semiconductors From thermal conductivity studies," *Phys. Rev.* 134(2A), A471–A480 (1964).

[46] A. Shore, A. Fritsch, M. Heim, A. Schuh, and M. Thoennessen, "Discovery of the arsenic isotopes," *At. Data Nucl. Data Tables* 96, 299–306 (2010).

[47] J. E. Parrott, "Heat conduction mechanisms in semiconducting materials," *Rev. Int. Hautes Temp. Refract.* 16(4), 393–403 (1979).

[48] M. T. Ramsbey, S. Tamura, and J. P. Wolfe, "Mode-selective scattering of phonons in a semi-insulating GaAs crystal: a case study using phonon imaging," *Phys. Rev. B* 46(3), 1358–1364 (1992).

[49] Z. Aksamija and I. Knezevic, "Thermal conductivity of $Si_{1-x}Ge_x/Si_{1-y}Ge_y$ superlattices: competition between interfacial and internal scattering," *Phys. Rev. B* 88(15), 155318 (2013).

[50] T. Yao, "Thermal properties of AlAs/GaAs superlattices," *Appl. Phys. Lett.*, 51, 1798–1800 (1987).

[51] X. Y. Yu, G. Chen, A. Verma and J. S. Smith, "Temperature dependence of thermophysical properties of GaAs/AlAs periodic structure," *Appl. Phys. Lett.* 67, 3554–3556 (1995).

[52] W. S. Capinski and H. J. Maris, "Thermal conductivity of GaAs/AlAs superlattices," *Physica B* 219–220, 699–701 (1996).

[53] A. Sood, J. A. Rowlette, C. G. Caneau, E. Bozorg-Grayeli, M. Asheghi, and K. E. Goodson, "Thermal conduction in lattice–matched superlattices of InGaAs/InAlAs," *Appl. Phys. Lett.* 105, 051909 (2014).

[54] G. R. Jaffe, S. Mei, C. Boyle, J. D. Kirch, D. E. Savage, D. Botez, L. J. Mawst, I. Knezevic, M. G. Lagally, and M. A. Eriksson, "Measurements of the thermal resistivity of InAlAs, InGaAs, and InAlAs/InGaAs superlattices," *ACS Appl. Mater. Interfaces* 11, 11970 (2019).

[55] W. S. Capinski, H. J. Maris, T. Ruf, M. Cardona, K. Ploog, and D. S. Katzer. Thermal conductivity measurements of GaAs/AlAs superlattices using a picosecond optical pump-and-probe technique, *Phys. Rev. B* 59, 8105–8113 (1999).

# Part II

## Active Research Topics

# 7 Quantum Cascade Laser Frequency Combs

Jérôme Faist and Giacomo Scalari

ETH Zurich

## 7.1 Optical Frequency Combs – Key Concepts

An optical frequency comb [1] is an optical source with a spectrum constituted by a set of modes which are perfectly equally spaced and have a well-defined phase relationship between each other. Frequency combs can be obtained by exploiting the natural phase locking mechanism arising when lasers operate in mode-locking regime producing ultrafast pulses. As a result, the ensemble of comb frequency lines at frequencies $f_n$, given by

$$f_n = f_{ceo} + n f_{rep},  \tag{7.1}$$

are spaced by the repetition rate of the laser $f_{rep}$, which can be made extremely stable and locked to an external microwave source. In addition, the ensemble of frequency lines can be shifted by the so-called carrier-envelope offset frequency $f_{ceo}$ [1]. As illustrated in Fig. 7.1, the difference between an array of single frequency lasers and an optical comb is the correlation in the noise of each individual line, immediately apparent when considering noise terms added to $f_{ceo}$ and to $f_{rep}$. Whereas the heterodyne beat between two independent single mode optical sources with similar linewidth $\delta f_n$ will yield a signal with a linewidth $\delta \nu_{RF} = \sqrt{2} \delta f_n$, the same experiment performed on a comb will yield a value that may be much below the one of the individual lines because of the correlation between the noise of the lines. This very peculiar relationship between the modes enables the concept of *self-referencing*. In a mode-locked laser broadened to more than an octave [1], by beating the second harmonic of a line in the red portion with a line in the blue portion of the spectrum, the offset frequency $f_{ceo}$ of the comb can be directly retrieved and stabilized [3]. As a result, the absolute optical frequency of each comb line is rigidly linked to the microwave reference frequency $f_{rep}$, allowing optical frequency combs to act as rulers in the frequency domain. By enabling extremely accurate frequency comparisons using a direct link between the microwave and optical spectral ranges, frequency combs have opened new avenues in a number of fields, including fundamental time metrology [1, 4], spectroscopy, and frequency synthesis. In addition, they also had a tremendous impact on many other fields such as astronomy, molecular sensing, range finding, optical sampling, and low phase noise microwave generation [5]. Their fundamental significance and impact on science was reflected by the attribution of the Nobel Prize in Physics in 2005 to Theodor W. Hänsch and John L. Hall [6].

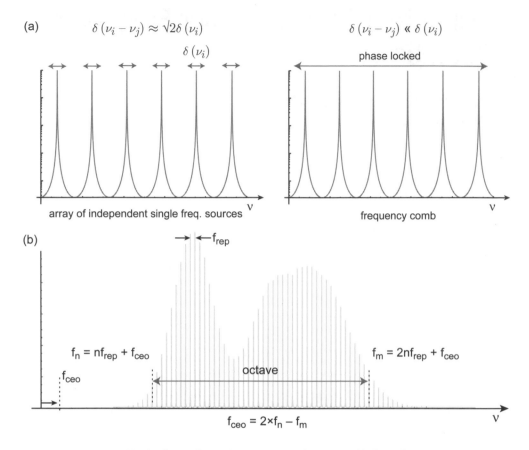

**Fig. 7.1** (a) The line to line noise $\delta(\nu_i - \nu_j)$ of an array of independent sources each one characterized by a noise $\delta(\nu_i)$ is much larger than that of a phase-locked frequency comb. (b) Octave-spanning frequency comb: the carrier envelope offset frequency $f_{ceo}$ can be extracted (and thus controlled) by frequency doubling the red part of the spectrum $f_n$ and extracting the beating with the high-frequency part $f_m$ as $2f_n - f_m = 2(n \times f_{rep} + f_{ceo}) - 2n \times f_{rep} - f_{ceo} = f_{ceo}$, implementing the so-called f-2f self-referencing scheme. Figure reproduced from [2], with the permission of AIP Publishing.

As shown schematically in Fig. 7.2, optical comb generation is relying on the combination of three elements. The first is an optical cavity, either a Fabry–Pérot as displayed or a ring cavity, that is engineered to provide nearly equidistant modes. Those modes are then made to reach threshold through a gain medium, and they are then locked together via the non-linearity provided by an element, usually a saturable absorbers. These three functions are neatly combined in traditional pulsed mode-locked lasers, which were the first approach to generate optical frequency combs. Besides pulsed mode-locked lasers, there is a great interest in miniaturizing the optical frequency comb generation, as shown schematically in Fig. 7.2(b). In a chip-based cavity, optimally the two functions, gain and non-linearity, are combined in the same materials. For instance, optical frequency combs have recently been generated using high-Q microcavity resonators pumped by narrow linewidth continuous-wave (CW)

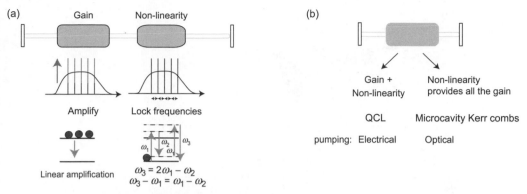

**Fig. 7.2** (a) Schematic functional building blocks of an optical frequency comb generator. The gain section brings the equidistant modes of a cavity above threshold while the fast non-linear section locks the mode separation via four-wave mixing. (b) In an integrated version, the gain and non-linearity can be combined together in a single element. In a QCL, the gain provides the non-linearity while in microresonator combs the non-linear section provides the gain via optical pumping

lasers [7, 8]. In this case the Kerr non-linearity is responsible for establishing the stable phase relationship between the laser modes and the gain for the comb modes. In contrast to mode-locked lasers, Kerr combs can exhibit complex phase relations between modes that do not correspond to the emission of single pulses while remaining highly coherent [9]. Operation of a Kerr comb in a pulsed regime with controlled formation of temporal solitons was recently demonstrated [10] and spurred intensive research efforts towards optimization and control of the soliton pulses [11, 12] and on-chip integration [13].

Among many others [14], an extremely appealing application of optical frequency combs is the so-called dual-comb spectroscopy, where multi-heterodyne detection is performed allowing Fourier transform spectroscopy with potentially high resolution, high sensitivity and no moving parts [15, 16]. As shown schematically in Fig. 7.3, two combs with slightly different repetition rates are used in a local oscillator-source configuration. As shown schematically in Fig. 7.3(b), by suitable choice of the repetition frequency and the offset of each comb, an image of the optical spectrum can be reproduced in the radio-frequency (RF) domain as each pair of lines is creating a separate beat-tone on the detector. For this reason, spectroscopy by means of optical frequency combs surpassing the precision and speed of traditional Fourier spectrometers by several orders of magnitude was recently demonstrated [17–19]. However, in this application the coherence property of the comb is of vital importance as the width of the optical comb lines is directly mapped also onto the RF domain. There, the phase locking of the modes by the optical non-linearity, as described in Fig. 7.1, is critical in maintaining a narrow enough optical linewdith $\delta v$ such that the frequency separation in the RF spectrum is much larger than the individual linewidths $\Delta = f_r - f_r' >> \delta v$. This can be achieved by actively locking each combs in both their repetition rates and

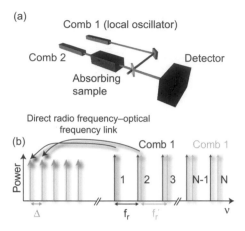

**Fig. 7.3** Principle of dual-comb spectroscopy. (a) Schematic diagram of a dual-comb spectroscopy setup. (b) Schematic diagram in the frequency domain.

offset frequencies. Adaptive techniques where the repetition rates and offset frequencies are either directly measured [20] or retrieved by a numerical techniques have also shown excellent results and are especially well adapted for combs with high repetition rates such as QCLs [21–23].

The use of dual-comb spectroscopy is especially interesting in the mid-infrared portion of the spectrum, the so-called "molecular fingerprint region" where most fundamental roto-vibrational absorption bands of light molecules can be found. Applications of mid-infrared spectroscopy are in the areas of environmental sensing, including isotopologues, medical, pharmaceutical, and toxicological measurements, as well as homeland security applications for molecules that are related to explosives. The THz region is also of high importance for non-invasive imaging, astronomy, security, and medical applications [24, 25]. For these reasons, there is a very strong demand to also create optical frequency combs centered in the mid-infrared and THz regions of the spectrum [26]. A possible approach is to downconvert the near-infrared emission by means of non-linear optics, still maintaining their short-pulse nature [27]. Direct mid-infrared emission, down to 6.2 μm wavelength, has been also achieved using direct pumping of an Optical Parametric Oscillator (OPO) using a Tm-doped fiber laser [28, 29]. Also very interesting are the results achieved using high-Q microresonators based on Si waveguides [30, 31] or microcrystalline resonators [32]. Nevertheless, one common drawback of these optical sources is that they consist of different optical elements that must be assembled together.

This chapter reviews an emerging new optical frequency comb technology based on QCLs. Section 7.2 reviews the basic physics, as well as the main operation characteristics of QCL combs. In Section 7.5, the techniques developed to characterize the comb operation are described. The topic of dispersion compensation is addressed in Section 7.4, together with some recent results on the topic. The application in spectroscopy of dual-combs systems based on QCL combs is discussed in Chapter 14 of this text. Conclusions and outlook are reported in Section 7.6.

## 7.2          Basic Principles of Comb Generation in Fast Gain Media

### 7.2.1          QCL Broadband Technology

QCLs [33] are semiconductor injection lasers emitting throughout the mid-infrared (3–24 µm) and THz (50–250 µm) [34, 35] regions of the electromagnetic spectrum. First demonstrated in 1994 [36] in the mid-infrared, they have undergone a tremendous development. Their capability to operate in a very wide frequency range makes them very convenient devices for optical sensing applications. In particular, single frequency emitter devices have demonstrated both very large continuous optical output power up to 2.4 W [37], high-temperature operation in CW [38] and low electrical dissipation below a watt [39]. The physics of QCLs is, in addition, very beneficial for their operation as broadband gain medium. First, intersubband transitions are transparent on either side of their transition energy. Interband transitions, in contrast, are transparent only on the low-frequency side of their gain spectrum and highly absorbing on the high-frequency side. Second, the cascading principle almost comes naturally because of the unipolar nature of the laser. These two features enable the cascading of dissimilar active region designs emitting at different wavelengths to create a broadband emitter. This concept, first demonstrated at cryogenic temperature [40] was further developed for applications with high performance, inherently broad gain spectra designs where a single upper state exhibits allows transitions to many lower states [41]. This technology was the base to the fabrication of external cavity QCLs with very large tunability [42].

### 7.2.2          Four-Wave Mixing in a Fast Gain Medium

It was suggested very early that broadband QCL could be mode-locked to provide ultrashort mid-infrared pulses [40]. In a mode-locked laser [43], the fixed phase relationship between the optical modes – needed for the formation of an optical comb spectrum as described by Eq. (7.1) – is obtained by having a single pulse propagating in the laser cavity. This mode of operation of the laser is generally obtained by either a subtle compensation between a negative dispersion in the cavity and the Kerr effect, giving rise to the propagation of a temporal soliton, or by a saturable absorber opening a time-window of low loss for a pulsed output. The short pulses that will be produced at the roundtrip frequency of the cavity will have, in well-designed lasers, an average power commensurate to the one of a CW laser. This is possible because the energy of the two-level system, stored in the inverted medium, is abruptly released in the optical pulse [44]. In contrast, QCLs, being based on intersubband transitions in quantum wells, exhibit very short upper state lifetime with, in high-performance devices operating at room temperature, a time in the sub-picosecond range ($\tau_2 \approx 0.6$ ps). This time is much shorter than the typical cavity round trip time $\tau_{rt}$ (64 ps for a 3 mm long device), such that the product $\omega\tau_2 \ll 1$, where $\omega = 2\pi/\tau_{rt}$ is the angular frequency corresponding to the longitudinal mode spacing. As a result, passive mode-locking in the sense described above is not possible.

For QCL devices operating in the THz region of the spectrum, however, lifetimes can be significantly longer, up to 40 ps [45]) such that the product $\omega\tau_2$ can be close to unity.

At the same time, the very short upper state lifetime, that strongly hinders fundamental mode-locking, is responsible for a very broadband four-wave mixing (FWM) process. FWM due to intersubband transition was measured in a multiquantum well system some years ago already [46]. In a QCL active region under operation, measurements of the FWM product was performed by injecting two single frequency sources and analyzing the output spectrum [47]. As shown schematically in Fig. 7.4(a), a single mode DFB QCL and the output of a tunable difference frequency generation source were injected in an anti-reflection coated QCL amplifier, driven in CW at room temperature close to its maximum current. The QCL amplifier consisted of a single stack of strain-compensated InGaAs/AlInAs active region operating at 4.3 μm wavelength [48]. For these operation conditions, the computed upper state lifetime is $T_1 = 0.29$ ps and dephasing time is $T_2 = 0.14$ ps. As in Fig. 7.4(b), the output showed the expected mixing product at frequencies corresponding to $\omega_4 = 2\omega_1 - \omega_2$ and $\omega_3 = 2\omega_2 - \omega_1$. As shown in Fig. 7.4(c), the dependence of FWM signals as a function of detuning was in good agreement with the value of the effective non-linear susceptibility, $\chi^{(3)}$, computed using a two-level density matrix model given by

$$\chi^{(3)}(\Delta, \delta\omega) = \frac{2e^4 \Delta N z_{ij}^4}{3\epsilon_0 \hbar^3} \frac{(\delta\omega - \Delta - i/T_2)}{(\Delta - \delta\omega + i/T_2)}$$
$$\times \frac{(-\delta\omega + 2i/T_2)(\Delta + i/T_2)^{-1}}{(\delta\omega - i/T_1)(\delta\omega - \Delta - iT_2)(\delta\omega + \Delta - i/T_2)}, \qquad (7.2)$$

where $\Delta = \omega_1 - \omega_{12}$ is the detuning between the pump and the intersubband transition, $\delta\omega = \omega_2 - \omega_1$ the separation between the pump and probe, $z_{ij} = 1.6$ nm the dipole matrix element, and $\Delta N$ the population inversion at threshold. For a detuning $\Delta$ of 210 GHz, an effective $\chi^{(3)} = 2.5 \times 10^{-15}$ m$^2$V$^{-2}$ is predicted, close to the measured value of $\chi^{(3)} = (0.9 \pm 0.2) \times 10^{-15}$ m$^2$V$^{-2}$.

The important conclusion of these measurements was that the QCL active region presents a *large, resonant* $\chi^{(3)}$, arising from the *active medium itself*. In addition, the bandwidth of the FWM process is much larger than in the case of semiconductor lasers because of the much shorter upper state lifetime.

## 7.2.3    Maxwell–Bloch Models

As shown schematically in Fig. 7.5, the dynamic behavior of a multimode QCL laser is the result of a complex interplay between the intersubband polarization of the active region and the dynamics of the cavity modes. The key quantity relating the electron motion in the heterostructure with the cavity electromagnetic field is represented by the polarization of the medium, which is usually expressed as function of the off-diagonal elements of the density matrix. The dynamical behavior of a multi-

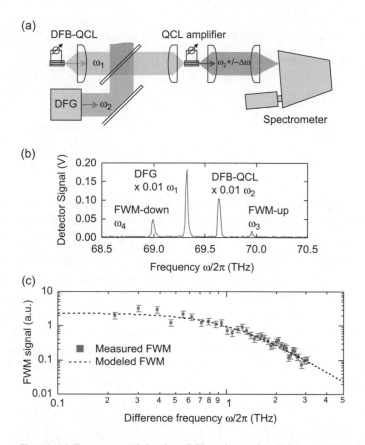

**Fig. 7.4** (a) Four-wave mixing in a QCL active region. Schematic diagram of the experimental setup. (b) Measured spectrum for a source separation $\delta\omega = \omega_2 - \omega_1$ of 314 GHz. (c) Response of the FWM as a function of detuning between the two sources (points), compared with the predictions of a two-level model (dashed lines). Reproduced from [47], with the permission of AIP Publishing.

level electronic system is driven by the optical Bloch equations for the density matrix describing the two-level system, while the Maxwell equations describe then the motion of the electromagnetic wave subjected to the matter polarization and losses in the cavity. The resulting Maxwell–Bloch coupled equations are the basis for the explanation of laser dynamics. Such sets of coupled equations exhibit a very rich set of solutions and are, in general, solvable only using numerical techniques.

While, in general, these equations are governing all laser types and have been studied extensively [49] in the past the case of the QCL, with its combination of fast relaxation time and wide bandwidth was relatively unexplored. In fact, the discussion of the form that these solutions take relies on the hierarchy between the different lifetimes: a system in which the cavity loss rate is smaller than both the transverse *and* longitudinal loss rate of the population is called a "class A" laser. Classical examples of such class A lasers are He-Ne gas lasers. One important feature of these type of lasers is that the electron population is able to "follow" the photon populations because of its very fast scattering time.

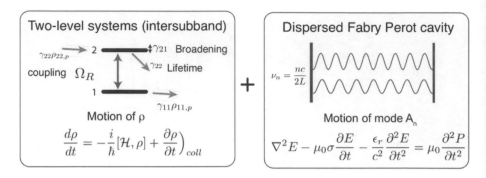

**Fig. 7.5** Schematic drawing showing the coupling between the optical Bloch equations, describing the motion of the density matrix and the Maxwell equation for the optical modes.

Initial work on solving the Maxwell–Bloch system for QCL [50, 51] was hampered by the assumption of a saturable absorber intrinsic to the device. A first important theoretical insight came from the work of Khurgin et al. [52] who solved the system in the frequency domain, using the modes of the laser cavity as a basis. This approach was already used by Lamb in 1964 to study the "three-mode laser," in which a third mode is locked by the beating of two other modes through four-wave mixing [53].

The method followed by Khurgin et al. [52] uses this approach and applies it to an arbitrary set of modes $N$. The final equation for the time evolution of the (slowly varying and normalized) amplitude $A_n$ of mode $n$ is given by [52, 54]

$$\dot{A}_n = \left\{ \underbrace{G_n - 1}_{\text{Net gain}} + i \underbrace{\left( \frac{\omega_n^2 - \omega_{nc}^2}{2\omega_n} \right)}_{\text{Cavity dispersion}} \right\} A_n - G_n \sum_{k,l} \underbrace{C_{kl} B_{kl} A_m A_l A_k^* \kappa_{n,k,l,m}}_{\text{FWM term}} \cdot$$

In the above equation, $G_n$ is the transition net gain, normalized to the cavity losses:

$$G_n = \frac{1}{1 - in\omega/\gamma_{12}}, \tag{7.3}$$

while

$$C_{kl} = \frac{\gamma_{22}}{\gamma_{22} - i(l - k)\omega} \tag{7.4}$$

defines the amplitude of the coherent population oscillations, and

$$B_{kl} = \frac{\gamma_{12}}{2i} \left( \frac{1}{-i\gamma_{12} - l\omega} - \frac{1}{i\gamma_{12} - k\omega} \right) \tag{7.5}$$

is the term driving the width of the FWM gain. In these equations, $\omega_{nc}$ is the angular frequency of the empty cavity mode, $\omega_n = n\omega$ is the angular frequency of one specific mode $n$. The broadening of the transition is $\gamma_{12}$ and the scattering rate of the upper state is $\gamma_{22}$. The result of the simulation, assuming for the QCL an upper state lifetime of 1 ps, a roundtrip of 65 ps, and neglecting at this stage the effect of the dispersion, is reported in Fig. 7.6. The results show that the short upper state lifetime prevents the formation of optical pulses; as a result the instantaneous intensity is almost constant

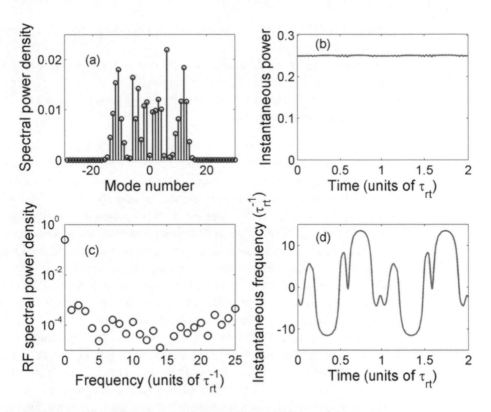

**Fig. 7.6** (a) Maxwell–Bloch model of a mid-infrared QCL comb. Optical spectrum. (b) Instantaneous intensity as a function of time. (c) RF spectrum of the intensity. (d) Instantaneous optical frequency. Figure reproduced from [52], with the permission of AIP Publishing.

(Fig. 7.6(b)) while the instantaneous frequency swings widely over a significant portion of the output spectrum (Fig. 7.6(d)). Note that the optical Bloch equations used in the derivation of Eq. (7.3) are the same as those used for the derivation of $\chi^{(3)}$ in Eq. (7.2), resulting in the expression (in normalized units) $\chi^{(3)} = G_n \sum_{k,l} B_{kl} C_{kl}$. This initial approach clarified a couple of important points:

- It showed that spatial hole burning can be seen as another term in $\chi^{(3)}$ non-linearity in a medium with a fast recovery time promoting multimode operation.
- It also showed that, in this regime, the optical intensity tends to remain constant while the instantaneous frequency exhibits large frequency swings.

At the same time, the approach had a number of limitations:

- The first results did neglect the waveguide dispersion. This was, however, added in a subsequent study using the same approach [54].
- The assumption of perfect mirrors at the end of the cavity is not realistic in semiconductor laser devices, which are often terminated by cleaved mirrors with a

reflectivity of about 30%. This effectively changes the profile of the intensity inside the laser cavity.

- By reducing the electronic states to a single two-level system, the treatment neglected the contribution of other transitions.
- Finally, this set of equations do not lend themselves easily to derive a semi-analytical approximation.

Most critically, however, was the need to predict experimental observations that were not accounted for by the model. One important observation was that, in specific experimental circumstances, the comb state became harmonic, such that the spectrum was consisting of modes separated by many times the round trip frequency. In [55], the study of the simple case of three modes was conducted, showing that the parametric gain was peaking at multiple of the round trip frequency, therefore explaining the appearance of the parametric gain. Another important experimental fact was the observation that QCL combs operate in a state characterized by a linear frequency chirp. Indeed, as shown later in the text, the frequency modulation of the output is not semi-random or sinusoidal but is characterized by a linear increase as a function of time while the intensity remained essentially constant [56].

Recently, a theoretical work solving the Maxwell–Bloch system of equations in time and space has been developed [57]. Compared to previous work, it included the real optical profile in the cavity and the dispersion, as well as the linewidth enhancement factor. The latter can be interpreted as a Kerr effect, since it corresponds to a change of refractive index as a function of the light intensity. Very interestingly, this model predicted the linear chirp and demonstrated the need to balance the LEF and dispersion to obtain stable combs. Such prediction was also obtained separately from similar Maxwell–Bloch computation developed originally for quantum dot devices but with parameters adapted to the QCL situation [58].

Another important result from Opacak et al. [57] was the derivation of the master equation for the comb state. This enables a much faster numerical treatment and also allows a better insight into the physics of the device. In fact, recent work has shown the intimate connection between the comb generation in QCLs with the Ginzburg–Landau equation [59] and with phase solitons [60].

In hindsight, these results are to be expected since QCL are so-called class A lasers, where the medium population inversion and polarization can follow the optical field dynamics. Indeed, FM mode-locking was observed in He-Ne gas lasers, which belong to the same class A [61]. In that experiment, however, the FM mode-locking was driven by an external phase modulator.

## 7.3        Comb Operation of Broadband QCL

Experimentally, it was shown that broadband QCLs can achieve frequency comb operation by using FWM as a phase-locking mechanism [62]. In a simplified view shown schematically in Fig. 7.7(a), degenerate and non-degenerate FWM processes induce

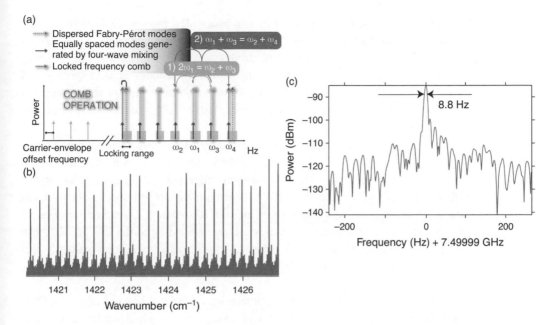

**Fig. 7.7** (a) The principle of operation of a QCL comb: the dispersed Fabry–Pérot resonator modes are injection-locked by the equally spaced modes generated by four-wave mixing processes. (b) Magnified view of the optical spectrum of a QCL comb operating in the mid-infrared, at a center wavenumber of $1430\,\text{cm}^{-1}$ ($7\,\mu\text{m}$) covering $60\,\text{cm}^{-1}$. (c) Radio-frequency spectrum of the intensity near the repetition frequency $f_{\text{rep}}$, showing a beatnote with a full-width at half-maximum of $< 10\,\text{Hz}$. The measurement is taken at the onset of the multimode emission and the resolution bandwidth of the spectrum analyzer is set to $10\,\text{Hz}$. Results are reproduced with permission from [62].

a proliferation of modes over the entire laser spectrum. The dispersed Fabry–Pérot modes – which are present in the case of a free-running multimode laser and are not phase-locked – can be injection-locked by the modes generated by the FWM process, thus creating a frequency comb. This phase-locking mechanism is very similar to the one that occurs in ring microresonator-combs [7, 8]. In addition, due to the short gain recovery of intersubband transitions, the laser output is not pulsed and resembles that of a frequency modulated laser [62].

Figure 7.7(b) shows a magnified view of the optical spectrum (measured with a Fourier transform infrared spectrometer, FTIR) of a QCL comb operating in the mid-infrared, at a center wavenumber of $1430\,\text{cm}^{-1}$ ($7\,\mu\text{m}$) covering $60\,\text{cm}^{-1}$. We observe a single set of longitudinal modes with no measurable dispersion within the resolution of the measurement ($0.0026\,\text{cm}^{-1}$, $78\,\text{MHz}$). This per se is not really a proof of comb operation as this measurement does not give any information about the relative stability of the phases of the modes.

Nevertheless, as a first indication, the radio-frequency spectrum of the laser intensity near the repetition frequency $f_{\text{rep}}$ can be measured. High-bandwidth detectors capable of detecting radio-frequency signals up to tens of GHz are needed for this type of characterization. Such a spectrum would show a beatnote with a large linewidth (5 to 10 MHz) in the case of two uncoupled modes (such as those generated by two independent lasers). In contrast, such a measurement would show a narrow beatnote in the case where the modes are locked in phase. The measurement of this radio-frequency spectrum, also called intermode beatnote spectrum, is shown in Fig. 7.7(c) for a laser biased just above the onset of multimode operation. It shows a beatnote with a full-width at half-maximum of 8.8 Hz, characteristic of comb operation.

Frequency comb operation was not observed over the entire laser operation range in the original work of [62], however, and these observations were also confirmed in QCL combs operating in the THz spectral range [63, 64]. Figure 7.8(a) shows different RF spectra together with their corresponding optical spectra for different values of laser current. In addition to the narrow RF beatnote characteristic of comb operation, broad beatnotes are also observed, corresponding to the loss of coherence of the optical comb. Figure 7.8(b) shows the light intensity–current–voltage characteristics of this same device, showing the range where QCL comb operation is observed.

Figure 7.9 shows an experimental confirmation of the FM-like behavior of a QCL comb. A sheet of polyethylene, which has strong wavelength-dependent absorption (see dashed line in Fig. 7.9(a)), is inserted between the laser output and a high-bandwidth detector. This sheet acts as an optical discriminator that can convert the frequency modulation of the laser output to an amplitude-modulated signal. We observe an amplification (16 dB in this example) of the measured RF beatnote at the round trip frequency (see Fig. 7.9(b)), as expected from a frequency-modulated signal.

## 7.3.1     THz QCL Comb Operation

In addition to mid-infrared combs, QCLs can be used to generate combs at THz frequencies, as the main driving mechanism is still FWM. THz QCLs can be driven in the comb regime by careful dispersion compensation [63], by engineering a broadband QCL with a flat gain curve [64] and by a combination of both elements. It was recently shown that it is possible to achieve octave-spanning lasers at THz frequencies using QCLs [64]. This very wide frequency coverage is possible due to the discussed gain engineering capability typical of intersubband transitions coupled to the extremely broadband nature of the double-metal waveguide. THz QCLs operate in double-metal cavities where the waveguide claddings are constituted by two metallic layers [35, 65], as in a microwave microstrip resonator. In TM polarization this cavity does not present any cutoff, making ultra-broadband operation possible. The heterogeneous cascade laser used for achieving octave-spanning operation is constituted by three different active regions stacked together in the same waveguide. Its emission spans from 1.64 THz to 3.35 THz when operated in CW operation at a temperature of 25 K, as shown in Fig. 7.10(a) [64]. Such a broad gain medium is beneficial for comb operation. As discussed below, the broad gain results in a low and flat group velocity dispersion (GVD). Furthermore, the double metal waveguides introduce only a small

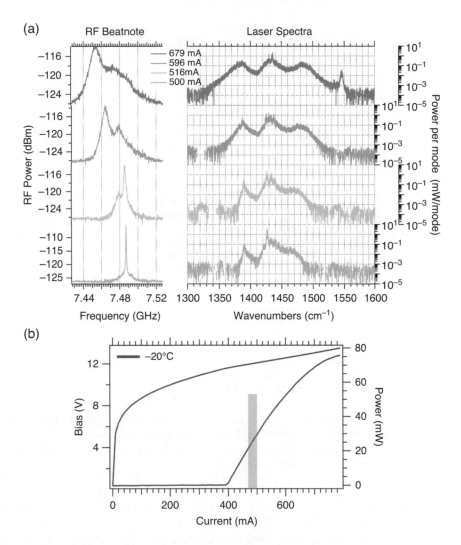

**Fig. 7.8** (a) QCL comb operation regimes. Left side: RF beatnote spectra measured using a fast quantum well photoconductor, together with their respective optical spectra (right graphs) of a QCL designed for comb operation. Above a certain value of current (510 mA in this example), the RF beatnote splits into two distinctive beats. (b) Corresponding light intensity–current–voltage characteristics of the device. The shaded area corresponds to the region where the narrow beatnote is observed. Results are reproduced with permission from [62].

amount of GVD to the laser. Therefore these lasers can act as frequency combs similar to mid-infrared QCL combs. The characteristics of such a THz QCL comb are displayed in Fig. 7.10(b). The shown comb has 1.55 mW of output power while spanning over a spectral bandwidth of 507 GHz. The beatnote has a linewidth of 800 Hz, which is still jitter-limited since the laser is not actively stabilized. The latest developments of wide bandwidth THz QCL combs based on multi-stack active cores show an impressive bandwidth coverage of 1.1 THz at 30 K ($\frac{\Delta f}{f} = 36\%$) [66] (Fig. 7.10(c)

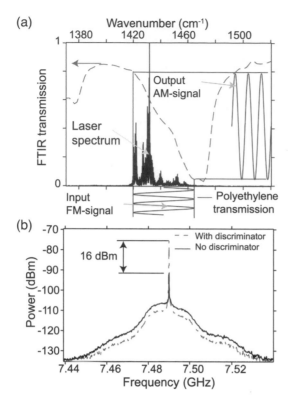

**Fig. 7.9** Frequency-modulated QCL comb. (a) Optical spectrum (black) with the polyethylene transmission spectrum (dashed line) acting as an optical discriminator. (b) RF spectrum showing the beatnote at the round trip frequency. An increase of the RF beatnote is characteristic of an FM to AM conversion.

and (d)) and 700 GHz at 80 K under RF injection from an homogeneous structure [67, 68]. Further bandwidth extension towards octave spanning combs is hindered by the presence of the strong dispersion coming from phonon bands at 8–10 THz.

## 7.4    Cavity Dispersion: Measurement and Compensation

The ability of FWM to generate mode proliferation – and ultimately comb operation – depends critically on the GVD of the cavity. If one considers a single active region with a gain bandwidth of $100 \, \text{cm}^{-1}$, the latter corresponds to about $N_m = 200$ cavity modes for a 3 mm long device. The finesse $\mathcal{F}$ of a typical QCL Fabry–Pérot cavity is very limited, but even assuming a value of $\mathcal{F} = 5$ requires that the fractional change of the group effective index is not larger than $\delta(\omega)/\omega = \delta n_g/n_g = 1/(N_m \mathcal{F}) = 10^{-3}$ over the gain bandwidth. The spectral change of the group refractive index is quantified by GVD, defined as

$$\text{GVD} = \frac{\partial}{\partial \omega} \frac{1}{v_g} = \frac{1}{c} \frac{\partial n_g}{\partial \omega}. \tag{7.6}$$

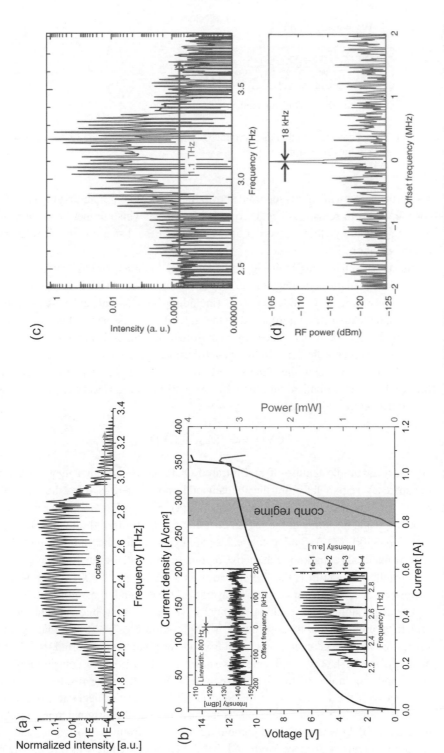

**Fig. 7.10** (a) Octave-spanning spectrum of a 2 mm × 150 μm long laser measured at 25 K in CW operation. Results are reproduced with permission from [64]. (b) LIV curve measured at 25 K in CW operation. The shaded area indicates the current range where the laser is a comb. The upper inset displays the intermode beatnote with a linewidth of 800 Hz at a driving current of 0.9 A. The lower inset shows the corresponding optical spectrum with a bandwidth of 507 GHz. (c) CW emission and (d) corresponding beatnote at T=30 K for a four-stack broadband comb. (Results are reproduced with permission from [66].)

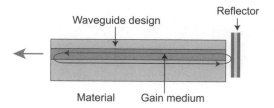

**Fig. 7.11** Schematic drawing of a mid-infrared QCL cavity, displaying the various components that can modify the dispersion.

Using the numerical values reported above, the GVD must remain below 560 fs$^2$/mm for the modes to remain efficiently coupled by FWM. This number is of course only a rough estimate yielding the order of magnitude of the dispersion that is relevant for comb operation.

As shown schematically in Fig. 7.11, the GVD in an active QCL waveguide consists of three main components: the dispersion of the material GVD$_{mat}$, the modal dispersion of the waveguide GVD$_{mod}$ (including both lateral and vertical confinement), and the dispersion introduced by the end reflectors. As shown below, a non-zero value of the GVD can be balanced by a controlled amount of dispersion introduced by specially designed reflectors. To these contributions one might also want to introduce the effect on the GVD of the gain itself. However, it is better taken into account by the Maxwell–Bloch formalism and therefore we will keep the GVD here as expressing one of the *empty* waveguides:

$$GVD = GVD_{mat} + GVD_{mod}. \tag{7.7}$$

One very favorable feature of mid-infrared QCLs is the fact that they operate in a transparency region far from the fundamental gap as well as from the reststrahlen band. For this reason, as shown in Fig. 7.12(a), the GVD of InP has a very low value ($< 1000$ fs$^2$/mm) in the region covering the mid-infrared range of 3.3–10 μm ($\approx 1000$–3000 cm$^{-1}$). Similar low values are expected for the undoped InGaAs and AlInAs materials constituting the active region. In contrast, the situation at the telecom wavelength of 1.55 μm is that InP already has a GVD$_{mat} \approx 2000$ fs$^2$/mm while the narrower gap confinement waveguide layers will have even larger values of GVD$_{mat}$.

The dispersion of the resulting "empty" waveguide, taking into account the effect of vertical and horizontal confinement, is displayed in Fig. 7.12(b). In that figure, the vertical waveguide geometry was taken from the original work reported [62]. Not surprisingly, strong lateral confinement is shown to add a positive contribution to the GVD, and the choice of the final waveguide width will be the result of a compromise between the need to keep the total GVD low and the necessity to guarantee a single transverse mode emission. Note also that in the original report [62] the waveguide GVD was computed combining the GVD from the horizontal and vertical confinements separately. We found that the latter procedure was a relatively poor approximation for tightly confined waveguides.

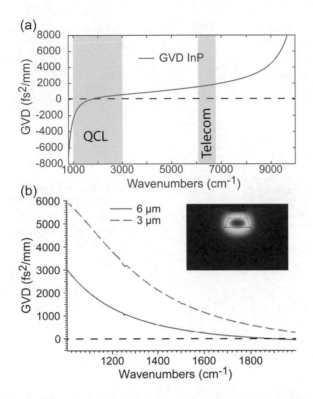

**Fig. 7.12** Contributions to the waveguide dispersion. (a) GVD for bulk InP material. (b) GVD due to the mode confinement, computed by performing a two-dimensional simulation, showing the impact of the ridge width on the GVD. Inset: mode profile of a typical buried heterostructure mid-infrared QCL.

In order to balance the effects of the materials and modal dispersions to produce near-zero GVD, it is crucial to have a knowledge of the exact amount of dispersion present in the laser cavity. Different methods exist for measuring the dispersion of a QCL. In general, one will perform this measurement *below* threshold, such that the the only difference between the measurement and the computation is the contribution from the gain itself.

As shown in Fig. 7.13(a), in the mid-infrared, the electroluminescence from the device is powerful enough such that the group delay dispersion (GDD, derivative of the group delay with respect to $\omega$) of the cavity can be extracted from a high-resolution measurement of the electroluminescence using a Fourier Transform interferometer. As discussed in [69], the GDD can be extracted directly by performing a low-resolution Fourier transform on the first satellite peak of the interferogram corresponding to the cavity optical delay and extracting its phase, as shown in Fig. 7.13(b). In this measurement, the group delay dispersion is encoded by the interference between the light that has crossed the cavity with the one that has performed an additional round trip. In Fig. 7.13(c), the effect of a dispersive coating (in that case a Gires–Tournois interferometric (GTI) mirror, further discussed below) deposited on the facet is shown. The

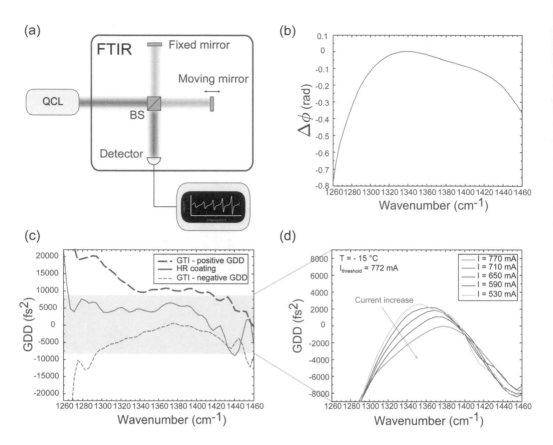

**Fig. 7.13** Dispersion measurements of QCL combs. (a) Set-up used to acquire the interferogram generated by the QCL comb biased below threshold on a FTIR. This interferogram is used to retrieve the relative phase accumulated through a round trip on the device. (b) Relative phase accumulated through a round trip on a QCL coated with a GTI mirror introducing negative dispersion ($T = -15\,^\circ$C, $I = 770$ mA corresponding to a value $\sim 2\%$ below threshold). (c) Measurement of the GDD of QCL combs. Three different coatings were evaporated on the back-facet of three different devices (4.5 mm long devices cleaved together, $T = -15\,^\circ$C, current being set to $\sim 2\%$ below threshold for the three devices). (d) Measurement of the GDD of the QCL showing negative GDD (the short-dash curve of (c)) as a function of the laser current ($T = -15\,^\circ$ C). Figure adapted with permission from [70] © The Optical Society.

modification of the total GVD by the gain of the active region is shown in Fig. 7.13(d) by comparing measurements performed at different current levels.

This technique is difficult to apply in the THz spectral range, however, because of the weakness of the spontaneous emission signal from the devices and the high detection noise. For this reason, very good results were achieved using time-domain techniques, as first shown in the work of David Burghoff et al. [63]. They measured a dispersion as high as $10^5$ fs$^2$ mm$^{-1}$, dominated by the material properties of GaAs that was then corrected using a chirped mirror. An example of such a measurement is

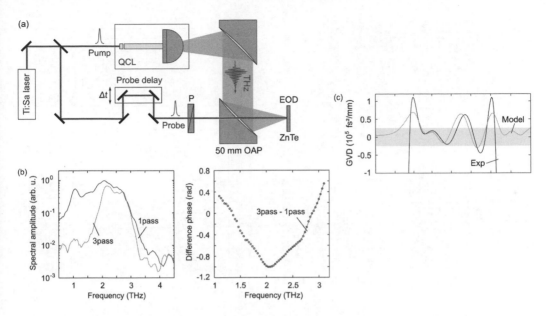

**Fig. 7.14** Time domain spectroscopy measurement of the dispersion in a broadband THz QCL using an integrated emitter. (a) Schematic drawing of the experimental setup. (b) Group delay after one and three passes through the active region. (c) GVD of the active region compared to the predictions. Figure reproduced from [71], with the permission of AIP Publishing.

shown in Fig. 7.14, performed on a high-bandwidth, three-stack active region of the kind demonstrated in [64].

In general, broadband comb operation can only occur if the dispersion is engineered to be within specified bounds and to a first order sufficiently low. As shown schematically in Fig. 7.11, this goal can be achieved in practice by the use of either a special waveguide design or by a dispersive coating at the end of the cavity.

As discussed in the previous sections, low materials GVD and a broad gain of the QCL ensures a sufficiently low device waveguide GVD to have comb operation. Nevertheless, dispersion is still large enough to prevent the comb from working over the full dynamical range of the laser, as can be seen, for example, in Figs. 7.8 and 7.10, and is typical in QCL comb formation [21, 62–64, 72, 73]. It is therefore of a major interest to compensate for this dispersion to get comb operation over the entire spectral bandwidth of the laser.

A well-known technique in ultrafast physics to compensate for positive dispersion over broad frequency ranges and achieve octave-spanning combs are double-chirped mirrors (DCM) [74]. The working principle of a DCM is to delay different frequencies with respect to each other, adding different phases for different frequencies. By careful engineering of the cavity one can introduce a negative dispersion which exactly compensates for the intrinsic dispersion of the laser. All that is needed is a sequence of two layers with different refractive index. The thickness of one period fulfills the Bragg condition as in a Bragg mirror. The layer thickness is then chirped in a way

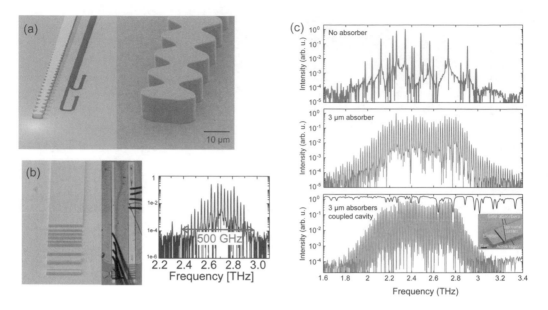

**Fig. 7.15** Double-chirped mirror (DCM) and mode control schemes for THz QCLs. (a) Scheme reported in [63] and reproduced with permission: the waveguide width is varied to achieve a contrast in refractive index. Both period and amplitude are chirped to introduce anomalous dispersion over a large frequency range. The longer the corrugation length the more dispersion is added. The right figure shows a scanning electron microscope (SEM) picture of a fabricated structure from [63]. (b) The DCM mirror is realized by alternating sections of BCB and active region (MQW) to get a high refractive index contrast. On the left a DCM grating before planarization is shown. On the right planarized and wire bonded devices are shown with the resulting single beatnote 500 GHz wide spectrum. (c) Mode control for QCLs to prevent high-order transverse modes from lasing. Lossy absorbers are introduced in the cavity showing greatly improved and uniform spectra although not combs over the full bandwidth. In the coupled cavity case there is also a GTI effect from the integrated emitter. Figure reproduced with permission from [76] © The Optical Society.

that longer wavelengths penetrate deeper into the DCM and are therefore delayed with respect to shorter wavelengths. In addition, the duty cycle is also chirped to impedance match the DCM and avoid unnecessary oscillation in the introduced group delay [74]. Recent work introduced DCMs to compensate the dispersion in THz QCLs [63]. Since the dimension of DCMs is given by the Bragg condition such structures can be fabricated at THz frequencies by using standard microfabrication techniques. Instead of using two materials with different refractive index, the double metal waveguide width has been corrugated and tapered in the implementation reported in [63]. Since the effective index of the waveguide is strongly dependent on the width, this approach is equivalent to using two media with different refractive indexes. This scheme is shown in Fig. 7.15(a). The starting and stopping period of the tapering define the frequency range covered by the DCM. The corrugation length defines the amount of dispersion introduced by the DCM. The optical spectrum of this device is shown in Fig. 7.19(b) [63] and discussed later in this chapter.

A similar approach is shown in Fig. 7.15(b) [76]. In order to fully exploit the technique of DCMs, sections of benzocyclobutene (BCB) and semiconductor active material have been alternated to get a higher refractive index contrast than is possible by just corrugating the waveguide as in [63]. The top metallization is kept continuous over the entire structure to get a good mode confinement. A narrow beatnote is observed on such devices, indicating comb operation over a bandwidth of 500 GHz [63, 76]. Further optimization on the design of the DCM should allow to also compensate for a full octave in frequency. An important design aspect in the THz QCL comb cavity is the control of transverse modes. To enforce a broadband QCL to emit light in the fundamental transverse waveguide mode across the entire octave-spanning bandwidth, the implementation of side absorbers on top of the waveguide is critical, as shown in Fig. 7.15(c). The setback in top metallization combined with lossy boundary conditions allows for the demonstration of octave-spanning spectra with regular mode distribution [75], even though the strong material dispersion still prevents such devices from octave spanning comb operation. Another concept that can be implemented to compensate for dispersion is the use of a GTI [77] directly integrated into the QCL comb. In contrast to DCMs, it can only compensate a limited frequency range but gives more flexibility on the amount of dispersion that can be introduced. This method was recently employed for the dispersion compensation of mid-infrared QCL combs, by directly evaporating the GTI on the back-facet of a QCL comb device [70]. By controlling the dispersion, the range where the comb operates significantly increases, effectively suppressing the high-phase noise regime usually observed in QCL combs [21, 62, 63, 72, 73]. In particular, the comb regime was observed over the full dynamical range of operation of the device up to powers larger than 100 mW. In the THz the implementation of the GTI is more difficult due to the high reflectivity of the double metal facet. A coupled-cavity GTI has been introduced in [71] to measure the dispersion of a multi-stack THz QCL and was successively optimized to decrease the pulse length in seeded THz QCLs [78]. Another possibility to compensate the dispersion is the one offered by a tunable GTI [79], realized by placing an object (a mirror) in close proximity to the facet of a ridge laser. When the distance of the mirror to the facet is increased, the GDD is modified, passing through a minimum value in correspondence of the GTI resonance. Thus, the GDD induced by the GTI at a fixed wavenumber can be changed from negative to positive and vice versa by tuning the mirror distance. As a consequence, the presented external GTI offers the possibility to tune the GDD of a QCL without permanently modifying it. This enables a systematic study of the influence of the GDD on the generation of frequency combs within one device. It is possible to tune to the offset frequency of the comb with the GTI while the repetition frequency is almost unaffected. Such an approach, implemented via a piezo-element [79, 80] or MEMS actuators [81], allows for the complete control of the state of a QCL comb. The position of the mirror, together with the current, form a couple of non-orthogonal but linearly independent actuators [81].

## 7.4.1          Coupled Waveguides

Generally, dispersion compensation by waveguide engineering is preferable in terms of its capability to correct for large amounts of dispersion, as well as its reliability. In a coupled-waveguide system, the wavelength dependence of the inter-waveguide coupling through the evanescent tails of the individual modes may lead to a significant change of its propagation vector. As shown in Fig. 7.16(a)–(c) one simple possibility is to exploit the coupling that arises naturally between the waveguide mode and the surface plasmon polariton that propagates at the top contact metalization [82, 83]. By changing the thickness of the top cladding, the group velocity of the waveguide can be modified as shown in Fig. 7.16(d). Such coupling was recognized early as a possible source of loss in the QCL waveguide and is unique to those devices that operate in the TM waveguide mode. The performance of a device implemented with such a waveguide dispersion scheme is shown in Fig. 7.16(e)–(f). Optical output powers up to one watt with a comb bandwidth up to $100\,\mathrm{cm}^{-1}$ with a narrow beatnote were achieved. Nevertheless, the design of surface plasmon polariton dispersion compensation is more challenging in the short wavelength region of the mid-infrared because of the need to achieve extremely high doping levels to control the propagation constant of the mode. Another possibility is then to couple the optical mode to a passive dielectric waveguide engineered in the device top cladding. Figure 7.17 shows the waveguide layout and the scanning electron microscope (SEM) cross-section of such a device, as well as the details of the symmetric and antisymmetric modes produced by coupling of the fundamental modes of the active and passive waveguides and the experimental results obtained with the device showing a narrow beatnote comb spanning over $40\,\mathrm{cm}^{-1}$ at $4.6\,\mu\mathrm{m}$ (data from [84]). One very attractive feature of such dispersion compensation design is that it does not lead to higher waveguide losses and is also feasible at near-infrared frequencies.

## 7.5          Assessing Comb Coherence

When discussing the coherence properties of a frequency comb, two different quantities have to be distinguished. The *relative coherence* between the different comb modes describes the relative phase stability between any two lines of a comb. This coherence property is intrinsically linked to the comb repetition frequency $f_{\mathrm{rep}}$, and a frequency comb with high degree of relative coherence shows perfectly spaced modes. On the other hand, the *absolute coherence* of a frequency comb characterizes the global phase stability of all modes. This second coherence property is linked to the comb offset frequency $f_{\mathrm{ceo}}$. As QCL combs are based on a different phase-locking mechanism as compared to frequency combs emitting pulses, new characterization techniques had to be developed to assess their coherence properties. Indeed, traditional characterization techniques – such as interferometric autocorrelation, FROG, or others [85] – are all based on non-linear effects induced by the short optical pulse emitted by traditional frequency combs and cannot be used for the characterization of QCL combs.

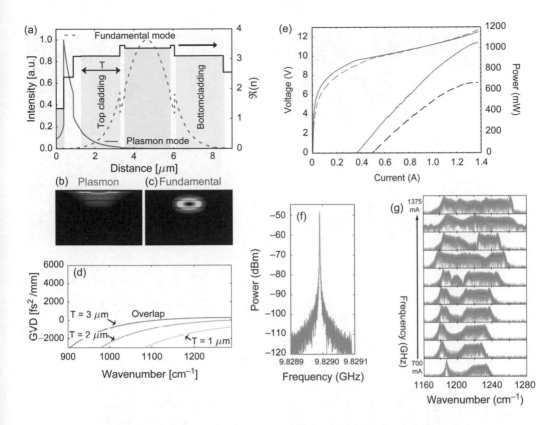

**Fig. 7.16** (a) Refractive index profile along the growth axis (right vertical axis) and 1D computed waveguide fundamental mode at 1250 cm$^{-1}$ and plasmon mode at 700 cm$^{-1}$. The shaded area on the left indicates the highly doped InP layer, the black area indicates the top cladding layer and the shaded area on the right area indicates the active region. (b) and (c) 2D profile of the plasmon and fundamental mode. (d) GVD of the fundamental mode as a function of the optical frequency for a top cladding thickness varying from 1.0 to 3.0 μm. High-power operation of a dispersion compensated QCL comb using a coupling with a surface-plasmon polariton. (e) Light versus current at 258 K (full lines) and 293 K (dashed line). (f) Measured beatnote at 1375 mA and 258 K with a span of 200 KHz and a bandwidth resolution of 100 Hz. (g) Measured optical spectra for driving currents of 700, 800, 900, 1000, 1075, 1125, 1200, 1290, 1330, and 1375 mA. Figures are reproduced with permission from [82] © The Optical Society and from [83], with the permission of AIP Publishing.

For this reason, characterization techniques which did not require pulse emission were developed. The first technique developed for the characterization of the relative coherence of QCL combs is an interferometric technique where the autocorrelation of the beatnote at the comb repetition frequency is measured using a FTIR [62].

## 7.5.1 Intermode Beat Spectroscopy and SWIFTS

The principle of *intermode beat spectroscopy* is schematically described in Fig. 7.18(a). The FTIR acts as an optical filter used for the spectral separation of the

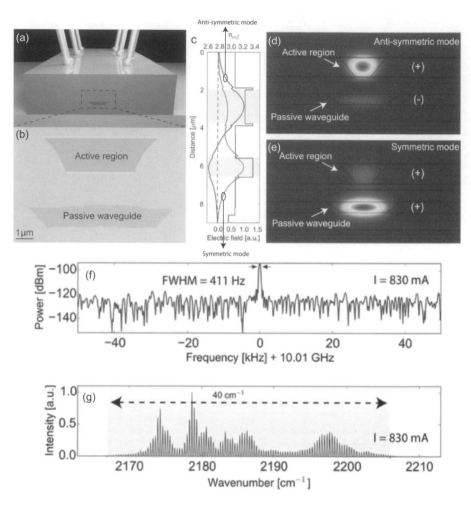

**Fig. 7.17** Device cross-section. (a) Schematic view of a mounted device. (b) SEM picture of the fabricated device. Passive waveguide and active region are highlighted. (c) Cut in the vertical plane of the electric field profile of the anti-symmetric and symmetric modes and refractive index profiles. (d) Intensity profile of the anti-symmetric mode at $2220\,\mathrm{cm}^{-1}$. (e) Intensity profile of the symmetric mode at $2220\,\mathrm{cm}^{-1}$. (f) Intermode beatnote measured at $-15°$ for a current of $830\,\mathrm{mA}$. (g) Optical spectrum measured in CW at $-15°$ for a current of $830\,\mathrm{mA}$. Data is reproduced with permission from [84].

comb modes, and the relative coherence is measured by investigating the beatnote at the comb repetition frequency as a function of the interferometric delay. This is done by using a high-bandwidth detector, sufficiently fast to measure the comb repetition frequency. The quantity measured is given by

$$I_{\mathrm{BS}}(\nu, \tau) \equiv \left| \mathcal{F}\left\{ \left| E(t) + E(t + \tau) \right|^2 \right\} \right| (\nu), \tag{7.8}$$

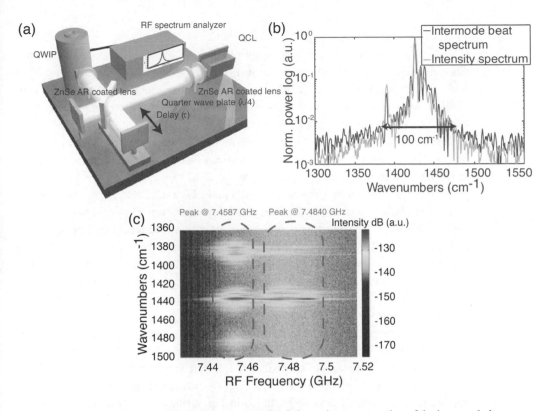

**Fig. 7.18** Intermode beat spectroscopy. (a) Schematic representation of the intermode beat spectroscopy characterization system. The output of a FTIR is sent to a high-bandwidth detector, in order to measure the intermode beatnote spectrum with a spectrum analyzer. (QWIP: quantum-well infrared photodetector; AR: anti-reflection). (b) The intermode beat spectrum together with the normal intensity spectrum of the laser, showing that the entire spectrum is contributing to the generation of the beatnote at the comb repetition frequency. (c) The intermode beat spectrum acquired over a large span of RF frequencies (100 MHz), in the case where the QCL is not operating as a comb, yielding important information about the mechanisms that destabilize the comb. Two broad beatnotes are observed on the RF spectrum. The peak at $\nu = 7.4840$ GHz involves mostly modes which are near the center of the optical spectrum ($\omega \simeq 1430\,\text{cm}^{-1}$), while the peak at $\nu = 7.4587$ GHz corresponds to modes locked for a wider bandwidth (from $\omega = 1360\,\text{cm}^{-1}$ to $\omega = 1500\,\text{cm}^{-1}$). Results are reproduced with permission from [62].

which is the absolute value of the Fourier component of the intensity at the detector at frequency $\nu$ over a chosen resolution bandwidth of the spectrum analyzer. The quantity $E$ denotes the time-dependent amplitude of the electric field of the light output from the QCL combs source and $\mathcal{F}$ stands for the Fourier transform that is applied to the detected intensity. The traditional intensity interferogram measured in a FTIR is given by $I_{\text{BS}}(0, \tau)$. The intermode beat interferogram $I_{\text{BS}}(\nu, \tau)$ is sensitive to the relative phase of the modes and can be used to measure the relative coherence of QCL combs. In analogy with Fourier-transform spectroscopy, one can calculate the Fourier transform of $I_{\text{BS}}(\nu, \tau)$ as

$$\mathcal{I}_{BS}(v,\omega) \equiv \mathcal{F}(I_{BS}(v,\tau)), \tag{7.9}$$

where $\omega$ is an optical frequency. The quantity $\mathcal{I}_{BS}(v_{rep},\omega)$ evaluated at the comb repetition frequency, called the intermode beat spectrum, is shown in Fig. 7.18(b), together with the normal intensity spectra of the laser. One can see that both spectrum are very similar, showing that the entire laser spectrum is contributing to the generation of the beatnote at the comb repetition frequency.

Another important characteristic of intermode beat spectroscopy is that one does not necessarily need to impose $v = v_{rep}$ and can therefore measure the entire intermode beat spectrum around $v_{rep}$. This can be used to yield important information about the mechanisms that destabilize the comb. Figure 7.18(c) shows an example of such characteristics, where the intermode beat spectrum $\mathcal{I}_{BS}(v,\omega)$ was acquired over a span of 100 MHz around the comb repetition frequency. In this measurement, two broad beatnotes are observed on the RF spectrum. The intermode beat spectrum $\mathcal{I}_{BS}(v,\omega)$ shows that the peak at $v = 7.4840$ GHz involves mostly modes which are near the center of the optical spectrum ($\omega \simeq 1430$ cm$^{-1}$), while the peak at $v = 7.4587$ GHz corresponds to modes locked for a wider bandwidth (from $\omega = 1360$ cm$^{-1}$ to $\omega = 1500$ cm$^{-1}$).

Although intermode beat spectroscopy can give access to the relative coherence of the modes, it cannot give direct access to the relative phases between two modes. For solving this issue, shifted wave interference Fourier transform spectroscopy (SWIFTS) has been introduced [63, 72]. A schematic description of SWIFTS applied for mid-infrared devices is shown in Fig. 7.19(a). The RF beatnote detected by a high-bandwidth quantum well infrared photodetector (QWIP) and the reference RF beatnote taken via a bias tee from the current feed of the laser are, after downmixing, are demodulated by a I/Q demodulator. This measurement yields the in-phase and quadrature signals $S_I(\tau,\omega_0)$ and $S_Q(\tau,\omega_0)$, respectively. The optical field of the QCL comb will induce current modulations on the QWIP detector at the comb repetition rate $\omega_r$, and harmonics thereof. The amplitudes of these modulations are proportional to

$$\langle E^*(t)E(t+\tau)\rangle_{(\omega_r)} = \sum_n A_n \left[ A_{n-1}^* e^{i\omega_r t} + A_{n+1}^* e^{-i\omega_r t} \right] e^{i(\omega_0 + n\omega_r)\tau}, \tag{7.10}$$

with $\tau$ being the time delay introduced by the two arms of the interferometer, which, for a rapid acquisition as in our case, becomes $\tau(t)$; $n$ denotes the mode index, $\omega_0$ the carrier envelope offset frequency in radians per second, and $\phi_n$ the average phase of mode $n$. The angle brackets indicate a narrowband filtering around $\omega_r$. Both quadratures of this signal, as retrieved by the lock-in amplifier, are recorded as a function of $\tau$. These quadratures are given by

$$x(\tau) = \frac{1}{2} \sum_n A_n \left[ A_{n-1}^* + A_{n+1}^* \right] e^{i(\omega_0 + n\omega_r)\tau}, \tag{7.11}$$

$$y(\tau) = \frac{1}{2i} \sum_n A_n \left[ A_{n-1}^* - A_{n+1}^* \right] e^{i(\omega_0 + n\omega_r)\tau}. \tag{7.12}$$

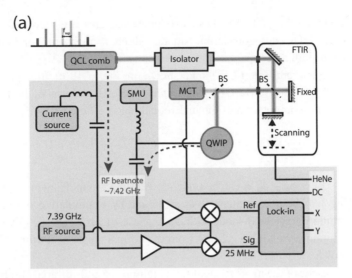

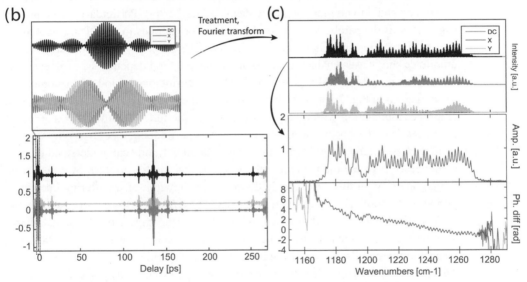

**Fig. 7.19** Shifted wave interference Fourier transform spectroscopy (SWIFTS). (a) Schematic representation of SWIFTS characterization system. The output of a FTIR is sent to a high-bandwidth detector. The detected signal and the reference, after downmixing, are sent to an I/Q demodulator in order to get access to the in-phase and quadrature signals as a function of delay of the interferometer arm. (b) Example of measured DC and demodulated (SWIFTS) interferogram traces (left); zoom about centerburst (right), showing that, when the interferometer arms are balanced, the RF signal is at a minimum, a characteristic of FM combs. (c) Fourier transform of the individual interferograms, as well as the extracted amplitude and phases of the lines. Figure adapted with permission from [56] © The Optical Society.

As shown in Fig. 7.19(b), these two interferograms are recorded in parallel with the normal DC autocorrelation signal, which is measured on a more sensitive, low-bandwidth detector.

As clear from Eq. (7.13) and shown in Fig. 7.19(c), the phasors $|A_n||A_{n-1}|$ $e^{i(\omega_r t + \phi_n - \phi_{n-1})}$ are retrieved by a simple Fourier transform of both interferograms, allowing the computation of the intermodal phase differences:

$$X(\omega) + iY(\omega) = \sum_n |A_n||A_{n-1}|e^{i(\phi_n - \phi_{n-1})}\delta\left[\omega - (\omega_0 + n\omega_r)\right]. \quad (7.13)$$

$\angle(X + iY) = \phi_n - \phi_{n-1}$ can be understood as the group delay multiplied by the mode spacing $\omega_r$. The phases themselves are found by a cumulative summation, assuming one arbitrarily selected mode to have a phase of zero. The positive amplitudes are found by simply taking a $\sqrt{|A_n|^2}$ from the normal power spectrum, which is acquired in parallel.

The results of the measurement for a high-power mid-infrared comb are shown after integrating and unwrapping the group delay in Fig. 7.20(a). Particularly prominent in Fig. 7.20(a) is the parabolic phase profile which corresponds to a GDD of $-6.4\,\mathrm{ps}^2$.

As a very interesting byproduct of the SWIFTS spectroscopy, knowledge of the amplitude and phases enabled us to compute the temporal dependence of intensity as well as the instantaneous frequency of the device during one period of the comb. The results shown in Fig. 7.20(b) show that the laser operates with a significant amount of amplitude modulation with a very short characteristic time. The most striking aspect of the result is the linear chirp, where the emission frequency is linearly increasing from the lower frequencies to the higher ones. Intuitively, one can understand the correspondence between the parabolic phase profile and the linear chirp by considering the group delay. In this case, the field sweeps a bandwidth of $f_{BW}$ during one single period of the comb $f_{rep}^{-1}$, yielding a group delay of $GD(\omega) = (\omega - \omega_0)/f_{BW}f_{rep} + \Psi$, with $\Psi$ representing an arbitrary delay offset, with no impact on the signal shape. The expression is linear in frequency; the phase, which is the integral of this with respect to the circular frequency, will then naturally yield a parabolic phase profile. This parabolic phase profile is then characterized by a group delay dispersion given by

$$GDD = \frac{1}{2\pi f_{rep} f_{BW}}. \quad (7.14)$$

This simple argument yields a predicted GDD of $-7.1\,\mathrm{ps}^2$, close to the value measured $(-6.4\,\mathrm{ps}^2)$.

Finally, the SWIFTS analysis also enables the computation of the degree of relative coherence between two modes [72], thus quantifying the coherence of the comb. It is worth mentioning that a technique analogous to the SWIFTS spectroscopy was basically already developed in [86] where a pulsed mode-locked laser was employed in combination with a Fourier transform interferometer and the phase measurement was used to spectrally characterize the sample, demodulating the signal at the roundtrip frequency of the fs laser.

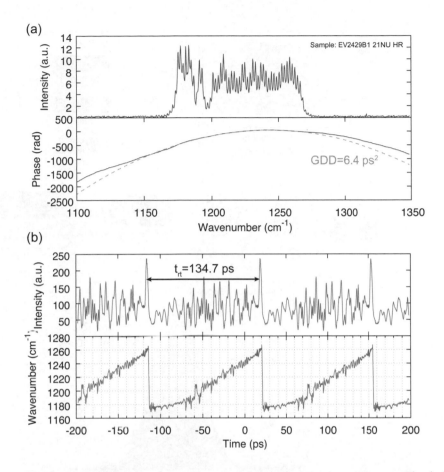

**Fig. 7.20** (a) Amplitude spectrum and computed phase for one measurement set. The dashed line indicates the quadratic fit to the phase, yielding an estimated GDD of 6.4 ps$^2$. (b) Instantaneous intensity and frequency of the field, simulated for the measurement state in (a). Both have been integrated to 1 ps to reflect the response times of the active region. Figure adapted with permission from [56] © The Optical Society.

An interesting variant of this technique is based on self-mixing of the laser output, which is fed back into the laser cavity after being filtered by a FTIR [87]. At the time of writing, however, the theoretical understanding of that technique is still incomplete and, being based on optical feedback, the laser state can change during the measurement.

## 7.5.2    Comb Line Equidistance Measurement Using a Dual-Comb Technique

Another method for quantifying the relative coherence of frequency combs is to verify the uniformity of the spacing between the comb lines. A direct verification of this uniformity can be done using a multi-heterodyne setup (also called dual-comb setup), as shown schematically in Fig. 7.21(a) [21]. By employing two frequency combs with

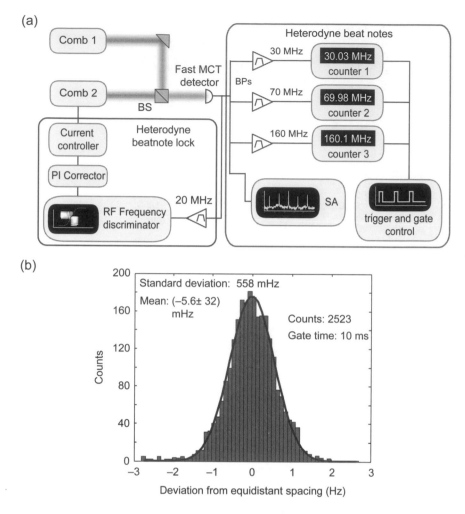

**Fig. 7.21** Measurement of the deviation from equidistant mode spacing. (a) Schematic representation of the setup used for measuring the deviation from equidistant mode spacing $\epsilon$ based on a multi-heterodyne measurement. SA: spectrum analyzer; BPs: band-pass filters; BS: beam splitter; PI: proportional integral; MCT: mercury cadmium telluride. (b) Distribution of the deviation from equidistance mode spacing $\epsilon$. Total measurement time = 25.23 s. Gate time = 10 ms. The distribution shows a Gaussian distribution with an average value of $(-5.6 \pm 32)$ mHz and a standard deviation of 558 mHz. Results published in [21].

slightly different repetition frequencies, this setup effectively maps the frequency comb from the optical domain down to the RF frequencies, where established, low-noise counting techniques can be employed. The uniformity of the mode spacing can be quantified by evaluating the deviation from equidistant mode spacing $\epsilon$, defined as [9]

$$\epsilon = \frac{f_M - f_0}{M} - \frac{f_N - f_0}{N},$$

(7.15)

where $f_0$, $f_N$, and $f_M$ are the frequencies of three multi-heterodyne beatnotes. After detecting the multi-heterodyne beat signal, three electrical bandpass filters are used to isolate the beatnotes at frequencies $f_0$, $f_N$, and $f_M$. After amplification, three frequency counters are used to measure the frequency oscillations of the beatnotes.

In this experiment, an active stabilization is implemented (by acting on the current of one device) in order to correct the slow frequency drifts of the multi-heterodyne beat spectrum, thus preventing each beatnote from drifting out of the bandwidth of the electrical filters. A precise control of both trigger and gate time of all the counters was implemented. It is particularly important as any drift of $f_{ceo}$ will be seen as a non-uniformity of the comb line spacing if the counters are not perfectly synchronized. Even though the comb is not fully stabilized and $f_{rep}$ drifts, $\epsilon$ should be zero at any point in time for a perfectly synchronized setup and a perfectly spaced comb. It is important to observe that this method offers a direct way to measure the uniform comb spacing using counting techniques even if the frequency comb is not fully stabilized.

Figure 7.21(b) shows the distribution of $\epsilon$ over the entire measurement (gate time = 10 ms, 2523 counts), showing a mean value of $\bar{\epsilon} = (-5.6 \pm 32)$ mHz. Normalized to the optical carrier frequency (42.8 THz or 7 μm), this gives an accuracy of the equidistance of $7.5 \times 10^{-16}$. This measurement definitely proves the equidistance of the mode spacing of a QCL frequency comb, confirming the high degree of relative coherence of QCL combs [21].

Access to the different phases of a QCL comb is given also by using a dual-comb technique where the sampling comb is a downconverted and mixed together with the QCL under investigation. By locking the two combs, in a similar fashion to [88] the phases of mid-IR and THz comb and their time stability have been measured [89].

### 7.5.3   Schawlow–Townes–Limited Linewidth of QCL Combs

As QCL combs are attractive for applications such as high-resolution molecular spectroscopy and optical metrology in the mid-infrared or THz, it is important to assess the frequency noise characteristics of these type of sources.

With this in mind, the investigation of the frequency noise power spectral density (FNPSD) of mid-infrared QCL combs was performed [73]. A high-finesse optical cavity is used to resolve the laser spectrum and to detect the frequency fluctuations of the laser, acting as frequency-to-amplitude (FA) converter, as shown in Fig. 7.22(a). The distance between the two mirrors is chosen in order to set the free spectral range (FSR) of the cavity close to comb repetition frequency $f_{rep}$. To utilize the cavity as a FA converter, a piezoelectric actuator controlling the cavity length is used and the temperature of the laser is precisely controlled in order to set the FSR to match exactly the round trip frequency of the comb, while simultaneously let the comb offset frequency $f_{ceo}$ be equal to that of the optical cavity. In this way, the comb modes and the optical cavity resonances are perfectly matched. Successfully achieving this alignment requires an independent control of the $f_{ceo}$ and of the $f_{rep}$ of the QCL comb. As a consequence, in these conditions and *only* in these conditions of temperature and driving current of the laser, all the comb modes are transmitted by the cavity. The cavity can thus be used

(a)

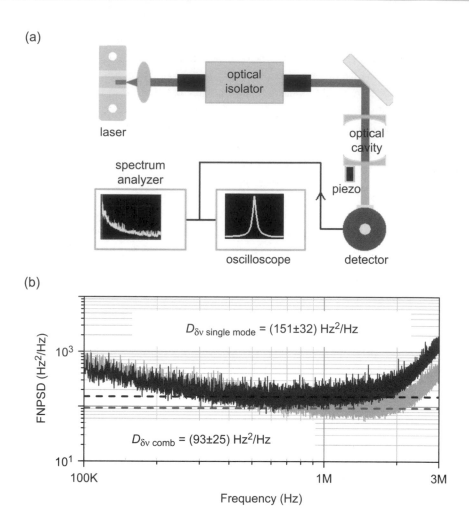

(b)

Fig. 7.22 Schawlow–Townes linewidth of QCL combs. (a) Experimental setup used to measure the FNPSD of the QCL comb. The main optical components include the laser (a multistack InGaAs/InAlAs QCL), the optical isolator, the high-finesse optical cavity, and the high-sensitivity MCT detector. The signal is processed by a high-sampling-rate oscilloscope. (b) Zoom of the flattening portion of the FNPSD of the QCL comb around 1 MHz, corresponding to the Schawlow–Townes limit. The spectra are compensated for the FA converter cutoff. The spectra are related to the two operating conditions of the laser: single-mode with $P = 15$ mW (black) and comb regime (with all the modes in resonance with the cavity) with $P = 25$ mW (gray). Figure adapted with permission from [73] © The Optical Society.

as a *multimode* frequency-to-amplitude converter to collect the frequency fluctuations of all the modes at the same time [90]. The laser emits a power of 25 mW when the comb modes are exactly matched to the cavity resonances. A spectrum retrieved with the laser in single-mode operation ($P = 15$ mW) can also be acquired.

The FNPSD measured on the single-mode and comb regimes can therefore be measured by employing this multimode FA. Figure 7.22(b) shows a magnified view of the FNPSDs in both regimes (shown from 100 kHz to 3 MHz). Around 1 MHz, a flattening characteristic of a white frequency noise is observed, corresponding to the intrinsic quantum noise level $D_{\delta\nu}$ due to the spontaneous emission, the so-called *Schawlow–Townes* limit [91]. One can also compare these levels of $D_{\delta\nu}$ to those expected for single-mode emission with the same characteristics, given by the Schawlow–Townes limit [92]:

$$\delta\nu = \frac{h\nu}{P}\frac{\alpha_{\text{tot}}c^2}{4\pi n_{\text{g}}^2}\alpha_m n_{\text{sp}}(1+\alpha_{\text{e}}^2).\qquad(7.16)$$

Taking $\nu = 42.2\,\text{THz}$ as the central frequency, $\alpha_m = 2.2\,\text{cm}^{-1}$ as the mirror losses, $\alpha_{\text{tot}} = 7.2\,\text{cm}^{-1}$ as the total losses (the relatively high waveguide losses are due to the residual *cross-absorption* given by the multistack structure), $n_{\text{g}} = 3.4$, $n_{\text{sp}} = 2$ as the spontaneous emission factor and $\langle\alpha_{\text{e}}^2\rangle = 0.0023$ as the squared *Henry linewidth enhancement factor* averaged over the laser spectrum, we can compute the Schawlow–Townes limit relative to the single-mode emission ($P = 15\,\text{mW}$) and to the comb emission ($P = 25\,\text{mW}$). The two values are $\delta\nu = 383\,\text{Hz}$ and $\delta\nu = 230\,\text{Hz}$ respectively. These values are consistent with those obtained from the spectra $\delta\nu = \pi D_{\delta\nu}$, which are $(474 \pm 100)\,\text{Hz}$ for the single-mode emission and $(292 \pm 79)\,\text{Hz}$ for the comb emission (see Fig. 7.22(b)).

The measurement of the FNPSD in comb regime shows that the quantum fluctuations of the different modes are correlated. In fact, it is observed that the FNPSD – in particular the portion limited by the quantum noise – is identical when measured with one comb mode and with all comb modes simultaneously. This quantum limit, which is given by the Schawlow–Townes expression, would be at least a factor of 6 larger than the one shown in Fig. 7.22(b), assuming that the quantum fluctuations of each comb mode were uncorrelated. This factor is outside the uncertainty of the measurement. This experimental work demonstrate that the four-wave mixing process – at the origin of the comb operation in QCLs – correlates the frequency fluctuations between the modes until the quantum limit. As a consequence, instruments using the spectral multiplexing of dual combs or multi-heterodyne spectrometers hold an inherent noise advantage compared to similar systems using arrays of single-mode lasers [73].

## 7.6     Dual Comb on a Chip

One very attractive feature of the QCL comb is the possibility to integrate two combs on the same chip, to have the basis for a compact dual-comb spectrometer. Such a device could bring broadband high-precision measurements to practical applications, such as trace gas or breath analysis, just to name a few. In [93], the authors demonstrated an on-chip dual-comb source based on MIR QCL combs. They showed that control of the offset and repetition frequencies of both combs is possible by integrating micro-heaters [94] close to the QCL combs, demonstrating that this device is ideal for compact dual-comb spectroscopy.

The concept consists of two QCL combs fabricated on a single chip, into which two independent micro-heaters [94] are integrated next to each QCL comb. A schematic representation of a QCL comb together with its micro-heater is shown in Fig. 7.23(a). The micro-heater consists of a small resistor created by etching a thin layer of Si-doped InP, previously utilized for achieving spectral tuning of single frequency QCLs [94]. The QCL comb and the micro-heater are biased by two independent current sources. The source driving the QCL is used for achieving laser action, but also for controlling the offset and repetition frequencies of the QCL-comb. In addition to this first way of controlling the comb parameters, the source driving the micro-heater controls the amount of heat generated by the micro-heater. By using the temperature tuning of the refractive index, both the offset and repetition frequencies of the QCL comb can also be controlled by the micro-heater. These two ways of controlling the comb parameters are not strictly equivalent as the laser current controls the amount of carriers in the structure, and thus its gain and refractive index. In addition to this effect, the laser current also increases the temperature of the structure and therefore induces a modification of the refractive index. Only this second effect is utilized by the micro-heater. Further control of the comb parameters is demonstrated by measuring the radio-frequency (RF) spectrum of the laser current [21]. The RF spectrum contains the beatnotes at the roundtrip frequency of the two combs $f_{rep,1}$ and $f_{rep,2}$, as they are biased in parallel by a unique current driver. One can therefore precisely measure both comb repetition frequencies with a spectrum analyzer, as shown in Fig. 7.23(c).

We show the control of $f_{rep}$ of a single comb by using one micro-heater to tune $f_{rep,1}$ while the second micro-heater is not utilized, therefore not acting on $f_{rep,2}$. The RF spectra acquired at different values of micro-heater current $I_{heater,1}$ (all other parameters being kept fixed) are shown in Fig. 7.23(f). When increasing the current in the micro-heater, we observe that one RF beatnote is shifted to lower frequencies while the other RF beatnote is not shifted, effectively decreasing $\Delta f_{rep}$ (observed from $I_{heater,1} = 80$ mA up to 214 mA). Around one precise value of the micro-heater current (around 214 mA in this example), both repetition frequencies are similar. By further increasing the value of the micro-heater current, one RF beatnote is further shifted to lower frequencies and $\Delta f_{rep}$ starts to increase. For important values of micro-heater current, we observe that both RF beatnotes start to shift to lower frequencies, one being more sensitive to the micro-heater current than the other. This is explained by the fact that at these values of the micro-heater current, the heat dissipated starts to influence the second comb. All these observations are summarized in Fig. 7.23(e), where $f_{rep,1}$ and $f_{rep,2}$ are represented as a function of $I_{heater,1}$. This experiment finally demonstrates the control of the repetition frequency of a single comb by employing the micro-heater. Although the control of $f_{ceo}$ and $f_{rep}$ of a single comb is possible with our system, these two quantities are yet not controllable independently. It is worth mentioning that the ultrafast lifetimes can make the QCL as an efficient fast detector. Such possibility can be exploited in order to integrate both elements of a dual-comb spectrometer on the same chip. In the THz, thanks to the presence of metallic layers, this allows for multi-heterodyne detection on-chip [95].

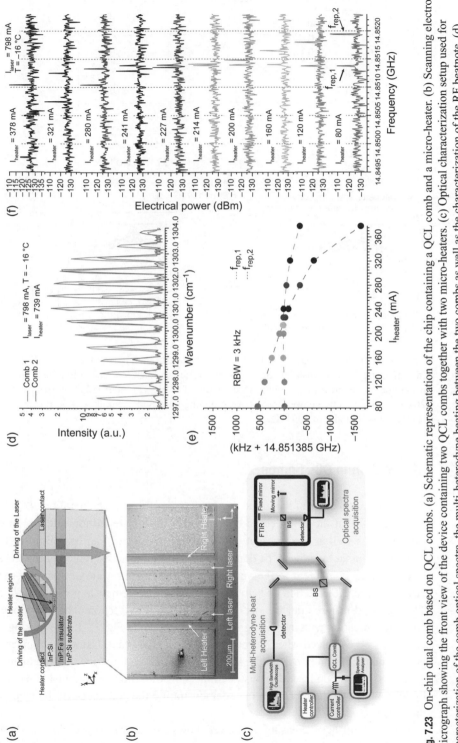

**Fig. 7.23** On-chip dual comb based on QCL combs. (a) Schematic representation of the chip containing a QCL comb and a micro-heater. (b) Scanning electron micrograph showing the front view of the device containing two QCL combs together with two micro-heaters. (c) Optical characterization setup used for characterization of the comb optical spectra, the multi-heterodyne beating between the two combs as well as the characterization of the RF beatnote. (d) Control of the comb parameters by using the integrated micro-heaters. Optical spectra of the two combs when both the laser current $I_{laser}$ and one of the micro-heater currents $I_{heater,1}$ were optimized in order to minimize the $\Delta f_{ceo}$. (e) Value of the frequencies $f_{rep,1}$ and $f_{rep,2}$, showing the independent control of the repetition frequency. (f) Set of RF spectra acquired for different values of $I_{heater,1}$, showing the two RF beatnotes corresponding to $f_{rep,1}$ and $f_{rep,2}$ (resolution bandwidth (RBW) = 3 kHz, span = 5 MHz, sweep time = 35 ms). Figure reproduced from [93], with the permission of AIP Publishing.

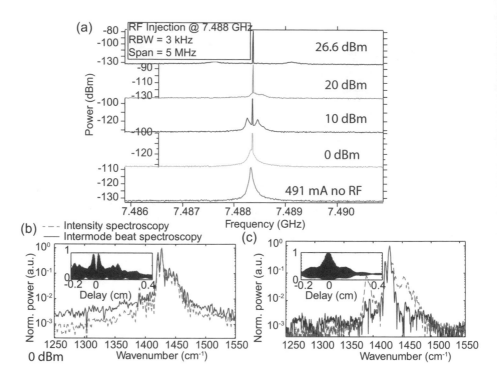

**Fig. 7.24** RF injection in QCL. (a) Intermode beat measurements with increasing RF power. (b) Intermode beat spectrum at 10 dBm RF injection power. Inset: corresponding intermode beat interferogram taken with a resolution bandwidth of 5 kHz. (c) Intermode beat spectrum at 26.6 dBm RF injection power. Inset: corresponding intermode beat interferogram taken with a resolution bandwidth of 100 kHz.

## 7.7    RF Injection in QCL Combs

One important feature of the initial demonstration of the comb operation in QCL [62] was to show that the comb could be controlled both in repetition rate and in offset frequency, as well as the fact that the round trip frequency can be controlled by RF injection. The fact that it can be done by modulating the active region current is again a feature that is inherent to the very short upper state lifetime of quantum cascade lasers. Indeed, RF injection had already been used for attempts at pulse-mode-locking THz and mid-infrared devices, as discussed in the next section. The intermode beat under RF injection with increasing RF power is shown in Fig. 7.24. The locking of the comb can be enhanced when injecting a 10 dBm RF signal. Using such injected RF power, the repetition rate could be shifted by 65 kHz. Heterodyne measurements with a stable single mode laser demonstrated the shift of a mode by about 11 GHz caused by this change of the repetition rate, demonstrating the coherence of the comb up to the edges.

However, further increasing the power shows a clear transient region, as shown by the contrast between an intermode beat spectroscopy at 10 dBm shown in Fig. 7.24(b) and at 26.6 dBm (Fig. 7.24(c)). The insets show the measured intermode beat interferograms. At 10 dBm RF injection power the frequency-modulated nature of the laser output is preserved, since the intermode beat interferogram shows a minimum at zero path difference. On the other hand, at 26.6 dBm RF injection power, the intermode beat interferogram starts to resemble the one of a fundamentally mode-locked laser. But when comparing the intermode beat spectra to the intensity spectra for the two cases in Figs. 7.24(b) and (c), it becomes obvious that not all laser modes are locked at high RF injection power.

Further experiments have greatly expanded these initial experimental results. In a first set of experiments, a mid-infrared QCL was embedded in a microwave microstrip cavity. The very good overlap between the microwave and the optical fields enabled to extend the range over which the optical beat-tone was locked to 1.4 MHz [96], although the coherence of the spectra was not discussed in this initial work.

Recent work have shown that, in devices that are optimized for RF injection, control of both the amplitude and the frequency of the RF injection could change the state of the comb.

The THz devices are naturally more adapted to the RF injection and control since the waveguide is in fact a segment of a microwave stripline. Early work analyzed the RF propertied of double metal waveguides [97] showing the possibility to injection lock the beatnote of the QCL (also in single plasmon waveguide) emitting over relatively narrow bandwidths ($<$ 200 GHz) [98]. Recent developments in high-efficiency, superdiagonal active regions show injection locking for RF powers as low as $-55$ dBm [67] reaching bandwidths wider than 900 GHz.

## 7.8    Pulsed Mode-Locked QCLs

Strong RF injection in QCL devices has also been used in the context of active pulsed mode-locking.

To that end, typically, one short section of the QCL active region is modulated at the laser round trip frequency (Fig. 7.25(a)). However, to mitigate the gain saturation in the active region, it is necessary to work with the active regions that provide upper state lifetimes in the tens of picoseconds, such that the condition $\omega\tau_2 \approx 1$ is fulfilled. Based on this principle, active mode-locking has been achieved in the mid-infrared with active regions based on photon-assisted tunneling transitions (as shown in Fig. 7.25(b)) [99] where very long intersubband lifetimes in the order of 20 ps can be achieved [100]. At a temperature of 77 K, a pulse length of 3 ps with a bandwidth of 15 cm$^{-1}$ was achieved [99] (see Fig. 7.25(b)–(d)). Nevertheless, the gain recovery time of 50 ps still restricted the operation of the device to a driving current range not much higher than 10% above threshold, as predicted by numerical simulations [51]. In recent years, a careful RF engineering of the laser packaging together with advances in the understanding of the mode-locking mechanisms allowed the demonstration of room temperature mode-locking in ICLs [101] and QCLs [102] with pulse lengths of less than 7 ps at 8 μm, with a corresponding spectral bandwidth of 8 cm$^{-1}$.

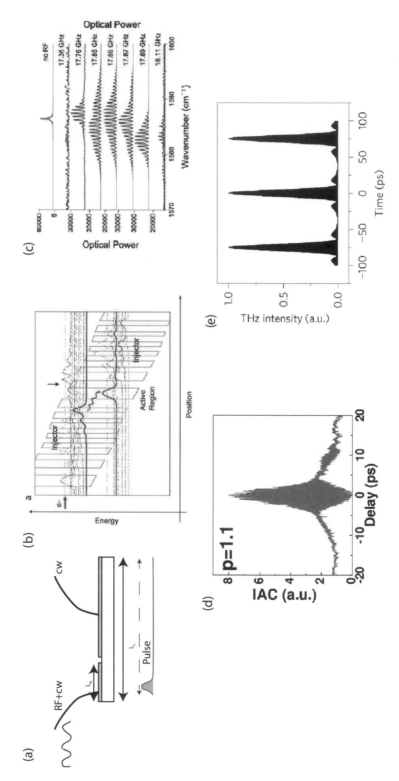

**Fig. 7.25** Mode-locking QCLs. (a) Schematic diagram of an actively mode-locked QCL: the modulated section opens a window in time, allowing a single pulse to propagate in the cavity. (b) Active region based on a photon-assisted tunneling transition, with very long upper state lifetime. Results are reproduced with permission from [99]. (c) Emission spectra of the device presented in (b) under modulation with different RF frequencies. (d) Characterization of the laser pulses using an autocorrelation setup with a two-photon quantum well photodetector that provides output quadratic to the incident light intensity. The resulting interferogram, measured at a bias current 10% above threshold, shows the ratio of 8:1 between the center and the wings, expected for single pulses. Results are reproduced with permission from [99] ©The Optical Society. (e) Electro-optic sampling, using a femto laser comb of the output of an actively mode-locked THz QCL. Results are reproduced with permission from [88].

In the THz, relatively long upper state lifetimes ($> 35\,\mathrm{ps}$) are achieved at cryogenic temperatures with state-of-the-art active regions based on hybrid bound-to-continuum transitions plus phonon depopulation [103]. Active mode-locking by RF injection was achieved at 3 THz with QCLs providing pulsed output with pulses lengths of about 10 ps and a bandwidth of 100 GHz [88, 104, 105], as shown in Fig. 7.25(i). One very interesting aspect of these experiments was the possibility of locking the THz QCL to a near-infrared fiber comb and using the latter to perform electro-optic sampling of the THz pulse (see Fig. 7.25(e)), as reviewed in [106]. Lately, such scheme has been extended to provide full stabilization against a metrological source [107]. Finally, mode-locking was also achieved in the THz by a combination of injection seeding and RF gain switching [108, 109]. In the latter case, in contrast to the conventional mode-locking by RF injection, the phase of the THz pulse is fixed by seeding the THz QCL laser with the output of a gated photoconductive antenna. These kinds of experiments benefited from broadband QCL gain engineering showing pulses as short as 2.5 ps [75] and similar results have been obtained by dispersion compensation schemes based on a Gires–Tournois interferometer [78].

However, compared to passive mode-locking in which the pulse length can routinely reach the inverse of the gain bandwidth $\Omega_g$, 1 active mode-locking leads to much longer pulses since the pulse length $\tau$ is now a geometrical average of the gain bandwidth and the round trip frequency $\omega$ [43]. In addition, as mentioned above, the output power is limited by the necessity to remain close to the laser threshold.

Further work in mode-locking QCLs include the use of a ring external cavity. This arrangement simplifies the mode selection because it minimizes the spatial hole burning effect, and it also allows the use of modulation frequencies low enough to enable complete switching of the laser on and off. This geometry has been analyzed both theoretically [110] and experimentally [111, 112].

Recent work using insight from more advanced modeling of the active region [58] made significant progress in addressing the limited mode-locking bandwidth and low optical powers observed in the initial work. Another potentially fruitful approach is also the self mode-locking in *integrated* ring QCLs [59, 113].

## 7.9     Conclusion and Outlook

As compared to mode-locked monolithic semiconductor lasers, QCL combs benefit from features specific to the physics of their intersubband gain medium: the capability to achieve very wide gain bandwidths, low material GVD, as well as the FM-like characteristics of the emission that allows large average powers without high peak powers incompatible with tightly confined waveguides. The possibility to build broadband spectrometers with chip-sized, electrically pumped optical frequency comb sources has a great potential for applications in sensing and spectroscopy. QCL combs have, in addition, some additional potential features that have not yet been fully exploited: because of the ultrafast transport in the active region, QCLs are also RF detectors that are in principle capable of detecting a heterodyne beatnote. The recent progress in broadband gain active regions in the mid-infrared and THz show that an octave-spanning QCL comb is feasible. In addition, the active region can also be engineered to provide a $\chi^{(2)}$ susceptibility, and therefore internally generate its own second harmonic. This capability of the QCL comb to integrate gain and non-linearity as well as detection is unique and offers great potential for device and system integration.

## Acknowledgement

The authors want to acknowledge the support from the Swiss National Science Foundation, the ERC CoG CHIC (724344), and the EU Flagship project QOMBS.

## References

[1]  T. Udem, R. Holzwarth, and T. Hansch, "Optical frequency metrology," Nature **416**, 233 (2002).

[2]  G. Scalari, J. Faist, and N. Picqué, "On-chip mid-infrared and THz frequency combs for spectroscopy," Applied Physics Letters **114**, 150401 (2019).

[3]  H. R. Telle, G. Steinmeyer, A. E. Dunlop, J. Stenger, D. H. Sutter, and U. Keller, "Carrier-envelope offset phase control: A novel concept for absolute optical frequency measurement and ultrashort pulse generation," Applied Physics B-Lasers and Optics **69**, 327 (1999).

[4]  S. T. Cundiff and J. Ye, "Colloquium: Femtosecond optical frequency combs," Reviews of Modern Physics **75**, 325 (2003).

[5]  S. A. Diddams, "The evolving optical frequency comb [Invited]," Journal of the Optical Society of America B–Optical Physics **27**, B51 (2010).

[6]  The Nobel Prize in Physics 2005, www.nobelprize.org, (2013).

[7]  P. Del'Haye, A. Schliesser, O. Arcizet, T. Wilken, R. Holzwarth, and T. J. Kippenberg, "Optical frequency comb generation from a monolithic microresonator," Nature **450**, 1214 (2007).

[8]  T. J. Kippenberg, R. Holzwarth, and S. A. Diddams, "Microresonator-based optical frequency combs," Science **332**, 555 (2011).

[9]  P. Del'Haye, K. Beha, S. B. Papp, and S. A. Diddams, "Self-injection locking and phase-locked states in microresonator-based optical frequency combs," Physical Review Letters **112**, 043905 (2014).

[10]  T. Herr, V. Brasch, J. D. Jost, I. Mirgorodskiy, G. Lihachev, M. L. Gorodetsky, and T. J. Kippenberg, "Mode spectrum and temporal soliton formation in optical microresonators," Physical Review Letters **113**, 123901 (2014).

[11]  T. J. Kippenberg, A. L. Gaeta, M. Lipson, and M. L. Gorodetsky, "Dissipative Kerr solitons in optical microresonators," Science **361**, e8083 (2018).

[12]  J. Riemensberger, A. Lukashchuk, M. Karpov, W. Weng, E. Lucas, J. Liu, and T. J. Kippenberg, "Massively parallel coherent laser ranging using a soliton microcomb," Nature **581**, 164 (2020).

[13]  A. L. Gaeta, M. Lipson, and T. J. Kippenberg, "Photonic-chip-based frequency combs," Nature Photonics **13**, 158–169 (2019).

[14] N. R. Newbury, "Searching for applications with a fine-tooth comb," Nature Photonics **5**, 186–188 (2011).

[15] F. Keilmann, C. Gohle, and R. Holzwarth, "Time-domain mid-infrared frequency-comb spectrometer," Optics Letters **29**, 1542 (2004).

[16] I. Coddington, W. C. Swann, and N. R. Newbury, "Coherent multiheterodyne spectroscopy using stabilized optical frequency combs," Physical Review Letters **100**, 013902 (2008).

[17] E. Baumann, F. R. Giorgetta, W. C. Swann, A. M. Zolot, I. Coddington, and N. R. Newbury, "Spectroscopy of the methane $\nu$ {3} band with an accurate midinfrared coherent dual-comb spectrometer," Physical Review A **84**, 062513 (2011).

[18] G. Rieker, F. R. Giorgetta, W. C. Swann, I. Coddington, L. C. Sinclair, C. L. Cromer, E. Baumann, A. Zolot, and N. R. Newbury, "Open-path dual-comb spectroscopy of greenhouse gases," in *CLEO: 2013*, OSA Technical Digest (online) (Optical Society of America, 2013), p. CTh5C.9.

[19] N. Leindecker, A. Marandi, R. L. Byer, K. L. Vodopyanov, J. Jiang, I. Hartl, M. Fermann, and P. G. Schunemann, "Octave-spanning ultrafast opo with 2.6–6.1 µm instantaneous bandwidth pumped by femtosecond TM-fiber laser," Optics Express **20**, 7046 (2012).

[20] T. Ideguchi, A. Poisson, G. Guelachvili, N. Picque, and T. W. Hansch, "Adaptive real-time dual-comb spectroscopy," Nature Communications **5** (2014).

[21] G. Villares, A. Hugi, S. Blaser, and J. Faist, "Dual-comb spectroscopy based on quantum-cascade-laser frequency combs," Nature Communications **5**, 5192 (2014).

[22] D. Burghoff, Y. Yang, and Q. Hu, "Computational multiheterodyne spectroscopy," Science Advances **2** (2016).

[23] L. A. Sterczewski, J. Westberg, and G. Wysocki, Computational coherent averaging for free-running dual-comb spectroscopy, Optics Express **27**, 23875 (2019).

[24] M. Tonouchi, "Cutting-edge terahertz technology," Nature Photonics **1**, 97 (2007).

[25] V. Spagnolo, P. Patimisco, R. Pennetta, A. Sampaolo, G. Scamarcio, M. S. Vitiello, and F. K. Tittel, "THz quartz-enhanced photoacoustic sensor for h(2)s trace gas detection," Optics Express **23**, 7574 (2015).

[26] A. Schliesser, N. Picqué, and T. W. Hänsch, "Mid-infrared frequency combs," Nature **6**, 440 (2012).

[27] F. Adler, P. Masłowski, A. Foltynowicz, K. C. Cossel, T. C. Briles, I. Hartl, and J. Ye, "Mid-infrared Fourier transform spectroscopy with a broadband frequency comb," Optics Express **18**, 21861 (2010).

[28] N. Leindecker, A. Marandi, R. L. Byer, K. L. Vodopyanov, J. Jiang, I. Hartl, M. Fermann, and P. G. Schunemann, "Octave-spanning ultrafast OPO with 2.6–6.1 µm instantaneous bandwidth pumped by femtosecond TM-fiber laser," Optics Express **20**, 7046 (2012).

[29] Q. Ru, T. Kawamori, P. G. Schunemann, S. Vasilyev, S. B. Mirov, and K. L. Vodopyanov, "Two-octave-wide (3–12 µm) subharmonic produced in a minimally dispersive optical parametric oscillator cavity," Optics Letters **46**, 709 (2021).

[30] A. G. Griffith, R. K. W. Lau, J. Cardenas, Y. Okawachi, A. Mohanty, R. Fain, Y. H. D. Lee, M. Yu, C. T. Phare, C. B. Poitras, A. L. Gaeta, and M. Lipson, "Silicon-chip mid-infrared frequency comb generation," Nature Communications **6** (2015).

[31] B. Kuyken, T. Ideguchi, S. Holzner, M. Yan, T. Haensch, J. Van Campenhout, P. Verheyen, S. Coen, F. Leo, R. Baets, G. Roelkens, and N. Picqué, "An octave-spanning mid-infrared frequency comb generated in a silicon nanophotonic wire waveguide," Nature Communications **6** (2015).

[32] C. Y. Wang, T. Herr, P. Del'haye, A. Schliesser, J. Hofer, R. Holzwarth, T. W. Hänsch, N. Picqué, and T. J. Kippenberg, "Mid-infrared optical frequency combs at 2.5 mum based on crystalline microresonators," Nature Communications **4**, 1345 (2013).

[33] J. Faist, *Quantum cascade lasers* (Oxford University Press, 2013).

[34] R. Kohler, A. Tredicucci, F. Beltram, H. Beere, E. Linfield, A. Davies, D. Ritchie, R. Iotti, and F. Rossi, "Terahertz semiconductor-heterostructure laser," Nature **417**, 156 (2002).

[35] G. Scalari, C. Walther, M. Fischer, R. Terazzi, H. Beere, D. Ritchie, and J. Faist, "THz and sub-THz quantum cascade lasers," Laser & Photonics Reviews **3**, 45 (2009).

[36] J. Faist, F. Capasso, D. L. Sivco, C. Sirtori, A. L. Hutchinson, and A. Y. Cho, "Quantum cascade laser," Science **264**, 553 (1994).

[37] Q. Y. Lu, Y. Bai, N. Bandyopadhyay, S. Slivken, and M. Razeghi, "2.4 W room temperature continuous wave operation of distributed feedback quantum cascade lasers," Applied Physics Letters **98**, 181106 (2011).

[38] A. Wittmann, Y. Bonetti, M. Fischer, J. Faist, S. Blaser, and E. Gini, "Distributed-feedback quantum-cascade lasers at 9 mu m operating in continuous wave up to 423 K," IEEE Photonics Technology Letters **21**, 814 (2009).

[39] A. Bismuto, S. Blaser, R. Terazzi, T. Gresch, and A. Müller, "High performance, low dissipation quantum cascade lasers across the mid-IR range," Optics Express **23**, 5477 (2015).

[40] C. Gmachl, D. Sivco, R. Colombelli, F. Capasso, and A. Cho, "Ultra-broadband semiconductor laser," Nature **415**, 883 (2002).

[41] A. Hugi, R. Terazzi, Y. Bonetti, A. Wittmann, M. Fischer, M. Beck, J. Faist, and E. Gini, "External cavity quantum cascade laser tunable from 7.6 to 11.4 μm," Applied Physics Letters **95**, 061103 (2009).

[42] A. Hugi, R. Maulini, and J. Faist, "External cavity quantum cascade laser," Semiconductor Science and Technology **25**, 083001 (2010).

[43] H. Haus, "Mode-locking of lasers," IEEE Journal of Selected Topics in Quantum Electronics **6**, 1173 (2000).

[44] A. E. Siegman, *Lasers* (University Science Books, 1986).

[45] C. G. Derntl, G. Scalari, D. Bachmann, M. Beck, J. Faist, K. Unterrainer, and J. Darmo, "Gain dynamics in a heterogeneous terahertz quantum cascade laser," Applied Physics Letters **113**, 181102 (2018).

[46] D. Walrod, S. Y. Auyang, P. A. Wolff, and M. Sugimoto, "Observation of third order optical nonlinearity due to intersubband transitions in AlGaAs/GaAs superlattices," Applied Physics Letters **59**, 2932 (1991).

[47] P. Friedli, H. Sigg, B. Hinkov, A. Hugi, S. Riedi, M. Beck, and J. Faist, "Four-wave mixing in a quantum cascade laser amplifier," Applied Physics Letters **102**, 222104 (2013).

[48] B. Hinkov, A. Bismuto, Y. Bonetti, M. Beck, S. Blaser, and J. Faist, "Singlemode quantum cascade lasers with power dissipation below 1 W," Electronics Letters **48**, 646 (2012).

[49] Y. Khanin, *Principles of lasers dynamics* (North Holland, 1995).

[50] A. Gordon, C. Y. Wang, L. Diehl, F. X. Kaertner, A. Belyanin, D. Bour, S. Corzine, G. Hoefler, H. C. Liu, H. Schneider, T. Maier, M. Troccoli, J. Faist, and F. Capasso, "Multimode regimes in quantum cascade lasers: from coherent instabilities to spatial hole burning," Physical Review A **77**, 053804 (2008).

[51] V. M. Gkortsas, C. Wang, L. Kuznetsova, L. Diehl, A. Gordon, C. Jirauschek, M. A. Belkin, A. Belyanin, F. Capasso, and F. X. Kaertner, "Dynamics of actively mode-locked quantum cascade lasers," Optics Express **18**, 13616 (2010).

[52] J. B. Khurgin, Y. Dikmelik, A. Hugi, and J. Faist, "Coherent frequency combs produced by self frequency modulation in quantum cascade lasers," Applied Physics Letters **104**, 081118 (2014).

[53] W. E. Lamb Jr, "Theory of an optical maser," Physical Review **134**, A1429 (1964).

[54] G. Villares and J. Faist, "Quantum cascade laser combs: effects of modulation and dispersion," Optics Express **23**, 1651 (2015).

[55] T. S. Mansuripur, C. Vernet, P. Chevalier, G. Aoust, B. Schwarz, F. Xie, C. Caneau, K. Lascola, C.-e. Zah, D. P. Caffey, T. Day, L. J. Missaggia, M. K. Connors, C. A. Wang, A. Belyanin, and F. Capasso, "Single-mode instability in standing-wave lasers: the quantum cascade laser as a self-pumped parametric oscillator," Physical Review A **94**, 063807 (2016).

[56] M. Singleton, P. Jouy, M. Beck, and J. Faist, "Evidence of linear chirp in mid-infrared quantum cascade lasers," Optica **5**, 948 (2018).

[57] N. Opačak and B. Schwarz, "Theory of frequency-modulated combs in lasers with spatial hole burning, dispersion, and kerr nonlinearity," Physical Review Letters **123**, 243902 (2019).

[58] J. Hillbrand, D. Auth, M. Piccardo, N. Opačak, E. Gornik, G. Strasser, F. Capasso, S. Breuer, and B. Schwarz, "In-phase and anti-phase synchronization in a laser frequency comb," Physical Review Letters **124**, 023901 (2020).

[59] M. Piccardo, B. Schwarz, D. Kazakov, M. Beiser, N. Opačak, Y. Wang, S. Jha, J. Hillbrand, M. Tamagnone, W. T. Chen, A. Y. Zhu, L. L. Columbo, A. Belyanin, and F. Capasso, "Frequency combs induced by phase turbulence," Nature **582**, 360 364 (2020).

[60] D. Burghoff, "Unraveling the origin of frequency modulated combs using active cavity mean-field theory," Optica **7**, 1781 (2020).

[61] S. E. Harris and O. P. McDuff, "Theory of FM laser oscillation," IEEE Journal of Quantum Electronics **1**, 245 (1965).

[62] A. Hugi, G. Villares, S. Blaser, H. C. Liu, and J. Faist, "Mid-infrared frequency comb based on a quantum cascade laser," Nature **492**, 229 (2012).

[63] D. Burghoff, T.-Y. Kao, N. Han, C. W. I. Chan, X. Cai, Y. Yang, D. J. Hayton, J.-R. Gao, J. L. Reno, and Q. Hu, "Terahertz laser frequency combs," Nature Photonics **8**, 462 (2014).

[64] M. Rösch, G. Scalari, M. Beck, and J. Faist, "Octave-spanning semiconductor laser," Nature Photonics **9**, 42 (2015).

[65] B. S. Williams, "Terahertz quantum-cascade lasers," Nature Photonics **1**, 517 (2007).

[66] M. Roesch, M. Beck, M. J. Suess, D. Bachmann, K. Unterrainer, J. Faist, and G. Scalari, "Heterogeneous terahertz quantum cascade lasers exceeding 1.9 THz spectral bandwidth and featuring dual comb operation," Nanophotonics **7**, 237 (2018).

[67] A. Forrer, M. Franckie, D. Stark, T. Olariu, M. Beck, J. Faist, and G. Scalari, "Photon-driven broadband emission and frequency comb RF injection locking in THz quantum cascade lasers," ACS Photonics **7**, 784 (2020).

[68] A. Forrer, L. Bosco, M. Beck, J. Faist, and G. Scalari, "RF injection of THz QCL combs at 80 K emitting over 700 GHz spectral bandwidth," Photonics, **7**, 9 (2020).

[69] D. Hofstetter, J. Faist, M. Beck, A. Muller, and U. Oesterle, "Demonstration of high-performance 10.16 μm quantum cascade distributed feedback lasers fabricated without epitaxial regrowth," Applied Physics Letters **75**, 665 (1999).

[70] G. Villares, S. Riedi, J. Wolf, D. Kazakov, M. J. Süess, P. Jouy, M. Beck, and J. Faist, "Dispersion engineering of quantum cascade laser frequency combs," Optica **3**, 252 (2016).

[71] D. Bachmann, M. Rösch, G. Scalari, M. Beck, J. Faist, K. Unterrainer, and J. Darmo, "Dispersion in a broadband terahertz quantum cascade laser," Applied Physics Letters **109**, 221107 (2016).

[72] D. Burghoff, Y. Yang, D. J. Hayton, J.-R. Gao, J. L. Reno, and Q. Hu, "Evaluating the coherence and time-domain profile of quantum cascade laser frequency combs," Optics Express **23**, 1190 (2015).

[73] F. Cappelli, G. Villares, S. Riedi, and J. Faist, "Intrinsic linewidth of quantum cascade laser frequency combs," Optica **2**, 836 (2015).

[74] N. Matuschek, F. X. Kärtner, and U. Keller, "Theory of double-chirped mirrors," IEEE Journal of Selected Topics in Quantum Electronics **4**, 197 (1998).

[75] D. Bachmann, M. Rösch, M. J. Süess, M. Beck, K. Unterrainer, J. Darmo, J. Faist, and G. Scalari, "Short pulse generation and mode control of broadband terahertz quantum cascade lasers," Optica **3**, 1087 (2016).

[76] J. Faist, G. Villares, G. Scalari, M. Rösch, C. Bonzon, A. Hugi, and M. Beck, "Quantum cascade laser frequency combs," Nanophotonics **5** (2016).

[77] F. Gires and P. Tournois, "Interferometre utilisable pour la compression d'impulsions lumineuses modulees en frequence," Comptes Rendus de l'Académie des Sciences, Paris **6112** (1964).

[78] F. Wang, H. Nong, T. Fobbe, V. Pistore, S. Houver, S. Markmann, N. Jukam, M. Amanti, C. Sirtori, S. Moumdji, R. Colombelli, L. Li, E. Linfield, G. Davies, J. Mangeney, J. Tignon, and S. Dhillon, "Short terahertz pulse generation from a dispersion compensated mode-locked semiconductor laser," Laser & Photonics Reviews **11** (2017), 10.1002/lpor.201700013.

[79] J. Hillbrand, P. Jouy, M. Beck, and J. Faist, "Tunable dispersion compensation of quantum cascade laser frequency combs," Optics Letters **43**, 1746 (2018).

[80] F. P. Mezzapesa, V. Pistore, K. Garrasi, L. Li, A. G. Davies, E. H. Linfield, S. Dhillon, and M. S. Vitiello, "Tunable and compact dispersion compensation of broadband THz quantum cascade laser frequency combs," Optics Express **27**, 20231 (2019).

[81] D. Burghoff, N. Han, F. Kapsalidis, N. Henry, M. Beck, J. Khurgin, J. Faist, and Q. Hu, "Microelectromechanical control of the state of quantum cascade laser frequency combs," Applied Physics Letters **115**, 021105 (2019).

[82] Y. Bidaux, I. Sergachev, W. Wuester, R. Maulini, T. Gresch, A. Bismuto, S. Blaser, A. Müller, and J. Faist, "Plasmon-enhanced waveguide for dispersion compensation in mid-infrared quantum cascade laser frequency combs," Optics Letters **42**, 1604 (2017).

[83] P. Jouy, J. M. Wolf, Y. Bidaux, P. Allmendinger, M. Mangold, M. Beck, and J. Faist, "Dual comb operation of λ ∼ 8.2 μm quantum cascade laser frequency comb with 1 W optical power," Applied Physics Letters **111**, 141102 (2017).

[84] Y. Bidaux, F. Kapsalidis, P. Jouy, M. Beck, and J. Faist, "Coupled-waveguides for dispersion compensation in semiconductor lasers," Laser & Photonics Reviews **12** (2018).

[85] R. Trebino, K. W. DeLong, D. N. Fittinghoff, J. N. Sweetser, M. A. Krumbügel, B. A. Richman, and D. J. Kane, "Measuring ultrashort laser pulses in the time-frequency domain using frequency-resolved optical gating," Review of Scientific Instruments **68**, 3277 (1997).

[86] J. Mandon, G. Guelachvili, and N. Picqué, "Fourier transform spectroscopy with a laser frequency comb," Nature Photonics **3**, 99 (2009).

[87] M. Wienold, B. Röben, L. Schrottke, and H. T. Grahn, "Evidence for frequency comb emission from a Fabry-Pérot terahertz quantum-cascade laser," Optics Express **22**, 30410 (2014).

[88] S. Barbieri, M. Ravaro, P. Gellie, G. Santarelli, C. Manquest, C. Sirtori, S. P. Khanna, E. H. Linfield, and A. G. Davies, "Coherent sampling of active mode-locked terahertz quantum cascade lasers and frequency synthesis," Nature Photonics **5**, 306 (2011).

[89] F. Cappelli, L. Consolino, G. Campo, I. Galli, D. Mazzotti, A. Campa, M. S. de Cumis, P. C. Pastor, R. Eramo, M. Rosch, M. Beck, G. Scalari, J. Faist, P. De Natale, and S. Bartalini, "Retrieval of phase relation and emission profile of quantum cascade laser frequency combs," Nature Photonics **13**, 562 (2019).

[90] I. Galli, F. Cappelli, P. Cancio, G. Giusfredi, D. Mazzotti, S. Bartalini, and P. De Natale, "High-coherence mid-infrared frequency comb," Optics Express **21**, 28877 (2013).

[91] A. L. Schawlow and C. H. Townes, "Infrared and optical masers," Physical Review **112**, 1940 (1958).

[92] C. H. Henry, "Theory of the linewidth of semiconductor lasers," IEEE Journal of Quantum Electronics **18**, 259 (1982).

[93] G. Villares, J. Wolf, D. Kazakov, M. J. Suess, A. Hugi, M. Beck, and J. Faist, "On-chip dual-comb based on quantum cascade laser frequency combs," Applied Physics Letters **107**, 251104 (2015).

[94] A. Bismuto, Y. Bidaux, C. Tardy, R. Terazzi, T. Gresch, J. Wolf, S. Blaser, A. Müller, and J. Faist, "Extended tuning of mid-IR quantum cascade lasers using integrated resistive heaters," Optics Express **23**, 29715 (2015).

[95] M. Rösch, G. Scalari, G. Villares, L. Bosco, M. Beck, and J. Faist, "On-chip, self-detected terahertz dual-comb source," Applied Physics Letters **108**, 171104 (2016).

[96] M. R. St-Jean, M. I. Amanti, A. Bernard, A. Calvar, A. Bismuto, E. Gini, M. Beck, J. Faist, H. C. Liu, and C. Sirtori, "Injection locking of mid-infrared quantum cascade laser at 14 GHz, by direct microwave modulation," Laser & Photonics Reviews **8**, 443 (2014).

[97] W. Maineult, L. Ding, P. Gellie, P. Filloux, C. Sirtori, S. Barbieri, T. Akalin, J. Lampin, I. Sagnes, and H. Beere, "Microwave modulation of terahertz quantum cascade lasers: a transmission-line approach," Applied Physics Letters **96**, 021108 (2010).

[98] P. Gellie, S. Barbieri, J.-F. Lampin, P. Filloux, C. Manquest, C. Sirtori, I. Sagnes, S. P. Khanna, E. H. Linfield, A. G. Davies, H. Beere, and D. Ritchie, "Injection-locking of terahertz quantum cascade lasers up to 35GHz using RF amplitude modulation," Optics Express **18**, 20799 (2010).

[99] C. Y. Wang, L. Kuznetsova, V. M. Gkortsas, L. Diehl, F. X. Kaertner, M. A. Belkin, A. Belyanin, X. Li, D. Ham, H. Schneider, P. Grant, C. Y. Song, S. Haffouz, Z. R. Wasilewski, H. C. Liu, and F. Capasso, "Mode-locked pulses from mid-infrared quantum cascade lasers," Optics Express **17**, 12929 (2009).

[100] J. Faist, F. Capasso, C. Sirtori, D. Sivco, A. Hutchinson, and A. Cho, "Laser action by tuning the oscillator strength," Nature **387**, 777 (1997).

[101] J. Hillbrand, M. Beiser, A. M. Andrews, H. Detz, R. Weih, A. Schade, S. Höfling, G. Strasser, and B. Schwarz, "Picosecond pulses from a mid-infrared interband cascade laser," Optica **6**, 1334 (2019).

[102] J. Hillbrand, N. Opacak, M. Piccardo, H. Schneider, G. Strasser, and B. Schwarz, "Mode-locked short pulses from an 8 mu m wavelength semiconductor laser," Nature Communications **11**, 5788 (2020).

[103] C. G. Derntl, G. Scalari, D. Bachmann, M. Beck, J. Faist, K. Unterrainer, and J. Darmo, "Gain dynamics in a heterogeneous terahertz quantum cascade laser," Applied Physics Letters **113**, 181102 (2018).

[104] S. Barbieri, P. Gellie, G. Santarelli, L. Ding, W. Maineult, C. Sirtori, R. Colombelli, H. Beere, and D. Ritchie, "Phase-locking of a 2.7-THz quantum cascade laser to a mode-locked erbiumdoped fibre laser," Nature Photonics **4**, 636 (2010).

[105] M. Ravaro, P. Gellie, G. Santarelli, C. Manquest, P. Filloux, C. Sirtori, J.-F. Lampin, G. Ferrari, S. P. Khanna, E. H. Linfield, H. E. Beere, D. A. Ritchie, and S. Barbieri, "Stabilization and mode locking of terahertz quantum cascade lasers," IEEE Journal of Selected Topics in Quantum Electronics **19**, (2013).

[106] C. Sirtori, S. Barbieri, and R. Colombelli, "Wave engineering with THz quantum cascade lasers," Nature Photonics **7**, 691–701 (2013).

[107] L. Consolino, M. Nafa, F. Cappelli, K. Garrasi, F. P. Mezzapesa, L. Li, A. G. Davies, E. H. Linfield, M. S. Vitiello, P. De Natale, and S. Bartalini, "Fully phase-stabilized quantum cascade laser frequency comb," Nature Communications **10** (2019).

[108] D. Oustinov, N. Jukam, R. Rungsawang, J. Madéo, S. Barbieri, P. Filloux, C. Sirtori, X. Marcadet, J. Tignon, and S. Dhillon, "Phase seeding of a terahertz quantum cascade laser," Nature Communications **1**, 6 (2010).

[109] J. R. Freeman, J. Maysonnave, N. Jukam, P. Cavalié, K. Maussang, H. E. Beere, D. A. Ritchie, J. Mangeney, S. S. Dhillon, and J. Tignon, "Direct intensity sampling of a mode-locked terahertz quantum cascade laser," Applied Physics Letters **101**, 181115 (2012).

[110] Y. Wang and A. Belyanin, "Active mode-locking of mid-infrared quantum cascade lasers with short gain recovery time," Optics Express **23**, 4173 (2015).

[111] P. Malara, R. Blanchard, T. S. Mansuripur, A. K. Wojcik, A. Belyanin, K. Fujita, T. Edamura, S. Furuta, M. Yamanishi, P. De Natale, and F. Capasso, "External ring-cavity quantum cascade lasers," Applied Physics Letters **102**, 141105 (2013).

[112] A. K. Wojcik, P. Malara, R. Blanchard, T. S. Mansuripur, F. Capasso, and A. Belyanin, "Generation of picosecond pulses and frequency combs in actively mode locked external ring cavity quantum cascade lasers," **103**, 231102 (2013).

[113] B. Meng, M. Singleton, M. Shahmohammadi, F. Kapsalidis, R. Wang, M. Beck, and J. Faist, "Mid-infrared frequency comb from a ring quantum cascade laser," Optica **7**, 162–167 (2020).

# 8 Frequency Noise and Frequency Stabilization of QCLs

Miriam Serena Vitiello,[1] Luigi Consolino,[2] and Paolo De Natale[2]

[1] Institute of Nanoscience (CNR-NANO) and Scuola Normale Superiore, Pisa, Italy
[2] National Institute of Optics (CNR-INO) and European Laboratory for Non-Linear Spectroscopy (LENS), Florence, Italy

## 8.1 Frequency Noise and Intrinsic Laser Linewidth

Quantum cascade lasers (QCLs) represent a masterpiece of electronic band structure engineering, exploiting the spectacular progress of semiconductor growth techniques to provide single atomic layer control [1]. This peculiar functionality endows QCLs with tailored properties, above all a tunable emission wavelength no more dependent on a fixed energy bandgap, as in bipolar lasers.

After their demonstration as key devices in many molecular sensing applications, QCLs recently emerged as ideal laser sources for a plethora of sophisticated applications, such as high-resolution and high-precision spectroscopy, frequency metrology, as well as for groundbreaking applications in quantum science [2], such as laser cooling and interrogation of ultracold molecules.

To address such applications, high-frequency stability is required. In this context, the knowledge of the intrinsic laser linewidth (LW) due to quantum noise plays a key role, ultimately determining the achievable spectral resolution and the coherence length.

The intrinsic LW is, by definition, a measure of the quantum-limited fluctuations ruled by the uncertainty principle. It is also called Schawlow–Townes linewidth [3] and in bipolar intraband semiconductor lasers it is typically affected by the variations of the refractive index of the semiconductor material by carrier density fluctuations, which contribute to induce a significant linewidth broadening. This effect is usually accounted for in the so-called Henry or linewidth enhancement factor ($\alpha_e$), which adds to the Schawlow–Townes linewidth inducing excess line-broadening [4, 5]. Conversely, the intersubband nature of the QCLs leads to a linewidth enhancement factor which is usually very low [6].

The QCL LW can be expressed by a slightly modified version of the Schawlow–Townes formula [3] including $\alpha_e$, which takes into account the refractive index variations with gain, caused by electron density fluctuations [5]:

$$\delta v = \frac{1}{4\pi} \frac{(1+\alpha_e^2)\gamma^2 \alpha_m (h\omega)}{2 P_{out}\alpha} n_{sp}, \tag{8.1}$$

where $\gamma = v_g \alpha$ is the cold cavity linewidth, $\alpha$ represents the total cavity losses, $\alpha_m$ is the mirror losses, $v_g = c/n_{eff}$ is the group velocity, $c$ is the light velocity in the vacuum, $n_{\text{eff}}$ is the effective refractive index, $n_{\text{sp}}$ is the population inversion, and $P_{\text{out}}$ is the output power of the laser.

Beyond its importance towards the development of ultra-narrow, metrological-grade sources, the LW is highly correlated with the carrier transport and relaxation dynamics inside the heterostructure, which can be particularly complex in QCLs, especially when their emission frequency falls in the far-infrared. Indeed, at tera-hertz frequencies, the electron energy distribution arises from the detailed balance between the input power associated with tunneling injection [7] and several energy relaxation channels, i.e., inter- and intra-subband $e$-$e$, $e$–phonon, $e$-impurity, and interface-roughness scattering [8, 9]. The interplay between the above processes can lead to a huge hot-electron distribution and to different temperatures ($T_e$) for the electronic subbands [10]. Furthermore, owing to the small energy separation ($< k_b T_e$, where $k_b$ is the Boltzmann constant) between the laser subbands, in addition to the normal non-radiative relaxation processes characterized by relaxation rates $1/\tau_{ij}$, also the reverse processes should be accounted for (the backfilling effect).

The QCL intrinsic LW is therefore deviating from the ideal delta-Dirac like function (Fig. 8.1(a)) and has to be evaluated including, on equal footing, all the mean

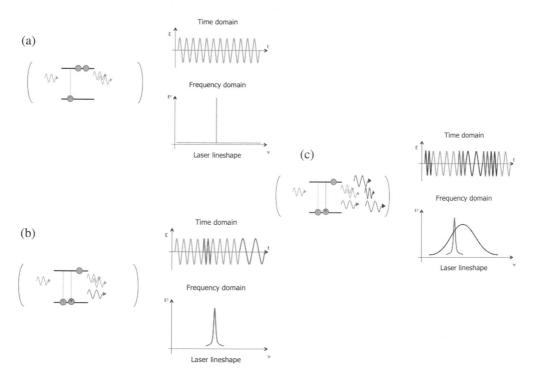

**Fig. 8.1** Laser linewidth. (a) Schematics of the ideal stimulated emission laser linewidth in the time domain and in the frequency domain; (b) deviations to the laser linewidth induced by spontaneous emission; (c) deviations to the laser linewidth induced by residual noise to external fluctuations (temperature, current, mechanical vibrations).

square fluctuations of relevant processes. One can identify, in the more general picture, five main noise sources: (i) spontaneous emission; (ii) stimulated emission of thermal (black-body) photons; (iii) absorption of thermal photons; (iv) waveguide loss of the thermal photons; and (v) thermal photon generation as the back action for the wave-guide loss (Fig. 8.1(b)). By schematizing the QCL as a three-level laser system, the intrinsic LW can be easily expressed by means of the following equation, in agreement with the prediction by a recently developed theoretical model [4]:

$$\delta\nu = \frac{1}{4\pi} \frac{(1 + \alpha_e^2)\lambda^2 \hbar\omega}{(P_{int})}\{(1 + N_{bb})[N_3/(N_3 - N_2)_{th}] + N_{bb}\}, \qquad (8.2)$$

where $N_3$ and $N_2$ represent the population of the upper and lower states, respectively, and $P_{int} = (\alpha/\alpha_m) \times 2P_{out}$ represents the photon number in the laser cavity. Here, the thermal photon population at the lasing mode, $N_{bb} = \left(\exp\left(\frac{\hbar\omega}{kT_b}\right) - 1\right)^{-1}$, is in turn highly dependent on the actual electron temperature, since the thermal photon temperature $T_b$ can be assumed equal to $T_e$ at quasi thermal equilibrium. Just above threshold one can assume that the population associated with spontaneous emission is $n_{sp} = N_3/(N_3 - N_2)_{th.} \approx 1$, meaning that the main LW broadening contribution comes from the factor $\approx 1 + 2N_{bb}$.

Other critical parameters affecting the QCL linewidth are all the environmental effects such as temperature, bias-current fluctuations, or mechanical oscillations. This means that any experimental LW measurement is typically mainly dominated by extrinsic noise (Fig. 8.1(c)). Environmental effects can be minimized by using frequency-stabilization or phase locking techniques, resulting in reduced laser LWs, mostly limited by the loop bandwidth or by the reference width of the specific experimental system [11, 12].

A clean and complete way to assess the inherent spectral purity of a QCL is the measurement of its frequency noise power spectral density (FNPSD). The latter provides, for each frequency, the amount of noise contributing to the spectral width of the laser emission by deriving the laser emission spectrum over any accessible time scale [13, 14]. It also allows calculation of the LW reduction achievable using a frequency-locking loop, once its gain/bandwidth characteristics are known, while enabling spurious noise sources to be singled out. By performing direct intensity measurements, this technique allows to retrieve information in the frequency domain by converting the laser frequency fluctuations into detectable intensity (amplitude) variations.

There are different possible approaches to convert laser frequency fluctuations in a detectable intensity signal. One option is to use, as a discriminator, the side of a Doppler-broadened molecular transition (Fig. 8.2) while the QCL frequency is stabilized at the center of the discriminator slope [2, 14, 15]. In combination with fast detection and low-noise fast-Fourier-transform acquisition, this technique can easily enable spectral measurements over seven frequency decades (from 10 Hz to 100 MHz) and 10 amplitude decades [15]. This approach has been exploited to report the first measurement of intrinsic linewidth (<1 kHz) on a cryogenically cooled mid-IR QCL [2] and soon confirmed by several other experiments [16].

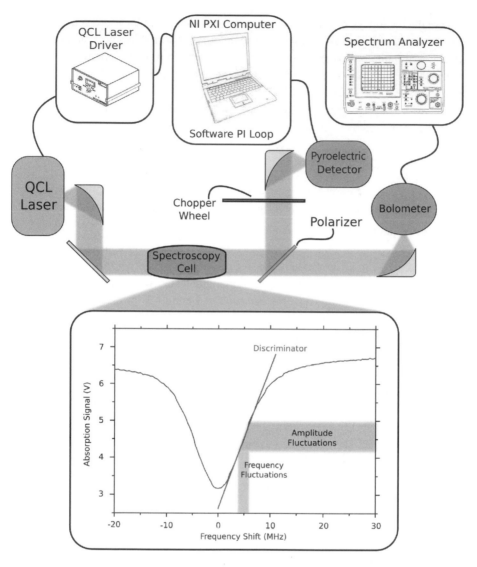

**Fig. 8.2** Schematic diagram of the experimental setup for measuring the LW using a molecular discriminator. The collimated THz QCL beam is sent to the gas cell for spectroscopy experiments. It is then split by a wire grid polarizer: the reflected beam is chopped and sent to a pyroelectric detector for the acquisition of the line profile and the frequency stabilization; the transmitted beam is acquired by means of two distinctive detectors (depending on the required bandwidth) and used for the frequency noise measurement. The discriminator function of the absorption molecular line is also shown, evidencing how frequency fluctuations are converted into detectable intensity fluctuations. Adapted from [15].

Following the first results, extensive studies have been then performed on the frequency noise of mid-infrared QCLs and its related dependence on operating conditions, opening new perspectives for design and processing of QCLs. In particular, a flicker $(1/f)$ contribution to the frequency noise has been singled out [17], and its

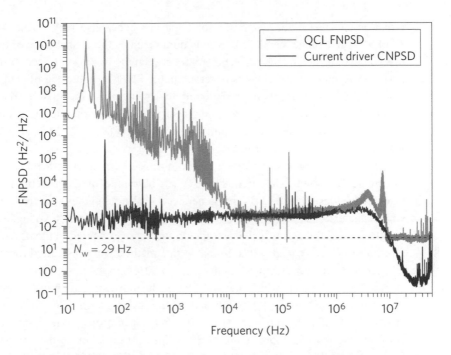

**Fig. 8.3** Frequency noise power spectral density. Experimental FNPSD of the THz QCL (gray trace), compared with the contribution to the frequency noise of the CNPSD of the current driver (black trace, starts below the QCL FNPSD trace at lower frequencies). The dashed line marks the white noise level. Adapted from [15].

dependence on the operating temperature was observed [16, 18, 19]. The presence of this noise broadens the linewidth of QCLs up to few MHz, and even larger values can be measured if no precautions on the driving current noise are taken [20]. An important step forward in the quest for narrower-linewidth QCLs was, indeed, the development of ultra-low-noise current drivers able to deliver currents of more than 1 A (and compliance voltages up to 15 V) with current-noise densities lower than $1 \, \text{nA/Hz}^{1/2}$ (relative stability better than $10^9$).

A prototypical frequency noise power spectral density measurement performed in a far-infrared QCL is shown in Fig. 8.3, which compare the FNPSD spectrum with the current-noise power spectral density (CNPSD) of the current driver, converted to the same units by using the current tuning coefficient.

Although residual external noise gives rise to the sharp peaks visible throughout the trace, it is possible to clearly recognize three distinct domains: (i) in the $f = 10 \, \text{Hz}$–10 kHz range the FNPSD is dominated by a noise not arising from the current driver and therefore ascribed to the QCL itself [14, 21]. It is worth noticing that, in a QCL, electric field fluctuations can lead to a non-negligible frequency noise contribution through the gain Stark shift [22] and the cavity mode pulling effect [23]. The prevailing frequency noise with respect to the CNPSD, absent at larger frequencies, can likely be attributed to this mechanism, as confirmed by the similar dependence observed in the amplitude noise. Additional spurious contributions of low-frequency background radiation signals and/or electronic noise may probably also play a role. (ii) In the 10 kHz–5 MHz range the FNPSD is fully dominated by the contribution of the current driver, as confirmed by the perfect overlap between the two noise traces. (iii) Above 8 MHz, an asymptotic flattening is observed in the FNPSD, with a significant deviation from CNPSD, thus suggesting a flattening to a white noise level ($N_w$). According to frequency noise theory, the laser power spectrum corresponding to the white component $N_w$ of the frequency noise is purely Lorentzian, with a FWHM $\delta\upsilon = \pi\, N_w$, from which we extracted an intrinsic LW, $\delta\upsilon = 90 \pm 30$ Hz [15].

An alternative procedure to extract the intrinsic laser linewidth relies on heterodyning the laser emission frequency with a harmonic of the repetition rate of a near-infrared laser comb [24]. This generates a beat-note in the radio frequency range whose fluctuations are then converted into a voltage signal using a voltage-controlled oscillator (VCO) that tracks the beat-note frequency. Compared to the use of a molecular transition as frequency discriminator, this method is intrinsically broadband, and allowed to retrieve comparable results ($\delta\upsilon = 250$ Hz) in the far-infrared.

Another technique, which cannot access the intrinsic LW but can be used to observ1e short-timescale LWs, is based on the investigation of the beat-note signal arising from the mixing of the QCL and a free space propagating optical frequency comb synthesizer. Such procedure involves direct observation of the spectrum of the laser optical field, after having down-converted it to a radio-frequency via heterodyne beating with a second, more stable, laser [25]. The latter approach had been recently exploited to determine the LW of a THz QCL based on intra-cavity difference frequency generation (THz DFG-QCL). The experimental set-up schematic is shown in Fig. 8.4(a) [26].

The THz frequency comb is generated in a MgO-doped lithium niobate waveguide by optical rectification in the Cherenkov configuration using a femtosecond mode-locked fiber laser emitting around 1.5 μm at a repetition rate ($f_{rep}$) adjustable around 250 MHz. The optical rectification process produces a zero-offset free-space THz comb with a spectral content broader than the THz QCL tunability range [27]. The frequency ($f_n$) of each tooth of the THz FCS can be parameterized as

$$f_n = n \times f_{rep}, \tag{8.3}$$

where $n$ is the comb tooth order.

Both the THz DFG-QCL and the reference THz comb beams were collected and collimated by a set of 90° off-axis parabolic mirrors, combined on a mylar film which acts as a THz beam splitter and are then sent to a hot electron bolometer (HEB) whose electrical signal is sent to a fast Fourier transform real-time spectrum analyzer. Within the HEB bandwidth, it is possible to detect both the signal generated by the intermodal beating of the THz comb, and the pair of beat-notes generated by heterodyning the THz DFG QCL with the closest right and left comb teeth.

Figure 8.4(b) shows one of these beat-note signals. In this case, the beating is with the right (higher-frequency) comb tooth, and the beat-note frequency ($f_b$) is given by

$$f_b = Nf_{rep} - f_{QCL}, \tag{8.4}$$

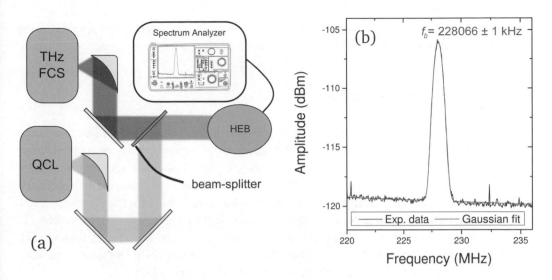

**Fig. 8.4** Experimental setup for retrieving the LW using a free space optical frequency comb synthesizer. (a) The schematic of the experimental setup: The THz frequency comb is generated in a MgO-doped lithium niobate waveguide by optical rectification with Cherenkov phase-matching of a femtosecond mode-locked fiber laser, working at 1.5 μm wavelength. The optical rectification process produces a zero-offset free-space THz comb with the repetition rate of approximately 250 MHz of the pump femtosecond laser. The THz frequency comb and the emission from the DFG-QCL were overlapped on the sensor element of the HEB with a 250 MHz electrical bandwidth. The beat-note signals of the THz DFG-QCL emission with the nearby frequency comb lines were analyzed by a FFT spectrum analyzer, having a 40 MHz real time bandwidth. (b) A typical beat-note spectrum observed on a spectrum analyzer for 2 ms integration time. The value of $f_b$ obtained by fitting is used in Eq. (8.2) for the determination of the QCL emission frequency $f_{QCL}$. Adapted from [26].

where $f_{QCL}$ is the THz DFG-QCL frequency and N is the order of the closest THz FCS tooth involved. The beat-note spectrum in Fig. 8.4(b) can be described by a Gaussian function, whose profile has been used to fit each line-shape. Since the LW of the THz comb tooth involved in the beating process (~130 Hz @ 1 s, as experimentally demonstrated) [27, 28] is negligible with respect to the DFG-QCL THz emission LW, the full width at half maximum (FWHM) of the Gaussian profiles provides an accurate quantitative estimation of the intrinsic emission LW of our device, whose upper limit was fixed to 125 kHz.

This procedure inherently implies that direct information on the laser linewidth can be retrieved for specific timescales. Conversely, the use of a molecular-lineshape discriminator is more general and contains also information about the spectral distribution of the laser's frequency fluctuations. As a consequence, it is always possible to retrieve the laser power spectrum in a specific timescale from its frequency noise power spectrum [13], while the inverse operation is not allowed.

Both the approaches described above can be used not only for retrieving quantum noise or intrinsic linewidth, but also for active stabilization of either QCL frequency or phase.

In the following, we will provide an overview on the experimental approaches for frequency stabilizing QCLs, limiting our discussion to terahertz frequencies.

## 8.2        Frequency Stabilization

The possibility to realize frequency-referenced, narrow-linewidth, high-power, terahertz frequency sources offers great perspectives for applications in many different fields. Besides their straightforward use for high-resolution and high-sensitivity molecular gas spectroscopy [14, 29] or far-infrared astronomy [30], a number of disciplines could benefit from their actual availability. In biophysics, for example, narrow and stable terahertz radiation can help understand complex dynamics problems, in principle addressing high-precision molecular recognition and protein folding [31]. Similarly, absolute-frequency-controlled, high-power sources at terahertz frequencies can be a key tool for controlling cold molecular ensembles [32].

Terahertz-emitting metrological sources, widely and continuously frequency-tunable, date back to the early 1980s [33, 34], and have recently evolved to systems based on photomixing techniques [35, 36] or difference frequency generation (DFG) [37]. They are based on either microwave or infrared standards, ensuring a very high spectral purity and an absolute frequency reference, but a drawback is their typically low emitted power, especially above 1 THz. In this context, the development of metrological-grade THz QCLs can represent an important breakthrough.

Despite their inherently high spectral purity [15, 24], which qualifies them as ideal metrological sources, free running QCLs have shown full width at half maximum linewidths in the tens of kHz range for short timescales (few milliseconds) [38, 39], while for longer integration times the linewidth can be as large as several MHz, even with stabilized temperature and low noise current drivers. For this reason, to fully exploit their potential and control QCL frequency and linewidth, frequency or phase stabilization to an external reference is needed.

The first report on a frequency stabilized THz QCL dates back to 2005, and made use of a molecular gas laser transition as local oscillator [12]. Since then, many different techniques and setups have been developed. In the following sections, we will limit our analysis to single-frequency THz QCLs.

### 8.2.1        Locking to a Molecular Transition

Soon after the invention of the laser [40], frequency stabilization techniques based on an atomic or molecular resonance serving as frequency reference, were developed [41, 42]. Since then, molecular referencing has become a well-established procedure in most parts of the electromagnetic spectrum, and has been applied to a variety of lasers operating in all regions of the electromagnetic spectrum, including the mid-infrared [43, 44]. Frequency-locking of a THz QCL to a molecular reference has been demonstrated in a first-derivative, direct-absorption spectroscopy configuration in 2010, achieving a linewidth of 300 kHz (FWHM) [45]. Soon after, linewidths of

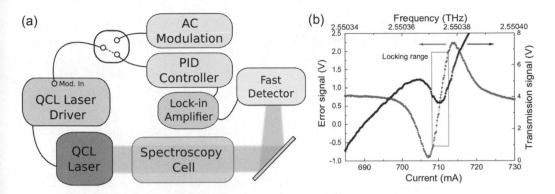

**Fig. 8.5** Locking to a molecular transition. (a) Schematic representation of the experimental setup for locking a THz QCL to a molecular transition. (b) Acquisition of a methanol absorption close to 2.55 THz in direct absorption spectroscopy ("transmission signal" data) and first derivative of the signal acquired by current modulation ("error signal" data). The gray rectangle represents the useful locking range. Adapted from [45].

18 kHz [46] and 240 kHz [47] have been demonstrated for a 3.5 THz and a 3.1 THz QCL, respectively. In all these experimental setups, the transition used as reference for locking belonged to low-pressure methanol gas, usually chosen for the large number of absorption lines around the QCL emission frequency. A schematic of the setup for laser frequency locking to a molecular transition is shown in Fig. 8.5(a). The QCL radiation is sent through a spectroscopy cell, which contains a reference gas with known spectrum, and is then detected with a fast THz detector, e.g., liquid He cooled detectors [45, 46], or a room temperature Schottky diode [47]. The QCL is supplied with a small sinusoidal current modulation superimposed on the DC driving current, so that, when the QCL frequency is swept through a molecular absorption, the first derivative of the absorption signal can be detected with a lock-in amplifier, as shown in Fig. 8.1(b). A proportional-integral-derivative (PID) controller is used to actively lock this derivative signal at zero value that corresponds to the center of the molecular transition. This is done by replacing the AC modulation of the driving current with a compensatory signal, properly adjusted by the PID parameters. Variations of the error signal are translated into frequency variations using the molecular transition as discriminator in the locking range around the center of the absorption line, where the error signal is almost linear (see Fig. 8.5(b)).

With respect to other stabilization approaches, frequency locking to a molecular absorption has got some advantages, such as the simplicity of the implementation, as it only requires a small absorption cell and an additional fast detector and the applicability for the whole THz domain, due to the rich absorption spectra of molecules. However, such a locking technique has provided spectral linewidths not better than tens of kHz, far away from the QCL intrinsic one (100 Hz) [15, 24]. Moreover, the uncertainty of the QCL absolute frequency is determined by the frequency accuracy of the reference line considered. There are not so many simple gases having transition frequencies that are known with an accuracy better than 10 kHz, e.g., CO molecular transitions in the 0.1–5.2 THz range [48], that are broadly spaced by 115 GHz. More commonly, molecules, such as, for example, the widely studied $CH_3OH$, provide a dense absorption spectrum [49, 50] with transitions having typical accuracies of a few MHz.

## 8.2.2    Locking to a Master Oscillator

A different approach for frequency stabilization relies in stabilizing a "slave" laser source via a phase lock loop (PLL) against a reference "master" oscillator, whose metrological properties are transferred to the locked laser. A PLL is a negative-feedback control system where the phase and frequency of the slave oscillator are forced to track those of the master [51]. This approach was pioneered a few years after the first demonstration of the laser, and it has been applied to a large variety of laser sources along the whole electromagnetic spectrum. In the case of THz QCLs slave lasers, the PLL loop is usually closed on the QCL driving current, for optimal fast performances, and a variety of master oscillators have been investigated, achieving dissimilar performances that will be briefly analyzed in the following paragraphs. Among the most important master oscillators we can include:

- molecular far-infrared (FIR) lasers;
- frequency stabilized IR carbon-dioxide lasers;
- frequency standards from microwave sources;
- frequency standards from optical or near-infrared sources.

### 8.2.2.1    PLL Basics

Figure 8.6 schematically shows a PLL setup that is a negative feedback system enabling electronic control of the frequency and phase of the output of a slave laser. The fields of the master laser or oscillator (MO) and the slave laser (SL) are mixed in a fast photodetector/mixer (PD), whose output signal is a down-converted beat-note of the two oscillators. This beat-note can be conveniently used to monitor the performances of the PLL. The signal is then amplified, mixed down with an "offset" radio frequency (RF) signal, filtered by a PID controller and fed back to the SL to close the loop.

If the SL and the MO have outputs $a_{SL} \cos(\omega_{SL}t + \phi_{SL}(t))$ and $a_{MO} \cos(\omega_{MO}t + \phi_{MO}(t))$ respectively, and if the RF frequency is written as $a_{RF} \sin(\omega_{RF}t + \phi_{RF}(t))$, then the output of the mixer can be written as

$$\delta\omega_{SL} = -K sin[(\omega_{MO} - \omega_{SL} \pm \omega_{RF})t + (\phi_{MO} - \phi_{SL} \pm \phi_{RF})]. \qquad (8.5)$$

This signal is sent to a phase detector and PID controller, and K contains all the gains of the various elements of the loop, from the PD responsivity to the PID parameters. This output has been written as $\delta\omega_{SL}$ because this will be the SL shift in frequency when the lock loop is activated. Once the loop is activated, in steady-state conditions, this variation will be equal to zero, leading to

$$\omega_{SL} = \omega_{MO} + \omega_{RF}, \qquad (8.6)$$

$$\phi_{SL} = \phi_{MO} + \phi_{RF} + \phi_e, \qquad (8.7)$$

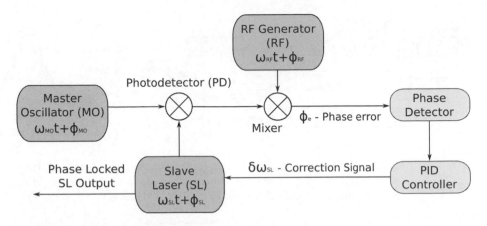

**Fig. 8.6** Schematic representation of a PLL setup.

where we have kept only the sign "+" for the RF oscillator, without loss of generality. The parameter $\phi_e$ is the steady-state phase error in the loop, and it is a consequence of the feedback current keeping locked the loop. In fact, substituting these last equations in the previous one, we get

$$\delta\omega_{SL} = -K \sin(\phi_e). \qquad (8.8)$$

The feedback loop produces a frequency shift $\delta\omega_{SL}$ that compensates the free-running frequency difference between the slave and master lasers, and its maximum value is limited by the gain of the loop. The steady-state phase error is given by

$$\phi_e \propto \sin^{-1}\left(\frac{\delta\omega_{SL}}{K}\right). \qquad (8.9)$$

It is important, therefore, that PLLs have sufficiently large gains to minimize the phase error. However, too high gains can produce self-oscillations in the SL emission frequency and should be avoided. For a properly designed PLL, $\phi_e$ is usually very small, and can be ignored in steady-state condition, meaning that the phase and frequency of the locked slave laser exactly follow those of the master oscillator, offset by the RF oscillator. For the sake of clarity, it should be noted that a perfect phase lock loop of the SL on the MO produces a very narrow beat-note signal, resolution bandwidth (RBW) limited, even at 1 Hz RBW of the spectrum analyzer instrument. This, however, does not provide information on the actual emission linewidth of the SL, indicating only that the SL spectral emission is as narrow as the MO one (supposing that the RF linewidth is much smaller than both the SL and MO ones). Moreover, the locking circuit can track the QCL frequency to the reference, meaning that the frequency of the locked SL can only be as accurate as the reference.

### 8.2.2.2 Locking to a Molecular Gas Laser
The first report on a phase-locked QCL exploited an FIR laser as MO [12]. A 3.06 THz QCL was actively referenced to the 3.1059368 THz line of a methanol gas laser, optically pumped by a $CO_2$ laser. The outputs of the QCL and the FIR laser were

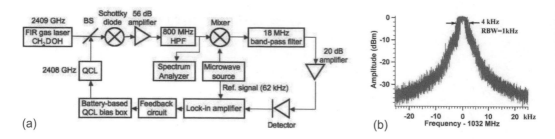

**Fig. 8.7** Locking to a molecular gas laser. (a) Experimental setup used in [52] to lock a THz QCL to the emission of a molecular FIR laser. (b) Performances of the lock loop in terms of locked beat-note signal width, with 60 s scan time (4 kHz FWHM). Adapted from [52].

combined by a wire grid polarizer and focused onto a room temperature GaAs–Schottky-diode mixer. This work actually proved that a THz QCL could be frequency controlled with conventional locking techniques, achieving a 65 kHz full width at half maximum (FWHM) long-term LW and a frequency accuracy limited by that of the reference oscillator (∼ tens of kHz). The rather large residual linewidth of the QCL was attributed to the 10 kHz bandwidth limit of the feedback loop. A similar approach, with improved signal-to-noise, was used for the frequency stabilization of a single-mode THz QCL against the 2.409293 THz line of a $CH_2DOH$ gas laser [52]. The experimental setup is reported in Fig. 8.7(a). A CW free-running THz QCL was mixed with the CW FIR laser, lasing on the 2409.293 GHz line of $CH_2DOH$, while the photodetector used was a whisker-contacted, corner-reflector-mounted Schottky barrier diode. The phase-locked beat-note (Fig. 8.7(b)), acquired for 60 seconds, shows a 4 kHz FWHM, narrower than that of the FIR reference laser, whose linewidth of 20–30 kHz is mainly due to mechanical and acoustic-induced laser cavity perturbations. This result proves that the lock circuit was able to track the reference frequency jitters, transferring the same accuracy to the THz QCL.

### 8.2.2.3   Locking to Microwave Standards

Phase locking to a reference line from a THz FIR gas laser is neither a very practical solution nor a very versatile technique since gas lasers are relatively bulky and appropriate laser lines become spare above 3 THz. Moreover, the stability of the FIR laser emission, providing tens of kHz accuracy can be much improved for other master oscillators. For these reasons, the possibility to phase lock a THz QCL to a microwave-driven harmonically generated multiplier chain source can provide significant advantages. An important advantage, in this case, is the simplicity of referencing to the primary frequency standard, which is traditionally distributed worldwide as a global positioning system (GPS) signal, and can be easily used to discipline chains of oscillators. At the same time, the linewidth of these multiplied oscillators in the THz range can be as narrow as few hundred hertz, thus significantly improving the accuracy performance of molecular FIR lasers.

The first attempts of using microwave radiation as MO were successful in 2009, when PLLs were demonstrated for QCLs emitting at 1.5 THz [53] and 2.7 THz [54].

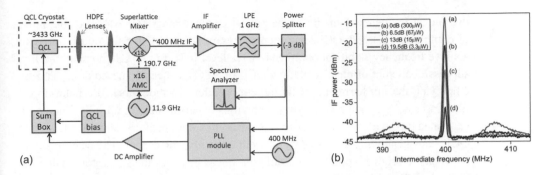

**Fig. 8.8** Locking to a microwave standard. (a) Experimental setup used in [56] to lock a THz QCL to a microwave standard. (b) Performances of the lock loop in terms of locked beat-note signal width, RBW = 100 kHz. Dependence of the lock signal on the QCL power used in the loop is also shown and it is obtained with increasing optical attenuation. Adapted from [56].

Despite the high complexity, the high electrical power consumption and the very low efficiency with a few pW of MO radiation power, such sources provided a very narrow and absolute-frequency reference, suitable for mixing with THz QCLs onto sensitive cryogenic detectors, like HEBs.

More recently, great efforts have been spent to use harmonic mixers for THz-QCL phase locking. The great advantage of these setups, apart from their room-temperature operation, is that the MO frequency does not fall in the QCL THz range, but rather in the hundreds GHz region. In fact, starting from this microwave signal, harmonic mixers are capable of internally generate a THz-frequency-stable reference signal that simultaneously down-convert the QCL frequency to the RF regime for electronic processing and QCL frequency/phase control. This approach, initially used for frequency-locking of a 2.32 THz QCL [55], was soon turned to phase-lock devices at progressively higher frequencies, as 3.4 THz and 4.7 THz [56, 57]. These two setups used a superlattice room-temperature mixer, capable of phase locking a THz QCL using only a small fraction of its emitted power. Figure 8.8(a) shows the experimental setup adopted in [56] for QCL phase locking. Note that the QCL emits at 3.433 THz, while the MO oscillates at 190.7 GHz, and therefore the 18th harmonic is used for beat-note retrieval. Figure 8.8(b) shows the performance of the PLL for different optical attenuation of the SL signal, proving that a good phase-lock is obtained also with a negligible amount (3.3 μW) of QCL power.

More recently, Schottky diodes have been also exploited as harmonic mixers for PLLs. In [58], phase locking of a 2.518 THz QCL was demonstrated using 10 μW of power, while in [59] two QCLs at 2.324 and 2.959 THz were phase locked using an optical power, delivered to the mixer, with a power as low as few hundreds nW, from a QCL.

### 8.2.2.4   Locking to a Frequency Comb

Frequency metrology has been revolutionized by the invention of Optical Frequency Comb synthesizers (FC), capable of providing a stable phase-coherent link between the optical-frequency domain and the radio-frequency range [60, 61]. The possibility of combining FCs with QCLs has been successfully exploited in the mid-infrared, to achieve frequency stabilization at tens-of-Hz level [62]. In the THz domain, the first successful attempt of phase locking a QCL to a FC has been achieved at 2.5 THz by employing the $n$th harmonic of the repetition rate of a mode-locked erbium-doped fiber laser [11], by slightly varying a previous experimental set-up [63]. The comb generated by the mode-locked laser was mixed in a nonlinear crystal with a CW THz QCL, thus generating THz sidebands around the near-IR carrier. The beating signal between the original comb and its shifted replicas provides the signal for closing the phase-lock loop. A similar scheme, but using a photoconductive antenna instead of electro-optic detection, was subsequently reported [64]. Despite the clear advantage of room-temperature detection of the beat-note, these two approaches are based on low-efficiency up-conversion processes, and they inherently require a CW THz power in the mW range. This can represent a major limitation in many practical cases, when enough output power is not available or when it is needed for the specific applications (e.g., spectroscopy).

One possible solution relies in acquiring the beat-note signal between the SL THz QCL and one mode of a free-standing THz FC, as MO [27]. This can be done by using a HEB photodetector/mixer with a very high efficiency, therefore involving only a very small fraction of the overall emitted QCL power, in the order of 100 nW. Of course, this detection involves a free standing and air-propagating THz FC, generated by optical rectification in Cherenkov configuration [65], of a femtosecond mode-locked laser in a single-mode waveguide. The waveguide can be fabricated on a MgO-doped LiNbO$_3$ crystal plate [66], and the generated radiation consists of a train of coherent THz pulses, carrying a very large spectral content (from a few GHz up to more than 6 THz). Since the pulses are identical, the comb-like spectrum of the infinite train has a perfectly zero offset, and a spacing corresponding to the repetition rate of the pump laser, which was stabilized against a Rb-GPS disciplined 10 MHz quartz oscillator. Stability in the mHz range was obtained for the repetition rate, therefore ensuring a proper stability of each tooth of the THz comb at the 100 Hz level, depending on the order of the tooth. However, it is worth mentioning that the possibility to use an optical frequency standard can, in principle, lower this value by several orders of magnitude.

Figure 8.9(b) shows the experimental setup therein used, while Fig. 8.9(b) shows the performances obtained for the active PLL, overall resulting in a $4 \times 10^{-11}$ relative accuracy in the determination of the QCL frequency. The experimental setup used for the beat-note detection in this approach is more complicated than the ones presented with microwave oscillators as MOs, requiring the use of a femtosecond mode-locked laser and of a cryogenically cooled detector. Nevertheless, it is the only one that has provided, up to now, a series of important scientific milestone applications, such as high-accuracy QCL-based spectroscopy [28], characterization of new-generation room-temperature sources [26], and characterization of THz QCL frequency combs.

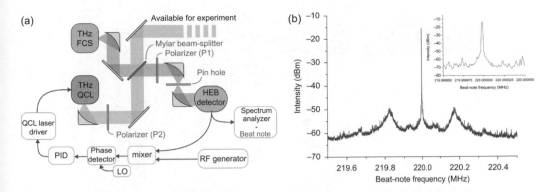

**Fig. 8.9** Locking to a frequency comb. (a) Experimental setup used in [27] to lock a THz QCL to one mode of a free-standing THz frequency comb. (b) Performances of the lock loop in terms of locked beat-note signal width, at 1 MHz span and 100 Hz RBW. The two sidebands indicate a phase-lock electronic bandwidth of about 200 kHz. (inset) Phase-locked beat-note signal acquired with a 100 Hz span and 1 Hz RBW. Adapted from [27].

### 8.2.3    Injection Locking

All the PLL loops described so far make use of a negative feedback loop with a limited loop-bandwidth to provide a feedback voltage onto the QCL. A completely different technique to achieve phase locking of a laser source is injection locking that has a simpler implementation, does not require a sensitive detector, and can compensate phase fluctuations over a wide bandwidth. Conceptually, the injection locking mechanism consists of seeding the electromagnetic field of the MO into the active cavity of the SL until their relative phase difference is constant. Injection-locked oscillators were initially studied by Adler in electronic circuits [67]. Optical injection locking was demonstrated for the first time in 1966 by Stover and Steier [68], and was extensively studied during the 1980s. In the THz region, phase locking of a QCL has been demonstrated very recently [69], using the experimental setup shown in Fig. 8.10. The QCL operates at 1.97 THz, while the MO radiation is retrieved by a FC. A wavelength selective switch selects two distinct comb lines, then individually used to injection lock a separate tunable laser, emitting at telecom wavelength. The signals from these two lasers are then combined, amplified and used both for generation and detection on a fiber-coupled CW THz photoconductive transmitter and receiver. This approach allows transferring the same high levels of frequency precision and accuracy available in the microwave region to the THz frequency domain.

### 8.3    Phase and Frequency Control of Quantum Cascade Frequency Combs

Frequency comb operation in QCLs naturally occurs via intra-cavity four-wave mixing (FWM) [70] that tends to homogenize the mode spacing and, consequently, acts as the main mode proliferation and comb generation mechanism in a free running QCL [70].

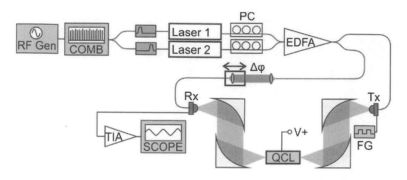

**Fig. 8.10** Experimental setup used for injection locking the THz QCL. RF signal generator (RF Gen); the wideband comb source (COMB); tunable DBR lasers, laser 1 and laser 2; polarization controllers (PC); erbium-doped fiber amplifier (EDFA); continuous-wave THz transmitter and receiver (Tx and Rx); a function generator (FG); the quantum cascade laser (QCL); transimpedance amplifier (TIA); oscilloscope (SCOPE) for data acquisition. Adapted from [69].

According to the theoretical setting in [71], the total complex field of a FC can be written as

$$E = \sum_n E_n e^{i[2\pi (f_s n + f_0)t + \phi_n]},$$

where $E_n$ is the amplitude of the $n$th comb mode, $f_s$ is the modes spacing, $f_0$ the frequency offset, and $\phi_n$ is the Fourier phase of the $n$th comb mode. The most important characteristic of a FC, which makes it different from a standard multi mode laser, is the fixed (i.e., constant over time) phase relation between the emitted optical modes, which translates into fixed Fourier phases for all the emitted modes. In a traditional passive mode-locked comb source, the presence of sharp pulses corresponds to a linear phase relation among the modes [71], but this specific trend is not guaranteed under different circumstances.

One option to assess the Fourier phases of a QCL-FC is the shifted-wave interference Fourier-transform spectroscopy (SWIFTS) [72, 73], which gives access to the phase domain by measuring the phase difference between adjacent FC modes. Such a procedure allows retrieving the phase relation of continuous portions of the FC spectrum by a cumulative sum on the phases. In SWIFTS, the need for a mechanical scan does not allow for a simultaneous analysis of the phases but, provided that the entire measurement frequency chain is properly calibrated, this technique can assess if the phases are actually stable during the measurement time.

In a more recent technique, the modal phases of a THz QCL comb are extracted through a Fourier analysis of the comb emission (FACE) [71, 74]. Relying on a multi-heterodyne detection scheme, this approach allows real-time tracing of the phases of the FC emitted modes, thus being insensitive to possible comb spectral gaps. A phase-stable reference comb overlapping with the sample comb under analysis is required for FACE, resulting in a more sophisticated set-up but providing a more general and accurate assessment of FC operation.

Phase measurements, performed with both SWIFT and FACE to assess a genuine FC operation regime in THz QCLs, proved that FWM establishes a tight phase relation among the modes emitted by the analyzed devices, resulting in stable Fourier modal phases [71–74]. The retrieved phase relations among the modes are never trivial (i.e., not simply with a linear trend, like in pulsed mode-locked combs), leading to a frequency and amplitude modulated laser emission different from that obtained by passive mode-locking pulsed operation. Nevertheless, these measurements confirm that FWM-based mode-locking is ensured at least in a limited operation range, usually very close to the laser threshold, and that QCL-based FCs are definitely well suited for most metrological grade applications. Indeed, this can only be done by tight phase referencing, to a well-suited frequency standard, both the QCL-FC modes spacing and frequency offset. These two degrees of freedom usually require two different orthogonal actuators to be stabilized while, to date, the only available fast actuator, acting simultaneously on spacing and offset, is the QCL driving current. It is worth noticing that the use of visible light as additional actuator on QCL-FC's degrees of freedom is under investigation by small optical frequency tuning (SOFT) technique [75]. SOFT has already proven its usefulness to phase stabilize QCL comb modes spacing, though non-orthogonality with respect to driving current is an issue that needs to be fixed, and that did not allow a complete phase stabilization. However, these difficulties have been overcome by exploiting the driving current actuator in two very different frequency ranges, recently [74]. Thanks to this technique, full phase stabilization of a THz QCL-FC has definitely been achieved, the emission linewidth of each comb mode being narrowed down to $\sim$2 Hz in 1 s, while metrological-grade tuning of their individual modes frequencies has been demonstrated, as shown in Fig. 8.11.

## 8.4 Perspectives

The stabilization setups described in the previous paragraphs prove that QCLs can be actively stabilized in frequency, and their emission can be narrowed down, approaching their intrinsic linewidth. However, present applications of frequency stabilized QCLs are still lagging behind with respect to other spectral regions. In fact, QCL-based spectroscopy has reached a relative accuracy of $4 \times 10^{-9}$ [28], while metrology measurements performed on atoms in the visible region can achieve a stability of few parts in $10^{-18}$ [76], and even measurements performed in the mid-IR on cold molecules can go below $10^{-15}$ [77]. This gap is partly technological in nature, and continuous efforts are being made by the scientific community to lower the present limit, with new and improved tools. For example, optical resonators are well-established tools commonly used in spectroscopy [78, 79]. They find widespread applications over the whole electromagnetic spectrum, from microwaves [80] to UV [81], while record-level optical finesse was achieved in the visible/near-IR [82], especially thanks to the advent of whispering gallery mode resonators [83]. In this context, the THz portion of the electromagnetic spectrum is still lacking such tools. Design and fabrication of cavities efficiently resonating at THz frequencies is challenging, due to the technological gap of THz materials and optical components, compared to

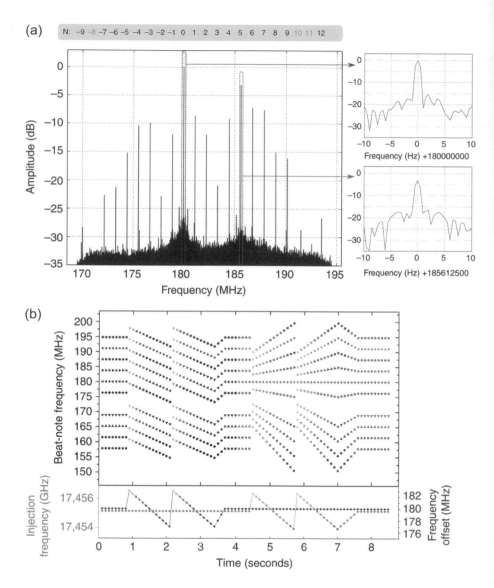

**Fig. 8.11** High-resolution comb spectra and frequency tuning. (a) RF-comb spectrum measured with 0.5 Hz RBW, limited by the acquisition time of 2 s. Insets: zoom on two RBW-limited peaks. (b) Fine tuning of the QCL-comb frequencies. First, the $f_{LO}$ frequency of the PL-loop that stabilizes the comb-offset is tuned, in steps of 500 kHz, between 183 and 177 MHz. Then, the frequency of the injecting RF-signal ($f_{RF}$) is tuned by 2.4 MHz, in steps of 200 kHz. Simultaneously, the RF-comb spectrum is acquired and the frequencies of individual modes are measured with a 100 Hz RBW.

other spectral regions. The best results have been obtained using a wire-grid-polarizer as input–output coupler, showing Q-factors in the order of $10^5$ [84, 85]. At the same time, these cavities have been for the first time used to influence the QCL emission, paving the way to cavity linewidth narrowing.

Absolute frequency references in the THz region generally rely on the Global Positioning System (GPS) for dissemination, which is ubiquitous at present, providing the most widespread time and frequency reference for the majority of industrial and research applications worldwide. On the other hand, the ultimate limits of GPS presently hinder further advances in high-precision, scientific, and industrial applications relying on this dissemination scheme. For this reason, more accurate standards are becoming the key for further progress. As a matter of facts, all the presented QCL PLL setups relying on optical- or near-infrared-based MOs would greatly benefit from a frequency standard directly delivered in the optical domain.

In this context, new optical atomic clocks [86] have already shown fractional accuracy capabilities of $10^{-18}$ in a few hours of measurement [87–89]. This is the main reason why the development and characterization of long-haul optical fiber links (OFLs), rather than microwave satellite links, recently featured a substantial thrust [90–94]. Indeed, in the last years, OFLs have shown undoubted effectiveness in comparing remote atomic clocks beyond the ultimate limit of GPS-based comparisons, obtaining precision levels not limited by the frequency transfer method [95–100], and are also envisioned as key elements for the development of a new kind of high-accuracy relativistic geodesy [99, 101].

Finally, quantum technologies represent another key area of development for QCL frequency combs. Indeed, QCLs are "quantum lasers" by-design, with a relatively simple quantum-well-based architecture, which is now the subject of analog simulation by Bose–Einstein condensates platforms, to fully exploit their quantum properties, which cannot be tailored by classical design [102]. This could be a really disruptive area of application for QCL-FCs, including novel quantum sensors, quantum imaging devices, and, in principle, q-bits made by entangled teeth for photonic-based quantum computation.

## Acknowledgements

We acknowledge financial support from the EC Project 665158 (ULTRAQCL), and the ERC Project 681379 (SPRINT).

## References

[1] Vitiello, M. S.; Scalari, G.; Williams, B.; De Natale, P. Quantum cascade lasers: 20 years of challenges. *Opt. Express* **2015**, *23*, 5167, doi:10.1364/OE.23.005167.
[2] Bartalini, S.; Borri, S.; Cancio, P.; Castrillo, A.; Galli, I.; Giusfredi, G.; Mazzotti, D.; Gianfrani, L.; De Natale, P. Observing the intrinsic linewidth of a quantum-cascade laser: Beyond the Schawlow-Townes limit. *Phys. Rev. Lett.* **2010**, *104*, 083904, doi:10.1103/PhysRevLett.104.083904.
[3] Schawlow, A. L.; Townes, C. H. Infrared and optical masers. *Phys. Rev.* **1958**, *112*, 1940–1949, doi:10.1103/PhysRev.112.1940.
[4] Yamanishi, M.; Edamura, T.; Fujita, K.; Akikusa, N.; Kan, H. Theory of the intrinsic linewidth of quantum-cascade lasers: Hidden reason for the narrow linewidth and

line-broadening by thermal photons. *IEEE J. Quantum Electron.* **2008**, *44*, 12–29, doi:10.1109/JQE.2007.907563.

[5]  Henry, C. Line broadening of semiconductor lasers. In *Coherence, Amplification, and Quantum Effects in Semiconductor Lasers*; Yamamoto, Y., Ed.; Wiley, 1991; pp. 5–76.

[6]  Green, R. P.; Xu, J. H.; Mahler, L.; Tredicucci, A.; Beltram, F.; Giuliani, G.; Beere, H. E.; Ritchie, D. A. Linewidth enhancement factor of terahertz quantum cascade lasers. *Appl. Phys. Lett.* **2008**, *92*, 071106, doi:10.1063/1.2883950.

[7]  Sirtori, C.; Capasso, F.; Faist, J.; Hutchinson, A. L.; Sivco, D. L.; Cho, A. Y. Resonant tunneling in quantum cascade lasers. *IEEE J. Quantum Electron.* **1998**, *34*, 1722–1729, doi:10.1109/3.709589.

[8]  Lee, S. C.; Wacker, A. Nonequilibrium Green's function theory for transport and gain properties of quantum cascade structures. *Phys. Rev. B – Condens. Matter Mater. Phys.* **2002**, *66*, 245314, doi:10.1103/PhysRevB.66.245314.

[9]  Callebaut, H.; Kumar, S.; Williams, B. S.; Hu, Q.; Reno, J. L. Importance of electron-impurity scattering for electron transport in terahertz quantum-cascade lasers. *Appl. Phys. Lett.* **2004**, *84*, 645–647, doi:10.1063/1.1644337.

[10]  Vitiello, M. S.; Scamarcio, G.; Spagnolo, V.; Williams, B. S.; Kumar, S.; Hu, Q.; Reno, J. L. Measurement of subband electronic temperatures and population inversion in THz quantum-cascade lasers. *Appl. Phys. Lett.* **2005**, *86*, 111115, doi:10.1063/1.1886266.

[11]  Barbieri, S.; Gellie, P.; Santarelli, G.; Ding, L.; Maineult, W.; Sirtori, C.; Colombelli, R.; Beere, H.; Ritchie, D. Phase-locking of a 2.7-THz quantum cascade laser to a mode-locked erbium-doped fibre laser. *Nat. Photonics* **2010**, *4*, 636–640, doi:10.1038/nphoton.2010.125.

[12]  Betz, A. L.; Boreiko, R. T.; Williams, B. S.; Kumar, S.; Hu, Q.; Reno, J. L. Frequency and phase-lock control of a 3 THz quantum cascade laser. *Opt. Lett.* **2005**, *30*, 1837, doi:10.1364/OL.30.001837.

[13]  Elliott, D. S.; Roy, R.; Smith, S. J. Extracavity laser band-shape and bandwidth modification. *Phys. Rev. A* **1982**, *26*, 12–18, doi:10.1103/PhysRevA.26.12.

[14]  Hübers, H. W.; Pavlov, S. G.; Richter, H.; Semenov, A. D.; Mahler, L.; Tredicucci, A.; Beere, H. E.; Ritchie, D. A. High-resolution gas phase spectroscopy with a distributed feedback terahertz quantum cascade laser. *Appl. Phys. Lett.* **2006**, *89*, 061115, doi:10.1063/1.2335803.

[15]  Vitiello, M. S.; Consolino, L.; Bartalini, S.; Taschin, A.; Tredicucci, A.; Inguscio, M.; De Natale, P. Quantum-limited frequency fluctuations in a terahertz laser. *Nat. Photonics* **2012**, *6*, 525–528, doi:10.1038/nphoton.2012.145.

[16]  Tombez, L.; Schilt, S.; Di Francesco, J.; Thomann, P.; Hofstetter, D. Temperature dependence of the frequency noise in a mid-IR DFB quantum cascade laser from cryogenic to room temperature. *Opt. Express* **2012**, *20*, 6851, doi:10.1364/OE.20.006851.

[17]  Borri, S.; Bartalini, S.; Cancio, P.; Galli, I.; Giusfredi, G.; Mazzotti, D.; Yamanishi, M.; De Natale, P. Frequency-noise dynamics of mid-infrared quantum cascade lasers. *IEEE J. Quantum Electron.* **2011**, *47*, 984–988, doi:10.1109/JQE.2011.2147760.

[18]  Tombez, L.; Di Francesco, J.; Schilt, S.; Di Domenico, G.; Faist, J.; Thomann, P.; Hofstetter, D. Frequency noise of free-running 4.6 µm distributed feedback quantum cascade lasers near room temperature. *Opt. Lett.* **2011**, *36*, 3109, doi:10.1364/OL.36.003109.

[19]  Bartalini, S.; Borri, S.; Galli, I.; Giusfredi, G.; Mazzotti, D.; Edamura, T.; Akikusa, N.; Yamanishi, M.; De Natale, P. Measuring frequency noise and intrinsic linewidth

of a room-temperature DFB quantum cascade laser. *Opt. Express* **2011**, *19*, 17996, doi:10.1364/OE.19.017996.

[20] Tombez, L.; Schilt, S.; Di Francesco, J.; Führer, T.; Rein, B.; Walther, T.; Di Domenico, G.; Hofstetter, D.; Thomann, P. Linewidth of a quantum-cascade laser assessed from its frequency noise spectrum and impact of the current driver. *Appl. Phys. B Lasers Opt.* **2012**, *109*, 407–414, doi:10.1007/s00340-012-5005-x.

[21] Wiseman, H. M. Light amplification without stimulated emission: Beyond the standard quantum limit to the laser linewidth. *Phys. Rev. A - At. Mol. Opt. Phys.* **1999**, *60*, 4083–4093, doi:10.1103/PhysRevA.60.4083.

[22] Dunbar, L. A.; Houdré, R.; Scalari, G.; Sirigu, L.; Giovannini, M.; Faist, J. Small optical volume terahertz emitting microdisk quantum cascade lasers. *Appl. Phys. Lett.* **2007**, *90*, 141114, doi:10.1063/1.2719674.

[23] Walther, C.; Scalari, G.; Beck, M.; Faist, J. Purcell effect in the inductor-capacitor laser. *Opt. Lett.* **2011**, *36*, 2623, doi:10.1364/OL.36.002623.

[24] Ravaro, M.; Barbieri, S.; Santarelli, G.; Jagtap, V.; Manquest, C.; Sirtori, C.; Khanna, S. P.; Linfield, E. H. Measurement of the intrinsic linewidth of terahertz quantum cascade lasers using a near-infrared frequency comb. *Opt. Express* **2012**, *20*, 25654, doi:10.1364/OE.20.025654.

[25] Weidmann, D.; Joly, L.; Parpillon, V.; Courtois, D.; Bonetti, Y.; Aellen, T.; Beck, M.; Faist, J.; Hofstetter, D. Free-running 9.1 μm distributed-feedback quantum cascade laser linewidth measurement by heterodyning with a $C_{18}O_2$ laser. *Opt. Lett.* **2003**, *28*, 704–706.

[26] Consolino, L.; Jung, S.; Campa, A.; De Regis, M.; Pal, S.; Kim, J. H.; Fujita, K.; Ito, A.; Hitaka, M.; Bartalini, S.; De Natale, P.; Belkin, M. A.; Vitiello, M. S. Spectral purity and tunability of terahertz quantum cascade laser sources based on intracavity difference-frequency generation. *Sci. Adv.* **2017**, *3*, e1603317, doi:10.1126/sciadv.1603317.

[27] Consolino, L.; Taschin, A.; Bartolini, P.; Bartalini, S.; Cancio, P.; Tredicucci, A.; Beere, H. E.; Ritchie, D. A.; Torre, R.; Vitiello, M. S.; De Natale, P. Phase-locking to a free-space terahertz comb for metrological-grade terahertz lasers. *Nat. Commun.* **2012**, *3*, 1040, doi:10.1038/ncomms2048.

[28] Bartalini, S.; Consolino, L.; Cancio, P.; De Natale, P.; Bartolini, P.; Taschin, A.; De Pas, M.; Beere, H.; Ritchie, D.; Vitiello, M. S.; Torre, R. Frequency-comb-assisted terahertz quantum cascade laser spectroscopy. *Phys. Rev. X* **2014**, *4*, 21006, doi:10.1103/PhysRevX.4.021006.

[29] Bellini, M.; Catacchini, E.; De Natale, P.; Di Lonardo, G.; Fusina, L.; Inguscio, M. Coherent FIR spectroscopy of molecules of atmospheric interest. *Infrared Phys. Technol.* **1995**, *36*, 37–44, doi:10.1016/1350-4495(94)00051-L.

[30] Graf, U. U.; Honingh, C. E.; Jacobs, K.; Stutzki, J. Terahertz heterodyne array receivers for astronomy. *J. Infrared, Millimeter, Terahertz Waves* **2015**, *36*, 896–921, doi:10.1007/s10762-015-0171-7.

[31] Tonouchi, M. Cutting-edge terahertz technology. *Nat. Photonics* **2007**, *1*, 97–105, doi:10.1038/nphoton.2007.3.

[32] Carr, L. D.; DeMille, D.; Krems, R. V.; Ye, J. Cold and ultracold molecules: Science, technology and applications. *New J. Phys.* **2009**, *11*, 055049, doi:10.1088/1367-2630/11/5/055049.

[33] Evenson, K. M.; Jennings, D. A.; Petersen, F. R. Tunable far-infrared spectroscopy. *Appl. Phys. Lett.* **1984**, *44*, 576–578, doi:10.1063/1.94845.

[34] Zink, L. R.; De Natale, P.; Pavone, F. S.; Prevedelli, M.; Evenson, K. M.; Inguscio, M. Rotational far infrared spectrum of $^{13}$CO. *J. Mol. Spectrosc.* **1990**, *143*, 304–310, doi:10.1016/0022-2852(91)90094-Q.

[35] Hindle, F.; Mouret, G.; Eliet, S.; Guinet, M.; Cuisset, A.; Bocquet, R.; Yasui, T.; Rovera, D. Widely tunable THz synthesizer. *Appl. Phys. B Lasers Opt.* **2011**, *104*, 763–768, doi:10.1007/s00340-011-4690-1.

[36] Mouret, G.; Hindle, F.; Cuisset, A.; Yang, C.; Bocquet, R.; Lours, M.; Rovera, D. THz photomixing synthesizer based on a fiber frequency comb. *Opt. Express* **2009**, *17*, 22031–40, doi:10.1364/OE.17.022031.

[37] De Regis, M.; Consolino, L.; Bartalini, S.; De Natale, P. Waveguided approach for difference frequency generation of broadly-tunable continuous-wave terahertz radiation. *Appl. Sci.* **2018**, *8*, 2374, doi:https://doi.org/10.3390/app8122374.

[38] Barkan, A.; Tittel, F. K.; Mittleman, D. M.; Dengler, R.; Siegel, P. H.; Scalari, G.; Ajili, L.; Faist, J.; Beere, H. E.; Linfield, E. H.; Davies, A. G.; Ritchie, D. A. Linewidth and tuning characteristics of terahertz quantum cascade lasers. *Opt. Lett.* **2004**, *29*, 575, doi:10.1364/OL.29.000575.

[39] Barbieri, S.; Alton, J.; Beere, H. E.; Linfield, E. H.; Ritchie, D. A.; Withington, S.; Scalari, G.; Ajili, L.; Faist, J. Heterodyne mixing of two far-infrared quantum cascade lasers by use of a point-contact Schottky diode. *Opt. Lett.* **2004**, *29*, 1632, doi:10.1364/OL.29.001632.

[40] Maiman, T. H. Stimulated optical radiation in Ruby. *Nature* **1960**, *187*, 493–494, doi:10.1038/187493a0.

[41] White, A. D.; Gordon, E. I.; Labuda, E. F. Frequency stabilization of single mode gas lasers. *Appl. Phys. Lett.* **1964**, *5*, 97, doi.org/10.1063/1.1754080.

[42] Lee, P. H.; Skolnick, M. L. Saturated neon absorption inside a 6238-A laser. *Appl. Phys. Lett.* **1967**, *10*, 303–305, doi:10.1063/1.1754821.

[43] Williams, R. M.; Kelly, J. F.; Hartman, J. S.; Sharpe, S. W.; Taubman, M. S.; Hall, J. L.; Capasso, F.; Gmachl, C.; Sivco, D. L.; Baillargeon, J. N.; Cho, A. Y. Kilohertz linewidth from frequency-stabilized mid-infrared quantum cascade lasers. *Opt. Lett.* **1999**, *24*, 1844, doi:10.1364/OL.24.001844.

[44] Borri, S.; Bartalini, S.; Galli, I.; Cancio, P.; Giusfredi, G.; Mazzotti, D.; Castrillo, A.; Gianfrani, L.; de Natale, P. Lamb-dip-locked quantum cascade laser for comb-referenced IR absolute frequency measurements. *Opt. Express* **2008**, *16*, 11637, doi:10.1364/OE.16.011637.

[45] Richter, H.; Pavlov, S. G.; Semenov, A. D.; Mahler, L.; Tredicucci, A.; Beere, H. E.; Ritchie, D. A.; Hübers, H. W. Submegahertz frequency stabilization of a terahertz quantum cascade laser to a molecular absorption line. *Appl. Phys. Lett.* **2010**, *96*, 71112, doi:10.1063/1.3324703.

[46] Ren, Y.; Hovenier, J. N.; Cui, M.; Hayton, D. J.; Gao, J. R.; Klapwijk, T. M.; Shi, S. C.; Kao, T. Y.; Hu, Q.; Reno, J. L. Frequency locking of single-mode 3.5-THz quantum cascade lasers using a gas cell. *Appl. Phys. Lett.* **2012**, *100*, 041111, doi:10.1063/1.3679620.

[47] Hübers, H. W.; Eichholz, R.; Pavlov, S. G.; Richter, H. High resolution terahertz spectroscopy with quantum cascade lasers. *J. Infrared, Millimeter, Terahertz Waves* **2013**, *34*, 325–341, doi:10.1007/s10762-013-9973-7.

[48] Varberg, T. D.; Evenson, K. M. Accurate far-infrared rotational frequencies of carbon monoxide. *Astrophys. J.* **1992**, *385*, 763, doi:10.1086/170983.

[49] Moruzzi, G.; Strumia, F.; Carnesecchi, P.; Carli, B.; Carlotti, M. High resolution spectrum of $CH_3OH$ between 8 and $100cm^{-1}$. *Infrared Phys.* **1989**, *29*, 47–86, doi:10.1016/0020-0891(89)90008-0.

[50] Moruzzi, G.; Riminucci, P.; Strumia, F.; Carli, B.; Carlotti, M.; Lees, R. M.; Mukhopadhyay, I.; Johns, J. W. C.; Winnewisser, B. P.; Winnewisser, M. The spectrum of $CH_3OH$ between 100 and 200 $cm^{-1}$: Torsional and "forbidden" transitions. *J. Mol. Spectrosc.* **1990**, *144*, 139–200, doi:10.1016/0022-2852(90)90313-F.

[51] Enloe, L. H.; Rodda, J. L. Laser phase-locked loop. *Proc. IEEE* **1965**, *53*, 165–166, doi:10.1109/PROC.1965.3585.

[52] Danylov, A. A.; Goyette, T. M.; Waldman, J.; Coulombe, M. J.; Gatesman, A. J.; Giles, R. H.; Goodhue, W. D.; Qian, X.; Nixon, W. E. Frequency stabilization of a single mode terahertz quantum cascade laser to the kilohertz level. *Opt. Express* **2009**, *17*, 7525, doi:10.1364/OE.17.007525.

[53] Rabanus, D.; Graf, U. U.; Phillips, M. C.; Ricken, O.; Stutzki, J.; Vowinkel, B.; Wiedner, M. C.; Walther, C.; Fischer, M.; Faist, J. Phase locking of a 1.5 Terahertz quantum cascade laser and use as a local oscillator in a heterodyne HEB receiver. *Opt. Express* **2009**, *17*, 1159, doi:10.1364/OE.17.001159.

[54] Khosropanah, P.; Baryshev, A.; Zhang, W.; Jellema, W.; Hovenier, J. N.; Gao, J. R.; Klapwijk, T. M.; Paveliev, D. G.; Williams, B. S.; Kumar, S.; Hu, Q.; Reno, J. L.; Klein, B.; Hesler, J. L. Phase locking of a 2.7 THz quantum cascade laser to a microwave reference. *Opt. Lett.* **2009**, *34*, 2958, doi:10.1364/OL.34.002958.

[55] Danylov, A. A.; Light, A. R.; Waldman, J.; Erickson, N. R.; Qian, X.; Goodhue, W. D. 2.32 THz quantum cascade laser frequency-locked to the harmonic of a microwave synthesizer source. *Opt. Express* **2012**, *20*, 27908, doi:10.1364/OE.20.027908.

[56] Hayton, D. J.; Khudchencko, A.; Pavelyev, D. G.; Hovenier, J. N.; Baryshev, A.; Gao, J. R.; Kao, T. Y.; Hu, Q.; Reno, J. L.; Vaks, V. Phase locking of a 3.4 THz third-order distributed feedback quantum cascade laser using a room-temperature superlattice harmonic mixer. *Appl. Phys. Lett.* **2013**, *103*, doi:10.1063/1.4817319.

[57] Khudchenko, A. V.; Hayton, D. J.; Pavelyev, D. G.; Baryshev, A. M.; Gao, J. R.; Kao, T. Y.; Hu, Q.; Reno, J. L.; Vaks, V. L. Phase locking a 4.7 THz quantum cascade laser using a super-lattice diode as harmonic mixer. In *International Conference on Infrared, Millimeter, and Terahertz Waves, IRMMW-THz*; 2014.

[58] Bulcha, B. T.; Hesler, J. L.; Valavanis, A.; Drakinskiy, V.; Stake, J.; Dong, R.; Zhu, J. X.; Dean, P.; Li, L. H.; Davies, A. G.; Linfield, E. H.; Barker, N. S. Phase locking of a 2.5 THz quantum cascade laser to a microwave reference using THz Schottky mixer. In *IRMMW-THz 2015 – 40th International Conference on Infrared, Millimeter, and Terahertz Waves*; 2015.

[59] Danylov, A.; Erickson, N.; Light, A.; Waldman, J. Phase locking of 2.324 and 2.959 terahertz quantum cascade lasers using a Schottky diode harmonic mixer. *Opt. Lett.* **2015**, *40*, 5090, doi:10.1364/OL.40.005090.

[60] Udem, T.; Holzwarth, R.; Hänsch, T. W. Optical frequency metrology. *Nature* **2002**, *416*, 233–237, doi:10.1038/416233a.

[61] Maddaloni, P.; Cancio, P.; De Natale, P. Optical comb generators for laser frequency measurement. *Meas. Sci. Technol.* **2009**, *20*, 20, 1–19, doi:10.1088/0957-0233/20/5/052001.

[62] Argence, B.; Chanteau, B.; Lopez, O.; Nicolodi, D.; Abgrall, M.; Chardonnet, C.; Daussy, C.; Darquié, B.; Le Coq, Y.; Amy-Klein, A. Quantum cascade laser

frequency stabilization at the sub-Hz level. *Nat. Photonics* **2015**, *9*, 456–460, doi:10.1038/nphoton.2015.93.

[63]  Löffler, T.; May, T.; Weg, C. A.; Alcin, A.; Hils, B.; Roskos, H. G. Continuous-wave terahertz imaging with a hybrid system. *Appl. Phys. Lett.* **2007**, *90*, 091111, doi:10.1063/1.2711183.

[64]  Ravaro, M.; Manquest, C.; Sirtori, C.; Barbieri, S.; Santarelli, G.; Blary, K.; Lampin, J.-F.; Khanna, S. P.; Linfield, E. H. Phase-locking of a 2.5 THz quantum cascade laser to a frequency comb using a GaAs photomixer. *Opt. Lett.* **2011**, *36*, 3969, doi:10.1364/OL.36.003969.

[65]  Askar'yan, G. A. Cerenkov and transition radiation from electromagnetic waves. *Sov. Phys. JETP* **1962**, *15*, 943–946.

[66]  Bodrov, S. B.; Stepanov, A. N.; Bakunov, M. I.; Shishkin, B. V.; Ilyakov, I. E.; Akhmedzhanov, R. A. Highly efficient optical-to-terahertz conversion in a sandwich structure with $LiNbO_3$ core. *Opt. Express* **2009**, *17*, 1871–1879, doi:10.1364/OE.17.001871.

[67]  Adler, R. A Study of locking phenomena in oscillators. *Proc. IRE* **1946**, *34*, 351–357, doi:10.1109/JRPROC.1946.229930.

[68]  Steier, W. H.; Stover, H. L. Locking of laser oscillators by light injection. *IEEE J. Quantum Electron.* **1966**, *2*, 111–112, doi:10.1109/JQE.1966.1073970.

[69]  Freeman, J. R.; Ponnampalam, L.; Shams, H.; Mohandas, R. A.; Renaud, C. C.; Dean, P.; Li, L.; Giles Davies, A.; Seeds, A. J.; Linfield, E. H. Injection locking of a terahertz quantum cascade laser to a telecommunications wavelength frequency comb. *Optica* **2017**, *4*, 1059, doi:10.1364/OPTICA.4.001059.

[70]  Khurgin, J.; Dikmelik, Y.; Hugi, A.; Faist, J. Coherent frequency combs produced by self frequency modulation in quantum cascade lasers. *Appl. Phys. Lett.* **2014**, *104*, 081118, doi:10.1063/1.4866868.

[71]  Cappelli, F.; Consolino, L.; Campo, G.; Galli, I.; Mazzotti, D.; Campa, A.; Siciliani de Cumis, M.; Cancio Pastor, P.; Eramo, R; Rösch, M.; Beck, M.; Scalari, G.; Faist, J.; De Natale, P.; Bartalini, S. Retrieval of phase relation and emission profile of quantum cascade laser frequency combs. *Nat. Photonics* **2019**, *13*, 562–568, doi:10.1038/s41566-019-0451-1.

[72]  Burghoff, D.; Yang, Y.; Hayton, D. J.; Gao, J.-R.; Reno, J. L.; Hu, Q. Evaluating the coherence and time-domain profile of quantum cascade laser frequency combs. *Opt. Express* **2015**, *23*, 1190, doi:10.1364/OE.23.001190.

[73]  Han, Z.; Ren, D.; Burghoff, D. Sensitivity of SWIFT spectroscopy, *Opt. Express* **2020**, *28*, 6002, doi:10.1364/OE.382243.

[74]  Consolino, L.; Nafa, M.; Cappelli, F.; Garrasi, K.; Mezzapesa, F. P.; Li, L.; Davies, A. G.; Linfield, A. G.; Vitiello, M. S.; De Natale, P.; Bartalini, S. Fully phase-stabilized quantum cascade laser frequency comb. *Nat. Commun.* **2019**, *10*, doi:10.1038/s41467-019-10913-7.

[75]  Consolino, L.; Campa, A.; De Regis, M.; Cappelli, F.; Scalari, G.; Faist, J.; Beck, M.; Rösch, M.; Bartalini, S.; De Natale, P. Controlling and phase-locking a THz quantum cascade laser frequency comb by small optical frequency tuning. *Laser Photonics Rev.* **2021** *15*, 2000417, doi:10.1002/lpor.202000417.

[76]  Nicholson, T. L.; Campbell, S. L.; Hutson, R. B.; Marti, G. E.; Bloom, B. J.; McNally, R. L.; Zhang, W.; Barrett, M. D.; Safronova, M. S.; Strouse, G. F.; Tew, W. L.; Ye, J.

Systematic evaluation of an atomic clock at 2 × 10-18 total uncertainty. *Nat. Commun.* **2015**, *6*, 6896, doi:10.1038/ncomms7896.

[77]  Borri, S.; Santambrogio, G. Laser spectroscopy of cold molecules. *Adv. Phys. X* **2016**, *1*, 368–386, doi:10.1080/23746149.2016.1203732.

[78]  Haroche, S. Controlling photons in a box and exploring the quantum to classical boundary. *Ann. Phys.* **2013**, *525*, 753–776.

[79]  Gagliardi, G.; Loock, H.-P. *Cavity-Enhanced Spectroscopy and Sensing*; Springer: Berlin, 2014.

[80]  Kuhr, S.; Gleyzes, S.; Guerlin, C.; Bernu, J.; Hoff, U. B.; Deléglise, S.; Osnaghi, S.; Brune, M.; Raimond, J. M.; Haroche, S.; Jacques, E.; Bosland, P.; Visentin, B. Ultrahigh finesse Fabry-Pérot superconducting resonator. *Appl. Phys. Lett.* **2007**, *90*, doi:10.1063/1.2724816.

[81]  Gary, G. A.; West, E. A. E.; Rees, D.; McKay, J. A.; Zukic, M.; Herman, P. Solar CIV vacuum-ultraviolet Fabry-Pérot interferometers. *Astron. Astrophys.* **2007**, *461*, 707–722, doi:10.1093/ije/26.1.224.

[82]  Rempe, G.; Lalezari, R.; Thompson, R. J.; Kimble, H. J. Measurement of ultralow losses in an optical interferometer. *Opt. Lett.* **1992**, *17*, 363, doi:10.1364/OL.17.000363.

[83]  Savchenkov, A. A.; Matsko, A. B.; Ilchenko, V. S.; Maleki, L. Optical resonators with ten million finesse. *Opt. Express* **2007**, *15*, 6768, doi:10.1364/OE.15.006768.

[84]  Campa, A.; Consolino, L.; Ravaro, M.; Mazzotti, D.; Vitiello, M. S.; Bartalini, S.; De Natale, P. High-Q resonant cavities for terahertz quantum cascade lasers. *Opt. Express* **2015**, *23*, 3751, doi:10.1364/OE.23.003751.

[85]  Consolino, L.; Campa, A.; Mazzotti, D.; Vitiello, M. S.; De Natale, P., Bartalini, S. Bow-tie cavity for terahertz radiation. *Photonics* **2018**, *6*, 1, doi:10.3390/photonics6010001.

[86]  Ludlow, A. D.; Boyd, M. M.; Ye, J.; Peik, E.; Schmidt, P. O. Optical atomic clocks. *Rev. Mod. Phys.* **2015**, *87*, 637.

[87]  Ushijima, I.; Takamoto, M.; Das, M.; Ohkubo, T.; Katori, H. Cryogenic optical lattice clocks. *Nat. Photonics* **2015**, *9*, 185–189, doi:10.1038/nphoton.2015.5.

[88]  Bloom, B. J.; Nicholson, T. L.; Williams, J. R.; Campbell, S. L.; Bishof, M.; Zhang, X.; Zhang, W.; Bromley, S. L.; Ye, J. An optical lattice clock with accuracy and stability at the 10-18 level. *Nature* **2014**, *506*, 71–75, doi:10.1038/nature12941.

[89]  Hinkley, N.; Sherman, A.; Phillips, N. B.; Schioppo, M.; Lemke, N. D.; Beloy, K.; Pizzocaro, M.; Oates, C. W.; Ludlow, A. D. An atomic clock with $10^{-18}$ instability. *Science (80-. ).* **2013**, *341*, 1215–1218, doi:10.1126/science.1240420.

[90]  Calonico, D.; Inguscio, M.; Levi, F. Light and the distribution of time. *Epl* **2015**, *110*, 40001, doi:10.1209/0295-5075/110/40001.

[91]  Droste, S.; Ozimek, F.; Udem, T.; Predehl, K.; Hänsch, T. W.; Schnatz, H.; Grosche, G.; Holzwarth, R. Optical-frequency transfer over a single-span 1840 km fiber link. *Phys. Rev. Lett.* **2013**, *111*, 110801, doi:10.1103/PhysRevLett.111.110801.

[92]  Calonico, D.; Bertacco, E. K.; Calosso, C. E.; Clivati, C.; Costanzo, G. A.; Frittelli, M.; Godone, A.; Mura, A.; Poli, N.; Sutyrin, D. V.; Tino, G.; Zucco, M. E.; Levi, F. High-accuracy coherent optical frequency transfer over a doubled 642-km fiber link. *Appl. Phys. B Lasers Opt.* **2014**, *117*, 979–986, doi:10.1007/s00340-014-5917-8.

[93]  Lopez, O.; Haboucha, A.; Chanteau, B.; Chardonnet, C.; Amy-Klein, A.; Santarelli, G. Ultra-stable long distance optical frequency distribution using the Internet fiber network. *Opt. Express* **2012**, *20*, 23518, doi:10.1364/OE.20.023518.

[94]  Fujieda, M.; Kumagai, M.; Nagano, S.; Yamaguchi, A.; Hachisu, H.; Ido, T. All-optical link for direct comparison of distant optical clocks. *Opt. Express* **2011**, *19*, 16498, doi:10.1364/OE.19.016498.

[95]  Yamaguchi, A.; Fujieda, M.; Kumagai, M.; Hachisu, H.; Nagano, S.; Li, Y.; Ido, T.; Takano, T.; Takamoto, M.; Katori, H. Direct comparison of distant optical lattice clocks at the $10^{-16}$ uncertainty. *Appl. Phys. Express* **2011**, *4*, 082203, doi:10.1143/APEX.4.082203.

[96]  Matveev, A.; Parthey, C. G.; Predehl, K.; Alnis, J.; Beyer, A.; Holzwarth, R.; Udem, T.; Wilken, T.; Kolachevsky, N.; Abgrall, M.; Rovera, D.; Salomon, C.; Laurent, P.; Grosche, G.; Terra, O.; Legero, T.; Schnatz, H.; Weyers, S.; Altschul, B.; Hänsch, T. W. Precision measurement of the hydrogen 1S-2S frequency via a 920-km fiber link. *Phys. Rev. Lett.* **2013**, *110*, 230801, doi:10.1103/PhysRevLett.110.230801.

[97]  Clivati, C.; Costanzo, G. A.; Frittelli, M.; Levi, F.; Mura, A.; Zucco, M.; Ambrosini, R.; Bortolotti, C.; Perini, F.; Roma, M.; Calonico, D. A coherent fiber link for very long baseline interferometry. *IEEE Trans. Ultrason. Ferroelectr. Freq. Control* **2015**, *62*, 1907–1912, doi:10.1109/TUFFC.2015.007221.

[98]  Droste, S.; Grebing, C.; Leute, J.; Raupach, S. M. F.; Matveev, A.; Hänsch, T. W.; Bauch, A.; Holzwarth, R.; Grosche, G. Characterization of a 450 km baseline GPS carrier-phase link using an optical fiber link. *New J. Phys.* **2015**, *17*, 083044, doi:10.1088/1367-2630/17/8/083044.

[99]  Lisdat, C.; Grosche, G.; Quintin, N.; Shi, C.; Raupach, S. M. F.; Grebing, C.; Nicolodi, D.; Stefani, F.; Al-Masoudi, A.; Dörscher, S.; Häfner, S.; Robyr, J. L.; Chiodo, N.; Bilicki, S.; Bookjans, E.; Koczwara, A.; Koke, S.; Kuhl, A.; Wiotte, F.; Meynadier, F.; Camisard, E.; Abgrall, M.; Lours, M.; Legero, T.; Schnatz, H.; Sterr, U.; Denker, H.; Chardonnet, C.; Le Coq, Y.; Santarelli, G.; Amy-Klein, A.; Le Targat, R.; Lodewyck, J.; Lopez, O.; Pottie, P. E. A clock network for geodesy and fundamental science. *Nat. Commun.* **2016**, *7*, 12443, doi:10.1038/ncomms12443.

[100]  Morzynski, P.; Bober, M.; Bartoszek-Bober, D.; Nawrocki, J.; Krehlik, P.; Sliwczynski, L.; Lipinski, M.; Maslowski, P.; Cygan, A.; Dunst, P.; Garus, M.; Lisak, D.; Zachorowski, J.; Gawlik, W.; Radzewicz, C.; Ciurylo, R.; Zawada, M. Absolute measurement of the 1S0-3P0 clock transition in neutral 88Sr over the 330 km-long stabilized fibre optic link. *Sci. Rep.* **2015**, *5*, 17495, doi:10.1038/srep17495.

[101]  Bjerhammar, A. On a relativistic geodesy. *Bull. Géodésique* **1985**, *59*, 207–220, doi:10.1007/BF02520327.

[102]  Trombettoni, A.; Scazza, F.; Minardi, F.; Roati, G.; Cappelli, F.; Consolino, L.; Smerzi, A.; De Natale, P. Quantum simulating the electron transport in quantum cascade laser structures, *Adv. Quantum Technol.* **2021**, *4*, 2100044, doi:10.1002/qute.202100044.

# 9 Distributed-Feedback and Beam Shaping in Monolithic Terahertz QCLs

Yuan Jin and Sushil Kumar

Lehigh University, Pennsylvania

## 9.1 Introduction

Terahertz quantum cascade lasers (QCLs), after 20 years of research and development, are now widely considered as the most promising compact solid-state sources of coherent terahertz radiation with reasonable optical output power, albeit with the requirement of cryogenic or, more recently, thermoelectric-cooler operating temperatures [1–3]. High-power terahertz QCLs are highly desired for a multitude of applications in non-destructive imaging [4–7], and spectroscopic sensing [8–11] in fields as diverse as astronomy, medicine, security, pharmaceuticals, and so forth. Especially, some applications such as heterodyne spectroscopy [12–14] for molecular sensing inherently require high-frequency stability and extremely narrow linewidths, which are only available from narrow-band sources such as terahertz QCLs.

The best performance at practical operating temperatures, output powers, and spectral coverage for single-mode terahertz QCLs is realized with metal–metal cavities, due to strong confinement of the optical mode within the QCL active-core regions of such cavities [15, 16]. However, the subwavelength cross-section of the radiating apertures in metallic Fabry–Pérot resonant cavities leads to highly non-directional far-field radiation patterns and poor radiative efficiencies from such QCLs [17]. Additionally, conventional Fabry–Pérot terahertz QCLs with metallic cavities easily excite multiple longitudinal and lateral cavity modes leading to multi-mode operation that is difficult to control by choosing cavities of different dimensions. Hence, the development of distributed-feedback (DFB) techniques to improve spectral as well as modal properties becomes indispensable for terahertz QCLs, in order to address targeted applications that typically require single-mode operation, frequency stability and specificity as well as optimal far-field beam quality with single-lobed profile and low angular divergence. This chapter describes the theory, design methodologies, and key results from a sampling of a wide variety of DFB techniques that have been implemented in literature for monolithic terahertz QCLs with metallic cavities in both edge-emitting and surface-emitting configurations, either of which have their specific application areas and advantages much like those for infrared diode lasers.

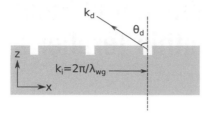

**Fig. 9.1** Distributed-feedback (DFB) with periodic gratings in a one-dimensional optical waveguide. The schematic depicts Bragg diffraction due to periodic gratings implemented along the $x$ dimension in a ridge-type waveguide. $k_i$ is the wavevector of an incident guided wave and $k_d$ is the wavevector of a wave that is generated due to diffraction from the periodic grating. Reproduced from [53].

## 9.2     Edge-Emitting Single-Mode Terahertz QCLs

DFB can be established in an optical cavity by introducing a periodic perturbation in the active medium and/or the claddings of the waveguide in the cavity, which causes Bragg diffraction of the guided optical modes in the waveguide up to multiple orders. Diffraction optically couples counter-propagating waves in the waveguide and establishes a specific resonant mode spectrum in the cavity, which is also referred to as the photonic bandstructure of the optical cavity. For a ridge-type cavity in which the mode propagates in a single (longitudinal) dimension, Eq. (9.1) describes the momentum conservation relation between the wavevectors of the incident guided wave inside the cavity, $k_i \approx 2\pi/\lambda_{wg} = 2\pi n_{eff}/\lambda$ (where $\lambda_{wg}$ is the wavelength inside the waveguide, $\lambda$ is the free-space wavelength, and $n_{eff}$ is the effective index of propagation along the direction of propagation in the longitudinal cavity), and that of the diffracted wave, $k_d$, which could be outside or inside the cavity at any angle $\theta_d$ (as defined with respect to the surface-normal direction of the cavity). This is also represented schematically in Fig. 9.1.

$$n\frac{2\pi}{\Lambda} = k_i + k_d \sin(\theta_d). \qquad (9.1)$$

Here $\Lambda$ is the periodicity of the grating in the cavity, $2\pi/\Lambda$ is the grating wavevector, and $n$ is an integer ($= \pm1, \pm2, \pm3 \ldots$) that specifies the diffraction order. From this equation, it can be concluded that there can be back reflection that results in distributed feedback such that $\theta_d = \pi/2$ and $k_d = k_i$, due to $n$th-order diffraction when $n(\pi/\Lambda) = k_i$. This condition could also be written as $\Lambda = n\lambda/(2n_{eff})$, which is the grating equation for the so-called $n$th-order one-dimensional DFB grating in an optical cavity.

### 9.2.1     First-Order DFB QCLs: Robust Single-Mode Operation but Divergent Beams

For mid-infrared QCLs single-mode operation is typically accomplished using DFB gratings that introduce index or gain/loss modulation in the waveguides [18–20]. This is typically achieved by etching into the dielectric cladding layers and/or the super-lattice gain region, sometimes in combination with a grating patterned into the top

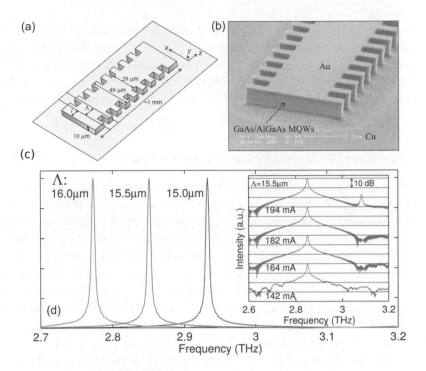

**Fig. 9.2** First-order DFB implemented with lateral corrugations in a ridge-cavity for a terahertz QCL [21]. (a) Schematic showing a metallic QCL cavity with lateral periodic corrugations for first-order DFB. (b) Scanning electron micrograph of a fractional length of the DFB cavity with dry-etched longitudinal facets and lateral corrugations. (c) Spectra from different DFB QCLs of different grating period $\Lambda$, taken at ∼8 K. The inset shows spectra taken at multiple biases from a specific QCL. Reprinted with permission from [21] © The Optical Society.

metal-contact layer of the cavity. For terahertz QCLs with metallic cavities, lateral gratings in the form of corrugations on sidewalls were utilized to implement first-order DFB, which allows for easy lithographic implementation when using dry-etching techniques for QCL fabrication [21]. The metal–metal waveguides were fabricated using the Cu-Cu thermocompression wafer-bonding technique. Following wafer bonding and substrate removal, corrugated Ti/Au metal contacts (20/500 µm) were defined on the 10-µm-thick epitaxial active region using contact lithography. The corrugated ridge waveguides were then defined by dry etching by using the top metallic contact as a self-aligned etch mask, as shown in Figs. 9.2(a) and (b). Spectra under continuous-wave (CW) operation taken at moderate injection currents at a heat-sink temperature of 8 K, from three devices with $\Lambda = $ 15.0 µm, 15.5 µm, and 16.0 µm, respectively, are shown in Fig. 9.2(c). These QCLs radiated at 2.93 THz, 2.85 THz, and 2.78 THz, respectively, which scale well with the grating period. While this technique allowed for ease of implementation of DFB with robust single-mode operation at the desired frequencies, the power output was low, even when compared to the already low output power from Fabry–Pérot cavity terahertz QCLs. This is due to negligible intensity of the electric field in the standing wave corresponding to the resonant DFB optical mode at the end-facets of the cavity, from where the cavity radiates. Second, and perhaps more important, the radiation pattern from such DFB cavities is highly divergent, similar to that from edge-emitting, metal–metal Fabry–Pérot cavity terahertz QCLs [22]. These poor radiative characteristics render such a first-order DFB technique less useful, which prompted further research on new DFB techniques that could improve the far-field radiation patterns and enhance the output power of single-mode terahertz QCLs.

## 9.2.2    Third-Order DFB Method for Low-Divergence Beams

In order to simultaneously achieve a low beam divergence and single-mode operation of edge-emitting terahertz QCLs with metallic cavities, an intuitive idea is to establish a strong surface-plasmon-polariton (SPP) mode at the top-metal layer that decays in vacuum/air (the surrounding medium of the cavity). The effective size of the wavefront of the resonant-optical mode for the DFB cavity would thus be considerably enhanced. This leads to a much narrower far-field beam profile, which is the Fourier transform of the source field as per the Fraunhofer-diffraction condition. Such an idea is unique to plasmonic lasers, and practical implementation is possible for terahertz QCLs due to their very long wavelengths. Two such schemes that effectively realize such an operating condition for terahertz DFB QCLs to achieve narrow-beam emission, namely the third-order DFB [23] and the antenna-feedback methods [24, 25], will now be described. The third-order DFB method is discussed first, for which the description of its operation is different from the treatment offered in [23] and is arguably easier to understand.

By considering a waveguide mode of the laser cavity with a propagation constant $k_i$, the wavevector of the SPP mode propagating in free space or above the top metallic layer, $k_{air}$ would be $\sim 2\pi/\lambda \sim k_i/n_{eff}$ (where $n_{eff}$ represents the effective propagation index inside the optical cavity) since the SPP wave travels in air. For a third-order DFB, as per Eq. (9.1), the grating period is set to be $\Lambda = 3\pi/k_i$. From Eq. (9.1), it can be calculated that for such a periodicity, the third-order Bragg diffraction for $n = 3$ results in the diffracted wavevector being the same as the incident wavevector in the opposite direction inside the cavity, $k_d = k_i$, which results in establishing a resonant optical mode due to DFB action. However, the second-order Bragg diffraction for $n = 2$ can satisfy a momentum conservation relation for $k_d = k_i/3$. If the optical cavity could be designed to achieve $n_{eff} = 3$ for the propagation index, which is possible through innovative cavity designs as will be shown next, a phase condition for which second-order Bragg diffraction is outside the cavity with $k_d = k_{air}$ is possible, and will establish a strong SPP mode in air on top of the cavity, as is desired for narrow-beam emission. Since the index of the GaAs-based active medium in terahertz QCLs is $\sim 3.6$, the use of slit-like gratings in the top-metal cladding by itself is not enough to satisfy the required phase-matching condition of $n_{eff} \sim 3$ since the refractive index of the active core is very different ($\sim 3.6$).

Finite-element (FEM) simulations are performed to comprehensively understand the characteristics of such metallic third-order Bragg gratings for terahertz QCLs. Figure 9.3 shows how the electric-field profiles look for lower and upper band-edge

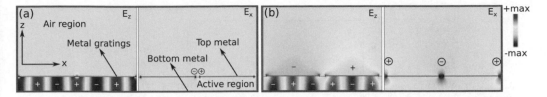

**Fig. 9.3** Finite-element modeling of a metallic terahertz QCL cavity with third-order DFB gratings: (a) and (b) show the electric fields for the resonant eigenmodes of the lower and upper band edge, respectively, for a parallel-plate metallic cavity in which third-order DFB gratings are implemented as slits in the top-metal cladding. Results from a two-dimensional simulation of a cavity of infinite width (in the $y$ dimension) and 10 μm height are shown, where the length of the cavity is along the $x$ dimension. A refractive index of 3.6 is used for the active-core (same as that of GaAs). Here, $\Lambda = 42.5$ μm and the modes are resonant around $\sim 3.5$ THz for the simulated structure.

resonant modes, respectively, for a structure in which the grating is realized in the form of periodic slits in the metal cladding. In this case, the radiative outcoupling for the upper band-edge mode (the symmetric in-plane $E_x$ field) is larger than that for lower band-edge mode (the antisymmetric in-plane field) due to $n_{\text{eff}}$ for the former being closer to 3 than that for the latter. The lower $n_{\text{eff}}$ for upper band-edge mode is due to a lower mode confinement factor of the resonant mode since a fraction of the optical intensity propagates outside the cavity in air. A low confinement factor also means the threshold gain to excite the upper band-edge mode will be high; therefore, it is not possible to preferentially excite the desired upper band-edge mode in such a simplistic implementation of third-order DFB gratings.

To excite a band-edge mode with strong outcoupling of radiation on top of the metallic cavity, such that an SPP wave with a large wavefront can be established, it is necessary to lower the effective propagation index inside the waveguide. The first successful demonstration of such an idea utilized gratings formed via deep dry-etched grooves inside the active region of the cavity to lower the $n_{\text{eff}}$ value [23, 26]. By alternating the active region and air with a duty cycle of $\sim 10\%$, a strong Bragg diffraction up to higher orders resulted in a collimated far-field radiation pattern for the resonantly excited optical DFB mode. An understanding of this type of DFB structure can be obtained through FEM simulations. The simulation results in Fig. 9.4 for such a cavity with deep-etched gratings show enhanced radiative losses for the different eigenmodes of the DFB cavity, compared to those in Fig. 9.3, which are effectively due to the large outcoupling of radiation outside the cavity in its surrounding medium.

The radiative emission loss per unit length is a very good indicator of the outcoupling efficiency and hence the output power of terahertz QCLs. The power out-coupled from a terahertz QCL with uniform distribution of the electromagnetic field in the cavity can be expressed as follows:

$$P_{\text{out}} = \hbar \omega n_{\text{ph}} \alpha_r \frac{c}{n_r} = \frac{N_p \hbar \omega}{|e|} \frac{\alpha_r}{\alpha_w + \alpha_r} [(I - I_{\text{th}}) \chi], \qquad (9.2)$$

where $N_p$ is the number of repeated stages in the QCL; $c/n_r$ is the phase-velocity of light inside the cavity of index $n_r$; and $\alpha_r$ and $\alpha_w$ are optical losses due to radiation

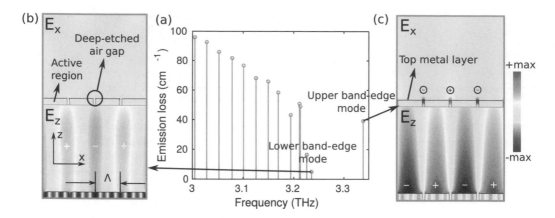

**Fig. 9.4** FEM modeling of third-order DFB gratings formed by deep-etched gratings as in [23]. Diagram shows the mode spectrum for a 1−mm-long and infinitely wide metallic terahertz QCL cavity with third-order DFB deep-etched gratings ($\Lambda = 42.5$ µm, alternating the active region and air with a duty cycle of ~10%) computed with the FEM method. Radiative losses for various resonant modes for such a cavity are shown in the step plot in (a). (b) and (c) show the electric-field profiles for the lower and upper band-edge modes, respectively.

(including that from edge facets and the top surface) and those in the waveguide (due to free-carrier and intersubband absorption), respectively. Further, $n_{ph}$ represents the number of photons in the lasing mode above the threshold, which is proportional to $(I - I_{th})\chi$, where $I$ is current flowing through the QCL, $I_{th}$ is the threshold current, and $\chi$ is the internal differential efficiency of the QCL superlattice. For a given type of active medium (such as the one based on resonant-phonon depopulation) and at a given frequency, $\chi$ and $\alpha_w$ are relatively constant across small variations in the superlattice designs and also do not depend significantly on the DFB design of the cavity. Hence, $\alpha_r/(\alpha_w + \alpha_r)$, defined as out-coupling efficiency [27, 28], is a good representation of the optical power output from the QCL. By differentiating Eq. (9.2) with respect to $I$, an expression for the slope efficiency, $dP_{out}/dI_{out}$, that is directly measurable is obtained:

$$\frac{dP_{out}}{dI} = \frac{N_p \hbar \omega}{|e|} \frac{\alpha_r}{\alpha_w + \alpha_r} \chi. \tag{9.3}$$

This equation suggests that the slope efficiency, for a given internal differential efficiency value, is an intuitive indicator of the out-coupling efficiency of the QCL. Since the radiation loss of edge-emitting terahertz QCLs contains the radiative losses from both its end-facets, and due to the difficulty in implementing high-reflectivity layers for terahertz QCLs, the optical power from one facet cannot be typically collected, and hence the measured slope efficiency is approximately halved.

For the specific case of a third-order DFB, shown in Fig. 9.4, due to a reduced effective mode index, there is a strong SPP mode established, on top of the metal cavity, for the band-edge modes. This effect does reshape the emission pattern into a single-lobed beam. The theory of phase-locked antenna arrays that operate at microwave frequencies can also explain this phenomenon: the periodic apertures mimic

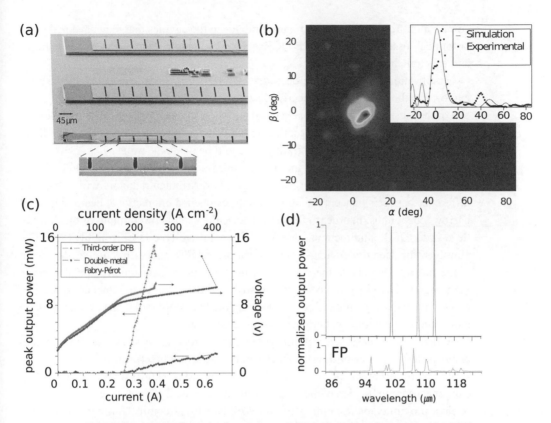

**Fig. 9.5** Experimental results for terahertz QCL with deep-etched third-order DFB gratings [23]. (a) SEM images of DFB structures with an enlarged view of the grating region. (b) Measured far-field pattern for a typical DFB QCL; the inset shows comparison of the calculated far-field for two-dimensional simulations with measurement. (c) Light–current–voltage (L–I–V) curves at 10 K for a typical third-order DFB QCL and for a Fabry–Pérot cavity QCL of similar dimensions. (d) Spectra for different third-order DFB QCLs with different grating periods and that from a Fabry–Pérot QCL. Used with permission of Springer Nature BV, from [23]; permission conveyed through Copyright Clearance Center, Inc.

a series of radiation sources consecutively phase-shifted by $\pi$ along the waveguide. Constructive interference from the different emitters can then be observed only in a narrow spot in the far field in the longitudinal direction along the array.

Figure 9.5 shows the experimental results from terahertz DFB QCLs with third-order gratings formed by deep-etching in the QCL superlattice medium, designed to achieve gain at ~3.2 THz. The QCLs were measured at 10 K and the typical peak output power from a DFB QCL was 15 mW, a factor of seven higher than that from a standard Fabry–Pérot cavity QCL of similar dimensions. The slope efficiency was also improved, reaching a value of 130 mW/A, which is due to the combined effect of improved outcoupling as well as collection of radiation from the cavities. All QCLs worked in single-mode operation with a 30 dB sideband suppression ratio, and their emission frequency was tuned over 0.17 THz by changing the grating period. A single-lobed narrow beam with a $10° \times 10°$ divergence was also observed. However, owing to competition from different-order lateral modes in the cavities that may have had similar optical losses, not all the DFB QCLs tested had similar far-field radiation patterns and spectral characteristics.

Modifications on this kind of deep dry etching-formed third-order DFB have been explored in the literature. A perfectly phase-matched third-order DFB structure [29] is one of the most promising approaches and it is shown in Fig. 9.6(a) [29]. In this design, by varying the duty cycle of alternating active-core and air-gap regions, the periodicity of the radiating elements of the grating, $\Lambda$ could be made the same as $\lambda_0/2$, where $\lambda_0$ is the free-space wavelength of the desired third-order DFB mode. The QCL can be modeled as a phase-locked laser array of evenly distributed radiators with a $\pi$-phase difference between adjacent radiators, which results in a bidirectional beam pattern of narrow lobe widths. In order to further increase the useful power output of this kind of third-order DFB, unidirectional operation from a ridge-type laser is highly desirable. However, for deep dry etching-formed third-order DFB, a large fraction of the lasing mode propagates outside the solid core; thus a conventional unidirectional scheme via a high-reflectivity (HR) coating or distributed Bragg reflector attached to the rear facet will have a negligible effect on blocking the backward radiation, which results in a bidirectional emission pattern even with an HR coating.

A deliberately designed asymmetric-element structure to skew the bidirectional beam pattern from a third-order DFB cavity into a unidirectional one was demonstrated more recently [30]. A schematic is shown in Fig. 9.6(b) where reflectors that can reflect a large fraction of the propagating mode outside the active region are placed in close proximity to the emitting apertures of the DFB structure. Unidirectionality can be achieved by placing the reflectors at asymmetric locations of the DFB grating. Specifically, the air gap of the antenna-coupled DFB structure can be approximated as an omnidirectional source that emits symmetrically in both forward and backward directions. The reflector at a distance $\delta$ from the aperture, when $\delta < \lambda/8$ and $\lambda$ being the free-space wavelength, reflects the aperture field in a way that destructively interferes with the backward traveling radiation and causes preferential radiation in the forward direction. Quantitatively, the reflection coefficient of the field is $r = (1-n)/(1+n)$ for a transverse-electric plane-wave, where $n$ is the index of the dielectric. Hence, $r < 0$, which causes a $180°$ phase-shift at the interface. The additional phase-shift from a round-trip travel with a distance of $\delta < \lambda/8$ which is $<90°$, results in a total phase-shift within the $180°$–$270°$ range for the reflected wave, which leads to destructive interference with the backward radiation from the aperture. Not obvious, but verified through simulations, there is also constructive interference in the forward direction because of double reflections from the reflectors closest to the aperture (a double-negative reflection coefficient leads to a positive coefficient). FEM simulations indicate that for constructive interference the optimal condition is $\delta \sim 8.5\,\mu m\ <\ \lambda/8\ (=\ \Lambda/4)$, where $\Lambda$ is the grating period for a perfectly phase-matched third-order DFB, and $\lambda = 80\,\mu m$. The simulation results are also shown in Fig. 9.6(b) where a strong asymmetry in the radiation pattern can be observed because of the reflectors. Insertion of partial reflectors in the DFB structure reduces the radiation loss from the cavity

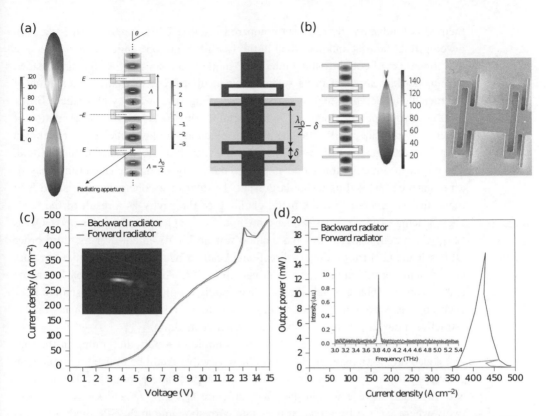

**Fig. 9.6** Perfectly phase-matched third-order DFB for terahertz QCLs [30]. (a) Simulated $E_z$ field (normal to the page; in arbitrary units) for a third-order DFB QCL cavity designed with a perfect phase-matching condition (right) and the corresponding far-field radiation pattern (left). (b) Schematic of a DFB structure modified for unidirectional emission with an antenna length interacting with adjacent reflectors (left), simulated $E_z$ field and the far-field radiation pattern (middle), and SEM image of the fabricated QCL (right). (c) V–I curves from DFB QCLs with backward and forward radiators respectively; inset: the single-lobe beam pattern measured with a terahertz camera from the QCL with forward radiator. (d) L–I curve of both QCLs; inset: the spectrum of the lasers. Used with permission of Springer Nature BV, from [30]; permission conveyed through Copyright Clearance Center, Inc.

slightly and larger air gaps are selected to compensate for this reduction. The measurement results from this design are shown in Fig. 9.6(c) and (d). The maximum power from a single-mode QCL was ~16 mW at 10 K.

## 9.2.3    Antenna-Feedback Scheme for Narrow-Beam Radiation from Plasmonic Lasers

Plasmonic lasers utilize metallic cavities and sustain long-range SPP resonant optical modes in subwavelength dimensions. They typically have highly divergent far-field radiation patterns. Recently, a new technique was developed to implement DFB in a plasmonic laser that leads to narrow far-field emission. The scheme has been termed as the antenna-feedback scheme [24] and was implemented for terahertz QCLs with metallic cavities to experimentally realize ultra-narrow far-field beams. This DFB method is fundamentally different compared to other DFB methods that have been developed in the past and described in the vast literature on semiconductor lasers, in that it consists of coupling the guided SPP mode in a laser's cavity to a single-sided SPP mode that can exist in its surrounding medium via a periodic perturbation of the metallic cladding in the cavity. Such a coupling is possible by choosing a Bragg grating of appropriate periodicity in the metallic film. This leads to the excitation of coherent single-sided SPPs atop the metallic cladding of plasmonic lasers, which, via coupling to the guided SPP, results in radiation with a low-divergence in the far field. The narrow beam emission is due, in part, to the cavity acting like an end-fire phased-array antenna, as well as to the large spatial extent of a coherent single-sided SPP mode that is generated on the metal cladding of the cavity as a result of the feedback scheme. The antenna-feedback method was implemented for terahertz QCLs [24], which demonstrated that the method was an improvement over the third-order DFB scheme [23] for producing directional beams since it does not require any specific design considerations for phase-matching [29]. The antenna-feedback scheme is automatically phase-matched and hence could be utilized for any plasmonic laser even at other wavelengths without any restrictions or specific requirements on the refractive-index or propagation index in the active medium.

The schematic in Fig. 9.7(a) shows an example of a periodic grating in the top-metal cladding for a parallel-plate metallic cavity that could be utilized to implement conventional $p$th-order DFB by choosing the appropriate periodicity. The incident and diffracted wavevectors in the waveguide are $k_i \approx 2\pi n_a/\lambda$ and $k_d = k_i$ respectively, where $\lambda$ is the free-space wavelength corresponding to the SPP mode and $n_a$ is the effective propagation index in the active medium (approximately the same as the refractive index of the medium), such that the so-called Bragg mode with $\lambda = 2n_a \Lambda/p$ is resonantly excited. The Bragg mode is, by design, the lowest-loss mode in the DFB cavity within the gain spectrum of the active medium, where $\Lambda$ is the grating period, and $p$ is an integer ($p = \pm 1, \pm 2, \pm 3 \ldots$) that specifies the diffraction order. The two SPP waves propagating in the opposite directions couple with each other to establish a standing-wave for the resonant-cavity mode inside the cavity. For conventional DFBs, the grating period always satisfies the condition $\Lambda = p\lambda/(2n_a)$, as shown in Fig. 9.7(a). Note that there is a phase mismatch for SPP waves on either side of the metal claddings and destructive interference between successive apertures, as shown in Fig. 9.7(b) for propagating SPP waves. However, in contrast to conventional DFB methods for which periodic gratings couple forward and backward propagating waves inside the active medium itself, the antenna-feedback scheme is implemented by using a slotted metal film that allows the guided SPP mode to diffract outside the cavity. By utilizing the unique ability of a metal film to support single-sided SPP modes propagating in air, a grating period could be chosen that leads to first-order Bragg diffraction in the opposite direction, but in the surrounding medium outside the cavity rather than inside the active medium itself. Similarly, the single-sided SPP mode in surrounding medium undergoes first-order Bragg diffraction to couple with the guided SPP wave inside the cavity, as illustrated in Fig 9.7(c). The SPP wave inside the active medium with

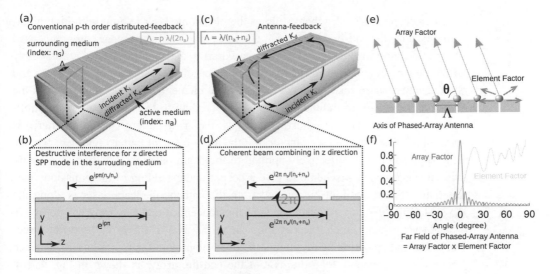

**Fig. 9.7** The antenna-feedback scheme for low-divergence radiation from plasmonic lasers. (a) The general principle of conventional DFB, which can be implemented in a THz QCL by introducing periodic slits or holes in its metallic cladding. A parallel-plate metallic cavity is illustrated. (b) With the periodicity of conventional DFBs in (a), there is a phase-mismatch between successive apertures for SPP waves on either side of the cladding. (c) The antenna-feedback scheme uses a grating period that effectively couples a single-sided SPP wave that travels in the surrounding medium with the SPP wave traveling inside the active medium. (d) The antenna-feedback scheme leads to a buildup of an in-phase condition at each aperture between counter-propagating SPP waves on either side of metal-cladding. Radiation from each aperture combines constructively to couple to far-field radiation in the end-fire ($z$) direction. (e) Linear phased-array antenna model to explain the far-field beam pattern for the antenna-feedback scheme. (f) Computed array factor and simulated far-field beam pattern from a single array element; multiplication of array factor and element factor results in the far-field beam pattern from the DFB cavity. Reproduced with permission from [25].

incident wavevector $k_i \approx 2\pi n_a/\lambda$ is diffracted in the opposite direction in the surrounding medium with wavevector $k_d \approx 2\pi n_s/\lambda$. Using Eq. (9.1), the first-order diffraction grating ($p = 1$) leads to excitation of a DFB mode for which

$$\lambda = (n_a + n_s)\Lambda, \qquad (9.4)$$

which is different from any of the $p$th-order DFB modes that occur at $\lambda = 2n_a\Lambda/p$. Hence, the antenna-feedback mode could always be selectively excited by just selecting the appropriate grating period $\Lambda$ such that the wavelength occurs close to the peak-gain wavelength in the active medium. The antenna-feedback gratings are neither completely periodic for the mode propagating inside the cavity, nor for the mode propagating on top of the metal cladding in free space. However, the combined phase-shift becomes $2\pi$, which suggests that the contribution from each grating aperture adds coherently in free space in the $z$-direction on top of the cavity [24]. The far-field radiation is therefore analogous to that from an end-fire phased-array antenna that produces a narrow beam in both $x$- and $y$-directions.

The antenna-feedback scheme allows the plasmonic laser to emit with high directionality along the end-fire direction, much like phased-array antennae commonly used at radio or microwave frequencies. A phased-array antenna consists of an assembly of single emitters. The array factor is a function of the geometry of the array and the excitation phase relation between individual elements. The element factor is the radiative pattern of an individual element. According to the beam pattern multiplication theorem of the phased-array antennae, the array pattern (far-field pattern) is the multiplication of the array factor and the array-element pattern (element factor). This approach can be effectively applied to the beam pattern estimation of a cavity with antenna-feedback scheme to give a deeper understanding of the emission characteristics. Since the light outcoupled from the antenna-feedback laser is along the cavity-length direction, it corresponds to the concept of end-fire array at microwave frequencies, which directs the radiation along the axis of the array (end-fire direction). The grating period in the antenna-feedback configuration represents the distance between each element along the axis of the end-fire array, as shown in Fig. 9.7(e), which determines the progressive phase between each successive element. The element factor is treated as the beam pattern corresponding to the radiation from a single emitting aperture, calculated from 2D simulation. The antenna-feedback scheme is engineered such that it can effectively provide the feedback for the mode inside and outside the cavity to establish DFB. A narrow far-field beam is produced since the radiation adds in phase only in the end-fire direction, while destructive interference happens in other directions as can be seen from the calculated array factor in Fig. 9.7(f) that has a single peak at the $0°$ (or $180°$), corresponding to the end-fire direction. The full-width half-maximum (FWHM) of the array factor is $\sim 5°$ calculated for a total length $\sim 1.4$ mm from a simple point-source model for the element factor. The FWHM of the beam pattern after multiplication of the array factor and element factor is $\sim 5°$, and the beam is peaked slightly off normal as a result of this simplified analytical calculation model assuming identical radiating elements and the effect of the finite lateral dimensions of the optical cavity.

For the antenna-feedback scheme to work effectively, optically absorbing boundaries near the two end-facets of the cavity are necessary so that the desired mode can be excited [24]. Figure 2(b) in [24] shows the eigenmode spectrum and the dominant electric field of a terahertz QCL cavity with antenna-feedback gratings. Radiation loss occurs through diffraction from apertures. The major fraction of electromagnetic (EM) energy for the resonant modes exists in a TM-polarized ($E_z$) electric field. The antenna-feedback grating excites a strong single-sided SPP standing wave on top of the metallic grating (in the air) [24]. The overall optical loss for the lowest-loss resonant mode is $\sim 10.6$ cm$^{-1}$, which is the sum of the loss due to radiation as well as that introduced by the absorbing boundaries.

Figure 9.8 shows experimental results from terahertz QCLs implemented with antenna-feedback gratings. The schematic in Fig. 9.8(a) shows how longitudinal and lateral absorbing boundaries were implemented by leaving 0.1-μm-thick highly doped ($5 \times 10^{18}$ cm$^{-3}$ $n$-doping) GaAs layers exposed without a metal cladding, which results in a highly lossy propagation of the SPP modes in those regions [31]. While the

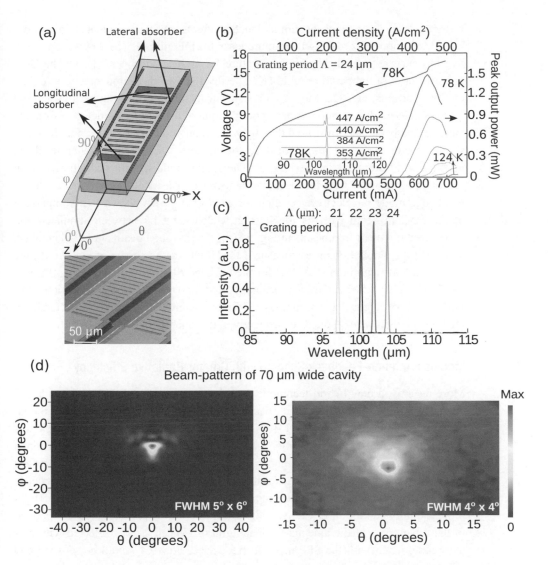

**Fig. 9.8** Experimental results for terahertz DFB QCLs implemented with antenna-feedback [24]. (a) The schematic shows the antenna-feedback gratings implemented in the top-metal cladding of the QCL cavity as well as the orientation and definition of angles. The QCL active medium is 10 μm thick with gain centered at ∼3 THz. A scanning electron microscope image of the fabricated QCLs is shown below the schematic. (b) Experimental L–I–V characteristics of a representative QCL of dimensions 1.4 mm × 100 μm at different heat-sink temperatures in pulsed mode of operation. Inset shows the lasing spectra at different bias where the spectral linewidth is limited by the instrument's resolution. The emitted optical power is measured without any cone collecting optic inside the cryostat. (c) Measured spectra for four different antenna-feedback QCLs with varying grating periods Λ, but similar overall cavity dimensions, at 78 K. (d) Simulated (left) and experimental (right) far-field radiation patterns where FWHM for the beams are ∼5° × 6° and ∼4° × 4° respectively. The plots are for QCLs with cavities ∼1.4 mm long and 70 μm wide, with a Λ = 21 μm grating, designed to excite a resonant DFB mode at ∼3.1 THz. Adapted with permission from [24] ©The Optical Society.

former helps in DFB-mode discrimination, the latter is useful to eliminate higher-order lateral SPP modes by making them more lossy [32]. Figure. 9.8(b) shows representative L–I curves versus heat-sink temperature for a QCL with $\Lambda = 24\,\mu m$. The QCL operated up to a temperature of 124 K. The inset shows measured spectra at different bias currents, at 78 K. Most QCLs tested with different grating periods showed robust single-mode operation. A peak power output of $\sim$1.5 mW was detected from the antenna-feedback QCL measured directly without any collecting optics. Figure 9.8(c) shows spectra measured from four different terahertz QCLs with antenna-feedback gratings of different grating periods, $\Lambda$. The single-mode spectra scale linearly with $\Lambda$, which confirms that the feedback mechanism works as expected and that the lower band-edge mode is selectively excited for each QCL. Also shown is a comparison of the far-field beam patterns for antenna-feedback QCLs between simulated and experimentally measured beams in Fig. 9.8(d). Single-lobed beams in both lateral $(x)$ and vertical $(y)$ directions were measured for all QCLs. As shown in Fig. 9.8(d), the full-width half-maximum (FWHM) for the QCL of 70 μm width and $\Lambda = 21\,\mu m$ is $\sim 4° \times 4°$, which was the narrowest beam reported for a terahertz QCL at the time [24]. Further improvements have been achieved for the radiative characteristics of terahertz QCLs since then, as explained subsequently in this chapter.

### 9.2.4  Modified Antenna-Feedback Scheme with Greater Radiative Efficiency

More recently, a new hybrid technique combining conventional DFB (with sidewall corrugation) and antenna-feedback has been demonstrated to realize higher output power edge-emitting terahertz DFB QCLs [33]. For that scheme, the feedback was provided by a sinusoidal corrugation (first-order DFB) of the cavity that determined the operating frequency, while light extraction was ensured by an array of surface holes based on an antenna-feedback scheme. The feedback wavevector $k_{fb}$ was determined by the first-order DFB scheme, which is equal to the momentum of light propagating inside the laser cavity, while the extraction wavevector $k_e$ corresponded to the wavevector of the antenna-feedback grating. When $k_e$ was selected with perfect phase-match with the SPP mode in free space, Eq. (9.4) could be satisfied and a low-divergence radiation pattern was realized. The perfect-matching condition was achieved by tuning the extraction wavevector via spanning a 10% range of the periodicity of the surface-hole array that was lithographically defined to achieve a sequence of different extraction wavevectors $\eta k_e$ (with $\eta$ in the range $0.9 < \eta < 1.1$ in steps of $\Delta\eta = 0.05$). This allowed the investigation of the optimal scattering condition. When $\eta = 0.95$, the peak output power exceeded 40 mW at 20 K, which translated to a perfect phase-matching condition. A maximum CW optical power of 6 mW and a slope efficiency of 100 mW/A were also measured at a heat-sink temperature of 10 K.

### 9.2.5  Phase-Locked Array of Edge-Emitting, Third-Order DFB QCLs

A perfect phase-matched third-order DFB has a large fraction of the mode propagating outside the active region [29, 30], which allows phase locking of multiple wire lasers in the lateral direction of laser ridges. Recently, a scheme for phase locking of several ($\leq 5$) wire lasers to form a coupled cavity was presented, which was called $\pi$-coupling [34]. The development of the $\pi$-coupling scheme was inspired by $\pi$ conjugation in chemistry. For a photonic-wire laser like the perfect phase-matched third-order DFB, the radiation from each subwavelength aperture is divergent. It can be concluded that by attaching several individual laser cavities together in the lateral direction, not only does the mirror loss increase, but also the lateral-field distribution in the intermediate zone is significantly enhanced. The strong lateral-field distribution of a third-order DFB opens up the possibility for the $\pi$-coupling of two or even more edge-emitting DFBs. Since the coupling exists between all in-phase antenna radiation gaps, this scheme will be far more effective compared to earlier schemes utilizing the facet radiation. By coupling several third-order DFBs, high power levels can be achieved. Figure 9.9 shows the results for $\pi$-coupled several third-order DFBs with each having 35 periods and a 5.5 µm-wide slot gap size between neighboring cavities. The coupling mechanism is sufficiently strong to maintain robust phase-locking at all biases. Narrow beam patterns ($10° \times 15°$) up to $\sim 50$ mW from two $\pi$-coupled third-order DFBs, and $\sim 90$ mW from five $\pi$-coupled DFBs with $6° \times 8°$ half-power beamwidth were demonstrated at 30 K.

## 9.3 Surface-Emitting, Single-Mode Terahertz QCLs

Compared to edge-emitting semiconductor lasers, their counterparts, high-power surface-emitting (SE) semiconductor lasers [35, 36], have significant advantages in multiple aspects related to coupling and alignment optics, testing and packaging, power-scaling through arrays, wavelength stability, and immunity to facet damage among others. The radiative efficiency of grating-coupled SE lasers has been increased by using different techniques, including the use of curved and chirped gratings for near-infrared diode lasers [37, 38], implementation of a central grating $\pi$-phase shift for short-cavity devices [39–41], plasmon-assisted mode selection for mid-infrared (IR) QCLs [27, 42, 43], and graded photonic structures for terahertz QCLs [44]. Grating-coupled, SE mid-IR QCLs are treated in Chapter 2. At near-IR wavelengths, such high-brightness lasers are predominantly realized as vertical-cavity SE lasers (VCSELs) [45]. However, some of the highest surface-emitted CW output powers for single-mode near-IR lasers have been demonstrated with second-order DFB gratings [37, 38]. For QCLs, which emit in the mid-IR and THz spectral ranges [28, 46–48], vertical cavities are not possible owing to the transverse-magnetic polarization of the intersubband optical field, hence second-order gratings are used for surface emission [42, 43, 49–52].

### 9.3.1 Terahertz QCLs with Second-Order Bragg Gratings

A SE DFB THz laser with second-order gratings typically excites an antisymmetric mode that has relatively low radiative efficiency and a double-lobed far-field beam pattern [52]. Conventional second-order gratings are used to realize broad-area

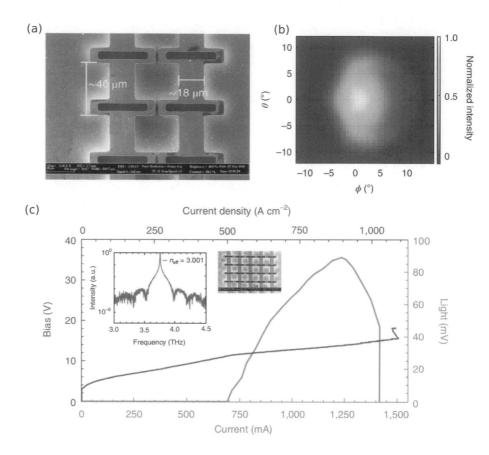

**Fig. 9.9** Phase-locked array of third-order DFB QCLs by π-coupling [34]. (a) SEM image of a fabricated laterally phase-locked array of two optical cavities. Here, three periods are shown (from a total of 35 periods) in each cavity. (b) Far-field beam pattern from a five-cavity edge-emitting phase-locked array. (c) Experimental spectrum, L–I–V curves, and SEM image of the five-cavity phase-locked array, which radiated single-mode with peak output power of 90 mW at 30 K. Used with permission of Springer Nature BV, from [34]; permission conveyed through Copyright Clearance Center, Inc.

surface (or substrate) normal emission from the laser cavity since the first-order Bragg diffraction is perpendicular to the guided-wave direction in the cavity as per Eq. (9.1) [35]. Coupled-mode [36] or finite-element modeling of the periodic structures are used to compute the photonic band structure. Typically, the band-edge modes in a longitudinal cavity are termed as symmetric and antisymmetric with high and low radiative efficiencies, respectively, where the latter, due to the lower threshold-current density, is excited for a SE-DFB laser, thereby limiting its useful output power. The radiative efficiency depends on the precise amplitude and phase of the standing-wave (resonant-mode) with respect to the periodic unit-cell of the grating. For the metallic cavities of THz QCLs, FEM simulations were carried out to reveal the radiative surface losses of the resonant modes, their frequencies, and electric-field profiles for the

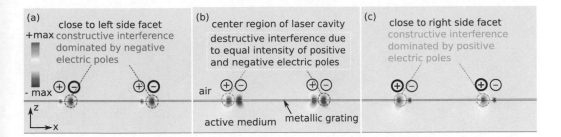

**Fig. 9.10** Field profiles for the antisymmetric band-edge mode for a cavity with second-order DFB. The plots illustrate the in-plane electric field ($E_x$) for the lower band-edge mode in a metallic THz QCL cavity with second-order DFB gratings implemented as slits in the top-metal cladding. The plot in (b) shows the $E_x$ field around the center of the laser cavity, whereas the plots in (a) and (c) show the $E_x$ field near each of the longitudinal end-facets in the cavity.

band-edge modes of finite-length cavities with gratings implemented in the top-metal cladding. Since the radiative loss itself is a very intuitive and direct indicator of the out-coupling efficiency of terahertz QCLs, it is intuitive to conclude that the radiative loss for the upper band-edge mode (the symmetric in-plane ($E_x$) field) is considerably larger than for the lower band-edge mode (the antisymmetric in-plane ($E_x$) field) [52] for a conventional second-order DFB laser.

Also, since conventional second-order DFB QCLs always show two-lobed far-field beam patterns it has been argued that this behavior is due to imperfect boundary conditions. Figure 9.10 shows the $E_x$ field of the anti-symmetric mode (lower band-edge mode) at three different locations along the length of a conventional second-order DFB based THz-QCLs cavity. Figure 9.10(b) shows that at the central region of the cavity, due to strong coupling of the guided waves, two equally strong electric poles with opposite polarity can be observed in each period of metallic grating, which, in turn, would lead to destructive interference in the far-field domain. However, due to the fact THz-QCLs have large end-facet reflectivities, the electric-field distribution is altered near the end-facets, which could cause a lateral shift of the electric field around the periodic slits due to a phase-mismatch of Fabry–Pérot and DFB effects. Figure 9.10(a) and (c) show that one of the pair of electric poles with opposite polarities would take a dominant position in the regions close to the two end facets. In that case, the vulnerable dipoles would be compressed. However, according to the fact that the two sets of constructive interference are caused by opposite electric dipoles, they could not combine with each other in the far-field domain. Therefore, conventional second-order DFB gratings lead to a double-lobed far-field beam for the antisymmetric mode.

One technique to achieve a single-lobed far-field beam pattern from a second-order DFB laser is introducing a $\pi$-phase shift in the grating center by means of a $\Lambda/2$ defect [39, 40]. This is essentially equivalent to flipping the phase of emission from all apertures in one half of the grating length by $\pi$ with respect to the other half. The mode spectrum, energy-density mode-shapes, and grating behavior for different end-lengths essentially remain the same as that of a grating without a $\Lambda/2$ defect.

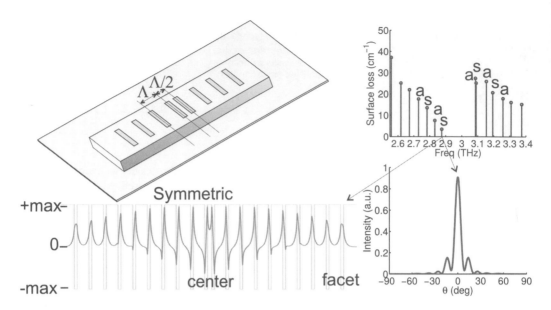

**Fig. 9.11** FEM results for a second-order DFB THz QCL cavity with a central defect [52], showing a resonant-mode spectrum, and vertically averaged field profile along its longitudinal direction, with the corresponding far-field pattern for the lower band-edge mode. The metallic cavity has a second-order DFB grating in the top-metal cladding with a $\Lambda/2$ defect at its center. Reprinted with permission from [52] © The Optical Society.

Figure 9.11 shows one such case when due to a $\Lambda/2$ central defect, the entire grating emits in-phase, and a single-lobed far-field pattern is obtained with a maximum in the center. Such a radiation pattern is highly desirable for practical applications.

Actually, all even-order diffractions could be operating in the SE configuration. Take a fourth-order DFB emitter as an example. Equation (9.1) shows that fourth-order DFB gratings ($n = 4$) would result into vertical diffraction ($\theta_d = 0$, for the first-order diffraction, where $k_d = k_i/n_w$, $n_w$ represents the refractive index of waveguide) or backward reflection ($\theta_d = \pi/2$, $k_d = k_i$, which is the second-order diffraction). It is assumed that $k_d = k_i/n_w$ if any off-normal-direction lobes exist. For a terahertz QCL with fourth-order gratings $2\pi/\Lambda = k_i/2$ ($\Lambda = 2\lambda_w$, where $\lambda_w$ is the wavelength inside active medium) and $n_w$ is about ~3.6. Equation (9.1) variants to $nk_i/2 = k_i + k_i \sin(\theta_d)/3.6$. Considering odd-order diffractions like $n = 3, \ldots, |\sin(\theta_d)|$ should be 1.8, which means that no viable solution exists for off-normal diffracted beams in the air. This fact reveals that there do not have to be any off-normal-direction lobes for metal–metal waveguide based terahertz QCLs with fourth-order DFB gratings.

Figure 9.12 shows the radiative surface losses of the resonant modes, their frequencies, and electric-field features for the band-edge modes of finite-length cavities with fourth-order Bragg gratings implemented in the top-metal layers. The lower- and upper-band-edge modes have similar characteristics of electric-field distribution as standard second-order DFBs, which respectively show two-lobe and single-lobe beam profiles, as well as low out-coupling and high out-coupling efficiency, respectively. The surface loss of the upper band-edge mode (symmetric mode) of the fourth-order

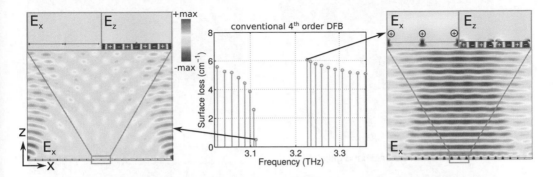

**Fig. 9.12** FEM modeling of metallic terahertz QCL cavities with a fourth-order DFB grating. The graph shows the mode spectrum for an infinitely wide cavity with fourth-order Bragg gratings implemented as slits in the top-metal cladding ($\Lambda = 54\,\mu$m, slit-width $\sim$4 $\mu$m, cavity-length $\sim$1.3 mm) computed by FEM simulations. The two pictures show the electric-field profiles for the eigenmodes of lower and upper band-edge modes respectively. The low-loss antisymmetric mode results in a dual-lobed far-field beam; conversely, the high-loss symmetric mode results in a single-lobed far-field beam. Reproduced from [53].

DFB is much lower than that of a second-order DFB, which could be explained by noting that half of the slits into the metal-cladding layer of the second-order DFB are closed. That is, the number of effective radiating apertures is halved.

### 9.3.2  Modified Second-Order DFB with Graded Photonic Heterostructure

The radiative efficiency and far-field beam pattern of second-order DFB-based terahertz QCLs can be improved by using different techniques. One of the most effective demonstrated methods is the so-called graded photonic heterostructure (GPH) for THz QCLs [44], which led to a high-power output emitted in a single-lobed far-field beam pattern. Figure 9.13(a) shows a schematic diagram of a device with a GPH resonator. Different from the conventional second-order DFB, the top metallic grating of the GPH structure is not periodic. Instead, its lattice spacing is symmetrically and gradually decreased from the center, towards each end of the laser ridge, according to the recursive formula: $a_{i+1} = a_i \times 0.99$ ($i = 0, 1, 2, \ldots$), where $a_i$ is the periodicity of the *ith* slit from the center. As the bandgap frequency scales with the periodicity, this structure features a position-dependent photonic bandgap, which smoothly shifts upwards in energy from the center to each end of the device. As a result, a V-shaped well for photons is formed, as shown at the bottom of Fig. 9.13(a). The total number of periods for the GPH laser is 29, which translates to a total laser-cavity length of about 700 $\mu$m, and it is argued to be sufficient to confine the symmetric mode. This technique realizes a localization of the upper band-edge mode with large out-coupling efficiency at the ridge center, while pushing the non-radiative mode close to the laser end facets. This spatial separation allows one to manipulate the modal loss of each mode differently. An absorbing boundary condition was implemented via a high-loss, n-doped layer located near the laser end facets. The net effect is to increase the modal loss of the antisymmetric mode with negligible out-coupling efficiency, without affecting the symmetric mode. Thus, by correctly shaping the mode profile, a GPH structure

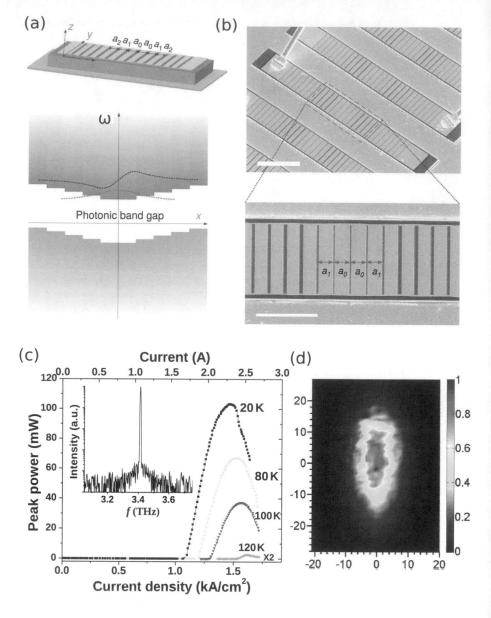

**Fig. 9.13** THz QCLs with a graded-photonic heterostructure (GPH) second-order DFB [44]. (a) Schematic illustration of surface-emitting THz QCL with GPH resonator. The top metallic grating is not periodic and its lattice spacing ($a_i$) is symmetrically and gradually decreased from the center towards each end of cavity. The bottom figure shows the schematic real-space, photonic-band diagram for the cavity. The band structure is position-dependent, and smoothly shifts up in energy from the center to each end of the device. Such a photonic-band structure behaves simultaneously as a localizing potential for the radiative modes and as a barrier for the non-radiative modes. The dotted curves are envelope functions of the first two resonant modes. (b) SEM images of GPH QCLs with an expanded view of a GPH resonator. The scale bars in the upper and lower panels are 200 μm and 100 μm long respectively. (c) and (d) Experimental L–I curves and far-field radiation pattern from a 214-μm-wide GPH cavity QCL with $a_0 = 27.4$ μm. The inset in (c) is a single-mode QCL spectrum at 20 K. Adapted by permission from Springer Nature Customer Service Centre GmbH: Springer Nature, *Nat. Comm.* vol. 3, p. 952 © 2012.

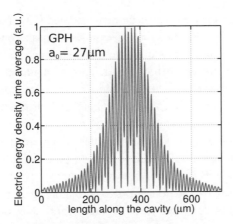

**Fig. 9.14** Mode profile in a terahertz QCL cavity with GPH [53]. Graph shows the time-averaged electric energy-density along the length of a metallic terahertz QCL GPH resonator cavity for its lowest-loss resonant mode Adapted from [53].

permits selective excitation of the desired radiative mode and control of its radiative efficiency.

Figure 9.13(c) and (d) show the pulsed spectral and light–current characteristics of a typical 214-μm-wide GPH laser, as well as its far-field beam pattern. The GPH laser exhibited a high-peak single-mode output power of 103 mW at 20 K, and the highest operating temperature was 120 K. The peak output power at liquid nitrogen temperature was still 67 mW, which is very appealing for practical applications. The full width at half maximum of the beam patterns are 9° and ~20° in the directions along and perpendicular to the ridge axes, respectively.

The design of GPH structure offers the advantage that the radiative symmetric mode can be confined inside the cavity and push the nonradiative mode to the absorbing boundaries. However, this is actually achieved with the restriction of a relatively short cavity. Otherwise, the antisymmetric mode might also be excited due to its extremely low surface loss. The normalized time-averaged electric energy-density profile for the resonant band-edge mode of the GPH is shown in Fig. 9.14. Considerable large non-uniformity of the electric-field distribution can be seen for the presented GPH design. The full-width half maximum of the electric energy-density profile is only ~180 μm, which indicates that the laser-cavity gain is not fully utilized for providing output power. Due to the internal defect of the GPH, the large non-uniformity in the electric-field distribution of its resonant mode would make it difficult to further enhance the radiative efficiency of GPH-based SE terahertz lasers due to the small volume of usable pumped active region. This indicates that there is still a large scope of improvement as far as the power performance of terahertz SE lasers.

### 9.3.3 High-Power Terahertz QCLs with Hybrid Second- and Fourth-Order DFB Gratings

Recently, a new hybrid-grating scheme that uses a superposition of second- and fourth-order Bragg gratings that excite a symmetric mode with much greater radiative efficiency into a single-lobed beam pattern has been demonstrated [53]. The scheme is implemented for terahertz QCLs with metallic waveguides. A peak power output of 170 mW is detected with robust single-mode, single-lobed emission for a 3.4 THz QCL operating at 62 K in a cryogen-free Stirling cooler. This is the highest reported power from any single-mode terahertz QCL to date. Moreover, a slope efficiency of 993 mW/A and a differential quantum efficiency of 71 photons/electron are realized, which represent records for terahertz QCLs (including those from multi-moded Fabry–Pérot QCLs). The hybrid-grating scheme is arguably simpler to implement compared to the aforementioned DFB schemes in literature and could potentially be used to increase the output power for SE-DFB lasers at any wavelength.

The concept of the hybrid second- and fourth-order DFB gratings for increased radiative efficiency of SE lasers is described next. Figure 9.15(a) shows conceptually how superposing a fourth-order grating structure on an existing second-order grating serves to enhance the overall radiative efficiency. Due to weaker Bragg diffraction for feedback, the fourth-order grating does not alter the mode shapes of band-edge modes significantly, which could only slightly change the electric-magnetic field distribution along the length of the cavity; that is, the maximum point of $E_x$ would be slightly offset the slit center and thus slightly reduce the radiative efficiency of the symmetric mode. Furthermore, an array of antenna-like dipoles with opposite polarity is introduced at each pair of metallic gratings to cause destructive interference, which, in turn, reduces the out-coupling efficiency of the symmetric mode. However, since the second-order Bragg diffraction is in the surface-normal direction, the gratings could be located so as to increase the radiative loss for the original antisymmetric mode of the second-order DFB structure. Such a hybrid-grating configuration enhances the radiative efficiency of the antisymmetric mode and reduces that of the symmetric mode, and hence, either of those band-edge modes could possibly be excited as per the precise implementation of the fourth-order grating. In either case, the radiative efficiency is enhanced when compared to the excited mode for the original second-order DFB structure.

Results from finite-element simulations are shown in Fig. 9.15(c). The radiative surface losses of the resonant modes, their frequencies, and electric-field profiles for the band-edge modes of finite-length cavities with gratings implemented in the top-metal cladding are shown. The cavity with the hybrid second- and fourth-order gratings greatly enhances the surface loss for the antisymmetric lower band-edge mode due to additional radiation from the slit corresponding to the fourth-order superposed grating. The loss for the symmetric upper band-edge mode is reduced from that of the second-order DFB structure. For the simulated case of $d/\Lambda = 3/8$, in Fig. 9.15(c), the upper band-edge mode is of lower loss for the hybrid-DFB grating and will be excited in a lasing cavity. More importantly, the surface loss for the excited mode in the cavity with hybrid DFB can be enhanced by an order of magnitude compared to that of a second-order DFB cavity.

The hybrid-DFB grating offers flexibility in design by simply altering the offset d at which the fourth-order grating is implemented, as shown in the schematic in Fig. 9.15(b). The relative radiative losses of the two band-edge modes could be modulated by adjusting $d/\Lambda$, which is useful to design for selective excitation of a lower or

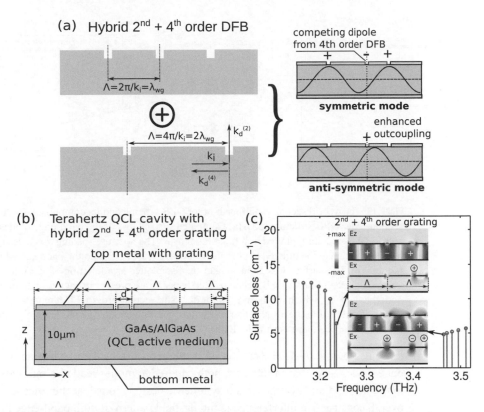

**Fig. 9.15** The hybrid second- and fourth-order grating scheme and its implementation for surface-emitting terahertz QCLs [53]. (a) The hybrid-grating concept that combines second- and fourth-order Bragg gratings. The second-order grating provides stronger DFB action and leads to establishing resonant optical modes with similar phase relationships to the grating as in conventional second-order DFBs. An added fourth-order grating serves to enhance the outcoupling of the antisymmetric mode and reduce that of the symmetric mode. By adjusting the design parameters, either of those modes could be excited in a DFB-laser cavity that has a greater radiative efficiency compared to that for the excited mode for a second-order DFB cavity. (b) Illustration of a metallic cavity for terahertz QCLs in which slits are opened in the top metal cladding to implement a periodic grating [52]. A fourth-order grating is superimposed at an offset of length $d$ to the original second-order grating with periodicity $\Lambda$, to realize a hybrid-grating structure as in (a). (c) The mode spectrum for a 1.4-mm-long and infinitely wide cavity with DFB gratings ($\Lambda = 27$ μm, slit-width ∼1 μm) computed with the finite-element modeling method. Radiative surface losses for various resonant modes for a cavity with hybrid grating ($d/\Lambda = 3/8$) in lines. The insets show electric-field profiles for the lower and upper band-edge modes, respectively, of the photonic bandstructure for each type of gratings. The radiative loss is effectively determined by the amplitude and phase of the in-plane electric field ($E_x$) at the slits. Reproduced from [53].

upper band-edge mode. This is shown in the plot of computed losses of the band-edge modes of an infinitely wide, but finite-length terahertz QCL cavity in Fig. 9.16(a), where the simulation is done with the same parameters as for the grating in Fig. 9.15. For a certain range of $d/\Lambda$ values, the loss of the upper band-edge mode could be low-ered and hence such a mode could be selectively excited. For the experimental results

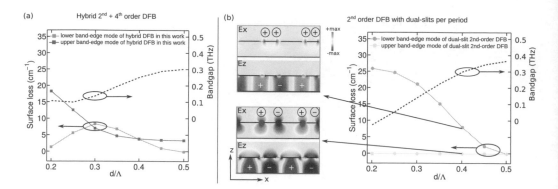

**Fig. 9.16** The hybrid second- and fourth-order DFB scheme and its comparison to a dual-slit, second-order DFB structure [53]. (a) Computed surface loss of the band-edge modes for the hybrid DFB structure, plotted as a function of $d/\Lambda$. The aperture spacing $d$ is a design parameter that can be utilized to alter the respective losses and also the bandgap, which is also plotted. (b) The surface loss and bandgap are also plotted as a function of aperture spacing for a dual-slit, second-order DFB structure (i.e., two slits are present in each grating period at the periodicity of $\Lambda$), which has been used for dual-slit, second-order DFB terahertz QCLs [54]. Electric-field profiles near the center of cavities, for both band-edge modes, are plotted for the case of $d/\Lambda = 0.4$, as an example. Reproduced from [53].

presented below $d/\Lambda = 3/8$ was implemented to selectively excite an upper band-edge mode. Note that in this case both band-edge modes lead to single-lobed beams in the far-field and symmetric/antisymmetric designations lose significance due to an overall constructive interference in the far-field pattern of both band-edge modes.

Given that the hybrid-DFB for terahertz QCLs is implemented with dual slits in the metal cladding in every alternate period of length $\Lambda$, it is intuitively compelling to consider whether the radiative loss for a second-order DFB for such QCLs could also be increased by introducing dual slits in each period, in order to avoid the nulls of the radiative field $E_x$ in one of the two slits. Such a DFB structure was indeed experimentally investigated in [54]. However, it turns out, such a DFB structure also has an antisymmetric mode for one of its band-edge modes, which has negligible radiative coupling. In this case, the role of symmetry is reversed for the band-edge modes, and the upper band-edge mode now becomes the low-loss mode with destructive interference for radiative coupling into the far-field pattern. The behavior of the surface losses for the two band-edge modes and the representative electric-field profiles for the dual-slit second-order grating structure are shown in Fig. 9.16(b). This type of structure can be considered as two offset second-order gratings, and can potentially be used to increase the output power if a large $d/\Lambda$ is chosen to make the photonic bandgap large, such that the upper band-edge modes lie outside the gain region of the active medium. This was indeed the strategy employed in [54], in order to increase the output power from SE-DFB THz QCLs. While the dual-slit structure may appear to offer somewhat similar functionality for the hybrid-DFB structure, that occurs at the cost of an inevitable trade-off between the out-coupling efficiency of the resonant mode and a large enough bandgap so that the upper band-edge modes could be located outside the gain region.

The coupling coefficient $\kappa$ is also calculated to determine the optimum total length of the DFB-laser cavity, which is a key factor for describing the coupling strength of the forward and backward traveling guided waves, and the absolute value of $\kappa$ should ideally not be less than the reciprocal of the cavity's length for a uniform field distribution as determined from the coupled-mode theory of grating SE-lasers [36, 55]. For a cavity with one-dimensional periodic gratings, the wave vectors of eigenfrequencies $k_{\pm}$ of the two band-edge modes are related to the complex coupling coefficient through the following relationships:

$$\mathrm{Re}(\kappa) \approx \frac{n_g}{2}\,(k_+ - k_-)\,, \tag{9.5}$$

$$\mathrm{Im}(\kappa) \approx \frac{\alpha_r}{2}\,, \tag{9.6}$$

where $n_g$ denotes the group refractive index. Using simulation results of the hybrid second- and fourth-order DFB grating from Fig. 9.15(b) with $d/\Lambda = 3/8$, $\Lambda = 27\,\mu\mathrm{m}$ and assuming no waveguide dispersion, Eqs. (9.5) and (9.6) lead to $|\kappa| \sim 12.07\,\mathrm{cm}^{-1}$. This relatively large value translates to the determination of the optimum length of the laser cavity to be $\sim 1\,\mathrm{mm}$ (not including the length of bonding pads at the longitudinal ends of the cavity). A slightly longer length of 1.3 mm was chosen for implementation, to minimize the optical losses due to longitudinally absorbing boundaries.

Finite-element simulations were performed to comprehensively understand the characteristics of the hybrid second- and fourth-order DFB scheme. Figure 9.17(b) shows the electric-field distribution as well as the energy-density profile for the resonant band-edge modes of the cavities with hybrid-DFB scheme, along the entire length ($\sim 1.3\,\mathrm{mm}$) of the cavity, with gratings implemented in the top-metal cladding. The energy density is estimated along the vertical center of the cavity from 2D simulations that effectively model cavities of infinite width. The eigenmode spectra and electric-field distributions for the cavities show non-uniform envelope shapes. In contrast to the GPH device, the FWHM of the electric energy-density profile is as large as $\sim 620\,\mu\mathrm{m}$, hence the stimulated emission could be generated over a large fraction of the laser cavity, which is significantly increased (by approximately three times) vs. that for the GPH-device cavity. Full-wave 3D FEM simulations were carried out for THz QCL cavities of the hybrid-DFB scheme. Figure 9.17(c) shows the computed far-field radiation pattern for a band-edge DFB mode of a THz QCL cavity implemented with hybrid-DFB gratings in the top-metal cladding, which demonstrates single-lobed beam operation for the upper band-edge mode, with narrow divergence in the $\theta_x$-direction and large divergence in the $\theta_y$-direction. The lower band-edge mode of the hybrid-DFB scheme would also show a single-lobed far-field beam pattern due to constructive interference along the entire length of the cavity. In this case, both band-edge modes could be selectively excited and provide enhanced output power compared to the case of conventional second-order DFBs.

Experimental results from representative SE THz QCLs implemented with hybrid-DFB gratings, in pulsed mode of operation and mounted inside a Stirling cooler are shown in Fig. 9.18. The results are from QCLs of dimensions $10\,\mu\mathrm{m} \times 200\,\mu\mathrm{m} \times 1.5\,\mathrm{mm}$. The hybrid-DFB grating in the form of slits is implemented in the top-metal

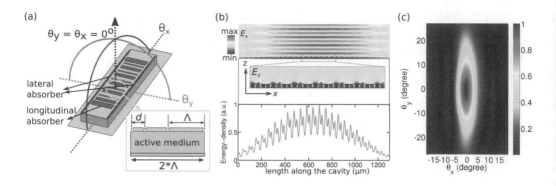

**Fig. 9.17** FEM modeling of a terahertz QCL cavity with hybrid second- and fourth-order DFB gratings [53]. (a) The schematic on the left shows the QCL's metallic cavity with a hybrid-DFB grating, implemented in the top-metal cladding, indicating the orientation of the QCL and the definition of angles. (b) The electric fields for the eigenmode of the lowest-loss (upper band-edge mode), calculated by FEM simulations of parallel-plate metallic cavities, and the energy density in the cavity along the ~1.3-mm-long metallic cavity (in the $x$-direction). (c) Full-wave 3D simulation result of the far-field radiation pattern of a 1.5 mm × 200 μm × 10 μm cavity with hybrid-DFB gratings. The full-width half-maximum (FWHM) beamwidth is ~ 7.5° × 30°. Reproduced from [53].

cladding. Figure 9.18(b) shows the light–current (L–I) curves versus heat-sink temperature, current–voltage (I–V) curve at 62 K, and also the spectra as a function of bias level at ~ 60 K. The QCL emits in a single-mode at all bias conditions at ~3.39 THz, and operates up to a maximum temperature of 105 K. Figure 9.18(c) shows the measured far-field radiation pattern, which is single-lobed and characteristic of symmetric-like mode excitation for the resonant-mode of the DFB structure. The full-width half-maximum divergence is ~ 5° × 25°. Also, the robustness of the DFB scheme in exciting the desired mode, based on lithographically defined periodicity is exemplified in Fig. 9.18(d) from the lasing spectra of three QCLs of different $\Lambda$ value. All QCLs showed single-mode operation over the entire dynamic range and the lasing frequencies scaled with $\Lambda$ with an effective propagation index $n_{\mathrm{eff}}$ ~ 3.16 for the guided modes. This relatively low $n_{\mathrm{eff}}$ value certifies that the upper band-edge mode is excited for these QCLs, as designed.

The hybrid second- and four-order DFB scheme significantly increases the radiative efficiency over that of surface-emitting semiconductor lasers based on the widely used second-order Bragg gratings. It excites a symmetric radiative mode with large radiative outcoupling in a single-lobed beam, which is also simpler to implement compared to published modifications to second-order gratings in the literature. The hybrid-DFB scheme has been implemented for terahertz QCLs to realize a single-mode QCL with peak optical power output of 170 mW, which is the highest reported value to date for monolithic terahertz DFB QCLs. In principle, the hybrid-DFB technique should also be applicable to semiconductor lasers at near-IR and mid-IR wavelengths to increase the output power for single-mode surface-emitting lasers at those wavelengths.

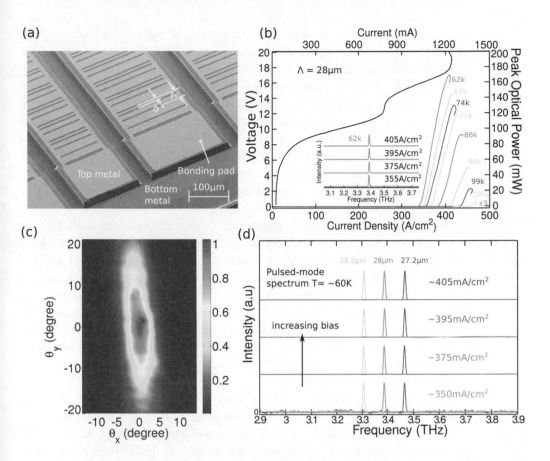

**Fig. 9.18** Experimental results: high-power surface-emitting terahertz QCLs with hybrid gratings [53]. (a) SEM image of the fabricated terahertz QCLs with hybrid-DFB gratings. (b) Current–voltage (I–V), and spectral intensity at different electric-bias levels at a heat-sink temperature of 62 K, and light–current characteristics at different temperatures measured in pulsed-mode operation. The device is of dimensions: 10 μm × 200 μm × 1.5 mm, grating period $\Lambda = 28$ μm, and $d/\Lambda = 3/8$. (c) Far-field radiation pattern (optical intensity) measured at 62 K, close to the peak-bias ($\sim$405 A/cm$^2$) at a distance of 40 mm from the QCL, in the surface-normal direction. $\theta_x$ and $\theta_y$ are the angles with respect to the surface normal along the longitudinal and lateral dimensions of the QCL cavity, respectively. (d) Spectral characteristics of three different QCLs, located next to each other on the wafer, of different grating periods, and $d/\Lambda = 3/8$ for each QCL. Reproduced from [53].

### 9.3.4    Phase-Locked Arrays of Terahertz SE-DFBs

The optical power performance of SE terahertz QCLs has been significantly improved during the past decade via many photonic structures like hybrid-DFB and GPH. In order to further improve the output power from SE terahertz QCLs, it is intuitive to laterally increase the size of SE-laser cavities; however, this is accompanied by an increased risk of exciting undesired spatial modes. Another strategy to achieve enhanced power and narrower far-field beam pattern is to use phase-locked arrays, a term that first emerged for device implementations at radio frequencies, and then was utilized for significantly scaling the coherent power of diode lasers and mid-IR QCLs by creating large-scale coherent modes distributed over an array of lasers by global coupling the array elements through the use of resonant leaky-wave coupling [56, 57]. Global coupling in the in-phase array mode was achieved by making the array interelement regions equal to an odd number of projected wavelengths, thus obtaining long-range coupling and, in fact, the phase-locked laser array becomes a high-index contrast photonic-crystal laser [61]. In addition, in-phase mode operation was ensured by suppressing nonresonant-mode operation by inserting high losses in the interelement regions [58]. This approach, applied to mid-IR QCL phase-locked arrays that select in-phase mode operation, is treated in Chapter 2.

However, resonant leaky-wave coupling via laterally propagating traveling waves is not possible for globally coupling arrays of individual ridge-type THz QCLs with metal–metal waveguide. Instead adjacent (i.e., parallel) ridge-type lasers (of same emission frequency) were connected via phase sectors corresponding to an even number of half wavelengths in the material [59, 60]. This is because, unlike in lateral resonant leaky-wave coupling, the coupling using a connecting section between two parallel lasers adds a phase factor of $\pi$ (i.e., one-half wavelength) as a result of the 180° rotation. Then the optical length of the phase sector should be an even number of half wavelengths (i.e., an integer number of wavelengths) of the propagating wave.

The scheme of terahertz SE QCLs with a central $\pi$-phase shift [52] was utilized to realize the first experimental demonstration of a phase-locked array of terahertz SE QCLs [59]. In-phase and out-of-phase coupling of two ridge guides was demonstrated by choosing the above-described proper phase-sector length. In-phase mode lasing in a diffraction-limited beam (i.e., in a single central lobe of 10° FWHM lobewidth) was obtained from a six-ridge array, albeit to relatively low power since means for suppressing nonresonant-array oscillation do not appear to have been introduced.

Similarly, phase-locked laser arrays involving coherent coupling of GPH terahertz semiconductor lasers via half-ring phase sectors have also been demonstrated [60], allowing the relative phase between the sources to be fixed by design. Unambiguous in-phase mode operation was obtained for two- and three-element arrays. However, for four- and five-element arrays the beam patterns revealed the excitation of other array modes such as the out-of-phase and adjacent array modes, most likely, as for the previous array architecture, due to lack of discrimination against nonresonant array-mode oscillation.

It has been confirmed that the phase-locked scheme for SE terahertz QCLs is an effective method to narrow the far-field beam profile of SE lasers, although the coherent power was scaled only for devices of up to only three elements. That is, to further improve the power performance together with achieving narrow far-field beam patterns, it is highly desired to develop techniques for suppressing nonresonant array-mode lasing, as has been done for diode lasers [58] and mid-IR QCLs [61].

## 9.3.5    Phase-Locked Arrays of Terahertz SE Lasers Through Traveling Plasmon Waves

Different mechanisms for phase-locking of multicavity arrays of metal cavities have been reported for terahertz QCLs toward improvement in their output power, radiative efficiency, and far-field radiation patterns [59, 60]. However, prior phase-locking schemes have not yielded better results for output power when compared to DFB methods.

Recently, a new scheme for phase-locking subwavelength metal cavities in a single-spatial-mode, terahertz plasmonic QCL, to achieve a large radiative efficiency and high power output in a narrow single-lobed far-field beam, was demonstrated [62]. An array of metallic microcavities is longitudinally coupled through traveling plasmon waves, which leads to radiation in a single spectral mode and a narrow far-field beam in the surface normal direction. The parallel metallic plates in the cavities sandwich a 10 μm-thick quantum-cascade GaAs/AlGaAs superlattice active medium with spectral gain at $\nu \sim 3.2$–3.4 THz ($\lambda \sim 88$–94 μm). Hence, the cavities are subwavelength in the dimension orthogonal to the metal plates of the resonant optical cavities, but are kept larger in the other two dimensions. Whereas previously reported phase-locking schemes for semiconductor lasers in the literature primarily relied on lateral coupling of cavities, the new method implements a novel longitudinal coupling mechanism by way of single-sided SPP waves that are established on top of the metal claddings of the cavities, and propagate in the surrounding medium as surface waves. The nature of surface waves thus generated is similar to that in [24]. However, the mechanism of generation of such surface waves and their role in the operation of the QCL differs significantly compared to that in [24] since they are not the primary contributor to the radiation from the phase-locked array.

The specific design of the phase-locked array and its operational principle are shown with an illustrative diagram in Fig. 9.19(a). The length of each microcavity is chosen to be $\sim 3\,\lambda_{\mathrm{SPP}}$, where $\lambda_{\mathrm{SPP}}$ is the wavelength of the hybrid SPP surface-wave in the surrounding medium that is determined through finite-element (FEM) simulations. For a resonant-mode excited at a free-space wavelength $\lambda$, $\lambda_{\mathrm{SPP}} < \lambda$ for the multicavity array surrounded by air or vacuum. $\lambda_{\mathrm{SPP}}$ can vary slightly based on the shape of the hybrid SPP mode that differs from one phase-locked resonant mode to another; hence, it is not an implicitly deterministic parameter for a given geometry of the microcavities. For a 3.3 THz-laser cavity design, $\lambda \sim 91$ μm, and $\lambda_{\mathrm{SPP}} \sim 81$ μm is deduced from experimental data for the measured QCL. The distance between neighboring cavities, $d_{\mathrm{air}} \sim \lambda_{\mathrm{SPP}}$, ensures that the hybrid SPP mode is periodic from one microcavity to another, that allows for phase-locking of the hybrid SPP surface waves to guided SPP modes within the cavities. Slit-like apertures are left open in the top-metal cladding at locations where the vertical electric-field component, $E_z$, has a null for the standing wave inside cavities; hence, the in-plane field component, $E_x$, has a maximum, leading to large diffractive outcoupling from the apertures [37]. The periodicity of aperture locations is kept at $\sim 3\lambda/n_{\mathrm{eff}}$, which is approximately an integer multiple of the wavelength of the guided SPPs within the semiconductor ($= \lambda/n_{\mathrm{eff}}$) where $n_{\mathrm{eff}}$ is effective propagation index of guided SPPs ($< 3.6$ in GaAs). The hybrid

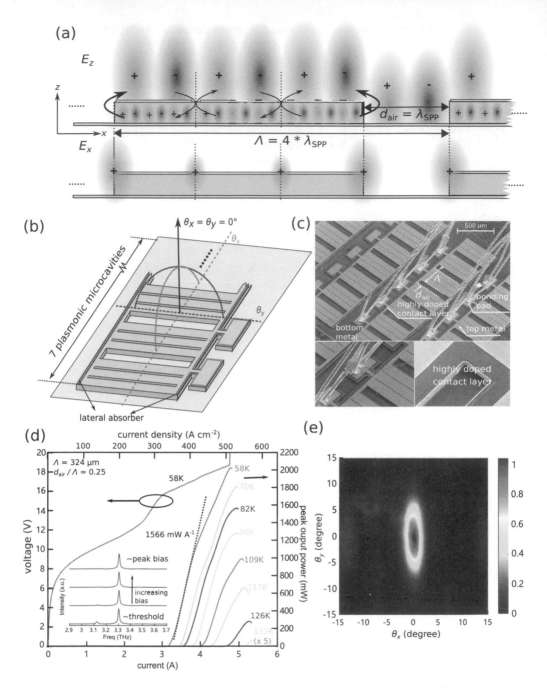

**Fig. 9.19** Longitudinal phase-locking scheme for subwavelength metallic cavities [62]. (a) The scheme that allows for phase-locked operation for multiple parallel-plate subwavelength metallic cavities. Single-sided SPPs traveling in the surrounding medium of the active region can be established due to a strong out-coupling effect from the end facets of each microcavity, which leads to a strong phase-locking effect. The distance between neighboring microcavities is equal to one wavelength ($\lambda_{SPP}$) of the single-sided SPPs, and the length of each individual microcavity is $3 \times \lambda_{SPP}$; two new apertures are thus introduced to the top metallic layer. The newly introduced apertures equally separate the top metal layer into three parts; the length of each part is equal to $\lambda_{SPP}$. A fixed-phase condition at each facet between counter-propagating SPP waves on either side of the metal-cladding can thus be established, and the dominant electric-field ($E_z$) distribution of the SPPs is shown.

SPP mode on top of the metal cladding automatically adjusts its shape to comply with such a periodicity such that $\lambda_{\text{SPP}} \sim 3\lambda/n_{\text{eff}}$ for the global phase-locked, resonant optical mode in the multicavity array. In that sense, there is no single parameter that exclusively determines the wavelength $\lambda$ of the resonant mode of the multicavity array. The length of the individual microcavities, the intercavity distance $d_{\text{air}}$, or the spacing between adjacent apertures could be each changed individually in the array for lithographically tuning the $\lambda$ value. Also, the length of microcavities in the multicavity array is chosen on the basis of two considerations. First, the length is such that the in-plane $(E_x)$ electric field has the same phase at the locations of both end-facets where it is most intense. This allows for constructive interference and radiation in the surface-normal direction for the array. Second, the cavities are kept short enough to keep $\alpha_{\text{rad}}$ large, with a goal of realizing $\alpha_{\text{rad}} \sim 10\,\text{cm}^{-1}$ for the desired lowest-loss resonant optical mode of the phase-locked array.

Experimental results from a representative terahertz QCL implemented with this phase-locking scheme, in pulsed-operation mode, are shown in Fig. 9.19 as well. The microcavities are combined on both lateral sides so they could be electrically biased simultaneously with few bonding pads. There is no optical coupling in those narrow connection regions. A scanning-electron-microscope (SEM) image of the fabricated and mounted QCL chip in Fig. 9.19(c) shows several QCLs of varying dimensions located side by side. Fig. 9.19(d) shows light–current (L–I) curves versus heat-sink temperature for the best performing QCL in terms of peak output power. A current–voltage (I–V) curve measured at 58 K is also shown. The results presented are from a QCL of overall dimensions: $10\,\mu\text{m} \times 500\,\mu\text{m} \times 2.2\,\text{mm}$, consisting of seven microcavities. The QCL emits in a single mode at $\sim 3.3$ THz at most bias conditions (except close to threshold when it also excites a second weaker spectral mode). The spectra as a function of bias are plotted in the inset of the figure. This QCL operated up to a maximum temperature of 132 K, whereas the best Fabry–Pérot QCLs fabricated from the same wafer operated up to 159 K. A peak optical power of $2030 \pm 10$ mW was

---

**Fig. 9.19** (*continued*) The specific arrangement of the multicavity array leads to a strong in-plane field of $E_x$ with the same phase at all locations of slit-like apertures and facets, as shown in the bottom half of (a). (b) Schematic of the QCL's metallic cavity with the phase-locking scheme, the orientation of the QCL, and the definition of angles. (c) Scanning-electron-microscope image of the fabricated QCLs. Insets show that a $10$-$\mu$m $\sim 15$-$\mu$m-wide highly doped GaAs layer below the top-metal cladding was left unetched at regions close to lateral facets, serving as the lateral absorbing boundary that ensures the excitation of the desired mode as the lowest-loss lasing mode. The same absorbing boundary was also implemented to surround the bonding pads to prevent the excitation of other undesired modes. (d) Experimental lasing characteristics (L–I and I–V curves) of a representative seven-microcavity phase-locked QCL array with dimensions of $7 \times 10\,\mu\text{m} \times 243\,\mu\text{m} \times 500\,\mu\text{m}$, at different heat-sink temperatures, and the spectra (inset) at different bias currents and a 58 K heat-sink temperature. (e) Far-field beam pattern of representative THz QCL measured close to peak-bias current. The measured beamwidth FWHM is $3.2° \times 11.5°$. Reprinted with permission from [62] © The Optical Society.

detected for the QCL at 58 K, which leads to a peak wall-plug efficiency of 2.3% for the corresponding bias condition. A slope efficiency of $1566\pm10$ mW/A is estimated from the slope of approximately the first half of the dynamic range at 58 K, over which the L–I curve is linear. The slope efficiency, in turn, corresponds to a differential quantum efficiency of 115 photons reaching the detector per electron transported through the QCL superlattice, which is 52% of the maximum theoretical limit for the QCL. This result makes the device the first single-mode QCL (including mid-IR QCLs) for which more photons are radiated than those that are absorbed as optical losses within the cavity.

## References

[1]  B. S. Williams, "Terahertz quantum-cascade lasers," *Nature Photonics*, vol. 1, p. 517, 2007.

[2]  S. Kumar, "Recent progress in terahertz quantum cascade lasers," *IEEE J. Sel. Topics Quant. Electron.*, vol. 17, pp. 38–47, 2011.

[3]  A. Khalatpour, A. K. Paulsen, C. Deimert, Z. R. Wasilewski, and Q. Hu, "High-power portable terahertz laser systems," *Nat. Photonics*, vol. 15, no. 1, pp. 16–20, 2021.

[4]  B. B. Hu and M. C. Nuss, "Imaging with terahertz waves," *Opt. Lett.*, vol. 20, pp. 1716–1718, 1995.

[5]  R. W. McMillan, "Terahertz imaging, millimeter-wave radar," in *Advances in Sensing with Security Applications*, J. Bymes, Ed. New York: Springer Verlag, 2006, pp. 243–268.

[6]  W. L. Chan, J. Deibel, and D. M. Mittleman, "Imaging with terahertz radiation," *Rep. Prog. Phys.*, vol. 70, p. 1325, 2007.

[7]  A. W. Lee, Q. Qin, S. Kumar, B. S. Williams, Q. Hu, and J. L. Reno, "Real-time terahertz imaging over a standoff distance (> 25 meters)," *Appl. Phys. Lett.*, vol. 89, p. 141125, 2006.

[8]  D. M. Mittleman, R. Jacobsen, R. Neelamani, R. G. Baraniuk, and M. C. Nuss, "Gas sensing using terahertz time-domain spectroscopy," *Applied Physics B: Lasers and Optics*, vol. 67, pp. 379–390, 1998.

[9]  J. F. Federici, B. Schulkin, F. Huang, D. Gary, R. Barat, F. Oliveira, and D. Zimdars, "THz imaging and sensing for security applications – explosives, weapons, and drugs," *Semicond. Sci. Technol.*, vol. 20, p. S266, 2005.

[10] H. H. V. Mendoza, "Mercury gas sensing using terahertz time-domain spectroscopy," 2013, US Patent Appl. 13/555,591.

[11] L. Consolino, S. Bartalini, H. E. Beere, D. A. Ritchie, M. S. Vitiello, and P. D. Natale, "THz QCL-based cryogen-free spectrometer for in situ trace gas sensing," *Sensors*, vol. 13, p. 3331, 2013.

[12] Y. Ren, J. Hovenier, R. Higgins, J. Gao, T. Klapwijk, S. Shi, B. Klein, T.-Y. Kao, Q. Hu, and J. Reno, "High-resolution heterodyne spectroscopy using a tunable quantum cascade laser around 3.5 THz," *Appl. Phys. Lett.*, vol. 98, p. 231109, 2011.

[13] N. Karpowicz, J. Dai, X. Lu, Y. Chen, M. Yamaguchi, H. Zhao, X.-C. Zhang, L. Zhang, C. Zhang, M. Price-Gallagher *et al.*, "Coherent heterodyne time-domain spectrometry covering the entire terahertz gap," *Appl. Phys. Lett.*, vol. 92, p. 011131, 2008.

[14] H.-W. Hübers, R. Eichholz, S. G. Pavlov, and H. Richter, "High resolution terahertz spectroscopy with quantum cascade lasers," *J. Infrared Milli. Terahz. Waves*, vol. 34, p. 325, 2013.

[15] B. S. Williams, S. Kumar, Q. Hu, and J. L. Reno, "Operation of terahertz quantum-cascade lasers at 164 K in pulsed mode and at 117 K in continuous-wave mode," *Opt. Express*, vol. 13, pp. 3331–3339, 2005.

[16] M. A. Belkin, J. A. Fan, S. Hormoz, F. Capasso, S. P. Khanna, M. Lachab, A. G. Davies, and E. H. Linfield, "Terahertz quantum cascade lasers with copper metal-metal waveguides operating up to 178 K," *Opt. Express*, vol. 16, pp. 3242–3248, 2008.

[17] A. Adam, I. Kašalynas, J. Hovenier, T. Klaassen, J. Gao, E. Orlova, B. Williams, S. Kumar, Q. Hu, and J. Reno, "Beam patterns of terahertz quantum cascade lasers with subwavelength cavity dimensions," *Appl. Phys. Lett.*, vol. 88, p. 151105, 2006.

[18] J. Yu, S. Slivken, S. Darvish, A. Evans, B. Gokden, and M. Razeghi, "High-power, room-temperature, and continuous-wave operation of distributed-feedback quantum-cascade lasers at λ = 4.8 μm," *Appl. Phys. Lett.*, vol. 87, p. 041104, 2005.

[19] S. Golka, C. Pflügl, W. Schrenk, and G. Strasser, "Quantum cascade lasers with lateral double-sided distributed feedback grating," *Appl. Phys. Lett.*, vol. 86, p. 111103, 2005.

[20] F. Xie, C. G. Caneau, H. P. LeBlanc, N. J. Visovsky, S. Coleman, L. C. Hughes, and C.-e. Zah, "High-temperature continuous-wave operation of low power consumption single-mode distributed-feedback quantum-cascade lasers at λ = 5.2 μm," *Appl. Phys. Lett.*, vol. 95, p. 091110, 2009.

[21] B. S. Williams, S. Kumar, Q. Hu, and J. L. Reno, "Distributed-feedback terahertz quantum-cascade lasers with laterally corrugated metal waveguides," *Opt. Lett.*, vol. 30, pp. 2909–2911, 2005.

[22] A. J. L. Adam, I. Kašalynas, J. N. Hovenier, T. O. Klaassen, J. R. Gao, E. E. Orlova, B. S. Williams, S. Kumar, Q. Hu, and J. L. Reno, "Beam patterns of terahertz quantum cascade lasers with subwavelength cavity dimensions," *Appl. Phys. Lett.*, vol. 88, p. 151105, 2006.

[23] M. I. Amanti, M. Fischer, G. Scalari, M. Beck, and J. Faist, "Low-divergence single-mode terahertz quantum cascade laser," *Nature Photonics*, vol. 3, pp. 586–590, 2009.

[24] C. Wu, S. Khanal, J. L. Reno, and S. Kumar, "Terahertz plasmonic laser radiating in an ultra-narrow beam," *Optica*, vol. 3, pp. 734–740, 2016.

[25] Y. Jin, C. Wu, J. L. Reno, and S. Kumar, "Terahertz plasmonic lasers with narrow beams and large tunability," *Proc. SPIE* 10123, 1012312 (12pp), 2017.

[26] M. I. Amanti, G. Scalari, F. Castellano, M. Beck, and J. Faist, "Low divergence terahertz photonic-wire laser," *Opt. Express*, vol. 18, pp. 6390–6395, 2010.

[27] C. Sigler, J. Kirch, T. Earles, L. Mawst, Z. Yu, and D. Botez, "Design for high-power, single-lobe, grating-surface-emitting quantum cascade lasers enabled by plasmon-enhanced absorption of antisymmetric modes," *Appl. Phys. Lett.*, vol. 104, p. 131108, 2014.

[28] D. Botez, C.-C. Chang, and L. J. Mawst, "Temperature sensitivity of the electro-optical characteristics for mid-infrared (λ= 3–16 μm)-emitting quantum cascade lasers," *J. Phys. D: Appl Phys*, vol. 49, no. 4, p. 043001, 2015.

[29] T.-Y. Kao, Q. Hu, and J. L. Reno, "Perfectly phase-matched third-order distributed feedback terahertz quantum-cascade lasers," *Opt. Lett.*, vol. 37, pp. 2070–2072, 2012.

[30] A. Khalatpour, J. L. Reno, N. P. Kherani, and Q. Hu, "Unidirectional photonic wire laser," *Nature Photonics*, vol. 11, p. 555, 2017.

[31] O. Demichel, L. Mahler, T. Losco, C. Mauro, R. Green, A. Tredicucci, J. Xu, F. Beltram, H. E. Beere, D. A. Ritchie, and V. Tamošinuas, "Surface plasmon photonic structures in terahertz quantum cascade lasers," *Opt. Express*, vol. 14, p. 5335, 2006.

[32] Y. Chassagneux, R. Colombelli, W. Maineult, S. Barbieri, H. E. Beere, D. A. Ritchie, S. P. Khanna, E. H. Linfield, and A. G. Davies, "Electrically pumped photonic-crystal terahertz lasers controlled by boundary conditions," *Nature*, vol. 457, p. 174, 2009.

[33] S. Biasco, K. Garrasi, F. Castellano, L. Li, H. E. Beere, D. A. Ritchie, E. H. Linfield, A. G. Davies, and M. S. Vitiello, "Continuous-wave highly-efficient low-divergence terahertz wire lasers," *Nature Communications.*, vol. 9, p. 1122, 2018.

[34] A. Khalatpour, J. L. Reno, N. P. Kherani, and Q. Hu, "Phase-locked photonic wire lasers by $\pi$ coupling." *Nature Photonics*, vol. 13, p. 47, 2019.

[35] R. Kazarinov and C. Henry, "Second-order distributed feedback lasers with mode selection provided by first-order radiation losses," *IEEE J. Quantum Electron.*, vol. 22, p. 144, 1985.

[36] R. J. Noll and S. H. Macomber, "Analysis of grating surface emitting lasers," *IEEE J. Quantum Electron.*, vol. 26, p. 456, 1990.

[37] S. H. Macomber, J. S. Mott, B. D. Schwartz, R. S. Setzko, J. J. Powers, P. A. Lee, D. P. Kwo, R. M. Dixon, and J. E. Logue, "Curved-grating surface-emitting DFB lasers and arrays," *Proc. SPIE*, vol. 3001, p. 42, 1997.

[38] M. Kanskar, J. Cai, D. Kedlaya, D. Olson, Y. Xiao, T. Klos, M. Martin, C. Galstad, and S. H. Macomber, "High-brightness 975-nm surface-emitting distributed feedback laser and arrays," *Proc. SPIE*, vol. 7686, p. 76860J, 2010.

[39] G. Witjaksono and D. Botez, "Surface-emitting, single-lobe operation from second-order distributed-reflector lasers with central grating phaseshift," *Appl. Phys. Lett.*, vol. 78, no. 26, pp. 4088–4090, 2001.

[40] G. Witjaksono, S. Li, J. J. Lee, D. Botez, and W. K. Chan, "Single-lobe, surface-normal beam surface emission from second-order distributed feedback lasers with half-wave grating phase shift," *Appl. Phys. Lett.*, vol. 83, no. 26, pp. 5365–5367, 2003.

[41] S. Li, G. Witjaksono, S. Macomber, and D. Botez, "Analysis of surface-emitting second-order distributed feedback lasers with central grating phaseshift," *IEEE J. Sel. Topics Quant. Electron.*, vol. 9, pp. 1153–1165, 2003.

[42] C. Boyle, C. Sigler, J. D. Kirch, D. F. Lindberg, T. Earles, D. Botez, and L. J. Mawst, "High-power, surface-emitting quantum cascade laser operating in a symmetric grating mode," *Appl. Phys. Lett.*, vol. 108, p. 121107, 2016.

[43] D. H. Wu and M. Razeghi, "High power, low divergent, substrate emitting quantum cascade ring laser in continuous wave operation," *APL Materials*, vol. 5, p. 035505, 2017.

[44] G. Xu, R. Colombelli, S. P. Khanna, A. Belarouci, X. Letartre, L. Li, E. H. Linfield, A. G. Davies, H. E. Beere, and D. A. Ritchie, "Efficient power extraction in surface-emitting semiconductor lasers using graded photonic heterostructures," *Nature Comm.*, vol. 3, p. 952, 2012.

[45] D. Zhou, J.-F. Seurin, G. Xu, P. Zhao, B. Xu, T. Chen, R. V. Leeuwen, J. Matheussen, Q. Wang, and C. Ghosh, "Progress on high-power high-brightness VCSELs and applications," *Proc. SPIE*, vol. 9381, p. 93810B, 2015.

[46] Y. Yao, A. J. Hoffman, and C. F. Gmachl, "Mid-infrared quantum cascade lasers," *Nature Photonics*, vol. 6, p. 432, 2012.

[47] R. Köhler, A. Tredicucci, F. Beltram, H. E. Beere, E. H. Linfield, A. G. Davies, and F. Rossi, "Terahertz semiconductor-heterostructure laser," *Nature* vol. 417, pp. 156–159, 2002.

[48] M. S. Vitiello, G. Scalari, B. Williams, and P. D. Natale, "Quantum cascade lasers: 20 years of challenges," *Opt. Express*, vol. 23, p. 5167, 2015.

[49] D. Hofstetter, J. Faist, M. Beck, and U. Oesterle, "Surface-emitting 10.1 μm quantum cascade distributed feedback lasers," *Appl. Phys. Lett.*, vol. 75, p. 3769, 1999.

[50] W. Schrenk, N. Finger, S. Gianordoli, L. Hvozdara, G. Strasser, and E. Gornik, "Surface-emitting distributed feedback quantum-cascade lasers," *Appl. Phys. Lett.*, vol. 77, p. 2086, 2000.

[51] J. A. Fan, M. A. Belkin, F. Capasso, S. Khanna, M. Lachab, A. G. Davies, and E. H. Linfield, "Surface emitting terahertz quantum cascade laser with a double-metal waveguide," *Opt. Express*, vol. 14, p. 11672, 2006.

[52] S. Kumar, B. S. Williams, Q. Qin, A. W. M. Lee, Q. Hu, and J. L. Reno, "Surface-emitting distributed feedback terahertz quantum-cascade lasers in metal-metal waveguides," *Opt. Express*, vol. 15, p. 113, 2007.

[53] Y. Jin, L. Gao, J. Chen, C. Wu, J. L. Reno, and S. Kumar, "High power surface emitting terahertz laser with hybrid second-and fourth-order Bragg gratings," *Nature Comm.*, vol. 9, p. 1407, 2018.

[54] L. Mahler, A. Tredicucci, F. Beltram, C. Walther, J. Faist, H. E. Beere, and D. A. Ritchie, "High-power surface emission from terahertz distributed feedback lasers with a dual-slit unit cell," *Appl. Phys. Lett.*, vol. 96, p. 191109, 2010.

[55] N. Finger, W. Schrenk, and E. Gornik, "Analysis of TM-polarized DFB laser structures with metal surface gratings," *IEEE J. Quantum Electron.*, vol. 36, no. 7, pp. 780–786, 2000.

[56] D. Botez, L. J. Mawst, and G. Peterson, Electron. Lett. 24, 1328 (1988). "Resonant leaky-wave coupling in linear arrays of antiguides," *Electron. Lett.*, vol. 24, no. 16, pp. 1328–1330, 1988.

[57] D. Botez, L. Mawst, G. Peterson, and T. Roth, "Resonant optical transmission and coupling in phase-locked diode laser arrays of antiguides: The resonant optical waveguide array," *Appl. Phys. Lett.*, vol. 54, no. 22, pp. 2183–2185, 1989.

[58] D. Botez, L. J. Mawst, G. Peterson, and T. J. Roth, "Phase-locked arrays of antiguides: modal content and discrimination," *IEEE J. Quantum Electron.*, vol. 26, no. 3, pp. 482–495, 1990.

[59] T.-Y. Kao, Q. Hu, and J. L. Reno, "Phase-locked arrays of surface-emitting terahertz quantum-cascade lasers," *Appl. Phys. Lett.*, vol. 96, no. 10, p. 101106, 2010.

[60] Y. Halioua, G. Xu, S. Moumdji, L. Li, J. Zhu, E. H. Linfield, A. G. Davies, H. E. Beere, D. A. Ritchie, and R. Colombelli, "Phase-locked arrays of surface-emitting graded-photonic-heterostructure terahertz semiconductor lasers," *Opt. Express*, vol. 23, no. 5, pp. 6915–6923, 2015.

[61] J. D. Kirch, C.-C. Chang, C. Boyle, L. J. Mawst, D. Lindberg, T. Earles, and D. Botez, "5.5 W near-diffraction-limited power from resonant leaky-wave coupled phase-locked arrays of quantum cascade lasers," *Appl. Phys. Lett.*, vol. 106, no. 6, p. 61113, 2015.

[62] Y. Jin, J. L. Reno, and S. Kumar, "Phase-locked terahertz plasmonic laser array with 2 W output power in a single spectral mode," *Optica*, vol. 7, pp. 708–715, 2020.

# 10 Metasurface-based THz Quantum Cascade Lasers

Benjamin S. Williams and Christopher A. Curwen

University of California Los Angeles, Los Angeles, CA 90095

## 10.1 Introduction

Since its first demonstration in 2002 [1], the THz QC-laser has faced two major challenges that have limited its practical value. The first is the issue of operating temperature – the maximum operating temperature remains below room temperature – approximately 250 K for pulsed operation at present [2, 3]. We will not dwell further on this issue here, since it is mostly a challenge related to the intersubband physics of the active region. Furthermore, for many applications cryogenic cooling is acceptable, and the electrical cost of a cryogen-free cooling system can simply be folded into the overall wall plug efficiency of the laser. The second major challenge for THz QC-lasers has been how to efficiently couple the THz output power into a high-quality beam. In other words, one usually desires a beam pattern that is circular, symmetric, directive, diffraction-limited, and that can be obtained without throwing away power using after-the-fact beam conditioning. Achieving a high-quality high-power beam is a challenge for many types of semiconductor lasers – however, the problem is particularly acute for THz QC-lasers, for which the wavelength is long (on the order of 100 microns) compared to the active region thickness (on the order of 10 microns). Due to this mismatch in length scales, terahertz QC-lasers almost exclusively use metallic and/or plasmonic waveguides, which help to concentrate or even fully confine the mode within the sub-wavelength sized active region [1, 4–6]. However, unless special measures are taken, emission from sub-wavelength sized facets leads to highly divergent beams often with large side-lobes and interference fringes.

In this chapter, we provide an overview of a new class of approaches for THz QC-lasers: large-area surface-emitting structures that we will refer to as "metasurface" lasers. As we will use it, the term metasurface loosely refers to two-dimensional arrays of sub-wavelength surface radiating antenna elements, in which the antennas are loaded with the QC-laser gain material. These elements cover a large surface area (transverse dimensions of several wavelengths or more), so that the large radiating aperture can produce a more directive beam. Due to the sub-wavelength size of the antenna elements, it is convenient to use reflectarray antenna formalism to design and describe their operation. Properly speaking, a metasurface should involve structures with periodicities less than the free-space wavelength so that Bragg scattering doesn't play a role, and the structure is treated as an effective medium; however, as we will see, that constraint is not strictly followed for all the devices described in this chapter. We will discuss two main schemes in which these antenna elements can be coaxed to radiate in a phase-locked manner. First, we introduce the vertical-external-cavity surface-emitting-laser (VECSEL) in which the metasurface is placed within an external cavity. Second, we introduce the antenna-coupled metasurface (AC-MS) laser, in which antenna elements couple via mutual radiative coupling to form a phase-locked laser array that will spontaneously oscillate without external feedback.

### 10.1.1   Metal–Metal Waveguides – Building Blocks of THz QC Metasurfaces

The building block for these metasurface QC-lasers is the so-called "metal–metal" (MM) waveguide, also known as a "double-metal" or "double-plasmon" waveguide [6, 7]. In this structure, the waveguide mode is tightly confined between metal cladding placed immediately above and below the 5–10 μm thick epitaxial active region. Such a structure is usually fabricated using copper-to-copper or gold-to-gold thermocompression wafer bonding, followed by a selective wet etch to remove the substrate [8]. Following this step, one is left with the epitaxial active region layers directly bonded to a metal ground plane, ready for patterning via photolithography and etching into ridges (or any other geometry) using conventional microfabrication processes. Since any heavily doped semiconductor contact layers are usually quite thin, waveguide losses are dominated by absorption in the metal, and any re-absorption originating from quasi-free carriers within the active region itself (which is often not negligible). The structure resembles a THz version of a microstrip transmission line, a fact that can be leveraged to (a) analyze and design structures using transmission-line formalism and (b) design radiating structures based upon microstrip patch antenna concepts [9–14].

Metal–metal waveguides have certain advantages over surface-plasmon (SP) waveguides (the other type of waveguide used for THz QC-lasers). Metal–metal waveguides tend to have lower absorption and radiative losses compared to the SP waveguide – particularly at higher temperatures. As a result, they tend to have the best high-temperature performance, both in pulsed as well as in continuous-wave mode. For example, the highest operating temperature in pulsed mode in a MM waveguide is approximately 250 K, compared to approximately 120 K in a SP waveguide [2]. Furthermore, the field is strongly confined within the MM waveguide with no low-frequency cutoff for its $TM_{00}$ mode; this allows both the vertical and lateral dimensions to be made smaller than the wavelength. This in turn reduces the total thermal dissipation and required cooling power, which is reflected in the higher record temperature for continuous wave (cw) operation (129 K) in a MM waveguide, compared to approximately 80 K in a SP waveguide [8, 15].

The main advantage of the SP waveguide has been the fact that it exhibits a larger transverse mode size as a result of the mode extending into the semi-insulating substrate. This leads to more efficient radiation from the waveguide facet, and a somewhat more directive and better quality beam. For example, watt-level powers from SP devices have been observed by several groups [16, 17], which is about 10 times larger than observed from edge-emitting MM waveguide lasers. It turns out to be quite difficult to get the THz power out of a MM waveguide. If a MM ridge waveguide is simply cleaved to form a Fabry–Pérot cavity, it performs poorly as an edge-emitting laser. The cleaved facet radiates as a sub-wavelength sized aperture, which exhibits an extremely divergent beam [18]. Even worse, the emitting aperture is poorly imped-ance matched to free space; as a result the effective facet reflectivity becomes large, $R \approx 0.6$–$0.9$ (depending on the waveguide dimensions relative to the wavelength). This is much higher than the expected Fresnel value of $R \approx 0.32$ calculated from the GaAs/vacuum index mismatch [5, 9]. For realistic dimensions, the radiation losses are often an order of magnitude smaller than absorption losses, which implies an optical coupling efficiency of 10% or less, with total output powers up to tens of milli-watts at best. The ideal solution would be a structure that retains the better thermal performance of MM waveguide QC-lasers, while also achieving high optical cou-pling efficiency into a high-quality free-space beam. A large variety of approaches have been investigated, and some have been very successful at improving the beam quality, improving the coupling efficiency, and sometimes both. An incomplete list includes facet-mounted lenses [19] and horn antennas [20], surface-emitting second-order distributed feedback (DFB) [21, 22] or end-fire antenna DFB cavities [23–25], surface-emitting photonic crystals [26], and arrays of patch antennas [12]. This is the landscape into which THz metasurface QC-lasers emerge.

## 10.2    Metasurface External Cavity Lasers

### 10.2.1    Overview of the Metasurface VECSEL Concept

One very successful solution for semiconductor lasers in the visible and near-IR to the problem of achieving high output power from a laser in a high-quality beam, is the vertical-external-cavity surface-emitting-laser (VECSEL). The VECSEL can be thought of as a semiconductor version of a diode-pumped solid-state disk laser [27]. In a typical VECSEL configuration, a quantum well (or dot) semiconductor active medium is grown monolithically with a distributed Bragg reflector on the bottom side only (i.e., the lower half of a VCSEL). The semiconductor chip forms an amplifying mirror that is inserted into an external laser cavity; lasing occurs when it is optically pumped in a spot that overlaps with the cavity mode. The cavity can be readily engi-neered to support only the fundamental Gaussian mode with near diffraction-limited beam quality. The geometry is favorable for heat sinking and readily allows scaling up of the output powers [28, 29].

It is not possible to naïvely implement the VECSEL geometry for QC-lasers. This is because the intersubband polarization selection rule ensures that a cavity mode only experiences gain if its electric field is polarized perpendicular to the plane of the quantum wells. This is incompatible with the natural in-plane polarization for sur-face incident waves in a VECSEL cavity. It is for this reason that QC-lasers have had to use gratings or photonic-crystal structures to couple surface radiating modes to the intersubband transitions.

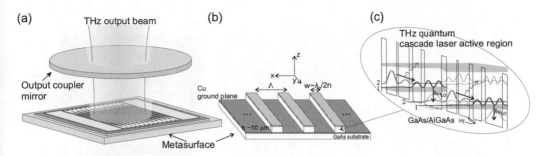

**Fig. 10.1**  (a) Schematic of metasurface QC-VECSEL configuration. (b) Close view of amplifying metasurface in its simplest form, which is made up of periodic arrays of sub-cavity antennas implemented in MM waveguide ridges of width $w$, height $h$, and period $\Lambda$, each loaded with QC-laser gain material. (c) Band diagram of typical GaAs/AlGaAs THz QC-laser active region. Figure partially adapted from [30] with the permission of AIP publishing.

In [30], Xu *et al.* addressed this problem by introducing the concept of an active reflectarray metasurface, which consisted of an array of MM waveguide sub-cavities loaded with QC gain material; each sub-cavity radiates as an antenna so as to efficiently couple in THz radiation, amplify it via stimulated emission, and re-radiate it into free space. The amplifying metasurface reflector is then paired with an output coupler mirror to create the external laser cavity. The schematic of the device is shown in Fig. 10.1(a). The simplest type of metasurface is made up of MM waveguides resonant in their first higher-order lateral mode ($TM_{01}$) at a cutoff wavelength approximately determined by $w \approx \lambda_0/2n$, where $n$ is the index of the quantum cascade semiconductor material as shown in Fig. 10.1(b) and (c). In this scenario each sub-cavity essentially acts as an elongated patch antenna and strongly radiates in the surface direction [9, 31, 32]. The MM waveguide sub-cavities are spaced with a period $\Lambda$ sufficiently smaller than the free-space wavelength to prevent higher-order Bragg scattering or excitation of lossy surface waves ($\Lambda < 0.8\lambda_0$ seems to be a sufficient criterion). While the period $\Lambda$ does have a second-order influence, the primary origin of the resonance is the dimensions of each sub-cavity – in other words the metasurface is not operating like a diffraction grating, but rather as an effective medium. Furthermore, in the VECSEL configuration, the metasurface is intentionally designed with strong radiative losses so that it acts as a low-quality-factor ($Q$) resonator and will not self-oscillate. For these reasons, the metasurface VECSEL approach is distinct from approaches that involve phase-locking of individual laser oscillators, such as arrays of second-order DFBs [33, 34], or the antenna-coupled metasurface (AC-MS) lasers discussed below [13]. It is the large-scale mode of the VECSEL cavity, rather than the sub-wavelength sized modes of the individual metallic sub-cavities on the metasurface, that exhibits lasing and shapes the beam to a near-Gaussian profile.

The amplifying metasurface can be readily understood by considering a finite-element (FEM) electromagnetic simulation of the reflectance spectrum. Results for a typical metasurface with a period $\Lambda = 70\,\mu m$ and ridge width $w = 11.5\,\mu m$ are shown in Fig. 10.2(a), including the effects of losses in the metal and semiconductor.

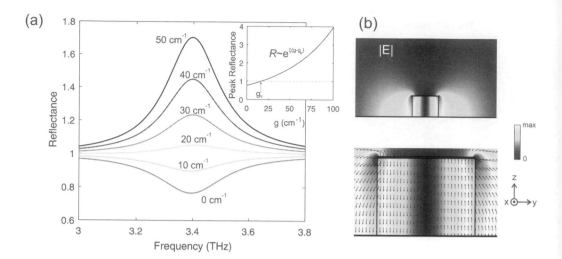

**Fig. 10.2** Finite element electromagnetic simulation of a typical active metasurface reflector. (a) Simulated reflectance spectra for different material gain coefficients $g$ for an active metasurface with MM waveguide ridge width of 11.5 µm and height of 10 µm, spaced in a period of $\Lambda = 70$ µm. The inset shows the peak reflectance vs. $g$. (b) Electric field intensity at resonance (2.94 THz). Lower panel shows close-up with electric field direction.

The results shown exhibit a characteristic resonant response peaked near 3.4 THz. As the QC material gain coefficient $g$ is increased above 18 cm$^{-1}$ – to simulate the effect of current injection – net amplification is observed. Figure 10.2(b) shows the electric field magnitude along with its vector direction, which is characteristic of the MM waveguide $TM_{01}$ mode at its cutoff frequency. The $TM_{01}$ sub-cavity antenna is similar in form to an elongated patch antenna, which is often analyzed according to the so-called cavity antenna model [35], in which one considers the radiation to originate from the fringing transverse fields at the waveguide sidewall surface. In the case of the ridge, the transverse magnetic fields are near zero at the sidewall, so one must only consider the transverse electric fields. The radiation from these fields can be considered to originate from equivalent magnetic current sources $\overrightarrow{M}_s = 2\hat{n} \times \vec{E}$, where $\hat{n}$ is the surface normal, and the factor of two reflects the presence of a ground plane [9, 35]. Although the $E_y$ field for the $TM_{01}$ mode is odd, $\hat{n}$ is also odd so that the sidewall magnetic currents $\overrightarrow{M}_s$ are in-phase for efficient surface normal radiation.

The choice of ridge width and period have a significant effect on the metasurface response. Figure 10.3(a) shows a set of simulated metasurface reflectance curves (including 40 cm$^{-1}$ of material gain) for various periods (with the width slightly adjusted to keep the resonance frequency constant). Figure 10.3(b) shows a set of reflectance curves for a fixed period ($\Lambda = 70$ µm) and various ridge widths. The resonance frequency is primarily determined by the width $w$, since the $TM_{01}$ cutoff resonance is approximately set by the condition that $w \approx \lambda_0/2n$. This gives a straightforward method to tune the response. The period $\Lambda$ has less direct effect on the resonant frequency, except when the period approaches the free-space wavelength

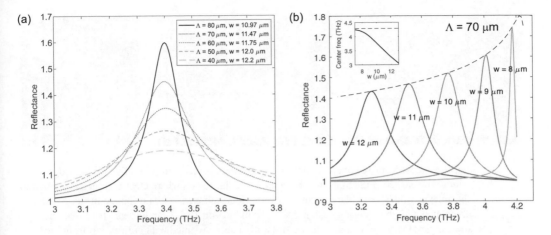

**Fig. 10.3** FEM simulation of metasurface reflectance for a set of metasurfaces with varying period $\Lambda$. The ridge width $w$ is varied slightly to keep the resonance frequency fixed at 3.4 THz. A fixed frequency-independent material gain coefficient of $40\,cm^{-1}$ is applied to the active material within the metasurface.

$\lambda_0$ at which an anticrossing occurs as the localized sub-cavity resonances mixes with surface waves. This effect can be seen in the inset of Fig. 10.3(b) for $w < 9\,\mu m$. Even away from this condition, the period $\Lambda$ significantly affects the quality factor of the resonance. As $\Lambda$ is reduced, the quality factor decreases due to increased coupling between adjacent sub-cavities leading to stronger radiative loss. This provides a mechanism to increase the bandwidth, albeit at the cost of a larger fill-factor. For further information, a full theory of metasurface QC-VECSEL performance is given in [36].

Several SEM images of a typical metasurface are shown in Fig. 10.4 at various levels of magnification. This particular metasurface is a "focusing" metasurface (discussed below in Section 10.2.2) in which the widths of the ridge sub-cavities are modulated as a function of position; for a uniform metasurface the ridge widths are identical. Figure 10.4(c) shows a close-up of a single MM waveguide ridge, with Ti/Au top contact, and Cu ground plane. The dotted circle in Fig. 10.4(a) represents the central circular bias area in which current is injected into the QC material. Outside this region, the ridges, tapered regions, and wire-bonding regions have a dielectric layer underneath the top metallization, so that no current is injected, and they exhibit loss rather than gain. This ensures that only the center of the metasurface provides gain, so that the fundamental Gaussian mode is preferentially pumped. A further consideration is the prevention of undesirable high-Q modes from lasing – for example, the fundamental confined $TM_{00}$ waveguide mode will lase unless losses are deliberately introduced at the ends of the waveguides to reduce feedback from reflections. Hence, the purpose of the tapered sections is to adiabatically transform the ridge waveguide to minimize reflections, and terminate traveling-wave waveguide modes in a large unbiased area where wire bonding occurs. The $TM_{01}$ antenna mode on which the

**Fig. 10.4** (a) SEM image of $2 \times 2 \, mm^2$ focusing metasurface with spatially varying ridge width. The dashed circle indicates the central circular region where electrical bias is applied to the antenna ridges. (b) Enlargement of several ridge sub-cavities. The change in ridge width associated with the focusing design discussed in Section 10.2.2 can be seen. (c) Enlargement of a single MM ridge with dry etched sidewall visible. Adapted with permission from [36], [OSA].

metasurface is based does not propagate longitudinally along the axis, and thus does not incur extra loss due to the terminations.

Compared to conventional THz QC-lasers, the advantages of the metasurface QC-VECSEL can be summarized as: (a) the beam quality is primarily determined and well-shaped by the external cavity and not individual sub-cavities, (b) the output power is scalable by electrically pumping a larger active area on the metasurface, (c) it is straightforward to achieve the optimum coupling condition and maximize the output power by optimizing the reflectance of the output coupler, and (d) the sparse arrangement of the sub-cavities reduces the power dissipation density for improved cw performance. The advantages (a)–(d) are somewhat related to the external cavity configuration; however, there is a further advantage (e) which is specific to the metasurface, in that one can engineer the phase, amplitude, and polarization response, in both the spatial and spectral domains.

## 10.2.2    Focusing Metasurface VECSEL

We begin by discussing the focusing metasurface QC-VECSEL introduced in [37] by Xu *et al.*, which serves as an example of the ability to engineer the metasurface phase. One of the most interesting functions realized in passive metasurfaces is focusing by spatially dependent phase shift imparted on a transmitted (or reflected) wave so that the action of a lens or a curved mirror is mimicked by a flat and thin equivalent structure [38–40]. This concept originates in the microwave regime in the form of the reflectarray antenna, in which a flat structure can be used to replace space-fed parabolic reflectors [41–43]. In its most common realization, a reflectarray comprises arrays of resonant patch antennas, which are used to engineer a spatially dependent reflection phase by varying a critical dimension of the patch. It is natural to implement this concept in an amplifying QC-metasurface for inclusion in a VECSEL cavity.

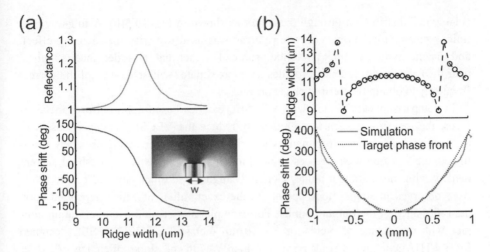

**Fig. 10.5** (a) Simulated reflectance and reflection phase from a uniform metasurface at a fixed frequency of 3.4 THz as a function of the ridge width w for $\Lambda = 70\,\mu m$. (b) Designed ridge width as a function of position (top) in order to achieve the target parabolic phase shift (bottom). The gray line gives the reflection phase extracted from a full-wave FEM simulation. Adapted with permission from [37], [OSA].

To date, QC-VECSELs have used flat output couplers. When paired with QC-metasurfaces with spatially uniform antenna elements a plano–plano Fabry–Pérot cavity is created. However, such a cavity lies at the edge of geometric stability, and therefore is intolerant to misalignment due to walk-off losses. This issue is particularly important for cavities longer than a few millimeters, or when using QC-metasurfaces with low peak-gain. The natural solution would be to use a concave mirror output coupler to form a stable hemispherical cavity. However, such components are not readily available in the THz range; planar OC components are far more convenient and can be easily manufactured using lithographic techniques [44]. By modifying the QC-metasurface to introduce a spatially dependent phase shift in reflection, it can act as an amplifying concave mirror to form a stable hemispherical cavity.

Control of the reflection phase at a fixed frequency is achieved by spatially modulating the ridge width both along and transverse to the ridges (see Fig. 10.4 and Fig. 10.5(b)). Figure 10.5(a) shows the reflection phase from a uniform metasurface at 3.4 THz as the ridge width $w$ is changed; one is able to vary the phase over a range of 311° by changing the width from 9 to 14 μm. However, detuning from the resonance condition also results in a reduction in the reflectance gain provided by that antenna element. A focusing metasurface designed for 3.4 THz was made up of 29 tapered MM waveguide ridges spaced with a period of $\Lambda = 70\,\mu m$. The modulation in ridge width is designed to achieve the target parabolic phase profile (for paraxial focusing) of $2\pi r^2 / R\lambda_0$, where $r$ is the radial distance to the metasurface center and $R$ is the effective radius of curvature (i.e., twice the desired focal length). As an example, a focusing metasurface designed with $R = 10\,mm$ at 3.4 THz has its transverse ridge width distribution through the center as shown in Fig. 10.5(b). A full-wave 2D finite element simulation of the entire array was used to verify the focusing effect, and matches with the target parabolic profile. The fact that the reflectance is highest near the resonance frequency provides an approximate "self-selection" of the correct frequency to obtain the desired phase profile.

In comparison with plano-plano VECSEL cavities based upon uniform metasurfaces, the focusing is seen to significantly reduce the VECSEL cavity's sensitivity to misalignment. In an experiment where the output coupler was deliberately misaligned, the focusing metasurfaces showed much slower increase in threshold current density with misalignment angle. Generally high slope efficiency and power levels were observed in single-mode operation, the exact values of which depended upon the reflectance of the output coupler. For example, at 77 K and using an output coupler with reflectance of ~60%, a $R = 10$ mm metasurface QC-VECSEL designed for 3.4 THz generated a peak power of 46 mW with the slope efficiency $dP/dI = 413$ mW/A. Since the active region used here has 163 periods, this corresponds to a slope efficiency of 18% (0.18 photons emitted per electron per period above threshold). At 6 K, the pulsed peak power increases to 78 mW, with $dP/dI = 572$ mW/A (25%) and a peak wall-plug efficiency reaching 1.15%. Continuous wave lasing is achieved at 10 K with peak power of 40 mW, $dP/dI = 339$ mW/A (15%), and wall-plug efficiency of 0.6%.

The focusing metasurface QC-VECSELs were observed to have excellent beam quality. The far-field beams were observed to be circular, very directive, close to Gaussian beam profile, as shown in Fig. 10.6(a) and (b). For example, a QC-VECSEL was tested with a $R = 20$ mm focusing metasurface and a circular 1 mm diameter bias area; its cavity length was 9 mm, with the output coupler external to the cryostat using a 3-mm-thick high-resistivity silicon cryostat window. This device produced a beam with 3.5° × 3.6° FWHM angular divergence. A similar QC-VECSEL built with a $R = 10$ mm metasurface produced a beam with 4.8° × 4.3° divergence. The measurement results agree with the expected divergence behavior of Gaussian modes in hemispherical resonators – the smaller value of $R$ produces a smaller spot on the output coupler, with consequent faster divergence in the far-field.

To further assess the beam quality, the beam propagation factor $M^2$ was measured using a knife-edge method through the focus of the beam along the propagation direction [45, 46]. The $M^2$ factor is the ratio of the angle of divergence of a laser beam to that of a fundamental Gaussian $TEM_{00}$ mode with the same beam waist diameter; it has a value of unity for a fundamental Gaussian beam. The beam waist evolution along the optical axis is shown in Fig. 10.7, with parameter fitting results. The best value measured is $M^2 = 1.3$ in both the $x$ and $y$ directions for a $R = 20$ mm metasurface QC-VECSEL. The peak power associated with this beam is 27 mW at 77 K, which leads to a high value of brightness $B_r = 1.86 \times 10^6$ Wsr$^{-1}$m$^{-2}$ given by $B_r = P/(M_x^2 M_y^2 \lambda^2)$, where $P$ is the output power. However, it must be noted that the ultimate beam quality is sensitive to cavity alignment. It is also noteworthy that only providing electrical bias to the center circular area with diameter of 1 mm is important in achieving high beam quality. By pumping only the center of the metasurface,

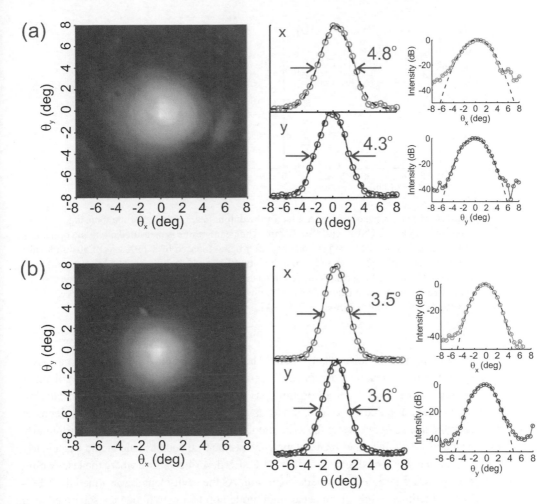

**Fig. 10.6** (a) The measured beam pattern from a focusing metasurface QC-VECSEL with $R = 10$ mm. (b) The measured beam pattern from a focusing metasurface QC-VECSEL with $R = 20$ mm. The angular resolution in measurement is $0.5°$. Black dashed lines are Gaussian curve fits to the 1D beam cuts through the beam center. 1D beams cuts are also plotted on a dB scale. Beams are measured at 77 K. Reprinted with permission from [37], [OSA].

the fundamental Gaussian mode exhibits the highest overlap, and is selectively excited. Several focusing devices were tested where the bias area was larger (1.5 mm diameter); for similar cavity conditions, these VECSELs were to exhibit beams with large side-lobes, indicating the presence of higher-order Hermite-Gaussian components in the cavity mode.

## 10.2.3 Intra-Cryostat Cavity VECSEL

The initial demonstrations of THz QC-VECSELs were based upon external cavities where the output coupler is placed external to the cryostat. This setup allows ease of alignment and changing of output couplers; however, it also prevents the cavity

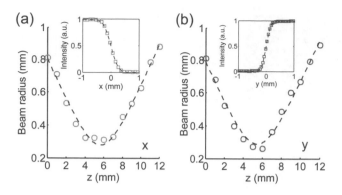

**Fig. 10.7** $M^2$ factor measurement results for the output beam directly from a focusing metasurface QC-VECSEL with $R = 20$ mm. The beam radius is measured along the optical axis ($z$-axis) in both the $x$- and $y$-directions after being focused by a TPX lens of 50 mm focal length which is placed 17 cm away from the VECSEL, and is represented by circles in (a) and (b), with the curve fitting results plotted in dashed line. The inset shows the knife-edge measurement raw data at beam waist position with curve fitting shown in dashed curve. Reprinted with permission from [37], [OSA].

from being made shorter than ~1 cm. The cryostat window introduces some loss – even a window made of a very-low-loss material such as high-resistivity silicon is estimated to exhibit 2–3% absorption loss per pass. Furthermore, if the cryostat window is not antireflection coated, it acts as an intra-cavity etalon filter, which tends to lock the lasing frequency to the axial modes of the window. This effect is particularly pronounced for windows made of high-index material such as silicon and was clearly observed in the focusing metasurface VECSEL described above where the laser would nearly always lase in single-mode operation. As the cavity length was tuned the VEC-SEL sometimes exhibited an occasional mode hop to a new frequency separated by a multiple of the free-spectral-range of the 3 mm silicon window.

In [47], Xu *et al.* reported an important development for improving the performance of THz QC-VECSELs: the introduction of the intra-cryostat cavity where the output coupler is placed inside of the cryostat. This allowed cavity lengths to be as short as several hundred microns. The primary disadvantage of this scheme is the inability to perform alignment once cold. However, if the output coupler is well aligned before cooldown, this disadvantage does not appear to be a serious impediment to good laser performance – at least for cavity lengths up to a few millimeters. Cavity lengths over 1 cm where misalignment might impose a greater penalty have not yet been explored in the intra-cryostat VCSEL configurations. As we will see in a later section, the removal of the window as an intra-cavity etalon also allows one to continuously tune the laser wavelength.

The intra-cryostat cavity configuration has yielded some of the best performing QC-VECSEL devices to date. One example is a 3.4 THz QC-VECSEL with a $2 \times 2$ mm$^2$ metasurface with a uniform ridge width of 11.5 μm and periodicity of 70 μm and a center circular bias area of 1 mm diameter. An output coupler was used

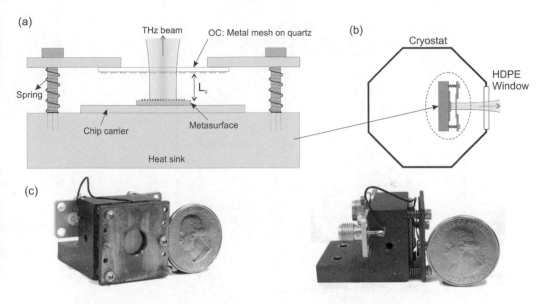

**Fig. 10.8** (a) Schematic of the intra-cryostat cavity design. (b) Schematic of the intra-cryostat QC-VECSEL mounted inside of a cryostat. (c) Front and side views of the actual intra-cryostat cavity setup. Figure adapted from [47], with the permission of AIP publishing.

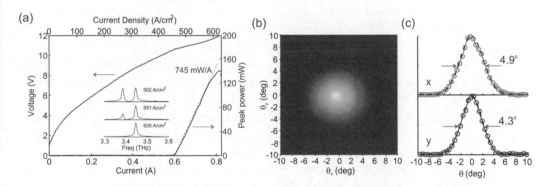

**Fig. 10.9** (a) Pulsed P–I–V for an intra-cryostat cavity QC-VECSEL based on a uniform metasurface paired with a metal mesh OC mounted inside the cryostat measured at 77 K. The inset shows the spectra change with the current injection level. (b) Measured 2D beam pattern for the intra-cryostat cavity QC-VECSEL. (c) 1D profiles of the beam in the x- and y-directions. Black dashed lines are the Gaussian curve fits. Figure © 2017 IEEE. Reprinted, with permission, from [36].

with a transmittance of 18–20% in the lasing frequency range. The best reported result is a high peak power of 140 mW in pulsed mode at 77 K with a slope efficiency of 745 mW/A, which corresponds to 0.33 photons emitted per electron per stage above threshold. The peak wall-plug efficiency (WPE) was 1.5%. The measured spectra show lasing in two neighboring longitudinal modes at low biases, which gradually evolves to a dominant high-frequency mode at a higher bias voltage. From the modal

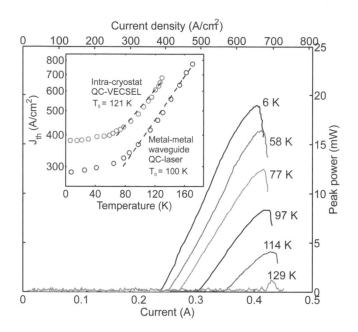

**Fig. 10.10** Pulsed P–I plots of the intra-cryostat QC-VECSEL measured at different cryostat temperatures. The inset plots the measured threshold current density, $J_{th}$, of the VECSEL and a reference ridge-waveguide QC-laser versus cryostat temperature, and the curve fits in black dashed line. Reprinted from [47], with the permission of AIP publishing.

separation in the lasing spectra, the cavity length is inferred to be 2.5 mm. The power is delivered in a near-Gaussian circular beam pattern with a FWHM divergence angle of $4.9° \times 4.3°$.

A similar intra-cryostat VECSEL device (focusing metasurface with a 10 mm effective radius of curvature, 0.7 mm bias area, 5% output coupler transmittance) was used to measure the P–I characteristics and thresholds as a function of temperature. These measurements allow one to assess the penalty of the VECSEL configuration on high-temperature performance compared to a MM ridge waveguide devices. In general, the threshold current density, $J_{th}$, is always higher for the VECSEL devices, indicating a higher total loss and threshold material gain. This data is provided in Fig. 10.10 and it shows that the VECSEL lases in pulsed mode up to a maximum temperature of 129 K. For comparison, a 50-μm-wide, 1.47-mm-long MM waveguide ridge fabricated from the same active material lased up to 170 K as shown in the inset to Fig. 10.10.

This loss penalty is a consequence of multiple effects. First, the MM sub-cavities that comprise the metasurface operate in the $TM_{01}$ mode at cutoff resonance; due to large fringing fields at the edges of the ridge, this mode intrinsically has a higher loss than the $TM_{00}$ mode of a MM waveguide. The simulated transparency gain (at which the metasurface gives unity reflectance) is 17.7 cm$^{-1}$ at the $TM_{01}$ mode cutoff frequency of ~3.4 THz, compared to 15.9 cm$^{-1}$ for a $TM_{00}$ mode (material parameters as in [47]). Second, the MM waveguide does not include losses due to cavity diffraction and absorption loss of the metal mesh output coupler that are present in a

QC-VECSEL cavity. Third, based upon the estimated Gaussian spot size and the bias area diameter of 0.7 mm, this QC-VECSEL has a transverse mode confinement factor of 0.76, which increases the estimated transparency gain from 17.7 cm$^{-1}$ to 23.3 cm$^{-1}$ and therefore further increases the threshold gain [36]. Despite the penalty in $J_{th}$ and $T_{max}$ for a VECSEL, in many cases the improved power output and beam pattern will still result in a net benefit.

The intra-cryostat cavity configuration is particularly beneficial for cw operation, which is essential for local-oscillator applications. In cw operation, thermal dissipation becomes very important, which requires efforts to minimize the total current consumed. In fact, the sparse arrangement of the active sub-cavities on the metasurface is a very favorable geometry for cw operation, since it reduces the power dissipated per unit area, so that one can maintain a large radiating aperture without excessive current draw. The sparsity is limited, however, since the period of the sub-cavities must remain smaller than the free-space wavelength ($\Lambda < 0.8\lambda_0$). Therefore, to further reduce the current consumption, one must reduce the active bias area on the metasurface. However, as the active area becomes smaller, the cavity length should also become smaller to allow a smaller modal spot size to be maintained without excessive diffraction losses. Such short cavities can only be obtained with an intra-cryostat setup.

The same device shown in in Fig. 10.10 was observed to lase in cw mode above liquid nitrogen temperature. The metasurface consumes a total power of less than 5 W. Measured power–current–voltage (P–I–V) curves in cw and pulsed modes are shown in Fig. 10.11(a). Steady cw operation with an output power of 5.1 mW is obtained from this intra-cryostat QC-VECSEL at 82 K, which improves to 13.5 mW at 16 K

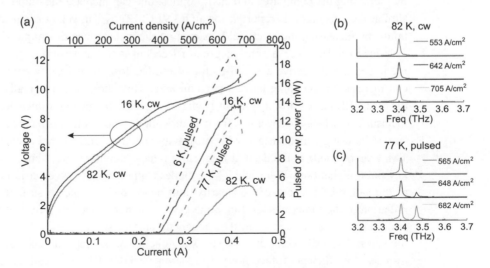

**Fig. 10.11** (a) cw and pulsed P–I–V curves at 6 K and 77 K for the intra-cryostat QC-VECSEL. (b) and (c) Lasing spectra at 82 K for cw mode (d) and 77 K for pulsed mode (c) at different current injection levels. Reprinted from [47], with the permission of AIP publishing.

[8, 48, 49]. The cw-mode slope efficiency at 82 K is 53 mW/A. In pulsed mode, the peak power reaches 12.5 mW at 77 K and 19 mW at 6 K. The slope efficiencies from this cw QC-VECSEL device are lower than those of the device shown in Fig. 10.9 because a much higher reflectivity output coupler (5% transmittance) is being used to keep the threshold of the cw QC-VECSEL as low as possible. The pulsed and cw lasing spectra at different biases at 77 K are shown in Fig. 10.11(b) and (c), which do not vary much at liquid-helium temperature. In cw mode, single-mode lasing at 3.403 THz is observed unchanged over the entire bias range (within the 7.5 GHz resolution of the FTIR spectrometer). In pulsed mode, lasing in two longitudinal cavity modes separated by 70 GHz is observed, from which the cavity length of 2.1 mm is deduced.

The 1D beam profiles are also measured to be identical in both cw mode (at two bias points) and in pulsed mode. This suggests that thermal lensing effects have a negligible impact on THz QC-VECSEL beams. This is attributed to the fact that the beam shape is primarily determined by the effective curvature radius of the focusing metasurface, which is primarily built-in by lithographic determination of the sub-cavity dimensions that determine the resonant phase response. Furthermore, the refractive index of GaAs does not change strongly with temperature for low lattice temperatures [50] (fractional changes on the order of $10^{-3}$ or less [51]), which means the metasurface resonances will shift very little as they are heated.

## 10.2.4    Power Scaling in VECSELs – Watt-Level Power

An advantage of conventional visible/infrared VECSELs is power scalability – simply by increasing the pump area and cavity optics, one can increase the output power. A similar demonstration is reported for a QC-VECSEL by Curwen *et al.* in [52]. Certainly, by increasing the electrical bias area on the metasurface the output power can be scaled up. However, another degree-of-freedom is the metasurface design itself. By increasing the active area fill-factor of the QC-loaded sub-cavity antennas, one can increase the pumping current per unit area. This allows one to greatly increase the VECSEL output power without having to increase the overall dimensions of the metasurface, which is convenient from a fabrication standpoint. This has been demonstrated by using a metasurface with sub-cavity antenna elements that resonate in the $TM_{03}$ lateral mode at the desired frequency, rather than the conventional $TM_{01}$. The key point is that both $TM_{01}$ and $TM_{03}$ modes support transverse electric fields ($E_y$) at the sidewalls that are of odd parity, and hence they couple to surface incident radiation in the same manner. Since the cavity antenna model predicts that the radiation takes place through equivalent magnetic current sources $\vec{M}_s = 2\hat{n} \times \vec{E}$, ($\hat{n}$ is the surface normal vector at each sidewall), the sidewall magnetic currents $\vec{M}_s$ are in-phase for efficient surface normal radiation for both the $TM_{01}$ and $TM_{03}$ modes (see Fig. 10.12(a) and (b)) [9, 35]. Using the $TM_{02}$ mode for example would not work, as it would produce a null in the far-field beam pattern at surface normal – it can be considered a "dark-mode" of the metasurface. Since the width of the $TM_{03}$

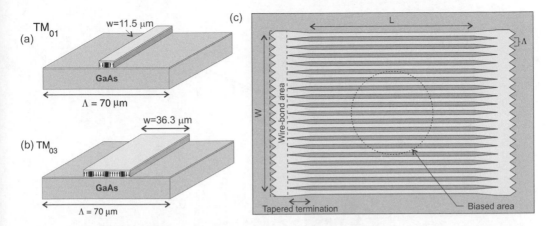

**Fig. 10.12** Cross-sectional schematic of metasurface unit cell that illustrates the difference between (a) $TM_{01}$ and (b) $TM_{03}$ metasurface designs. (c) Areal view of $TM_{03}$ metasurface. $W = L = 3$ mm for fabricated devices and bias area diameter = 1.5 mm. The top contacts of the laser ridges and terminations are shown in light gray and the ground plane is in dark gray. The tapered terminations are 300 μm long, and each wire bond area is another 300 μm in length, making the total length of the metasurface 4.2 mm. Reprinted from [52], with the permission of AIP publishing.

antenna is approximately three half-wavelengths (within the semiconductor) at resonance, the $TM_{03}$ ridges are approximately three times as wide as the $TM_{01}$ ridges for a given frequency. Hence for metasurfaces with identical periods $\Lambda$, the amount of gain material (and injection current) per unit area is likewise tripled.

Results from a $TM_{03}$ metasurface QC-VECSEL designed for 3.4 THz operation is shown in Fig. 10.13. The metasurface itself was $3 \times 3$ mm$^2$ with tapered terminations leading to wire-bonding areas, with a central circular bias area of diameter 1.5 mm. The VECSEL cavity was intra-cryostat, with an inductive metal mesh output coupler with 18% transmittance, and a cavity length of approximately 2 millimeters. The lasing frequency was measured to be 3.38 THz. In pulsed mode (550 ns pulses at a 2 kHz repetition rate), a maximum peak power of 1.35 W and peak slope efficiency of 767 mW/A (0.33 photons emitted per electron per period above threshold) is observed in a single mode with a narrow Gaussian-shaped beam. Absolute power levels were measured using a calibrated thermopile detector. A circular, high-quality beam with a full-width half-maximum divergence angle of 4° is consistently observed. The peak wall-plug efficiency at 6 K is just under 2%.

## 10.2.5 Tunable and Broadband VECSELs

### 10.2.5.1 Short-Cavity Tuning

As discussed above, the resonant frequency of the $TM_{01}$ metasurface is determined by the width of the individual MM ridges (sub-cavity resonators) that make up the metasurface. In most cases, this provides a low-quality-factor resonance ($Q \leq 20$) that allows the metasurface to provide gain over hundreds of gigahertz when bias is

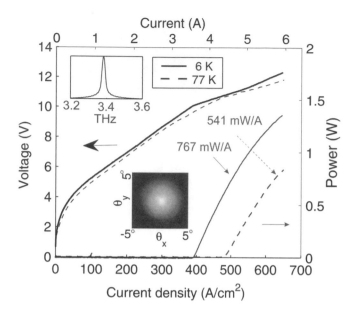

**Fig. 10.13** P–I–V data from $TM_{03}$ metasurface VECSEL using a low reflective output coupler with a measured transmission of ~18%. Inset shows emission spectrum and far-field beam measured at 77 K. Reprinted from [52], with the permission of AIP publishing.

applied (provided that the underlying quantum cascade material has sufficient gain bandwidth). However, the specific frequency at which the VECSEL lases is determined by the length of the external cavity, such that the roundtrip phase change is equal to an integer multiple of $2\pi$. Therefore, it is natural to consider building a frequency tunable VECSEL by adding mechanical tunability to the output coupler mirror position. In theory, the tuning range of such a construction is limited by either the bandwidth of the metasurface resonance, or the bandwidth of the quantum cascade gain material (whichever is narrower); however, extra consideration must be given to the possibility of multi-moding. If the free-spectral range (FSR) of the external cavity (FSR $= c/2L_{cav}$) is smaller than the gain bandwidth of the metasurface, multiple longitudinal modes of the external cavity may lase simultaneously. Therefore, to maximize the single-mode tuning range, the cavity must be made as short as possible – at least less than 1 mm. For a simple Fabry–Pérot cavity lasing in a longitudinal of order $m$, it can be shown that the maximum fractional tuning is given by the relation $\delta v/v = 1/(m + 1/2)$.

For practical purposes, the tunable VECSEL must be built entirely within the cryostat in order to avoid the etalon filtering effect of the cryostat window (which would prevent continuous tuning), and to allow a sufficiently short cavity length. A tunable setup was constructed by mounting the VECSEL output coupler on a piezoelectric translational stage, as is shown in Fig. 10.14. It has been observed that as the cavity length is reduced, the bias diameter must also be reduced in order to maintain single-mode behavior. For long cavity lengths (several mm or longer), metasurfaces with

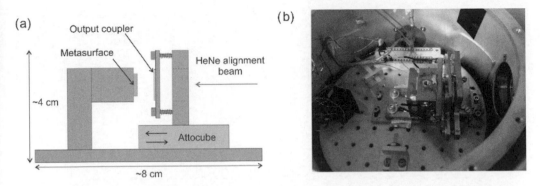

**Fig. 10.14** (a) Schematic of the piezoelectric tuning scheme for short-cavity QC-VECSEL. (b) Image of the piezoelectric tuning setup within a cryostat. Reprinted from [52], with the permission of AIP publishing.

small bias areas are unlikely to lase due to the diminishing overlap of the cavity mode with the active area; however, for short cavities, diffraction loss of the terahertz mode is reduced, allowing for intra-cavity modes with smaller transverse size. Furthermore, when a large bias area device is used with a very short cavity, it has been observed that multi-mode lasing is more likely to occur. This is believed to result from multiple small, independent cavity modes that simultaneously lase at different locations on the metasurface. Therefore, to maintain single-mode behavior at shorter cavity lengths, a smaller metasurface was used with overall dimensions of $1.5 \times 1.5\,mm^2$ with a $500\,\mu m$ diameter circular bias area. The metasurface antenna was $TM_{01}$ resonant type targeted at 3.4 THz. Additionally, for this experiment, the period of the metasurface was reduced to $\Lambda = 42\,\mu m$ (compared to the typical period of 60–70 μm). Reducing the period of the metasurface increases the radiative loss, leading to a lower-quality-factor resonance with a more broadband response (see Fig. 10.3).

Results from the piezoelectric setup are plotted in Fig. 10.15. Continuous, single-mode tuning was observed over a bandwidth of 650 GHz as the laser is tuned from 3.15 to 3.8 THz – approximately 19% of the fractional tuning range. For the most part, the emission spectrum is single-mode across the entire bias range. However, it is noted that at the extremes of the tuning range, the lasing frequency becomes bias dependent, and the single-mode behavior is only observed in certain bias ranges. The tuning range is limited by the free-spectral range of the external cavity; i.e., as the laser is tuned above 3.8 THz, it mode hops back to 3.1 THz. This implies that both the bandwidths of the QC-gain material and the metasurface resonance are broader than 650 GHz. Although the exact cavity length was not directly measured, from the observed free-spectral range, and the modeled reflection phase of the metasurface, we can infer that the cavity length was actuated from 177 to 130 μm (maintaining an optical cavity length of approximately two wavelengths). The measured beam patterns are plotted in Fig. 10.16. Good quality beams are observed throughout the tuning range; however, the divergence angle of the beam is larger (15°) than that of the previously studied devices (~5°). This is expected as the smaller bias diameter

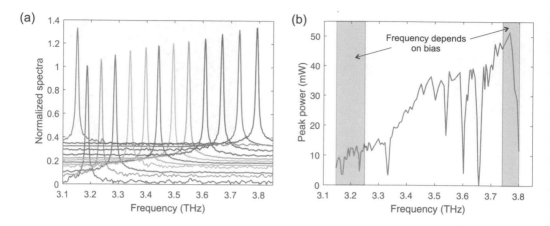

**Fig. 10.15** (a) Series of emission spectra (vertically offset for clarity) from a short-cavity tunable QC-VECSEL operated in pulsed mode at 77 K with a single-resonator metasurface as length of cavity is changed. (b) Measured peak emission power from the device in (a) as it is tuned. The dips are a result of atmospheric absorption related to the distance from the device to the detector. Figure adapted from [53].

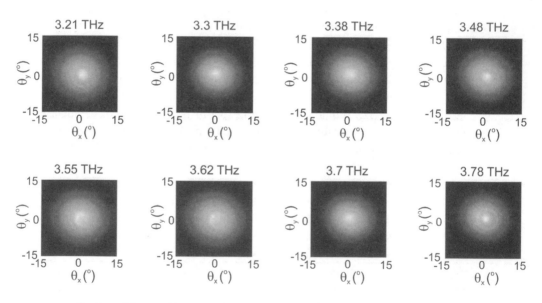

**Fig. 10.16** Measured beam patterns from short-cavity VECSEL (data taken with spectra in Fig. 10.15) as the laser is tuned. Figure adapted from [53].

forces the VECSEL to lase with smaller spot sizes that are more divergent. The peak output power varies from 10 to 40 mW across the tuning range in a non-trivial fashion, depending on a combination of factors such as the variation of the output coupler reflectance with wavelength, the laser gain lineshape, and the metasurface spectral response.

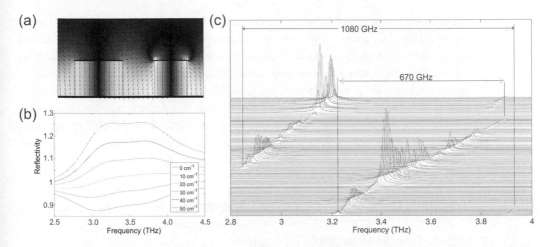

**Fig. 10.17** (a) FEM simulation of coupled resonator metasurface. Plot of E-field intensity with E-field direction given by arrows. (b) Simulated reflectance spectra as QC material gain is added to the sub-cavities. (c) Set of emission spectra (vertically offset for clarity) from short-cavity QC-VECSEL with a broadband coupled resonator metasurface as the cavity length is tuned. Spectra are for the most-part multi-mode, spanning an overall bandwidth of 1080 GHz.

## 10.2.5.2   Coupled Resonator Metasurfaces for Broadband Amplification

The ultimate tuning range for a short-cavity QC-VECSEL will be limited by a combination of the cavity (i.e., the maximum mirror displacement), and the gain bandwidth which is determined by both the QC gain material as well as the metasurface design, and the reflection phase of the metasurface itself. For a metasurface based upon a single sub-cavity per period, reducing the quality factor (e.g., by changing the MM waveguide height or period $\Lambda$) is the primary method of increasing the gain bandwidth. However, the resonance lineshape will remain close to Lorentzian. One can greatly increase the flexibility of the QC-VECSEL design by including multiple inhomogeneous sub-cavities within a unit cell of the metasurface. An example of this is shown in Fig. 10.17(a), where the unit cell is made up of two MM sub-cavity ridges with different widths; together they form a coupled resonator system which produces a considerably broader and flatter gain spectrum. This coupled resonator metasurface has been tested in a short-cavity tunable VECSEL configuration, and the observed spectra are plotted in Fig. 10.17(b). First, it is interesting to note that the tuning bandwidth of a single mode in this configuration was 670 GHz – almost the same as in the single ridge metasurface case shown in Fig. 10.15(a). Instead, two modes are observed to lase simultaneously at many cavity lengths; together lasing over very broad range of frequencies is observed (2.8–3.9 THz). This implies that the coupled resonator design does in fact broaden the metasurface response considerably to more than 1 THz bandwidth. Attempts to achieve a larger FSR by further decreasing the cavity length were not successful – the device ceased to lase. While offering no particular advantage for single-mode tuning in the measured configuration, single-mode lasing might be established over the full bandwidth in a different cavity configuration – for example, with the use of an external grating as a frequency selective feedback element (Littrow cavity configuration). Additionally, the relative flat gain spectrum of the coupled resonator metasurface corresponds to a significant reduction in the group delay dispersion of the reflected waves compared to the single-ridge design. This suggests that coupled resonator metasurface designs may be a promising approach for implementing frequency combs and related mode-locking schemes in QC-VECSELs, particularly when paired with a very broadband active material [54–56].

## 10.3     Monolithic Metasurface Lasers and Amplifiers

### 10.3.1     Antenna-Coupled Metasurface QC-Laser

While metasurface lasers based upon external cavities have many attractive features, the requirement for an external cavity also brings disadvantages – e.g., a VECSEL cavity requires alignment, stability, larger physical size, etc. However, it is possible to design a quantum cascade metasurface laser that does not require external feedback. The most notable example of this is a THz QC-lasers based upon a 2D array of sub-wavelength laser elements that are phase-locked via global antenna mutual coupling, as demonstrated by Kao et al. [13]. For convenience, we will refer to this as an antenna-coupled metasurface (AC-MS) QC-laser. Like the aforementioned metasurface VECSELs, the AC-MS laser is made up an array of sub-wavelength sized MM waveguide antenna elements loaded with QC-material. However, a critical difference is this: the metasurface (i.e., a collective mode of all mutually coupled antenna elements) has a sufficiently high quality factor to lase on its own with the available intersubband gain. Hence there is no need for an output coupler to reach lasing threshold. The challenge lies in how to ensure that the 2D array reaches a phase locked condition in the in-phase supermode that will produce a high-quality beam. How to phase lock large numbers of laser oscillators is a well-considered challenge in laser physics. Typical techniques of phase locking laser diode arrays include nearest neighbor coupling via the evanescent fields of the cavity mode (including antiguiding) [57], diffracted feedback from an external cavity (i.e., Talbot feedback) [58], connecting multiple laser waveguides to a single-mode waveguide (i.e., Y-coupling, end-coupling [33, 34, 59, 60]), or using leaky-wave coupling within the cladding between adjacent ridge waveguides [61–63]. The AC-MS laser introduces a new method of coupling, termed global antenna mutual coupling, where interaction of the radiated waves between each sub-wavelength sized laser array element mediates the coupling. Such a mechanism allows a more sparse distribution of laser sub-cavities compared an evanescent coupling scheme, since the evanescent field decays within several microns outside of MM waveguides. Due to the propagating-nature of the coupling waves, the phase relations between two elements can be adjusted via their spacing. The global antenna mutual coupling scheme has the most in common with leaky-wave coupling, since both are based upon radiated waves. However, leaky-wave coupling is typically

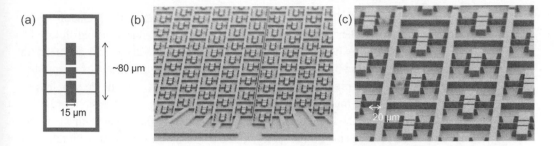

**Fig. 10.18** (a) Schematic of a three-section surface-emitting second-order DFB laser element. (b) and (c) SEM pictures of the antenna-coupled metasurface array. Air-bridge electrical contacts are visible in (c). Figure courtesy of T.-Y. Kao.

performed using 1D arrays of waveguides that support propagating modes, and the leaky-wave radiation is into the cladding layers. On the other hand, the AC-MS laser naturally supports 2D arrays of standing-wave cavities where the radiation occurs into free space.

In [13], the individual laser array element was chosen to be an ultra-short-cavity second-order distributed feedback laser with a deep-etched grating and a central $\pi$ phase defect so that it will emit in a strongly radiating mode. The basic design is shown in Fig. 10.18. For a design near 3 THz ($\lambda_0 = 100\,\mu m$), the element is approximately 10 $\mu$m high, 10–15 $\mu$m wide, and 70–80 $\mu$m long (approximately 2.5 wavelengths long within the semiconductor). Even such a small cavity supports multiple higher-order modes; however, one can ensure by design that the unwanted modes are sufficiently detuned from the gain spectrum that they are less likely to lase.

With this basic framework in place, the challenge lies in how to force the array of laser elements to phase-lock in the in-phase supermode, so that a far-field beam pattern with a single directive spot is produced. The problem is common across many laser phase locking schemes – one must ensure that the desired supermode exhibits the lowest total loss (absorption + radiative loss) at the frequency in which the gain medium produces the largest gain. This task is not trivial, since supermodes with overall odd-parity in their phases tend to cancel in the far-field, which reduces the overall radiative loss, and creates multi-lobed beam patterns. This is similar to the set of problems associated with second-order DFB lasers and photonic crystals, which have been addressed using a variety of techniques including introducing $\pi$ phase defects within the center of the cavity to change the overall parity of the mode, and forcing the laser to operate in the symmetric parity mode which strongly radiates by selectively introducing loss to de-select the non-radiative mode [48].

In [13], the issue of modeling radiative losses is addressed using an antenna array formalism which describes the mutual admittance between multiple electromagnetic emitters. Because of the sub-wavelength profile of the waveguide, each laser element can be modeled as radiating from a set of slot antennas. Such an approximation allows one to consider the total radiated power from an arbitrary array of laser elements (assuming a given set of phase relations) and quantify it as a mutual admittance. A large positive value of admittance corresponds to a reduction of the radiation resistance, which is equivalent to reduced radiative loss and a reduced lasing threshold. Such a simplified formalism allowed the optimization of the array grid geometry to ensure that the in-phase mode had the lowest losses among the various supermodes. From this point, more precise finite element electromagnetic eigenmode simulations can be used to finalize the design. When the correct sets of grid dimensions are used, the spatial mode with the lowest lasing threshold (at least at the wavelength where the active material provides maximum gain) will be the in-phase supermode. Simulations show that the expected far-field beam divergence narrows considerably with increasing number of elements. Another notable feature in the simulation is the reduction in the gain required to bring the laser array to threshold as the number of laser elements in increased, i.e., while the lasing threshold gain for a single element is simulated to be $g_{th} = 36 \, \text{cm}^{-1}$, for a nine-element array $g_{th} = 26 \, \text{cm}^{-1}$, and for a 16-element array $g_{th} = 22 \, \text{cm}^{-1}$. The larger the array size, the greater the ability of the collective system to harvest and re-emit radiation from nearby emitters, which increases the overall quality factor of the collective system.

AC-MS lasers were fabricated such that the bias area was separated into four separately biased quadrants. The individual laser elements were surrounded by an unbiased "pattern grid" that electrically connects to the top of each element via an air-bridge. Each quadrant contained between 28 and 37 individual laser elements. Figure 10.19 shows an image taken using a microbolometer THz camera of the metasurface when each of the four quadrants are sequentially biased. These images show a near-field spot consisting of a single spatial mode without nulls that moves depending upon which quadrant is biased. Pulsed voltage–current (V–I) and peak optical power–current (P–I) curves are shown in Fig. 10.20(a), where each quadrant is biased separately. From the V–I curves, it is seen that the threshold voltage for lasing is nearly the same for each array, and that it is significantly smaller than that for the single-element three-section DFB laser. This is an experimental confirmation of the reduction in radiative losses as the array size is scaled up. The threshold currents are different for each quadrant due to the fact that each quadrant contains a different number of elements. The current above threshold also increases at a faster rate with bias for an array vs. a single element. This is an indirect indication of an increased intra-cavity "circulating" THz intensity among the array elements; this stronger modal intensity within the active material leads to increased stimulated emission assisted transport that increases the differential conductance of the device. The spectral measurements (Fig. 10.20(b)) show single-mode emission from all four quadrants near 2.95–2.96 THz. The laser array from quadrant C1 delivers ~6.5 mW pulsed power (4% duty cycle) with a maximum slope efficiency of ~450 mW/A (20%) at 10 K. Such a large slope efficiency is a good indication that the radiative loss of the lasing mode is quite strong – it is possible that it could be optimized to further increase the efficiency and output power. While cw lasing was not reported, the sparse arrangement of small laser elements is likely to present a favorable geometry for heat removal. Far-field measurements show narrow and symmetric beam patterns were observed from the laser arrays, with beam divergence (full-width at half-maximum) less than $10 \times 10°$ for all cases (Fig. 10.20(c)).

Considering both the spectral and beam-pattern measurements, it is evident that these subwavelength confined lasers are strongly coupled and form a larger laser array with a single global phase and coherent emission. The array spans over a size of $\sim 8\lambda_0$ and includes 37 elements. The limits for scaling the size of the array are not clear, but it seems likely that much larger arrays may be possible, particularly since the far-field nature of the coupling mechanism allows some degree of global coupling and not just nearest neighbor.

## 10.3.2 Free-Space Metasurface Amplifier

An interesting application of the AC-MS laser is to bias it slightly below threshold, and to use it as a narrowband free-space THz amplifier. The THz range is lacking for convenient low-noise amplifiers. While QC-laser material has been used as a traveling-wave amplifier in the past, the typical configuration involves monolithic integration of the amplifier section with a THz laser source, i.e., a master-oscillator power-amplifier (MOPA) configuration [64]. QC-amplifiers for free-space radiation are difficult to develop, however, since it is very difficult to couple radiation in-and-out of the metallic/plasmonic waveguides without significant insertion loss. Indeed, the difficulty of fabricating MM waveguides with anti-reflective coatings means it is difficult to bias a putative QC-amplifier without it beginning to self-lase.

In [65], Kao *et al.* report an alternate approach that bears a resemblance to a resonant Fabry–Pérot amplifier, where the feedback from the cavity is used to create multi-pass amplification within the gain material, albeit with a much narrower bandwidth than a traveling-wave amplifier. The AC-MS acts as an amplifying grid reflectarray surface, which (a) eases coupling with free-space THz radiation without the need for extreme focusing and alignment, and (b) increases the volume of gain material that interacts with the radiation to increase saturation level. The laser element used is similar to that used in [13], except with a $3\lambda_0/2n$ length central section rather than a $\lambda_0/2n$ central section; 120 antenna elements are used covering a $2 \times 2 \, \mathrm{mm^2}$ area, with an overall areal fill factor of 8%.

Measurement takes place in the reflection mode configuration shown in Fig. 10.21(a), where single frequency emission from a separate DFB THz QC-laser is used as the source excitation. The metasurface amplifier is biased continuously at a temperature of 14–16 K, and the source laser is operated in pulsed mode at a 1–2% duty cycle. The THz signal reflected from the amplifier is measured using a liquid helium-cooled Ge:Ga photodetector using lock-in detection. The spectral characteristics of the source laser, along with the metasurface (biased above threshold) are shown in Fig. 10.21(b).

Figure 10.21(c) shows a set of plots of the reflected THz power at different amplifier bias voltages. As the bias on the amplifier is increased, so does the material gain of the embedded QC material. A clear amplification peak is observed. To measure the system-level gain, i.e., the ratio of the power measured at the output port after and before the introduction the THz light amplifier, the amplifier is replaced with a high-reflective mirror. A detailed source bias scan is then performed to obtain the reference

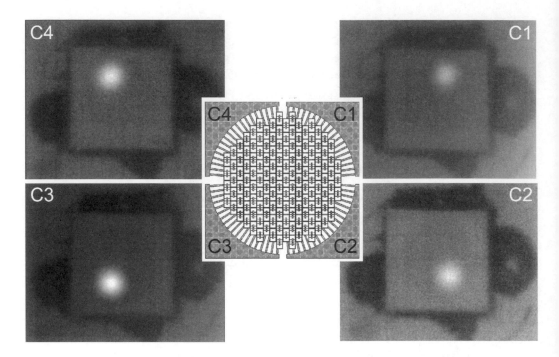

**Fig. 10.19** Terahertz images taken using a microbolometer camera of the emission from the laser array when different quadrants are biased. The images show narrow single-lobe symmetric terahertz emissions from the laser arrays. The lasers are only biased slightly above the lasing threshold to prevent the terahertz emissions from saturating the camera, so that the outline of the laser array chip could be visible in the images. Figure reproduced courtesy of T.-Y. Kao.

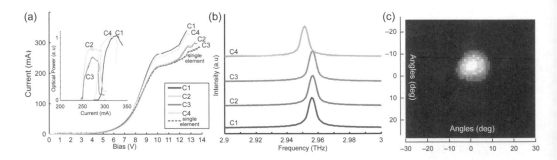

**Fig. 10.20** (a) Pulsed I–V curves and P–I curves (inset) taken when biasing each of the four quadrants of the metasurface (indicated by C1, C2, C3, C4). For comparison, the I–V curve of a single element laser is plotted. The thresholds can be observed in the I–V curves as a "kink" near 12 V associated with the change in differential resistance at threshold. The single element design lases at a significantly higher voltage. The current thresholds are different for the different quadrants because the number of elements is different. (b) Lasing spectra measured when each of the four quadrants are biased. (c) Typical far-field beam pattern when single quadrant is biased, with FWHM divergence of approximately 10 degrees. Figure courtesy of T.-Y. Kao.

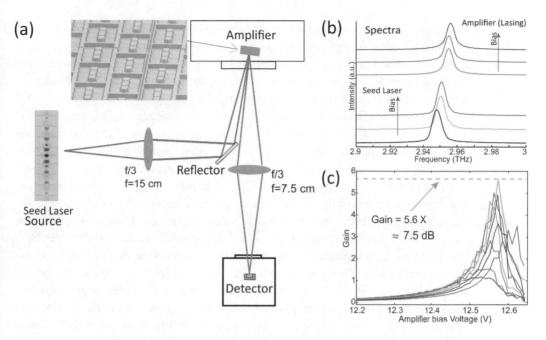

**Fig. 10.21** (a) Schematic of optical setup for testing the metasurface free-space amplifier under reflection mode with another THz QC-laser as the excitation source. (b) Pulsed spectra data from an amplifier device at 10 K. Spectra from seed laser are also shown. Both devices have single frequency emission at all bias. (c) Calibrated system-level gain curves of amplified signals under various amplifier/seed laser bias points. Figure adapted with permission from [65], [OSA].

power level at every source bias. The reference scheme also takes into account the power loss along the optical path due to strong atmosphere absorption at terahertz frequency range, which eliminates the need to purge the whole measurement setup. The amplification increases rapidly near the lasing threshold, and also strongly depends on the excitation frequency. A peak system-level gain of $5.6\times$ (7.5 dB) is obtained just before the lasing threshold bias is reached. From the photodetector and the lock-in amplifier output, the peak amplified signal is estimated to have $\sim$1 mW peak power. The bandwidth (3 dB) of the THz light amplifier is estimated to be only $\sim$500 MHz from the spectral data of the source laser. The limited bandwidth might be suitable or even desirable for applications such as preamplifiers for heterodyne detection systems targeting a specific spectral line or single frequency THz imaging. Larger gain levels may be achievable by increasing the fill factor of the antenna elements, by optimizing the coupling efficiency of incident radiation, and more precise control of the source laser frequency with respect to the amplifier resonance.

## 10.4     Conclusion

The relatively new development of metasurface based lasers is very promising for addressing one of the most limiting problems for THz quantum cascade lasers – the difficulty of obtaining excellent beam patterns with high wall-plug efficiencies and medium to high output powers. Results so far show that both the metasurface VEC-SELs and AC-MS lasers exhibit the ability to achieve high slope efficiencies of many hundreds of mW/A in pulsed mode, equivalent to roughly 0.1–0.33 photons emitted per injected electron per stage above threshold. Comparing exact values is tricky, since the same active material is not used in all cases. Nonetheless, the very fact that such high efficiencies are possible points to the value of being able to optimize the level of desirable radiative loss relative to other losses; i.e., output power can be maximized by engineering the correct balance between a large slope efficiency without overly increasing the lasing threshold. These numbers are the more notable for the fact that they are generally single-mode output in a high-quality beam pattern. For the AC-MS, the radiative loss is intrinsically tied into design of the monolithic metasurface itself (via engineering of the mutual admittance), whereas for the VECSEL the out-coupling is separately controlled by the choice of the output coupler transmittance/reflectance. Furthermore, although only a limited data in cw operation has been reported so far, both schemes share the advantage of having sparse distributions of active material, which generally helps to distribute the thermal load over a large area. The record-high reported power output of 19 mW in cw at 80 K from a VECSEL lends credence to this idea [53]. We believe it is likely that significant further improvements in fundamental performance metrics are possible. For example, the output power can be further increased by scaling up the metasurface size, smaller metasurfaces can lead to low-power consumption devices, new metasurface designs can potentially improve the slope and wall-plug efficiency, and improved heat sinking may improve cw performance.

For both QC-VECSELs and AC-MS lasers, there is a wide open design space to engineer the phase, amplitude, and polarization response, in both the spatial and spectral domains. This provides a platform to leverage the extensive past research in reflectarray antennas and emerging research in metasurfaces and flat optical components [38, 66]. One particular example of this approach is a polarimetric VECSEL, in which a metasurface was designed to electrically switch between orthogonal polarization states, depending upon how a certain sub-set of metasurface antennas was electrically biased [67]. In the future, QC-VECSELs and AC-MS lasers might include designs that directly generate specific (or even arbitrary) beams, such as vector beams, vortex beams, multi-lobed beams. What is more, dynamic electrical control of gain on the unit cell level can be incorporated into the metasurface itself.

The value of the QC-VECSEL external cavity and low-quality-factor metasurface is particularly evident when it comes to broadband tuning of single-mode lasers. The value of 650 GHz continuous tuning (19% of center frequency) is a record for a THz QC-laser, but it seems unlikely to be the limit – particularly since octave spanning THz QC-laser gain media has recently been demonstrated [56]. Widely tunable QC-VECSELs are particularly important, since spectroscopy is a major application for THz QC-lasers, and the various integrated Bragg gratings (i.e., DFB or distributed Bragg reflector) strategies for THz QC-laser beam control are not suitable for broadband tuning. A few demonstrations of THz QC-external cavity lasers have been reported, with the most successful demonstrations based upon surface-plasmon waveguides for either tuning [68] or detection via self-mixing [69]. However, it is nearly impossible to implement an external cavity over a large bandwidth based upon an edge-emitting MM waveguide, due to the difficulty of creating an antireflective coating at the sub-wavelength sized facet; a smoothly tapered horn antenna is needed to transform the waveguide impedance to that of free space. The metasurface *eliminates the facet entirely* and presents a large emitting aperture to ease the coupling of external optics. Furthermore, there is a wide design space available to create metasurfaces with a large gain bandwidth; when this is paired with a broadband QC-laser gain material extremely broad tuning may be possible. It is unclear whether the AC-MS approach can be adapted for broadband operation. Initial speculation suggests it cannot be, since the phase locking mechanism built into the spacing of the array grid, which is wavelength dependent. However, it is clear that the AC-MS is more suitable as a narrowband resonant amplifier due to its ability to reach lasing threshold.

On a related note, an open question is the possibility of developing a broadband frequency comb within a QC-VECSEL cavity. So far, THz QC-laser frequency combs have been based only on monolithic MM waveguides. However, it may be possible that the metasurface can be engineered to not only have extremely large gain bandwidth, but also to provide built-in dispersion compensation. However, it is believed that emergence of a comb state within QC-lasers is strongly related to the phenomena of spatial hole burning, which plays a key role in driving the multi-mode instability [70, 71]. A metasurface VECSEL experiences spatial hole burning in a very different manner than a MM waveguide ridge cavity. For example, in a VECSEL based upon a uniform metasurface made up of a single size of sub-cavity, even if multiple longitudinal cavity modes are lasing, their interaction with the active medium is mediated through the same $TM_{01}$ resonance. In other words, all longitudinal modes will burn the same spatial hole in the active region, which may suppress the onset of multi-mode lasing. This issue will require more study.

In the opinion of the authors, the metasurface laser approach is perfectly suited for the THz QC-laser. The THz range is unique in the photonic world in that using metallic/plasmonic waveguides/antennas is not just tolerated, but preferable from a loss perspective. This allows a ready combination of reflectarray concepts from microwave/mm-wave, metasurface photonic concepts, and quantum-electronic and laser cavity concepts. Moreover, in some sense, the world of THz QC-lasers has some of the greatest need for these ideas. Due to the wavelength/metal-waveguide size mismatch, the THz QC-lasers have struggled much more with beam quality compared to their mid-infrared cousins, and the facet reflectivity has been nearly impossible to eliminate for MM waveguides over a significant bandwidth. The metasurface approach now allows us to tap into the extensive library of external cavity techniques that have been widely developed for shorter wavelength lasers of all types. The surface has just been scratched, and we look forward to what the future holds.

## Acknowledgements

The authors would like to particularly thank Dr. Tsung-Yu Kao (Longwave Photonics/MIT) and Professor Qing Hu (MIT) for providing information and figures on the antenna-coupled metasurface lasers, and Dr. Luyao Xu (UCLA) for providing data and figures on VECSELs. The QC active material used in all lasers reported in this chapter was grown by molecular beam epitaxy by Dr. John Reno from Sandia National Laboratory Center for Integrated Nanotechnologies (CINT).

## References

[1]  R. Köhler, A. Tredicucci, F. Beltram, H. E. Beere, E. H. Linfield, A. G. Davies, D. A. Ritchie, R. C. Iotti, and F. Rossi, "Terahertz semiconductor-heterostructure laser," *Nature*, vol. 417, p. 156, 2002.

[2]  A. Khalatpour, A. K. Paulsen, C. Deimert, Z. R. Wasilewski, and Q. Hu, "High power portable terahertz laser systems," *Nature Photon.*, vol. 15, pp. 16–20, 2021.

[3]  L. Bosco, M. Franckié, G. Scalari, M. Beck, A. Wacker, and J. Faist, "Thermoelectrically cooled THz quantum cascade laser operating up to 210 K," *Appl. Phys. Lett.*, vol. 115, p. 010601, 2019.

[4]  B. S. Williams, "Terahertz quantum cascade lasers," *Nature Photon.*, vol. 1, pp. 517–524, 2007.

[5]  S. Kohen, B. S. Williams, and Q. Hu, "Electromagnetic modeling of terahertz quantum cascade laser waveguides and resonators," *J. Appl. Phys.*, vol. 97, p. 053106, 2005.

[6]  B. S. Williams, S. Kumar, H. Callebaut, Q. Hu, and J. L. Reno, "Terahertz quantum-cascade laser at $\lambda \sim 100\,\mu m$ using metal waveguide for mode confinement," *Appl. Phys. Lett.*, vol. 83, p. 2124, 2003.

[7]  K. Unterrainer, R. Colombelli, C. Gmachl, F. Capasso, H. Y. Hwang, A. M. Sergent, D. L. Sivco, and A. Y. Cho, "Quantum cascade lasers with double metal-semiconductor waveguide resonators," *Appl. Phys. Lett.*, vol. 80, p. 3060, 2002.

[8]  B. S. Williams, S. Kumar, Q. Hu, and J. L. Reno, "Operation of terahertz quantum-cascade lasers at 164 K in pulsed mode and at 117 K in continuous-wave mode," *Opt. Express*, vol. 13, p. 3331, 2005.

[9]  P. W. C. Hon, A. A. Tavallaee, Q.-S. Chen, B. S. Williams, and T. Itoh, "Radiation model for terahertz transmission-line metamaterial quantum-cascade lasers," *IEEE Transactions on Terahertz Science and Technology* vol. 2, no. 3, pp. 323–332, 2012.

[10] A. A. Tavallaee, P. Hon, K. Mehta, T. Itoh, and B. S. Williams, "Zero-index terahertz quantum-cascade metamaterial lasers," *IEEE J. Quantum Electron.*, vol. 46, pp. 1091–1098, 2010.

[11] C. Bonzon, I. C. B. Chelmus, K. Ohtani, M. Geiser, M. Beck, and J. Faist, "Integrated patch and slot array antenna for terahertz quantum cascade lasers at 4.7 THz," *Appl. Phys. Lett.*, vol. 104, p. 161102, 2014.

[12] M. Justen, C. Bonzon, K. Ohtani, M. Beck, U. Graf, and J. Faist, "2D patch antenna array on a double metal quantum cascade laser with >90% coupling to a Gaussian beam and selectable facet transparency at 1.9 THz," *Opt. Lett.*, vol. 41, pp. 4590–4592, 2016.

[13] T.-Y. Kao, J. L. Reno, and Q. Hu, "Phase-locked laser arrays through global antenna mutual coupling," *Nature Photon.*, vol. 10, pp. 541–546, 2016.

[14] T.-Y. Kao, X. Cai, A. W. M. Lee, J. L. Reno, and Q. Hu, "Antenna coupled photonic wire lasers," *Opt. Express*, vol. 23, p. 17091, 2015.

[15] M. Wienold, B. Röben, L. Schrottke, R. Sharma, A. Tahraoui, K. Biermann, and H. T. Grahn, "High-temperature, continuous-wave operation of terahertz quantum-cascade lasers with metal-metal waveguides and third-order distributed feedback," *Opt. Express*, vol. 22, pp. 3334–3348, 2014.

[16] L. Li, L. Chen, J. Zhu, J. Freeman, P. Dean, A. Valavanis, A. G. Davies, and E. H. Linfield, "Terahertz quantum cascade lasers with >1 W output powers," *Electron. Lett.*, vol. 50, pp. 309–311, 2014.

[17] M. Brandstetter, C. Deutsch, M. Krall, H. Detz, D. C. MacFarland, T. Zederbauer, A. M. Andrews, W. Schrenk, G. Strasser, and K. Unterrainer, "High power terahertz quantum cascade lasers with symmetric wafer bonded active regions," *Appl. Phys. Lett.*, vol. 103, p. 171113, 2013.

[18] A. J. L. Adam, I. Kašalynas, J. N. Hovenier, T. O. Klaassen, J. R. Gao, E. E. Orlova, B. S. Williams, S. Kumar, Q. Hu, and J. L. Reno, "Beam patterns of terahertz quantum cascade lasers with subwavelength cavity dimensions," *Appl. Phys. Lett.*, vol. 88, p. 151105, 2006.

[19] A. W. M. Lee, Q. Qin, S. Kumar, B. S. Williams, Q. Hu, and J. L. Reno, "High-power and high-temperature THz quantum-cascade lasers based on lens-coupled metal-metal waveguides," *Opt. Lett.*, vol. 32, pp. 2840–2842, 2007.

[20] M. I. Amanti, M. Fischer, C. Walther, G. Scalari, and J. Faist, "Horn antennas for terahertz quantum cascade lasers," *Electron. Lett.*, vol. 43, p. 573, 2007.

[21] S. Kumar, B. S. Williams, Q. Qin, A. W. M. Lee, Q. Hu, and J. L. Reno, "Surface-emitting distributed feedback terahertz quantum-cascade lasers in metal-metal waveguides," *Opt. Express*, vol. 15, p. 113, 2007.

[22] J. A. Fan, M. A. Belkin, F. Capasso, S. Khanna, M. Lachab, A. G. Davies, and E. H. Linfield, "Surface emitting terahertz quantum cascade laser with a double-metal waveguide," *Opt. Express*, vol. 14, p. 11672, 2006.

[23] M. I. Amanti, M. Fischer, G. Scalari, M. Beck, and J. Faist, "Low-divergence single-mode terahertz quantum cascade laser," *Nature Photon.*, vol. 3, pp. 586–590, 2009.

[24] T.-Y. Kao, Q. Hu, and J. L. Reno, "Perfectly phase-matched third-order distributed feedback terahertz quantum-cascade lasers," *Opt. Lett.*, vol. 37, pp. 2070–2072, 2012.

[25] C. Wu, S. Khanal, J. L. Reno, and S. Kumar, "Terahertz plasmonic laser radiating in an ultra-narrow beam," *Optica*, vol. 3, pp. 734–740, 2016.

[26] Y. Chassagneux, R. Colombelli, W. Maineult, S. Barbieri, S. P. Khanna, E. H. Linfield, and A. G. Davies, "Graded photonic crystal terahertz quantum cascade lasers," *Appl. Phys. Lett.*, vol. 96, p. 031104, 2010.

[27] A. C. Tropper, H. D. Foreman, A. Garnache, K. G. Wilcox, and S. H. Hoogland, "Vertical-external-cavity semiconductor lasers," *J. Phys. D: Appl. Phys.*, vol. 37, pp. R75–R85, 2004.

[28] M. Kuznetsov, F. Hakimi, R. Sprague, and A. Mooradian, "High-power (0.5-W CW) diode-pumped vertical-external-cavity surface-emitting semiconductor lasers with Circular TEM Beams," *IEEE Photon. Tech. Lett.*, vol. 9, pp. 1063–1065, 1997.

[29] B. Rudin, A. Rutz, M. Hoffmann, D. J. H. C. Maas, A.-R. Bellancourt, E. Gini, T. Südmeyer, and U. Keller, "Highly efficient optically pumped vertical-emitting semiconductor laser with more than 20 W average output power in a fundamental transverse mode," *Opt. Lett.*, vol. 33, pp. 2719–2721, 2008.

[30] L. Xu, C. A. Curwen, P. W. C. Hon, Q.-S. Chen, T. Itoh, and B. S. Williams, "Metasurface external cavity laser," *Appl. Phys. Lett.*, vol. 107, p. 221105, 2015.

[31] A. A. Tavallaee, B. S. Williams, P. W. C. Hon, T. Itoh, and Q.-S. Chen, "Terahertz quantum-cascade laser with active leaky-wave antenna," *Appl. Phys. Lett.*, vol. 99, p. 141115, 2011.

[32] Y. Todorov, L. Tosetto, J. Teissier, A. M. Andrews, P. Klang, R. Colombelli, I. Sagnes, G. Strasser, and C. Sirtori, "Optical properties of metal-dielectric-metal microcavities in the THz frequency range," *Opt. Express*, vol. 18, pp. 13886-13907, 2010.

[33] Y. Halioua, G. Xu, S. Moumdji, L. Li, J. Zhu, E. H. Linfield, A. G. Davies, H. E. Beere, D. A. Ritchie, and R. Colombelli, "Phase-locked arrays of surface-emitting graded-photonic-heterostructure terahertz semiconductor lasers," *Opt. Express*, vol. 23, p. 6915, 2015.

[34] T.-Y. Kao, Q. Hu, and J. L. Reno, "Phase-locked arrays of surface-emitting terahertz quantum-cascade lasers," *Appl. Phys. Lett.*, vol. 96, p. 101106, 2010.

[35] C. A. Balanis, *Antenna Theory – Analysis and Design*. Hoboken, NJ: Wiley-Interscience, 2005.

[36] L. Xu, C. A. Curwen, D. Chen, J. L. Reno, T. Itoh, and B. S. Williams, "Terahertz metasurface quantum-cascade VECSELs: theory and performance," *IEEE J. Sel. Topics Quantum Electron.*, vol. 23, p. 1200512, 2017.

[37] L. Xu, D. Chen, T. Itoh, J. L. Reno, and B. S. Williams, "Focusing metasurface quantum-cascade laser with a near diffraction-limited beam," *Optics Express*, vol. 24, pp. 24117-24128, 2016.

[38] N. Yu and F. Capasso, "Flat optics with designer metasurfaces," *Nature Materials*, vol. 13, pp. 139–150, 2014.

[39] A. V. Kildishev, A. Boltasseva, and V. M. Shalaev, "Planar photonics with metasurfaces," *Science*, vol. 339, p. 1232009, 2013.

[40] F. Aieta, P. Genevet, M. A. Kats, N. Yu, R. Blanchard, Z. Gaburro, and F. Capasso, "Aberration-free ultrathin flat lenses and axicons at telecom wavelengths based on plasmonic metasurfaces," *Nano Lett.*, vol. 23, pp. 4932–4936, 2012.

[41] J. Huang and J. A. Encinar, *Reflectarray Antennas*. Hoboken, NJ: Wiley-IEEE, 2007.

[42] Y. Rahmat-Samii, "Reflector Antennas," in *Antenna Engineering Handbook*, J. L. Volakis, Ed., New York: McGraw-Hill, 2007.

[43] D. Berry, R. Malech, and W. Kennedy, "The reflectarray antenna," *IEEE Trans. Antennas and Propagat.*, vol. 11, pp. 645–651, 1963.

[44] R. Densing, A. Erstling, M. Gogolewski, H.-P. Gemund, G. Lunderhausen, and A. Gatesman, "Effective far infrared laser operation with mesh couplers," *Infrared Phys.*, vol. 33, pp. 219–226, 1992.

[45] A. E. Siegman, M. W. Sasnett, and T. F. Johnston, "Choice of clip levels for beam width measurements using knife-edge techniques," *IEEE J. Quantum Electron.*, vol. 27, pp. 1098–1104, 1991.

[46] ISO 11146-1: Lasers and laser-related equipment – Test methods for laser beam widths, divergence angles and beam propagation ratios, 2005.

[47] L. Xu, C. A. Curwen, J. L. Reno, and B. S. Williams, "High performance terahertz metasurface quantum-cascade VECSEL with an intra-cryostat cavity," *Appl. Phys. Lett.*, vol. 111, p. 101101, 2017.

[48] G. Xu, R. Colombelli, S. P. Khanna, A. Belarouci, X. Letartre, L. Li, E. H. Linfield, A. G. Davies, H. E. Beere, and D. A. Ritchie, "Efficient power extraction in surface-emitting

semiconductor lasers using graded photonic heterostructures," *Nature Commu.*, vol. 3, p. 952, 2012.

[49] M. I. Amanti, G. Scalari, F. Castellano, M. Beck, and J. Faist, "Low divergence terahertz photonic-wire laser," *Opt. Express*, vol. 18, pp. 6390–6395, 2010.

[50] J. S. Blakemore, "Semiconducting and other major properties of gallium arsenide," *J. Appl. Phys.*, vol. 53, p. R123, 1982.

[51] B. S. Williams, S. Kumar, Q. Hu, and J. L. Reno, "Distributed-feedback terahertz quantum-cascade lasers with laterally corrugated metal waveguides," *Opt. Lett.*, vol. 30, p. 2909, 2005.

[52] C. A. Curwen, J. L. Reno, and B. S. Williams, "Terahertz quantum cascade VECSEL with watt-level output power," *Appl. Phys. Lett.*, vol. 113, p. 011104, 2018.

[53] C. A. Curwen, J. L. Reno, and B. S. Williams, "Broadband continuous single-mode tuning of a short-cavity quantum-cascade VECSEL," *Nature Photon.*, vol. 13, pp. 855–859, 2019.

[54] D. Burghoff, Y. Yang, J. L. Reno, and Q. Hu, "Dispersion dynamics of quantum cascade lasers," *Optica*, vol. 3, pp. 1362–1365, 2016.

[55] D. Burghoff, T.-Y. Kao, N. Han, C. W. I. Chan, X. Cai, Y. Yang, D. J. Hayton, J.-R. Gao, J. L. Reno, and Q. Hu, "Terahertz laser frequency combs," *Nature Photon.*, vol. 8, pp. 462–467, 2014.

[56] M. Rösch, G. Scalari, M. Beck, and J. Faist, "Octave-spanning semiconductor laser," *Nature Photon.*, vol. 9, pp. 42–47, 2015.

[57] D. E. Ackley, "Single longitudinal mode operation of high power multiple-stripe injection lasers," *Appl. Phys. Lett.*, vol. 42, p. 152, 1983.

[58] J. Katz, S. Margalit, and A. Yariv, "Diffraction coupled phase-locked semiconductor laser array," *Appl. Phys. Lett.*, vol. 42, p. 554, 1983.

[59] K. L. Chen and S. Wang, "Single-lobe symmetric coupled laser arrays," *Electron. Lett.*, vol. 21, pp. 347–349, 1985.

[60] W. Streifer, D. Welch, P. Cross, and D. Scifres, "Y-Junction semiconductor laser arrays: part I – theory," *IEEE J. Quantum Electron.*, vol. 23, pp. 744–751, 1987.

[61] D. Botez, "High-power monolithic phase-locked arrays of antiguided semiconductor diode lasers," *IEE Proc. J.*, vol. 139, pp. 14–23, 1992.

[62] D. Botez and G. Peterson, "Modes of phase-locked diode-laser arrays of closely spaced antiguides," *Electron. Lett.*, vol. 24, pp. 1042–1044, 1988.

[63] J. D. Kirch, C.-C. Chang, C. Boyle, L. J. Mawst, D. Lindberg, T. Earles, and D. Botez, "5.5 W near-diffraction limited power from resonant leaky-wave coupled phase-locked arrays of quantum cascade lasers," *Appl. Phys. Lett.*, vol. 106, p. 061113, 2015.

[64] H. Zhu, F. Wang, Q. Yan, C. Yu, J. Chen, G. Xu, L. He, L. Li, L. Chen, A. G. Davies, E. H. Linfield, J. Hao, P.-B. Vigneron, and R. Colombelli, "Terahertz master-oscillator power-amplifier quantum cascade lasers," *Appl. Phys. Lett.*, vol. 109, p. 231105, 2016.

[65] T.-Y. Kao, J. L. Reno, and Q. Hu, "Amplifiers of free-space terahertz radiation," *Optica*, vol. 4, pp. 713–716, 2017.

[66] P. Genevet and F. Capasso, "Holographic optical metasurfaces: a review of current progress," *Rep. Prog. Phys.*, vol. 78, p. 024401, 2015.

[67] L. Xu, D. Chen, C. A. Curwen, M. Memarian, J. L. Reno, T. Itoh, and B. S. Williams, "Metasurface quantum-cascade laser with electrically switchable polarization," *Optica*, vol. 4, pp. 468–475, 2017.

[68]  A. W. M. Lee, B. S. Williams, S. Kumar, Q. Hu, and J. L. Reno, "Tunable terahertz quantum cascade lasers with external gratings," *Opt. Lett.*, vol. 35, pp. 910–912, 2010.

[69]  Y. Ren, R. Wallis, D. S. Jessop, R. Degl'Innocenti, A. Klimont, H. E. Beere, and D. A. Ritchie, "Fast terahertz imaging using a quantum cascade amplifier," *Appl. Phys. Lett.*, vol. 107, p. 011107, 2015.

[70]  J. B. Khurgin, Y. Dikmelik, A. Hugi, and J. Faist, "Coherent frequency combs produced by self frequency modulation in quantum cascade lasers," *Appl. Phys. Lett.*, vol. 104, p. 081118, 2014.

[71]  A. Gordon, C. Y. Wang, L. Diehl, F. X. Kartner, A. Belyanin, D. Bour, S. Corzine, G. Hofler, H. C. Liu, H. Schneider, T. Maier, M. Troccoli, J. Faist, and F. Capasso, "Multimode regimes in quantum cascade lasers: From coherent instabilities to spatial hole burning," *Physical Review A*, vol. 77, p. 053804, 2008.

# 11 Terahertz Quantum Cascade Laser Sources Based on Intra-Cavity Difference-Frequency Generation

Mikhail A. Belkin

Technical University of Munich

## 11.1 Introduction

As discussed in Chapter 4, despite dramatic progress since their first demonstration, terahertz (THz) QCLs have not yet been able to meet the key desired characteristics for a semiconductor laser – room-temperature operation. A terahertz QCL source can also be produced using a nonlinear optical process of difference-frequency generation (DFG) inside of a dual-wavelength mid-infrared (mid-IR) QCL. These devices, known as THz DFG-QCLs, rely on a mid-IR QCL active region engineered to provide both laser gain for mid-IR pumps and giant intersubband nonlinear susceptibility $\chi^{(2)}$ for THz DFG between the mid-IR pumps.

THz DFG-QCLs are monolithic QCL devices and, from a user perspective, they look and operate similar to mid-IR QCLs. Upon application of a bias current, THz DFG-QCLs produce two mid-IR pump frequencies as well as the THz frequency output created via a DFG process inside of the laser cavity, as shown schematically in Fig 11.1(a). Since the optical nonlinearity for the THz DFG process does not require population inversion across THz transitions and since mid-IR QCLs can operate with high optical power at and above room temperature, THz DFG-QCLs can provide THz output at and above room temperature. This device technology is currently the only approach to produce a monolithic semiconductor laser source of coherent THz output that is operable at and above room temperature.

The timeline of the improvements in the power output of room-temperature THz DFG-QCLs is given in Fig. 11.1(b). Also shown in the figure is the timeline of improvements in the THz nonlinear conversion efficiency in these devices, which is defined as the ratio of the THz power output of the device to the product of its mid-IR pump power outputs. The nonlinear conversion efficiency is a convenient, although not a strictly defined, parameter that is used to characterize the efficiency of mid-IR frequency conversion in THz DFG-QCLs. By convention, it is computed using mid-IR powers output through the uncoated front QCL facet (corrected for the collection efficiency, if necessary) and the collected THz power output from the device.

Besides room-temperature operation, another important favorable feature of THz DFG-QCLs is their wide spectral tunability. Since the tuning bandwidth of the mid-IR QCLs can exceed 10 THz and the optical nonlinearity in THz DFG-QCLs remains significant across the entire THz spectral range, the tuning bandwidth of a single THz DFG-QCL device can span 1–6 THz, limited by the optical phonon absorption of the

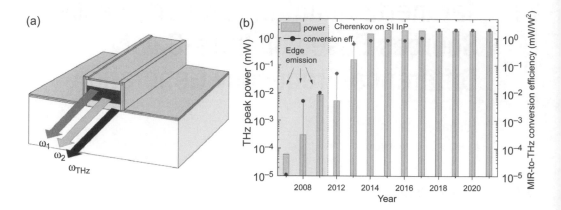

**Fig. 11.1** (a) Schematic of an edge-emitting ridge-waveguide THz DFG-QCL. (b) Timeline of the peak THz power output (bars and right axis) and the mid-IR-to-THz nonlinear conversion efficiency (circles and left axis) of THz DFG-QCLs at room temperature. Prior to 2012, THz DFG-QCLs were designed to confine the THz DFG wave in the laser waveguide and provide THz emission from the front facet of the waveguide. These devices are referred to as edge-emitting THz DFG-QCLs. Starting 2012, high-performance THz DFG-QCLs were all designed based on the Cherenkov emission into their semi-insulating (SI) InP substrates. These devices are referred to as Cherenkov THz DFG-QCLs.

III-As and InP materials on the high frequency end of the THz spectrum and by the reduction of the THz DFG efficiency, and increase in free-carrier losses on the low-frequency end [1, 2].

THz power output of room-temperature DFG-QCLs has been improving steadily since their initial demonstration in 2007 [3], as shown in Fig. 11.1(b). Record-performing devices with peak power output of nearly 2 mW and continuous-wave power output in excess of 14 μW at room temperature were reported at 3.4 THz in [4]. THz DFG-QCLs have a significant potential for further improvements in the THz generation efficiency and power as discussed below.

## 11.2    Operating Principles of THz DFG-QCLs

DFG is a nonlinear optical process in which two beams at frequencies $\omega_1$ and $\omega_2$ interact in a medium with nonlinear response described by the second-order nonlinear susceptibility tensor $\overset{\leftrightarrow}{\chi}^{(2)}$ to produce radiation at frequency $\omega = \omega_1 - \omega_2$ [5]. For the case of a DFG process between plane waves in a boundless uniform nonlinear crystal, the DFG intensity is given as [5, 6]

$$I(\omega) = \frac{\omega^2}{8\varepsilon_0 c^3 n_1 n_2 n} \left| \chi_{\text{eff}}^{(2)} \right|^2 \times I(\omega_1) I(\omega_2) \, l_{\text{eff}}^2, \tag{11.1}$$

where $l_{\text{eff}}^2 = ((k_1 - k_2 - k)^2 + (\alpha/2)^2)^{-1}$ is the effective length of the nonlinear inter-action; $I(\omega_i)$, $k_i$, and $n_i$ are the intensity, propagation constant, and refractive index of the nonlinear crystal, respectively, for the beam at frequency $\omega_i$; here we assumed that the nonlinear crystal is transparent for the pump waves and have the intensity loss coefficient $\alpha$ for the DFG wave. The value of the effective nonlinear susceptibility $\chi_{eff}^{(2)}$ in Eq. (11.1) is given by the product of the second-order nonlinear susceptibility tensor of the crystal $\overset{\leftrightarrow}{\chi}^{(2)}$ and the polarization unit vectors of the input and output beams, $\chi_{eff}^{(2)} = \hat{e} \times \overset{\leftrightarrow}{\chi}^{(2)} : \hat{e}_1 \hat{e}_2$, where $\hat{e}_i$ is the polarization unit vector of the beam at frequency $\omega_i$ [5, 6].

We can use Eq. (11.1) to estimate the value of $\left| \chi_{eff}^{(2)} \right|$ required to convert 1% of the mid-IR pump intensity into THz radiation in a QCL, assuming the same intensities for both mid-IR pumps. The intensity of the laser radiation in the current state-of-the-art mid-IR QCLs is limited to $\sim 10\,\mathrm{MW/cm^2}$ and $l_{eff}$ is smaller than 0.2 mm, even in the case of perfect phase-matching, $(k_1 - k_2 - k)^2 = 0$, due to free-carrier absorption of THz radiation in QCL waveguides, as shown in Fig. 11.2(a). Assuming a DFG frequency of 3 THz, $I(\omega_2) = I(\omega_2) = 10\,\mathrm{MW/cm^2}$, and $l_{eff} = 0.2$ mm, we obtain that $\left| \chi_{eff}^{(2)} \right|$ of approximately $2 \times 10^4$ pm/V is required to attain 1% of mid-IR to THz power conversion. Traditional nonlinear crystals have second-order nonlinear susceptibility values in the range of 1–500 pm/V [5], which would prevent us from attaining THz frequency conversion with sufficiently high efficiency in mid-IR QCLs. To solve this problem, the active regions in THz DFG-QCL devices are designed to provide giant optical nonlinearity of over $10^4$ pm/V associated with engineered resonant intersubband nonlinearity.

Intersubband transitions in multi-quantum-well semiconductor heterostructures can be designed to produce strong resonantly enhanced nonlinear optical response [12]. For the case of a THz DFG process, resonant coupled-quantum-well semiconductor heterostructures with electron subband configurations shown in Fig. 11.2(b) [9] and Fig. 11.2(c) [10] were reported in both InGaAs/AlInAs/InP and GaAs/AlGaAs materials systems. Optical nonlinearities over $10^6$ pm/V for the THz DFG process between mid-IR pumps have been experimentally measured [9, 10]. However, it would be difficult to integrate these nonlinear elements into mid-IR QCLs directly as these structures produce very large resonant intersubband absorption at the mid-IR pump frequencies that prevents high-performance room-temperature QCL operation [13]. The problem is solved by designing a QCL active region that simultaneously provides laser gain and resonant optical nonlinearity with population inversion for THz DFG. Two schematic examples of electron subband configurations that can provide giant nonlinear response for DFG with population inversion are shown in Fig. 11.2(d) and (e), after the QCL structures reported in [3] and [11], respectively. Figures 11.3(a) and (b) show the actual QCLs bandstructures [3, 11], which inevitably produce a more complex energy levels configurations than the idealized DFG diagrams shown in Fig. 11.2(d) and (e).

The contribution of intersubband transitions to the second-order nonlinear susceptibility can be computed using the quantum mechanical expressions for $\chi_{zzz}^{(2)}$ obtained with the time-dependent perturbation formalism in the electric-dipole approximation [5, 6]. Considering the subband configurations and energy states labeling shown in

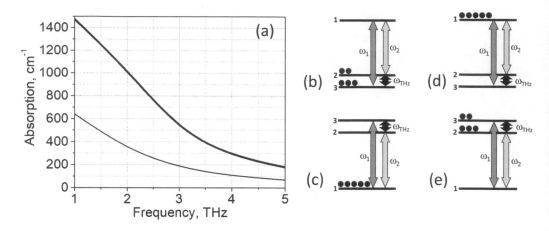

**Fig. 11.2** (a) Power absorption coefficient ($\alpha$) for THz radiation in InP n-doped to $2 \times 10^{16}$ cm$^{-3}$ (thin line) and $5 \times 10^{16}$ cm$^{-3}$ (thick line) due to free-carrier absorption. Calculations use a bulk Drude–Lorentz approximation with a scattering time constant of $\tau = 0.1$ ps [7] and neglect THz absorption of intrinsic InP [8]. Other doped materials in the QCL heterostructure, such as InGaAs and AlInAs ternary alloys, have similar (within a factor of ~2) values of THz loss for a given doping. (b) A schematic showing resonant DFG process between electron subbands in a passive multi-quantum-well heterostructure without population inversion, after [9]. (c) Another scheme for a passive resonant THz DFG process between electron subbands, after [10]. (d) A diagram showing resonant DFG process between electron subbands in an active electrically pumped QCL heterostructure with population inversion, after [3]. (e) Another scheme for the resonant DFG process between the electron subbands in an active electrically pumped QCL heterostructure with population inversion, after [11]. Black circles at different energy levels refer to electron population.

Fig. 11.2(d) and (e) and keeping only the resonant or close-to-resonance terms in the $\chi^{(2)}$ expression, one obtains [10, 15, 16]:

$$\chi_{zzz}^{(2)}(\omega_1, \omega_2) \approx -\frac{e^3 \Delta N_e}{\hbar^2 \varepsilon_0}$$

$$\times \left( \frac{z_{12}z_{23}z_{31}}{(\omega - \omega_{23} + i\Gamma_{23})(\omega_1 - \omega_{13} + i\Gamma_{31})} + \frac{z_{12}z_{23}z_{31}}{(\omega - \omega_{23} + i\Gamma_{23})(-\omega_2 + \omega_{12} + i\Gamma_{21})} \right.$$

$$\left. + \frac{z_{13}^2(z_{33} - z_{11})}{(\omega_{13} - \omega_2 + i\Gamma_{13})(\omega_1 - \omega_{13} + i\Gamma_{13})} \frac{\omega + i2\Gamma_{13}}{\omega + i\Gamma} + \frac{z_{12}^2(z_{22} - z_{11})}{(\omega_{12} - \omega_2 + i\Gamma_{12})(\omega_1 - \omega_{12} + i\Gamma_{12})} \frac{\omega + i2\Gamma_{12}}{\omega + i\Gamma} \right),$$

$$(11.2)$$

where $\omega_1$ and $\omega_2$ are the frequencies of the mid-IR pumps; $\omega = \omega_1 - \omega_2$ is the THz difference-frequency; $\Delta N_e$ is the population inversion density (assuming all the states with similar energies, i.e., states 2 and 3 in Fig. 11.2(d) and (e), have the same populations); $e$, $\hbar$, $\varepsilon_0$ are, respectively, the electron charge, the reduced Planck constant, and the permittivity in vacuum; and $ez_{ij}$, $\omega_{ij}$, and $2\Gamma_{ij}$ are, respectively, the transition dipole moment, the transition frequency, and the transition linewidth (full width at half maximum) between states $i$ and $j$. The value of $1/\Gamma$ can be approximated as the lifetime of the lower laser states (which is several times smaller than that of the upper laser

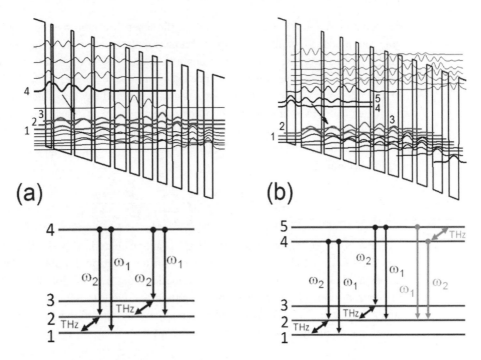

**Fig. 11.3** (a) Bandstructure (top panel) and energy level schematic (bottom panel) of a bound-to-continuum active region design for a THz DFG–QCL [3]. State 4 is the upper laser state, while states 3, 2, and 1 represent the lower laser states manifold. Relevant triplets of states that produce strong nonlinearity for THz DFG are shown in the bottom panel. (b) Bandstructure (top panel) and energy level schematic (bottom panel) of a dual-upper-state (DAU) active region design for a THz DFG–QCL [11]. States 5 and 4 are the upper laser states, while states 3, 2, and 1 represent the lower laser states manifold. Relevant triplets of states that produce strong nonlinearity for THz DFG are shown in the bottom panel. © IOP Publishing. Reproduced with permission from [14]. All rights reserved.

states). The $z$-direction is taken to be perpendicular to the semiconductor layers. The first two and last two terms in the parentheses of Eq. (11.2) correspond to the doubly resonant DFG process [9, 10] and the optical rectification process [10, 17], respectively. We note that only the first two terms were initially considered in the analysis of the intersubband nonlinearity for DFG [3]; however, the last two terms in Eq. (11.2) can be as significant as the first two terms, particularly for THz DFG below 3 THz [10, 15, 16].

Due to the additional requirements of electron transport and the maintenance of population inversion, the energy levels configurations in the actual DFG-QCL active region designs shown in Fig. 11.3 are significantly more complex than the idealized ones shown in Fig. 11.2. However, their nonlinear susceptibility can still be estimated analytically using Eq. (11.2) by first calculating the transition dipole moments $ez_{ij}$ between electron states from the computed QCL bandstructure (cf. Fig. 11.3), taking experimentally measured values of the intersubband transition linewidths, and, finally, deducing the population inversion $\Delta N_e$ at the laser threshold by equating the modal gain and modal loss and assuming the same populations for all the upper and, separately, all the lower laser states. The transition linewidths for mid-IR transitions can be estimated from the electroluminescence spectrum of the device, while the linewidths of the intersubband transition between states separated by THz frequencies can be estimated using two-dimensional nonlinear spectroscopy as described in [18].

Assuming the same populations in all of the upper and, separately, all of the lower laser states considered in Eq. (11.2), the expression for the optical gain in a QCL active region becomes

$$g(\omega) \approx \frac{e^2 \Delta N_e \omega}{\hbar \varepsilon_0 nc} \times \sum_{i,j} \frac{|z_{ij}|^2 \Gamma_{ij}}{(\omega - \omega_{ij})^2 + \Gamma_{ij}^2}, \tag{11.3}$$

where $n$ is the refractive index of the QCL active region, $\omega$ is the light frequency, c is the speed of light, $\Delta N_e$ is a difference in the electron populations between the upper and lower laser states, and the summation goes over all possible combinations of electron subbands $i$ and $j$ that correspond, respectively, to the upper and lower laser states. The value of $\Delta N_e$ can then be computed by specifying that the modal loss equals the modal gain at threshold:

$$g\Gamma = \alpha_{\text{tot}} = \alpha_{\text{wg}} + \alpha_{\text{m}}, \tag{11.4}$$

where $\alpha_{\text{wg}}$ and $\alpha_{\text{m}}$ are, respectfully, the waveguide loss and the mirror loss for the lasing mode, $\alpha_{\text{tot}}$ is the total loss, and $\Gamma$ is the modal overlap with the gain section. For typical DFG-QCL active regions, such as that shown in Fig. 11.3, this analysis yields the values of $\chi_{zzz}^{(2)}$ in the range 10–30 nm/V [1, 3, 11, 19, 20]. We note that these values of the nonlinearity result in a reasonably good agreement of theoretical calculations with the experimentally measured THz generation efficiency in DFG-QCLs as discussed in the following [1, 20].

The active region of THz DFG-QCLs can contain up to 90 quantum cascade (QC) stages [21, 22], with a typical number of 30–60 [23]. The stages could be the same or slightly different to broaden the mid-IR gain bandwidth of a device to be able to simultaneously generate mid-IR pumps separated by as much as 6 THz frequency difference. Starting from the initial demonstration [3], the active regions in the majority of THz DFG-QCL devices reported to date contain two stacks of QC-stages designed to provide slightly different mid-IR gain peaks to broaden the mid-IR gain bandwidth of the laser. However, there are also THz DFG-QCL devices with the active region based on multiple repetitions of only one QC-stage design [19, 21, 22].

To enable narrow-line THz emission from THz DFG-QCLs, two single-frequency mid-IR pumps are selected using distributed feedback (DFB) and/or distributed Bragg mirror (DBR) gratings etched in the laser cavity [1, 24] or by employing an external cavity (EC) setup with external diffraction gratings [2, 18]. Broadly tunable THz DFG-QCL sources, described in the latter sections of this chapter, typically employ a DFB grating to select one mid-IR pump frequency and an EC diffraction grating to tune the frequency of the second mid-IR pump to produce THz DFG frequency tuning [1, 2].

## 11.3 Edge-Emitting THz DFG-QCLs Based on Modal Phase-Matching

The first THz DFG-QCLs to be developed were designed with their waveguides supporting confined optical modes at both mid-IR pump and THz DFG frequencies [3, 20, 24, 25]. These devices are grown on doped InP substrates and are processed similarly to regular ridge-waveguide or buried heterostructure mid-IR QCLs. Mid-infrared modes are confined in the QCL active region by dielectric (InP) cladding layers, while the THz mode is confined between the device top metal contact layer and the doped substrate. It is also possible to implement THz DFG-QCLs with double-metal waveguides for THz mode confinement, similar to the double-metal waveguide of the THz QCLs described in Chapter 4. In this case, mid-IR laser modes are still confined in the waveguide core by the InP cladding layers. A typical waveguide configuration for THz DFG-QCLs based on modal phase-matching is described in Fig. 11.4, after the device reported in [20]. The ridge waveguides of the edge-emitting THz DFG-QCLs are often tapered towards the output facet to reduce THz impedance mismatch [7] between the subwavelength QCL waveguide and free space.

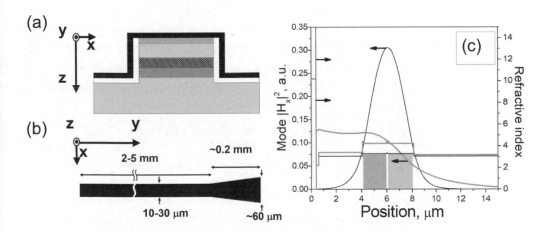

**Fig. 11.4** (a) Waveguide structure of a typical THz DFG-QCL based on modal phase-matching: the gold layer is shown in black, the insulating $Si_3N_4$ layer is in white, and the semiconductor is in gray (for low-doped InP layers) and dark gray (for a high-doped InP layer near the metal contact and the two active region sections in the waveguide core). (b) Ridge waveguide shape, viewed from the top: the back facet (left side) may have a high-reflection coating. Optional waveguide tapering may be done to facilitate THz radiation outcoupling to free space. (c) The magnetic field intensity in the $TM_{00}$ waveguide mode (right axis) and the refractive index profile (left axis) for a THz wave (thick gray line, data for $\lambda = 60\,\mu m$ ($\nu = 5\,THz$) is shown) and a mid-IR pump (thin black line; data for the $\lambda \approx 9\,\mu m$ mode is shown). Note that both mid-IR pumps have similar mode structure owing to their similar wavelengths. Also shown in gray are the two stacks of QC-stages in the active region, with each stack having a gain peak at a slightly different mid-IR wavelength. Reproduced from [20], with the permission of AIP Publishing.

Equation (11.1) needs to be modified in the case of a DFG process between TM-polarized optical modes in a QCL waveguide with the intersubband optical nonlinearity $\chi_{zzz}^{(2)}$ situated in the waveguide core. In this case, the power of the difference-frequency wave in the waveguide is given as [3]

$$W(\omega) \approx \frac{\omega^2}{8\varepsilon_0 c^3 n_1 n_2 n} \frac{\left|\chi_{eff}^{(2)}\right|^2}{S_{eff}} \times W(\omega_1) W(\omega_2) l_{eff}^2, \tag{11.5}$$

where $l_{eff}^2 = ((\beta_1 - \beta_2 - \beta)^2 + (\alpha/2)^2)^{-1}$ is the effective length of the nonlinear interaction; $W(\omega_i)$, $n_i$, and $\beta$ are the power, effective refractive index, and propagation constant of the optical mode at frequency $\omega_i$; $\alpha$ is the intensity loss of the mode at difference-frequency; and we have assumed no waveguide loss for the pump waves at $\omega_1$ and $\omega_2$.

Referring to the coordinate system shown in Fig. 11.4 and considering waveguide modes propagating in the $y$-direction, the expression for the term $\frac{\left|\chi_{eff}^{(2)}\right|^2}{S_{eff}}$ in Eq. (11.5) is given as [3]

$$\frac{\left|\chi_{eff}^{(2)}\right|^2}{S_{eff}} = \frac{\left|\int H_{x,\omega}^*(x,z)\,\chi_{zzz}^{(2)}(x,z)\,H_{x,\omega_1}(x,z)\,H_{x,\omega_2}(x,z)\,dxdz\right|^2}{\int \left|H_{x,\omega}(x,z)\right|^2 dxdz \int \left|H_{x,\omega_1}(x,z)\right|^2 dxdz \int \left|H_{x,\omega_2}(x,z)\right|^2 dxdz}, \tag{11.6}$$

where $H_{x,\omega_i}(x,z)$ is the magnetic field amplitude distribution across the waveguide in a TM-polarized mode at frequency $\omega_i$. One may separate the optical nonlinearity, $\chi_{eff}^{(2)}$, and the effective area of modal overlap $S_{eff}$ in Eq. (11.6) if one takes the value of $\chi_{eff}^{(2)}$ to be the maximum value of $\chi_{zzz}^{(2)}(x,z)$ in a QCL waveguide core. The expression for $S_{eff}$ then becomes

$$S_{eff} = \frac{\int \left|H_{x,\omega}(x,z)\right|^2 dxdz \int \left|H_{x,\omega_1}(x,z)\right|^2 dxdz \int \left|H_{x,\omega_2}(x,z)\right|^2 dxdz}{\left|\int H_{x,\omega}^*(x,z) \frac{\chi_{zzz}^{(2)}(x,z)}{\chi_{eff}^{(2)}} H_{x,\omega_1}(x,z) H_{x,\omega_2}(x,z)\,dxdz\right|^2}. \tag{11.7}$$

Typical values of $\chi_{eff}^{(2)}$ and $S_{eff}$ for THz DFG-QCLs based on modal phase-matching are 10–30 nm/V (corresponding to the maximum values of $\chi_{zzz}^{(2)}$ in the DFG-QCL active regions) and 500–1500 $\mu m^2$, respectively, where we assumed the QCL ridge width of 10–30 $\mu m$ for the calculations of $S_{eff}$ in Eq. (11.7) [20].

Phase-matching, $(\beta_1 - \beta_2 - \beta)^2 = 0$, is required to maximize $l_{eff}$ and the THz power output in Eq. (11.5). Fortunately, the values of refractive indices of intrinsic As-based III-V semiconductor compounds in the terahertz frequency range are significantly higher than that in the mid-IR range, due to the presence of strong optical phonon absorption region positioned between the two bands. This is shown for the case of semi-insulating InP in Fig. 11.5(a) [8]. By controlling the doping and the thickness of the waveguide cladding layers as well as by changing the waveguide width, it is possible to effectively phase-match the THz DFG process in the DFG-QCL waveguide, $\beta = \beta_1 - \beta_2$ [20, 26]. As a result, the value of $l_{eff}$ in Eq. (11.4) is determined principally by the THz loss $\alpha$ in a QCL waveguide. Note that the THz

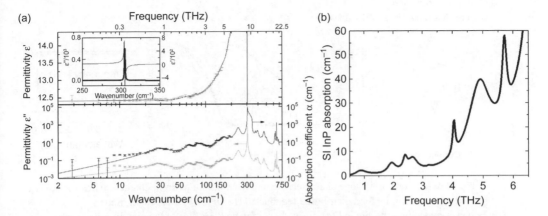

**Fig. 11.5** (a) Room-temperature terahertz-infrared spectra of the real (top panel) and imaginary (bottom panel, left vertical axis) parts of the dielectric permittivity of Fe-doped semi-insulating (SI) InP. Also shown is the power absorption coefficient $\alpha$ (bottom panel, right vertical axis) of SI InP. The inset in the top panel shows the dielectric permittivity near the transverse optical phonon resonance [8]. (b) Power absorption in SI InP in the THz spectral range [8]. Panel (a) is reproduced from [8] under Creative Commons Attribution International License (http://creativecommons.org/licenses/by-nc/4.0/).

abortion in semi-insulating InP (shown in Fig. 11.5(b)) is much smaller than the free-carrier absorption in doped InP (shown in in Fig. 11.2(a)). Other doped materials in the QCL heterostructure, such as InGaAs and AlInAs ternary alloys, have the values of THz loss within a factor of 2 of that of a similarly doped InP.

Because of the strong free-carrier loss at lower THz frequencies, the sweet spot for edge-emitting THz DFG-QCLs based on modal phase-matching is in the 3–5 THz range. The value of $l_{\text{eff}}$ is strongly limited by free-carrier absorption below $\sim$3 THz and by the tails of the optical phonon absorption bands above $\sim$5 THz in intrinsic InP (as shown in Fig. 11.5(b)) as well as in InGaAs/AlInAs semiconductor compounds.

Assuming THz DFG at 4 THz and taking the typical value of $\chi_{zzz}^{(2)} \approx 20$ nm/V, $S_{\text{eff}} \approx 1000 \, \mu\text{m}^2$, and $l_{\text{eff}} \approx 100 \, \mu\text{m}$ (cf. Fig. 11.2(a)), we can use Eq. (11.5) to obtain an order-of-magnitude estimate of the nonlinear conversion efficiency $\eta_{\text{int}}$ in the DFG-QCL waveguide:

$$\eta_{\text{int}} = \frac{W(\omega)}{W(\omega_1)W(\omega_2)} \approx 40 \frac{\mu W}{W^2}. \tag{11.8}$$

Since the mid-IR pump powers can be as high as few watts each, Eq. (11.8) predicts that THz output in the range of 10–100 $\mu$W may be expected from DFG-QCLs based on modal phase-matching and this value is in a good agreement with experiments.

Figure 11.6(a)–(c) shows some of the best performance attained by an edge-emitting THz DFG-QCL based on modal phase-matching at room temperature. The data is taken from [24]. The device was processed as a 3-mm-long double-channel ridge-waveguide QCL with a ridge width of 16 $\mu$m tapered into a 60-$\mu$m-wide output section with the taper angle of 1°. A dual-period mid-IR DFB grating was fabricated

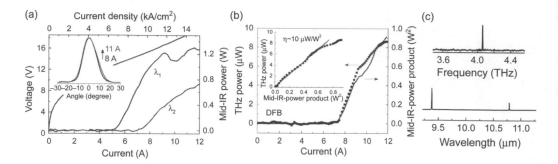

**Fig. 11.6** (a) Current–voltage characteristic and light output–current characteristics for the two mid-IR wavelengths for a distributed feedback THz DFG-QCL based on modal phase-matching reported in [24]. Inset shows far-field profiles of the mid-IR emission at different pump currents that indicate that mid-IR pumps operate in the $TM_{00}$ mode. (b) THz DFG power output vs device pump current. Inset shows THz power output plotted against the product of the mid-IR pump powers. (c) THz (top panel) and mid-IR (bottom panel) emission spectra of the laser. The laser is operated at room temperature with 60 ns pulses at 250 kHz repetition frequency. Both mid-IR and THz powers were corrected for the collection efficiency of the setup. Reproduced from [24], with the permission of AIP Publishing.

along the waveguide ridge to select two mid-IR pump frequencies for a single-frequency THz DFG output as shown in Fig. 11.6(c). Over 8 μW of THz peak power output was recorded at room temperature as described in Fig. 11.6(b) with the mid-IR-to-THz conversion efficiency of approximately 10 μW/W², in a reasonable agreement with the theoretical predictions shown in Eq. (11.8).

The outcoupling efficiency for THz radiation in edge-emitting devices based on modal phase-matching is very poor as THz radiation generated further than ~10–100 μm away from the waveguide facet is absorbed in the QCL waveguide due to high free-carrier losses at THz frequencies as shown in Fig. 11.2(a). To overcome this problem, devices with surface-grating outcouplers for extracting THz radiation in the vertical [27] or forward direction [22] along the entire waveguide length were demonstrated. However, the most successful method to overcome the problem of terahertz extraction efficiency was the development of THz DFG-QCLs based on the Cherenkov phase-matching scheme. Unlike grating-based outcouplers [22, 27], the Cherenkov DFG scheme allows for broadband THz radiation extraction.

## 11.4     THz DFG-QCLs Based on Cherenkov Phase-Matching

To enable efficient and broadband extraction of THz radiation along the whole length of the QCL waveguide, THz DFG-QCLs based on the Cherenkov phase-matching scheme were developed [28]. Cherenkov emission occurs when the phase velocity of the nonlinear polarization wave in an optically thin slab of nonlinear optical material is higher than the phase velocity of the generated radiation in the medium surrounding the slab [29]. In this case, the radiation at a frequency generated in a nonlinear process is emitted into the medium surrounding the slab at the emission angle $\theta_C$ as shown schematically in Fig. 11.7(a). For THz DFG, the Cherenkov emission angle $\theta_C$ is given as

$$\theta_C = \cos^{-1}\left(k_{nl}/k_{THz}\right), \tag{11.9}$$

where $k_{nl} = \beta_1 - \beta_2$ is the propagation constant of the nonlinear polarization wave with $\beta_1$ and $\beta_2$ being the propagation constants of the two pumps in the DFG process and $k_{THz}$ is the propagation constant of the terahertz wave in the medium surrounding the nonlinear polarization slab. Referring to the coordinate system shown in Fig. 11.7(a), we can write an expression for the nonlinear polarization wave at $\omega_{THz} = \omega_1 - \omega_2$ in the slab waveguide approximation as

$$P_z^{(2)}(x,z) = \varepsilon_0 \chi_{zzz}^{(2)}(z) E_z^{\omega_1}(z) E_z^{\omega_2}(z) e^{i(\beta_1 - \beta_2)x}, \tag{11.10}$$

where the $z$-direction is normal to the QCL layers and the $x$-direction is along the waveguide, $E_z^{\omega_1}(z)$, and $E_z^{\omega_2}(z)$ are the $z$-components of the E-field of the mid-IR pump modes, and $\chi_{zzz}^{(2)}(z)$ is the giant intersubband optical nonlinearity for DFG in the QCL active region defined in Eq. (11.2). The propagation constant of the nonlinear polarization wave is thus given as $\beta_1 - \beta_2$. Since the two mid-IR pump frequencies are close, $\omega_1 \approx \omega_2$, one can write [26]

$$|\beta_1 - \beta_2| \approx \frac{n_g \omega_{THz}}{c} \tag{11.11}$$

where $n_g = n_{eff}(\omega_1) + \omega_1 \left.\frac{\partial n_{eff}}{\partial \omega}\right|_{\omega=\omega_1}$ is the group effective refractive index for the mid-IR pumps. In order to produce Cherenkov DFG emission into the substrate, the substrate refractive index at $\omega_{THz}$ must be larger than $n_g$ and this condition is satisfied throughout the 0.1–6 THz spectral range for InP/GaInAs/AlInAs QCLs grown on semi-insulating (SI) InP. In particular, the value of $n_g$ is typically in the range 3.3–3.4 for mid-IR pumps [30] and the refractive index of InP in the 0.1–6 THz range is 3.5–3.7 [8]. Thus, the value of the Cherenkov angle in a typical THz DFG-QCL on an SI InP substrate is expected to be in the range of 20±7 degrees, depending on the waveguide design and operating frequencies. Since the SI InP substrates are non-conducting, lateral current extraction is implemented in all Cherenkov THz DFG-QCLs.

In the slab-waveguide approximation, the Cherenkov THz emission may be modeled analytically as a leaky slab-waveguide mode produced by the polarization source described in Eq. (11.10) following the formalism outlined in [31] and generalized for the multi-layer waveguide structure of DFG-QCLs. In a physical picture, two Cherenkov waves are generated by the nonlinear polarization wave: one propagates towards the top contact and the other one towards the substrate as shown in Fig. 11.7(a). These two waves are partially reflected by various waveguide layers and may pass through the active region multiple times and interfere with each other before finally exiting to the substrate. The calculated squared magnitude of the H-field for the TM-polarized Cherenkov wave ($|H_y|^2$) for a representative Cherenkov THz DFG-QCL device reported in [1] are shown in Fig. 11.7(b).

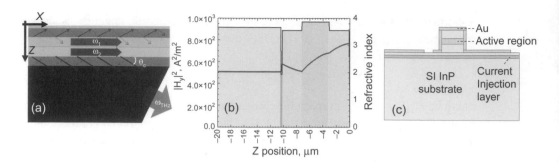

**Fig. 11.7** (a) Schematic of Cherenkov THz DFG emission in QCLs. Mid-IR pumps $\omega_1$ and $\omega_2$ are confined in the laser active region (light gray) by the waveguide cladding layers (gray), THz DFG emission occurs from the active region into the SI InP substrate (black) at the Cherenkov angle $\theta_C$. (b) Calculated square of the H-field in the TM-polarized THz Cherenkov wave (thick line, left axis) and the waveguide refractive index profile (thin line, right axis) at 4 THz for the devices reported in [1]. A slab waveguide model was used for calculations. The gold contact layer is positioned at $z = 0$, cladding layers (InP doped $1.5 \times 10^{16}$ cm$^{-3}$) extend from approximately 0 to $-3$ μm and from $-7$ to $-10$ μm, the active region is positioned between $-3$ and $-7$ μm, the current injection layer (InGaAs doped $7 \times 10^{17}$ cm$^{-3}$) is shown in gray at approximately $-10$ μm, and the SI InP substrate extends from approximately $-10.5$ μm. Calculations are performed assuming 56 mW/μm slab-waveguide linear pump power density in each mid-IR pump [1]. (c) Schematic of a ridge-waveguide Cherenkov DFG-QCL with a side contact for current injection. Panels (a) and (b) are reproduced from [1] under Creative Commons Attribution International License (http://creativecommons.org/licenses/by-nc/4.0/).

To avoid total internal reflection of the Cherenkov wave at the InP/air interface of the QCL cleaved facet, the front facet of the substrate has to be polished to form a wedge as shown in Fig. 11.7(a). We found that a 30-degree substrate polishing results in the Cherenkov THz wave outcoupling in the forward direction at 3.5 THz for typical THz DFG-QCLs. The angle varies for other THz frequencies and depends on the QCL waveguide design due to variations in $n_g$ (see Eq. (11.11)) and on the InP refractive index dispersion [2]. The use of semi-insulating InP makes it necessary to inject current laterally from under the laser active region. A schematic of a typical ridge-waveguide Cherenkov THz DFG-QCL with a side contact for lateral current injection is shown in Fig. 11.7(c).

Typical pulsed-mode room-temperature device performance of ridge-waveguide THz DFG-QCLs is shown in Fig. 11.8, after [1]. Depending on the mid-IR pump spacing, THz emission in these devices can be varied in the entire 1–6 THz range and beyond, limited only by the materials losses and the reduction of the THz DFG efficiency, which falls down with the square of the THz frequency as shown in Eq. (11.5). The active region of the devices in [1] was made of two stacks of QCL stages designed for emission at $\lambda_1 = 8.2$ μm and $\lambda_2 = 9.2$ μm, based on the bound-to-continuum active region design similar to that shown in Fig. 11.3(a). The waveguide structure is shown schematically in Fig. 11.7(b).

Processing the Cherenkov THz DFG-QCL into buried heterostructure (BH) lasers using SI InP overgrowth on the sides of the ridge-waveguide results in substantial

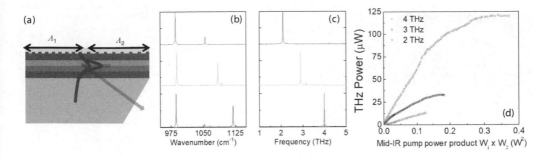

**Fig. 11.8** (a) Schematic of a dual-period surface DFB grating Cherenkov THz DFG-QCL, after [1]. The schematic shows the gold contact layer (top layer, light gray), cladding layers (dark gray), active regions with nonlinearity (light gray and gray, for the two stacks), and a substrate (bottom, light-gray). Cherenkov THz radiation is emitted into the substrate. The mid-IR pump modes are also shown. Dual-color lasers in [1] had an approximately equal length of DFB gratings sections $\Lambda_1$ and $\Lambda_2$. (b)–(d) Performance of 2 THz, 3 THz, and 4 THz Cherenkov DFG-QCL sources operated in pulsed mode at room temperature with 100 ns current pulses at a 5 kHz repetition frequency: (b) mid-infrared and (c) THz emission spectra taken with a 0.2 cm$^{-1}$ resolution, (d) terahertz peak power output versus the product of the mid-IR pump powers. All devices were 2-mm-long, 25-μm-wide ridge-waveguide lasers. Reproduced from [1] under Creative Commons Attribution International License (http://creativecommons.org/licenses/by-nc/4.0/).

improvements in the THz power output compared to that of the ridge-waveguide devices depicted in Fig. 11.7(c) due to a significant reduction of the waveguide loss for both mid-IR and THz radiation and an increase in the mid-IR pump powers. For efficient thermal management and to reduce series resistance in the current extraction layer, BH Cherenkov THz DFG-QCLs are mounted episide-down on diamond submounts and processed into the configuration with ground contacts on both sides of the waveguide. An image of a typical BH Cherenkov THz DFG-QCL is shown in Fig. 11.9(a), after [32].

THz DFG-QCLs processed in such a way and with a buried DFB grating to select two mid-IR pump frequencies achieved continuous-wave (CW) operation with over 14 μW of power output at 3.4 THz at room temperature [4], which represents the highest room-temperature CW power output of all THz DFG-QCLs reported to date. Their active region consists of 40 repetitions of the active stages that are conceptually similar to that shown in Fig. 11.3(b) but were designed and grown using a strain-compensated heterostructure. Compared to lattice-matched designs, strain-compensated THz DFG-QCL active regions provide both higher mid-IR pump powers and higher DFG nonlinearity for the same doping and current density [4]. This is principally due to the reduction of the electron effective mass in indium-rich wells that leads to the increase of transition dipole moments, cf. Eq. (11.2) [33]. Figures 11.9(b) and (c) show the performance of one of these devices, after [4]. The room-temperature wall-plug efficiency (WPE) for CW THz generation in the devices shown in Fig. 11.2 was as high as $2.6 \times 10^{-6}$.

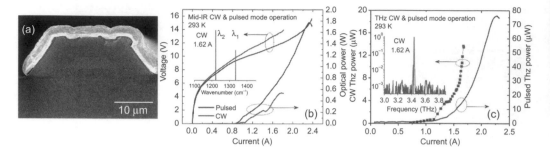

**Fig. 11.9** (a) Scanning electron microscope image of the buried-ridge Cherenkov THz DFG-QCL. (b) Current–voltage and mid-IR light output–current characteristics of a 4-mm-long buried-ridge Cherenkov THz DFG-QCL with a composite DFB grating operated in continuous-wave (smaller dynamic range and lower powers) and pulsed (higher current range and higher powers) regimes at room temperature. Inset shows the mid-IR spectrum of the device. After [4]. (c) THz performance of the device reported in [4] in pulsed and continuous-wave operation. Inset shows its THz emission spectrum. Panel (a) is reproduced from [32], with the permission of AIP Publishing. Panels (b) and (c) are reproduced from [4] under Creative Commons Attribution International License.

Lower doping in the DFG-QCL active region, which is necessary to achieve room-temperature CW operation, results in a decrease of both the optical nonlinearity and the mid-IR peak pump powers. Both of these parameters scale linearly with doping in the first approximation (cf. Eq. (11.2)). The mid-IR-to-THz conversion efficiency of the state-of-the-art room-temperature CW devices in Fig. 11.9 is about $100\,\mu W/W^2$. In comparison, state-of-the-art Cherenkov THz DFG-QCLs optimized for pulsed operation can provide significantly higher mid-IR-to-THz conversion efficiency and THz power output since more QCL stages and higher active region doping can be used in these devices. Currently, the highest THz peak power output of 1.9 mW with a conversion efficiency of $0.8\,mW/W^2$ is achieved at 3.5 THz from an HR-coated, 3-mm-long, 23-μm-wide DFB laser epilayer-down mounted on a patterned diamond submount [34]. The performance of this device is shown in Fig. 11.10, after [34]. Note that the measured THz power is not corrected for collection efficiency. The active region of these lasers was made of 60 stages with an average active region doping of $7 \times 10^{16}\,cm^{-3}$, which is a factor 2.5 higher doping density compared to that of the CW devices shown in Fig. 11.9 [34]. A relatively thin InP cladding layer of 3 μm for strong surface DFB coupling was implemented, similar to that of the devices shown in Fig. 11.8. The WPE for THz generation of the record devices reported in [34] is as high as $0.7 \times 10^{-5}$ at 3.5 THz. The results shown in Figs. 11.9 and 11.10 represent the current state of the art in the THz DFG-QCL technology in terms of CW and pulsed THz power output and WPE at room temperature.

## 11.5     THz Emission Characteristics of the Cherenkov THz DFG-QCLs

Our theoretical discussion of the Cherenkov emission has been focused on a slab waveguide model, cf. Fig. 11.7(a). In real devices, however, Cherenkov THz DFG emission

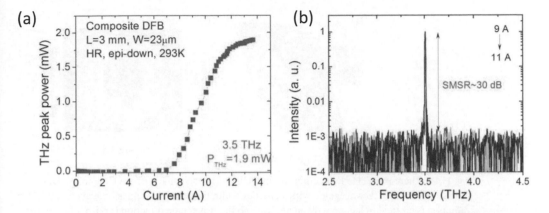

**Fig. 11.10** (a) THz performance of the device reported in [34] in pulsed operation at room temperature. (b) Emission spectra of the device in (a) in pulsed operation at different pump currents. Reproduced from [34] under Creative Commons Attribution International License.

is produced from a QCL waveguide with typical transverse dimensions (ridge width) of 8–30 μm, which are smaller than the wavelength of the emitted THz radiation. As a result, THz radiation is emitted into the SI InP substrate in a conical beam with an angle $\theta_C$ to the waveguide direction, see Fig. 11.11(a). Despite substrate polishing, only the bottom portion of the cone (cf. Fig. 11.11) has a high transmission at the substrate/air interface, while the lateral portions of the THz beam undergo total internal reflection and are not outcoupled to free space. COMSOL simulations shown in Fig. 11.11(b) confirm this understanding and demonstrate that only about 30% of the THz radiation power propagating towards the polished substrate facet is outcoupled to free space in a typical Cherenkov THz DFG-QCL device. For a normal plane wave propagation, this number is expected to be about 70% [35]. The THz far-field emission profile of a typical THz DFG-QCL is shown in Fig. 11.11(c), after [23]. Unlike traditional QCLs, in THz DFG-QCLs the THz beam divergence along the fast axis (along the crystal growth direction) is smaller than that along the slow axis because the entire length of the waveguide operates as a coherent THz emitter.

The terahertz emission linewidth (LW) of CW Cherenkov THz DFG-QCLs has so far only been characterized at cryogenic temperatures, owing to difficulties in producing devices operable CW at room temperature. It is expected, however, to be not very different for room-temperature devices. The measurements were reported in [36]. To characterize the THz LW of a DFG-QCL, a heterodyne detection system was used to measure a beat note between the THz emission of a free-running DFG-QCL and a reference comb tooth of a free-space THz frequency comb synthesizer (FCS) mixed in a hot electron bolometer (HEB) as shown in Fig. 11.12(a). The RF beat note was sent from the HEB to a fast Fourier-transform (FFT) spectrum analyzer that was used to retrieve the beat-note spectra over different integration times and therefore to evaluate the THz DFG-QCL emission LW at different observation times (ranging from $t = 20\,\mu s$ to $t = 20\,ms$).

(a)　　　　　(b)　　　　　(c)

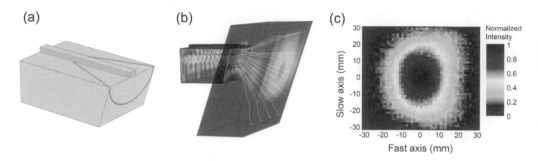

**Fig. 11.11** Cherenkov THz DFG emission in QCLs. (a) A schematic of the Cherenkov DFG emission from a QCL waveguide with a smaller width than the THz emission wavelength. THz radiation is emitted in a cone with an angle $\theta_C$ to the waveguide direction (cf. Eq. (11.9)). (b) Three-dimensional COMSOL simulation of the THz emission from a 200-μm-long section of a 22-μm-wide ridge waveguide THz DFG-QCL at 3.5 THz. The lines are the power streamlines indicating the propagation direction of THz power outcoupled to the points on the air monitor on the right. (c) Far-field THz emission profile for a 3-mm-long, 14-μm-wide Cherenkov THz DFG-QCL on a SI InP substrate operating at 2.7 THz. The image is obtained by two-dimensional scanning of a Golay cell 5.9 cm away from the laser facet. The fast axis is along the crystal growth direction, the slow axis is along the QCL layers. Panel (b) is reproduced with permission from [35]. © The Optical Society. Panel (c) is reproduced with permission from [23] under a CC BY License.

The CW Cherenkov DFG-QCL used for the THz LW characterization was a 2-mm-long, 12-μm-wide buried-heterostructure dual-mid-IR-wavelength DFB laser with the mid-IR pump wavelengths fixed at approximately 8.5 μm and 9.1 μm. The LW of the 2.58 THz DFG emission line of the device was measured as a function of the observation time $t$ at two different heat sink temperatures ($T_H = 45$ K and $T_H = 78$ K) as shown in Fig. 11.12(b). The laser was operated with a pump current of 450 mA for both measurements and provided approximately 0.2 μW of THz power output. An ultra-low-noise current driver (ppqSense model QubeCL05) was used to drive the laser. The data in Fig. 11.12(b) shows that the LW of the THz DFG emission from the laser was reduced as the operating temperature increased. This trend is consistent with the previously observed narrowing of the emission LW of mid-IR QCLs at higher temperatures [37, 38].

The logarithmic dependence of the LW with $t$, indicates the presence of a $1/f$ noise component in the frequency noise spectrum between 1 Hz and 1 kHz that likely comes from the typical "pink" frequency noise of the mid-IR pumps [37, 39]. This is also confirmed by the noise decrease at increasing temperatures (see Fig. 11.12(b)), in agreement with previous experiments on the $1/f$ noise of mid-IR QCLs [37, 38, 40] that show a similar trend. A quantitative comparison with previous LW measurement reports of mid-IR QCLs can help to understand the possible correlations between the emission LW of the mid-IR pumps and the THz DFG. To this purpose, the inverse method proposed in [41] was applied to the data in Fig. 11.12(b) in order to retrieve the frequency noise power spectral density (FNPSD) of the devices. The FNPSD of THz DFG-QCLs can be directly compared with the FNPSDs of a THz QCL operating

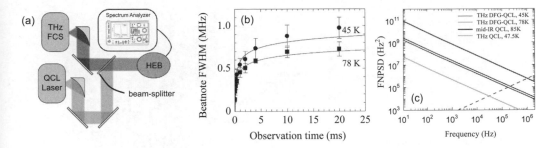

**Fig. 11.12** Linewidth measurements of THz emission from a Cherenkov THz DFG-QCL. (a) Schematic of an optical setup used for the measurements. QCL laser refers to a Cherenkov THz DFG-QCL device, THz FCS is the THz frequency comb synthesizer, and HEB stands for a hot electron bolometer (HEB). (b) The width of the beat note at different time scales measured at two different operating temperatures of the device. Solid lines are fits with a logarithmic function. (c) Reconstruction of the frequency noise power spectral density (FNPSD) of the THz DFG-QCL emission. Given the smallest time scale of 20 μs of the setup, the measurements do not include frequencies higher than 50 kHz. Reproduced from [36]. © The Authors, some rights reserved; exclusive licensee AAAS. Distributed under a CC BY-NC 4.0 License (http://creativecommons.org/licenses/by-nc/4.0/).

at 47.5 K [42] and of a mid-IR QCL operating at 85 K [39], as shown in Fig. 11.12 (c). The plot clearly unveils that the FNPSD of the DFG-QCL, even in the upper edge of the investigated temperature range ($T_H = 45$–85 K), is almost one order of magnitude lower than that of mid-IR QCLs. The reduction on the emission LW of THz DFG can be explained by the correlation between the phase/frequency noises of the two mid-IR pumps in DFG-QCLs that are partly compensated by the DFG process. The FNPSD of a single-mode bound-to-continuum THz QCL, exploiting a single plasmon waveguide, shows a significantly ($\sim$ a factor of seven) narrower LW. Nevertheless, the measurements indicate that the LW of THz DFG-QCLs is already suitable for heterodyne THz detection, and one may expect that it can be further reduced with frequency/phase stabilization.

## 11.6   Widely Tunable Cherenkov THz DFG-QCLs

Relaxed phase-matching conditions in Cherenkov THz DFG-QCL devices enable broad spectral tunability of THz output in these lasers. An example of a broadly tunable external cavity (EC) THz DFG-QCL system is presented in Fig. 11.13, after [2]. To enable THz DFG emission tuning, the system is designed to continuously tune one mid-IR pump frequency by an external grating in a Littrow-type EC configuration, while having the second mid-IR pump frequency fixed by a DFB grating etched into the QCL waveguide as shown schematically in Fig. 11.13(a). Due to spatial hole burning, gain competition between the two mid-IR pumps is reduced and simultaneous two-color emission can be achieved.

Figure 11.13(b) shows a photograph of the EC system. Experimental results obtained with a 2.3-mm-long ridge waveguide Cherenkov THz DFG-QCL that

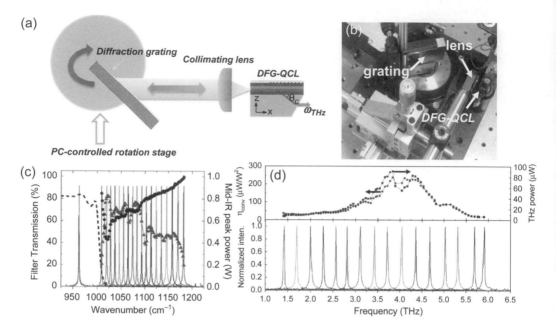

**Fig. 11.13** Schematic (a) and a photograph (b) of a widely tunable external-cavity (EC) THz DFG-QCL system, after [2]. (c) Mid-IR emission spectra and power output of the two mid-IR pumps for the EC THz DFG-QCL system described in (a) at different EC diffraction grating positions taken at a current density of $8.0\,\mathrm{kA/cm^2}$. Also shown is the mid-IR power for the fixed-frequency $\omega_1$ pump (circles) and the tunable $\omega_2$ pump (triangles) as a function of $\omega_2$ pump wavenumber. The dashed line shows the transmission spectrum of mid-IR long-pass filter used for power measurements. (d) THz emission spectra of the EC THz DFG-QCL system in (a)–(c) taken at a current density of $8.0\,\mathrm{kA/cm^2}$. Also plotted are the THz peak power (squares and right axis) and mid-IR-to-THz conversion efficiency (circles and left axis) as a function of THz frequency. Panels (a) and (c) are reproduced with permission from [2]. © IOP Publishing. All rights reserved.

contains a 1.6-mm-long DFB grating section and a back facet anti-reflection coating are shown in Fig. 11.13(c) and (d). The device structure was identical to the devices discussed in Fig. 11.8 and [1]. The DFB grating was designed to fix $\omega_1$ pump at $963\,\mathrm{cm^{-1}}$. The substrate front facet of the device was polished at a 30° angle. Figure 11.13(c) shows the emission spectra and power output of the two mid-IR pumps at different EC grating positions at a current density of $8.0\,\mathrm{kA/cm^2}$ through the device, close to the rollover point. A long pass filter was used to separate the two mid-IR pumps for the measurements. The spectral data shows that, as expected, the DFB pump frequency $\omega_1$ stays fixed at $963\,\mathrm{cm^{-1}}$ while the EC pump frequency $\omega_2$ is tuned continuously from $1009\,\mathrm{cm^{-1}}$ to $1182\,\mathrm{cm^{-1}}$, resulting in the available difference-frequency tuning from 1.4 THz to 5.9 THz. Due to the gain competition, the mid-IR emission switches from dual-frequency ($\omega_1$ and $\omega_2$) to single-frequency ($\omega_1$) output as the EC grating tunes pump frequency $\omega_2$ towards the DFB frequency $\omega_1$ as shown in Fig. 11.13(c).

The THz emission spectra and THz peak power for different EC grating positions taken at a current density of $8.0\,\mathrm{kA/cm^2}$ through the device are displayed in

Fig. 11.13(d). Also shown is mid-IR-to-THz nonlinear conversion efficiency at a current density of $8.0 \, \text{kA/cm}^2$. THz peak power of $90 \, \mu\text{W}$ and the mid-IR-to-THz conversion efficiency of nearly $250 \, \mu\text{W/W}^2$ were observed at 3.8 THz. The conversion efficiency peaks in the 3.7–4.5 THz range of frequencies and falls off at both high- and low-frequency ends of the tuning curve, which is consistent with the results obtained with other THz DFG-QCL devices. At the high-frequency end, efficiency of THz generation is limited by the onset of high optical losses in InGaAs/AlInAs/InP materials due to the tails of LO-phonon absorption bands (Reststrahlen band). At the low-frequency end, THz generation efficiency is principally limited by high free-carrier absorption in the QCL waveguide and the $\omega_{\text{THz}}^2$ dependence of the DFG efficiency, see Eq. (11.1). Additional factors include spectral dependence of the intersubband optical nonlinearity and a residual absorption in the SI InP substrate through which the THz radiation is extracted, see Fig. 11.5(b).

Besides EC systems, a variety of monolithic THz DFG-QCL sources have been demonstrated based on thermal tuning of mid-IR pumps (using direct thermal tuning or the Vernier tuning mechanism) [4, 43], a dual-grating EC setup for EC tuning of both mid-IR pumps [18], and/or a QCL array concept [44].

## 11.7 Frequency Comb Generation in THz DFG-QCLs

As discussed in Chapter 7, mid-IR QCLs can generate frequency combs if their active regions and waveguides are configured to possess low group velocity dispersion. Similar mid-IR active regions and waveguides can be implemented in the Cherenkov THz DFG-QCLs grown on SI InP substrates. Intra-cavity frequency mixing can then be used to down-convert mid-IR frequency combs lines to the THz spectral region while Cherenkov phase-matching scheme enables extraction of the generated THz radiation through the laser substrate.

Two approaches may in principle be used to down-convert mid-IR comb lines to the THz spectral region. The first approach is to rely on wave mixing between the mid-IR comb lines themselves. Because individual comb lines are phase-locked, this approach is expected to produce THz comb lines with very narrow linewidths that are comparable to the linewidths of the radio-frequency beat notes observed in the mid-IR frequency comb QCLs ($\sim$1–10 kHz). Despite promising THz characteristics, there are several factors that make this approach relatively difficult to implement experimentally. As discussed in Chapter 9, measurements of QCL frequency combs indicate that the phase relationship between the individual comb lines in frequency comb mid-IR QCLs is close to that of a frequency-modulated signal. This property is expected to make THz down-conversion of these combs relatively inefficient due to the destructive interference of the THz outputs produced by mixing of different pairs of mid-IR comb lines separated by the same THz frequency. Additionally, the bandwidth of the mid-IR QCL frequency combs is currently below $\sim$3 THz ($100 \, \text{cm}^{-1}$) and the most efficient difference-frequency mixing in THz DFG-QCLs has so far been demonstrated at frequencies above 3 THz.

The second approach to generate THz frequency combs in DFG-QCLs is to use a strong single-frequency mid-IR pump emission line that is spectrally detuned from the

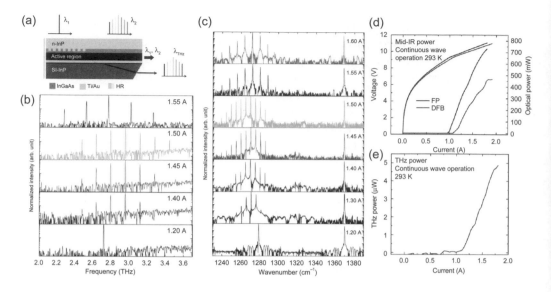

**Fig. 11.14** Demonstration of room-temperature THz frequency comb generation in a Cherenkov THz DFG-QCL, after [45]. (a) Schematic of the device and the process of THz DFG frequency comb generation in the device. (b) and (c) Room-temperature continuous-wave THz and mid-IR emission spectra of a 4-mm-long, 8–9-μm-wide buried heterostructure device shown schematically in panel (a). (d) Room-temperature continuous-wave current–voltage and mid-IR light output–current characteristics of the "DFB" device shown in (a)–(c) and a Fabry–Pérot "FP" device without a DFB grating. The FP device was not reported to produce a measurable THz emission. (e) THz power output of the frequency comb device shown in (a)–(d). Reproduced from [45] under Creative Commons Attribution International License (http://creativecommons.org/licenses/by-nc/4.0/).

group of lines forming the frequency comb. The separate line is then used to down-convert every line of the mid-IR comb to the THz spectral range. This approach has recently been used to produce THz DFG-QCL frequency comb sources [23, 45] and is shown schematically in Fig. 11.14(a). The only room-temperature THz DFG-QCL frequency comb devices were reported in [45] and we focus on their description in Fig. 11.14. The laser active region, buried heterostructure processing, and mounting was similar to that of CW THz DFG-QCLs shown in Fig. 11.9. A buried DFB grating with high coupling strength ($\kappa L \sim 3$–5, where $\kappa$ and $L$ are the coupling constant and the length of the grating) was introduced in the laser waveguide to select a single-mode QCL pump wavelength separated from that of a frequency comb as shown schematically in Fig. 11.14(a). The mid-IR frequency comb emission occurred in a high-order harmonic state [46] and was down-converted by a single-line mid-IR pump into the THz frequency comb as shown in Fig. 11.14(b) and (c). The mid-IR power output of the device described in Fig. 11.14(a)–(c) and that of a reference Fabry–Pérot device without a DFB grating are shown in Fig. 11.14(d). As expected, the Fabry–Pérot device did not produce a detectable THz power output. Fig. 11.14(d) shows the THz power output vs pump current of the device described in Fig. 11.14(a)–(c).

The frequency comb nature of the observed mid-IR emission was confirmed by mixing the laser output with the emission of a reference Fabry–Pérot mid-IR comb device (fabricated from the same material) on a fast mid-IR quantum well infrared photodetector. The radio-frequency spectrum of the detector output showed a train of narrow beat notes as expected for a frequency combs mixing. THz emission of the laser shown in Fig. 11.14(b) has only been characterized with relatively low spectral resolution using a Fourier-transform infrared spectrometer. The true emission linewidths of the THz comb lines and their phase-locking currently remain undetermined for now.

## 11.8    THz Radiation Extraction Efficiency

The SI InP substrates are used for Cherenkov THz DFG-QCLs because InP is a native substrate for high performance InGaAs/AlInAs mid-IR QCLs. However, SI InP has both considerable losses and significant refractive index dispersion in the THz spectral range as shown in Fig. 11.5. As follows from Eq. (11.9), the substrate refractive index dispersion leads to beam steering in broadly tunable Cherenkov THz DFG-QCLs and this effect was reported in [2]. More importantly, SI InP has relatively high THz absorption, which prevents efficient THz extraction from the device. Assuming devices are operated on sufficiently thick InP substrates and the front facet is polished normal to the Cherenkov wave direction, one can estimate the extraction efficiency $\eta_{\text{ext}}$ of the generated THz radiation to free space using the following analytical function [35]

$$\eta_{\text{ext}} = \frac{1}{2L} T \int_0^{2L} e^{-\alpha_{\text{sub}} \cos(\theta_c)x} dx = \frac{T}{2L\alpha_{\text{sub}} \cos(\theta_c)} \left( 1 - e^{-\alpha_{\text{sub}} \cos(\theta_c)2L} \right). \quad (11.12)$$

Here $L$ is the length of the laser, $T$ is the polished facet power transmission for the Cherenkov wave ($\sim$30% for SI InP as discussed earlier), $\alpha_{\text{sub}}$ is the THz power absorption of the substrate, $\theta_C$ is the Cherenkov angle, and we assumed a uniform intensity of the Cherenkov THz generation along the device waveguide. Integration from 0 to $2L$ is used because both forward- and backward-going THz Cherenkov waves are generated in the device and we assume that the backward going wave reflected from the unpolished back facet of the device can be outcoupled through the front polished facet.

Figure 11.15(a) plots the computed values of $\eta_{\text{ext}}$ as a function of the laser cavity length for different values of $\alpha_{\text{sub}}$ in the range 1–20 cm$^{-1}$ (cf. Fig. 11.5(b)). We considered a Cherenkov THz DFG-QCL on a SI InP substrate with the front facet polished normal to the Cherenkov wave and used $\theta_C = 20°$. Terahertz extraction efficiency is computed for the case of a device on a very thick SI InP substrate that sets no geometrical limits on the THz extraction and for the case of a device on a 0.5-mm-thick SI InP substrate, which is in the middle of the typical range of substrate thicknesses used for THz DFG-QCLs (typically 0.35–0.7-mm-thick SI InP substrates are used for the Cherenkov devices growth [1, 35]). The finite substrate thickness sets geometrical limits on the region from which the Cherenkov radiation can be extracted through the polished device facet, see Fig. 11.7(a). Geometric analysis shows that, for the case of a 0.5-mm-thick SI InP substrate, the Cherenkov forward-going wave generated only within approximately 1.6 mm of the device facet reaches the polished facet and have a chance to be outcoupled to free space. Given that we include both forward- and

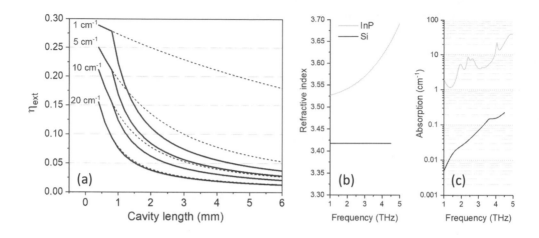

**Fig. 11.15** (a) Extraction efficiency of the generated Cherenkov THz radiation in THz DFG-QCLs on SI InP as a function of SI InP loss and the device cavity length. Computations are shown for the case of a 0.5-mm-thick substrate (solid lines) and that of an infinitely thick substrate (dashed lines). (b) Refractive index and (c) intensity absorption in SI InP and high-resistivity silicon.

backward-going waves in our analysis, we obtain that for the case of a 0.5-mm-long SI InP substrate, the geometrical limits affect the THz extraction efficiency for the device length exceeding 0.8 mm as shown in Fig. 11.15(a).

The results in Fig. 11.15(a) indicate that only about 5% of the generated THz Cherenkov power is outcoupled to free space for the record room-temperature CW and pulsed THz devices shown in Figs. 11.9 and 11.10 that were grown in 0.5-mm-long SI InP substrates, operated in the 3.4–3.5 THz frequency range where $\alpha_{sub} \approx 5\,cm^{-1}$ for SI InP (cf. Fig. 11.5(b)), and had a cavity length of 3–4 mm.

One can avoid beam steering in broadly tunable Cherenkov THz DFG-QCLs and improve the outcoupling efficiency of THz radiation in these devices by transferring the QCLs from their native InP substrates to high-resistivity silicon (HR Si) substrates that have very low THz refractive index dispersion and losses compared to SI InP as shown in Fig. 11.15(b) and (c) [8, 35].

Figure 11.16 shows the performance of the Cherenkov THz DFG-QCLs on 1-mm-thick HR Si substrates in comparison with that of a similar device on 0.66-mm-thick SI InP. A transfer-printing method was used to transfer completely processed devices from SI InP to HR Si. Details of the processing are given in [35]. More than a factor of five improvement in the THz power generation and more than a factor of eight improvement in the mid-IR-to-THz conversion efficiency have been achieved experimentally for the devices on SI InP operating at a frequency around 3.5 THz [35].

While the THz power output and the mid-IR-to-THz conversion efficiency of the devices on HR Si have shown considerable improvement, the question of thermal management of such devices still remains to be addressed. Alternative methods of

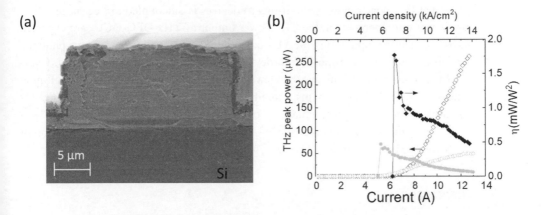

**Fig. 11.16** (a) Scanning electron microscope image of a THz DFG-QCL epi-transferred to HR Si substrate using adhesive bonding with a 100-nm-thick layer of SU-8 epoxy at the interface between the Si substrate and the QCL III-V epilayer. The laser facets are created by dicing with a dicing saw. (b) Comparison of representative THz performance of a Cherenkov THz DFG-QCL devices on HR Si (black diamonds) and a reference device on SI InP (gray circles). Both devices are processed as ridge-waveguide lasers with 22-μm-wide and 4.2-mm-long ridges. Testing is performed at room temperature with 40 ns current pules at 15 kHz repetition rate. Open symbols refer to THz power output (left axis) and solid symbols refer to mid-IR-to-THz conversion efficiency (right axis) of the devices. Reproduced with permission from [35]. © The Optical Society.

overcoming the limitations of the SI InP substrate extraction efficiency include the development of THz DFG-QCLs that employ double-metal THz waveguides and use gratings to outcouple THz radiation from the laser waveguide, instead of a Cherenkov scheme [22]. The development of these grating-based THz DFG-QCL devices, however, is still at its infancy and they also lack the ability to provide broadband tuning and broadband frequency comb generation. Overall, the computations shown in Fig. 11.15(a) indicate significant room for improvements of the power output of THz DFG-QCL systems beyond the current state of the art.

## 11.9 Conclusion and Outlook

Since their first demonstration in 2007, the power output of THz DFG-QCLs has increased dramatically, boosted by the employment of the Cherenkov phase-matching scheme. State-of-the-art results include demonstration of room-temperature 3.5 THz DFG-QCLs with THz peak power output of 1.9 mW and the THz WPE of $0.7 \times 10^{-5}$ as well as room-temperature CW devices with over 14 μW of THz power output at 3.4 THz. Cherenkov THz DFG-QCLs have also been shown to provide broadly tunable THz output that spans a range of nearly 1–6 THz employing the external cavity setup or using various monolithic tuner configurations. Cherenkov THz DFG-QCLs can be operated as room-temperature frequency comb sources with the potential of spanning the entire 1–6 THz spectral range and beyond. It is estimated that less than 5% of the generated THz power in the state-of-the-art devices is outcoupled to free space. Thus, significant improvements in the THz power output of THz DFG-QCLs are expected to be produced by optimizing the THz outcoupling efficiency using various photonic structures, including epi-transferred devices and grating-outcoupled devices. Overall, Manley–Rowe relations may be viewed as the fundamental limit on the wall-plug efficiency for THz DFG systems. From this perspective, given the current values for room-temperature wall-plug efficiency of mid-IR QCLs in the range of 10–20% at $\lambda \approx 5\text{–}10\,\mu m$, one obtains the Manley–Rowe-limited wall-plug efficiency value of $\sim 1\%$ for 4 THz generation in THz DFG-QCLs at room temperature.

# References

[1] K. Vijayraghavan et al., "Broadly tunable terahertz generation in mid-infrared quantum cascade lasers," Nat. Commun., vol. 4, no. 1, p. 2021, 2013, doi: 10.1038/ncomms3021.
[2] Y. Jiang et al., "External cavity terahertz quantum cascade laser sources based on intracavity frequency mixing with 1.2–5.9 THz tuning range," J. Opt., vol. 16, no. 9, p. 094002, 2014, doi: 10.1088/2040-8978/16/9/094002.
[3] M. A. Belkin et al., "Terahertz quantum-cascade-laser source based on intracavity difference-frequency generation," Nat. Photonics, vol. 1, no. 5, pp. 288–292, 2007, doi: 10.1038/nphoton.2007.70.
[4] Q. Lu, D. Wu, S. Sengupta, S. Slivken, and M. Razeghi, "Room temperature continuous wave, monolithic tunable THz sources based on highly efficient mid-infrared quantum cascade lasers," Sci. Rep., vol. 6, no. 1, p. 23595, 2016, doi: 10.1038/srep23595.
[5] R. W. Boyd, Nonlinear Optics, 3rd ed. New York: Academic Press, 2008.
[6] Y. R. Shen, The Principles of Nonlinear Optics. Hoboken, NJ: John Willey & Sons, 2003.
[7] S. Kohen, B. S. Williams, and Q. Hu, "Electromagnetic modeling of terahertz quantum cascade laser waveguides and resonators," J. Appl. Phys., vol. 97, no. 5, p. 053106, 2005, doi: 10.1063/1.1855394.
[8] L. N. Alyabyeva, E. S. Zhukova, M. A. Belkin, and B. P. Gorshunov, "Dielectric properties of semi-insulating Fe-doped InP in the terahertz spectral region," Sci. Rep., vol. 7, no. 1, p. 7360, 2017, doi: 10.1038/s41598-017-07164-1.
[9] C. Sirtori, F. Capasso, J. Faist, L. N. Pfeiffer, and K. W. West, "Far-infrared generation by doubly resonant difference frequency mixing in a coupled quantum well two-dimensional electron gas system," Appl. Phys. Lett., vol. 65, no. 4, pp. 445–447, 1994, doi: 10.1063/1.112328.
[10] E. Dupont, Z. R. Wasilewski, and H. C. Liu, "Terahertz emission in asymmetric quantum wells by frequency mixing of midinfrared waves," IEEE J. Quantum Electron., vol. 42, no. 11, pp. 1157–1174, 2006, doi: 10.1109/JQE.2006.882877.
[11] K. Fujita et al., "Terahertz generation in mid-infrared quantum cascade lasers with a dual-upper-state active region," Appl. Phys. Lett., vol. 106, no. 25, p. 251104, 2015, doi: 10.1063/1.4923203.

[12]  E. Rosencher, A. Fiore, B. Vinter, V. Berger, P. Bois, and J. Nagle, "Quantum engi-
      neering of optical nonlinearities," *Science*, vol. 271, no. 5246, pp. 168–173, 1996,
      doi: 10.1126/science.271.5246.168.

[13]  R. W. Adams *et al.*, "Terahertz sources based on intracavity frequency mixing in mid-
      infrared quantum cascade lasers with passive nonlinear sections," *Appl. Phys. Lett.*,
      vol. 98, no. 15, p. 151114, 2011, doi: 10.1063/1.3579260.

[14]  M. A. Belkin and F. Capasso, "New frontiers in quantum cascade lasers: High perfor-
      mance room temperature terahertz sources," *Phys. Scr.*, vol. 90, no. 11, p. 118002, 2015,
      doi: 10.1088/0031-8949/90/11/118002.

[15]  B. A. Burnett and B. S. Williams, "Origins of terahertz difference frequency susceptibility
      in midinfrared quantum cascade lasers," *Phys. Rev. Appl.*, vol. 5, no. 3, p. 034013, 2016,
      doi: 10.1103/PhysRevApplied.5.034013.

[16]  S. Jung *et al.*, "Narrow-linewidth ultra-broadband terahertz sources based on difference-
      frequency generation in mid-infrared quantum cascade lasers," vol. 10123, p. 1012315,
      2017, doi: 10.1117/12.2256099.

[17]  E. Rosencher, P. Bois, J. Nagle, E. Costard, and S. Delaitre, "Observation of nonlinear
      optical rectification at 10.6 μm in compositionally asymmetrical AlGaAs multiquantum
      wells," *Appl. Phys. Lett.*, vol. 55, no. 16, pp. 1597–1599, 1989, doi: 10.1063/1.102248.

[18]  Y. Jiang *et al.*, "Spectroscopic study of terahertz generation in mid-infrared quantum
      cascade lasers," *Sci. Rep.*, vol. 6, no. 1, p. 21169, 2016, doi: 10.1038/srep21169.

[19]  F. Demmerle *et al.*, "Single stack active region nonlinear quantum cascade lasers
      for improved THz emission," *IEEE Photonics J.*, vol. 9, no. 3, pp. 1–9, 2017,
      doi: 10.1109/JPHOT.2017.2708423.

[20]  M. A. Belkin *et al.*, "Room temperature terahertz quantum cascade laser source based on
      intracavity difference-frequency generation," *Appl. Phys. Lett.*, vol. 92, no. 20, p. 201101,
      2008, doi: 10.1063/1.2919051.

[21]  K. Fujita, M. Hitaka, A. Ito, M. Yamanishi, T. Dougakiuchi, and T. Edamura,
      "Ultra-broadband room-temperature terahertz quantum cascade laser sources based
      on difference frequency generation," *Opt. Express*, vol. 24, no. 15, p. 16357, 2016,
      doi: 10.1364/OE.24.016357.

[22]  J. H. Kim *et al.*, "Double-metal waveguide terahertz difference-frequency generation
      quantum cascade lasers with surface grating outcouplers," *Appl. Phys. Lett.*, vol. 113,
      no. 16, p. 161102, 2018, doi: 10.1063/1.5043095.

[23]  K. Fujita *et al.*, "Recent progress in terahertz difference-frequency quantum cascade laser
      sources," *Nanophotonics*, vol. 7, no. 11, pp. 1795–1817, 2018, doi: 10.1515/nanoph-2018-
      0093.

[24]  Q. Y. Lu, N. Bandyopadhyay, S. Slivken, Y. Bai, and M. Razeghi, "Room tempera-
      ture single-mode terahertz sources based on intracavity difference-frequency generation
      in quantum cascade lasers," *Appl. Phys. Lett.*, vol. 99, no. 13, p. 131106, 2011,
      doi: 10.1063/1.3645016.

[25]  Q. Y. Lu, N. Bandyopadhyay, S. Slivken, Y. Bai, and M. Razeghi, "High perfor-
      mance terahertz quantum cascade laser sources based on intracavity difference frequency
      generation," *Opt. Express*, vol. 21, no. 1, p. 968, 2013, doi: 10.1364/OE.21.000968.

[26]  V. Berger and C. Sirtori, "Nonlinear phase matching in THz semiconductor waveg-
      uides," *Semicond. Sci. Technol.*, vol. 19, no. 8, pp. 964–970, 2004, doi: 10.1088/0268-
      1242/19/8/003.

[27] C. Pflügl *et al.*, "Surface-emitting terahertz quantum cascade laser source based on intra-cavity difference-frequency generation," *Appl. Phys. Lett.*, vol. 93, no. 16, p. 161110, 2008, doi: 10.1063/1.3009198.

[28] K. Vijayraghavan *et al.*, "Terahertz sources based on Čerenkov difference-frequency generation in quantum cascade lasers," *Appl. Phys. Lett.*, vol. 100, no. 25, p. 251104, 2012, doi: 10.1063/1.4729042.

[29] G. A. Askaryan, "Cerenkov radiation and transition radiation from electromagnetic waves," *J. Exp. Theor. Phys.*, vol. 15, no. 5, p. 943, 1962.

[30] A. Hugi, G. Villares, S. Blaser, H. C. Liu, and J. Faist, "Mid-infrared frequency comb based on a quantum cascade laser," *Nature*, vol. 492, no. 7428, pp. 229–233, 2012, doi: 10.1038/nature11620.

[31] N. Hashizume, T. Kondo, T. Onda, N. Ogasawara, S. Umegaki, and R. Ito, "Theoretical analysis of Cerenkov-type optical second-harmonic generation in slab waveguides," *IEEE J. Quantum Electron.*, vol. 28, no. 8, pp. 1798–1815, 1992, doi: 10.1109/3.142578.

[32] Q. Y. Lu, N. Bandyopadhyay, S. Slivken, Y. Bai, and M. Razeghi, "Continuous operation of a monolithic semiconductor terahertz source at room temperature," *Appl. Phys. Lett.*, vol. 104, no. 22, p. 221105, 2014, doi: 10.1063/1.4881182.

[33] E. Benveniste *et al.*, "Influence of the material parameters on quantum cascade devices," *Appl. Phys. Lett.*, vol. 93, no. 13, p. 131108, 2008, doi: 10.1063/1.2991447.

[34] Q. Lu and M. Razeghi, "Recent advances in room temperature, high-power terahertz quantum cascade laser sources based on difference-frequency generation," *Photonics*, vol. 3, no. 3, p. 42, 2016, doi: 10.3390/photonics3030042.

[35] S. Jung, J. H. Kim, Y. Jiang, K. Vijayraghavan, and M. A. Belkin, "Terahertz difference-frequency quantum cascade laser sources on silicon," *Optica*, vol. 4, no. 1, p. 38, 2017, doi: 10.1364/optica.4.000038.

[36] L. Consolino *et al.*, "Spectral purity and tunability of terahertz quantum cascade laser sources based on intracavity difference-frequency generation," *Sci. Adv.*, vol. 3, e160331, 2017, doi: 10.1126/sciadv.1603317.

[37] L. Tombez, S. Schilt, J. Di Francesco, P. Thomann, and D. Hofstetter, "Temperature dependence of the frequency noise in a mid-IR DFB quantum cascade laser from cryogenic to room temperature," *Opt. Express*, vol. 20, no. 7, p. 6851, 2012, doi: 10.1364/oe.20.006851.

[38] S. Bartalini *et al.*, "Measuring frequency noise and intrinsic linewidth of a room-temperature DFB quantum cascade laser," *Opt. Express*, vol. 19, no. 19, p. 17996, 2011, doi: 10.1364/oe.19.017996.

[39] S. Bartalini *et al.*, "Observing the intrinsic linewidth of a quantum-cascade laser: Beyond the Schawlow-Townes limit," *Phys. Rev. Lett.*, vol. 104, no. 8, p. 083904, 2010, doi: 10.1103/PhysRevLett.104.083904.

[40] S. Borri *et al.*, "Frequency-noise dynamics of mid-infrared quantum cascade lasers," *IEEE J. Quantum Electron.*, vol. 47, no. 7, pp. 984–988, 2011, doi: 10.1109/JQE.2011.2147760.

[41] G. Di Domenico, S. Schilt, and P. Thomann, "Simple approach to the relation between laser frequency noise and laser line shape," *Appl. Opt.*, vol. 49, no. 25, pp. 4801–4807, 2010, doi: 10.1364/AO.49.004801.

[42] M. S. Vitiello *et al.*, "Quantum-limited frequency fluctuations in a terahertz laser," *Nat. Photonics*, vol. 6, no. 8, pp. 525–528, 2012, doi: 10.1038/nphoton.2012.145.

[43] S. Jung *et al.*, "Broadly tunable monolithic room-temperature terahertz quantum cascade laser sources," *Nat. Commun.*, vol. 5, no. 1, p. 4267, 2014, doi: 10.1038/ncomms5267.

[44]  A. Jiang, S. Jung, Y. Jiang, K. Vijayraghavan, J. H. Kim, and M. A. Belkin, "Widely tunable terahertz source based on intra-cavity frequency mixing in quantum cascade laser arrays," *Appl. Phys. Lett.*, vol. 106, no. 26, p. 261107, 2015, doi: 10.1063/1.4923374.

[45]  Q. Lu, F. Wang, D. Wu, S. Slivken, and M. Razeghi, "Room temperature terahertz semiconductor frequency comb," *Nat. Commun.*, vol. 10, no. 1, pp. 1–7, 2019, doi: 10.1038/s41467-019-10395-7.

[46]  T. S. Mansuripur *et al.*, "Single-mode instability in standing-wave lasers: The quantum cascade laser as a self-pumped parametric oscillator single-mode instability in standing-wave lasers: the quantum cascade laser as a parametric oscillator," *Phys. Rev. A*, vol. 94, p. 063807, 2016, doi: 10.1103/PhysRevA.94.063807.

# Part III

## Applications

# 12 QCL Applications in Scientific Research, Commercial, and Defense and Security Markets

Jeremy Rowlette, Eric Takeuchi, and Timothy Day

DRS Daylight Solutions, San Diego

## Primus lux

In 1994, the quantum cascade laser (QCL) made first light in the basement of a renowned industrial research lab in Murray Hill, New Jersey [1]. Its untamed Fabry–Pérot emission spectrum was measured in microwatts, its viability maintained only at cryogenic temperatures and its potential impact on society entirely uncertain. Nearly 30 years later, we are just beginning to fully appreciate the transformative power of QCL technology as we watch it enable new fields of research and displace incumbent technologies in major commercial, medical, and defense industries. In this chapter, we will discuss applications of QCLs in scientific research, commercial, and defense and security markets.

## 12.1 Scientific Research Applications

### 12.1.1 Introduction

Since its first laboratory demonstration a quarter century ago, the QCL has become an invaluable research tool for the Physical and Life Sciences. Today, researchers in these and related fields are continuing to find innovative ways to exploit the unique properties of the QCL, some of which are outlined in Table 12.1, in order to unlock the potential of the mid-infrared (IR) and terahertz spectral bands. Chemical sensing based on the well-known IR vibrational fingerprint of chemical functional groups remains the primary application domain for QCLs today (Fig. 12.1). However, unlike in the early years, chemical sensing has extended well beyond monitoring of purified gases to include the analysis of extremely complex chemical mixtures such as tumor microenvironments.

There are dozens of application categories that the QCL has either enabled or advanced significantly. However, its use in micro and nanoscale imaging instrumentation is of particular interest as these platforms promise to greatly expand the application space and user base in the coming decade. These new QCL imaging modalities enable simultaneous spectral and spatial resolution of complex scenes and sample specimens and with this comes rich datasets to analyze. As QCL imaging and microscopy instrumentation have become more robust and readily available; the

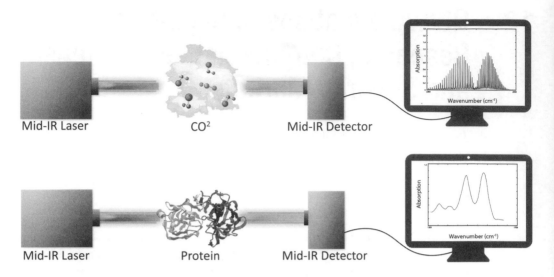

**Fig. 12.1** Two examples of chemical sensing using quantum cascade lasers. In the top portion of the figure, a vibrational spectrum is obtained of a gas species (e.g., $CO_2$) by passing a tunable QCL beam through a gas cloud and detecting the attenuation of the light with a mid-IR detector. In this case, the spectral features are narrow (<GHz) and may be comparable to the linewidth of the QCL. In the bottom portion of the figure, a vibrational spectrum is obtained of protein molecules in a similar manner to the gas-phase example. However, unlike in the gas-phase measurement, the spectral features are broad (>500 GHz) compared to the effective linewidth of the QCL. In the protein measurement example, broadly tunable external cavity lasers have significant advantages over narrowly tunable lasers such as distributed feedback (DFB) lasers.

sheer volume of high-quality spectral data has grown exponentially. With this trend, a significant fraction of QCL-based applied research today requires big data analytics, machine learning, or artificial intelligence to extract useful information from an experiment. It follows that many of the leading research groups today operate at the intersection of analytical chemistry, spectroscopy, biology, and machine learning. It is awe inspiring to see that in recent years, with the aid of modern computational tools, researchers using QCL technology have achieved an ability to routinely identify and track disease progression in human tissue on clinically relevant time scales. Because of its profound enabling force within the scientific community, we will devote much of the following sections to discussing key advancements in QCL-based microscopy at multiple length scales. Before doing so, let us briefly review the history and pace of QCL applied research.

## 12.1.2    History of QCL Applied Research

During the first 10 years after the birth of the QCL (1994–2004), applications using QCLs were relatively limited by today's standards and served mainly to highlight the advancements in the rapidly evolving core semiconductor technology. The first application demonstrated was gas sensing using a distributed feedback (DFB)

**Table 12.1** The QCL has a number of intrinsic advantages, which make them well suited for scientific research.

| Attribute | Research benefit |
| --- | --- |
| Mid-IR fingerprint coverage (3–14 μm) | Molecular sensing |
| High spectral brightness | More sensitive measurements in light-starved applications |
| High degree of coherence | Phase-sensitive measurements |
| Semiconductor | Wafer-scale packaging |
| Room temperature operation | Mobile and field-deployed applications (e.g., environmental monitoring and medical device) |
| High intrinsic reliability | Long lifetime research in instruments |
| Small étendue | Coupling of beams to fibers and small sample volumes (e.g., atomic-force microscopy) |
| Broad gain bandwidth | Broad tuning to support multi-analyte sensing, liquids and solid phase |
| Direct electrical amplitude modulation | Pulse rates adjustable to match mechanical resonances, Q-factor enhancements |
| Small linewidth confinement factor | Weak dependence on refractive-index modulation |
| Strong $\chi^{(3)}$ nonlinearity | Spontaneous frequency and harmonic comb generation |
| Short upper-state lifetime | Spatial hole burning |

QCL [2]. During this period, QCL pioneers, mostly within the quantum electronics communities, focused their research on pushing QCL-device technology to operate at higher temperatures, at higher optical powers, at higher wall-plug efficiencies, and over wider wavelength ranges. These advancements would be essential in making room for an expanded application space to take hold in the subsequent decade. The expanded application space would be accomplished with the placement of a QCL in a tunable external-cavity configuration (see Fig. 12.2).

During the second decade (2004–2014), the commercialization of turnkey scientific tunable QCL sources using improved QCL die ushered in an expansionary phase of applied research. These tunable QCLs incorporated into external-cavity configurations to fully exploit the broad QCL gain bandwidth. Rather than the QCL being the focus of the research, the QCL became an essential tool for spectroscopists to develop and advance a wide range of applications beginning with high-precision, environmental trace gas monitoring [3–6]. Quickly the application space expanded to include stand-off detection [7], breath diagnostics [8], microscopy instrumentation [9], combustion dynamics [10], silicon photonics, chemical mixing [11], and much more. Feedback from the spectroscopy community during this phase was critical for encouraging continued investment and maturation of commercial QCL sources with expanded capabilities and improved reliability.

Within the past five years, the commercialization of turnkey QCL-based microscopy, spectrometry, and analyzer instruments having user-friendly software interfaces

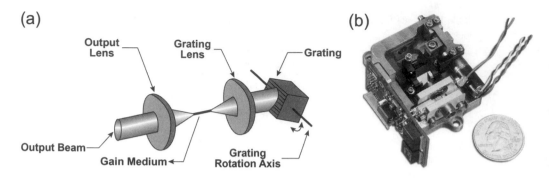

**Fig. 12.2** [Reproduced and adapted with permission from [12].] (a) Schematic representation of an external-cavity quantum cascade laser showing the semiconductor gain medium centered between two collimating lenses. The left lens enables light to propagate into free space and interact with a sample downstream of the laser. The right lens enables the light to propagate over a short distance from mm to cm and interact with a dispersive element such as a diffraction grating that can be rotated to select specific frequencies of light amplified by the laser gain medium. (b) a photograph of an early commercial tunable external cavity QCL source.

and built-in spectral data analysis capabilities, have enabled yet another wave of applied research to take place. With this higher degree of instrument integration, an expanded community of scientists no longer limited to the fields of photonics and spectroscopy, has been able to take advantage of QCL technology to accelerate their scientific discovery. The Materials and Life Sciences communities have particularly benefited from advancements in microscopy and imaging. Consequently, the balance of applied research is shifting towards condensed and liquid-phase applications which favor higher power and more broadly tunable QCL sources.

Figure 12.3 shows the trend in the overall number of QCL publications since 1994, along with a breakdown showing all peer-reviewed journal publications and the subset of those articles which are focused on the application of a QCL source. As of the end of 2018, about 1000 peer-reviewed articles have been written based on the QCL since 1994, of which 40% have been application focused. Interestingly, the data shows all three curves share the same basic trend, including an apparent slow-down in the rate of publications, reaching a plateau of about 100 journal articles per year.

Figure 12.4 shows the fractional trend of journal articles which use QCL technology in an application. Since the first demonstration of a QCL in an application [2] in the late 1990s, there has been a steady shift towards applied research of about 2% per year.

### 12.1.3    QCL Microscopy

Many applications, especially those within the Materials and Life Sciences, benefit greatly from the ability to visualize chemistry at multiple length scales. Since 2011 [13], QCLs have been deployed with great success in a variety of microscopy and imaging modalities to investigate chemical structure across seven orders of spatial resolution, from 10 nm to 10 cm. Figure 12.5 illustrates the relevant biological and

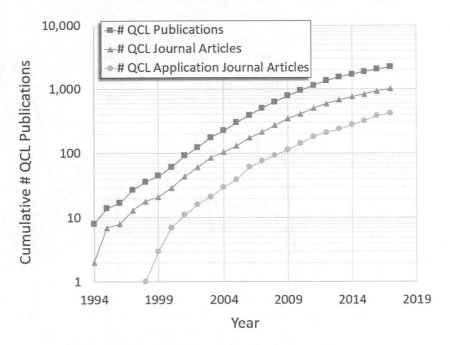

**Fig. 12.3** The cumulative number of journal publications written about QCL technology. The circle dataset shows the total cumulative number of publications whereas the square and datasets show peer-reviewed journal articles focused on QCL core technology and applications of QCLs respectively. From these trends, it becomes apparent that an overall slowing of the rate of journal articles published has plateaued at about 100 per year.

material length scales which QCL microscopy techniques currently address. In all these applications, the QCL is tunable to allow spectroscopic chemical sensing at each pixel of a two-dimensional (2D) image. One advantage of the external cavity QCL (EC-QCL) is that hyperspectral image cubes can be built up one wavelength at a time. That is, a 2D image can be formed at each tuned wavelength of the laser. The end result is a spectral response of the sample at each pixel location. This three-dimensional data structure can then be mined using a wide range of multi-dimensional chemometrics analysis tools.

It is interesting to note that the latest cutting-edge laser-based infrared microscopes actually bear a number of similarities to the earliest infrared microscopes developed in the late 1940s [13, 14], which used dispersive techniques. However, unlike the early days, when narrow bands of light were created from broadband incoherent sources through tunable dispersive elements, today the laser wavelength is tuned using an intra-cavity dispersive filter (e.g., a pivoting grating) thereby taking advantage of optical gain to produce orders of magnitude higher spectral density.

There are several advantages to building up hyperspectral image cubes one wavelength (frequency) at a time. First, the microscopist has the freedom to determine a priori the set of wavelengths used in the analysis. By reducing the number of wave-lengths in the dataset, faster data acquisitions can be achieved while simultaneously

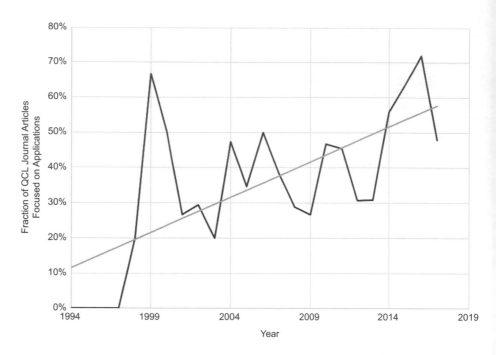

**Fig. 12.4** Fraction of journal articles written on QCL technology which were focused on the use of or application of a QCL source. Over the complete history of the QCL, there has been a long-term trend towards applications, as one might expect, at about 2% increase per year.

reducing the size of the dataset. This becomes particularly important when considering applications like cancer diagnostics in pathology where a single specimen (cm$^2$) typically requires roughly 10 million pixels of information (assuming ~5 μm pixel sampling size). If each pixel requires a full spectrum of 500 spectral points and each spectral point contains 4 bytes of memory, the hyperspectral cube per specimen would be of order 10 GB. This is not a terribly large file size by today's standards, but imagine a hospital processing hundreds or thousands of slides per day. This modality also opens up the possibility of fast survey screens using a single wavelength to help first identify regions of interest followed by a dense hyperspectral cube acquired over the limited region.

All QCL-based microscopy embodiments to date fit into two basic classes: (1) far-field and (2) near-field techniques. Far-field techniques are limited by Fraunhofer diffraction which is determined by the wavelength (λ) of light and the effective aperture limiting the spatial bandwidth of the system. In practice, the spatial resolving power is limited to ~λ/2. Since mid-IR QCLs generate light between 3 and 14 μm, the spatial resolution of far-field systems is expected to be wavelength dependent and limited to 1.5 μm at the blue end of the wavelength band. Exceptions to this rule will be discussed in future sections. However, for simplification, we shall equate the term far-field with microscale when discussing this application space.

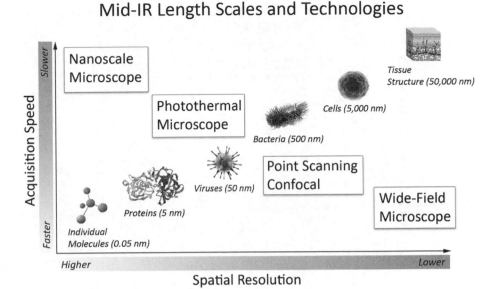

**Fig. 12.5** QCL-based microscope technology performance in the context of Life Sciences length scales.

In contrast, near-field techniques rely on a sharp tip with a radius of order 10 nm placed in close proximity to or in physical contact with a surface to localize and enhance the optical field. Such techniques have the benefit of achieving spatial resolving power on par with the tip radius, which is much smaller than the wavelength of light, thus circumventing the Fraunhofer limit. With a spatial resolving power on the order of 10 nm, we shall for simplicity equate the term near-field with nanoscale when discussing this application space.

Figure 12.6 shows the overall trend in peer-reviewed journal articles published on QCL-based microscopy techniques or their use since 2011, when QCL-based microscopy techniques first emerged. Of the approximately 1000 journal articles written on or using QCLs by the end of 2018, about 56 articles were associated with microscopy applications. Within this application space, approximately two thirds of these publications have been associated with near-field techniques, the remainder being far-field techniques. Both near-field and far-field technique and application publications are growing exponentially as they become more accepted as scientific workhorse tools and the user base expands into new fields.

Within these two major branches of QCL microscopy, a number of instrument variations have been conceived of and demonstrated to address a growing diversity of application needs. In the following sections, we shall explore how QCLs are expanding the applicability of these two branches separately.

## 12.1.4    QCL Microscopy – Nanoscale

We shall only summarize here the more salient developments relevant to QCL-based near-field microscopy and refer the reader to an excellent review by A. Centrone [15]

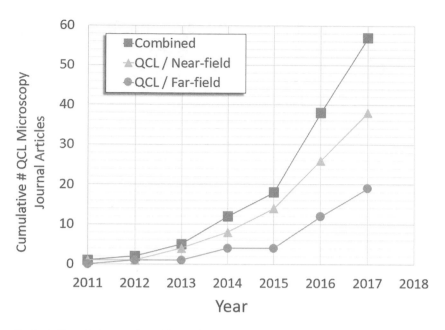

**Fig. 12.6** Cumulative number of journal publications associated with QCL based microscopy tools. Near-field (nanoscale) and far-field (microscale) techniques are broken out separately showing a roughly 2:1 split since 2011.

for a more comprehensive history, which began ca. 1999 when mid-IR lasers were first combined with AFM tips [16]. In essence, two major branches have evolved within the near-field (nanoscale) imaging and spectroscopy community. The first branch was based on detection of a tip-scattered signal measured in the far field and referred to as infrared scattering, scanning nearfield optical microscopy (IR-sSNOM). The second branch is based on detection of a photothermal response of the sample at the cantilever and is referred to as AFM-IR.

In 2011, Fei *et al.* [13] demonstrated the first successful use of a tunable QCL source for imaging a material at the nanoscale. As shown in Fig. 12.7, the high-contrast discrete-frequency imaging of graphene, $SiO_2$ and silicon substrate at four mid-IR frequencies was achieved showing clear boundary delineation between the three materials. The method employed was an extension of the already well-established infrared scattering, scanning near-field optical microscopy (IR s-SNOM) technique first introduced by Noll and Keilmann [16] in 1999 with a narrowly tunable $CO_2$ laser source.

The seminal work by Fei *et al.* employed an interferometric pseudo-heterodyne technique first discussed by Ocelic *et al.* [17] in 2006, and which is illustrated in Fig. 12.8 for a tunable-QCL setup [18]. The tunable QCL source (or sources) of light is injected into one arm of a Michelson interferometer and the AFM cantilever tip and sample are placed in the opposing arm. Unlike the standard FTIR configuration used in conjunction with a broadband light source (e.g., globar, synchrotron, fiber laser, etc.), the translating mirror is held at a nominal position while the

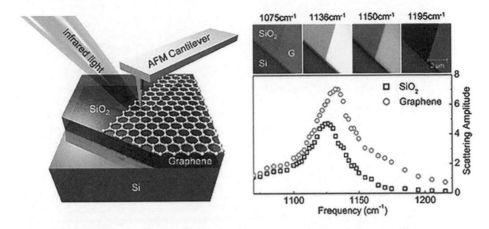

**Fig. 12.7** [Reproduced with permission from [13].] First demonstration of QCL-based nanoscale imaging. The example shown was of graphene on $SiO_2$.

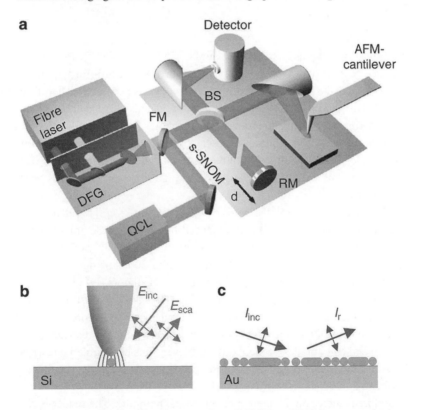

**Fig. 12.8** [Reproduced with permission from [18].] IR s-SNOM setup employing a tunable QCL. The light backscattered from the tip is analyzed with a Michelson interferometer that is operated in pseudo-heterodyne mode for s-SNOM imaging.

QCL-source wavelength is tuned to discrete mid-IR frequencies. In accordance with the interferometric pseudo-heterodyne technique, the translating mirror is dithered at

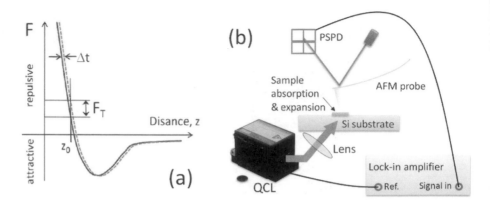

**Fig. 12.9** [Reproduced with permission from [19].] (a) Mechanism of AFM cantilever deflection during sample photoexpansion. The solid curve shows the dependence of the interaction force between the sample surface and the AFM cantilever tip on tip-surface distance.

a modulation frequency to provide higher SNR images [17]. Because the interferometric s-SNOM approach is highly influenced by phase information, continuous or quasi-continuous phase tuning of the laser source is necessary. Therefore, mode-hop-free (MHF) QCLs or CW operation, where islands of repeatable phase continuous regions can be present, are quite advantageous for this application. Furthermore, because the s-SNOM technique is sensitive to both the real and imaginary part of the complex index of refraction of the material, the technique tends to be preferred for harder materials with high reflectivity and high dielectric constants or strong optical resonances.

Shortly after the s-SNOM work by Fei *et al.*, Lu *et al.* [19] demonstrated the first use of a tunable QCL in a photothermal-based nanoscale imaging technique (see Fig. 12.9). This work was an extension of the previously well-established AFM-IR technique developed by Dazzi *et al.* in the mid-2000s using a tunable free-electron laser source and later using optical parametric oscillators (OPOs) [20].

In this groundbreaking work, Lu *et al.* [19] also demonstrated that the SNR of the photothermal signal could be greatly enhanced (by more than 100× in some cases) when the QCL pulse repetition rate was brought into resonance with one of the mechanical eigenmodes of the AFM cantilever. This highlighted one of the core benefits of the QCL over other mid-IR laser sources such as OPOs, specifically, the ability to directly modulate the light output from QCLs from DC to MHz or even higher repetition frequencies on demand using electrical pumping. The reader is referred to [21] for an excellent discussion of the many applications being enabled by this approach.

In 2016, Nowak *et al.* [22] described a variant of nanoscale IR imaging, referred to by the authors as photoinduced force microscopy (PiFM). The technique is schematically described in Fig. 12.10. Closely resembling that of the AFM-IR setup, the apparent distinguishing characteristic between AFM-IR and PiFM is that the pulse repetition rate of the QCL source in the latter is tuned to the difference between two

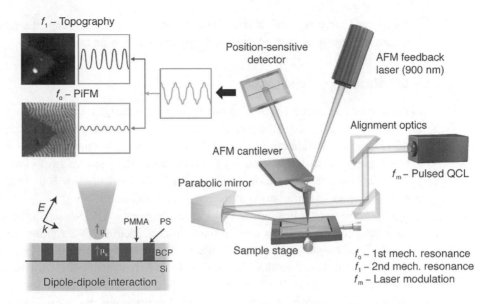

$f_1$ – Topography

$f_o$ – PiFM

Position-sensitive detector

AFM feedback laser (900 nm)

Alignment optics

AFM cantilever

$f_m$ – Pulsed QCL

Parabolic mirror

E

k

PMMA  PS

↑μ_t

BCP

↑μ_s

Si

Sample stage

Dipole-dipole interaction

$f_o$ – 1st mech. resonance
$f_1$ – 2nd mech. resonance
$f_m$ – Laser modulation

**Fig. 12.10** [Reproduced with permission from [22].] Schematic of IR PiFM experiment. The incident mid-IR laser is electrically triggered to pulse at $f_m = f_1 - f_0$, where $f_0$ and $f_1$ are the first and second mechanical eigenmode resonances of the cantilever. The topography of the sample is recorded by the AFM feedback system at $f_1$, and the PiFM is concurrently recorded at $f_0$ by the feedback laser position-sensitive detector and lock-in electronics. The sample is raster-scanned under the tip to generate the image. The incident light is polarized along the tip axis to maximize the signal coupling of the dipole-dipole force along the vertical direction of the cantilever vibration.

eigenmodes of the AFM cantilever as opposed to a particular eigenmode frequency. By doing so, additional gains in SNR were achieved while also rejecting common mode signal during top side illumination. It should be noted that the physical origin of the signal seen in PiFM is subject to discussion as both optical forces and sample photoexpansion forces were suggested as the source of the signal [23, 24]. For material identification and imaging this may not be an issue, but could lead to barriers in molecular quantitation.

Mathurin *et al.* [25] provided additional context to this exciting and dynamic subfield by describing the technique whereby the laser repetition rate is specifically tuned to the difference of the first and second cantilever eigen mode of a dynamically driven tip (non-contact) as Tapping Mode AFM-IR. In this mode, large signal enhancement factors could be observed for soft materials like biologic and organic polymer materials.

Within the infrared nanoscale microscopy community, there is a growing desire to move away from discrete frequency to hyperspectral imaging whereby a complete spectrum over the infrared fingerprint band is obtained at every point in the image. This trend is motivated by the desire to perform chemical identification of unknown materials or map out chemical processes or particular biological processes at the nanoscale. As a consequence, this requires QCL sources having much faster scan speeds and more repeatable wavelength tuning and beam pointing, the latter being essential in ensuring spatio-wavelength alignment across an image.

## 12.1.5    QCL Microscopy – Microscale

Around the same time that the first articles on nanoscale QCL imaging were first being published (c. 2011–2014), a parallel community was forming with the motivation of developing and advancing far-field QCL microscopy techniques. Rather than looking at samples with spatial resolution of tens of nanometers and fields of view (FOV) of a few microns, the application goal for far-field QCL microscopy is to measure large samples (up to 10 cm or more) at micron-scale resolution.

One of the key applications driving the development of QCL-based far-field micros-copy has been the desire to make infrared micro-spectroscopy a routine technique in the fields of biomedical research and clinical pathology where it is desired to ana-lyze large (cm) tissue sections in minutes. Prior to the advent of QCL-based far-field microscopy, a large body of scientific work by the infrared spectroscopy community, see e.g., [26], overwhelmingly demonstrated the intrinsic diagnostic value of infra-red spectroscopy using FTIR microscopes and machine learning. However, broad acceptance by the biomedical and clinical research communities was not achieved primarily because single-specimen data collection could take many hours or even days for typical clinical samples. QCL far-field microscopy has offered substantially higher ($>150\times$) sample collection speeds [27] over FTIR microscopes, bringing col-lection times down to clinically relevant time scales. QCL microscopes have also enabled smaller instruments which do not require cryogenic cooling. Though biomed-ical imaging has been the primary application driving its development, QCL far-field microscopy is enabling a multitude of new application areas, particularly in materials research and sample inspection.

Since their first demonstrations, multiple QCL-microscope strategies have formed to take advantage of the high brightness (Fig. 12.11) of the QCL. These strate-gies broadly fit into two classes: wide-field and point-scanning (or mapping)-type architectures. In point-scanning architectures, the photons are focused into a tight, diffraction-limited spot in order to either increase the signal magnitude, as in the case of photothermal-based imaging, or to overcome substantial optical losses from either weakly scattering samples in the reflection mode or from highly absorbing materials (e.g., borosilicate glasses or water) in the transmission mode. The ability to leverage all available photons in a measurement could, in principle, increase the shot-noise detection limit by a factor of 1000 or more compared to other mid-IR sources, pro-vided, of course, that all other noise sources can be sufficiently suppressed. However, in practice, most sample materials have an upper tolerance limit of optical intensity, which requires throttling of the source. Alternatively, the photons can be spread out over a larger area in order to perform wide-field imaging [28]. This approach can have significant (as much as $10^5$) throughput advantage over point-scanning or mapping instruments, particularly when it comes to measuring large samples. Furthermore, as there is no benefit to increasing the signal beam flux beyond the dynamic range

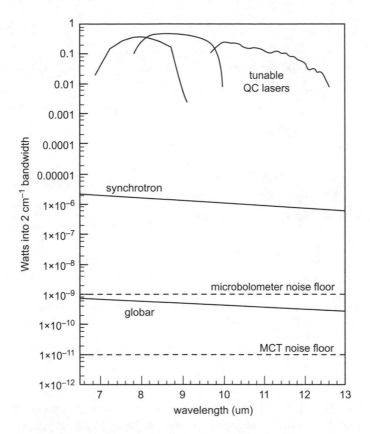

**Fig. 12.11** [Reproduced with permission from [29].] Comparison of signal level through a 10 μm pinhole for three different mid-IR sources. The MCT noise level is for liquid-nitrogen cooled single-element detector. The microbolometer noise level is for uncooled FPA with 100 mK NEP. Globar and synchrotron data from Reffner *et al.* [30].

of the detector, "excess photons" in a wide-field illumination mode can be gainfully employed by equally distributing them across the sensor array such that the linear dynamic range of each pixel is fully exercised.

In 2012, Kole *et al.* [9] presented the first peer-reviewed demonstration of the use of a QCL in a microscopy configuration by coupling a tunable QCL source with a 128 × 128 mercury cadmium telluride (MCT) focal plane array and a high-NA, single-element refractive optic. This seminal work demonstrated significant data-collection throughput advantages over an FTIR microscope, particularly for limited spectral-band ranges. However, it also highlighted some major engineering hurdles that would need to be overcome in order to achieve data quality parity with FTIR microscopes. In particular, using coherent illumination in a wide-field (or broad area illumination) instrument architecture required special attention. Kole *et al.* reported significant image quality reduction as well as excess noise well above the shot noise limit, both issues being attributed to source coherence.

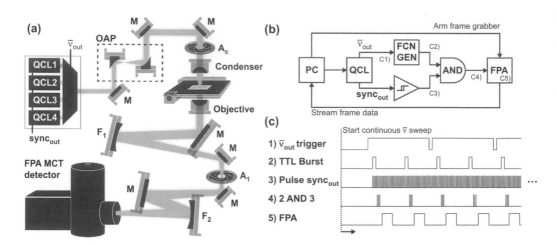

**Fig. 12.12** [Reproduced with permission from [31].] (a) Schematic of the QCL-based wide field-imaging microscope configured in the inverted transmission mode. The microscope is coupled to a four-chip widely tunable QCL and a $128 \times 128$ FPA MCT detector. (b) Wiring diagram showing the primary control components of the system (stage omitted). The timing structure on the annotated trigger lines are then shown (not to scale). (c) A generalized synchronization protocol. The real-time wavenumber-monitoring line output from the QCL controls an external oscillator that triggers the FPA depending on the specified spectral resolution and scan speed.

Several reports [31, 32] demonstrated some success in improving image quality using a spinning disk diffuser approach, the goal being to spatially scramble the diffraction pattern by pseudo randomization of the beam's angular (or k-space) distribution. Figures 12.12 and 12.13 show two embodiments of wide-field QCL microscopes incorporating a spatial-diffuser strategy. This approach has shown reduced diffraction effects around edges, thus leading to softer images which more closely resemble an incoherent, extended source. However, as Schönhals *et al.* [32] concluded, the improvement in image quality comes at the price of raising the overall baseline noise floor, while at the same time reducing light throughput. That is, the pseudo-random noise pattern generated by the laser illumination of the spinning disk is an additive noise source, and does not address the fundamental temporal noise source of the laser.

In 2014, the first peer-reviewed articles using commercial wide-field QCL microscopes were published [26, 33]. These papers highlighted, for the first time, the use of compound refractive imaging objectives, which enabled both high-NA (>0.7) and wide-FOV (650 μm × 650 μm) performance while maintaining sufficient achromaticity across the full molecular fingerprint band (900–1800 cm$^{-1}$). Prior to these seminal works, reflective based optics such as the Cassegrain (Schwarzschild) design, were found [34] to be unsuitable in wide-field QCL illumination architectures due in part to their central obstruction. Figure 12.14 demonstrated the significant throughput advantage enabled by this imaging advancement by showing that a breast tissue microarray containing 207 cores could be spectroscopically imaged in less than 20 minutes compared to over 19 hours using a state-of-the-art FTIR microscope.

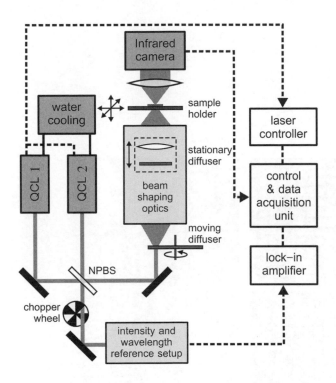

**Fig. 12.13** [Reproduced with permission from [32].] Schematic of the QCL-based infrared microscope. Two external cavity QCLs (1020–1097 cm$^{-1}$ and 1135–1319 cm$^{-1}$) are combined using a non-polarizing beamsplitter. One of the resulting beams is used for referencing of the laser power and wavelength. A collimated beam is directed at a moving diffuser disk. The scattered radiation is guided to a second, stationary diffuser disk, and then used for illumination of the sample by beam shaping optics. A microbolometer array detector with a microscope objective is used for detection of the transmitted radiation.

With the challenges presented by Kole *et al.* [9], several groups began investigating point-scanning alternatives to the wide-field architecture; the idea being that spatial coherence could be better managed by averaging over a single point detector rather than an array of detectors. The culmination of direct, laser-reflectance-based imaging is best represented by Mittal *et al.* [35] who, in 2018, demonstrated very high-quality QCL imaging and spectral classification of tissue using a reflection-based confocal mapping microscope architecture. As shown in Fig. 12.15, each pixel was added to the image in a sequential manner rather than in parallel, in a wide-field architecture.

The sensitivity improvement shown by [35] was achieved at the cost of data-acquisition speed. According to the authors, a tissue micro-array contained on a single slide could be analyzed using a limited set of pre-selected 12 mid-IR frequencies, enough to perform a 6-class stroma segmentation analysis, in about 3 hours. This should be compared to the results of Kuepper *et al.* [27] who demonstrated a wide-field QCL-microscopy technique possessing the ability to accurately analyze large tissue specimens (cm$^2$) in about 100 seconds using over 400 spectral frequencies.

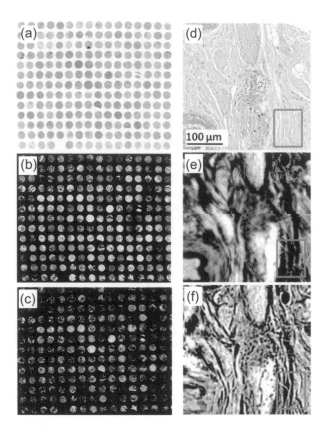

**Fig. 12.14** [Reproduced with permission from [33].] (a) Visible image of the 207 core breast tissue micro array (TMA). (b) and (c) are chemical images showing the absorbance of the amide I band at 1655 cm$^{-1}$ using FTIR- and QCL-imaging microscopes, respectively. (d) Expanded view from part (a) in the left column of core B4 (row 2, column 4 from top left). (e) FTIR chemical image of amide I, pixel size of $5.5 \times 5.5\ \mu m^2$. (f) QCL chemical image of amide I, pixel size of $1.36 \times 1.36\ \mu m^2$.

In 2016, Zhang *et al.* [36] introduced an alternative point-scanning far-field microscopy technique based on the well-known photothermal phenomenon. As schematically shown in Fig. 12.16, the photothermal technique builds up images point-by-point by detecting small changes in the far-field scattering distribution of a visible probe beam induced by sample absorption from an overlapping mid-IR QCL pump laser beam. As the QCL is tuned over a spectral band containing different absorption features, the amount of heating will vary according to the molecular absorption strength as well as the optical and thermal parameters of the sample and its local environment. The resultant signal is therefore a combination of multiple physical parameters including the absorption cross-section ($\sigma$), molecular concentration (N), thermal conductivity (k), thermal variation of the refractive index (dn/dT) and the product of the powers of the infrared pump $P_{IR}$ and probe beams $P_{pr}$ according to Eq. (12.1):

$$\Delta P_{pr} \propto \frac{\sigma N}{k C_p} \frac{\partial n}{\partial T} P_{pr} P_{IR}. \tag{12.1}$$

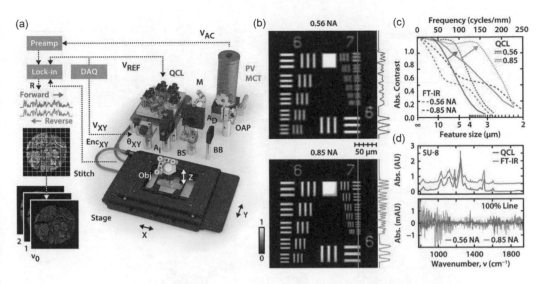

Fig. 12.15 [Reproduced with permission from [35].] Laser-based confocal IR microscopy. (a) The microscope consists of a quantum cascade laser (QCL) source, tunable for narrow-band emission across the mid-IR fingerprint region. A high-speed stage and focus unit raster scans the sample, while the detection is locked into the laser's pulse rate and each pixel is triggered by the stage encoder counter. (b) Images of USAF 1951 resolution test targets show diffraction-limited performance for absorbance at $1658\ cm^{-1}$ with both the 0.56 NA objective (top) at 2 μm pixel size and the 0.85 NA objective (bottom) at 1 μm pixel size. (c) The optical contrast of each set of bars is plotted as a function of spatial frequency. The bars are no longer resolvable if the contrast drops below 26%, which corresponds to the Rayleigh-criterion separation distance. The arrow indicates the deconvolution of the raw data (solid line) with the simulated PSF at the specified wavenumber to achieve a substantial resolution enhancement (dotted line). These results are compared with the simulated performance (dashed line) of a FTIR instrument with optimized Schwarzschild objectives used in FTIR imaging. (d) The unprocessed spectrum of a 5 μm layer of SU-8 epoxy, acquired by our QCL instrument at $1\ cm^{-1}$ resolution, shows accurate spectral features compared with a reference FTIR spectrometer. The 100% spectral profile lines show absorbance noise of $\sim 10^{-4}$ and $10^{-3}$ for the two objectives over most the fingerprint region.

The technique has significant advantages over other QCL-imaging approaches including spatial resolution that is limited by the probe beam wavelength, in this case, of order 0.5 μm. Furthermore, because the technique is directly proportional to the absorption cross-section of the material (Eq. (12.1)), reflectance spectra appear similar to typical transmission spectra. This makes sample identification based on vibrational-spectral fingerprint straightforward by using existing mid-IR spectral libraries that have been developed over decades using FTIR instruments. The technique currently suffers from slow speed; a full spectrum across the fingerprint band taking on the order of 10 seconds nowadays. However, in principle, this throughput of the technique could be increased with wide-field alternatives combined with higher power QCL sources. Quantitation can be challenging with this technique because the effective signal strength is a combination of multiple sample properties including thermal effusivity and its thermo-optic coefficient.

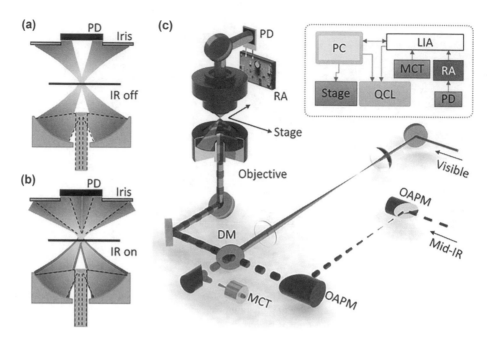

**Fig. 12.16** [Reproduced with permission from [36].] Principle and schematic of MIP imaging: (a) Probe-beam propagation through the sample via a dark-field objective (not to scale; condenser was omitted for simplicity). PD, photodiode; IR, infrared. (b) The probe-beam propagation is perturbed by the addition of an infrared-pump beam due to infrared absorption and the development of a thermal lens. (c) Setup: a pulsed mid-IR pump beam is provided by a QCL, and a continuous probe beam is provided by a visible laser, both of which are collinearly combined by a silicon dichroic mirror (DM) and sent into a reflective objective. The residual reflection of the infrared beam from the dichroic mirror is measured by a mercury cadmium telluride (MCT) detector. The probe beam is collected by a condenser with a variable iris and sent to a silicon PD connected to a resonant amplifier (RA). Inset: the photothermal signal is selectively amplified by the RA and detected by a lock-in amplifier (LIA). A computer is used for control and data acquisition. OAPM: off-axis parabolic mirror.

## 12.1.6    QCL Biomedical Microscopy

QCL-based microscopy is contributing to a major transformation within the biomedical-imaging field by offering cancer researchers, pathologists, and those working in translational medicine an entirely new imaging modality which is non-destructive, quantitative, and directly correlated with the specimen's native bio-chemistry. It follows that the QCL is helping to position mid-IR imaging as the perfect frontend screening tool within the histology and *omics*; e.g., proteomics, workflows.

In the 20 years preceding the advent of the QCL microscope, FTIR microscopy was able to successfully validate the science of infrared biomarkers. However, FTIR was not able to bridge the gap between the scientific spectroscopy community and the medical community largely because of the speed limitations. The QCL microscope has reduced the time scale for performing high-resolution IR imaging by more than two orders of magnitude, to the extent that whole tissue specimens can now be analyzed in minutes rather than hours or days.

In 2017, Schwaighofer *et al.* [37] presented an excellent review of the major uses of QCL-based instruments in biomedical research including developments in biomedical imaging. Since that publication, there have been three notable contributions to the field in the area of biomedical microscopy. The first contribution since the review by Schwaighofer *et al.* was from Mittal *et al.* [35] who demonstrated the ability to accurately segment epithelial and stromal tissue in cancerous breast tissue by using a limited set of discrete mid-IR frequencies. Figure 12.17 provides a summary of the imaging and tissue segmentation results achieved in this work. In this work, a point-scanning confocal QCL microscope was used to build up hyperspectral images of tissue microarrays point-by-point with excellent image quality and tissue-segmentation performance. Though speed was not a particular aim of this research, it was able to highlight the intrinsic spectral- and spatial-resolution capabilities of QCL microscopy in tissue analysis.

The second notable contribution was the work by Varma *et al.* [38] who demonstrated an ability to predict the progression of fibrosis in patients receiving kidney transplants using a commercially available wide-field QCL microscope. Fibrosis is a common issue that can be found in a wide range of organ types, often occurring as a byproduct of the organ's own intrinsic repair process in response to physical or chemical damage. As the summary of results shown in Fig. 12.18 support, the research group demonstrated not only that QCL imaging could be used to rapidly visualize the extent of fibrosis using a small number of mid-IR frequencies associated with collagen proteins, but could also be used to accurately predict the progression and remission of fibrosis in transplanted kidneys. This work is particularly interesting to the pathology community because it offers a capability not currently available by any other analytical techniques or immunohistochemical (IHC) staining protocols.

The third major contribution was the work by Kuepper *et al.* [27], which remains, arguably, the most significant step towards the clinical translation of infrared microscopy reported to date. In their work, Kuepper *et al.* achieved sensitivity and accuracy values of 100% and 95%, respectively, in detecting stage-II and -III colorectal cancers in a 110-randomized-patients study. Figure 12.19 summarizes the key results of this study, which culminated in the ability to automatically image and detect cancerous regions in under 5 minutes.

Although similar classification accuracies and sensitivities had been reported previously in FTIR and QCL tissue studies both by this group and others, this study was ground breaking in that it was the first large patient mid-IR tissue classification study which did not rely on the use of tissue microarrays. Tissue microarrays, typically containing several hundred small ($< 1$ mm$^2$) biopsies from multiple patients, are excellent research tools, but they tend to have limited use in clinical validation studies as the tissue cores are typically too small to observe larger scale tissue structure (which provide context to local tissue anomalies) and they do not support classifier robustness testing against sample preparation variation. In the study by Kuepper *et al.*, a large ($> 1$ cm$^2$) biopsied tissue slice for every patient sample was fixed to its own dedicated slide and

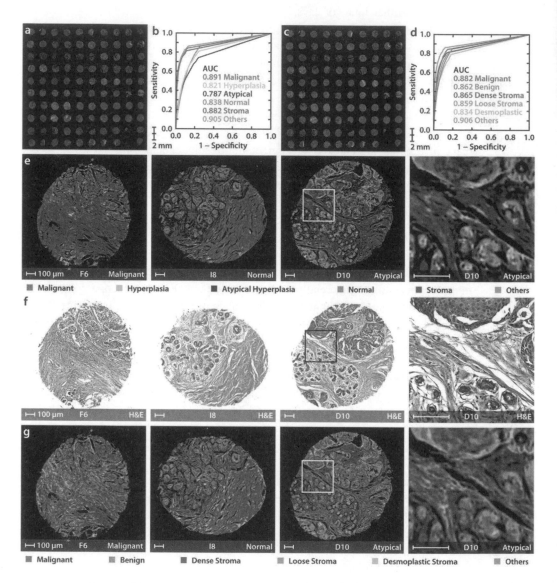

**Fig. 12.17** [Reproduced with permission from [35].] DF epithelial and stromal classification for rapid breast-cancer diagnosis: (a) TMA classified using the six-class epithelial (6E) model. (b) Receiver-operating-characteristic (ROC) curves representing the performance of each class in the 6E model. (c) TMA classified using the six-class stromal (6S) differentiation model. (d) ROC curves for the 6S model. (e) 6E-model classified images of three samples from the TMA with malignant, normal, and atypical hyperplasia states (left to right) along with their corresponding H&E-stained images in F. (g) Classification using the 6S model. A small region from the hyperplasia with atypia sample is also shown, along with its H&E stain, to demonstrate the spatial distribution of normal and malignant cells. The letter and numbers below each image correspond to the row and column of the TMA (A1 is the top-left sample), respectively. All scale bars: 100 μm.

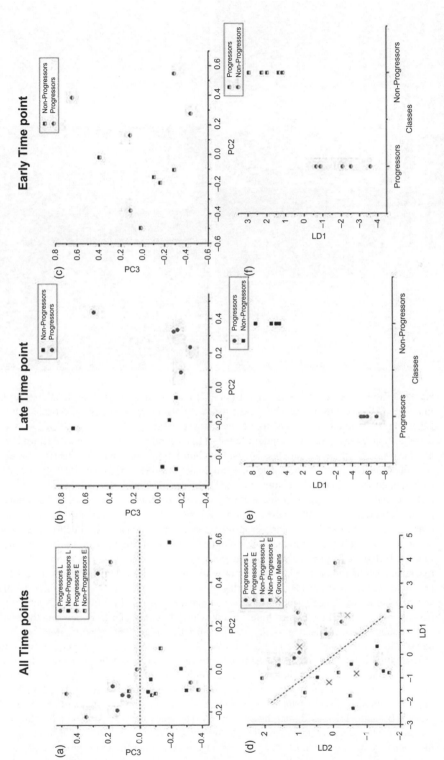

**Fig. 12.18** [Reproduced with permission from [38].] Multivariate time point analysis of biopsies for classification of progressors. Using principal component analysis (PCA) to reduce the data dimensionality (using the 1800–900 $cm^{-1}$) while maintaining the variance in an unsupervised fashion, it was possible to see segregation between all-time points (a–c). This shows alteration in the inherent biochemistry of fibrosis between the progressors and non-progressions. Furthermore, if linear discrimination analysis is used to classify the reduced dataset, it is possible to see near perfect segregation between progressors and non-progressors at all time points (d–f).

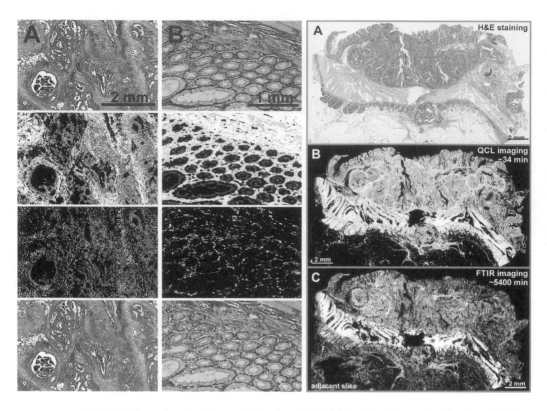

**Fig. 12.19** [Reproduced with permission from [27].] (a) Detailed tissue classification with IR imaging in 100 seconds. Colorectal cancer tissue annotation provided by QCL-based imaging: Cancerous (A) and tumor-free (B) tissue sections stained with H&E (top row), QCL-based classification, showing the first random forest (RF) (second row), the second RF (third row), and the overlay with the tumor class (A) and the infiltrating inflammatory cells (B) of the second RF analysis (fourth row). The QCL-imaging tissue classification takes 100 seconds for each sample shown here. The colors indicated below correspond to the colors in Fig. 2 of [27]. (A) The analysis of cancerous tissue sections, showing the tumor cells invading the connective tissue. Second row: red, pathological regions; yellow, infiltrating inflammatory cells; white, muscularis mucosae and muscularis propria; green, connective tissue. Cell debris and blood are sorted out and presented in the background color black. All regions marked in red using the first-level RF were further analyzed by the second-level RF. Third row: red, tumor cells; magenta, necrotic regions; yellow, infiltrating inflammatory cells. The inflammatory cells originate from the first-level RF. The overlay of tumor class (red) and HE underlines the high spatial accuracy of detection (bottom row). (B) The analysis of tumor-free tissue sections. Second row: cyan, healthy crypts; blue, lumen; white, thin muscular layer between the crypts and the submucosal connective tissue; green, connective tissue. The representative pathological, second-level RF analysis is presented in the third-row image: yellow, infiltrating inflammatory cells; magenta, necrosis; red, tumor cells. No tumor cells can be detected here. Following, the HE overlay is shown for the infiltrating inflammatory cells. (b) Comparison of the infrared-based tissue analysis. The colors indicated below correspond to the colors in Fig. 1 of [27]. Colorectal cancer tissue analysis using H&E staining as Gold standard (A), Spero QT system (B), and FTIR-based imaging system (C). The listed times illustrate the duration of the measurements. Red, pathological region comprising of tumorous regions and infiltrating inflammatory cells; white, muscles; green, connective tissue; cyan, crypts; and blue, lumen.

spectroscopically imaged by the QCL microscope with over 400 mid-IR frequencies across the molecular fingerprint band (950–1800 cm$^{-1}$) spaced apart by 2 cm$^{-1}$, and at 4 μm pixel resolution. In doing so, this study closely aligned to standard clinical procedures. With the tremendous speed advantage of the QCL wide-field microscope (each sample taking on average 30 minutes), Kuepper *et al.* were able to show that multi-thousand-patient studies were now feasible.

As Kuepper *et al.* pointed out, large patient datasets, made possible by high-throughput QCL microscopes, will be the key to developing statistically robust disease classifiers, which are usable in routine pathology labs. With the wide-field QCL microscope such a study could be conducted within a week. To achieve the same goal with a state-of-the-art FTIR-imaging microscope would take close to two years. With over one billion spectral data points collected per sample, the 110-patient study represents a dataset of well over 100 billion spectral points. Based on this pioneering work, it is conceivable that multi-thousand-patient, mid-IR spectral-imaging studies will soon become routine within the clinical translational community. As a final note, it should be pointed out that Kuepper *et al.* took the additional step of validating their results by using two QCL microscopes, built months apart and operated by different operators.

## 12.1.7    The Future of QCL Microscopy

Despite rapid progress in the field of QCL microscopy over the past five years, the field remains very much in its infancy. The capabilities demonstrated thus far are certainly impressive compared to those of the technologies being replaced, but do not come close to pushing the theoretical limits of QCL-imaging technology. As an example, data-acquisition speeds could be reduced by an additional two orders of magnitude; possibly more. It follows that, whole-slide, hyperspectral (hundreds of mid-IR frequencies) imaging at micron resolution could be achieved within seconds. QCL-based photothermal imaging techniques have already reduced the spatial resolution by an order of magnitude, but currently suffers from slow scanning speeds. Wide-field imaging modalities could alleviate the speed bottleneck and begin to address clinical biomedical imaging markets.

---

**Fig. 12.19** *(continued.)*
The comparison of the images demonstrates convincingly that the QCL-IR imaging results are in nice agreement with those obtained using the FTIR imaging. The observed deviations seem to be caused mainly by the use of an adjacent slice and the training of the FTIR classifier on samples of another study which could slightly differ in sample handling and processing. Furthermore, the previously FTIR classifier is performing the classification in one step which is less accurate by means of tumor detection. While the improved classifier for the QCL imaging recognizes infiltrating inflammatory cells in the first-level RF and cancerous regions in the second-level RF, which allows a much more accurate classification. (For color version, see www.nature.com/articles/s41598-018-26098-w.)

## 12.1.8    Commercial Applications

The QCL continues to represent significant commercial potential. One day, QCL sensors may be embedded in every smart phone to monitor everything from local $CO_2$ levels to your blood glucose level. However, to reach its full market potential, the cost of the QCL will need to fall by three orders of magnitude. The QCL has yet to benefit from a multi-billion-dollar investment in its supply chain, like the near-IR diode laser received in the 1990s during the telecom build-out. Hence, the QCL remains an expensive light source, typically commanding thousands of dollars for an uncoated, Fabry–Pérot chip. Therefore, the QCL is finding its niche in markets which benefit uniquely from mid-IR illumination, while also supporting price points of tens to hundreds of thousands of dollars. Recently, the latest generation of external cavity QCLs is fueling a surge in the commercialization of analytical instruments targeting the biopharmaceutical industry. Specifically, QCL-based spectroscopy tools for the analysis of protein-based therapeutics is experiencing good market adoption. Companies such as Redshift Bioanalytics [39] and Protein Dynamic Solutions are offering turnkey fluidic instruments designed to measure key attributes of aqueous proteins including aggregation, quantitation, structure, stability, and similarity.

Finally, mid-IR free-space communication (FSC) could be an important avenue for large market growth potential which could enable the level of investment required to bring QCL component costs down by multiple orders of magnitude. Delga and Leviandier [40] recently reviewed the strong technical motivation for deploying mid-IR-band FSC links to address "the last mile" in the continued expansion of communication networks as well as the current state of QCL-based FSC. Though the best QCL-based FSC link demonstration to date [41] has been limited to 3 Gbps over less than a meter, communication bandwidths exceeding Tbps over kilometer distances are fundamentally possible provided direct QCL modulation rates can reach 50 GHz at room temperature and low-loss, wavelength multiplexing solutions with excellent channel isolation are realized. Though the long-haul vacuum channel capacity is significantly lower for mid-IR than in the vicinity of the 1.55 μm band used in fiber-based communication networks, the robustness of mid-IR solutions to atmospheric dynamics by more than 20 dB/km makes mid-IR QCL based approach a very interesting architecture to be considered.

## 12.2    QCL Applications in Defense and Security Markets

### 12.2.1    Introduction

There are many applications in Defense and Security markets that are enabled by, and can utilize, laser sources operating directly in the mid-IR range. These include directed infrared countermeasures, standoff detection of chemical agents and explosives using molecular-detection techniques [42–45], and other emerging applications.

Due to their inherent simplicity of operation, systems based upon QCLs are particularly well suited to meet the demands of Defense and Security applications. These applications often require products that are ruggedized to meet military-specific requirements and/or operate in harsh environments to meet the typical use-case demands.

The physically compact and electrically efficient nature of QCLs naturally leads to system architectures that are relatively simple in design and operation. These characteristics typically yield product embodiments whose size, weight, and power consumption (SWaP) are minimized.

Applications in Defense and Security are driven primarily by the need to address a particular threat or group of threats. New technology capabilities are then integrated into product embodiments that can effectively address those threats to warfighters and civilians. It is also critical that products be flexible in design to provide scalability of new capabilities that are constantly and rapidly evolving. Providing wide spectral coverage, QCL technology lends itself to open architecture designs that allow for product design flexibility using common components throughout the mid-IR range.

Directional infrared countermeasures (DIRCM) applications benefit greatly from QCL technology. Aircraft protection from heat-seeking missiles requires DIRCM systems to operate effectively in harsh environmental conditions. High-brightness laser sources such as QCLs that can operate at wavelengths similar to the heat signatures of aircraft (i.e., in the mid-IR) provide an enabling capability. Furthermore, the inherent simplicity of QCL technology allows for system size, weight, and power consumption (SWaP) to be minimized, which is critical for integrating onto lightweight, rotary-wing and fixed-wing aircraft.

Standoff detection of explosives and other harmful agents has been demonstrated using QCL sources as an effective probe. By using spectroscopic analysis techniques, the wavelength coverage offered by QCL technology provides a common system to detect almost any chemical or explosive element of interest. Furthermore, the inherent high power/brightness offered by QCLs provides higher signal strength and longer standoff distances to be achieved.

As a relatively young technology, applications for QCLs continue to be identified. Emerging applications such as combat identification and sensor test and measurement also benefit from QCL capabilities.

## 12.2.2    Directional Infrared Countermeasures

Infrared countermeasures (IRCM) provide a defensive mechanism for protecting against infrared homing (or "heat-seeking") missiles. Simplistically, the IRCM system is designed to confuse the missile's guidance system so that it will miss the target. An excellent description of countermeasure systems has been compiled [45, 46].

For aircraft survivability applications, it is critical that a DIRCM system be able to match the aircraft's heat signature on which the missile is tracking. For typical jet engine plumes, this lies generally in the 3–5 micron wavelength range.

This range overlaps ideally with QCL technology [47]. As described in [44], QCLs represent the latest advance in DIRCM source technology, offering all-band (wavelength) coverage, as well as additional benefits in design simplicity and production scalability.

Early IRCM systems integrated a missile-warning-system (MWS) with an electronics processor and flares. Upon detection of an incoming missile by the MWS, the processor would cue the release of flares that masked the heat signature of the target; thereby confusing the incoming missile so that it would miss.

As heat-seeking missiles became more sophisticated, additional jamming energy became required to overcome the missile's ability to distinguish the target signal from the decoy signal.

Directional infrared countermeasure (DIRCM) systems were developed to generate and direct radiation towards the heat-seeking missile, thereby providing a significant increase in jamming energy (compared to flares) in order to defeat the incoming missile. By modulating the intensity of the output radiation, DIRCM systems are able to disrupt the missile guidance system and cause it to break lock from the target (see Fig. 12.20).

The threats from air-to-air and man-portable air-defense systems (MANPADS) against aircraft provide major technology drivers of DIRCM-system requirements. Minimizing size, weight, power consumption, and cost (SWaP-C) is also critical for viable DIRCM systems.

QCL technology offers many advantages for DIRCM systems. In many cases, these advantages are enabling.

QCLs operate throughout the mid-IR spectral range. In many ways, this range can be considered the "color of heat". The nature of QCL design lends them to be "wavelength engineered" to match the operating band(s) of heat-seeking missiles; making them an ideal technology platform to address legacy, current, and future threats.

QCLs can now generate multi-watt power levels throughout the mid-IR, and with near-diffraction-limited beam quality. Typical irradiance levels that are required from DIRCM systems have been discussed elsewhere [45].

For most military applications, operation (and storage) at extreme temperature ranges is required. QCL designs that enable high optical output at high heatsink temperatures are advantageous for such applications. For high-power applications, heat–dissipation requirements drive the size, weight and power (SWaP) performance of practical systems. It is highly desirable, therefore, to maximize the operating temperature of the laser device, so as to minimize the thermal management and system SWaP. As illustrated Fig. 12.21, QCL devices provide over 1 W of continuous–wave (cw) optical power while operating (device-heatsink temperature) at 75 °C.

For aircraft installations, SWaP is a critical parameter. QCL devices offer performance that can be measured in the several watts per ounce at the device level. While additional SWaP must be considered for the overall package (and that is largely dictated by particular requirements and packaging), this type of performance enables QCLs to be installed onboard ultra-light platforms such as unmanned aerial systems (UAS), as well as conventional fixed- and rotary-wing aircraft.

**Fig. 12.20** Directional infrared countermeasures (DIRCM) systems based on QCL technology are highly valuable for defeating heat-seeking missiles with low size, weight, and power consumption to enable their installation on lightweight rotary-wing aircraft.

For military aircraft installations, environmental ruggedization and reliability are also major drivers for adoption of DIRCM systems. The direct conversion from input electrical power to output optical power (in the mid-IR) without requiring frequency/wavelength conversion or optical pumping, is uniquely characteristic of QCL technology. This provides an inherent simplicity that translates directly into military-rugged system architectures with very high reliability.

The amplitude modulation characteristics of DIRCM lasers are critical to exploit the signal-processing electronics of the incoming missile threat to achieve optical breaklock (OBL); i.e., the process of causing the heat-seeking missile to lose "sight" of the target completely. The modulation flexibility of semiconductor sources such as QCLs provides inherent benefits to the system designer when developing advanced algorithms to achieve OBL. As true cw output-power sources, or with high modulation bandwidth (in excess of several MHz up to GHz), QCLs can provide very high bandwidth and cw output powers to address a wide variety of threats.

DIRCM systems will typically engage incoming missiles at the earliest possible time; i.e., typically the longest range to target. Over this range, the spectral characteristics of atmospheric transmission play a role in the wavelength coverage required from a DIRCM system. It is well known from imaging cameras that operation in the mid-wave infrared (MWIR), or roughly 3–5 micron wavelength range, is appropriate for maximizing atmospheric transmission (see Fig. 12.22). This wavelength range fortuitously overlaps with the capabilities of QCL technology, making it an ideal technology to address this range.

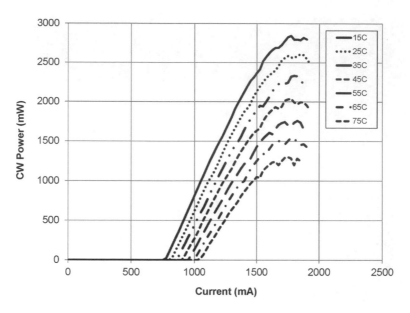

**Fig. 12.21** The operating temperature of QCL-based systems drives overall SWaP. Typical system-operating temperatures of ~70 °C are required for most military-grade applications. Being able to generate over 1 W cw optical power with device temperatures in excess of 70 °C is significant to minimize thermal-management-system requirements and SWaP.

As mentioned above, QCL technology can routinely reach multi-Watt output powers throughout the mid-IR spectral range and with near-diffraction-limited beam quality; thus, it has the same performance throughout the MWIR spectral range. This is another critical parameter, as high radiant intensity at the missile translates directly into jam effectiveness.

## 12.2.3    Standoff Detection

Terrorist threats continue to drive interest and requirements for explosives detection. Improvised explosive devices (IEDs) and vehicle-borne improvised explosive devices (VBIEDs) pose grave threat to personnel and assets. To minimize the impact and risk of damage to personnel and assets, it is desirable maximize the standoff-detection range.

Optical detection techniques based on mid-IR laser spectroscopy represent a promising approach as most chemical compounds exhibit strong characteristic absorbance patterns in this spectral range. These molecular "fingerprints" can be used to uniquely identify chemical compounds, and identify specific explosives types and reduce false alarm rates.

Absorption spectroscopy in the mid-IR (3–12 μm) is one of the most established analytical techniques for molecular detection, as almost all molecules have fundamental vibrational transitions in this region. A vibrational mode is "infrared active" if the change of states causes a change in the dipole moment of the molecule. In this

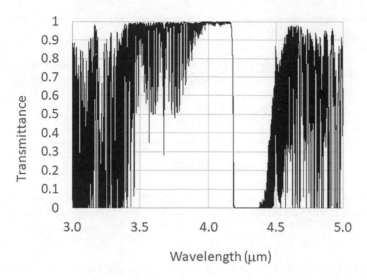

**Fig. 12.22** Atmospheric-transmission model based on HITRAN molecular constants. Under standard atmospheric conditions, the optical transmission varies dramatically throughout the MWIR spectral range. Atmospheric-constituent values used were taken from EPA data (https://www.epa.gov/climate-indicators/climate-change-indicators-atmospheric-concentrations-greenhouse-gases) at 1 km path length.

case, a strong absorption of light at the wavelength corresponding to a given vibration transition can be observed. The fundamental vibrations in this region of the spectrum are typically 30–1000 times stronger than harmonic "overtones" or "combination bands" which occur in the near-IR and visible portions of the spectrum. This "fingerprint" region is of particular interest because almost every compound produces its own unique pattern in this area (see Fig. 12.23).

The increased absorption strength in the mid-IR means that sensitivities can be much higher. At the same time, these fundamental transitions are unique to the particular molecular species, providing a high degree of specificity that can be useful with heterogeneous mixtures of analytes in a given sample.

Many, if not most molecular species of interest for military and homeland security applications, such as explosives, precursors, and chemical agents, possess distinctive, fundamental absorption features in the mid-IR. The combination of these absorption fingerprints along with pattern recognition algorithms allows for these targets to be unambiguously identified in a very robust way against a strongly interfering background matrix. Figure 12.23 illustrates example molecular species with absorption features in the mid-IR spectral range.

QCL technologies overlap ideally with detectors and imaging sensors that are used throughout the defense and security markets. Thermal-imaging technologies that include Indium Antimonide (InSb), Indium Gallium Arsenide (InGaAs), Mercury Cadmium Telluride (HgCdTe or MCT), Vanadium Oxide (VOx) bolometers, and Quantum-Well Infrared Photodetectors (QWIPs). However, only a few QWIPs operate

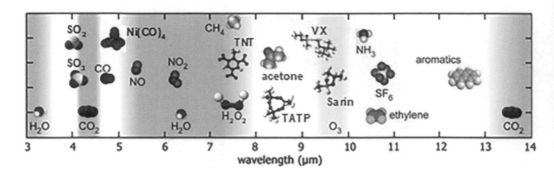

**Fig. 12.23** Example molecules with mid-IR "fingerprint" signatures. Examples of explosives, precursors, and chemical agents with features in the 7–10 micron wavelength range are also provided.

in the MWIR and LWIR; that is, have spectral sensitivities that overlap with QCL technologies.

High-explosive materials such as trinitrotoluene (TNT), hexahydro-1,3,5-trinitro-1,3,5 (RDX), pentaerythritol tetranitrate (PETN), and ammonium nitrate/fuel oil (ANFO) are but a few compounds that are well-known to the public, and have very strong, characteristic mid-IR fingerprints. Other volatile species such as ethylene glycol dinitrate (EGDN), nitroglycerine (NG), urea nitrate (UN), and triamino-trinitrobenzene (TATB) also possess unique, strong mid-IR absorption signatures that can be used for spectroscopic detection. The mid-IR spectra of many of these agents have been published previously [48–50]. A review of the spectroscopic techniques for explosives detection has been conducted [51, 52].

The application of QCLs for standoff explosives detection has been discussed previously. Members of Sandia National Laboratories, for example, have reviewed the requirements for QCLs in stand-off spectroscopic systems [53].

A generalized standoff-detection system geometry is illustrated in Fig. 12.24. The emission from a laser interacts with the explosives' vapor cloud created by the bulk explosives sample. The spectroscopic signature created by the interaction can be detected and analyzed using an optical receiver and data acquisition system. While schematically simple, this interaction results from several system-level considerations.

Theisen and Linker [53] generated a mathematical model (see Fig. 12.25) that captures the laser-vapor interaction in terms of photons that are available for detection. It is important to note that while several of these parameters can be addressed using QCLs, system-level considerations must be taken into account to optimize performance.

QCL technology has been demonstrated in several system geometries that provide non-contact detection of explosive residues [7, 54–56]. In each of these detection methodologies, the mid-IR light generated by QCL technology serves as the interrogation source for explosive materials of interest. Since mid-IR light is strongly absorbed by explosive residues, it can be used as a probe for variety detection technologies.

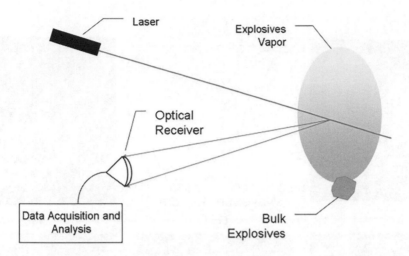

**Fig. 12.24** [Reproduced with permission from [53].] Generalized standoff-detection system geometry for explosives detection using a laser.

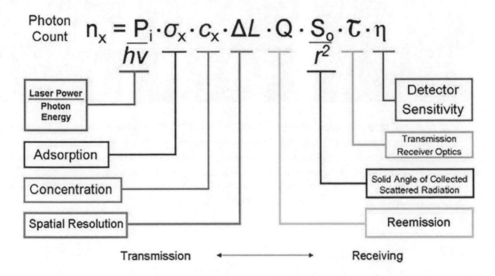

**Fig. 12.25** [Reproduced with permission from [53].] Mathematical model depicting laser-based detection.

Furstenberg *et al.* [7] described the technique for using a QCL to probe trace explosives by selectively heating the analyte, thereby creating an elevated-temperature change that can be detected by thermal-imaging technologies. Here, resonant absorption by the analyte at specific wavelengths emitted by the QCL generates selective heating. Detection using this technique of both trinitrotoluene (TNT) and cyclotrimethylenetrinitramine (RDX) analytes was demonstrated (Fig. 12.26).

Van Neste *et al.* [51] demonstrated standoff detection of tributyl phosphate (TBP), cyclotrimethylenetrinitramine (RDX), pentaerythritol tetrantitrate (PETN) and trinitrotoluene (TNT) using QCLs and photoacoustic detection. A detection limit of $100 \, ng/cm^2$ at a standoff distance of 20 meters was demonstrated using the technique.

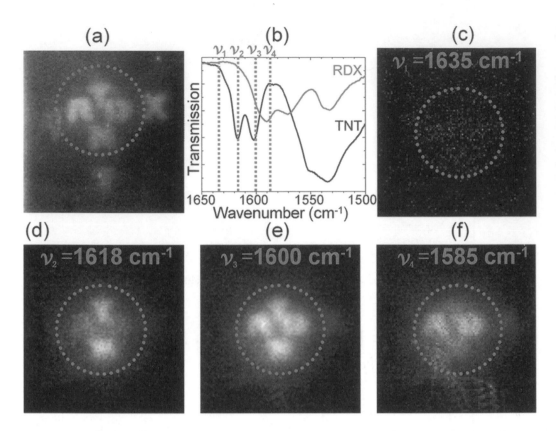

**Fig. 12.26** [Reproduced with permission from [7].] Illustration of resonant photothermal heating. Letters "RDX" and "TNT" were written on a polypropylene film using a solution of the explosives. (a) Result of nonselective heating using a heat gun. (b) IR spectrum of RDX and TNT showing resonances. (c) Laser heating with off-resonance wavelength. (d) Laser heating on resonance with TNT only. (e) Laser heating on resonance with RDX and TNT. (f) Laser heating on resonance with RDX only.

## 12.2.4　Emerging Applications

### 12.2.4.1　Combat Identification and Free-Space Communications

Compact, high-brightness QCL sources have shown their usefulness as beacon emitters when viewed by thermal cameras operating in the LWIR and MWIR spectral bands [57]. Wavelengths throughout these bands have been demonstrated in battery-operated, handheld and wearable form factors. Operation in the MWIR and/or LWIR provides capabilities to be detected at ranges in excess of 20 km, but only with non-proliferated detectors.

QCLs offer an inherent high modulation bandwidth capability that can provide high bandwidth free-space communication capabilities [40] for defense applications. Optical free-space communications provides line-of-sight communications that are difficult to intercept. Combined with operation in non-proliferated bands, this provides additional layers of security for military applications.

### 12.2.4.2 Test and Measurement

Numerous military applications are now employing mid-IR sensor technologies such as those listed previously. As such, there are significant opportunities for QCL sources to operate with such detectors as active illumination sources to test these emerging systems.

Infrared scene generation is used to test infrared cameras. Simulations conducted using infrared scenes allow the military to evaluate and test the effectiveness of imaging systems.

QCLs have been combined with spatial light modulators to generate infrared scenes. Such infrared scene projectors can be used to evaluate thermal-imaging cameras that operate in the MWIR and/or LWIR directly. High-resolution displays can be generated using state-of-the-art modulators (e.g., digital micro-mirrors), which are much more capable than their legacy resistor-based system counterparts.

Similarly, QCLs can be employed as compact, handheld sources to stimulate missile warning sensors to test operation on a flight line or in a laboratory environment. Missile warning sensors that detect the heat signature of a threat can be evaluated against the "thermal light" that is emitted from QCLs.

## 12.3 Conclusion

At 30 years *young*, QCL technologies have come a long way from their first-light demonstration at Bell Labs in 1994. With their unique ability to lase at wavelengths across the molecularly rich mid-IR and THz electromagnetic spectrum (4–25 μm and 75–300 μm), QCLs have enabled new capabilities across a wide range of applications. QCLs have evolved from laboratory demonstrations of ever increasing power and operating temperature (at an increasing range of wavelengths) to being used in applications that include applied research, materials analysis, combustion and gas-phase diagnostics, new approaches to microscopy, aircraft protection, standoff detection of explosives, and analytical instrumentation. In the decade to follow, we will see the further deployment of QCL-based solutions in cancer diagnostics equipment, compact sensors, communications devices, drug development, autonomous vehicles, aircraft protection, thermal imaging products, search and rescue systems, test and measurement systems, and likely much more. The future is indeed bright for the QCL and its myriad of potential applications.

### References

[1]  J. Faist, F. Capasso, D. L. Sivco, C. Sirtori, A. L. Hutchinson, and A. Y. Cho, "Quantum cascade laser," *Science*, vol. 264, no. 5158, pp. 553–556, 1994, doi: 10.1126/science.264.5158.553 %J Science.

[2]  K. Namjou *et al.*, "Sensitive absorption spectroscopy with a room-temperature distributed-feedback quantum-cascade laser," *Opt. Lett.*, vol. 23, no. 3, pp. 219–221, 1998, doi: 10.1364/OL.23.000219.

[3] R. F. Curl *et al.*, "Quantum cascade lasers in chemical physics," *Chemical Physics Letters*, vol. 487, no. 1, pp. 1–18, 2010, doi: https://doi.org/10.1016/j.cplett.2009.12.073.

[4] A. P. M. Michel *et al.*, "Quantum cascade laser open-path system for remote sensing of trace gases in Beijing, China," *Opt. Eng.*, vol. 49, no. 11, 111125, 2010.

[5] M. T. Parsons *et al.*, "Real-time monitoring of benzene, toluene, and p-xylene in a photoreaction chamber with a tunable mid-infrared laser and ultraviolet differential optical absorption spectroscopy," *Appl. Opt.*, vol. 50, no. 4, pp. A90-A99, 2011, doi: 10.1364/AO.50.000A90.

[6] M. Rezaei, K. H. Michaelian, and N. Moazzen-Ahmadi, "Nonpolar nitrous oxide dimer: Observation of combination bands of $(14N_2O)_2$ and $(15N_2O)_2$ involving the torsion and antigeared bending modes," *J. Chem. Phys.*, vol. 136, no. 12, p. 124308, 2012, doi: 10.1063/1.3697869.

[7] R. Furstenberg *et al.*, "Stand-off detection of trace explosives via resonant infrared photothermal imaging," *Phys. Rev. B*, vol. 93, no. 22, p. 224103, 2008, doi: 10.1063/1.3027461.

[8] T. H. Risby and S. F. Solga, "Current status of clinical breath analysis," *Appl. Phys. B*, vol. 85, no. 2, pp. 421–426, 2006, doi: 10.1007/s00340-006-2280-4.

[9] M. R. Kole, R. K. Reddy, M. V. Schulmerich, M. K. Gelber, and R. Bhargava, "Discrete frequency infrared microspectroscopy and imaging with a tunable quantum cascade laser," *Analytical Chem.* vol. 84, no. 23, pp. 10366–10372, 2012, doi: 10.1021/ac302513f.

[10] I. A. Schultz *et al.*, "Multispecies midinfrared absorption measurements in a hydrocarbon-fueled scramjet combustor," *J. Propulsion Power*, vol. 30, no. 6, pp. 1595–1604, 2014, doi: 10.2514/1.B35261.

[11] D. P. Kise, D. Magana, M. J. Reddish, and R. B. Dyer, "Submillisecond mixing in a continuous-flow, microfluidic mixer utilizing mid-infrared hyperspectral imaging detection," *Lab on a Chip*, vol. 14, no. 3, pp. 584–591, 2014, doi: 10.1039/C3LC51171E.

[12] D. Caffey *et al.*, "Performance characteristics of a continuous-wave compact widely tunable external cavity interband cascade lasers," *Opt. Express*, vol. 18, no. 15, pp. 15691–15696, 2010, doi: 10.1364/OE.18.015691.

[13] Z. Fei *et al.*, "Infrared nanoscopy of Dirac plasmons at the graphene–$SiO_2$ interface," *Nano Letters*, vol. 11, no. 11, pp. 4701–4705, 2011, doi: 10.1021/nl202362d.

[14] R. C. Gore, "Infrared spectrometry of small samples with the reflecting microscope," *Science*, vol. 110, no. 2870, p. 710, 1949., doi: 10.1126/science.110.2870.710

[15] A. Centrone, "Infrared imaging and spectroscopy beyond the diffraction limit," *Ann. Rev. Anal. Chem.* vol. 8, no. 1, pp. 101–126, 2015, doi: 10.1146/annurev-anchem-071114-040435.

[16] B. Knoll and F. Keilmann, "Near-field probing of vibrational absorption for chemical microscopy," *Nature*, vol. 399, p. 134, 1999, doi: 10.1038/20154.

[17] N. Ocelic, A. Huber, and R. Hillenbrand, "Pseudoheterodyne detection for background-free near-field spectroscopy," *Appl. Phys. Lett.*, vol. 89, no.10, 101124, 2006

[18] I. Amenabar *et al.*, "Structural analysis and mapping of individual protein complexes by infrared nanospectroscopy," *Nature Comm.*, vol. 4, p. 2890, 2013, doi: 10.1038/ncomms3890.

[19] F. Lu and M. A. Belkin, "Infrared absorption nano-spectroscopy using sample photoexpansion induced by tunable quantum cascade lasers," *Opt. Express*, vol. 19, no. 21, pp. 19942–19947, 2011, doi: 10.1364/OE.19.019942.

[20] A. Dazzi, R. Prazeres, F. Glotin, and J. M. Ortega, "Local infrared microspectroscopy with subwavelength spatial resolution with an atomic force microscope tip used as a photothermal sensor," *Opt. Lett.*, vol. 30, no. 18, pp. 2388–2390, 2005, doi: 10.1364/OL.30.002388.

[21] A. Dazzi and C. B. Prater, "AFM-IR: Technology and applications in nanoscale infrared spectroscopy and chemical imaging," *Chem. Rev.*, vol. 117, no. 7, pp. 5146–5173, 2017, doi: 10.1021/acs.chemrev.6b00448.

[22] D. Nowak *et al.*, "Nanoscale chemical imaging by photoinduced force microscopy," *Sci. Adv.*, 10.1126/sciadv.1501571 vol. 2, no. 3, 2016.

[23] B. T. O'Callahan, J. Yan, F. Menges, E. A. Muller, and M. B. Raschke, "Photo-induced tip-sample forces for chemical nano-imaging and spectroscopy," *Nano Lett.*, vol. 18, no. 9 , pp. 5499–5505, 2018.

[24] E. O. P. Junghoon Jahng and E. S. Lee, "Nanoscale spectroscopic origins of photoinduced tip–sample force in the midinfrared," *PNAS*, vol. 116, p. 52, 2019.

[25] A. D.-B. J. Mathurin and A. Dazzi, "Advanced infrared nanospectroscopy using photothermal induced resonance technique, AFMIR: New approach using tapping mode," *Acta Phys. Polonica Ser. A*, vol. 137, no. 1, pp. 29–32, doi: 10.12693/APhysPolA.137.29.

[26] M. J. Baker *et al.*, "Using Fourier transform IR spectroscopy to analyze biological materials," *Nature Protocols*, vol. 9, p. 1771, 2014, doi: 10.1038/nprot.2014.110.

[27] C. Kuepper, A. Kallenbach-Thieltges, H. Juette, A. Tannapfel, F. Großerueschkamp, and K. Gerwert, "Quantum cascade laser-based infrared microscopy for label-free and automated cancer classification in tissue sections," *Sci. Rep.*, vol. 8, no. 1, p. 7717, 2018, doi: 10.1038/s41598-018-26098-w.

[28] M. C. Phillips and N. Hô, "Infrared hyperspectral imaging using a broadly tunable external cavity quantum cascade laser and microbolometer focal plane array," *Opt. Express*, vol. 16, no. 3, pp. 1836–1845, 2008, doi: 10.1364/OE.16.001836.

[29] M. J. Weida and B. Yee, "Quantum cascade laser-based replacement for FTIR microscopy," *Proc. SPIE*, vol. 7902, 79021C, 2011.

[30] J. A. Reffner, P. A. Martoglio, and G. P. Williams, "Fourier transform infrared microscopical analysis with synchrotron radiation: The microscope optics and system performance (invited)," *Rev. Sci. Instruments*, vol. 66, no. 2, pp. 1298–1302, 1995, doi: 10.1063/1.1145958.

[31] K. Yeh, S. Kenkel, J.-N. Liu, and R. Bhargava, "Fast infrared chemical imaging with a quantum cascade laser," *Anal. Chem.*, vol. 87, no. 1, pp. 485–493, 2015 https://pubs.acs.org/doi/abs/10.1021/ac5027513 further permissions related to this figure should be directed to the ACS., doi: 10.1021/ac5027513.

[32] A. Schönhals, N. Kröger-Lui, A. Pucci, and W. Petrich, "On the role of interference in laser-based mid-infrared widefield microspectroscopy," *J. Biophotonics*, vol. 11, no. 7, p. e201800015, 2018, doi: 10.1002/jbio.201800015.

[33] P. Bassan, M. J. Weida, J. Rowlette, and P. Gardner, "Large scale infrared imaging of tissue micro arrays (TMAs) using a tunable quantum cascade laser (QCL) based microscope," *Analyst*, vol. 139, no. 16, pp. 3856–3859, 2014, doi: 10.1039/C4AN00638K.

[34] M. Weida, P. Buerki, M. Pushkarsky, and T. Day, "QCL-assisted infrared chemical imaging," *Proc. SPIE*, vol. 8031, 803127, 2011.

[35] S. Mittal, K. Yeh, L. S. Leslie, S. Kenkel, A. Kajdacsy-Balla, and R. Bhargava, "Simultaneous cancer and tumor microenvironment subtyping using confocal infrared

microscopy for all-digital molecular histopathology," *Proc. National Acad. Sci.*, 2018, doi: 10.1073/pnas.1719551115.

[36] D. Zhang, C. Li, C. Zhang, M. N. Slipchenko, G. Eakins, and J.-X. Cheng, "Depth-resolved mid-infrared photothermal imaging of living cells and organisms with sub-micrometer spatial resolution," *Science Adv.*, vol. 2, no. 9, 2016, doi: 10.1126/sciadv.1600521.

[37] A. Schwaighofer, M. Brandstetter, and B. Lendl, "Quantum cascade lasers (QCLs) in biomedical spectroscopy," *Chem. Soc. Rev.*, 46, no. 19, pp. 5903–5924, 2017, doi: 10.1039/C7CS00403F.

[38] V. K. Varma, A. Kajdacsy-Balla, S. Akkina, S. Setty, and M. J. Walsh, "Predicting fibrosis progression in renal transplant recipients using laser-based infrared spectroscopic Imaging," *Scientific Reports*, vol. 8, no. 1, p. 686, 2018, doi: 10.1038/s41598-017-19006-1.

[39] L. Wang, B. Kendrick, and E. Ma, "Enhanced protein structural characterization using microfluidic modulation spectroscopy," *Spectroscopy*, vol. 33, no. 7, pp. 46–52, 2018.

[40] A. Delga and L. Leviandier, "Free-space optical communications with quantum cascade lasers," *Proc. SPIE*, vol. 10926, 1092617, 2019.

[41] X. Pang *et al.*, "Gigabit free-space multi-level signal transmission with a mid-infrared quantum cascade laser operating at room temperature," *Opt. Lett.*, vol. 42, no. 18, pp. 3646–3649, 2017, doi: 10.1364/OL.42.003646.

[42] F. Fuchs *et al.*, "Remote sensing of explosives using mid-infrared quantum cascade lasers," *Proc. SPIE*, vol. 6739, 673904, 2007.

[43] A. A. Kosterev and F. K. Tittel, "Chemical sensors based on quantum cascade lasers," *IEEE J. Quantum Elec.*, vol. 38, no. 6, pp. 582–591, 2002, doi: 10.1109/JQE.2002.1005408.

[44] T. Risby and S. F. Solga, "Current status of clinical breath analysis," vol. 85, no. 2–3, pp. 421–426, 2006.

[45] D. H. Pollock, J. S. Accetta, and D. L. Shumaker, *The Infrared & Electro-Optical Systems Handbook. Countermeasure Systems*, Volume 7. Ann Arbor, MIL: Infrared Information and Analysis Center, 1993.

[46] D. H. Titterton, "Development of infrared countermeasure technology and systems," in *Mid-infrared Semiconductor Optoelectronics*, A. Krier Ed. London: Springer London, 2006, pp. 635–671.

[47] R. J. Grasso, "Source technology as the foundation for modern infra-red counter measures (IRCM)," *Proc. SPIE*, vol. 7836, 783604, 2010

[48] F. Pristera, M. Halik, A. Castelli, and W. Fredericks, "Analysis of explosives using infrared spectroscopy," *Anal. Chem.* vol. 32, no. 4, pp. 495–508, 1960, doi: 10.1021/ac60160a013.

[49] D. E. Chasan and G. Norwitz, "Qualitative analysis of primers, tracers, igniters, incendiaries, boosters, and delay compositions on a microscale by use of infrared spectroscopy," *Microchem. J.*, vol. 17, no. 1, pp. 31–60, 1972, doi: https://doi.org/10.1016/0026-265X(72)90035-5.

[50] Z. Iqbal, K. Suryanarayanan, S. Bulusu, and J. Autera, *Infrared and Raman Spectra of 1, 3, 5-trinitro-1, 3, 5-triazacyclohexane (RDX)*. Dover, NJ: Picatinny Arsenal, 1972.

[51] J. I. Steinfeld and J. Wormhoudt, "Explosives detection: A challenge for physical chemistry," *Annu. Rev. Phys. Chem.*, vol. 49, no. 1, pp. 203–232, 1998, doi: 10.1146/annurev.physchem.49.1.203.

[52] M. W. P. Petryk, "Promising spectroscopic techniques for the portable detection of condensed-phase contaminants on surfaces," *Appl. Spectro. Rev.*, vol. 42, no. 3, pp. 287–343, 2007, doi: 10.1080/05704920701293794.

[53] L. A. Theisen and K. L. Linker, "Quantum cascade lasers (QCLs) for standoff explosives detection: LDRD 138733 final report," technical report, 2009.

[54] C. W. Van Neste, L. R. Senesac, and T. Thundat, "Standoff spectroscopy of surface adsorbed chemicals," *Anal. Chem.*, vol. 81, no. 5, pp. 1952–1956, 2009, doi: 10.1021/ac802364e.

[55] F. Fuchs *et al.*, "Standoff detection of explosives with broad band tunable external cavity quantum cascade lasers," *Proc. SPIE*, vol. 8268, 82681N, 2012.

[56] A. K. Goyal *et al.*, "Active infrared multispectral imaging of chemicals on surfaces," *Proc. SPIE*, vol. 8018, 80180N, 2011.

[57] M. Pralle, I. Puscasu, E. Johnson, P. Loges, and J. Melnyk, "High-visibility infrared beacons for IFF and combat ID," *Proc. SPIE*, vol. 5780, pp 18–25, 2005.

# 13 QCL-based Gas Sensing with Photoacoustic Spectroscopy

Vincenzo Spagnolo, Pietro Patimisco, Angelo Sampaolo, and Marilena Giglio
Polytechnic University of Bari, Italy

## 13.1    Overview on Optical Gas Sensing

Gas detection has a great impact in a wide range of applications. For example, the use of high-sensitivity gas detectors is widespread in atmospheric science, to measure and study the profile and pathways of different gas species, including greenhouse gases. Gas sensing finds applications also in medicine and life sciences, mostly for breath analysis: various potential biomarker gases are under study for use in human breath diagnostics, including nitric oxide (NO), ethane ($C_2H_6$), ammonia ($NH_3$), carbon oxide (CO), and many others.

Quantitative detection of gases is traditionally performed by laboratory analytical equipment such as gas chromatographs and mass spectrometers, or by small low-cost devices such as pellistors, semiconductor gas sensors, or electrochemical devices. Pellistors are four-terminal devices based on a Wheatstone bridge circuit which includes two beads: one of these beads is treated with a catalyst, which lowers the ignition temperature of explosive gases. The heating coming from combustion events changes the conductivity of the catalyst-treated bead, resulting in an imbalanced voltage, from which the gas concentration can be retrieved [1]. However, pellistors suffer from drift at parts per million (ppm) levels and can be subjected to poisoning, i.e., a change (partially reversible or irreversible) in sensor response due to chemical reactions between certain gases with the catalyst surface. Semiconductor gas sensors possess an electrical resistance made with a porous assembly of tiny crystals of an n-type metal oxide semiconductor (MOS), typically $SnO_2$, $In_2O_3$, or $WO_3$. The resistance value changes sharply on contact with a small concentration of reducing or oxidizing gas, enabling the gas concentration to be related to the resistance change [2]. However, these types of sensors also suffer from drift, cross-talking with other gases, and from changes of humidity levels. Electrochemical gas sensors are based on the electro-adsorptive effect, where electrical fields applied on the gas-sensitive layer may alter the adsorption characteristics of the material. They can be specifically related to individual gases and are sensitive up to the parts per billion (ppb) concentration level; however, they have limited lifetimes and suffer from cross-response issues.

Gas sensors based on optical absorption offer fast responses times (usually below 1 second), minimal drift and high gas specificity, and, in most cases, have zero cross-response to other gases. Measurements can be performed in real time and in situ without altering the gas sample, which can be important for industrial process control. The detection is based on the direct or indirect measurement of the target molecules' absorption at a specific wavelength. Each gas species is characterized by specific absorption lines or bands, thus guaranteeing the high selectivity of single gas species detection [3]. Many molecular species exhibit strong absorption in the infrared region, where roto-vibrational transitions occur. The excitation of these transitions provides a highly sensitive tool for specific recognition of gas molecules. For mid-infrared optical sources, quantum cascade lasers (QCLs) are the optimal choice, due to their high output power, compactness, narrow spectral linewidth, and broad wavelength tunability. Among laser-based techniques, photoacoustic spectroscopy (PAS) is an indirect technique in that an effect of optical absorption is measured rather than the absorption itself [4]. When light at a specific wavelength is absorbed by the gas sample, the excited molecules will subsequently relax through non-radiative processes. These processes produce localized heating in the gas, which in turn results in an increase in the local pressure. PAS can detect weak acoustic waves and local heating in a gas sample, respectively, and is capable of reaching concentration levels down to the part-per trillion range and below [5] with a compact and robust detection module.

## 13.2 Introduction on the Photoacoustic Effect

The photoacoustic effect was observed by Graham Bell more than a century ago (1880) while he was working on the improvement of a photophone, in an accidental way. Bell realized that when a light beam is periodically interrupted by a chopper and subsequently focused on a layer of thin material (a diaphragm in connection with an acoustic horn) a sound wave is produced. In addition, the generated acoustic signal increased in intensity when the layer exposed to the beam was dark in color. Thus, Bell realized that this effect was related to the absorption of light by the thin layer. The photoacoustic effect occurs in all kind of materials (solids, liquids, and gases) and scientists began to study the phenomenon and its possible applications [6]. However, due to the lack of appropriate equipment (such as light sources, microphones), the photoacoustic effect was completely forgotten for more than half a century.

The photoacoustic effect for gas species can be divided into three main processes that can be analyzed separately:

- generation of heat in the gas as a consequence of the absorption of the modulated optical radiation;
- generation of the acoustic wave;
- detection of the photoacoustic signal.

In Fig. 13.1 the main processes that take place in the production of thermal effects are sketched.

In the following subsections, a detailed description of these three steps is presented.

### 13.2.1 Absorption of Modulated Light and Generation of Heat

When modulated light is absorbed by molecules via resonant absorption processes, the optically excited molecules may transfer the excess energy to the neighbor molecules by non-radiative collision processes. The absorption of a photon causes a transition

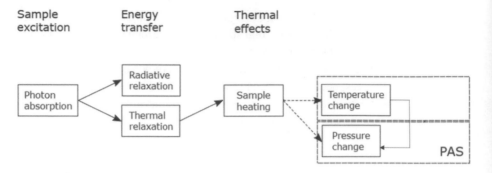

**Fig. 13.1**  Processes that involve the generation of thermal effects in a gas sample.

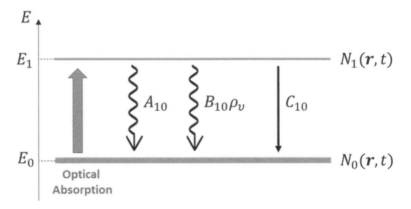

**Fig. 13.2**  Radiative and nonradiative transitions for a two-level system.

from the molecule fundamental energy state $E_0$ to the excited one $E_1$, being $E_1 - E_0 = h\nu$ the energy of the photon and $\nu$ its frequency. Then, the molecule relaxes to the fundamental state via different relaxation processes:

- radiative relaxation with the emission of a photon;
- photochemical processes;
- collisions with neighbor molecules of same species at the fundamental energy $E_0$ with transfer of the vibrational energy and their consequent excitation at the level $E_1$;
- collisions with other type of gas molecules with transfer of the vibrational energy of excited analytes into kinetic energy (translation) of the surrounding molecules (V-T relaxation), causing an increase of the local temperature.

When the optical absorption occurs in the infrared spectral range, the roto-vibrational energy states are involved. Typical non-radiative relaxation times at pressure below 1 bar are of the order of $10^{-6}$–$10^{-9}$ s, while radiative decay times are $10^{-1}$–$10^{-3}$ s. In addition, energies of the infrared photons are too small to induce chemical reactions. Thus, the absorbed optical energy is mainly released as heat because of the increase of kinetic energy of collided molecules.

Let us suppose that the absorption process in a gas can be schematized as a two-level system, as sketched in Fig. 13.2.

$N_0(r, t)$ and $N_1(r, t)$ are molecular density for the fundamental energy state $E_0$ and the excited energy state $E_1$, respectively, $\rho_v$ is the energy density of radiation, $A_{10}$ and $B_{10}$ are the Einstein coefficient for spontaneous emission and stimulated emission, respectively, and $C_{10}$ is the non-radiative transfer rate via collisions. The rate equation for the population $N_1(r, t)$ can be expressed as:

$$\frac{dN_1}{dt} = \rho_v B_{10} (N_0 - N_1) - (A_{10} + C_{10}) N_1. \tag{13.1}$$

The inverse of lifetime $\tau$ of the excited state can be expressed as $1/\tau = A_{10} + C_{10}$ and, in a weak optical absorption approximation, namely $N_1 \ll N_0$ and $N_0$ time-independent, the previous equation can be rewritten as:

$$\frac{dN_1}{dt} = \rho_v B_{10} N_0 - \frac{N_1}{\tau}. \tag{13.2}$$

The term $\rho_v B_{10}$ represents the absorption optical rate and can be expressed as the product between the photon flux and the absorption cross-section $\sigma$. When the photon flux is modulated at a frequency $\omega$, it can be expressed as:

$$F(r, t) = F_0(r) \left(1 + \delta e^{i\omega t}\right). \tag{13.3}$$

When $\delta \ll 1$, a small amplitude sinusoidal modulation is superimposed to a time-independent photon flux $F_0(r)$. By combing Eqs. (13.2) and (13.3), the differential equation can be solved, and the density of the excited molecules can be related to the optical absorption of modulated light:

$$N_1(r, t) = \frac{N_0 \sigma F_0(r) \delta}{\sqrt{1 + \omega^2 \tau^2}} \tau e^{i(\omega t - \vartheta)}, \tag{13.4}$$

where $\vartheta$ is the phase difference between the modulation of the population density and the photon flux. The heat rate generated by the optical absorption can be related with the population density:

$$H(r, t) = \frac{N_1(r, t) \cdot h\upsilon}{\tau} = \frac{N_0 \sigma F_0(r) \delta h\upsilon}{\sqrt{1 + \omega^2 \tau^2}} e^{i(\omega t - \vartheta)}, \tag{13.5}$$

where $h\upsilon = E_1 - E_0$ is the energy of the absorbed photon. The product $N_0 \sigma$ represents the absorption coefficient of the optical transition and it is proportional to the gas target concentration. This model requires two assumptions: (i) $F_0(r)\sigma$ should be low in order to avoid absorption saturation and neglect stimulated emission; and (ii) low-frequency modulation, $\omega \ll 1/\tau$. The first condition is verified for low concentrations of absorbing gas and the second one by keeping the modulation frequency in the kHz range.

## 13.2.2  Generation of Acoustic Waves

The generation of the acoustic wave is related to the local periodic variation of the pressure as a consequence of the periodic heating of the sample due to the non-radiative energy relaxation of absorbing gas. The theoretical model is based on a combination of fluid mechanics and thermodynamics laws. We consider the case of an ideal, uniform, and continuous fluid that is perfectly elastic and at the thermodynamic equilibrium except for the local variations produced by the pressure wave, which are small enough to neglect any non-linear effects. The effect of the gravitational force is also neglected, so that pressure and density at equilibrium can be considered constant throughout the fluid. Dissipative terms due to fluid viscosity are not considered in the model. The propagation of the acoustic wave produces local changes in pressure, density, and temperature in the gas, proportional to the acoustic wave amplitude. Pressure changes are easily measured by means of sound detectors. For this reason, the acoustic wave is usually described in terms of acoustic pressure, which is defined as the difference between the instantaneous pressure and the equilibrium pressure. The equations that rule the acoustic pressure $p$ are: (i) the ideal gas law

$$\frac{p}{\rho} = (\gamma - 1) \frac{C_V}{M} T, \tag{13.6}$$

where $\rho$ is the gas density, $\gamma$ is the ratio between the specific heat for the ideal gas at constant pressure ($C_p$) and at constant volume ($C_V$), $M$ is the molar mass, and $T$ is the temperature; (ii) the continuity equation

$$-\frac{1}{\rho}\frac{\partial p}{\partial t} = \nabla \cdot \mathbf{v}, \tag{13.7}$$

where $\mathbf{v}$ is the velocity field; (iii) the law of momentum conservation

$$\rho \frac{\partial v}{\partial t} = -\nabla p, \tag{13.8}$$

and (iv) the conservation of energy

$$\rho \frac{C_V}{M} \frac{\partial T}{\partial t} + p \nabla \cdot \mathbf{v} = H. \tag{13.9}$$

By combining Eqs. (13.6)–(13.9), it can be demonstrated that the acoustic pressure is governed by the D'Alembert wave equation:

$$\nabla^2 p(\vec{r}, t) - \frac{1}{c^2}\frac{\partial^2 p(\vec{r}, t)}{\partial t^2} = -\frac{\gamma - 1}{c^2}\frac{\partial H}{\partial t}, \tag{13.10}$$

where

$$c = \sqrt{\frac{\gamma R T}{M}} \tag{13.11}$$

is the speed of sound in the gas. Since $H$ is sinusoidally modulated, it is convenient to take the Fourier transform of Eq. (13.10), considering $\mathbf{r}$ as a parameter:

$$\left(\nabla^2 + \frac{\omega^2}{c^2}\right) p(r, \omega) = \frac{\gamma - 1}{c^2} i\omega H(r, \omega). \tag{13.12}$$

A general solution $p(r, \omega)$ can be expressed as infinite summation of acoustic modes $p_k(\mathbf{r})$:

$$p(r, \omega) = \sum_k A_k(\omega) p_k(r), \tag{13.13}$$

where $A_k(\omega)$ is the amplitude of the $k$th acoustic mode with angular frequency $\omega_k$ and $p_k(\mathbf{r})$ is the solution of the associated homogeneous equation.

## 13.2.3     Detection of Acoustic Waves

In PAS, the gas is enclosed within an acoustic cell where optical absorption takes place [4]. The calculation of the normal acoustic modes of the gas cell is determined by imposing boundary conditions, dependent on the geometry of the acoustic cell. Most common acoustic cell designs adopt the basic symmetry of the exciting light source and are cylindrically shaped. The excitation source is a small diameter light beam centered along the cylinder axis. Local pressure propagates radially outward perpendicular to the exciting beam. The resonances occur at frequencies $f_{\mathrm{res}}$ and for a cylindrical acoustic cell of length $L$ and radius $R$, it is given by [7]:

$$f_{\mathrm{res}} = \frac{c}{2}\sqrt{\left(\frac{k}{L}\right)^2 + \left(\frac{\alpha_{m,n}}{R}\right)^2},$$  (13.14)

where $k$, $m$, and $n$ are indices related to the longitudinal, azimuthal, and radial acoustic modes, and $\alpha_{m,n}$ are determined by boundary conditions. Numerical values for the lowest order roots are: $\alpha_{0,0} = 0$, $\alpha_{0,1} = 1.226$, and $\alpha_{1,0} = 0.5891$. For air ($c = 3.31 \times 10^4$ cm s$^{-1}$) in a cell with a diameter of 5 cm and a length of 15 cm, the lowest resonance frequencies are 1.1 kHz for the longitudinal ($k = 1$, $m = n = 0$) mode, 3.9 kHz for the azimuthal ($m = 1$, $k = n = 0$) mode, and 8.1 kHz for the radial ($n = 1$, $k = m = 0$) mode.

When the radiation modulation frequency matches the resonance frequency of the acoustic cell, i.e., $\omega = 2\pi f_{\mathrm{res}}$, a standing wave vibrational pattern is created within the acoustic cell which acts as an acoustic resonator. In other words, cell resonances enhance and amplify acoustic powers at the resonance frequency. The degree of amplification of an acoustic cell is described by a quantity called the quality factor $Q$. $Q$ values of more than 1000 have been reported and values of a few hundred are typical for a radial resonance mode. Due to the cylindrical symmetry, longitudinal (and azimuthal) modes are not as strongly excited and have somewhat lower $Q$ values compared to the radial mode. The amplitude of the acoustic mode at the resonance frequency $f_{\mathrm{res}}$ is proportional to:

$$A_k(f_{\mathrm{res}}) \propto \frac{\alpha PLQ}{f_{\mathrm{res}}V},$$  (13.15)

where $P$ is the radiation power and $V$ is the cell volume. Geometries other than cylindrical symmetry have also been employed. Cells that exploit Helmholtz resonances are constructed with two chambers, a sample chamber of volume $V_1$ and a microphone chamber of volume $V_2$, connected by a cylindrical tube of length $l$ and inside diameter $D$ [8]. The resonance frequency exhibited by such cells is given by the expression:

$$f_{\mathrm{res}} = \frac{Dc}{2\sqrt{\pi}}\sqrt{\frac{V_1 + V_2}{V_1 V_2 \left(l + \frac{\pi D}{2}\right)}}.$$  (13.16)

From Eq. (13.16), the resonance frequency is reduced if a long narrow tube is used between the two cell chambers.

Several devices can be used to detect the sound wave generated by sample absorption. Most photoacoustic experiments utilize pressure sensors, the most common of which are microphones. A condenser microphone produces an electrical signal when a pressure wave impinging on the diaphragm pushes the diaphragm closer to a fixed metal plate, thereby increasing the capacitance between these two surfaces. The capacitance change leads to a voltage signal which increases with bias voltage and the diaphragm area. Microphones generally have flat frequency response up to 15 kHz, have low distortion, are generally not sensitive to mechanical vibrations, and respond well to pressure impulses, which enables their use in pulsed applications [9]. The dielectric material between the condenser plates in this type of microphone is air. In contrast to condenser microphones, electret microphones are constructed using solid materials of high dielectric constant which are electrically polarized [10]. One side of the electret foil is metallized and the insulating side is placed on a fixed back plate. A sound wave impinging on the metallized side causes a change in the polarization characteristics of the electret material, which in turn produces a small voltage between the metallized front electret surface and the back plate of the microphone. Thus, an electret microphone requires no bias voltage, which simplifies the apparatus. Another striking difference between air-spaced condenser microphones and electret microphones is physical size. Due to the large capacitance per unit area possible from electret materials they can be made into miniaturized microphones.

The need for windows on the acoustic cell affects its performance in two ways. Brewster windows, for example, destroy the cylindrical shape, which tends to dampen the natural acoustic resonances of the cell. Consequently, cell Q factors are lowered somewhat by the presence of these windows. Windows also generate a background acoustic signal at the source modulation frequency which can interfere with the detection of weak acoustic signals. Background acoustic signals are due to light absorption or scattering by the window material itself or from dust or adsorbed sample molecules on the window surface. The resultant window heating is transferred to the gas sample, generating an acoustic signal. Several cell designs, such as the use of baffles, for example, have been partially successful in minimizing cell window backgrounds. Different designs for PAS cells have been proposed and implemented, including a resonant cell with acoustic buffers [11], windowless [12], and a differential cell [13]. A differential cell includes two acoustic resonators equipped with microphones having the same responsivity at the resonance frequency of the cell. Since the laser light excites only one of the two resonators, the difference between the two signals removes noise components that are coherent in both resonators.

Optical microphones based on an interferometric displacement concept have been also employed in PAS [14]. Their design utilizes an interferometric technique whereby a metal coated 50/50 transmission diffraction grating is integrated in the middle of the acoustic backcavity. Together with the metal-coated acoustic diaphragm, a Michelson interferometer is formed. This approach provides sub-picometer displacement resolution and pressure sensitivity in the low $\mu Pa/\sqrt{Hz}$ range. The acoustic resonant frequency is adjustable from 10 kHz to 100 kHz, depending on membrane stiffness (thickness) and diameter [15].

The highest sensitivity has been achieved by using a cantilever pressure sensor [5] that is over hundred times more sensitive compared to membrane microphones. An extremely thin cantilever portion moves like a flexible door due to the pressure variations in the surrounding gas. The movement of the free end of the cantilever can be about two orders of magnitude greater than the movement of the middle point of the tightened membrane under the same pressure variation. The displacement of the cantilever is measured optically [16].

PAS has been successfully applied in trace gas sensing applications, which include atmospheric chemistry, volcanic activity, agriculture, industrial processes, workplace surveillance, and medical diagnostics. For instance, PAS has been used to monitor nitric oxide (NO) from vehicle exhaust emissions, which contributes to respiratory allergic diseases, inflammatory lung diseases, bronchial asthma, and the depletion of ozone [17]. In medicine, PAS has been used to monitor drug diffusion rates in skin [18] and to detect trace concentrations of disease biomarkers, such as ethylene ($C_2H_4$), ethane ($C_2H_6$), and pentane ($C_5H_{12}$), which are emitted by UV-exposed skin [19]. Other applications include monitoring respiratory $NH_3$ emission from cockroaches as well as detecting the intake of prohibited substances by athletes [20]. Low-cost portable PAS sensors have been on the market, examples of which include smoke detectors, toxic gas monitoring, and oil sensors for monitoring hydrocarbons in water.

## 13.3    Focus on QEPAS

The main issue suffered by PAS sensors is the sensitivity of the employed microphone for external sound sources. This issue can be solved exploiting the quartz-enhanced PAS (QEPAS) approach. In QEPAS a quartz tuning fork (QTF) is used as a sharply resonant acoustic transducer to detect weak photoacoustic waves, avoiding the use of acoustic gas cells, thereby removing restrictions imposed on the gas cell design by the acoustic resonance conditions [21, 22]. The laser beam is focused between the prongs of the QTF and the sound wave generated between them causes an antisymmetric vibration of the prongs in the QTF plane, with the two prongs moving in opposite directions at the same time: as a result, a flexural mode is excited. This vibration mode is piezoelectrically active and thus electrical charges are generated, proportional to the sound wave intensity. The most important advantages of QEPAS are [23]: (1) a high resonance frequency of the tuning fork combined with narrow bandwidth, yielding a very high $Q > 10,000$; (2) applicability over a wide range of pressures, including atmospheric pressure; (3) sample volumes as low as a few cm$^3$; and (4) QTFs are not spectrally sensitive and practically unaffected by environmental noise. Indeed, external acoustic sources tend to apply a force in the same direction on the two QTF prongs (in-plane symmetric mode) and the resulting deflection of QTF prongs is not piezoelectrically active. Thus, in contrast with the antisymmetric vibration of QTF produced by the laser beam, most of the external environmental noise does not generate an electrical signal.

For a harmonic modulation at frequency $f$ of the photon flux, the frequency-dependent generated QEPAS signal at the modulation frequency $f$ can be obtained by revising Eq. (13.15) [24]:

$$S = KQ(P) P_L c\varepsilon(f, \tau) = \frac{KQ(P) P_L c}{\sqrt{1 + (2\pi f \tau)^2}}, \tag{13.17}$$

where $K$ is a sensor constant, $P_L$ is the laser power, $c$ is the analyte concentration, and $\varepsilon(f, P)$ is the radiation-to-sound conversion efficiency, given by [25]:

$$\varepsilon(f, \tau) = \frac{1}{\sqrt{1 + (2\pi f \tau)^2}}. \tag{13.18}$$

For instantaneous relaxation, i.e., $2\pi f \tau \ll 1$, $\varepsilon(f, P) = 1$ and the QEPAS signal will be not dependent on the modulation frequency. According to the Landau–Teller model [26], the dependence of the relaxation time on the gas pressure can be expressed by:

$$\tau = \frac{1}{kP}, \tag{13.19}$$

where $k$ is a constant.

A crucial aspect in QEPAS detection is the dependence of the QEPAS signal on the radiation-to-sound conversion efficiency, which in turn affects the acoustic wave generation within the gas. It is mainly determined by the transfer rate of the vibrational energy of excited analyte molecules into the kinetic energy (translation) of the surrounding molecules (V-T relaxation) [27]. The use for several years of standard QTFs operating at frequencies as high as 32.7 kHz in QEPAS sensors has divided the gas species into two categories: fast and slow relaxing gases. For slow relaxing gases, a laser modulation frequency of 32.7 kHz is higher than the effective analyte relaxation rate in the gas matrix, not ensuring a complete release of absorbed energy during each oscillation period. The realization of custom QTFs with resonance frequencies lower than 32.7 kHz paved the way for the detection of slow relaxing molecule concentrations with better sensitivity [28–31]. Conversely, for fast relaxing gases, the relaxation rate is significantly faster than that of the optical modulation and an efficient energy transfer occurs between absorbing analytes and surrounding molecules. Thus, fast relaxing gases (such as $H_2O$ and $SF_6$) have been used also as relaxation promoters for slow relaxing gases to enhance the detection sensitivity of the latter [32, 33]. In this case, the effective relaxation rate of the analyte depends on the gas matrix composition. Indeed, excited analyte molecules can relax on different channels through collisions with the different types of molecules composing the matrix, both the buffer and the promoter. The relaxation rate is then given by the sum of the relaxation rates of every possible energy transfer pathway, weighted by the concentration of each species in the mixture [34]. Since for slow relaxing gases, the QEPAS signal is dependent on the promoter concentration within the mixture, accurate measurements of the promoter concentration and additional sensor calibration are required. To overcome this issue, custom QTFs with low resonance frequencies (<32.7 kHz) have been realized for QEPAS sensing.

## 13.3.1 Quartz Tuning Forks

The QTF represents the core of any QEPAS sensor. A tuning fork can be considered as two cantilevers bars (prongs) joined at a common base. The in-plane flexural modes of vibrations of the QTFs can be classified into two groups: symmetric modes, in which case the prongs move along the same direction; and antisymmetric modes, in which case the two prongs oscillate along opposite directions. The in-plane antisymmetric modes will be the predominant modes when a sound source is positioned between the prongs, forcing them to move in the opposite directions. In QEPAS sensors, the light source (typically a laser) is focused between the QTF prongs and sound waves produced by the modulated absorption force the prongs to vibrate antisymmetrically back and forth. QTFs can be designed to resonate at any frequency in the 3–200 kHz range and beyond, since resonance frequencies are defined by the properties of the piezoelectric material and by its geometry [28, 35].

The interaction between the laser modulated beam and absorbing molecules leads to the generation of acoustic waves that mechanically bend the QTF prongs. Hence, the electrode pairs of the QTF will be electrically charged due to the quartz piezoelectricity. Piezoelectricity is the coupling between internal dielectric polarization and strain, and it is present in most crystals lacking a center of inversion symmetry [36]. When a stress is applied to these materials, it induces a displacement of charge and a net electric field. The effect is reversible: when a voltage is applied across a piezoelectric material, it is accompanied by strain. The resonance frequencies of a QTF flexural mode can be calculated in the approximation of an independent cantilever vibrating in the in-plane modes. Each prong of the tuning fork can be treated as a clamped beam. According to the Euler–Bernoulli approximation, the description of the vibration is given by the following fourth-order differential equation:

$$EI \frac{\partial^4 y(x,t)}{\partial z^4} + \rho A \frac{\partial^2 y(x,t)}{\partial t^2} = 0, \tag{13.20}$$

where $E$ and $\rho$ are the Young's modulus and the volume density of quartz, respectively, $A = LT$ is the cross-sectional area of the beam (where $L$ and $T$ are the prong length and thickness, respectively, see Fig. 13.3(a)), $I$ is the second moment of beam's cross-sectional area, $y(z,t)$ is the prong displacement function, $x$ and $y$ are the directions in the QTF plane, and $t$ is the time. The geometry of a QTF is sketched in Fig. 13.3.

The product $EI$ is usually referred as the flexural rigidity. The latter, together with the cross-sectional area $A$ and the volume density, are assumed to be constant along the whole beam. The Euler-Bernoulli equation can be solved by the separation of variables method. The displacement can be separated into two parts, one is dependent on position and the other one is a function of time. This leads to a simplified differential equation for the $x$-direction that can be solved by superimposing boundary conditions. The boundary conditions originate from the support of the QTF. The fixed end must have a zero displacement and zero slope due to the clamp, while the free end does not have a bending moment or a shearing force (free-clamped boundary conditions). The related eigenfrequencies $f_n$ are given by the following expression [24]:

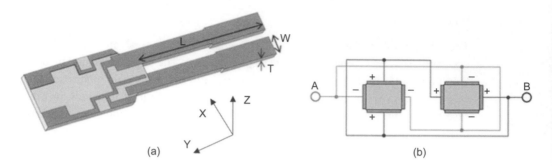

**Fig. 13.3** (a) Schematic of a QTF with the reference frame. (b) Top view of the tuning fork with the electrical configuration for the electrodes A and B. As shown, the charges cumulate in a quadrupole configuration.

$$f_n = \frac{\pi T}{8\sqrt{12}L^2}\sqrt{\frac{E}{\rho}}v_n^2, \tag{13.21}$$

where $v_n$ identifies the mode number, i.e., $n=0$ identifies the fundamental flexural mode, $n=1$ the first overtone mode, $n=2$ the second overtone mode, and so on.

## 13.3.2    Influence of Gas Properties

The dependence of the resonance frequency on the surrounding gas pressure can be determined by the assumption that the gas effect on prong vibration damping increases the inertia of the prong. In other words, the effect of the surrounding medium can be modeled as an additive mass for the vibrating prong. Thus, the Euler–Bernoulli equation in Eq. (13.20) requires an additional term corresponding to the reactive part, which attributes additional inertia to the vibrating prong [37]:

$$EI\frac{\partial^4 y(z,t)}{\partial z^4} + (\rho A + u)\frac{\partial^2 y(z,t)}{\partial t^2} = 0, \tag{13.22}$$

where $u$ is the added mass per unit length. The eigenfrequency $f_u$ of the fundamental mode changes as:

$$f_u = f_0\left(1 - \frac{1}{2}\frac{u}{\rho A}\right), \tag{13.23}$$

where $f_0$ is determined in vacuum by Eq. (13.21) when $n=0$. The exact derivation of the added mass $u$ is a complicated problem even for simple prong structures. In [38], the added mass per unit length of a thin prong has been found to be proportional to gas density $\rho_0$. By using the ideal gas law, the gas density can been be expressed by $\rho_0 = MP/RT$, where $M$ is molar mass, $P$ is the gas pressure (in torr), $R=62.3637\,\text{L·torr·K}^{-1}\text{·mol}^{-1}$ is the gas constant, and $T$ is the prong temperature (in K). Thus, the QTF resonance frequency shifts linearly with the pressure, if the temperature is assumed to be fixed. The Q-factor of a resonance mode is a measure of the energy loss of the prongs while they are vibrating. It has been demonstrated that the

QTF mainly loses energy via interaction with the surrounding viscous medium. Gas damping is the main loss mechanism for the fundamental and first overtone modes and is proportional to the viscosity and density of the gas. A good approximation for the dependence of the quality factor on the gas pressure is provided by the following equation [39, 40]:

$$Q(P) = \frac{Q_o}{1 + Q_0 b \sqrt{P}}, \tag{13.24}$$

where $Q_0$ is the quality factor at $P = 0$, which includes all pressure-independent loss mechanisms (support losses and thermoelastic damping), and $b$ is a fitting parameter. Both $Q_0$ and $b$ are mainly related to QTF geometry and gas viscosity, and thus vary for each QTF.

## 13.3.3  T-Shaped Grooved Tuning Forks

Moving to low resonance frequencies by varying the ratio $T/L$ following the Euler–Bernoulli equation for rectangular prongs, the Q-factor decreases, as demonstrated in [31]. In particular, QTFs with a resonance frequency lower than 10 kHz cannot ensure $Q > 10,000$ at atmospheric pressure. A further reduction of the resonance frequency without strongly affecting the overall quality factor requires a change of prong geometry. T-shaped prongs have been proposed to accomplish both requirements, with a further increase of stress field distribution along vibrating prongs with respect to rectangular prongs, which is beneficial for piezoelectric charge generation when prongs are deflected. [41–43]. In addition, applying rectangular grooves in prongs increases the electrical coupling between electrodes, leading to a reduction of the electrical resistance without affecting the quality factor [31]. To investigate the resonance properties of a T-shaped grooved QTF, a finite-element-analysis (FEA) using COMSOL Multiphysics was performed. Taking $T_1$ as the thickness of the T-shaped prong head and $T_2$ as the thickness of the T-shaped prong body (see Fig. 13.4(a)), the influence of the ratio $T_2/T_1$ on the resonance frequency was investigated in [44]. $T_1$, $L_1$, and $L_2$ were kept constant to 2.0 mm, 2.4 mm, and 7.0 mm, respectively, while $T_2$ was varied. Grooves 50 μm deep were carved on both sides of each prong, as sketched in Fig. 13.4(b). The simulation was performed in vacuum environment to avoid the influence of the surrounding medium and the results are shown in Fig. 13.4(c). $T_2/T_1 = 0.5$ was set as lower limit for the simulation.

The simulation results clearly indicate a linear trend of the fundamental resonance frequency with respect to the $T_2/T_1$ ratio, with a $R^2$ value of 0.9998. Ratios lower than 0.7, i.e., $T_1$ smaller than 1.4 mm, lead to a resonance frequency lower than 10 kHz. However, the lowering of the resonance frequency through reduction of the $T_2/T_1$ ratio is in competition with the mechanical stability of the QTF. A low $T_2/T_1$ ratio would not guarantee a stable oscillation of the QTF prong due to a low moment of inertia. Moreover, prong oscillations would be highly influenced by the surrounding medium damping leading to high viscous losses that in turn negatively affect the quality factor.

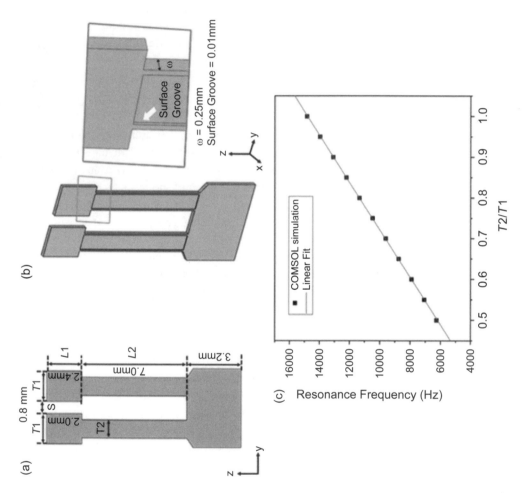

**Fig. 13.4** (a) Schematic diagram of the T-shaped grooved QTF, with geometrical parameters shown. (b) Schematic diagram of the grooved prong. (c) Results of COMSOL Multiphysics simulation of fundamental resonance frequency of a T-shaped grooved QTF as a function of the $T_2/T_1$ ratio when $T_1$, $L_1$, and $L_2$ are kept constant to 2.0 mm, 2.4 mm, and 7.0 mm, respectively.

### 13.3.4    Overtone Modes

The realization of custom-made QTFs optimized for QEPAS applications can significantly improve the sensitivity of the detection technique. With respect to the standard 32.7 kHz QTF, the resonance frequency of the fundamental mode can be reduced to a few kilohertz in order to better approach the typical energy relaxation time of targeted gases [45], while maintaining a high resonator quality factor. Both conditions can be simultaneously satisfied by an appropriate design of the QTF prong sizes [31]. However, lowering the fundamental resonance frequency also reduces the overtone frequencies, leading to their possible use in QEPAS sensor systems. This is not feasible with a 32.7 kHz QTF, since its first overtone mode occurs at frequencies higher than 190 kHz, which is impractical for QEPAS-based gas detection. By using Eq. (13.21), the first overtone mode exhibits a resonance frequency $(2.9882/1.1942)^2$ $\sim$ 6.26 times higher than the fundamental mode. This implies that in order to have the first overtone resonance frequency <30 kHz, the fundamental mode must vibrate at frequencies lower than 4.5 kHz. Lowering the operating frequency to values <3 kHz is not recommended in QEPAS since the sensor system would be more influenced by environmental acoustic noise. The fundamental and first overtone modes exhibit different quality factor values because the associated loss mechanisms depend on the related vibrational dynamics and on the geometry of the QTF prongs [39, 46]. Hence, the QTF geometry can be designed to provide an enhancement of the overtone mode resonance Q-factor and thereby a higher QEPAS SNR with respect to the fundamental mode can be expected. According to Hosaka's model [47], the air damping is significantly reduced when moving from the fundamental to overtone mode. However, by moving to higher modes support losses (due to the interaction of the prong with its support) can start to dominate, deteriorating the overall quality factor [48, 49]. An increase of a factor of $\sim$12 in support losses is expected for a QTF, when changing from the fundamental to the first overtone mode. The possibility to excite independently and simultaneously both the fundamental and the first overtone flexural mode of a QTF opened the way to simultaneous detection of two different gas species targeted by two different lasers.

### 13.3.5    Microresonator Tubes

The QTF is not used as a standalone in a QEPAS sensor but it is acoustically coupled with a pair of millimeter-size resonator tubes located on both sides of the QTF, as sketched in Fig. 13.5. This arrangement is known as the on-beam configuration. Conversely, in the off-beam configuration, a single resonator tube is positioned adjacent to the QTF: the QTF senses the pressure in the microresonator through a small aperture in its center [50, 51]. Resonator tubes act as an acoustic resonator: the standing wave vibrational pattern within the tubes enhances the intensity of the acoustic field between the QTF prongs up to 60 times [31]. The QTF coupled with a single or a pair of resonator tubes constitutes the QEPAS spectrophone and represents the core of any QEPAS detection module.

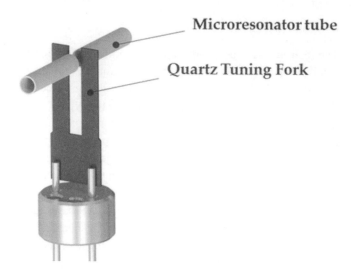

**Fig. 13.5** Schematic diagram of the QEPAS spectrometer.

In the on-beam configuration, when two tubes are acoustically coupled with a QTF, the internal diameter (ID) of tubes and their length influence the enhancement of the sound wave between the QTF prongs. The optimal tube diameter can be estimated by considering the theoretical model proposed in [52], in which the acoustic coupling between the two tubes is expressed in terms of the amount of the acoustic field transferred from one tube to the other one. As a result, the optimal tube radius strongly depends on the sound wavelength $\lambda_s$ and prong spacing $g$. As an example, for a QTF resonating at 8 kHz in pure $SF_6$ ($\lambda_s = v_s/f_0 = 16.7$ mm, with $v_s = 136$ m/s being the speed of sound in the $SF_6$ gas and $g = 800$ µm) the theoretical model predicts the optimal ID $= 1.0$ mm. The optimal tube length can be estimated by considering the open-end correction, which assumes that the antinode of a standing sound wave in an open-ended resonator is located outside the tube end, due to an impedance mismatch between the acoustic field inside and outside of the resonator. The optimal tube length $l$ depends on the tube ID and the sound wavelength by the relation [53]:

$$l = \frac{v_s}{2f_0} - \frac{8ID}{3\pi}. \tag{13.25}$$

Thus, the optimal tube length depends on the tube ID, the speed of sound, and the resonance frequency of the acoustic wave. The QTF acoustically coupled with a pair of resonator tubes composes the QEPAS spectrophone, which is the core of the detection module of any QEPAS sensor.

## 13.4     Overview of QCL-based QEPAS Gas Sensors

High selectivity and detection sensitivity, and short response time are the primary requirements for trace gas sensing. These characteristics, combined with no downtime between data sampling and analysis, no need for frequent recalibration, the absence of any saturable or degradable components, portability, and low cost, are the primary

**Table 13.1** The list of the main methods and techniques employed in gas analyzer technologies and their characteristics (shaded cells).

| Technique | High sensitivity | Short response time | No degradation | Small size | In-situ monitoring | Low cost |
|---|---|---|---|---|---|---|
| GC-MS | 👍 | 👍 | 👍 | 👎 | 👎 | 👎 |
| Chemi-luminescence | 👍 | 👍 | 👎 | 👎 | 👎 | 👎 |
| Electrochemical-based | 👎 | 👎 | 👎 | 👍 | 👎 | 👍 |
| Laser-based | 👍 | 👍 | 👍 | 👍 | 👍 | 👎 |

requirements for a gas sensor system to be particularly suitable for in-field applications like environmental monitoring, breath sensing, leak detection, industrial processes control, and oil and gas exploration. The characteristics of the commonly employed methods in gas analyzers, i.e., gas chromatography–mass spectrometry (GC-MS), chemiluminescence, and electrochemical- and laser-based techniques, are summarized in Table 13.1.

Although the peaks in the GC-MS technique allow unambiguous identification of gas components with a very high sensitivity, the gas chromatography columns and mass spectrometers are bulky, expensive, and do not allow for in-situ measurements [54]. When dealing with chemiluminescence-based analyzers, instability over time and the need for daily calibration must be added to these drawbacks [55]. Electrochemical-based analyzers are inexpensive and very small, making them suitable for compact devices. However, these sensors exhibit a lower response time and a lower sensitivity, compared to the other techniques, and a reliability degradation over time [56]. Finally, laser-based analyzers (LAs) can provide highly sensitive and selective detection, with fast response time [57]. Mid-infrared QCLs are perfect excitation sources for laser-based gas sensor systems due to their operation in the spectral range covering most of the fundamental gas absorption lines with single-mode relatively narrow linewidth emission, fulfilling the selectivity requirement [58]. Compared to other laser sources, QCLs have high optical power output which leads to higher achievable sensitivity in QEPAS sensors and can be implemented in a distributed-feedback or external-cavity configuration, allowing a wide, mode-hop free wavelength tunability. The ability to operate at room temperature and the overall laser compact size allow portable sensors to be developed.

While there are a number of gas sensing techniques employing laser sources, we will focus on gas sensors based on QEPAS, whose fundamentals and advantages have been discussed in Section 13.3.

## 13.4.1    Environmental Monitoring

Climate, human health, terrestrial and aquatic ecosystems, and agricultural productivity depend on atmospheric composition. Human activities have an increasing influence on atmospheric composition, with subsequent environmental impacts [59]. Therefore, environmental trace gas monitoring is fundamental to assess human health risks in polluted urban or industrial surroundings, and to mitigate climate change caused by greenhouse gas emissions [60]. Directive 2008/50/EC of the European Parliament and of the Council of May 21, 2008 on ambient air quality and cleaner air for Europe established threshold values for the following pollutants: sulfur dioxide ($SO_2$), nitrogen oxides ($NO_x$), particulate matter (PM), lead (Pb), benzene ($C_6H_6$), and carbon monoxide (CO), due to their contributions to photochemical smog, acid rains, ozone depletion in the atmosphere, and troposphere ozone formation [61]. In addition, carbon dioxide ($CO_2$), methane ($CH_4$), and nitrous oxide ($N_2O$) must be monitored as the most important greenhouse gases [62].

Most of the environmentally relevant gas molecules as well as explosives and toxic gases exhibit strong absorption features in the mid-infrared spectral range. For this reason, spectroscopic techniques employing mid-infrared QCLs have been developed, aiming at high-sensitivity trace gas detection. In particular, QEPAS has been demonstrated as a powerful technique for harmful pollutant or greenhouse gas sensing. For example, a QEPAS sensor approach for the detection of triacetone triperoxide (TATP), a peroxide-based explosive, was presented for the first time in [63]. C-O stretch bands near $1200 \, \mathrm{cm}^{-1}$ are detected by employing a widely tunable pulsed external cavity QCL (EC-QCL), at a duty cycle of 5%, producing an average output power of 5 mW. The sensor architecture employs a QEPAS spectrophone in on-beam configuration. Starting from an amount of 500 μg of TATP synthesized ad hoc for the experiment, a concentration of 65 ppm at ambient air is detected, with a detection limit of 1 part per million (ppm). Continuous wave EC-QCLs have been employed in off-beam QEPAS sensors for the detection of laboratory samples of nitrous acid in gas phase (HONO) and hydrogen sulfide ($H_2S$) in the $1254\text{--}1256 \, \mathrm{cm}^{-1}$ and $1233\text{--}1235 \, \mathrm{cm}^{-1}$ spectral regions, respectively, as reported in [64] and [65]. Interest on HONO detection rises from the central role it plays in the atmospheric oxidation capacity, which in turn significantly affects the regional air quality and the global climate change. With a laser output power of 50 mW and a lock-in integration time of 1 s, the authors report a minimum detection limit of 66 ppb by volume (ppbv). Hydrogen sulfide is a harmful gas and its occupational exposure limit is regulated. It is of essential importance to perform $H_2S$ monitoring in the field of petrochemical and biotechnological processes. The reported sensitivity for 160 mW of QCL optical power, at 1 s lock-in time constant, is 492 ppbv.

Another example of a QCL-based QEPAS sensor targeting a gas molecule relevant for environmental monitoring is reported in [66]. Here, an array of 32 pulsed distributed-feedback QCLs emitting in the $1190\text{--}1340 \, \mathrm{cm}^{-1}$ spectral range is employed as the excitation source of a quartz-enhanced photoacoustic sensor in on-beam configuration for nitrous oxide broadband detection. At a duty cycle of 0.75%, the highest average optical power measured between the 32-QCLs device is of 1.6 mW. With a 10 s lock-in integration time, a detection limit of less than 60 ppb is achieved, matching the sensitivity level required for nitrous oxide monitoring at global atmospheric levels.

The reviewed examples put solid basis on the development of QCL-based QEPAS sensors for environmental applications. Up to now, four in-field tested sensors have been developed, whose main characteristics are summarized in Table 13.2.

The first in-field test of a QEPAS sensor for environmentally relevant gas molecules monitoring was demonstrated in 2013 in [67]. Here, an on-beam spectrophone configuration and a high-power continuous wave distributed feedback (DFB) QCL emitting at $\sim 2169\,\text{cm}^{-1}$ have been selected to develop a QEPAS sensor for carbon monoxide (CO) and nitrous oxide ($N_2O$) detection. By adding 2.6% water vapor concentration in the gas line and using a lock-in integration time of 1 s, the achieved minimum detection limit is of 1.5 ppbv at atmospheric pressure for CO and of 23 ppbv at 100 torr for $N_2O$. The sensor was first optimized and calibrated in the laboratory, by employing gas cylinders containing certified concentration of the two investigated gas species. Then, a continuous monitoring of atmospheric CO and $N_2O$ concentration levels was performed by placing the inlet tube of the QEPAS sensor outside the laboratory, in the Rice University campus (Houston, TX, USA). The concentration levels of the two gases were monitored by locking the QCL emission wavelength to the CO absorption line for the first five and a half hours, and then to the $N_2O$ absorption line for the next five hours. The obtained results are shown in Fig. 13.6.

The measured $N_2O$ atmospheric concentration level is relatively stable, while peaks in the CO concentration over a background of $\sim 130$ ppbv are measured when people smoke close to the sensor inlet (concentration over 750 ppbv) or automobiles pass through the campus road $\sim 20$ meters away from the sensor sampling inlet. When pure nitrogen is flushed through the gas line, the signal drops to zero, as can be observed in the final part of both data plots. Reliable and robust operation of the QEPAS sensor was thus demonstrated.

In 2019 an improved QEPAS sensor for CO detection in air, at atmospheric pressure, was reported [41]. Compared to the previous work, the employed laser source targeted the CO absorption line at $2169.2\,\text{cm}^{-1}$ with an optical power $\sim 50$ times lower. However, the use of an optimized spectrophone allowed comparable sensitivity levels to be achieved. The sensor was calibrated in the laboratory and tested for in-field measurements. Seven-days continuous measurements were implemented by placing

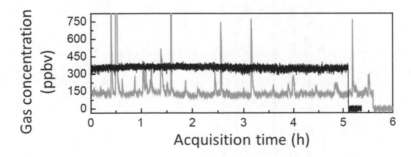

**Fig. 13.6** Atmospheric CO (gray line) and $N_2O$ (black line) concentration levels measured from air sampled at Rice University campus, Houston, TX, USA. The two tracks were acquired in succession.

**Table 13.2** List of gas molecules detected in air by employing a QCL-based QEPAS sensor for environmental monitoring purposes. For each molecule, the QCL configuration (DFB – distributed feedback), operation (cw – continuous wave), emission wavenumber, and optical power (P) are reported, as well as the QEPAS sensor configuration, operating pressure, achieved sensitivity at 1 s lock-in integration time (i.t.), and the sensor field-test location.

| Gas molecule | QCL config. and op. | Wavenumber $(cm^{-1})$ | $P$ (mW) | QEPAS config. | Pressure (torr) | Sensitivity at 1 s i.t. (ppbv) | Field-test location | Ref. |
|---|---|---|---|---|---|---|---|---|
| CO | DFB, cw | 2169.2 | 987 | On-beam | 760 | $(+H_2O\ 2.6\%)\ 1.5$ | Rice University campus, Houston, TX, USA | [67] |
| | DFB, cw | 2169.2 | 21 | On-beam | 700 | $(+H_2O\ 2.5\%)\ 7$ | Shanxi University, Taiyuan, China | [41] |
| | DFB, cw | 2169.6 | 987 | On-beam | 100 | $(+H_2O\ 2.6\%)\ 23$ | Rice University campus, Houston, TX, USA | [67] |
| $N_2O$ | DFB, cw | 1275.49 | 120 | On-beam | 130 | 6 | BFI McCarty landfill, Houston, TX, USA | [68] |
| $CH_4$ | DFB, cw | 1275.04 | 120 | On-beam | 130 | 13 | BFI McCarty landfill, Houston, TX, USA | [68] |
| NO | DFB, cw | 1900.07 | 65 | Off-beam | 760 | 120 | Chinese University of Hong Kong, Hong Kong | [69] |

the sensor out of the Yifu building of Shanxi University, China, to verify the reliability and validity of the developed system. The obtained data well-matched the one recorded by a China National Environmental Monitoring Center monitoring station, 7 km away from the campus.

In 2014 a compact QCL-based on-beam QEPAS sensor for environmental monitoring of greenhouse gases was developed [68]. A continuous wave DFB-QCL was employed to excite the $CH_4$ and $N_2O$ absorption lines at $1275.04 \, cm^{-1}$ and $1275.49 \, cm^{-1}$, respectively, with an optical power of 120 mW. For a working pressure of 130 torr and a lock-in integration time of 1 s, detection limits of 13 ppbv and 6 ppbv were achieved, respectively. Laboratory measurements aiming at proving the sensor stability and reproducibility were performed. The sensor was later installed in the Aerodyne Research, Inc. mobile laboratory for atmospheric $CH_4$ and $N_2O$ sensing at the BFI McCarty landfill, an urban solid waste disposal site in the Greater Houston area. Measurements revealed high $CH_4$ concentration levels, up to 27 ppm by volume (ppmv), confirmed by the values measured by a multi-pass absorption cell-based sensor installed in a van.

Another air pollutant sensor based on the QEPAS technique was reported in 2017 in [69]. A compact and sensitive mid-infrared nitric oxide (NO) sensor was designed by using a distributed-feedback QCL emitting at $1900.07 \, cm^{-1}$ with an optical power of 65 mW and a spectrophone in off-beam configuration. With a minimum detection limit of 120 ppbv, for a lock-in integration time of 1 s, the QEPAS NO sensor was deployed in real-time diesel-engine exhaust monitoring. Measurements were performed by sampling the exhaust of a ISUZU LT 134L bus diesel engine (6HK1-TC Euro-III Compliant Rear Engine) at the Chinese University of Hong Kong. A slight decrease of the QEPAS signal was observed due to soot scattering the laser light, highlighting the importance of integrating a soot filter at the sensor inlet. An increase of NO concentrations from 126 ppmv to 187 ppmv in the exhaust was measured when engine rotational speed increased from 500 to 2000 revolutions per minute.

## 13.4.2    Breath Analysis

Human breath is composed of a matrix of nitrogen, oxygen, water vapor, carbon dioxide and inert gases (Ar, Ne, He), where a mixture of as many as 500 molecules is diluted in traces concentrations. Such a composition is determined by the inspiratory air and the volatile organic compounds present in the blood, which are originated by molecules and their metabolites that are (i) inhaled, (ii) absorbed by skin, (iii) ingested, and (iv) produced by tissues and cells in the body [70]. The presence of some molecules or metabolites in the exhaled molecular profile, in elevated concentrations compared to the normal ones, is closely associated with inflammatory processes and diseases [71]. For example, kidney and liver dysfunction lead to high levels of ammonia ($NH_3$) in breath [72], diabetes causes an increase of acetone ($C_3H_6O$) [73], asthma and upper respiratory track inflammation cause high levels of nitric oxide (NO) [74], and neonatal icterus leads to a carbon monoxide (CO) concentration increase [75]. Therefore, breath analysis has an enormous potential for clinical applications aiming at early disease diagnosis and screening. Moreover, compared to other diagnostic techniques (e.g., histologic examination), breath sampling is non-invasive, safe for the patient, painless, and can be performed, if necessary, multiple times, at any site (laboratory, operation room, bedside) and over a large number of subjects.

Despite its undeniable potential, clinical breath analysis is still in its infancy and just a few molecules and metabolites breath tests have been approved by national public health institutions, mainly due to the stringent requirements on sampling procedures and measurements reproducibility [76].

In the last decade, the characteristics of laser-based spectroscopy and, in particular, of QEPAS have led researchers to develop QCL-based QEPAS sensors for potential breath sensing application. Indeed, the high-power (compared to diode lasers and interband cascade lasers) narrow linewidth emission of QCLs, combined with the performance of the QEPAS spectrophones and with the selection of the best operating pressure, provide the selectivity and sensitivity levels required to avoid false positives and false negatives in the breath analysis. For example, in [77] a QEPAS sensor for ethylene ($C_2H_4$) detection is reported, employing a continuous wave DFB-QCL, emitting at $967.40\,cm^{-1}$ with an optical power of $74.2\,mW$, and an optimized spectrophone in on-beam configuration. The best working pressure for interferents-isolated, high-sensitivity $C_2H_4$ detection is found to be 120 torr. At 10 s lock-in integration time, a sensitivity of 10 ppb is achieved, comparable to the $C_2H_4$ concentration levels in breath indicative of lipid peroxidation in lung epithelium. Another example is reported in [78], where a continuous wave DFB-QCL emitting at $2178.69\,cm^{-1}$ with an optical power of 75 mW is employed in an on-beam QEPAS sensor to detect carbon disulfide ($CS_2$), a biomarker of respiratory bacterial colonization in cystic fibrosis. Here, a minimum detection limit of 28 ppbv is achieved for a lock-in integration time of 1 s when the gas sample is moisturized with a water vapor concentration of 2.3 vol%, to improve the gas vibrational-translational relaxation process, at a pressure of 56 torr. In [33] a widely tunable, mode-hop-free EC-QCL is employed in a QEPAS sensor in on-beam configuration for nitric oxide detection at $1900.08\,cm^{-1}$ with 66 mW optical excitation power. Here, a NO detection sensitivity of 4.9 ppbv is achieved with a 1 s lock-in integration time, by adding a 2.5% of water vapor concentration, at 250 torr. In the abovementioned three examples, measurements are performed by employing cylinders containing a certified concentration of the target gas, later diluted with nitrogen (and, in some cases, also water vapor). A further step toward a clinical breath sensing application is proposed in [79], where the EC-QCL is replaced with a $\sim$100 mW DFB-QCL emitting at $1900.08\,cm^{-1}$, in order to reduce the setup size and realize a compact and portable sensor with a comparable sensitivity. By employing the sensor described in [79], a preliminary breath-test was performed by measuring the airway wall nitric oxide concentration exhaled by four volunteers. This test was performed in Rice University, Houston, TX, USA. In Fig. 13.7 the volunteers' exhaled NO QEPAS signal (labels: v.A, v.B, v.C, v.D) is reported and compared with the signal measured for a certified reference $N_2$ gas mixture containing 95 ppb of NO concentration and a water vapor concentration of 2.5 vol%. The whole measurement was performed by locking

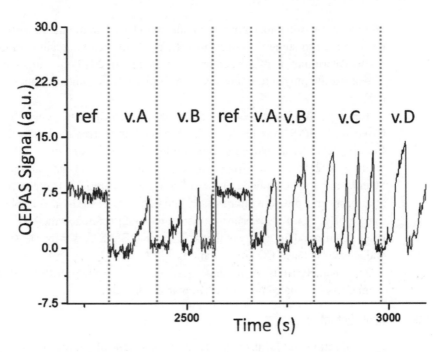

**Fig. 13.7** QEPAS signal amplitude of a mixture of 95 ppb of NO, 2.5% $H_2O$, rest $N_2$ (ref), and of breath samples from four volunteers (v.A, v.B, v.C, v.D), obtained by locking the QCL emission wavenumber to the NO absorption line.

the laser emission wavenumber to the nitric oxide absorption line at 1900.08 cm$^{-1}$. The gas line inlet allows for switching between the reference gas mixture and a breath collector.

Each time the inlet is switched from the mixture containing a known concentration of NO to the breath collector, the signal quickly drops to zero till the volunteer starts breathing. The NO breath signal changes for each volunteer, depending on breath duration, mouth pressure, and NO concentration, while the QEPAS signal measured between 2550 s and 2680 s is the same as the one measured in the first part of the acquired track. All these characteristics prove a complete absence of any memory or saturation effect. Moreover, the signal resembles the breath dynamic of the volunteer, due to the fast response time of the sensor.

The work presented in [79] provides a proof-of-concept on the potential of QCL-based QEPAS as a powerful breath sensing technique. The first clinical study of a QEPAS-based breath sensor was performed at the St. Luke's University Hospital of Bethlehem, PA, USA, in 2013 [80]. In this study, breath ammonia is detected with a fast, real-time monitor using a continuous wave DFB-QCL-based on-beam QEPAS sensor coupled to a breath sampling device that measures mouth pressure and the real-time concentration of carbon dioxide. The selected laser source allows the 967.35 cm$^{-1}$ $NH_3$ absorption line to be targeted with an optical power of 24 mW. At a working pressure of 120 torr a detection sensitivity as low as six ppbv is achieved with a 1 s lock-in integration time. The $NH_3$ QEPAS sensor was used to evaluate potential clinical utility of breath ammonia compared to blood ammonia. In particular, a proto-col was used to study the effects that the mode of breathing and the mouth pH have on the determination of exhaled breath ammonia. In 2015, the same breath ammonia sensor was employed to evaluate ammonia physiology in healthy and cirrhotic adults [81].

Recently, the first carbon monoxide sensor dedicated to human breath analysis, based on QEPAS, was tested at the Montpellier University Hospital, Montpellier, France [82]. The developed sensor employed a continuous wave Fabry–Pérot QCL emitting at $2103.26\,\mathrm{cm}^{-1}$ (targeting CO absorption line) with an optical power of $80\,\mathrm{mW}$ and a spectrophone in the off-beam configuration. The achieved sensitivity at atmospheric pressure, at a lock-in integration time of 1 s is 20 ppbv. The performance of the QEPAS sensor was compared with reference methods. The obtained results confirmed the ability of the QCL-based QEPAS sensor for measuring low-level endogenous exhaled CO concentrations.

The characteristics of the three described QCL-based QEPAS sensors targeting biomarkers in exhaled breath are summarized in Table 13.3.

### 13.4.3 Leak Detection

Leak detection and localization plays an important role in mechatronics, biotechnology, petrochemical and automotive industries, and all the industrial fields requiring manufacturing quality-control processes [83]. In contrast to other control tests, leak detection employs techniques which do not damage the characteristics or performances of components like vessels, valves, pipes, or rubber seals [84, 85].

In 2016, a leak detector based on QEPAS technique was developed to test mechatronics systems and components, such as vacuum-valves and diesel injectors, usually operating at high pressures [86]. Leak tightness requirements are extremely important in automotive industry to lower emissions and improve fuel consumption. The sensor employed sulfur hexafluoride ($SF_6$) as a leak tracer gas due to being an extremely stable, inert, non-flammable, and non-toxic, gas, making it particularly suitable and safe for working-environment use [87]. A QCL in external-cavity configuration, resonant with a strong $SF_6$ absorption band around $947\,\mathrm{cm}^{-1}$, was employed as the excitation source. The QEPAS sensor was integrated in a vacuum-sealed test station. Validation tests showed a minimum detection sensitivity of 2.75 ppb for a 1 s integration time, corresponding to a minimum detectable leak of $4.5 \cdot 10^{-7}\,\mathrm{mbar \cdot l/s}$, for a nitrogen gas carrier flow of $0.67\,\mathrm{mbar \cdot l/s}$. If pure $SF_6$ is used as leak test gas, the sensitivity can be improved down to $\sim 4.5 \cdot 10^{-9}\,\mathrm{mbar \cdot l/s}$. In terms of sensitivity, QEPAS is competitive with most of the techniques reported in literature [88], some of which are listed in Table 13.4. However, compared with these techniques, QEPAS is the only one also combining the characteristics of short response time, compactness, low weight and cost, and safety [86].

### 13.4.4 Multi-Gas Detection

Isotopes concentration ratio measurements, molecular relaxation dynamics studies, industrial process control, as well as the aforementioned environmental monitoring

**Table 13.3** The list of breath biomarkers detected by employing QCL-based QEPAS sensors. For each molecule, the QCL configuration (DFB – distributed feedback, FP – Fabry–Pérot), operation (cw – continuous wave), emission wavenumber, and optical power (P) are reported, as well as the QEPAS sensor configuration, operating pressure, and achieved sensitivity at 1 s lock-in integration time (i.t.), as well as the sensor field-test location.

| Gas molecule | QCL config. and op. | Wavenumber $(cm^{-1})$ | P (mW) | QEPAS config. | Pressure (torr) | Sensitivity at 1 s i.t. (ppbv) | Field-test location | Ref. |
|---|---|---|---|---|---|---|---|---|
| NO | DFB, cw | 1900.08 | 60 | On-beam | 250 | $(+H_2O\ 2.5\%)$ 4 | Rice University campus, Houston, TX, USA (proof of concept) | [79] |
| $NH_3$ | DFB, cw | 967.35 | 24 | On-beam | 120 | 6 | St. Luke's University Hospital, Bethlehem, PA, USA | [80] |
| CO | FP, cw | 2103.26 | 80 | Off-beam | 760 | 20 | Montpellier University Hospital, Montpellier, France | [82] |

**Table 13.4** List of the main leak detection methods and corresponding sensitivity levels.

| Leak detection technique | Sensitivity (mbar·l/s) |
|---|---|
| QEPAS | $10^{-9}$ |
| Bubble in water | $10^{-2}$ |
| Vacuum decay | $10^{-3}$ |
| Halogen detector | $10^{-6}$ |
| Radio-isotope | $10^{-10}$ |
| Mass spectroscopy | $10^{-11}$ |

and breath biomarkers detection are some of the sensing applications requiring detection and quantification of different components in a gas mixture, simultaneously or quasi-simultaneously [89]. With this purpose, several on-beam QEPAS sensors have been developed, whose characteristics, in terms of analyzed gas mixture, employed QCL source features, and possible applications, are summarized in Table 13.5.

In [90], the dynamic range of a continuous wave distributed-feedback QCL was exploited to selectively detect isotopes $^{12}CH_4$ and $^{13}CH_4$ in methane samples at the natural abundance, in the $1296.00$–$1296.15\,cm^{-1}$ spectral range. Isotopic ratio can be used as both fingerprint and indicator for many processes involving isotopic fractionations. The thermogenic or biogenic origin of natural gas, as well as its source and history, can be identified by measuring the methane isotopic ratio. Therefore, with a sensitivity at the parts-per-billion level, this sensor opens the way to the development of QEPAS-based gas spectrometers to be employed downhole for oil and gas exploration. The cw DFB-QCL-based QEPAS sensor demonstrated in [91] has a great potential for both breath pattern recognition and environmental monitoring. Indeed, this sensor detects ethylene and carbon dioxide, which are both biomarkers and pollutant gases, in the $949.1$–$949.6\,cm^{-1}$ spectral range, and provides a study of the influence of the $CO_2$ addition to the $C_2H_4/N_2$ mixture on the photoacoustic signal of ethylene, supported by the development of an energy transfer model in the analyzed gas mixture. Environmental monitoring is also the possible field of application of the dual-gas sensors described in [92] and [93]. In the first case, a $CO$-$N_2O$ sensor is developed by employing a cw DFB-QCL. In the second case, detection of chlorodifluoromethane ($CHClF_2$) and acetone ($C_3H_6O$) is demonstrated in the $1180$–$1345\,cm^{-1}$ spectral region by exploiting the tunability range of an external cavity QCL. In this last example, each target molecule exhibits unresolved rotational–vibrational absorption lines. However, the absorption features of the two molecules lay in different spectral regions. When the strong absorption features of the target molecules overlap, multi-gas sensing becomes challenging.

The first strongly overlapped, broadband absorbing molecules QEPAS detection is reported in [94]. A widely tunable mode-hop free external cavity QCL is used to target the absorption lines of pentafluorocthane (Freon 125, $C_2HF_5$) and acetone in the spectral range from $1127\,cm^{-1}$ to $1245\,cm^{-1}$. QEPAS reference spectra of the single-gas

**Table 13.5** List of gas mixtures detected by employing a QCL-based on-beam QEPAS sensor. For each mixture, the gas components concentration is reported and the spectral overlap (Ov.) of absorption features of different molecules is specified. The employed QCL configuration (DFB – distributed feedback, EC – external cavity), operation (cw – continuous wave), emission wavenumber range, and average optical power are reported, together with the sensing application (BS – breath sensing, EM – environmental monitoring).

| Gas mixture | Analyzed gas components concentration | Ov. | QCL config. and op. | Wavenumber range ($cm^{-1}$) | Optical power (mW) | Possible application | Ref. |
|---|---|---|---|---|---|---|---|
| $^{12}CH_4$–$^{13}CH_4$ | 49.41 ppm – 0.555 ppm | no | DFB, cw | 1296–1296.15 | ~112 | Oil and gas exploration | [90] |
| $C_2H_4$–$CO_2$ | 9 ppm – 1% | no | DFB, cw | 949.1–949.6 | ~16 | BS, EM | [91] |
| CO–$N_2O$ | 250 ppm – 750 ppm | no | DFB, cw | 2188.8–2191.2 | ~75 | EM | [92] |
| $CHClF_2$–$C_3H_6O$ | 1.32% – 0.96% | no | EC, pulsed | 1180–1345 | ~2 | EM | [93] |
| $C_2HF_5$–$C_3H_6O$ | 4.4 ppm – 47.2 ppm | yes | EC, cw | 1127–1245 | ~6.6 | Broadband spectroscopy | [94] |
| $CH_4$–$N_2O$ | 150 ppm – 510 ppm | yes | DFB QCL array, pulsed | 1190–1340 | ~1.6 | EM | [95] |
| $C_2H_2$–$CH_4$–$N_2O$ | 150 ppm – 3000 ppm – 550 ppm | yes | DFB, cw | 1295.5–1296.5 | ~112 | EM | [92] |

species are first acquired separately. Then, the QEPAS spectrum recorded for a gas mixture with an unknown concentration of both species is fitted by a linear combination of the reference spectra. A similar approach is adopted in [95], where broadband detection of mixtures of $N_2O$ and $CH_4$ in dry nitrogen is demonstrated by using an array of 32 DFB-QCLs having an emission in the spectral range between $1190 \, cm^{-1}$ and $1340 \, cm^{-1}$. Despite the significant overlap in the target gas molecules absorption spectra, determination of nitrous oxide and methane concentrations is demonstrated, representing a starting point for the development of sensors aiming at environmental monitoring of broadband greenhouse gases. However, linear regression is efficient as long as the experimental data are uncorrelated or at least weakly correlated; otherwise, a lack of precision and accuracy can occur [96]. For this reason, an algorithm based on partial least squares regression is employed in [92] to retrieve the components concentration in a three-gas mixture. Here, a cw DFB QCL is employed as the excitation source of a QEPAS sensor for environmentally relevant gas detection. Analyzed gas mixtures are composed of different amounts of acetylene ($C_2H_2$), methane, and nitrous oxide, showing a spectral overlap greater than 97% in the laser emission range ($1295.5$–$1296.5 \, cm^{-1}$). The calibration error reported for the three-gas components concentrations predicted by PLSR tool is up to five times better, compared with standard linear regression.

## 13.5     Conclusion

This chapter presents an overview of the gas sensors based on quartz-enhanced photoacoustic spectroscopy employing quantum cascade lasers as the excitation light source. Optical gas sensing techniques are sensitive, non-destructive tools to detect gas species in situ and in real time. In particular, photoacoustic spectroscopy and quartz-enhanced photoacoustic spectroscopy, whose physical principles are described in Sections 13.2 and 13.3, combined with the advantages of quantum cascade lasers, have been exploited to develop high-sensitivity, compact, robust, portable, multi-gas detectors. QCL-based QEPAS sensors have been demonstrated for several applications, such as environmental monitoring, breath analysis, leak detection, and multi-gas detection, as reviewed in Section 13.4, thus demonstrating the versatility and great potential of the QCL-based QEPAS technique in gas sensing.

## References

[1]  P. T. Moseley, "Solid state gas sensors," *Meas. Sci. Techn.*, vol. 8, no. 3, pp. 223–237, 1997.
[2]  N. Yamazoe and K. Shimanoe, "Receptor function and response of semiconductor gas sensor," *J. Sensors*, vol. 875704, 2009.
[3]  J. Hodgkinson and R. P. Tatam, "Optical gas sensing: A review," *Meas. Sci. Technol.*, vol. 24, no. 1, p. 012004, 2013.
[4]  A. Elia, P. M. Lugarà, C. Di Franco, and V. Spagnolo, "Photoacoustic techniques for trace gas sensing based on semiconductor laser sources," *Sensors*, vol. 9, pp. 9616–9628, 2009.

[5]  T. Tomberg, M. Vainio, T. Hieta, and L. Halonen, "Sub-parts-per-trillion level sensitivity in trace gas detection by cantilever-enhanced photo-acoustic spectroscopy," *Sci. Rep.*, vol. 8, p. 1848, 2018.

[6]  S. Manohar and D. Razansky, "Photoacoustics: a historical review," *Adv. Opt. Photon.*, Vol. 8, pp. 586–617, 2016.

[7]  L. Y. Hao *et al.*, "A new cylindrical photoacoustic cell with improved performance," *Rev. Sci. Instrum.*, vol. 73, no. 2 I, p. 404, 2002.

[8]  K. Song *et al.*, "Differential Helmholtz resonant photoacoustic cell for spectroscopy and gas analysis with room-temperature diode lasers," *Appl. Phys. B Lasers Opt.*, vol. 75, no. 2–3, pp. 215–227, 2002.

[9]  F. G. C. Bijnen, J. Reuss, and F. J. M. Harren, "Geometrical optimization of a longitudinal resonant photoacoustic cell for sensitive and fast trace gas detection," *Rev. Sci. Instrum.*, vol. 67, no. 8, pp. 2914–2923, 1996.

[10]  K. Wilcken and J. Kauppinen, "Optimization of a microphone for photoacoustic spectros-copy," *Appl. Spectrosc.*, vol. 57, no. 9, pp. 1087–1092, 2003.

[11]  D. C. Dumitras, D. C. Dutu, C. Matei, A. M. Magureanu, M. Petrus, and C. Popa, "Laser photoacoustic spectroscopy: Principles, instrumentation, and characterization," *J. Optoelectron. Adv. Mater.*, vol. 9, no. 12, pp. 3655–3701, 2007.

[12]  A. Miklós and A. Lörincz, "Windowless resonant acoustic chamber for laser-photoacoustic applications," *Appl. Phys. B Photophysics Laser Chem.*, vol. 48, no. 3, pp. 213–218, 1989.

[13]  Z. Yu *et al.*, "Aerosol Science and Technology Differential photoacoustic spectroscopic (DPAS)-based technique for PM optical absorption measurements in the presence of light absorbing gaseous species Differential photoacoustic spectroscopic (DPAS)-based tech-nique for PM optical absorption measurements in the presence of light absorbing gaseous species," *Aerosol Sci. Technol.*, vol. 51, pp. 1438–1447, 2017.

[14]  O. E. Bonilla-Manrique, J. E. Posada-Roman, J. A. Garcia-Souto, and M. Ruiz-Llata, "Sub-ppm-level ammonia detection using photoacoustic spectroscopy with an optical microphone based on a phase interferometer," *Sensors*, vol. 19, no. 13, p. 2890, 2019.

[15]  N. A. Hall, M. Okandan, R. Littrell, B. Bicen, and F. L. Degertekin, "Micromachined optical microphone structures with low thermal-mechanical noise levels," *J. Acoust. Soc. Am.*, vol. 122, no. 4, pp. 2031–2037, 2007.

[16]  T. Kuusela and J. Kauppinen, "Photoacoustic gas analysis using interferometric cantilever microphone," *Appl. Spectrosc. Rev.*, vol. 42, no. 5, pp. 443–474, 2007.

[17]  A. Berrou, M. Raybaut, A. Godard, and M. Lefebvre, "High-resolution photoacoustic and direct absorption spectroscopy of main greenhouse gases by use of a pulsed entangled cavity doubly resonant OPO," *Appl. Phys. B Lasers Opt.*, vol. 98, no. 1, pp. 217–230, 2010.

[18]  S. D. Campbell, S. S. Yee, and M. A. Afromowitz, "Applications of photoacoustic spec-troscopy to problems in dermatology research," *IEEE Trans. Biomed. Eng.*, vol. BME-26, no. 4, pp. 220–227, 1979.

[19]  F. J. M. Harren *et al.*, "On-line laser photoacoustic detection of ethene in exhaled air as biomarker of ultraviolet radiation damage of the human skin," *Appl. Phys. Lett.*, vol. 74, no. 12, pp. 1761–1763, 1999.

[20]  C. Fischer, R. Bartlome, and M. W. Sigrist, "The potential of mid-infrared photoacoustic spectroscopy for the detection of various doping agents used by athletes," *Appl. Phys. B Lasers Opt.*, vol. 85, no. 2–3, pp. 289–294, 2006.

[21] P. Patimisco, G. Scamarcio, F. K. Tittel, and V. Spagnolo, "Quartz-enhanced photoacoustic spectroscopy: A review," *Sensors*, vol. 14, pp. 6165–6206, 2014.

[22] P. Patimisco, A. Sampaolo, L. Dong, F. K. Tittel, and V. Spagnolo, "Recent advances in quartz enhanced photoacoustic sensing," *Appl. Phys. Rev.*, vol. 5, no. 1, p. 011106, 2018.

[23] A. A. Kosterev, F. K. Tittel, D. V. Serebryakov, A. L. Malinovsky, and I. V. Morozov, "Applications of quartz tuning forks in spectroscopic gas sensing," *Rev. Sci. Instrum.*, vol. 76, no. 4, p. 043105, 2005.

[24] F. K. Tittel *et al.*, "Analysis of overtone flexural modes operation in quartz-enhanced photoacoustic spectroscopy," *Opt. Express,* Vol. 24, pp. A682–A692, 2016.

[25] G. Wysocki, A. A. Kosterev, and F. K. Tittel, "Influence of molecular relaxation dynamics on quartz-enhanced photoacoustic detection of $CO_2$ at $\lambda = 2$ μm," *Appl. Phys. B Lasers Opt.*, vol. 85, no. 2–3, pp. 301–306, 2006.

[26] L. Landau, "Theory of sound dispersion," *Phys. Zeitschrift der Sowjetunion*, vol. 10, pp. 34–43, 1936.

[27] S. Schilt, J. P. Besson, and L. Thévenaz, "Near-infrared laser photoacoustic detection of methane: The impact of molecular relaxation," *Appl. Phys. B Lasers Opt.*, vol. 82, no. 2 SPEC. ISS., pp. 319–329, 2006.

[28] P. Patimisco *et al.*, "Analysis of the electro-elastic properties of custom quartz tuning forks for optoacoustic gas sensing," *Sensors Actuators, B Chem.*, vol. 227, pp. 539–546, 2016.

[29] A. Sampaolo *et al.*, "Improved tuning fork for terahertz quartz-enhanced photoacoustic spectroscopy," *Sensors (Switzerland)*, vol. 16, no. 4, p. 439, 2016.

[30] P. Patimisco, A. Sampaolo, H. Zheng, L. Dong, F. K. Tittel, and V. Spagnolo, "Quartz–enhanced photoacoustic spectrophones exploiting custom tuning forks: A review," *Adv. Phys. X*, vol. 2, no. 1, pp. 169–187, 2017.

[31] P. Patimisco *et al.*, "Tuning forks with optimized geometries for quartz-enhanced photoacoustic spectroscopy," *Opt. Express*, vol. 27, no. 2, p. 1401, 2019.

[32] A. A. Kosterev, Y. A. Bakhirkin, and F. K. Tittel, "Ultrasensitive gas detection by quartz-enhanced photoacoustic spectroscopy in the fundamental molecular absorption bands region," *Appl. Phys. B Lasers Opt.*, vol. 80, no. 1, pp. 133–138, 2005.

[33] L. Dong, V. Spagnolo, R. Lewicki, and F. K. Tittel, "Ppb-level detection of nitric oxide using an external cavity quantum cascade laser based QEPAS sensor," *Opt. Express*, vol. 19, no. 24, p. 24037, 2011.

[34] A. Elefante *et al.*, "Environmental monitoring of methane with quartz-enhanced photoacoustic spectroscopy exploiting an electronic hygrometer to compensate the $H_2O$ influence on the sensor signal," *Sensors*, Vol. 20, 2935, 2020.

[35] P. Patimisco *et al.*, "A quartz enhanced photo-acoustic gas sensor based on a custom tuning fork and a terahertz quantum cascade laser," *Analyst*, vol. 139, no. 9, pp. 2079–2087, 2014.

[36] P. Patimisco *et al.*, "Octupole electrode pattern for tuning forks vibrating at the first overtone mode in quartz-enhanced photoacoustic spectroscopy," *Opt. Lett.*, vol. 43, no. 8, p. 1854, 2018.

[37] M. Christen, "Air and gas damping of quartz tuning forks," *Sensors and Actuators*, vol. 4, no. C, pp. 555–564, 1983.

[38] W. K. Blake, "The radiation from free-free beams in air and in water," *J. Sound Vib.*, vol. 33, no. 4, pp. 427–450, 1974.

[39] P. Patimisco *et al.*, "Loss mechanisms determining the quality factors in quartz tuning forks vibrating at the fundamental and first overtone modes," *IEEE Trans. Ultrason. Ferroelectr. Freq. Control*, vol. 65, no. 10, pp. 1951–1957, 2018.

[40] M. Giglio *et al.*, "Damping mechanisms of piezoelectric quartz tuning forks employed in photoacoustic spectroscopy for trace gas sensing," *Phys. Status Solidi Appl. Mater. Sci.*, vol. 216, no. 3, p. 1800552, 2019.

[41] S. Li *et al.*, "Ppb-level quartz-enhanced photoacoustic detection of carbon monoxide exploiting a surface grooved tuning fork," *Anal. Chem.*, vol. 91, no. 9, pp. 5834–5840, 2019.

[42] F. Sgobba *et al.*, "Quartz-enhanced photoacoustic detection of ethane in the near-IR exploiting a highly performant spectrophone," *Appl. Sci.*, vol. 10, p. 2447, 2020.

[43] Y. Ma *et al.*, "Ultra-high sensitive trace gas detection based on light-induced thermoelastic spectroscopy and a custom quartz tuning fork," *Appl. Phys. Lett.*, vol. 116, no. 1, p. 011103, 2020.

[44] B. Sun *et al.*, "Mid-infrared quartz-enhanced photoacoustic sensor for ppb-level CO detection in a SF 6 gas matrix exploiting a T-grooved quartz tuning fork," *Anal. Chem.*, vol. 92, no. 16, pp. 13922–13929, 2020.

[45] M. Relaxation and W. H. Flygare, "Molecular relaxation," *Acc. Chem. Res.*, vol. 1, no. 4, pp. 121–127, 1968.

[46] H. Zheng *et al.*, "Double antinode excited quartz-enhanced photoacoustic spectrophone," *Appl. Phys. Lett.*, vol. 110, no. 2, p. 021110, 2017.

[47] H. Hosaka, K. Itao, and S. Kuroda, "Damping characteristics of beam-shaped micro-oscillators," *Sensors Actuators, A Phys.*, vol. 49, no. 1–2, pp. 87–95, 1995.

[48] Y. Jimbo and K. Itao, "Energy loss of a cantilever vibrator," *J. Horol. Inst. Japan*, no. 47, pp. 1–15, 1968.

[49] D. M. Photiadis and J. A. Judge, "Attachment losses of high Q oscillators," *Appl. Phys. Lett.*, vol. 85, no. 3, pp. 482–484, 2004.

[50] K. Liu, X. Guo, H. Yi, W. Chen, W. Zhang, and X. Gao, "Off-beam quartz-enhanced photoacoustic spectroscopy," *Opt. Lett.*, vol. 34, no. 10, p. 1594, 2009.

[51] S. Böttger, M. Köhring, U. Willer, and W. Schade, "Off-beam quartz-enhanced photoacoustic spectroscopy with LEDs," *Appl. Phys. B Lasers Opt.*, vol. 113, no. 2, pp. 227–232, 2013.

[52] S. Dello Russo *et al.*, "Acoustic coupling between resonator tubes in quartz-enhanced photoacoustic spectrophones employing a large prong spacing tuning fork," *Sensors (Switzerland)*, vol. 19, no. 19, 2019.

[53] N. Ogawa and F. Kaneko, "Open end correction for a flanged circular tube using the diffusion process," *Eur. J. Phys.*, vol. 34, no. 5, pp. 1159–1165, 2013.

[54] F. G. Kitson, B. S. Larsen, and C. N. McEwen, *Gas Chromatography and Mass Spectrometry*. Burlington, MA: Academic Press, 1996.

[55] R. E. Kirk, D. F. Othmer, M. Grayson, and D. Eckroth, *Kirk-Othmer Concise Encyclopedia of Chemical Technology*. Chichester: Wiley, 1985.

[56] X. Giraud, N. N. Le-Dong, K. Hogben, and J. B. Martinot, "The measurement of DLNO and DLCO: A manufacturer's perspective," *Respir. Physiol. Neurobiol.*, vol. 241, pp. 36–44, 2017.

[57] T. H. Risby and F. K. Tittel, "Current status of midinfrared quantum and interband cascade lasers for clinical breath analysis," *Opt. Eng.*, vol. 49, no. 11, p. 111123, 2010.

[58] J. Faist, F. Capasso, D. L. Sivco, C. Sirtori, A. L. Hutchinson, and A. Y. Cho, "Quantum cascade laser," *Science (80-. ).*, vol. 264, no. 5158, pp. 553–556, 1994.

[59] World Meteorological Organization, *WMO Global Atmosphere Watch (GAW) Implementation Plan: 2016–2023*. 2018.

[60] P. M. Hundt *et al.*, "Multi-species trace gas sensing with dual-wavelength QCLs," *Appl. Phys. B Lasers Opt.*, vol. 124, no. 6, p. 108, 2018.

[61] "Directive 2008/50/EC of the European Parliament and of the Council of 21 May 2008 on ambient air quality and cleaner air for Europe," *Off. J. Eur. Union*, 2008.

[62] B. V. Braatz and S. Barvenik, "National greenhouse gas emission inventories in developing countries and countries with economies in transition: Global synthesis," in *Greenhouse Gas Emission Inventories: Interim Results from the U.S. Country Studies Program*, B. V Braatz, B. P. Jallow, S. Molnár, D. Murdiyarso, M. Perdomo, and J. F. Fitzgerald, Eds. Dordrecht: Springer Netherlands, 1996, pp. 1–57.

[63] C. Bauer *et al.*, "A mid-infrared QEPAS sensor device for TATP detection," *J. Phys. Conf. Ser.*, vol. 157, no. 1, 2009.

[64] H. Yi, R. Maamary, X. Gao, M. W. Sigrist, E. Fertein, and W. Chen, "Short-lived species detection of nitrous acid by external-cavity quantum cascade laser based quartz-enhanced photoacoustic absorption spectroscopy," *Appl. Phys. Lett.*, vol. 106, no. 10, p. 101109, 2015.

[65] M. Helman, H. Moser, A. Dudkowiak, and B. Lendl, "Off-beam quartz-enhanced photoacoustic spectroscopy-based sensor for hydrogen sulfide trace gas detection using a mode-hop-free external cavity quantum cascade laser," *Appl. Phys. B Lasers Opt.*, vol. 123, no. 5, p. 141, 2017.

[66] M. Giglio *et al.*, "Nitrous oxide quartz-enhanced photoacoustic detection employing a broadband distributed-feedback quantum cascade laser array," *Appl. Phys. Lett.*, vol. 113, no. 17, p. 171101, 2018.

[67] Y. Ma, R. Lewicki, M. Razeghi, and F. K. Tittel, "QEPAS based ppb-level detection of CO and $N_2O$ using a high power CW DFB-QCL," *Opt. Express*, vol. 21, no. 1, pp. 1008–1019, 2013.

[68] M. Jahjah *et al.*, "A compact QCL based methane and nitrous oxide sensor for environmental and medical applications," *Analyst*, vol. 139, no. 9, pp. 2065–2069, 2014.

[69] C. Shi *et al.*, "A mid-infrared fiber-coupled QEPAS nitric oxide sensor for real-time engine exhaust monitoring," *IEEE Sens. J.*, vol. 17, no. 22, pp. 7418–7424, 2017.

[70] K. H. Kim, S. A. Jahan, and E. Kabir, "A review of breath analysis for diagnosis of human health," *TrAC – Trends in Analytical Chemistry*, vol. 33. pp. 1–8, 01, 2012.

[71] C. Wang and P. Sahay, "Breath analysis using laser spectroscopic techniques: Breath biomarkers, spectral fingerprints, and detection limits," *Sensors*, vol. 9, no. 10, pp. 8230–8262, 2009.

[72] S. Davies, P. Spanel, and D. Smith, "Quantitative analysis of ammonia on the breath of patients in end-stage renal failure," *Kidney Int.*, vol. 52, no. 1, pp. 223–228, 1997.

[73] C. N. Tassopoulos, D. Barnett, and T. R. Fraser, "Breath-acetone and blood-sugar measurements in diabetes.," *Lancet*, vol. 293, no. 7609, pp. 1282–1286, 1969.

[74] M. R. McCurdy, A. Sharafkhaneh, H. Abdel-Monem, J. Rojo, and F. K. Tittel, "Exhaled nitric oxide parameters and functional capacity in chronic obstructive pulmonary disease," *J. Breath Res.*, vol. 5, no. 1, p. 016003, 2011.

[75] H. Okuyama, M. Yonetani, Y. Uetani, and H. Nakamura, "End-tidal carbon monoxide is predictive for neonatal non-hemolytic hyperbilirubinemia," *Pediatr. Int.*, vol. 43, no. 4, pp. 329–333, 2001.

[76] A. Amann, P. Spanel, and D. Smith, "Breath analysis: the approach towards clinical applications," *Mini-Reviews Med. Chem.*, vol. 7, no. 2, pp. 115–129, 2007.

[77] M. Giglio *et al.*, "Quartz-enhanced photoacoustic sensor for ethylene detection implementing optimized custom tuning fork-based spectrophone," *Opt. Express*, vol. 27, no. 4, pp. 4271–4280, Feb. 2019.

[78] J. P. Waclawek, H. Moser, and B. Lendl, "Compact quantum cascade laser based quartz-enhanced photoacoustic spectroscopy sensor system for detection of carbon disulfide," *Opt. Express*, vol. 24, no. 6, p. 6559, 2016.

[79] R. Lewicki, L. Dong, Y. Ma, and F. K. Tittel, "A compact CW quantum cascade laser based QEPAS sensor for sensitive detection of nitric oxide," in *CLEO: Science and Innovations, CLEO_SI 2012*, May 2012, p. CW3B.4.

[80] S. F. Solga *et al.*, "Factors influencing breath ammonia determination," *J. Breath Res.*, vol. 7, no. 3, 2013.

[81] L. A. Spacek, M. Mudalel, F. Tittel, T. H. Risby, and S. F. Solga, "Clinical utility of breath ammonia for evaluation of ammonia physiology in healthy and cirrhotic adults," *J. Breath Res.*, vol. 9, no. 4, p. 047109, 2015.

[82] N. Maurin *et al.*, "First clinical evaluation of a quartz enhanced photo-acoustic CO sensor for human breath analysis," *Sensors Actuators, B Chem.*, vol. 319, p. 128247, 2020.

[83] L. Li, K. Yang, X. Bian, Q. Liu, Y. Yang, and F. Ma, "A gas leakage localization method based on a virtual ultrasonic sensor array," *Sensors (Switzerland)*, vol. 19, no. 14, 2019.

[84] O. M. Aamo, "Leak detection, size estimation and localization in pipe flows," *IEEE Trans. Automat. Contr.*, vol. 61, no. 1, pp. 246–251, 2016.

[85] Q. He, A. Li, Y. Zhang, S. Liu, Y. Guo, and L. Kong, "A study on mechanical and tribological properties of silicone rubber reinforced with white carbon black," *Tribol. – Mater. Surfaces Interfaces*, vol. 12, no. 1, pp. 9–16, 2018.

[86] A. Sampaolo *et al.*, "Highly sensitive gas leak detector based on a quartz-enhanced photoacoustic SF6 sensor," *Opt. Express*, vol. 24, no. 14, p. 15872, 2016.

[87] M. Maiss and C. A. M. Brenninkmeijer, "Atmospheric SF6: Trends, sources, and prospects," *Environ. Sci. Technol.*, vol. 32, no. 20, pp. 3077–3086, 1998.

[88] A. Pregelj, M. Drab, and M. Mozetic, "Leak detection methods and defining the sizes of leaks," *Proc. 4th International Conference of Slovenian Society for Nondestructive Testing*, Ljubljana, Slovenia, 1997.

[89] A. Elefante *et al.*, "Dual-gas quartz-enhanced photoacoustic sensor for simultaneous detection of methane/nitrous oxide and water vapor," *Anal. Chem.*, vol. 91, no. 20, pp. 12866–12873, 2019.

[90] A. Sampaolo *et al.*, "Quartz-enhanced photoacoustic spectroscopy for hydrocarbon trace gas detection and petroleum exploration," *Fuel*, vol. 277, p. 118118, 2020.

[91] Z. Wang, J. Geng, and W. Ren, "Quartz-enhanced photoacoustic spectroscopy (QEPAS) detection of the $\nu 7$ band of ethylene at low pressure with $CO_2$ interference analysis," *Appl. Spectrosc.*, vol. 71, no. 8, pp. 1834–1841, 2017.

[92] A. Zifarelli *et al.*, "Partial least-squares regression as a tool to retrieve gas concentrations in mixtures detected using quartz-enhanced photoacoustic spectroscopy," *Anal. Chem.*, vol. 92, no. 16, 2020.

[93] S. Zhou *et al.*, "External cavity quantum cascade laser-based QEPAS for chlorodifluoromethane spectroscopy and sensing," *Appl. Phys. B Lasers Opt.*, vol. 125, no. 7, p. 125, 2019.

[94] R. Lewicki *et al.*, "QEPAS based detection of broadband absorbing molecules using a widely tunable, cw quantum cascade laser at 8.4 μm," *Opt. Express*, Vol. 15, 7357–7366, 2007.

[95]　M. Giglio *et al.*, "Broadband detection of methane and nitrous oxide using a distributed-feedback quantum cascade laser array and quartz-enhanced photoacoustic sensing," *Photoacoustics*, vol. 17, p. 100159, 2020.

[96]　S. Wold, A. Ruhe, H. Wold, and W. J. Dunn, III, "The collinearity problem in linear regression: The partial least squares (PLS) approach to generalized inverses," *SIAM J. Sci. Stat. Comput.*, vol. 5, no. 3, pp. 735–743, 1984.

# 14 Multiheterodyne Spectroscopic Sensing and Applications of Mid-Infrared and Terahertz Quantum Cascade Laser Combs

Gerard Wysocki, Jonas Westberg, and Lukasz Sterczewski

Princeton University, New Jersey

## 14.1 Introduction

Laser spectroscopy has been increasingly the technology of choice in many applications that require non-destructive quantification of chemical compounds with high selectivity and high sensitivity. The mid-infrared (mid-IR) spectral region (wavelengths between $3\,\mu m$ and $16\,\mu m$, or in wavenumbers between $\sim 3300\,cm^{-1}$ and $600\,cm^{-1}$) is of particular interest because it gives access to the fundamental rovibrational bands of many molecules. This region is sometimes referred to as the molecular-fingerprint region, which is particularly suitable for sensitive chemical detection. At longer wavelengths, extending to the so-called THz range (1–10 THz or in wavenumber scale between $\sim 30\,cm^{-1}$ and $\sim 300\,cm^{-1}$), one can also perform chemical identification by accessing the molecular rotational bands and lattice vibrations in solid state samples.

One particular subset of chemical sensing applications focuses on detection, identification, and quantification of molecular samples in the gas phase. Given the narrow absorption lines of gases under atmospheric- and reduced-pressure conditions, gas-sensing applications have stringent requirements in terms of spectral resolution, which from the very early days of lasers particularly favored laser spectroscopy [1] over other spectrometer technologies that utilize broadband light sources and wavelength-selective detection systems based on conventional grating spectrometers [2], or Fourier-transform spectrometers [3]. Despite the spectral resolution advantage, historically, lasers have been lacking in terms of spectral coverage, which has resulted in difficulties in detection of large molecules with broad ro-vibrational spectra that require broadband optical-frequency coverage ($>50\,cm^{-1}$) in the infrared. Over the last 50 years, the steady innovation in infrared laser technologies has led to a significant revolution in laser spectrometer design, transitioning from early gas-laser systems (e.g., $CO_2$ or CO lasers) that could be tuned over a set of discrete frequencies defined by the molecular transitions of the gain medium [4], through solid- and liquid-state gain media, such as color-center lasers, that enabled continuous broadband tuning in the infrared at the expense of difficult operation and maintenance [5], to more recent continuously tunable semiconductor-laser technologies such as quantum cascade lasers (QCLs) [6] and interband cascade lasers (ICLs) [7].

Similarly to the visible and near-infrared semiconductor diode lasers that revolutionized areas of optical communications and everyday consumer electronics (e.g., DVD players, barcode scanners, and laser pointers), QCLs and ICLs have already shown to have a transformational impact on the field of mid-IR spectroscopy and chemical sensing [8]. These compact, electrically pumped semiconductor-laser sources provide direct emission of infrared radiation that is tunable in frequency by simply varying the injection current or device temperature. When used in external cavity (EC) configurations these semiconductor gain media can provide single-mode EC-QCLs and EC-ICLs tunable over a relatively broad >400 cm$^{-1}$ frequency range [9], which proficiently addresses the need for broadband high-resolution spectroscopy in laboratory environments [10]. The primary reason that EC lasers are less common outside well-controlled laboratory settings originates from their rather complex opto-mechanical construction, which makes them vibration-sensitive. The construction of robust and transportable spectroscopic systems based on EC-laser technology is a challenging engineering task. Therefore, there is a clear need for infrared semiconductor-laser technologies that can support the development of compact, high-resolution, and all-electronically controlled spectrometers with no optomechanical moving parts, simultaneously providing broadband optical coverage and narrow-linewidth emission.

Among all modern laser technologies, optical frequency comb (OFC) sources have emerged as the most promising for high-resolution spectrometers with broadband spectral coverage. The metrological application of OFC sources providing efficient frequency synchronization between the microwave standards and the optical domain was recognized with the Nobel Prize in Physics awarded to John L. Hall and Theodor W. Hänsch in 2005. The functional extension of this technology into the mid-IR [11] and THz [12] spectral regions required development of new gain media [13], leveraging various nonlinear optical processes such as difference-frequency generation [14], optical rectification [15], and four-wave mixing [16], or utilizing photoconductive generation [17]. While these technologies deliver very impressive results in terms of spectral coverage, with super-octave coverage [15], or output powers at multi-watt levels [13], their reliance on femtosecond laser pumps or continuous-wave (CW) optical-parametric oscillator pumps makes them less likely to serve as building blocks for compact, high-resolution spectrometers with monolithic/hybrid-integration potential. Therefore, recent advancements in the area of all-electrically pumped semiconductor-laser frequency combs operating in the mid-IR [18] and THz [19, 20] spectral regions represent an important step towards field applications with truly integrated and scalable frequency combs.

In this chapter we provide an overview of the spectroscopic capabilities provided by QCL and ICL frequency-comb laser sources. The multiheterodyne detection methodology patented in 1998 by Keilmann *et al.* for microwave sources [21], and later adopted at optical frequencies for dual-comb spectroscopy (DCS) [22, 23], has been used successfully with QCL/ICL-based DCS spectrometers. These DCS spectrometers offer fast chemical sensing without the need for optomechanical tuning or focal plane array detectors [24, 25]. The DCS technique offers efficient down-conversion of optical signals to the radio frequency (rf) domain through a multiheterodyne process that allows signal acquisition and processing with conventional rf electronics. When combined with intrinsically compact semiconductor frequency combs, DCS enables simple and compact broadband spectrometers that give access to the fundamental molecular absorption bands in the mid-IR while providing high optical resolution (<15 MHz, free-running). In the following sections we provide an overview of the DCS technique and present measurement approaches and experimental implementations of DCS for mid-IR and THz spectroscopy of chemicals using QCL and ICL frequency combs.

## 14.2    Multiheterodyne Detection

The main strength of dual-comb spectroscopy originates from the simultaneous down-conversion of the optical modes of a frequency comb to the radio-frequency domain [22]. This is obtained through a multi-parallel optical-heterodyne process, commonly referred to as multiheterodyne detection. The process makes use of the electrical non-linearity of a photodetector, which results in an output voltage proportional to the optical intensity rather than the electric field (commonly referred to as a square-law detection), thus enabling optical heterodyning. For convenience, we will consider a simple optical heterodyning process, where two electric fields with the same polarization direction, $E_{sig}$ and $E_{LO}$, oscillating at angular optical frequencies: $\omega_{LO} = 2\pi f_{LO}$ and $\omega_{sig} = 2\pi f_{sig}$ are simultaneously overlaid on a photodetector. Assuming the two fields differ in phase by $\varphi$, the intensity-proportional output of the photodetector $V_{PD}$ can be written as

$$
\begin{aligned}
V_{PD} \propto I \propto \ & \left[ E_{sig} \cos \left( \omega_{sig} t + \varphi \right) + E_{LO} \cos \left( \omega_{LO} t \right) \right]^2 \\
\propto \ & \frac{E_{sig}^2}{2} + \frac{E_{LO}^2}{2} + \frac{E_{sig}^2}{2} \cos \left( 2\omega_{sig} t + 2\varphi \right) + \frac{E_{LO}^2}{2} \cos \left( 2\omega_{LO} t \right) \\
& + 2 E_{sig} E_{LO} \cos \left( \omega_{sig} t + \varphi \right) \cos \left( \omega_{LO} t \right) .
\end{aligned}
\tag{14.1}
$$

As expected, the photodetector response is a constant DC term proportional to the sum of the intensities of the local oscillator (LO) and the signal, but more importantly, numerous frequency-mixing terms appear. The two terms following the DC terms are the second-harmonic products at optical frequencies far beyond the electrical bandwidth of the photodetector. Of larger interest is the last term, where the product-to-sum formula for cosine functions can be used:

$$2 \cos\left(\omega_{\text{sig}}t + \varphi\right) \cos\left(\omega_{\text{LO}}t\right) = \underbrace{\cos\left(\left(\omega_{\text{LO}} + \omega_{\text{sig}}\right)t + \varphi\right)}_{\text{sum frequency}} + \underbrace{\cos\left(\left(\omega_{\text{sig}} - \omega_{\text{LO}}\right)t + \varphi\right)}_{\text{difference frequency}}.$$

$$(14.2)$$

The sum frequency usually lies beyond the electrical bandwidth of the photodetector, while the difference frequency term, which oscillates at significantly lower frequencies than the optical carrier, performs down-conversion of both the amplitude and the phase of the optical fields to the rf domain. Consequently, the bandwidth-limited AC response of the photodetector can be written as

$$V_{\text{AC}} \propto E_{\text{sig}}E_{\text{LO}} \cos\left(\left(\omega_{\text{sig}} - \omega_{\text{LO}}\right)t + \varphi\right). \tag{14.3}$$

Now, let us expand this process to multiple frequency components, also known as multiheterodyne beating. The optical frequency of each comb tooth (mode) can be described using two parameters: the optical offset frequency $f_0$ and the optical repetition frequency $f_{\text{rep}}$, which implies that the optical frequency of the $n$th comb line can be expressed as

$$f_n = f_0 + n f_{\text{rep}}. \tag{14.4}$$

The electric field of such a comb consisting of $N$ lines with electric-field amplitudes $A_n$ and phases $\varphi_n = \varphi(f_n)$ is given by

$$E_{\text{comb}}\left(t\right) = \text{Re}\left\{\sum_{n=1}^{N} A_n e^{i\varphi_n} e^{i2\pi\left(f_0 + n f_{\text{rep}}\right)t}\right\}. \tag{14.5}$$

The multiheterodyne process requires a mismatch of the repetition rate of the signal comb with respect to the LO comb, i.e., $f_{\text{rep,LO}} = f_{\text{rep,sig}} - \Delta f_{\text{rep}}$. If the two sources are combined on a photodetector, after neglecting the DC, second-harmonic, and sum-frequency terms, we arrive at

$$V_{\text{DCS}}\left(t\right) \propto \text{Re}\left\{\sum_{n=1}^{N} A_n B_n e^{i\varphi_n} e^{i2\pi\left(\Delta f_0 + n\Delta f_{\text{rep}}\right)t}\right\}, \tag{14.6}$$

where $B_n$ denotes the amplitude of the $n$th LO-comb tooth. The down-converted spectrum is essentially a radio-frequency comb with the mixing products appearing at frequencies equal to multiples of the repetition rate difference, $\Delta f_{\text{rep}}$, offset by the difference in carrier-envelope offset frequencies, $\Delta f_0$. The $n$th rf-comb tooth appears at a frequency

$$f_{\text{rf},n} = \Delta f_0 + n\Delta f_{\text{rep}}. \tag{14.7}$$

A simplified illustration of the process is shown in Fig. 14.1. Note that, by convention, the repetition rate of the signal comb is greater than that of the LO.

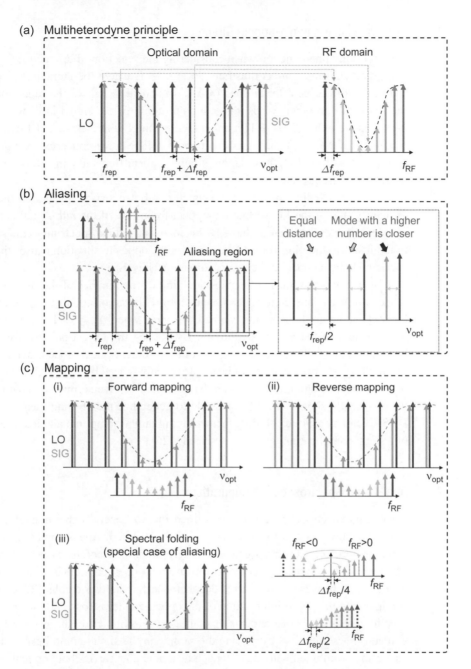

**Fig. 14.1** (a) Schematic of the optical down-conversion process in dual-comb spectroscopy. Two optical frequency combs with different repetition rates are combined on a fast photodetector, which maps the optical spectrum to the rf domain, creating a one-to-one link between the two domains. (b) Aliasing in DCS occurs when cumulative repetition-frequency difference after $m$ lines ($m\Delta f_{rep}$) is equal to half of the repetition rate of the LO comb. Above the mode beating at equal frequencies the mapping direction flips. Therefore, the rf beat notes mapping higher optical frequencies overlap with the lower frequency ones. (c) Illustrations of the forward (i) and reverse (ii) mapping of the optical modes to the rf domain. Spectral folding for doubling the number of down-converted modes within a given bandwidth is shown in (iii).

## 14.2.1   Spectral Aliasing and Mapping Direction

An interesting phenomenon when considering the selection of $\Delta f_0$ and $\Delta f_{rep}$ in DCS is spectral aliasing. This is illustrated in Fig. 14.1, where the cumulative repetition-frequency difference after $m$ lines $(m\Delta f_{rep})$ is equal to half of the repetition rate of the LO comb. In other words, the $m$th comb tooth of the signal laser will optically beat with the $m$th and $(m+1)$th comb teeth of the LO at the same rf frequency. For $n > m+1$ (if $\Delta f_0 \approx 0$), the signal comb tooth will be down-converted by higher-order comb modes of the LO, thus mapping higher optical frequency in reverse order with decreasing rf frequencies.

In DCS, the optical modes of the signal comb can also be placed on the higher-frequency side of the LO so that the optical frequency of the $m$th signal comb tooth is always greater than that of the LO. This maps the optical spectrum to the rf domain in the forward direction (Fig. 14.1), whereas the opposite situation causes the optical spectrum to be mapped in the reverse direction (Fig. 14.1).

A special case of aliasing enables doubling of the number of the mapped optical modes within a given photodetector bandwidth. This was originally proposed by Schiller [22] and its implementation can be considered when $\Delta f_{rep} \gg \Gamma_{rf,n}$, where $\Gamma_{rf,n}$ is the linewidth of the $n$th rf-comb tooth. By placing the lowest positive frequency rf beat note at $\Delta f_{rep}/4$, a reverse-mapping of every other optical mode is achieved. As a result, the rf spectrum consists of beat notes evenly spaced by $\Delta f_{rep}/2$, with half being forward-mapped and interlaced with the other half of reverse-mapped beat notes. As it was recently shown, it is also possible to generalize the aliasing idea to compress DCS spectra by arbitrary folding factors during digital sampling, albeit at the expense of a slight complication on the analysis side [26].

## 14.2.2   Dual-Comb Spectroscopy Configurations

Dual-comb spectroscopy can be performed in two general experimental configurations, as shown in the top two panels of Fig. 14.2. In the symmetric (collinear) configuration, the LO and signal combs are combined before passing through the sample, which is typically preferred for turbulent open-path environments or stand-off detection because it minimizes decoherence for the two combs [27, 28]. In this configuration the phase information of the optical modes is lost, and thus it cannot be used for dispersion measurements. In contrast, the asymmetric (dispersive) configuration adds phase retrieval by placing the sample in the signal-comb beam path before the beam is combined with the LO. Each beat note now carries both optical amplitude and phase information, which encode spectroscopic information. As shown in the bottom two panels of Fig. 14.2, the aforementioned configurations can also be combined with balanced detection that can further improve system performance by suppressing common mode amplitude noise. The merit of using balanced detection must be individually assessed for each DCS spectrometer, where specific system parameters, such as noise-equivalent power (NEP) of the photodetectors together with beat note stability, need to be carefully considered. In general, systems based on QCLs or ICLs

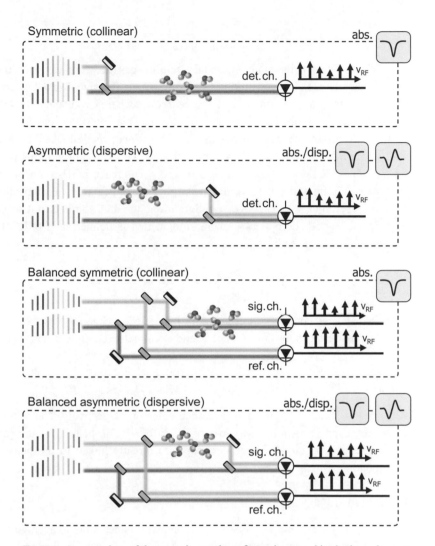

**Fig. 14.2** An overview of the experimental configurations used in dual-comb spectroscopy.

exhibit beat note instabilities that warrant the added complexity of a second detector, but more stable systems may, on the contrary, suffer from the added detector noise.

It is interesting to note how the DCS configurations differ in the retrieval of the optical-transmission spectrum from the beat note amplitudes. First, one can consider the digitized photodetector AC voltage response, $V_{AC,n}$, for a given beat note, $n$, which is proportional to the product of the electric fields of the signal and LO-comb teeth,

$$V_{AC,n} \propto E_{sig,n} \cdot E_{LO,n}. \tag{14.8}$$

Since the optical transmission is defined as the ratio of the transmitted and initial intensities, and the intensity is proportional to the square of the electric field, the expression for the optical transmission in the symmetric (collinear) configuration can be written as

$$T_n^{(s)} \propto \frac{\left[E_{\text{sig},n} \cdot E_{\text{LO},n}\right]_{\text{G}}}{\left[E_{\text{sig},n} \cdot E_{\text{LO},n}\right]_{\text{Z}}} = \frac{\left[V_{\text{AC},n}\right]_{\text{G}}}{\left[V_{\text{AC},n}\right]_{\text{Z}}}, \tag{14.9}$$

where G and Z denote the sample-gas and reference-gas (i.e., zero-gas) measurements, respectively. This expression implicitly assumes that the signal and LO combs are affected similarly by the sample, i.e., $\delta\nu_n > n\Delta f_{\text{rep}}$, where $\delta\nu_n$ is the linewidth of the absorption feature probed by signal comb mode $n$. This criterion is generally fulfilled for pressure-broadened gas-sensing using conventional mode-locked DCS systems [28], where $n\Delta f_{\text{rep}}$ is typically less than 100 MHz. However, for systems with larger $\Delta f_{\text{rep}}$, such as QCL- or ICL-based systems, $n\Delta f_{\text{rep}}$ can reach 1 GHz, in which case the asymmetric configuration is preferred to simplify the optical transmission calculation.

In the asymmetric (dispersive) configuration, the LO comb is not affected by the sample. Hence, the optical transmission, which is defined as a ratio of optical intensities, will be given by

$$T_n^{(a)} \propto \frac{\left[E_{\text{sig},n} \cdot E_{\text{LO},n}\right]_{\text{G}}^2}{\left[E_{\text{sig},n} \cdot E_{\text{LO},n}\right]_{\text{Z}}^2} = \frac{\left[V_{\text{AC},n}\right]_{\text{G}}^2}{\left[V_{\text{AC},n}\right]_{\text{Z}}^2}. \tag{14.10}$$

With the balanced configurations, an additional detector channel is introduced, which leads to transmissions for the symmetric and asymmetric configurations according to

$$T_n^{(bs)} \propto \frac{\left[V_{AC,n}\right]_{\text{GS}} / \left[V_{AC,n}\right]_{\text{GR}}}{\left[V_{AC,n}\right]_{\text{ZS}} / \left[V_{AC,n}\right]_{\text{ZR}}}, \qquad T_n^{(ba)} \propto \frac{\left[V_{AC,n}\right]_{\text{GS}}^2 / \left[V_{AC,n}\right]_{\text{GR}}^2}{\left[V_{AC,n}\right]_{\text{ZS}}^2 / \left[V_{AC,n}\right]_{\text{ZR}}^2}, \tag{14.11}$$

where S and R denote signal and reference detector, respectively. The optical phase can be retrieved in a similar way, where the difference in phases for the same beat notes measured by the two photodetectors are directly linked to the phases of the corresponding optical modes.

## 14.2.3    Coherent Averaging

To reduce the uncertainty in DCS measurements, it is desirable to average the repetitive interferogram signal over extended time scales. This is motivated by the fact that the dominating short-term sources of noise have a random-like character. Consequently, the value of the measured signal should be analyzed in statistical terms: the mean, variance, and other higher-order moments. If the underlying random process is weakly stationary, i.e., the mean value and the autocorrelation function do not vary with time, one can estimate the population mean (true noise-free intensity value) and the uncertainty can be lowered by performing multiple independent measurements. In other words, by averaging multiple interferograms arising from the optical beating between the two combs, one can retrieve the absorption and dispersion spectra of the probed specimen more accurately than that retrieved from a single interferogram. In the case of a Gaussian character of the fluctuations, the measurement uncertainty is understood as the standard deviation of the sample mean, which decreases with the square root of acquisition time. Unfortunately, in multiheterodyne spectroscopy the stationarity assumption is rarely fulfilled due to the presence of multiplicative phase noise, which necessitates alignment of the interferograms before averaging, so-called coherent averaging, either by very tight frequency stabilization of the combs [29], adaptive signal acquisition hardware [30], or by computational means [31–33].

A perfectly stable (noise-free) down-converted radio frequency comb with $N$ lines appears in the voltage on the photodetector as

$$V_{\text{DCS}}(t) \propto \text{Re} \left\{ e^{i2\pi \Delta f_0 t} \sum_{n=1}^{N} C_n e^{i\varphi_n} e^{i2\pi n \Delta f_{\text{rep}} t} \right\}. \tag{14.12}$$

For easier analysis, it is convenient to obtain this signal in a complex form using a quadrature demodulator or a Hilbert transform. We will also ignore any sum-frequency components in the signal, thus leaving only the signal of interest:

$$\tilde{V}_{\text{DCS}}(t) = e^{i2\pi \Delta f_0 t} \sum_{n=1}^{N} C_n e^{i\varphi_n} e^{i2\pi n \Delta f_{\text{rep}} t}. \tag{14.13}$$

Phase noise causes the beat notes in the radio-frequency spectrum to fluctuate in frequency. Generally, there are two kinds of radio-frequency instabilities that occur simultaneously: fluctuations of the repetition rate difference $\Delta f_{\text{rep}}$, and the offset frequency difference $\Delta f_0$. Since both now become a function of time, $\Delta f_0 = \Delta f_0(t)$, $\Delta f_{\text{rep}} = \Delta f_{\text{rep}}(t)$, the equation above takes the form

$$\tilde{V}_{\text{DCS}}^{(\text{VAR})}(t) = e^{i\left(2\pi \int_0^t \Delta f_0(\tau) d\tau + \varphi_0\right)} \sum_{n=1}^{N} C_n e^{i\Delta\varphi_n} e^{i2\pi n \int_0^t \Delta f_{\text{rep}}(\tau) d\tau}. \tag{14.14}$$

The first term including $\Delta f_0(t)$ shifts all multiheterodyne beat notes equally, while the second describes the effect of $\Delta f_{\text{rep}}(t)$, which causes the spectrum to stretch and compress by introducing a mode-number-dependent shift (resembling "breathing"). In the time domain, a variable rf repetition rate $\Delta f_{\text{rep}}(t)$ changes the duration of interferogram frames, while a time-varying rf offset $\Delta f_0(t)$ modifies the shape of the individual frames. Intuitively, the stringent requirement of stationary signal fails in such conditions, and coherent averaging of such a non-stationary signal without any signal manipulation is impossible. Any attempt to average a non-stationary multiheterodyne signal over extended time scales will yield a spectrum with dispersed beat notes, which, in extreme cases, can become unresolvable.

There are two ways to address this issue in semiconductor frequency-comb based DCS. The first involves an active stabilization of the radio-frequency spectrum beat notes using hardware frequency and phase-locked loops that instantaneously modulate the injection current of one or both lasers. This has the major advantage of being real-time without employing fast digitizers and dedicated computational techniques, but it cannot correct for both $\Delta f_{\text{rep}}(t)$ and $\Delta f_0(t)$ simultaneously because direct current modulation affects both parameters. Moreover, hardware stabilization techniques are limited in bandwidth by the slowest servo loop in the system. The second approach, acts directly on the digitized photodetector signal through computational means, which works with free-running lasers, corrects for both parameters, and does not show bandwidth limitations.

**14.2.3.1    Active Stabilization**

One of the first active stabilization schemes facilitating prolonged averaging in a QCL-based dual-comb setup was employed in 2014 by Villares *et al.* [34]. The authors isolated a strong low-frequency RF beat note using a band-pass filter and supplied it to a radio-frequency discriminator, which was essentially a linear slope converting frequency deviation into amplitude variations. In this configuration, often referred to as a frequency locked loop (FLL), the error signal from the discriminator was further fed into a slow (5 kHz bandwidth) proportional-integral-derivative (PID) controller acting on an injection current of one of the lasers. The inter-locking system stabilized the two lasers mutually, thus yielding an improvement in the stability of the radio-frequency spectrum; however, in the optical frequency domain both lasers continued to fluctuate. In other words, one of the QCL combs was made to follow the other.

A similar approach was developed by Hangauer *et al.* [35], where a frequency discriminator based on a logarithmic amplifier together with the frequency-dependent characteristics of an RC filter (later replaced by a high-order Butterworth low-pass filter in [36]) was used. This resulted in a power-independent error signal that could be supplied to a fast (100 kHz bandwidth) analog PID controller to stabilize the RF beat notes. Later, Westberg *et al.* [37] combined frequency-discriminator locking with a high bandwidth optical phase-locked loop (OPLL), schematically shown in Fig. 14.3, which resulted in a narrowing of the RF beat note linewidths accompanied by an increase in beat note amplitudes and effective improvement in the spectrometer's signal-to-noise ratio (SNR).

The OPLL was based on the Toptica mFalc 100 module, which comprises the two critical elements of the loop: the phase detector and the loop filter. A stable 40 MHz phase reference signal was provided from an external function generator. The last block of the OPLL is the QCL itself acting as a voltage-controlled (via injection current) oscillator. A change in the voltage and the laser driver modulation input causes variation in laser injection current that yields a locally proportional change in optical frequency of the emitted frequency comb. This OPLL arrangement allowed for continuous stabilization of the optical frequency difference between the signal and the LO combs resulting in narrower beat notes.

The performance of this hybrid locking system was characterized using an RF spectrum analyzer. The center position of the locked beat note, recorded over 600 s (Fig. 14.3(c)), becomes considerably more stable when the fastest loop is enabled. This can be observed in the Allan deviation plot (Figure 14.3(d)), where a $1\sigma$ deviation of the OPLL in combination with the FLL reaches 400 Hz for 10 s integration time, which is almost a two orders of magnitude improvement compared to the FLL alone. The fastest loop also provides improvements in 3 dB line widths of the RF beat notes, which get narrowed down to 11–220 kHz (with the lower values in the proximity of the locked beat note). The reduction in line widths is accompanied by up to 20 dB improvement in the signal-to-noise ratio (Fig. 14.3(e)) for the narrowest beat notes with the efficacy degrading for beat notes further away from the locked beat

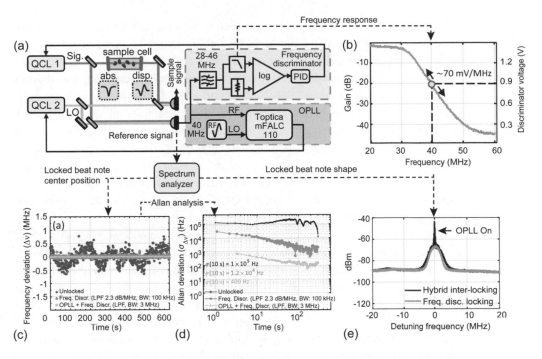

**Fig. 14.3** (a) Schematic of the hybrid locking system [37]. The RF signal from the photodetector feeds the frequency discriminator to pre-stabilize the system with FLL, and the OPLL is used in addition to FLL to further improve the quality of the lock. (b) Frequency characteristics of the frequency discriminator. (c) Center position of the beat note used for locking (d) Allan deviation of the instantaneous frequency of the beat note used for frequency locking measured in three modes of operation (unlocked, FLL, and FLL+OPLL). (e) Frequency spectrum of the beat note with the OPLL enabled and disabled. Adapted from *Applied Physics Letters*, vol. 110, p. 141108, 2017, with the permission of AIP Publishing.

note. It should be noted that this locking scheme still only provided mutual stability, i.e., locking one of the QCL combs to its free-running partner and absolute frequency calibration was performed using a known molecular spectrum.

### 14.2.3.2 Coherent Averaging with Free-Running Lasers

The main disadvantage of the active stabilization in QCL-based multiheterodyne spectrometers is the difficulty of addressing frequency fluctuations of both $\Delta f_{rep}$ and $\Delta f_0$, simultaneously. In the case of conventional injection current control, RF beat notes far away from the stabilized one will suffer from a gradual broadening due to the cumulative effect of the non-stationary $\Delta f_{rep}(t)$. For example, a typical intermode beat note linewidth, which indicates the stability of $f_{rep}(t)$ in a high-performance QCL comb, is on the order of kilohertz [38]. Modern QCL combs develop a spectrum of more than 100 lines [39], which implies that the multiheterodyne beat notes located at frequencies far away from the locked line would show close to megahertz linewidths even if the locking loop could perfectly correct for $\Delta f_0(t)$. In addition, the tunings of $f_0(t)$

and $f_{rep}$ $(t)$ in a QCL are not orthogonal, which is not an issue when using a computational method, either through a real-time DSP platform or post-processing routine, if the instantaneous frequencies of the two parameters can be extracted.

One of the first demonstrations of computational multiheterodyne spectroscopy with QCLs was performed in 2016 by Burghoff et al. [32], who used an extended Kalman filter to iteratively predict the time-varying phases and amplitudes of RF comb teeth detected using a superconducting hot-electron bolometer and a Schottky mixer pumped by two terahertz QCL combs. The algorithm effectively compensated for both time-varying parameters caused by severe vibrations of the pulsed tube cryostat housing the QCLs at 37 K. The procedure enabled the first demonstration of multiheterodyne spectroscopy with terahertz QCL combs [40], where the transmission spectrum of a submillimeter-thick GaAs wafer was measured.

While effective, the Kalman filter shows increasing computational cost with an increasing number of comb teeth, which limits its real-time application to sources with up to a couple of dozens of lines. Instead of the time-domain solution of the Kalman filter, a frequency-domain approach inspired by rf electronics can be used. This was demonstrated by Sterczewski et al. [41], who digitally filtered two arbitrary beat notes and mixed them to retrieve the instantaneous rf repetition rate difference, followed by tracking of an arbitrary beat note for global rf offset correction. In principle, this procedure can be implemented purely in hardware, where the repetition rate difference adaptively adjusts the sampling intervals, known as adaptive sampling [42], and the estimated frequency offset difference is supplied to an rf phase shifter in counterphase.

This basic methodology was further extended by adding a nonlinear operation to obtain pure repetition rate harmonics through digital difference-frequency generation (DDFG), and introducing a fast-frequency tracker to deal with frequency drifts beyond the rf line spacing. This methodology, referred to as CoCoA (computational coherent averaging) [33], has shown to be compatible with a wide range of phase-noisy comb sources including fiber lasers with tens of thousands of teeth, ICLs, and QCLs.

It should be noted that the main difficulty with multiheterodyne phase correction algorithms lies in the retrieval of the correction signals rather than the manipulation of the signal phase itself. It is for instance possible to perform coherent averaging with free-running sources without the computational extraction of $\Delta f_0(t)$ or $\Delta f_{rep}$ $(t)$, where the instantaneous repetition rates of the combs $f_{rep,1}$ $(t)$ and $f_{rep,2}$ $(t)$ are extracted electrically using microwave bias tees. After external electrical mixing, the repetition rate difference may be used to adaptively trigger the acquisition, similar to approaches used for fiber lasers [30] and successfully exported to QCL combs by Faist et al. [43].

The retrieval of the instantaneous offset frequency $f_0(t)$ is more challenging. There have been considerable efforts made to facilitate the use of the well-known $f$–$2f$ self-referencing scheme that employs optical beating between the frequency doubled lower end of the comb spectrum and the higher end to obtain an electrical signal located at $f_0$. The central requirement is that the optical spectrum of the comb covers an octave, rarely seen in present-day QCL combs. Consequently, a different scheme, utilizing an additional CW reference in the form of a single-mode laser must be used for absolute calibration of the frequency. This procedure may be omitted in harmonic frequency

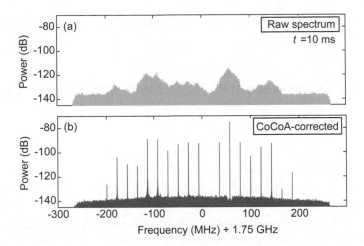

**Fig. 14.4** Results of the CoCoA algorithm in application to an extremely unstable THz multiheterodyne spectrum. (a) Raw uncorrected spectrum, and (b) after computational correction. Adapted with permission from [33] © The Optical Society.

combs [20] generated through difference-frequency generation (DFG) processes [44], which by nature are frequency offset-free.

No matter if the correction signals are extracted computationally or using external optoelectronics, the correction procedure follows the same steps. First, the variable repetition-rate difference is corrected to ensure an equal duration of the interferogram frames. This can be obtained either in hardware by synthesizing an adaptive clock signal compatible with the acquisition device [30], or in software by adaptive resampling [33], which is a linear interpolation on a nonlinear time grid. The time axis for resampling is given by

$$t'(t) = \int_0^t \frac{k\langle \Delta f_{rep}\rangle}{k\Delta f_{rep}(\tau)}d\tau, \tag{14.15}$$

where $\langle \Delta f_{rep}\rangle$ denotes the expected value of the repetition rate and $k$ is the number of the repetition rate difference harmonic extracted. As a result of this operation the signal takes the following form:

$$\tilde{V}_{DCS}^{(RCOR)}(t) = e^{i\left(2\pi \int_0^t \Delta f_0(\tau)d\tau + \varphi_0\right)} \sum_{n=1}^N C_n e^{i\Delta \varphi_n} e^{i2\pi n\langle \Delta f_{rep}\rangle t}. \tag{14.16}$$

It is clear that to obtain a phase-noise-free signal, one needs to apply a global phase shift in counterphase, yet preserving its expected value. Otherwise, the spectrum would unnecessarily shift towards DC.

$$\hat{\tilde{V}}_{DCS}(t) \approx e^{-i\left(2\pi \int_0^t \Delta \hat{f}_0(\tau) - \langle \Delta f_0\rangle d\tau\right)} \cdot \tilde{V}_{DCS}^{(RCOR)}(t). \tag{14.17}$$

An example demonstration of computational correction is shown in Fig. 14.4, which plots the results of the CoCoA algorithm applied to an extremely noisy free-running THz DCS system. All correction signals have been extracted purely computationally, which allowed for restoring the discrete comb nature of the multiheterodyne spectrum in the rf domain.

## 14.3    Dual-Comb Spectroscopy in the Mid-Infrared with QCLs and ICLs

### 14.3.1    Absorption Spectroscopy

In 2012, Hugi *et al.* [18] published the first results showing comb operation of mid-infrared quantum cascade lasers and proposed that the QCL had the potential to become a standard technology for broadband, compact, all-solid-state mid-infrared spectrometers. Shortly thereafter, the demonstrations of multiheterodyne or dual-comb spectroscopy using QCLs were given by Wang *et al.* [45, 46] and Villares *et al.* [34]. Since then, the technique has been expanded to cover wavelengths from the 3 μm range using interband cascade lasers [41] to the terahertz spectral region measuring both broadband [47] and narrowband absorbers [48]. An overview of the results published to date is given in the following sections.

#### 14.3.1.1    High-Resolution Spectroscopy

One of the unique and convenient capabilities of semiconductor lasers is their ability to be frequency tuned via modulation of the injection current. This feature was very early recognized to be of great utility for high-resolution spectroscopy by frequency tuning a single-mode diode laser [49]. The same feature also applies to semiconductor-laser frequency combs and enables continuous tuning of comb modes in the optical frequency domain. Such high-resolution capabilities of QCL-based multiheterodyne spectroscopy was first demonstrated in 2013 by Wang *et al.* [45, 46], who used two

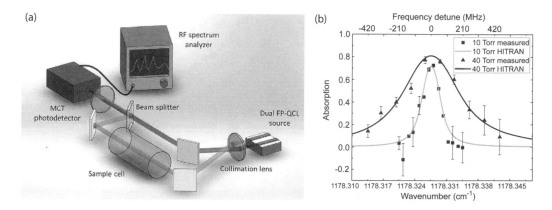

**Fig. 14.5** (a) Experimental configuration from Wang *et al.* [46]. The output beams from the two QCLs are collimated with a single lens. The beams are combined on a beamsplitter and focused on a fast photodetector. (b) High-resolution measurement of a $N_2O$ transition at low pressure. The triangular and square markers represent data measured at 10 torr and 40 torr, respectively. The black and gray lines represent HITRAN simulations. Adapted from *Applied Physics Letters*, vol. 104, pp. 0311141–5, 2014, with the permission of AIP Publishing.

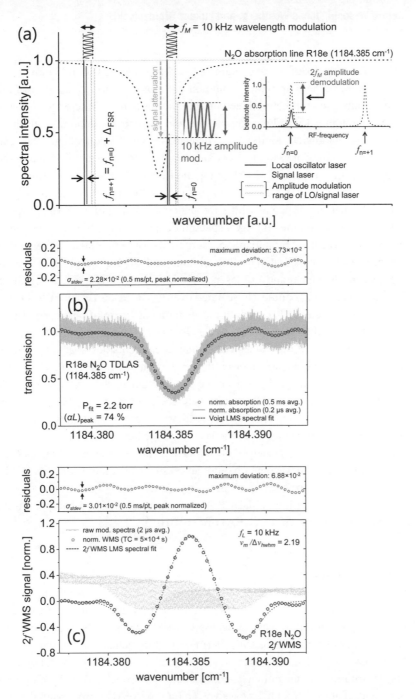

**Fig. 14.6** (a) Basic principle of wavelength-modulated DCS. The injection currents of the lasers are sinusoidally modulated at 10 kHz, which is converted into an intensity modulation by the absorption feature. The intensity modulation is directly mapped to the corresponding beat note and can be demodulated by a lock-in amplifier to obtain the WM-DCS spectrum. (b) Direct absorption measurement of the R18e $N_2O$ transition at ~8.5 μm. The gray trace represents the squared amplitude-vs-time measurement and the circles represent the time-windowed average (0.5 ms) of the same data. The black dotted line is a fit based on the HITRAN database. (c) Corresponding wavelength modulation measurements. The gray trace shows the squared amplitude-vs-time measurement and the circles show the digital lock-in signal. The dotted line shows a 2f-WMS fit. Adapted with permission from [35] © The Optical Society.

Fabry–Pérot QCLs at a wavelength of ~8.5 μm, cleaved to 1.23 mm in length, situated on the same chip with a common collimation lens, as shown in Fig. 14.5(a). The asymmetric multiheterodyne configuration was used, where the signal beam was guided through a 15 cm long sample cell and thereafter combined with the LO beam on a 1 GHz bandwidth (3 dB) MCT detector (VIGO Systems S.A.). The sample cell was filled with low pressure (10 torr and 40 torr) nitrous oxide ($N_2O$), and the absorption spectrum was acquired by stepping the frequency of the signal laser over the R11 $N_2O$ transition at $1178.3\,cm^{-1}$ (~8.5 μm) using the injection current. Figure 14.5(b) shows the measured spectra with a frequency resolution of ~15 MHz together with a simulation based on the HITRAN spectral database [50]. However, the large $\Delta f_{rep}$ of ~230 MHz precluded broadband measurements using more than a few optical modes.

An extension of the initial high-resolution work in [46] was given by Hangauer *et al.* [35], where the signal and LO lasers were inter-locked by a frequency-discriminator locking scheme and collinearly scanned over the R18e transition of $N_2O$ at ~8.5 μm. This approach removed the limitation in frequency resolution imposed by the step-scan, which resulted in a system whose frequency resolution was determined by the optical linewidths of the modes rather than their frequency spacing within the comb spectrum. In addition, Hangauer *et al.* [35] introduced wavelength modulation (WM) by superimposing a 10 kHz sinusoidal modulation to the linear scan of the lasers, which is known to improve the system sensitivity by reducing the influence of amplitude noise with $1/f$-characteristics. Modulation-based techniques, such as WM, are well suited for semiconductor diode lasers and are known to yield sensitivity improvements [51]. The basic principle of WM-DCS [35] is shown in Fig. 14.6(a) and the direct absorption and wavelength modulation spectra acquired with one DCS mode are shown in Fig. 14.6(b) and (c), respectively. A sensitivity improvement by a factor of ~3 was observed for the WM-DCS measurements reaching a bandwidth-normalized noise-equivalent absorption (NEA) of ~$5 \times 10^{-4}/\sqrt{Hz}$.

## 14.3.1.2    Broadband Spectroscopy

Compared to single-mode QCL and ICL devices, the broadband-frequency coverage of QCL and ICL combs represents a significant advantage in spectroscopic applications. This capability opens up access to molecular species with broad unresolved spectra without the need for slow thermal or optomechanical frequency tuning of the laser source. In 2014, Villares *et al.* [34] provided the first demonstration of high temporal resolution broadband QCL-based dual-comb spectroscopy by employing a pair of matched Fabry–Pérot QCLs based on an InGaAs/InAlAs broadband design for measurements of water vapor around $1400\,cm^{-1}$ (~7 μm). The devices were cleaved to 6 mm in length, which translated into mode spacings ($f_{rep}$) of ~7.5 GHz (~$0.25\,cm^{-1}$) with a mode spacing difference ($\Delta f_{rep}$) of 5 to 40 MHz, tunable by temperature or bias current. The mode spectra for the two QCLs are shown in Fig. 14.7(a). The balanced-asymmetric configuration was used (see Fig. 14.7(c)), which rendered a rf multiheterodyne beat note spectrum as shown in Fig. 14.7(b). As a gas-sensing proof of principle, a high-resolution (80 MHz) transmission spectrum of water vapor in air

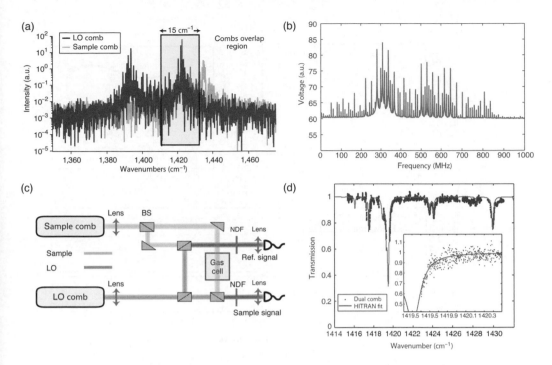

**Fig. 14.7** (a) Optical-mode spectra of two QCL combs. (b) rf multiheterodyne beat note spectrum generated by two QCL combs with slightly different comb repetition rates. (c) Schematic overview of the balanced-asymmetric DCS configuration used. BS: AR-coated beam splitter, NDF: neutral density filter. (d) Water-vapor optical transmission spectra in air ($P_{H2O} = 1.63$ kPa, total pressure $P_{tot} = 101.3$ kPa, $T = 20\,^\circ$C). The solid line is a HITRAN simulation, the inset shows an expansion of the measured transmission over $\sim1$ cm$^{-1}$ with a resolution of 80 MHz [34]. Adapted by permission from Springer Nature: *Nat. Comm.*, vol. 5, p. 5192, Copyright 2014.

at atmospheric pressure was measured by interleaving transmission spectra via stepping of the laser temperature. For each step, data was acquired using a total acquisition time of 100 ms and processed by segmentation into 3.2 μs bins followed by fast Fourier transform (FFT) to yield short-time rf spectra, which were aligned numerically to allow averaging beyond the mutual coherence time of the devices.

The optical transmission results from such a measurement, covering $\sim16$ cm$^{-1}$ (0.48 THz), is shown in Fig. 14.7(d), where a fit based on parameters from the HITRAN database [50] was added. The system achieved a peak short-term SNR ($1\sigma$) of $\sim118$ for 3.2 μs of acquisition time, which demonstrated the feasibility of precise broadband measurements with temporal resolutions down to the microsecond range.

To improve the mutual coherence of the two QCL comb devices, Westberg *et al.* [37] added an OPLL to a balanced-asymmetric dual-comb system by bandpass filtering a selected rf multiheterodyne beat note and phase-locking it to a stable local oscillator generated by a rf signal generator. The beat note center frequency stability was characterized by an Allan deviation plot, which is shown in Fig. 14.8(a). A noticeable improvement in frequency stability was observed for the OPLL system compared

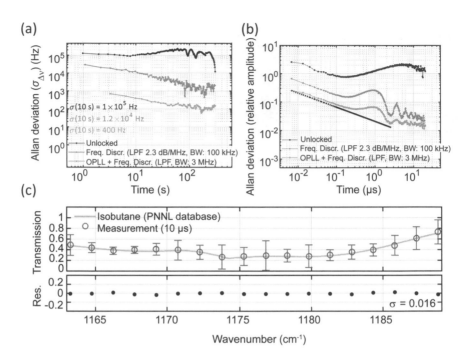

**Fig. 14.8** (a) Allan deviation plot of the center frequency of an individual rf multiheterodyne beat note for different modes of operation. The center frequency was evaluated by fitting a Gaussian profile to the beat note and extracting the center frequency from the fit. (b) Short-term relative amplitude stability using the different modes of operation. (c) Transmission spectrum measured with dual-comb system for ~15% isobutane in $N_2$ at atmospheric pressure (circles) together with a calculated spectrum based on the PNNL database (line). The lower panel shows the residuals with a standard deviation of 1.6% [37]. Adapted from *Applied Physics Letters*, vol. 110, p. 141108, 2017, with the permission of AIP Publishing.

to the same system operated in free-running mode. The short-term amplitude stability of the system is shown in Fig. 14.8(b), which shows relative beat note amplitude stability in the percent-level for acquisition times down to tens of microseconds. Broadband spectroscopic measurements of isobutane covering ~25 cm$^{-1}$ were used to demonstrate the system performance and good agreement with a reference spectrum from the PNNL database was obtained even with the short acquisition time of 10 μs (see Fig. 14.8(c)). The standard deviation of the residual was 1.6%, which translates to a bandwidth-normalized NEA of ~5 × 10$^{-5}$/$\sqrt{\text{Hz}}$.

The high brightness of QCL frequency combs was utilized by Hensley *et al.* [52] to perform close proximity stand-off detection from diffusely scattering targets for detection of hazardous compounds. Two dispersion-compensated QCLs with a peak optical output power of ~1 W and a bandwidth of ~100 cm$^{-1}$ around ~8 μm were used in the balanced-symmetric DCS configuration, where they were combined and scattered of a rough surface with deposited fluorinated silicon oil (FSO) and Krytox$^{\text{TM}}$. The scattered light was collected and focused on a 0.5 GHz bandwidth detector, see Fig. 14.9(a). The system achieved μg/cm$^2$ level detection limits using up

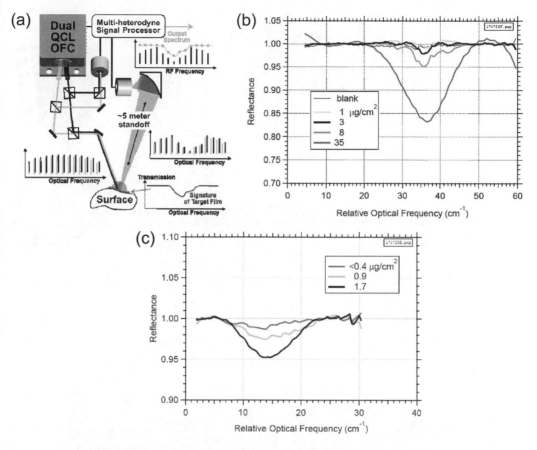

**Fig. 14.9** (a) Schematic overview of the close-proximity standoff-detection system based on QCL-DCS. (b) A series of reflectance measurements with different mass loadings of FSO with a third-order polynomial baseline removed. (c) A series of reflectance measurements of Krytox™ at 0.98 m standoff with a third-order polynomial baseline removed [52]. Adapted from J. Hensley, J. Brown, M. Allen, M. Geiser, P. Allmendinger, M. Mangold, *et al.*, "Standoff detection from diffusely scattering surfaces using dual quantum cascade laser comb spectroscopy," vol. 10638: SPIE, 2018 by permission from SPIE.

to 1 m of stand-off distance, as shown in the reflectance spectra of Fig. 14.9(b) and (c). The system sensitivity could be further improved by managing the optical feedback to the lasers without sacrificing the optical-power probing the sample, with these improvements stand-off detection at meter-scale distance will become feasible.

One of the most important advantages of QCL- or ICL-based DCS is its inherent ability to perform fast broadband spectroscopic measurements with sub-microsecond time resolution. This capability was recently demonstrated by Klocke *et al.* [53]. Their QCL-DCS spectrometer covered 55 cm$^{-1}$ (1185–1240 cm$^{-1}$) with a temporal resolution down to 320 ns and was used to measure the absorption spectra of the photocycle of bacteriorhodopsin exposed to short pulses of 532 nm light. The spectrometer was able to resolve all the characteristics of the intermediate states shown in Fig. 14.10. In

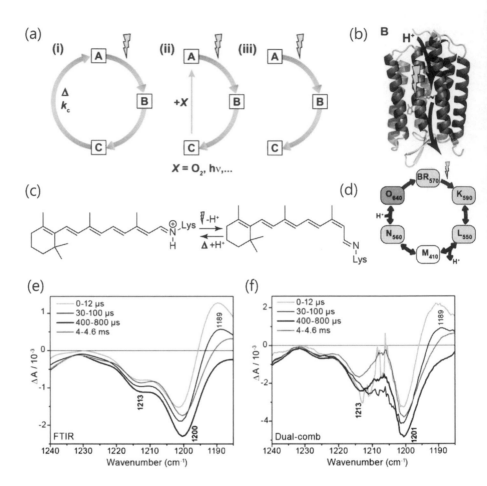

**Fig. 14.10** The authors of [53] studied photon-absorption induced reactions shown in (a) that may have a thermal back-reaction (i), stimulated back-reaction (ii), or may be irreversible (iii) and thereby challenging to study. Shown in (b) is bacteriorhodopsin, which binds a retinal covalently and acts as light-driven proton pump across the membrane. (c) Light-induced retinal isomerization from all-*trans* to 13-*cis* 15-*anti* and the subsequent deprotonation are fully reversible within milliseconds. In (d) authors showed a simplified photocycle of bacteriorhodopsin. Spectra obtained with step-scanned FTIR for states KL, L, M, and N in the photocycle (d) are shown in (e) and similar spectra obtained with QCL-based DCS are shown in (f) and show good agreement. QCL DCS spectra shown in (f) represent an average of 500 single spectra. Adapted with permission from J. L. Klocke, M. Mangold, P. Allmendinger, A. Hugi, M. Geiser, P. Jouy, *et al.*, "Single-shot sub-microsecond mid-infrared spectroscopy on protein reactions with quantum cascade Laser frequency combs," *Analytical Chemistry*, vol. 90, pp. 10494–10500, 2018/09/04 2018. Copyright (2018) American Chemical Society.

addition, the effect of pH on the photocycle was studied and compared to step-scan FTIR experiments with good agreement. The spectrometer, operating at a resolution of 4.5 cm$^{-1}$, achieved an RMS measurement noise of 15 μOD for 500 averages and even a single-shot measurement was sufficient to resolve the most important characteristics

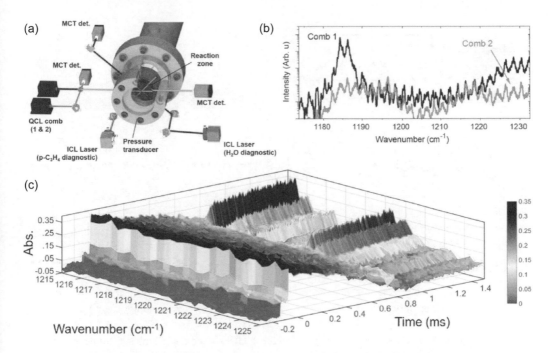

**Fig. 14.11** (a) 3D schematic of the shock-tube equipped with a QCL dual-comb spectrometer. (b) FTIR spectra of the QCL combs. (c) DCS data from 1215 to 1225 cm$^{-1}$ of propyne oxidation (2% p-C$_3$H$_4$ $T_0 = 1225$ K, $P_0 = 2.8$ atm) illustrating the arrival of the incident and reflected shock before time-zero. Figures from [57]. Adapted from https://doi.org/10.1088/1361-6501/ab6ecc © IOP Publishing. Reproduced with permission. All rights reserved.

in the mid-IR absorption spectrum. This indicates the potential usefulness of such an instrument to study irreversible reactions at microsecond time-scales, where step-scanned FTIRs or slowly tunable external-cavity QCLs would be impractical or result in very long acquisition times.

Other QCL-DCS examples include vibrational stark spectroscopy of fluorobenzene [54], attenuated total reflection spectroscopy [55], and multi-species measurements at high temperatures and pressures [56]. Recently, Pinkowski *et al.* [57] used QCL-DCS to study high temperature reactions between propyne and oxygen in a shock-tube using a pair of free-running QCLs. The lasers covered 1174 to 1233 cm$^{-1}$ with 179 modes spaced by 9.86 GHz and with a time resolution of 4 μs. This application demonstrates some of the most prominent features of QCL-DCS, fast response time and broad-band spectral coverage, which sets it apart from interferometric or frequency-swept techniques.

Figure 14.11(a) shows the shock-tube together with the QCL-DCS spectrometer. Separate laser diagnostics tools were also employed for validation of the DCS spectrometer. The spectrometer spectral coverage is shown by the FTIR measurements in Fig. 14.11(b). Figure 14.11(c) shows the time evolution of the reaction where the broadband absorption features of propyne can be seen from 0 to 0.6 ms and the finer structure of water vapor is visible after 0.8 ms.

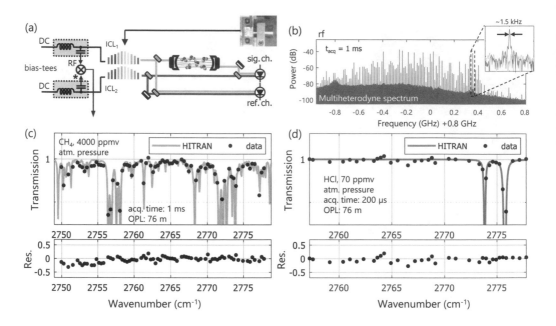

**Fig. 14.12** (a) Experimental configuration for the ICL-based dual-comb spectrometer [60]. The intermode beat notes are extracted using microwave probes (see inset photo) and bias-tees. (b) Coherently averaged multiheterodyne spectrum. (c) Optical transmission spectrum (log-scale) of 4000 ppmv of methane in $N_2$ at atmospheric pressure acquired during 1 ms with a pathlength of 76 m together with a fit based on parameters from the HITRAN database. (d) Optical transmission spectrum (log-scale) of 70 ppmv of hydrogen chloride in $N_2$ at atmospheric pressure acquired during 200 μs with a pathlength of 76 m together with a fit based on parameters from the HITRAN database. The lower panels show the residual of the fit.

### 14.3.1.3    High-Sensitivity Spectroscopy

Most work on gas-sensing using QCL-DCS has been performed using short absorption cells with bandwidth-normalized NEAs in the $10^{-3}$ to $10^{-5}$ range. This limits the achievable sensitivities to percentage-level concentrations for most gases, which is not sufficient for trace-gas sensing. An improvement to this modest sensitivity can be obtained by extending the optical pathlength via a multipass cell, which is a standard approach in the field of laser-spectroscopy [58]. Sensitivity enhancements of up to three orders of magnitude (10 cm to 100 m) can be achieved using commercially available multipass cells, given that the added optical fringes and imperfect reflectivity of the multipass mirrors do not severely degrade the beat note SNR.

To explore this sensitivity enhancement, Westberg and Sterczewski *et al.* [59, 60] used a balanced asymmetric system based on ICLs operating at ~3.6 μm, where the signal beam was directed through a 76 m astigmatic Herriott cell, after which it was combined with a matched local oscillator, as shown in Fig. 14.12(a). The multipass cell was filled with 4000 ppmv of methane in $N_2$ at atmospheric pressure. The resulting rf multiheterodyne beat note spectrum is shown in Fig. 14.12(b) and the corresponding methane spectrum together with a fit based on the HITRAN

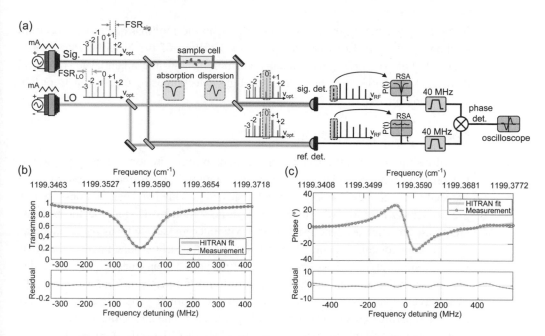

**Fig. 14.13** (a) Schematic of the balanced-asymmetric DCS configuration used. Two bandpass filters are used to isolate an individual beat note from the signal and reference channels and their phase difference is measured by a phase detector. (b) Measured transmission spectrum for the R35e $N_2O$ transition at $1199.3576\,cm^{-1}$ acquired with a 1 kHz triangular ramp. The solid line shows a HITRAN fit based on the Voigt absorption lineshape. (c) The corresponding phase measurement obtained by detection of the differential phase between the two beat notes. The solid line represents a HITRAN fit based on a Voigt dispersion lineshape. Adapted with permission from [36] © The Optical Society.

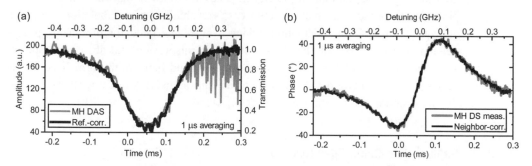

**Fig. 14.14** (a) High-resolution swept absorption measurement of 1.6 kPa ethylene at $3112.6\,cm^{-1}$ using a 1 kHz triangular ramp, which gives an acquisition time of 0.5 ms per spectrum. (b) High-resolution swept dispersion measurement using the same time-domain data as in (a). An adjacent beat note unaffected by the molecule was used as a phase reference. Adapted from [41].

database [50] is shown in Fig. 14.12(c). The system operated with ~28 $cm^{-1}$ of optical bandwidth and a minimum detection limit of 9 ppmv/$\sqrt{Hz}$ was achieved. To reach lower minimum detection limits (MDL), 70 ppmv of hydrogen chloride diluted in $N_2$ to atmospheric pressure was introduced in the multipass cell. The spectrum is shown in Fig. 14.12(d) and the standard deviation of the baseline equates to a MDL ($1\sigma$) of $\sim$60 ppb/$\sqrt{Hz}$. This demonstration shows the potential for ppb-level detection sensitivities for atmospheric trace-gas detection using on-chip QCL- or ICL dual-comb sources.

## 14.3.2    Molecular Dispersion Spectroscopy

An intrinsic property of the asymmetric dual-comb configuration is its ability to assess not only the sample-induced absorption via amplitude measurements, but also the induced phase change (sample-induced dispersion). This has been demonstrated by Sterczewski et al. [36], who measured high-resolution dispersion spectra by using QCL-based multiheterodyne spectroscopy with inter-locked QCLs using a frequency-discriminator locking scheme. The lasers were collinearly scanned across the R35e $N_2O$ transition at $1199.3576\,cm^{-1}$ ($\sim$8.34 µm) with a scan-rate of 1 kHz. The dispersion spectra were obtained by bandpass filtering the pertinent beat notes from the two detector channels (signal and reference) and supplying them to a phase detector. The experimental setup is shown in Fig. 14.13(a). The absorption and dispersion spectra obtained by this method are shown in Fig. 14.13(b) and (c), respectively. The bandwidth-normalized NEA was estimated to be $2.8\times10^{-4}/\sqrt{Hz}$, and the bandwidth-normalized noise-equivalent phase was estimated to be $6.6\times10^{-6}\,rad/\sqrt{Hz}$.

The first demonstration of interband cascade laser-based multiheterodyne spectroscopy by Sterczewski et al. [41] also assessed the dispersion properties of a molecular sample, in this case low-pressure ethylene was measured. The two ICLs used emitted around 30 mW of optical power while consuming less than 0.5 W each, and were operated in scan-mode, similar to [35]. A notable difference is that the retrieval of the phase and amplitude information in this case was obtained by computational in-phase and quadrature (IQ) demodulation, which avoided the need for the analog phase-detector and custom-made bandpass filters. The dispersion and absorption spectra are shown in Fig. 14.14(a) and (b), respectively, and bandwidth-normalized NEAs of $3.3\times10^{-4}/\sqrt{Hz}$ for the absorption and an equivalent NEA of $2.3\times10^{-4}/\sqrt{Hz}$ for the dispersion measurement were reported.

## 14.3.3    Gapless Dual-Comb Spectroscopy

In recent years QCL combs have been developed even further by incorporating sophisticated dispersion engineering into the device waveguides [61, 62], which has resulted in improved stability of the comb operation and significantly extended the tuning range of the frequency combs. These advancements enabled frequency tuning of the comb structure that could cover the frequency-comb mode spacing, resulting in gap-less coverage of the frequency domain over the entire optical bandwidth of the QCL frequency comb. Both absorption and dispersion measurement techniques discussed in prior sections can benefit from gap-less tuning and can provide broadband high-resolution spectra of both small molecules with well-resolved molecular

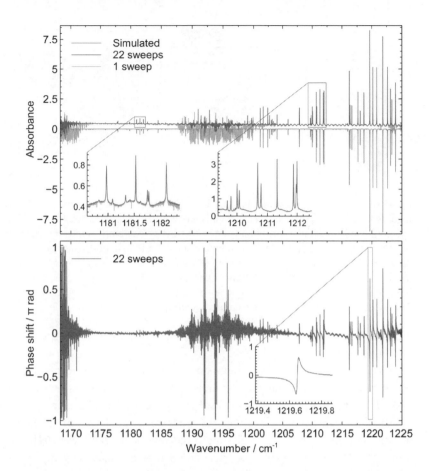

**Fig. 14.15** Gap-less DCS absorption (top) and dispersion (bottom) spectra of methane acquired with a high spectral resolution of $0.001\ \mathrm{cm}^{-1}$. The insets show the data in the corresponding boxes. Reprinted with permission from [63] © The Optical Society.

transitions as well as of large molecules with unresolved absorption bands. The demonstration of gap-less DCS in both absorption as well as dispersion spectroscopy modes was recently demonstrated by Gianella *et al.* [63] and an example of high-resolution gap-less DCS spectra presented in that work are reproduced in Fig. 14.15. These advancements in QCL comb performance are opening new avenues to perform selective and sensitive chemical detection in the mid-IR.

## 14.4 Dual-Comb Spectroscopy in the THz Using QCLs

The THz region has recently attracted much spectroscopic interest because of the large number of complex molecules that have well-resolvable and strong far-infrared absorption features enabling unambiguous species identification. THz spectroscopy is also intriguing for studies of solid-state systems with THz-driven dynamics, where, for instance, *in situ* polymorphic transitions of active pharmaceutical ingredients can be

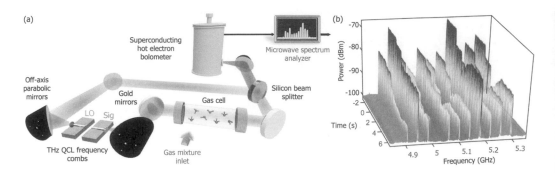

**Fig. 14.16** (a) 3D schematic of the THz-QCL DCS system. (b) Nine seconds of THz-DCS data recorded with the microwave spectrum analyzers with 1 millisecond temporal resolution (9000 spectra). Each 1 ms frame corresponds to 39 000 quasi-coherently averaged interferograms with a fundamental refresh rate defined by the spacing of the radio-frequency comb equal to $\Delta f_{\text{rep}} = 39$ MHz. Figure from [48]. Reprinted with permission from L. A. Sterczewski, J. Westberg, Y. Yang, D. Burghoff, J. Reno, Q. Hu, *et al.*, "Terahertz spectroscopy of gas mixtures with dual quantum cascade laser frequency combs," *ACS Photonics*, vol. 7, pp. 1082–1087, 2020. Copyright (2020) American Chemical Society.

characterized without using ionizing radiation, in contrast to the X-ray diffractometry (XRD) technique.

Prior to the first QCL-based dual-comb spectroscopy experiments, the THz dual-comb technique predominantly employed photoconductive switches or nonlinear crystals pumped by femtosecond lasers to generate THz radiation. However, the conversion efficiency of these sources leads to a rapid roll-off in emitted power above 1–2 THz down to the single-microwatt region or even lower [64]. The directly generated THz emission from a QCL comb provides a remedy to this issue with demonstrations of several milliwatts of optical power at wavelengths above 3 THz [19, 65].

## 14.4.1    Absorption Spectroscopy of Gases

The first demonstration of THz QCL multiheterodyne spectroscopy was achieved by Yang *et al.* in 2016 [40], where the transmission through a submillimeter GaAs wafer, over ∼250 GHz of optical bandwidth around 2.8 THz, was characterized. Shortly thereafter, gas-phase multiheterodyne measurements of water vapor, ammonia, and nitrogen dioxide were demonstrated [48]. Here, a pair of QCL combs was equipped with pre-collimating lenses and housed in a pulsed tube cryostat with a thermally insulating vacuum. The dispersion-compensated QCL devices [19] were biased through cryogenic coaxial cables and microwave bias tees to monitor the intermode beat notes at the laser-cavity roundtrip frequencies. The experimental setup together with the time evolution of the measured rf spectra are shown in Fig. 14.16.

The THz spectroscopy setup (Fig. 14.16(a)) was arranged in a single-photodetector asymmetric configuration, where the sample cell was placed before combining the attenuated beam with the LO on a high-resistivity float-zone silicon (HRFZ-Si) beam splitter. The off-axis parabolic mirrors (OAPM) ensured proper collimation of the

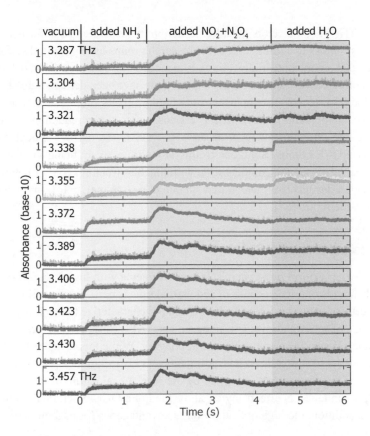

**Fig. 14.17** Time-resolved decadic (base-10) absorbance retrieved from the dual-comb spectrogram at the four phases of the experiment. Bold lines are 10-point moving averages. Figure from [48]. Reprinted with permission from L. A. Sterczewski, J. Westberg, Y. Yang, D. Burghoff, J. Reno, Q. Hu, *et al.*, "Terahertz spectroscopy of gas mixtures with dual quantum cascade laser frequency combs," *ACS Photonics*, vol. 7, pp. 1082–1087, 2020. Copyright (2020) American Chemical Society.

beam, and provided a tight focus on the surface of the superconducting hot electron bolometer. Through injection-current tuning, the lasers were biased in the comb regime with near-kHz intermode beat notes and mismatched in repetition rates leading to a down-converted $\sim$180 GHz-wide portion of THz spectrum centered around 3.4 THz.

The spectroscopic capabilities of the system were evaluated by studying evolving spectra from gas mixtures with 1 ms temporal resolution. Figure 14.17 shows the absorbance spectra of individual modes as a function of time.

Despite the many advantages of the chip-scale, electrically pumped THz-comb sources, widespread applications to molecular spectroscopy are currently hindered by the requirement of cryogenic operation of the THz-QCL sources. Therefore, the most recent developments in the area of novel THz QCL-gain designs [66] and nonlinear generation of THz-QCL combs via an internal DFG process [20] will have significant impact on the field of THz DCS and applications. In addition, the most sensitive THz photodetectors need liquid-helium baths, which adds further complexity to the systems. To address this, some research groups have proposed self-detection schemes that could potentially mitigate the need for sensitive THz detectors and early proof-of-concept spectroscopic experiments have been demonstrated by Li *et al.* [67]. Moreover, recent developments in Schottky mixers promise room-temperature photodetector operation, albeit at the expense of a slight reduction in sensitivity [40]. Another limitation of QCL-based THz-DCS is the narrow optical bandwidth of the THz-QCL combs. Solutions to this issue are actively being pursued through more effective dispersion compensation with tunable structures [68] or heterogeneous active-region designs [69], which may enable octave-spanning THz-frequency comb sources in the future [65].

## 14.4.2    Hyperspectral Imaging of Solids

The ability to perform broadband THz measurements with high signal-to-noise ratio within hundreds of microseconds of acquisition time is particularly well-suited for the pharmaceutical industry, where there is a need to characterize the quantity and spatial distribution of active pharmaceutical ingredients (API), various excipients, and protective coating layers, through hyperspectral imaging. While this idea has been explored in experiments using terahertz-pulsed-imaging (TPI) techniques [70–73], the acquisition times in such systems is limited by the mechanical nature of the spectroscopic scan. This limitation is effectively circumvented in a QCL-based dual-comb system, where, in addition, the high optical power is advantageous for probing optically thick samples.

To demonstrate this capability, a three-zone pellet was imaged in a raster-scan transmission experiment [47], as shown in Fig. 14.18. The experimental setup was similar to the gas-spectroscopy arrangement discussed in the prior section with the absorption cell replaced by two off-axis parabolic mirrors and a motorized sample holder placed at the focal point between them (Fig. 14.18(a)). To provide more power-per-mode with a resolution sufficient for solid spectroscopy, two-section THz chips with a shorter cavity length were used, capable of developing 200 GHz-wide combs centered around 3.3 THz with mode spacings of 17 GHz, similar to those in [68].

The imaged polyethylene-based sample contained two popular pharmaceutical excipients $\alpha$(-D-lactose monohydrate, $\alpha$-D-glucose monohydrate), and an amino acid (L-histidine monohydrochloride monohydrate) in a concentration of 10%, which were pressed together into a three-zone 30-mm-diameter disk with a thickness of approximately 3 mm. The sample image was recorded by stepping the XY-stage in 0.5 mm increments and recording a multiheterodyne spectrum at each position. Next, the acquired data were post-processed to obtain a false-color RGB composite image. Three comb modes (two on the side and one in the center) were selected for differential analysis on a logarithmic scale. The mean value was subtracted from the intensity of each beat note, thus yielding the differential absorption, as proposed in [74]. In contrast to the optical photography of the pellet (Fig. 14.18(b)), which shows no distinguishable regions, the terahertz hyperspectral image clearly reveals

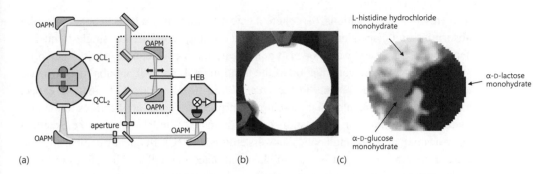

**Fig. 14.18** Hyperspectral imaging using THz QCLs. (a) Experimental setup wherein the imaged sample is raster-scanned at the focal point of the two parabolic mirrors. (b) Photography of a polyethylene pellet composed of three zones: α-D-glucose monohydrate, α-D-lactose monohydrate, and L-histidine monohydrochloride monohydrate. (c) THz RGB composite image obtained from three multiheterodyne beat notes.

the chemical composition difference (Fig. 14.18(c)) due to the wavelength-dependent absorptions of the different materials. The QCL comb probes the positive slope of the glucose monohydrate's absorption [75], whereas lactose shows the opposite trend. L-histidine on the contrary has a parabolic-like shape of absorption with a peak located approximately in the center of the comb (~3.37 THz) [76].

This demonstration is a first step towards real-time monitoring of pharmaceutical formulations, where further improvements can be made by replacing the raster scan with a single-pixel imaging technique that employs spatial light modulators and computational reconstruction. In principle, video-rate hyperspectral images are attainable with this technique.

## 14.5 Conclusions and Future Outlook

In a relatively short span of time quantum-cascade-laser frequency combs have proven to be useful spectroscopic lasers sources for high-resolution and broadband spectroscopy with microsecond time resolution both in the mid-infrared molecular-fingerprint region as well as in the THz frequency range. In this chapter we have discussed a number of molecular sensing technologies that utilize this laser technology to perform spectroscopy of gaseous and solid samples. Given the fast pace of innovation in this field, in the next few years significant advancements in the performance of QCL comb sources coupled with more mature and novel solutions in the areas of spectroscopic systems and signal processing are expected to yield a new generation of high-resolution spectrometers that will show increasingly broader spectral coverage and enable new chemical sensing applications. It has already been clear from the first demonstrations surveyed in this chapter that this technology has a unique potential for system integration and can offer unprecedented capabilities in terms of: (1) temporal resolution for monitoring irreversible reactions; (2) optical power suitable for optically dense media; (3) high spectral resolution, enabling sensing application of low-pressure gas transitions; (4) coherent detection that gives access to both molecular absorption and dispersion; and (5) broadband coverage compared to single-frequency semiconductor lasers, which opens up applications to characterize liquid- and solid-state samples. With a continuation of the trend in QCL and ICL combs' performance and the recent developments in the area of mid-IR photonics [77, 78], we expect significant advancements in lab-on-chip spectroscopic-sensor applications in the bio-medical, environmental, and health sensing domains. Similarly, due to the excellent laser-beam quality and high optical powers emitted by QCL frequency combs, applications in remote sensing and stand-off chemical detection will also benefit from future advancements in performance of mid-IR- and THz-laser frequency combs. This is a unique and rapidly growing technology that has the potential to make a significant impact in many areas of chemical sensing.

## References

[1]   R. R. Patty, G. M. Russwurm, W. A. McClenny, and D. R. Morgan, "$CO_2$ laser absorption coefficients for determining ambient levels of $O_3$, $NH_3$, and $C_2H_4$," *Applied Optics*, vol. 13, pp. 2850–2854, 1974.

[2]   C. H. Kenneth and C. P. George, "A rapid-scan infrared spectrometer: Flash photolytic detection of chloroformic acid and of CF2," *Applied Optics*, vol. 4, pp. 25–30, 1965.

[3]   P. R. Griffiths, "The early days of commercial FT-IR spectrometry: A personal perspective," *Applied Spectroscopy*, vol. 71, pp. 329–340, 2017.

[4]   W. Urban, "Physics and spectroscopic applications of carbon monoxide lasers, a review," *Infrared Physics & Technology*, vol. 36, pp. 465–473, 1995.

[5]   R. F. Curl, J. V. V. Kasper, P. G. Carrick, E. Koester, and F. K. Tittel, "High sensitivity color center laser spectroscopy," in *Laser Spectroscopy V*, Berlin, Heidelberg, 1981, pp. 346–350.

[6]   J. Faist, F. Capasso, D. L. Sivco, C. Sirtori, A. A. Hutchinson, and A. Y. Cho, "Quantum cascade laser," *Science*, vol. 264, p. 4, 1994.

[7]   R. Q. Yang, "Infrared laser based on intersubband transitions in quantum wells," *Superlattices and Microstructures*, vol. 17, pp. 77–83, 1995.

[8]   R. F. Curl, F. Capasso, C. Gmachl, A. A. Kosterev, B. McManus, R. Lewicki, *et al.*, "Quantum cascade lasers in chemical physics," *Chemical Physics Letters*, vol. 487, pp. 1–18, 2010.

[9]   A. Hugi, R. Terazzi, Y. Bonetti, A. Wittmann, M. Fischer, M. Beck, *et al.*, "External cavity quantum cascade laser tunable from 7.6 to 11.4 μm," *Applied Physics Letters*, vol. 95, p. 061103, 2009.

[10]   C. L. Strand, Y. Ding, S. E. Johnson, and R. K. Hanson, "Measurement of the mid-infrared absorption spectra of ethylene ($C_2H_4$) and other molecules at high temperatures and pressures," *Journal of Quantitative Spectroscopy and Radiative Transfer*, vol. 222–223, pp. 122–129, 2019.

[11]   A. Schliesser, N. Picque, and T. W. Hansch, "Mid-infrared frequency combs," *Nat Photon*, vol. 6, pp. 440–449, 2012.

[12]   H. Füser and M. Bieler, "Terahertz frequency combs," *Journal of Infrared, Millimeter, and Terahertz Waves*, vol. 35, pp. 585–609, 2014.

[13]  S. B. Mirov, V. V. Fedorov, D. Martyshkin, I. S. Moskalev, M. Mirov, and S. Vasilyev, "Progress in mid-IR lasers based on Cr and Fe-doped II–VI chalcogenides," *IEEE Journal of Selected Topics in Quantum Electronics*, vol. 21, pp. 292–310, 2015.

[14]  M. F. Seth, J. J. David, and Y. Jun, "Flexible and rapidly configurable femtosecond pulse generation in the mid-IR," *Opt. Lett.*, vol. 28, pp. 370–372, 2003.

[15]  S. Vasilyev, I. S. Moskalev, V. O. Smolski, J. M. Peppers, M. Mirov, A. V. Muraviev, *et al.*, "Super-octave longwave mid-infrared coherent transients produced by optical rectification of few-cycle 2.5-μm pulses," *Optica*, vol. 6, pp. 111–114, 2019.

[16]  A. G. Griffith, R. K. W. Lau, J. Cardenas, Y. Okawachi, A. Mohanty, R. Fain, *et al.*, "Silicon-chip mid-infrared frequency comb generation," *Nature Communications*, vol. 6, p. 6299, 2015.

[17]  T. Yasui, Y. Kabetani, E. Saneyoshi, S. Yokoyama, and T. Araki, "Terahertz frequency comb by multifrequency-heterodyning photoconductive detection for high-accuracy, high-resolution terahertz spectroscopy," *Applied Physics Letters*, vol. 88, p. 241104, 2006.

[18]  A. Hugi, G. Villares, S. Blaser, H. C. Liu, and J. Faist, "Mid-infrared frequency comb based on a quantum cascade laser," *Nature*, vol. 492, pp. 229–233, 2012.

[19]  D. Burghoff, T.-Y. Kao, N. Han, C. W. I. Chan, X. Cai, Y. Yang, *et al.*, "Terahertz laser frequency combs," *Nature Photonics*, vol. 8, p. 462, 2014.

[20]  Q. Lu, F. Wang, D. Wu, S. Slivken, and M. Razeghi, "Room temperature terahertz semiconductor frequency comb," *Nature Communications*, vol. 10, p. 2403, 2019.

[21]  D. van der Weide and F. Keilmann, "Coherent periodically pulsed radiation spectrometer," US Patent 5,748,309, 1998.

[22]  S. Schiller, "Spectrometry with frequency combs," *Optics Letters*, vol. 27, pp. 766–768, 2002.

[23]  F. Keilmann, C. Gohle, and R. Holzwarth, "Time-domain mid-infrared frequency-comb spectrometer," *Optics Letters*, vol. 29, pp. 1542–1544, 2004.

[24]  M. J. Thorpe, D. Balslev-Clausen, M. S. Kirchner, and J. Ye, "Cavity-enhanced optical frequency comb spectroscopy: application to human breath analysis," *Opt. Express*, vol. 16, pp. 2387–2397, 2008.

[25]  J. Mandon, G. Guelachvili, and N. Picque, "Fourier transform spectroscopy with a laser frequency comb," *Nat Photon*, vol. 3, pp. 99–102, 2009.

[26]  L. A. Sterczewski and M. Bagheri, "Subsampling dual-comb spectroscopy," *Optics Letters*, vol. 45, pp. 4895–4898, 2020.

[27]  K. C. Cossel, E. M. Waxman, F. R. Giorgetta, M. Cermak, I. R. Coddington, D. Hesselius, *et al.*, "Open-path dual-comb spectroscopy to an airborne retroreflector," *Optica*, vol. 4, pp. 724–728, 2017.

[28]  I. Coddington, N. Newbury, and W. Swann, "Dual-comb spectroscopy," *Optica*, vol. 3, pp. 414–426, 2016.

[29]  Z. Chen, M. Yan, T. W. Hänsch, and N. Picqué, "A phase-stable dual-comb interferometer," *Nature Communications*, vol. 9, p. 3035, 2018.

[30]  T. Ideguchi, A. Poisson, G. Guelachvili, N. Picqué, and T. W. Hänsch, "Adaptive real-time dual-comb spectroscopy," *Nature Communications*, vol. 5, p. 3375, 2014.

[31]  J. Roy, J.-D. Deschênes, S. Potvin, and J. Genest, "Continuous real-time correction and averaging for frequency comb interferometry," *Optics Express*, vol. 20, pp. 21932-21939, 2012.

[32]  D. Burghoff, Y. Yang, and Q. Hu, "Computational multiheterodyne spectroscopy," *Science Advances*, vol. 2, p. e1601227, 2016.

[33] L. A. Sterczewski, J. Westberg, and G. Wysocki, "Computational coherent averaging for free-running dual-comb spectroscopy," *Optics Express*, vol. 27, pp. 23875–23893, 2019.

[34] G. Villares, A. Hugi, S. Blaser, and J. Faist, "Dual-comb spectroscopy based on quantum-cascade-laser frequency combs," *Nature Communications*, vol. 5, p. 5192, 2014.

[35] A. Hangauer, J. Westberg, E. Zhang, and G. Wysocki, "Wavelength modulated multi-heterodyne spectroscopy using Fabry-Pérot quantum cascade lasers," *Optics Express*, vol. 24, pp. 25298–25307, 2016.

[36] L. A. Sterczewski, J. Westberg, and G. Wysocki, "Molecular dispersion spectroscopy based on Fabry–Pérot quantum cascade lasers," *Optics Letters*, vol. 42, pp. 243–246, 2017.

[37] J. Westberg, L. A. Sterczewski, and G. Wysocki, "Mid-infrared multiheterodyne spectroscopy with phase-locked quantum cascade lasers," *Applied Physics Letters*, vol. 110, p. 141108, 2017.

[38] P. Jouy, J. M. Wolf, Y. Bidaux, P. Allmendinger, M. Mangold, M. Beck, *et al.*, "Dual comb operation of $\lambda \sim 8.2$ μm quantum cascade laser frequency comb with 1 W optical power," *Applied Physics Letters*, vol. 111, p. 141102, 2017.

[39] J. Westberg, L. A. Sterczewski, F. Kapsalidis, Y. Bidaux, J. M. Wolf, M. Beck, *et al.*, "Dual-comb spectroscopy using plasmon-enhanced-waveguide dispersion-compensated quantum cascade lasers," *Optics Letters*, vol. 43, pp. 4522–4525, 2018.

[40] Y. Yang, D. Burghoff, D. J. Hayton, J.-R. Gao, J. L. Reno, and Q. Hu, "Terahertz multiheterodyne spectroscopy using laser frequency combs," *Optica*, vol. 3, pp. 499–502, 2016.

[41] L. A. Sterczewski, J. Westberg, C. L. Patrick, C. S. Kim, M. Kim, C. L. Canedy, *et al.*, "Multiheterodyne spectroscopy using interband cascade lasers," *Optical Engineering*, vol. 57, p. 12, 2017.

[42] T. Yasui, R. Ichikawa, Y.-D. Hsieh, K. Hayashi, H. Cahyadi, F. Hindle, *et al.*, "Adaptive sampling dual terahertz comb spectroscopy using dual free-running femtosecond lasers," *Scientific Reports*, vol. 5, p. 10786, 2015.

[43] J. Faist, G. Villares, G. Scalari, M. Rösch, C. Bonzon, A. Hugi, *et al.*, "Quantum cascade laser frequency combs," *Nanophotonics*, vol. 5, p. 272, 2016.

[44] M. A. Belkin, F. Capasso, F. Xie, A. Belyanin, M. Fischer, A. Wittmann, *et al.*, "Room temperature terahertz quantum cascade laser source based on intracavity difference-frequency generation," *Applied Physics Letters*, vol. 92, p. 201101, 2008.

[45] Y. Wang and G. Wysocki, "High resolution molecular spectroscopy using dual comb-like structures of Fabry-Perot quantum cascade lasers," in Advanced *Solid-State Lasers Congress*, M. Ebrahim-Zadeh and I. Sorokina, eds., OSA Technical Digest (online), Optica Publishing Group, 2013, paper MTh3C.5, available at: https://doi.org/10.1364/MICS.2013.MTh3C.5.

[46] Y. Wang, M. G. Soskind, W. Wang, and G. Wysocki, "High-resolution multi-heterodyne spectroscopy based on Fabry-Pérot quantum cascade lasers," *Applied Physics Letters*, vol. 104, pp. 0311141–5, 2014.

[47] L. A. Sterczewski, J. Westberg, Y. Yang, D. Burghoff, J. Reno, Q. Hu, *et al.*, "Terahertz hyperspectral imaging with dual chip-scale combs," *Optica*, vol. 6, pp. 766–771, 2019.

[48] L. A. Sterczewski, J. Westberg, Y. Yang, D. Burghoff, J. Reno, Q. Hu, *et al.*, "Terahertz spectroscopy of gas mixtures with dual quantum cascade laser frequency combs," *ACS Photonics*, vol. 7, pp. 1082–1087, 2020.

[49] E. D. Hinkley, "High-resolution infrared spectroscopy with a tunable diode laser," *Applied Physics Letters*, vol. 16, pp. 351–354, 1970.

[50] I. E. Gordon, L. S. Rothman, C. Hill, R. V. Kochanov, Y. Tan, P. F. Bernath, *et al.*, "The HITRAN2016 molecular spectroscopic database," *Journal of Quantitative Spectroscopy and Radiative Transfer*, vol. 203, pp. 3–69, 2017.

[51] P. Werle, "Spectroscopic trace gas analysis using semiconductor diode lasers," *Spectrochimica Acta Part A: Molecular and Biomolecular Spectroscopy*, vol. 52, pp. 805–822, 1996.

[52] J. Hensley, J. Brown, M. Allen, M. Geiser, P. Allmendinger, M. Mangold, *et al.*, "Stand-off detection from diffusely scattering surfaces using dual quantum cascade laser comb spectroscopy," *Proceedings SPIE*, vol. 10638, p. 1063820, 2018.

[53] J. L. Klocke, M. Mangold, P. Allmendinger, A. Hugi, M. Geiser, P. Jouy, *et al.*, "Single-shot sub-microsecond mid-infrared spectroscopy on protein reactions with quantum cascade laser frequency combs," *Analytical Chemistry*, vol. 90, pp. 10494–10500, 2018.

[54] U. Szczepaniak, S. H. Schneider, R. Horvath, J. Kozuch, and M. Geiser, "Vibrational stark spectroscopy of fluorobenzene using quantum cascade laser dual frequency combs," *Applied Spectroscopy*, vol. 74, pp. 347–356, 2019.

[55] E. Lins, S. Read, B. Unni, S. M. Rosendahl, and I. J. Burgess, "Microsecond resolved infrared spectroelectrochemistry using dual frequency comb IR lasers," *Analytical Chemistry*, vol. 92, pp. 6241–6244, 2020.

[56] G. Zhang, R. Horvath, D. Liu, M. Geiser, and A. Farooq, "QCL-based dual-comb spectrometer for multi-species measurements at high temperatures and High pressures," *Sensors*, vol. 20, p. 3602, 2020.

[57] N. H. Pinkowski, Y. Ding, C. L. Strand, R. K. Hanson, R. Horvath, and M. Geiser, "Dual-comb spectroscopy for high-temperature reaction kinetics," *Measurement Science and Technology*, vol. 31, p. 055501, 2020.

[58] D. Herriott, H. Kogelnik, and R. Kompfner, "Off-axis paths in spherical mirror interferometers," *Applied Optics*, vol. 3, pp. 523–526, 1964.

[59] L. A. Sterczewski, J. Westberg, M. Bagheri, C. Frez, I. Vurgaftman, C. L. Canedy, *et al.*, "Mid-infrared dual-comb spectroscopy with interband cascade lasers," *Optics Letters*, vol. 44, pp. 2113–2116, 2019.

[60] J. Westberg, L. A. Sterczewski, M. Bagheri, C. Frez, I. Vurgaftman, C. L. Canedy, *et al.*, "Interband cascade laser-based dual-comb spectroscopy for methane sensing," in *Light, Energy and the Environment 2018 (E2, FTS, HISE, SOLAR, SSL)*, Singapore, 2018, p. EW2A.6.

[61] Y. Bidaux, I. Sergachev, W. Wuester, R. Maulini, T. Gresch, A. Bismuto, *et al.*, "Plasmon-enhanced waveguide for dispersion compensation in mid-infrared quantum cascade laser frequency combs," *Optics Letters*, vol. 42, pp. 1604–1607, 2017.

[62] B. Yves, K. Filippos, J. Pierre, B. Mattias, and F. Jérôme, "Coupled-waveguides for dispersion compensation in semiconductor lasers," *Laser & Photonics Reviews*, vol. 12, p. 1870025, 2018.

[63] M. Gianella, A. Nataraj, B. Tuzson, P. Jouy, F. Kapsalidis, M. Beck, *et al.*, "High-resolution and gapless dual comb spectroscopy with current-tuned quantum cascade lasers," *Optics Express*, vol. 28, pp. 6197–6208, 2020.

[64] S.-H. Yang and M. Jarrahi, "Frequency-tunable continuous-wave terahertz sources based on GaAs plasmonic photomixers," *Applied Physics Letters*, vol. 107, p. 131111, 2015.

[65] M. Rösch, G. Scalari, M. Beck, and J. Faist, "Octave-spanning semiconductor laser," *Nature Photonics*, vol. 9, pp. 42–47, 2015.

[66] A. Khalatpour, A. K. Paulsen, C. Deimert, Z. R. Wasilewski, and Q. Hu, "High-power portable terahertz laser systems," *Nature Photonics*, vol. 15, pp. 16–20, 2021.

[67] H. Li, Z. Li, W. Wan, K. Zhou, X. Liao, S. Yang, *et al.*, "Toward compact and real-time terahertz dual-comb spectroscopy employing a self-detection scheme," *ACS Photonics*, vol. 7, pp. 49–56, 2020.

[68] Y. Yang, D. Burghoff, J. Reno, and Q. Hu, "Achieving comb formation over the entire lasing range of quantum cascade lasers," *Optics Letters*, vol. 42, pp. 3888–3891, 2017.

[69] M. Rösch, M. Beck, M. J. Süess, D. Bachmann, K. Unterrainer, J. Faist, *et al.*, "Heterogeneous terahertz quantum cascade lasers exceeding 1.9 THz spectral bandwidth and featuring dual comb operation," *Nanophotonics*, vol. 7, p. 237, 2018.

[70] T. Sakamoto, A. Portieri, D. D. Arnone, P. F. Taday, T. Kawanishi, and Y. Hiyama, "Coating and density distribution analysis of commercial ciprofloxacin hydrochloride monohydrate tablets by terahertz pulsed spectroscopy and imaging," *Journal of Pharmaceutical Innovation*, vol. 7, pp. 87–93, 2012.

[71] S. Zhong, Y.-C. Shen, L. Ho, R. K. May, J. A. Zeitler, M. Evans, *et al.*, "Non-destructive quantification of pharmaceutical tablet coatings using terahertz pulsed imaging and optical coherence tomography," *Optics and Lasers in Engineering*, vol. 49, pp. 361–365, 2011.

[72] L. Ho, R. Müller, M. Römer, K. C. Gordon, J. Heinämäki, P. Kleinebudde, *et al.*, "Analysis of sustained-release tablet film coats using terahertz pulsed imaging," *Journal of Controlled Release*, vol. 119, pp. 253–261, 2007.

[73] Y.-C. Shen, P. Taday, D. Newnham, M. Kemp, and M. Pepper, "3D chemical mapping using terahertz pulsed imaging" vol. 5727, pp. 24–31, 2005.

[74] P. Dean, N. K. Saat, S. P. Khanna, M. Salih, A. Burnett, J. Cunningham, *et al.*, "Dual-frequency imaging using an electrically tunable terahertz quantum cascade laser," *Optics Express*, vol. 17, pp. 20631–20641, 2009.

[75] M. Takahashi and Y. Ishikawa, "Terahertz vibrations of crystalline $\alpha$-D-glucose and the spectral change in mutual transitions between the anhydride and monohydrate," *Chemical Physics Letters*, vol. 642, pp. 29–34, 2015.

[76] S. Zong, G. Ren, S. Li, B. Zhang, J. Zhang, W. Qi, *et al.*, "Terahertz time-domain spectroscopy of l-histidine hydrochloride monohydrate," *Journal of Molecular Structure*, vol. 1157, pp. 486–491, 2018.

[77] M. Nedeljkovic, S. Stanković, C. J. Mitchell, A. Z. Khokhar, S. A. Reynolds, D. J. Thomson, *et al.*, "Mid-infrared thermo-optic modulators in SoI," *IEEE Photonics Technology Letters*, vol. 26, pp. 1352–1355, 2014.

[78] M. M. Milošević, M. Nedeljkovic, T. M. B. Masaud, E. Jaberansary, H. M. H. Chong, N. G. Emerson, *et al.*, "Silicon waveguides and devices for the mid-infrared," *Applied Physics Letters*, vol. 101, p. 121105, 2012.

# 15 Self-Mixing in Quantum Cascade Lasers: Theory and Applications

Paul Dean,[1] Jay Keeley,[1] Yah Leng Lim,[2] Karl Bertling,[2] Thomas Taimre,[2] Pierluigi Rubino,[1] Dragan Indjin,[1] and Aleksandar Rakic[2]

[1] University of Leeds
[2] University of Queensland

## 15.1 Introduction

Optical feedback (OF) occurs when a fraction of radiation emitted by a laser is reinjected into the laser cavity from an external reflector [1]. Its effects in laser systems have been studied since the first days of laser development, and are well documented to include the onset of many undesirable phenomenon such as increased intensity noise [2], chaotic behavior [3], and even coherence collapse [4]. Nevertheless, it was identified early on that OF could also be exploited to produce beneficial effects, for example the discrimination against unwanted Fabry–Pérot modes [5]. Indeed, the optical mixing ("self-mixing" (SM)) that occurs between the intra-cavity and reinjected fields under OF is now known under certain conditions to cause predictable and controllable perturbations to the laser operation that depend on both the amplitude and phase of the reinjected field [1, 6]. This realization led the first demonstrations of laser-feedback interferometry (LFI) [7, 8], a technique in which the optical properties of an external target forming an external cavity are transferred, through the SM effect, to measurable changes in laser operating parameters. In this way, a single laser device can be used as a compact interferometric sensor comprising the source, local oscillator, mixer, and shot noise-limited detector.

The first report of LFI demonstrated its application to displacement sensing of a remote target; in that case an external mirror situated 10 m from an optical maser [8]. However, the experimentally simple form of the LFI scheme, combined with its coherent sensing capability, has since motivated its application across a wide range of sensing modalities, including distance-ranging [9], vibrometry [10] and displacement sensing [11, 12], coherent imaging [13] and microscopy [14, 15], Doppler flow measurements [16], and materials analysis [17, 18]. Furthermore, the SM effect has more generally been exploited for the measurement of fundamental laser parameters including the emission spectrum [19], laser linewidth [20, 21], and linewidth enhancement factor (LEF) [22, 23]. The universality of the SM effect has also been reported within both class-A and class-B laser systems including gas lasers [8], semiconductor diode lasers [24], vertical-cavity surface emitting lasers (VCSELs) [25], mid-infrared [26] and terahertz (THz)-frequency [27] quantum cascade lasers (QCLs), interband cascade lasers [28], and fiber lasers [15].

Yet it is the THz region of the electromagnetic spectrum (conventionally taken to lie from ∼0.1 THz to 10 THz) in which the phenomenon of OF has sparked significant research interest in recent years. This interest stems primarily from the high output powers and low phase noise offered by THz QCL sources, which when combined with the high sensitivity afforded by coherent LFI schemes opens up enormous potential for the development of compact sensing and imaging systems at THz frequencies. Indeed, exploitation of the SM effect in THz QCLs offers an attractive solution to the lack of convenient, fast, and sensitive detector technologies that has hindered metrological applications in this region of the spectrum to date.

In this chapter, we will begin in Section 15.2 by presenting a simple theoretical framework that describes the SM effect in a generalized laser system: the three-mirror laser model. This model provides a quantitative description of the measurable changes to the operating parameters of a laser under OF in the quasi-static regime, and lays the basis for the application of LFI across a range of experimental implementations. The physics and experimental observations of the SM effect in THz QCLs will then be discussed in Section 15.3. In Section 15.4, recent research in this field will be presented. This will focus firstly on the use of the SM effect for the measurement of QCL parameters including the emission spectrum and LEF. Secondly, we will present application-driven research in the field of THz sensing, including coherent imaging, materials analysis, and near-field microscopy. We conclude with a summary and outlook of LFI utilizing THz QCLs in Section 15.5.

## 15.2    Theory of Lasers under Optical Feedback

The SM effect describes the perturbations to the electrical and optical properties of a laser that occur when a fraction of emitted radiation is reinjected into the laser cavity from an external reflector. The reflected radiation interferes with the field in the laser cavity, causing a change in optical gain (or equivalently the carrier density) that depends not only on the amplitude but also the phase of the reinjected field [1, 6]. In turn, this change in gain induces variations in the optical power, lasing frequency and, in the case of semiconductor lasers, the laser terminal voltage (referred to herein as the "SM voltage," but alternatively referred to as the "LFI signal"). Through measurement of these variations, a single laser device can be employed simultaneously as a radiation source and a coherent detector.

The foundational model describing a semiconductor laser subject to OF was presented in the seminal work of Lang and Kobayashi [29]. This model adopts a rate equation approach to describe the carrier density and complex field of a laser system, with the inclusion of a time-delayed field term to describe the presence of OF. Although this model fully captures the essence of laser dynamics under OF, here we will concern ourselves only with the steady state solutions for the threshold carrier density $N$ and laser frequency $\nu$ [1, 7]:

$$2\pi \tau_{\text{ext}} (\nu_0 - \nu) = C \sin (2\pi \nu \tau_{\text{ext}} + \arctan (\alpha)); \qquad (15.1)$$

$$N - N_0 = -\beta' \cos (2\pi \nu \tau_{\text{ext}}), \qquad (15.2)$$

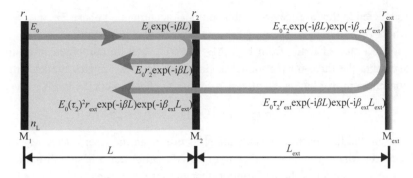

**Fig. 15.1** The three-mirror model of a laser under optical feedback. Light with electric field amplitude $E_0$ at laser mirror $M_1$ propagates in the laser cavity of length $L$ and undergoes a reflection at laser mirror $M_2$. A portion of the radiation is reinjected to the cavity after transmission through $M_2$ and reflection from the external mirror $M_{ext}$.

where the subscript 0 denotes parameters for the solitary laser, $\alpha$ is the LEF, and $\beta'$ is a parameter representing the coupling rate of feedback relative to the rate of carrier density gain. Unsurprisingly, these perturbations to the carrier density and laser frequency are also intrinsically linked to the physical parameters of the external cavity including the external cavity length $L_{ext}$, which gives rise to the round-trip delay $\tau_{ext} = 2L_{ext}/c$, and the reflectance of the external cavity mirror $R_{ext}$. This latter quantity is encapsulated in the dimensionless feedback parameter $C$ that defines the strength of OF and is written as

$$C = \varepsilon \frac{\tau_{ext}}{\tau_L} \sqrt{\left(1 + \alpha^2\right)} \sqrt{\frac{R_{ext}}{R_L}} \left(1 - R_L\right) \qquad (15.3)$$

where $R_L$ is the reflectance of the laser facet, $\tau_L$ is the laser cavity round-trip time, and $\varepsilon$ is a coupling constant that accounts for partial reinjection of radiation into the laser cavity and other optical losses.

It is particularly noteworthy that Eq. (15.1), known as the "excess phase equation," can also be obtained through entirely geometric considerations using a three-mirror laser model. This alternative theoretical description captures the simple architecture of the LFI scheme, and is outlined here. For further details the reader is referred to the derivations presented elsewhere [6, 7, 30].

The three-mirror model for a laser under OF is depicted in Fig. 15.1. The laser cavity is defined by two mirrors $M_1$ and $M_2$ with field reflection coefficients $r_1$ and $r_2$, respectively, which form a cavity of length $L$ and with internal refractive index $n_L$. The external reflector responsible for OF to this laser cavity is represented by a third external mirror $M_{ext}$, with field reflection coefficient $r_{ext}$, which forms an external cavity with the mirror $M_2$ of length $L_{ext}$. Light circulates within the laser cavity undergoing reflections at mirrors $M_1$ and $M_2$ with a round-trip time $\tau_L = 2n_L L/c$. The phase accumulated for each round-trip is $2\pi\nu\tau_L = 2\beta L$, where the wavenumber $\beta = 2\pi n_L \nu/c$. A fraction of light is also transmitted through the mirror $M_2$ with field transmission coefficient $t_2 = 1 - r_2$, reflected at the external mirror $M_{ext}$ and reinjected to the laser cavity with additional accumulated phase $2\pi\nu\tau_{ext} = 2\beta_{ext}L_{ext}$, where $\beta_{ext} = 2\pi\nu/c$ assuming the external cavity is in vacuum.

The reflected fields from both $M_2$ and $M_{ext}$ can be combined in a single term, such that the three-mirror model is simplified to a two-mirror model with a single, combined mirror having an effective complex reflection coefficient $r_{eff}$ given by

$$r_{eff} = r_2 + \left(1 - r_2^2\right) r_{ext} \exp\left(-i2\pi\nu\tau_{ext}\right). \tag{15.4}$$

Here the first term represents the reflection from the laser mirror $M_2$ whereas the second term represents the reflection from $M_{ext}$ after transmission through $M_2$. Inherent to this model is the assumption that the laser operates on a single longitudinal mode and that radiation experiences only a single round-trip in the external cavity. This latter assumption is valid under weak OF for which $r_2 \gg r_{ext}$, which is satisfied in the majority of experimental LFI schemes. Nevertheless, for stronger feedback, a summation of multiple reflected electric fields injected into the internal cavity can replace the second term in Eq. (15.4) [31, 32].

To proceed, the effective reflection coefficient $r_{eff}$ is rewritten as a single complex term

$$r_{eff} = |r_{eff}| \exp\left(-i\phi_{eff}\right), \tag{15.5}$$

where $|r_{eff}|$ and $\phi_{eff}$ are the amplitude and phase of the effective reflection coefficient. Under the assumption of weak OF the amplitude and phase of the reflection coefficient can be expressed as [6]

$$|r_{eff}| \approx \mathrm{Re}\left[r_{eff}\right] = r_2\left[1 + \kappa_{ext}\cos\left(2\pi\nu\tau_{ext}\right)\right], \tag{15.6}$$

$$\phi_{eff} \approx \frac{\mathrm{Im}\left[r_{eff}\right]}{r_2} = \kappa_{ext}\sin\left(2\pi\nu\tau_{ext}\right), \tag{15.7}$$

where the feedback coupling coefficient is

$$\kappa_{ext} = \varepsilon\sqrt{\frac{R_{ext}}{R_L}}\left(1 - R_L\right) = C\frac{\tau_L}{\tau_{ext}}\frac{1}{\sqrt{\left(1 + \alpha^2\right)}}, \tag{15.8}$$

in which the mirror reflection coefficients $r_2$ and $r_{ext}$ for field amplitudes have been replaced by their respective power reflectances $R_L$ and $R_{ext}$.

To obtain the perturbation to the threshold gain under OF we invoke the standard laser threshold condition for a two-mirror laser with facet reflection coefficients $|r_1|(= \sqrt{R_L})$ and $r_{eff}$, and cavity losses per unit length $\alpha_c$,

$$g_{th} = \alpha_c - \frac{1}{L}ln(|r_1 r_{eff}|). \tag{15.9}$$

By substituting Eq. (15.6), expanding the natural logarithm as a power series, and noting that $\kappa_{ext} \ll 1$ for weak OF, the change in threshold gain under OF becomes

$$\Delta g = -\frac{\kappa_{ext}}{L}\cos\left(2\pi\nu\tau_{ext}\right). \tag{15.10}$$

Since the change in gain can be assumed to be proportional to the change in carrier density for small perturbations, it can be seen that this equation is equivalent to

Eq. (15.2) obtained through the Lang and Kobayashi formalism. Equation (15.10) also elucidates the form of the LFI signal that can be observed by monitoring the emitted power or the laser voltage, both of which can be assumed proportional to the perturbation to the carrier density [6]. This leads to the following equation for the SM voltage applicable to a generic LFI scheme:

$$V_{SM} \propto \varepsilon \sqrt{R_{\text{ext}}} \cos \left( \frac{4\pi \, (L_{\text{ext}} + \partial L) \, \nu}{c} + \theta_R \right), \tag{15.11}$$

in which $\partial L$ represents a small change in external cavity length arising, for example, from the surface morphology or longitudinal displacement of the target, and $\theta_R$ accounts for any phase change occurring on reflection at the target.

To arrive at the excess phase equation (Eq. (15.1)) we first note the phase condition for the field propagating in the external cavity,

$$2\beta L + \phi_{\text{eff}} = 2\pi m, \tag{15.12}$$

where $m$ is an integer. At threshold and without OF (i.e., with $\phi_{\text{eff}} = 0$), this equation determines the laser emission frequency $\nu_0$. However, under OF both the emission frequency and cavity index experience perturbations $(\nu - \nu_0)$ and $\Delta n_L$, respectively, such that

$$\Delta(n_L \nu) = \nu_0 \Delta n_L + (\nu - \nu_0) \, n_L. \tag{15.13}$$

The change in the compound cavity phase accumulation (relative to a fixed phase $2\pi m$) due to OF can therefore be expressed as

$$\Delta\phi = \frac{4\pi L}{c} \left[ \nu_0 \Delta n_L + (\nu - \nu_0) \, n_L \right] + \phi_{\text{eff}}. \tag{15.14}$$

Noting that the cavity index is a function of both the carrier density and emission frequency, and making use of the relation [6],

$$\frac{\delta n_L}{\delta N} (N - N_0) = -\frac{\alpha c}{4\pi \nu_0} \Delta g. \tag{15.15}$$

Equation (15.14) can be rewritten as

$$
\begin{aligned}
\Delta\phi &= \frac{4\pi L}{c} \left[ \nu_0 \frac{\delta n_L}{\delta N} (N - N_0) + \nu_0 \frac{\delta n_L}{\delta \nu} (\nu - \nu_0) + (\nu - \nu_0) \, n_L \right] + \phi_{\text{eff}} \\
&= \frac{4\pi L}{c} \left[ -\frac{\alpha c}{4\pi \nu_0} \Delta g + \nu_0 \frac{\delta n_L}{\delta \nu} (\nu - \nu_0) + (\nu - \nu_0) \, n_L \right] + \phi_{\text{eff}}.
\end{aligned}
\tag{15.16}
$$

Substituting for the group index $n_g = n_L + \nu \frac{\delta n_L}{\delta \nu}$ and making use of Eqs. (15.7) and (15.10), we thereby arrive at the excess phase equation,

$$\Delta\phi = \frac{4\pi n_g L}{c} (\nu - \nu_0) + \kappa_{\text{ext}} \left[ \sin \left( 2\pi \nu \tau_{\text{ext}} \right) + \alpha \cos \left( 2\pi \nu \tau_{\text{ext}} \right) \right]. \tag{15.17}$$

Considering that $\Delta\phi = 0$ at the resonant frequency of the compound cavity and using Eq. (15.8), this equation reduces to the identical form of Eq. (15.1) obtained from the Lang and Kobayashi model.

At this stage it is worth briefly highlighting some noteworthy properties and physical consequences of this excess phase equation, which is central to the theoretical description of the SM effect. In particular, Eq. (15.1) is a transcendental equation that exhibits a unique solution only when $C \leq 1$, corresponding to the regime of weak feedback. Under this regime the perturbed laser emission frequency also remains close to that of the solitary laser ($\nu \approx \nu_0$), and the SM voltage varies approximately sinusoidally with a linear change in external cavity phase according to Eq. (15.11). For moderate and strong feedback ($C > 1$) multiple solutions to Eq. (15.1) exist, and the alternating stability of these solutions brings about phase-dependent hysteresis as $\nu_0$, $L_{ext}$, $C$, or $\alpha$ varies. Further discussion of how the feedback strength affects the morphology of SM signals obtained in LFI can be found in [7, 24]. For details of how to numerically solve the excess phase equation, the reader is referred to [33].

It is also worth noting here that five distinct and qualitatively different regimes of operation have been identified in interband lasers under different feedback levels [34, 35], including unstable regimes of coherence collapse for very strong feedback. However, recent research suggests that both mid-infrared (MIR) and THz QCLs can tolerate feedback levels far greater than those that would cause instability in interband devices [36, 37]. This absence of coherence collapse or other continuous-wave instabilities has been attributed to the high value of the photon to carrier lifetime ratio and the negligible linewidth enhancement factor of QCLs.

In this chapter we will focus on QCL devices operating under weak to moderate feedback. Under these regimes, the operating parameters of the laser remain dependent on the external cavity phase, and the resulting SM signals vary predictably with variation of the external cavity length as well as the optical properties of the external target.

## 15.3     The Self-Mixing Effect in Terahertz QCLs

Over the past decades there has been significant research interest in the investigation of the SM effect and the implementation of LFI systems utilizing visible and near-infrared lasers [24]. Nevertheless, despite the THz QCL being first demonstrated in 2002 [38] it is only within recent years that the phenomenon of OF in these devices has been studied and its unique potential for sensing schemes appreciated fully [23, 27]. One of the foremost advantages offered by the SM detection approach is the possibility of circumventing the reliance on traditional THz detectors, which are commonly either slow, unresponsive, bulky or require external cryogenic cooling. This opportunity arises from the remarkably large voltage perturbation observed in THz QCLs under OF, which can be two orders of magnitude greater than that observed in traditional diode lasers [27]. This contrasts the situation in diode lasers, for which monitoring the optical output power using a photodiode is preferred due to the superior SNR this approach offers compared to voltage sensing [39].

A typical experimental implementation of the SM voltage sensing scheme is shown in Fig. 15.2. In this example the THz QCL was based on a bound-to-continuum active

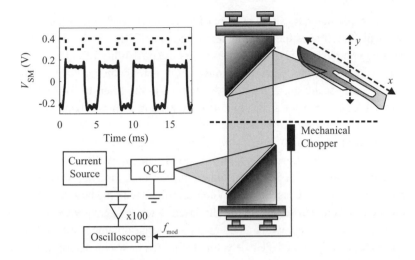

**Fig. 15.2** Experimental system for observing the voltage perturbation on a THz QCL under optical feedback from an external reflector. The QCL terminal voltage is monitored using an oscilloscope or a lock-in amplifier following amplification using an ac-coupled differential amplifier. In this scheme optical feedback to the laser is modulated at a frequency $f_{mod} = 215\,\text{Hz}$ using a mechanical chopper. Inset: the amplified QCL voltage $V_{SM}$ (solid line) and the reference waveform of the mechanical modulation (dashed line, offset). (Adapted with permission from [27], Copyright 2011 The Optical Society.)

region [40] emitting at $\sim$2.6 THz. The laser was processed into a surface plasmon ridge waveguide device with dimensions $3\,\text{mm} \times 140\,\mu\text{m}$. The QCL was cooled to a temperature of 25 K using a continuous-flow cryostat and driven with a dc current source. It is determined experimentally that the largest SM voltages are observed when the QCL is driven just above threshold [27] (in this particular case, at a current of 900 mA), which is also in accordance with the simple model described in [39]. Radiation from the QCL was collected using an F/2 parabolic reflector and focused onto the external target using a second identical reflector. These two parabolic reflectors also served to couple radiation reflected from the target back into the laser cavity. In this scheme the response to OF was monitored directly via the QCL terminal voltage following amplification using an ac-coupled differential amplifier providing 20 dB voltage gain.

In an LFI system the phase response described by Eq. (15.1) is not usually observed directly. Rather, it is the perturbation to the laser gain under OF that is responsible for the experimentally observable SM voltage. In order to facilitate detection of this voltage response it is common to modulate the optical field reinjected into the laser cavity and employ a lock-in detection at this modulation frequency. This is most simply achieved using a mechanical chopper as shown in Fig. 15.2, although modulation of the laser driving current can also be employed [17, 41, 42]. The inset to Fig. 15.2 shows the voltage perturbation observed directly on an oscilloscope, as well as the reference waveform for the mechanical light modulation. As can be seen, the voltage perturbation induced by OF is of the order of millivolts before amplification, far exceeding the microvolt level commonly observed in laser diodes.

The experimentally simple nature of this detection scheme is particularly notewor-thy, which facilitates the development of compact THz sensing systems. The voltage sensing approach in THz QCLs also offers the possibility of extremely fast detection bandwidths since the speed of response to OF is fundamentally determined by elastic and inelastic scattering dynamics of electrons in the laser. In THz QCLs these occur on the picosecond time scales [43, 44], suggesting response bandwidths on the order of $\sim$10–100 GHz should be achievable.

One further remarkable consideration concerning the voltage sensing scheme in THz QCLs is the extremely high detection sensitivity that it provides. The self-coherent nature of the SM effect inherently ensures the suppression of background radiation entering the laser cavity. When combined with the large voltage perturbations induced by coherent reinjection to the laser, detection sensitivities at the $\sim$pW/$\sqrt{\text{Hz}}$ level can be achieved, which is comparable to that obtained in commercially availa-ble cryogenically cooled THz detectors [45]. This has been determined recently using an experimental system similar to that shown in Fig. 15.2, with a planar gold mirror forming an external cavity of length $L_{\text{ext}} = 0.48$ m. The QCL used consisted of a 14-μm-thick GaAs/AlGaAs nine-well active region emitting at a frequency of $\sim$3.3 THz [46], which was processed into a waveguide of length $L = 2.9$ mm. At the driving current of 400 mA the emitted power measured at the external planar mirror using a calibrated THz power meter was found to be $\sim$22.4 μW. Interferometric variations on the terminal voltage, of the form described by Eq. (15.11), were recorded by extend-ing the external cavity over a distance of $\partial L = 200$ μm in steps of 4 μm, with $V_{\text{SM}}$ being recorded at each mirror position. In this experiment the total field attenuation factor can be expressed as $\varepsilon = \varepsilon_c \varepsilon_a$, in which $\varepsilon_c$ is a coupling constant representing losses inherent to the optical system arising due to imperfect collection of the emit-ted radiation, attenuation of the field propagating in the external cavity, and spatial mode mismatch between the reinjected and the cavity mode. The coupling constant $\varepsilon_a$ was controlled by introducing calibrated THz power attenuators in the external cavity, which provided additional double-pass (power) attenuation in the range $\varepsilon_a^2 = 0$–90 dB. This allowed the maximum tolerable additional cavity attenuation to be determined, as the level of attenuation where interferometric variations on the laser terminal voltage were reduced to the voltage noise level.

Figure 15.3(a) shows the SM voltage interferograms recorded with a detection bandwidth $\sim$0.025 Hz, for varying degrees of additional cavity attenuation. The non-sinusoidal form of these interferograms is a result of $\nu$ in Eq. (15.11) being dependent on the external cavity length, as described by the excess phase equation, Eq. (15.1). Figure 15.3(b) shows the peak-to-peak signal as a function of additional field atten-uation in the cavity; a linear relationship between $\varepsilon_a$ and $V_{\text{SM}}$ can be established, as expected from Eq. (15.11). Also shown is the voltage noise floor of the sys-tem, which was dominated by current noise in the laser driver and determined to be $\sim$1.5 μV/$\sqrt{\text{Hz}}$. As can be seen from this data, variations on the laser voltage can be resolved with up to $\sim$40 dB (80 dB) of additional field (power) attenuation introduced in the external cavity. Based on the power measured in the external cavity, and with-out accounting for reinjection losses, this measurement corresponds to a remarkably

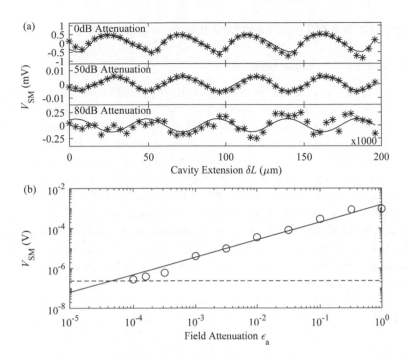

**Fig. 15.3** (a) Self-mixing voltage interferograms recorded over an external cavity extension of 200 μm for various levels of double-pass power attenuation $\varepsilon_a^2$ in the external cavity. Fits to Eq. (15.11) are shown as solid lines. The data in the bottom panel has been scaled by a factor of 1000. (b) Self-mixing signal amplitude as a function of an additional field attenuation introduced in the external cavity, $\varepsilon_a$. The dashed horizontal line indicates the voltage noise floor, at a level $\sim 1.5\,\mu V/\sqrt{Hz}$.

low detection sensitivity of $\sim 1.4\,pW/\sqrt{Hz}$. It should be noted, however, that the power coupled coherently to the laser active region is likely to be several orders of magnitude lower than this figure. For example, by inserting experimental values into Eq. (15.3) and assuming a value of $C = 1$ we can estimate collection and reinjection losses on the order of $\varepsilon_c^2 \sim -35\,dB$ for this system. Assuming collection losses of 10 dB this implies a reinjected power smaller than $\sim 10\,fW/\sqrt{Hz}$.

As a final note, it is interesting to evaluate the intrinsic limit for the maximum tolerable total field attenuation for this QCL source. It can be expected that OF induces a non-negligible effect on the operating parameters of a laser when the rate of photon reinjection to the cavity exceeds the rate of spontaneous emission in the cavity [47]. Using a recently reported reduced rate equation model of a THz QCL under OF [48] with typical parameter values for this QCL device, we predict an intrinsic limit for the maximum tolerable total power attenuation $\varepsilon^2 \sim 150\,dB$, which is in reasonable agreement with the experimentally determined value obtained above. Considering the intrinsic stability of QCLs under optical reinjection [36, 37], this experimental observation also suggests an extraordinarily large range of feedback strengths under which QCL operation remains both stable and measurable.

## 15.4        Applications of Laser Feedback Interferometry

### 15.4.1        Measurement of Laser Parameters

As we have seen from the theoretical framework presented in Section 15.2, the laser response to OF is dependent on parameters including the external cavity length, target reflectance, laser emission frequency, and LEF. The first two of these relate to the target properties and present opportunities for imaging and sensing applications, as outlined in Section 15.4.2. The latter two relate to optical parameters of the laser itself and can be determined through LFI measurements, provided the properties of the target and external cavity are known. In fact, the LFI approach provides a convenient means of measuring laser parameters electrically and without need for an external detector or spectrometer. One example of this is the measurement of RMS phase noise in a MIR QCL from the associated LFI spectrum [21]. However, LFI is perhaps most advantageous in the THz region of the spectrum which suffers from a lack of convenient instruments and detectors.

The LEF, $\alpha$, is a dimensionless parameter that describes the coupling between refractive index and gain in semiconductor lasers; a finite LEF indicates that the laser linewidth becomes broadened relative to the Schawlow–Townes limit [49]. One method used commonly to measure the LEF in semiconductor lasers relies on measurements of the sub-threshold emission spectrum and determination of the change in gain and wavelength as a function of pump current [50]. However, this method provides only the LEF below threshold. An alternative technique involves measuring amplitude and phase modulation of the laser but requires a fast detector [51]. The LFI approach overcomes both of these limitations.

For intersubband devices such as QCLs, the laser subbands responsible for the optical transition exhibit the same curvature in k-space, giving rise to a spectrally symmetric differential gain. As such, an LEF close to zero is expected [52], although LFI measurements undertaken with both MIR and THz QCLs have shown that $\alpha$ can deviate from zero in practical devices, and furthermore can vary with laser operating conditions. These measurement approaches rely implicitly on the fact that the shape and form of the SM signals obtained in LFI depend on $\alpha$, as described by Eq. (15.11). By measuring the laser response to a changing external cavity length, the value of $\alpha$ can thus be extracted through suitable analysis of these signals. Such analysis techniques include simple methods based on analysis of the shape of the SM signals [23, 53–55] and parameter fitting methods [56, 57]. For MIR QCLs, values of $\alpha$ in the range $-2$ to $+2.3$ have been reported using these measurement techniques [54, 56], and attributed to asymmetric gain arising from nonparabolicity of the laser subbands. In the case of THz QCLs, values up to $\alpha \sim 0.5$ have been reported and similarly explained by an asymmetric gain arising from specific absorbing transitions between states not directly involved in the laser transition [23].

This same realization that the laser response to OF inherently portrays the optical parameters of the laser itself has led to the recent demonstration of a new approach to measurement of the laser emission spectrum by LFI [19]. The success of this technique relies inherently on several useful qualities of the LFI scheme and the behavior of lasers under weak OF. The first of these is the ability of LFI schemes to operate with high sensitivity in the regime of weak OF; this has been demonstrated for the case of THz QCLs in Section 15.3. The second is the fact that the spectral characteristics of a laser under weak OF replicate those of the solitary laser (rather than those of the external cavity), as can be seen from inspection of the excess phase equation (Eq. (15.1)). Lastly it is the realization that under weak OF the SM signal can be approximated as a linear combination of individual signals arising from different lasing modes [58, 59]. Specifically, for a solitary laser emitting on $M$ longitudinal modes, the SM voltage signal can be expressed as

$$V_{\text{SM}} = \sum_{i=1}^{M} \beta_i \cos\left(2\pi \nu_i \tau_{\text{ext}}\right), \tag{15.18}$$

in which $\nu_i$ is the frequency of the $i$th mode and $\beta_i$ is a coefficient describing the intensity of that mode. Through monitoring the SM voltage as the cavity length is extended an interferogram can thereby be obtained in a means conceptually similar to that used in conventional Fourier transform infrared (FTIR) spectroscopy. Through Fourier analysis of this interferogram the emission spectrum of the laser can then be reproduced over a wide spectral bandwidth and with high resolution.

Figure 15.4(a) shows exemplar interferograms recorded for a cavity extension $\partial L = 200 \, \text{mm}$, for the cases of a THz QCL emitting both on a single longitudinal mode (top panel) and in multiple longitudinal modes (bottom panel) at $\sim 2.25 \, \text{THz}$, controlled through the driving current. For these measurements the feedback parameter was controlled using crossed polarizers positioned in the external cavity to provide a value $C \sim 0.6$. In the latter case the multiple frequencies propagating in the external cavity manifest as periodicities observed in the SM signal, as described by Eq. (15.18). Figure 15.4(b) shows the emission spectra obtained at two different laser driving currents by performing an FFT of the recorded interferometric signals. These spectra reproduce the expected single- and multiple-longitudinal mode emission, which were confirmed through a traditional FTIR measurement performed using radiation emitted from the opposite laser facet (also shown for comparison in Fig. 15.4(b)).

One advantage of this approach to THz spectral measurements is the comparative simplicity of the experimental arrangement. In addition to circumventing the need for a THz detector this technique readily permits the use of long-scan range motion hardware, thereby providing a high spectral resolution without recourse to an expensive FTIR system. For example, the cavity extension $\partial L = 200 \, \text{mm}$ used above translates to a spectral resolution $c/2\partial L = 750 \, \text{MHz}$. Another notable advantage is the ability to recover strong signals close to the laser threshold, at which the SM signal is largest but the emitted power is low.

## 15.4.2 Terahertz Imaging Using the Self-Mixing Effect in QCLs

In recent years there has been significant research activity in the realms of THz imaging and sensing using THz QCL sources. Much of this activity has been driven by

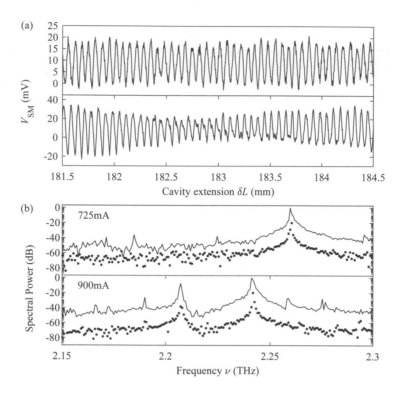

**Fig. 15.4** (a) Self-mixing interferograms recorded with a ~2.25 THz QCL lasing on a single (top panel) and multiple (bottom panel) longitudinal laser cavity modes. (b) Normalized emission spectra obtained from self-mixing interferograms (solid lines) and FTIR spectroscopy measurements (dotted lines; offset for clarity) for single-mode (top panel; driving current 725 mA) and multiple-mode (bottom panel; driving current 900 mA) operation. (Adapted with permission from [19] under a Creative Commons Attribution 4.0 International License.)

the diverse range of potential applications of THz radiation including biomedicine, microscopy, security controls, non-destructive analysis and inspection, and spectroscopic mapping of materials [60, 61]. These opportunities stem from the unique combination of properties of THz radiation, including: the ability to penetrate many non-polar and non-metallic materials that are opaque to visible/near-infrared radiation; the capability of sub-millimeter imaging resolution in the far-field; the ability to excite vibrational modes in organic and inorganic materials, providing sensitivity to both chemical composition and crystalline structure [62, 63]; and the non-ionizing nature and strong sensitivity to water content and hydration, of relevance to biomedical imaging applications [64, 65]. The THz QCL, in particular, offers significant benefits for imaging applications compared to alternative THz sources – namely: high optical powers [66]; narrow linewidths and high spectral purity [67]; emission at frequencies in the entire 1–5 THz range and beyond [68]; and compact size. For a detailed review of the use of QCLs for THz imaging the reader is referred to [41].

Given the benefits and convenience of THz QCL sources, as well as the experimental simplicity of the LFI sensing approach described in Section 15.3, it is perhaps not surprising that significant research activity has focused on the combination of these technologies for THz imaging and sensing applications. Of particular relevance to the development of stand-off imaging systems is the geometry of the LFI approach, which inherently lends itself to reflective imaging modalities. This has enabled the investigation of highly absorbing or geometrically thick samples without the need for additional optical components or the experimental complexity of coupling the reflected beam to an external detector [26, 27, 69–71]. As described in Section 15.2, the SM detection approach is also inherently sensitive to both the amplitude and phase of the reinjected THz field. This enables a departure from traditional QCL-based imaging approaches, which have until recently almost exclusively employed incoherent thermal detectors [41]. In contrast, the high-power spectral density of QCL sources combined with coherent detection may offer extremely high dynamic range and detection close to the shot-noise limit [72, 73]. But perhaps most significant is the sensitivity of coherent LFI approaches to changes in the optical phase in the external cavity. Such changes can arise from changes in the cavity length, the refractive index inside the cavity, the phase change on reflection at the target, or a combination of these effects (see Eq. (15.11)). This opens up many opportunities for the development of compact LFI systems for depth-resolved imaging [13, 74], biomedical imaging [69], materials analysis [17, 18], displacement/vibration sensing [57, 75], and gas spectroscopy [76] at THz frequencies. Some of these applications will be discussed in the next section.

### 15.4.2.1 Two-Dimensional Imaging

The experimental setup illustrated in Fig. 15.2 lends itself naturally to the development of compact reflective imaging systems; by raster-scanning the target in the plane perpendicular to the incident beam and concurrently recording the SM voltage, a two-dimensional (2D) image of the target can be acquired readily. This experimental geometry is also inherently confocal due to the requirement to couple reflected radiation back into the small laser aperture, which enables high imaging resolutions to be attained. Figure 15.5(a) shows such an exemplar image of a UK two-pence coin with a diameter of 25.9 mm, obtained at a frequency of ~2.6 THz. In this case, the pixel size was 100 µm and the image resolution of the system was determined using reflective resolution targets to be better than 250 µm [27]. Similar sub-millimeter resolutions have also been reported elsewhere [26, 70]. In the case of Ren et al. [70], an adaptation to the traditional approach of exploiting OF in a laser was implemented through the use of a quantum cascade *amplifier* providing gain at ~2.9 THz. The amplifier was formed by coupling a high-refractive-index lens with an anti-reflective coating to the facet of a QCL, which has the effect of enhancing the mirror losses and thereby suppressing lasing action. In the presence of an external reflector the introduction of OF to the amplifier initiates lasing action and enhances photon-assisted transport through the structure, which in turn reduces the voltage across the device. Through the use of fast electrical pulses supplied to the quantum cascade amplifier an acquisition rate of 20 000 pixels per second was demonstrated. An image obtained using this system

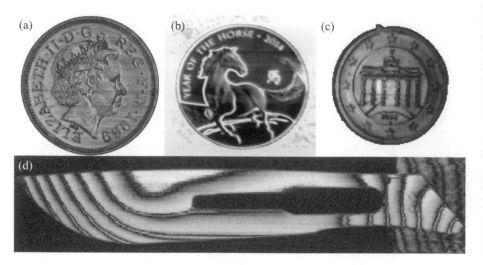

**Fig. 15.5** (a) Image of a British two-pence coin (25.9 mm diameter) obtained via the self-mixing voltage signal with a $\sim$2.6 THz QCL. (Adapted with permission from [27], Copyright 2011 The Optical Society.) (b) Image of a Lunar Year of the Horse 2014 Gold Coin (16.5 mm diameter) obtained with a $\sim$2.9 THz quantum cascade amplifier. (Reproduced from [70], with the permission of AIP Publishing.) (c) Image of a German 50-cent coin (24.2 mm diameter) obtained with a $\sim$3.3 THz QCL at a frame rate of 1 Hz. (Reproduced from [71], with the permission of AIP Publishing.) (d) Image of a scalpel blade obtained with a $\sim$2.6 THz QCL over a round-trip beam path of 21 m in unpurged air. The feature on the right-hand side arises due to reflections from the sample mount. (Copyright 2013 IEEE. Reprinted, with permission, from [77].)

is shown in Fig. 15.5(b). One further notable extension of the experimental arrangement depicted in Fig. 15.2 has been the implementation of beam-steering approaches enabling fast 2D imaging [71]. In this case a fast-scanning mirror is employed to translate the beam waist in the object plane, rather than mechanically scanning the target. The fast scanning capability provided by the small deflecting mirror, exploited in conjunction with the fast electrical response inherent to the SM effect, has enabled impressive frame rates up to 2 Hz (2200 pixels per second) to be demonstrated through this approach. An image obtained using this system at a frame rate of 1 Hz is shown in Fig. 15.5(c).

The high detection sensitivity afforded by the SM phenomenon in THz QCLs (see Section 15.3) also enables imaging over extended distances, provided care is taken to select a suitable QCL source emitting within a window of relatively high atmospheric transmission [78]. Figure 15.5(d) shows an image of a scalpel blade obtained at a frequency of $\sim$2.6 THz with a round-trip beam path of 21 m through air. The well-defined interference fringes evident in this figure arise from the dependence of the SM voltage on the length of the external cavity formed by the target (Eq. (15.11)); unless the target is optically flat and aligned perpendicular to the beam axis the length of the external cavity may vary as the target is scanned. In this case the long beam path was achieved by folding the collimated THz beam using multiple planar mirrors positioned between the parabolic reflectors. Even over this large distance a suitably large SNR is obtained in this image, suggesting that imaging over much greater distances may be possible. In fact, assuming a free-running linewidth $\Delta \nu_{LW} \sim 20\text{--}30\,\text{kHz}$ [79–81] for the QCL source in the presence of thermally and electrically induced frequency noise, the typical coherence length can be estimated to be $L_c = c/\Delta \nu_{LW} \sim 10\,\text{km}$ in vacuum. As such, the main practical limitations for long range stand-off sensing based on this approach can be identified as arising only from the practicalities of efficiently collecting and reinjecting radiation back-scattered from the remote target.

### 15.4.2.2 Three-Dimensional Imaging

The mechanical modulation scheme employed to obtain the images shown in Fig. 15.5(a) modulates only the amplitude of the reinjected field. The image contrast therefore arises from a combination of the spatial variations in both the target reflectance (through changes to the feedback parameter), as well as the surface morphology of the target (through the associated change in external cavity length). It is this latter effect that gives rise to the well-defined interference fringes evident in these figures. In order to untangle both the phase and amplitude variations embedded in the SM voltage, and therefore resolve both the reflectance and the surface morphology of the target, an alternative modulation scheme is required.

One experimental implementation of LFI enabling three-dimensional (3D) image reconstruction adopts the system architecture shown in Fig. 15.2, but in which the target is also mounted on a computer-controlled translation stage allowing longitudinal scanning along the beam path [13]. A complete SM voltage interferogram of the form described by Eq. (15.11) can thus be captured at each imaging pixel by synchronizing the longitudinal scanning with the 2D raster-scan of the target. This scheme has been demonstrated using a GaAs test target comprising three stepped regions, with a nominal step height $\sim 10\,\mu\text{m}$, formed by wet chemical etching. Half the surface of each of these three steps was also coated with 125 nm of gold in order to provide regions of differing reflectance. For each transverse ($x$–$y$) position the target was scanned longitudinally ($z$-direction) through a distance of 0.5 mm in steps of 0.5 $\mu$m, and the SM voltage captured at each $z$-position. With a lock-in time constant of 10 ms the resulting acquisition rate was 20 s/pixel. Figure 15.6 shows SM voltage interferograms recorded from two representative sampling positions on the target surface. These positions correspond to gold-coated and uncoated regions on different steps of the structure, which gives rise to relative variations in both the phase and amplitude of the SM interferograms. Also shown in the figure are fits to Eq. (15.11) obtained by treating $\varepsilon\sqrt{R_{ext}}$, $C$, $\alpha$, and $\partial L$ as free parameters (with the known target displacement $z$ being subsumed within the cavity length $L_{ext}$). From these fits the surface depth $\partial L$ profile of the sample could be determined, and showed good agreement with the profile measured using a non-contact optical profilometer.

The requirement for longitudinal scanning of the target, combined with the slow ($\sim 200\,\text{Hz}$) mechanical modulation employed for lock-in detection of the SM voltage, imposes a slow acquisition rate for the non-optimized system described above. A far superior approach to coherent 3D image reconstruction becomes apparent by

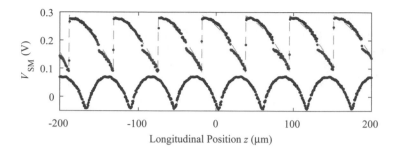

**Fig. 15.6** Self-mixing interferograms recorded from two positions on the surface of a stepped GaAs structure using a ~2.6 THz QCL, revealing different phases and amplitudes corresponding to different surface heights and reflectances. The top trace corresponds to a gold-coated region of the surface. Fits to Eq. (15.11) are shown as dashed lines. (Reproduced from [13], with the permission of AIP Publishing.)

considering that, according to Eq. (15.11), variation of the phase accumulated in the external cavity can also be induced through *frequency* modulation of the laser source. By applying a linear modulation to the laser driving current the SM signal can then be observed as a sequence of periodic perturbations superimposed on the (directly) modulated laser voltage. This electrically modulated LFI scheme in essence constitutes a swept-frequency delayed self-homodyning approach that shares similarity with swept-source optical coherence tomography [82]. In addition to the considerably faster acquisition that this approach provides, far superior depth resolutions can also be achieved through the reduced effects of slow thermal drift of the QCL emission frequency.

In this scheme the laser was modulated by a sawtooth current modulation with frequency $f_{mod} = 1$ kHz (90% duty cycle) and an amplitude of 100 mA, which was superimposed on a dc driving current. The THz QCL was based on a bound-to-continuum active region [40] emitting at ~2.65 THz ($\lambda$~113 μm), for which the modulation coefficient was $-8.5$ MHz/mA, providing $-850$ MHz frequency modulation. Unlike the previous scheme that employed lock-in detection the amplified QCL terminal voltage was here sampled at 500 kS/s using a 16-bit digital acquisition board. For each sampling position on the target the QCL voltage signal was averaged over $N = 200$ modulation periods and the direct voltage modulation, recorded in the absence of OF, was subtracted. Figure 15.7(a) shows the SM interferograms recorded from two representative sampling positions on the target, which again exhibit different amplitudes and phases in an equivalent fashion to Fig. 15.6.

The interferometric phase of the SM signal in this swept-frequency scheme can be expressed as [33, 74]

$$\phi(t) = \frac{4\pi L_0}{c}\gamma t + \frac{4\pi L_0}{c}v_0 = 2\pi f_c t + \varphi, \tag{15.19}$$

where $v_0$ is the laser frequency without feedback at $t = 0$, $L_0 = L_{ext} + \delta L$ is the external cavity length, $\gamma = -945$ GHz/s is the laser frequency modulation rate, and $f_c$ is the carrier frequency of the SM signal given by $f_c = 2\gamma L_0/c$. The quantity $\varphi$ in

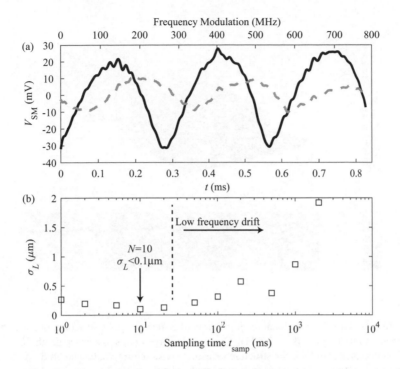

**Fig. 15.7** (a) Self-mixing interferograms recorded from two positions on the surface of a stepped GaAs structure, obtained by frequency modulation of a ∼2.6 THz QCL. The solid trace corresponds to a gold-coated region of the surface. (b) Depth resolution $\sigma_L$ plotted as a function of sampling time per pixel $t_{samp}$. For $N = 10$ waveform averages a resolution <0.1 μm is determined. (Adapted with permission from [74], Copyright 2015 The Optical Society.)

Eq. (15.19) is the initial phase of the SM fringe, which can be obtained from the phase of the complex fast Fourier transform (FFT) of the SM signal evaluated at the carrier frequency $f_c$, and is related to $L_0$ (and hence the surface morphology $\partial L(x, y)$) through the relation

$$L_0(x, y) = \frac{c}{4\pi\nu_0}\varphi(x, y). \tag{15.20}$$

The depth resolution achievable through this swept-frequency approach is determined by SM voltage noise, which can be reduced through averaging multiple interferograms, but also by frequency instability of the QCL that arise from driving current and temperature fluctuations. This latter effect can cause frequency drifts on the order of several MHz over time-scales of several seconds [80, 81]; the maximum frequency drift in this experiment was estimated to be ∼10 MHz based on a ±100 mK instability of the heat-sink temperature. To quantify the depth resolution, 50 successive SM interferograms were recorded for a fixed cavity length. The cavity length $L_0$ was then determined for each of these measurements using the analysis described above, and the standard deviation $\sigma_L$ calculated. This procedure was repeated for different

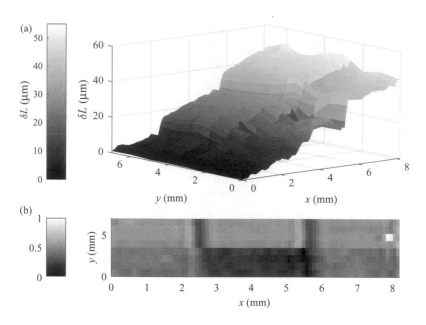

**Fig. 15.8** (a) Three dimensional reconstruction of a stepped GaAs structure, obtained by frequency modulation of a ∼2.6 THz QCL. Scale corresponds to surface depth $\partial L$.
(b) Reflectance profile of the structure revealing GaAs (bottom half) and gold-coated (top half) regions. The gray scale corresponds to normalized reflectance. (Adapted with permission from [74], Copyright 2015 The Optical Society.)

amounts of waveform averaging in the range $N = 1\text{–}2000$, corresponding to total sampling times per pixel $t_{samp} = N/f_{mod}$ in the range 1–2000 ms. The results of this analysis are shown in Fig. 15.7(b). For short sampling times $t_{samp} < 10$ ms, increased waveform averaging causes a reduction in depth resolution. An optimum resolution of better than 0.1 μm (corresponding to a phase change $\Delta\varphi \sim 0.6°$, or equivalently $\sim \lambda/1000$) is achieved for a sampling time of 10 ms/pixel. However, for sampling times $t_{samp} > 10$ ms a degradation of depth resolution is observed, which can be attributed to low-frequency drift of the laser emission frequency.

By applying Eq. (15.20) to the measured phase of the SM interferograms on a pixel-by-pixel basis, a full 3D image of the surface geometry can be constructed. This is shown in Fig. 15.8(a), which reproduces well the 10 μm steps in the target as well as the presence of a small (∼0.36°) tilt of the target surface. A spatial map of the target reflectance can also be obtained from the amplitudes of the SM interferograms, which is expected to scale with $\varepsilon\sqrt{R_{ext}}$ in accordance with Eq. (15.11). Figure 15.8(b) shows the variation of $\varepsilon^2 R_{ext}$ obtained from the magnitude of the complex FFT evaluated at the carrier frequency for each pixel. Regions of higher reflectance are clearly visible in the top rows of pixels, which correspond to the gold-coated half of the target. Nevertheless, the reconstructed amplitude is not homogenous across these regions. This can be explained by variations in the sample geometry and the presence of surface defects arising from the etching process, which cause spatial variation in the coupling constant $\varepsilon$.

### 15.4.2.3 Imaging for Materials Analysis

In the above examples of coherent imaging based on the LFI technique, the phase information encoded in the SM signals was used to reconstruct the target geometry. For optically flat samples the same approach can also be employed to spatially map the change in the phase of the THz field upon reflection at the target. This information, along with the field amplitude, can in turn be related to the complex refractive index of the target material. This frequency-domain approach can be considered analogous to the well-established technique of THz time-domain reflection spectroscopy [83].

The interferometric phase under frequency modulation of the laser follows the form given by Eq. (15.19), but in which $L_0 = L_{ext}$ is considered fixed and an additional term is included that accounts for the phase change on reflection, $\theta_R$ [17, 84]. For 2D imaging applications this implies the inherent assumption that the target surface is perpendicular to the beam axis. In practice this can be achieved most easily on small targets by adjusting the target alignment until no fringes due to sample tilt can be observed in the image; in the case of flat targets, such fringes are aligned parallel to each other and cover the entire sample surface, and can therefore be distinguished easily from fringes arising due to spatial variation of $\theta_R$. The phase change $\theta_R$ (and hence the interferometric phase) depends predominantly on the complex part of the refractive index (extinction coefficient $k$) and to a lesser extent the real part; the latter predominantly affects the shape of the SM interferogram through the feedback parameter $C$. By fitting the SM signal to the three-mirror laser model (see Section 15.2) the complex reflection coefficient of the target can thereby be deduced on a pixel-by-pixel basis. Using calibration standards of known complex refractive index, the complex indices of unknown materials can then be calculated with high accuracy. For a detailed description of the experimental method and the relationship between the SM signal parameters and the optical properties of the target the reader is referred to [17, 84]. One notable variation on this analysis arises in the case of spatially inhomogeneous targets such as granular materials or mixtures of materials; one example being plastic explosives that consist of a powdered explosive embedded in a matrix [85]. This challenge can be overcome by acquiring SM signals over multiple sampling positions to produce a "cloud" of data points in amplitude-phase space. A central point in this parameter space is then selected using the mean-shift algorithm, and is taken as representative of the ensemble characteristics of the target. Table 15.1 summarizes the refractive index and extinction coefficient for a range of polymers and plastic explosives measured using this materials analysis technique at a frequency of 2.6 THz. The excellent agreement with values obtained from other techniques demonstrates the high accuracy of this approach. One obvious and valuable extension to this technique would be the use of broadly tunable laser sources [86] or multiple sources emitting at targeted frequencies. This would potentially provide a powerful and compact technique for fast and high-resolution spectroscopic imaging at THz frequencies.

One particular field that may benefit from the capability to coherently probe THz properties of inhomogeneous materials using a reflective sensing geometry is the realm of biomedical imaging. Indeed, THz biomedical imaging has drawn much interest over recent decades for applications including discriminating common basal cell carcinoma [64], breast tumors [65], and metastasis liver [87] from surrounding healthy

**Table 15.1** Refractive index *n* and extinction coefficient *k* of exemplar polymers and plastic explosive samples obtained by self-mixing (SM) measurements and time-domain spectroscopy (TDS) [17, 85]: SX2 (1,3,5-trinitroperhydro-1,3,5-triazine (RDX) based); Metabel (1,3-dinitrato-2,2-bis(nitratomethyl)propane (PETN) based); Semtex (RDX and PETN based); POM (polyoxymethylene); PVC (polyvinyl chloride); and PA6 (polycaprolactam).

| Material | $n_{SM}$ | $n_{TDS}$ | $k_{SM}$ | $k_{TDS}$ |
|---|---|---|---|---|
| SX2 | 1.76 | 1.75 | 0.09 | 0.09 |
| METABEL | 1.66 | 1.66 | 0.06 | 0.07 |
| SEMTEX | 1.56 | 1.55 | 0.07 | 0.06 |
| POM | 1.65 | 1.66 | 0.011 | 0.012 |
| PVC | 1.66 | 1.66 | 0.063 | 0.062 |
| PA6 | 1.66 | 1.67 | 0.11 | 0.11 |

**Fig. 15.9** (a) THz amplitude and (b) THz phase images of a porcine skin sample obtained by frequency modulation of a ~2.6 THz QCL. The image size is 101 × 101 pixels with a pixel size of 200 μm. The inset of (a) is a high-resolution (51 × 301 pixels) image with a pixel size of 50 μm. The bright diagonal feature is an aluminum separator dividing samples oriented in cross-section (above) and *en face* (below) to the incident THz beam. The tissue types in the tissue cross-section are (from top to bottom): muscle; sub-dermal fat; and skin. (Adapted with permission from [69], Copyright 2014 The Optical Society.)

tissue, as well as for in vivo imaging and assessment of skin burns [88]. Figure 15.9 shows amplitude and phase images obtained using an LFI scheme with a QCL source at a frequency ~2.6 THz. In this case the exemplar targets were excised samples of porcine tissue oriented in both cross-section (top part of images) and *en face* (bottom part of images) geometries, which were fixed between a 2-mm-thick polymethylpen-tene (TPX) window and a backing plate. In the *en face* images several features are

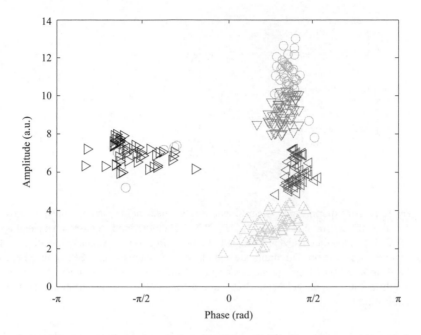

**Fig. 15.10** Amplitude-phase map of SM signals obtained from the images shown in Fig. 15.9, and corresponding to different types of tissue: epidermis (circles); upper dermis (down arrow); lower dermis (up arrow); sub-dermal fat (left arrow); and muscle tissue (right arrow). (Adapted with permission from [69], Copyright 2014 The Optical Society.)

present that are not evident by visual inspection. These have been identified through stained visible microscopy of the microtomed sample as arising from blood vessels, an arteriole and venule pair, and a sebaceous gland that are below the tissue surface (see [69] for details). Clear contrast between the different types of tissue (from top to bottom: muscle; sub-dermal fat; sub-dermal fascia; dermis; and epidermis) can also be seen in the cross-section part of the target. Furthermore, when the individual pixels corresponding to these tissue types are represented in amplitude–phase space, they form distinct clusters as shown in Fig. 15.10. This clustering of signatures suggests that THz-frequency LFI could find application to the non-invasive classification of skin tissue types.

### 15.4.2.4 Near-Field Microscopy

One fundamental limit imposed by the far-field LFI imaging approaches discussed above arises from the relatively long wavelength of THz frequency radiation. A simple calculation of the diffraction-limited beam size for a $TEM_{00}$ (Gaussian) mode focused by an F/1 optic reveals a resolution limit $\sim$64 μm for radiation at a frequency 3 THz. It is therefore clear that far-field approaches precludes the microscopic and nanoscopic investigation of samples in the THz region. In order to overcome this limitation and enable the investigation of samples exhibiting sub-wavelength features a number of

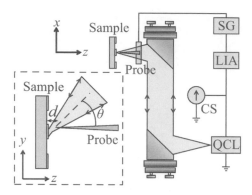

**Fig. 15.11** An experimental system for apertureless near-field scanning optical microscopy based on the self-mixing effect in a THz QCL. THz radiation is focused on a needle probe (or AFM tip) that is modulated mechanically in the $z$-direction, causing a modulation of the field scattered from the tip. The amplitude and phase of the demodulated SM voltage depend on the local (near-field) dielectric properties of the sample. CS – current source; LIA – lock-in amplifier; SG – signal generator. (Reproduced from [14], with the permission of AIP Publishing.)

near-field imaging techniques have been adopted from the visible/infrared regions [89–91]. Within recent years these techniques have found a range of THz applications, including the mapping of charge carriers in semiconductors and nanostructures [92, 93], the microscopic investigation of quantum dots [94] and nanowires [95], the investigation of metamaterials [96, 97], and microscopy of biomedical samples [98].

One powerful near-field imaging approach demonstrated in the THz range is apertureless (scattering-type) near-field scanning optical microscopy (s-SNOM) [99]. In this technique an atomic force microscope (AFM) probe equipped with a sub-micron scattering tip apex is positioned in close proximity to the surface of a sample and illuminated by a THz source. The near-field interaction between the tip apex and the sample surface introduces a change of the field scattered from the tip, which is typically collected in the far-field using an external detector. Crucially, this change depends on the local dielectric properties of the sample allowing its optical properties to be probed with sub-wavelength resolution. In fact, the spatial resolution achievable by s-SNOM is determined principally by the tip apex size and tip-sample distance rather than the radiation wavelength. However, the scattering efficiency of the tip is small, particularly at THz frequencies, which demands the use of a sensitive detector. It is perhaps therefore not surprising that the benefits afforded by LFI have recently been exploited for s-SNOM using QCL sources at both MIR [100] and THz frequencies [14, 101].

The experimental apparatus for one such implementation based on a ∼2.5 THz QCL is illustrated in Fig. 15.11. The QCL was operated at a temperature of 25 K and driven with a dc current just above the lasing threshold. Radiation was focused on a platinum-iridium needle with a sub-micron tip apex that was orientated at an angle $\theta = 50°$ from the beam axis. The sample was positioned at a small separation $d$ from the tip apex, and the field scattered from the needle and reinjected to the

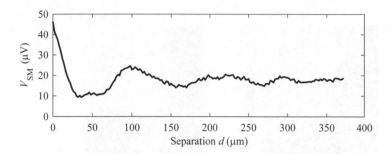

**Fig. 15.12** Self-mixing voltage $V_{SM}$ recorded as a function of tip-sample separation $d$ for the case of a flat metallic target. $V_{SM}$ is obtained by demodulation of the QCL voltage at the first harmonic of the modulation frequency. A near-field enhancement of the signal is observed for small separations. (Reproduced from [14], with the permission of AIP Publishing.)

laser cavity was measured via the QCL terminal voltage. This reinjected field can be expressed as

$$E_T = E_s + E_b + E_{NF}. \tag{15.21}$$

Here $E_s$ represents the background field arising from radiation scattered directly from the sample without interaction with the needle probe. This component was discriminated against by mechanically modulating the needle in the $z$-direction with amplitude $\sim 1$ um and employing lock-in detection of the SM voltage at this modulation frequency ($f_{mod} = 90\,\text{Hz}$). The second term, $E_b$, accounts for radiation scattered either directly from the needle or from the needle via a reflection from the sample surface. This term is not suppressed fully by lock-in detection, although demodulation at higher harmonics of $f_{mod}$ can be employed to further mitigate against this unwanted signal [102–104]. The final term $E_{NF}$ represents the enhanced scattered signal due to interaction between the excited needle probe and the dielectric sample. These remaining two contributions each exhibit a different dependence on $d$, and can be distinguished by observing the SM signal as the sample approaches the needle apex. Figure 15.12 shows an approach curve recorded in this way using a flat metallic target. The growing oscillations observed in the signal as the separation $d$ decreases can be attributed to the "background" term, $E_b$, in Eq. (15.21) [102, 103]. As the needle apex approaches the sample ($d \rightarrow 0$), a sharp increase in the signal is observed; this represents the near-field enhancement of the scattering cross-section and carries information on the local (near-field) dielectric properties of the sample.

In order to obtain a 2D image using this s-SNOM approach, the tip apex was positioned in proximity to the sample surface ($d \approx 0$), and the SM voltage recorded as the sample was raster scanned in the $x$–$y$ plane. Figure 15.13(a) shows an exemplar THz image of a rectangular region of gold (thickness 115 nm) lithographically defined on a quartz substrate. In this case a pixel size of 1 μm was used, which was limited by the motion control hardware. Clear contrast can be observed between the gold and quartz regions of the sample, and the THz image shows excellent agreement with the topographic AFM image of the same region shown in Fig. 15.13(b). By analyzing the

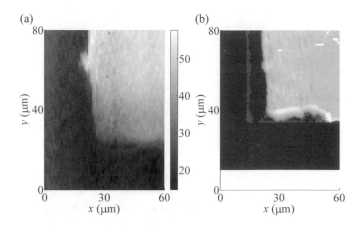

**Fig. 15.13** (a) Two-dimensional near-field image of the corner of a rectangular region of 115-nm-thick gold deposited on a quartz substrate. The gray scale corresponds to the magnitude of $V_{SM}$. (b) Topographic image of the same region of the sample. (Reproduced from [14], with the permission of AIP Publishing.)

signals measured across the edges of the gold region, the spatial resolutions obtained for the $x$- and $y$-directions are $\sigma_x = 1\,\mu\text{m}$ and $\sigma_y = 7\,\mu\text{m}$, respectively. This value of $\sigma_x$, which corresponds to $< \lambda/100$, is consistent with the expected resolution limit of the s-SNOM system imposed by the diameter of the tip apex. In the case of $\sigma_y$ the inferior resolution could be the result of the sample surface not being perfectly normal to the tip axis, causing the tip-sample separation to vary with $y$-position. An alternative explanation could be the signal component $E_b$ arising from radiation illuminating the tip via a surface reflection.

More recently, this initial demonstration of THz-s-SNOM based on LFI has been improved significantly through the use of an AFM tip acting as a sensitive nanopositioning element. By virtue of the ability to maintain nanometric separations between the scattering tip and sample surface, far superior resolution can be achieved, down to the order of tens of nanometers. This enhanced capability, when combined with fully coherent measurement of the scattered field, has seen its utilization across a range of applications including nanoscale mapping of material parameters [105, 106], coherent mapping of field distribution in THz micro-resonators [107], and the mapping of hyperbolic plasmon–phonon-polariton modes in 2D materials [108].

## 15.5   Summary and Outlook

LFI based on the SM effect that occurs in QCLs under optical feedback offers enormous potential for the future development of sensing applications in the MIR and THz regions of the spectrum. In the THz region of the spectrum, the LFI approach provides an opportunity for circumventing the reliance on traditional THz detectors, which are commonly either slow, unresponsive, bulky, or require external cryogenic cooling. The coherent nature of the LFI scheme opens up opportunities across a broad range of applications. Notable examples include: the measurement of fundamental laser parameters including the emission spectrum [19] and linewidth enhancement factor [23] of QCLs; fast coherent THz imaging in real-time [71]; three-dimensional imaging with sub-micron depth resolution [74]; materials analysis enabling the spatial mapping of plastic explosives [85] and free carriers in semiconductors [18]; biomedical imaging [69]; and high-resolution gas spectroscopy [76]. Whilst the majority of the demonstrate LFI systems employ continuous-wave operation of the QCL, it is noteworthy to mention that pulsed operation would facilitate higher operating temperatures as well as the potential of time-gating. Emerging models for understanding the dynamic behavior of QCLs under OF [48] will be of future importance in this respect.

LFI using THz QCLs has also been demonstrated to provide extremely high detection sensitivities, at the level $\sim$ pW/$\sqrt{}$Hz. This has recently facilitated a new approach for probing structural information far below the diffraction limit, based on the NSOM techniques that exploit the SM effect in THz QCLs [14, 101]. This emerging field promises to unlock a host of exciting new research opportunities for probing nanostructures, plasmonic resonators and metamaterials, and surface plasmons in two-dimensional materials.

## References

[1]  D. M. Kane and K. A. Shore, *Unlocking Dynamical Diversity: Optical Feedback Effects on Semiconductor Lasers*. New York: Wiley, 2005.

[2]  G. Acket, D. Lenstra, A. D. Boef, and B. Verbeek, "The influence of feedback intensity on longitudinal mode properties and optical noise in index-guided semiconductor lasers," *IEEE J. Quantum Electron*, vol. 20, no. 10, pp. 1163–1169, 1984.

[3]  J. Mork, B. Tromborg, and J. Mark, "Chaos in semiconductor lasers with optical feedback: theory and experiment," *IEEE J. Quantum Electronics*, vol. 28, no. 1, pp. 93–108, 1992.

[4]  D. Lenstra, B. Verbeek, and A. D. Boef, "Coherence collapse in single-mode semiconductor lasers due to optical feedback," *IEEE J. Quantum Electronics*, vol. 21, no. 6, pp. 674–679, 1985.

[5]  D. A. Kleinman and P. P. Kisliuk, "Discrimination against unwanted orders in the Fabry-Pérot resonator," *The Bell System Technical Journal*, vol. 41, no. 2, pp. 453–462, 1962.

[6]  K. Petermann, *Laser Diode Modulation and Noise*, 3rd ed., Dordrecht: Kluwer Academic, 1991.

[7]  T. Taimre, M. Nikolić, K. Bertling, Y. L. Lim, T. Bosch, and A. D. Rakić, "Laser feedback interferometry: a tutorial on the self-mixing effect for coherent sensing," *Adv. Opt. Photonics*, vol. 7, no. 3, pp. 570–631, 2015.

[8]  P. G. R. King and G. J. Steward, "Metrology with an optical maser," *New Sci.*, vol. 17, pp. 180, 1963.

[9]  E. Gagnon and J. F. Rivest, "Laser range imaging using the self-mixing effect in a laser diode," *IEEE Trans. Instrum. Meas.*, vol. 48, no. 3, pp. 693–699, 1999.

[10]  G. Guido, B.-P. Simone, and D. Silvano, "Self-mixing laser diode vibrometer," *Meas. Sci. Technol.*, vol. 14, no. 1, pp. 24, 2003.

[11]  S. Donati, G. Giuliani, and S. Merlo, "Laser diode feedback interferometer for measurement of displacements without ambiguity," *IEEE J. Quantum Electronics*, vol. 31, no. 1, pp. 113–119, 1995.

[12]  F. P. Mezzapesa, L. L. Columbo, G. D. Risi, M. Brambilla, M. Dabbicco, V. Spagnolo, and G. Scamarcio, "Nanoscale displacement sensing based on nonlinear frequency mixing in quantum cascade lasers," *IEEE J. Sel. Top. Quantum Electron.*, vol. 21, no. 6, pp. 107–114, 2015.

[13]  P. Dean, A. Valavanis, J. Keeley, K. Bertling, Y. Leng Lim, R. Alhathlool, S. Chowdhury, T. Taimre, L. H. Li, D. Indjin, S. J. Wilson, A. D. Rakić, E. H. Linfield, and A. Giles Davies, "Coherent three-dimensional terahertz imaging through self-mixing in a quantum cascade laser," *Appl. Phys. Lett.*, vol. 103, no. 18, pp. 181112, 2013.

[14]  P. Dean, O. Mitrofanov, J. Keeley, I. Kundu, L. Li, E. H. Linfield, and A. Giles Davies, "Apertureless near-field terahertz imaging using the self-mixing effect in a quantum cascade laser," *Appl. Phys. Lett.*, vol. 108, no. 9, pp. 091113, 2016.

[15]  S. Blaize, B. Bérenguier, I. Stéfanon, A. Bruyant, G. Lérondel, P. Royer, O. Hugon, O. Jacquin, and E. Lacot, "Phase sensitive optical near-field mapping using frequency-shifted laser optical feedback interferometry," *Opt. Express*, vol. 16, no. 16, pp. 11718–11726, 2008.

[16]  Y. L. Lim, R. Kliese, K. Bertling, K. Tanimizu, P. A. Jacobs, and A. D. Rakic, "Self-mixing flow sensor using a monolithic VCSEL array with parallel readout," *Opt. Express*, vol. 18, no. 11, pp. 11720–11727, 2010.

[17]  A. D. Rakic, T. Taimre, K. Bertling, Y. L. Lim, P. Dean, D. Indjin, Z. Ikonic, P. Harrison, A. Valavanis, S. P. Khanna, M. Lachab, S. J. Wilson, E. H. Linfield, and A. G. Davies, "Swept-frequency feedback interferometry using terahertz frequency QCLs: a method for imaging and materials analysis," *Opt. Express*, vol. 21, no. 19, pp. 22194–22205, 2013.

[18]  F. P. Mezzapesa, L. L. Columbo, M. Brambilla, M. Dabbicco, M. S. Vitiello, and G. Scamarcio, "Imaging of free carriers in semiconductors via optical feedback in terahertz quantum cascade lasers," *Appl. Phys. Lett.*, vol. 104, no. 4, pp. 041112, 2014.

[19]  J. Keeley, J. Freeman, K. Bertling, Y. L. Lim, R. A. Mohandas, T. Taimre, L. H. Li, D. Indjin, A. D. Rakić, E. H. Linfield, A. G. Davies, and P. Dean, "Measurement of the emission spectrum of a semiconductor laser using laser-feedback interferometry," *Sci. Rep.*, vol. 7, no. 1, pp. 7236, 2017.

[20]  G. Giuliani and M. Norgia, "Laser diode linewidth measurement by means of self-mixing interferometry," *IEEE Photon. Technol. Lett.*, vol. 12, no. 8, pp. 1028–1030, 2000.

[21]  M. C. Cardilli, M. Dabbicco, F. P. Mezzapesa, and G. Scamarcio, "Linewidth measurement of mid infrared quantum cascade laser by optical feedback interferometry," *Appl. Phys. Lett.*, vol. 108, no. 3, pp. 031105, 2016.

[22]  Y. Yanguang, G. Giuliani, and S. Donati, "Measurement of the linewidth enhancement factor of semiconductor lasers based on the optical feedback self-mixing effect," *IEEE Photon. Technol. Lett.*, vol. 16, no. 4, pp. 990–992, 2004.

[23]  R. P. Green, J.-H. Xu, L. Mahler, A. Tredicucci, F. Beltram, G. Giuliani, H. E. Beere, and D. A. Ritchie, "Linewidth enhancement factor of terahertz quantum cascade lasers," *Appl. Phys. Lett.*, vol. 92, no. 7, pp. 071106, 2008.

[24] G. Giuliani, M. Norgia, S. Donati, and T. Bosch, "Laser diode self-mixing technique for sensing applications," *J. Opt. A: Pure Appl. Opt.*, vol. 4, no. 6, pp. S283–S294, 2002.

[25] Y. L. Lim, M. Nikolic, K. Bertling, R. Kliese, and A. D. Rakić, "Self-mixing imaging sensor using a monolithic VCSEL array with parallel readout," *Opt. Express*, vol. 17, no. 7, pp. 5517–5525, 2009.

[26] F. P. Mezzapesa, M. Petruzzella, M. Dabbicco, H. E. Beere, D. A. Ritchie, M. S. Vitiello, and G. Scamarcio, "Continuous-wave reflection imaging using optical feedback interferometry in terahertz and mid-infrared quantum cascade lasers," *IEEE Trans. Terahertz Sci. Technol.*, vol. 4, no. 5, pp. 631–633, 2014.

[27] P. Dean, Y. L. Lim, A. Valavanis, R. Kliese, M. Nikolić, S. P. Khanna, M. Lachab, D. Indjin, Z. Ikonić, P. Harrison, A. D. Rakić, E. H. Linfield, and A. G. Davies, "Terahertz imaging through self-mixing in a quantum cascade laser," *Opt. Lett.*, vol. 36, pp. 2587–2589, 2011.

[28] K. Bertling, Y. L. Lim, T. Taimre, D. Indjin, P. Dean, R. Weih, S. Höfling, M. Kamp, M. von Edlinger, J. Koeth, and A. D. Rakić, "Demonstration of the self-mixing effect in interband cascade lasers," *Appl. Phys. Lett.*, vol. 103, no. 23, pp. 231107, 2013.

[29] R. Lang and K. Kobayashi, "External optical feedback effects on semiconductor injection laser properties," *IEEE J. Quantum Electron.*, vol. QE-16, pp. 347–355, 1980.

[30] J. R. Tucker, "A self-mixing imaging system based on an array of vertical-cavity surface-emitting lasers (VCSELs)," PhD thesis, School of Information Technology and Electrical Engineering, The University of Queensland, 2007.

[31] J. Y. Law and G. P. Agrawal, "Effects of optical feedback on static and dynamic characteristics of vertical-cavity surface-emitting lasers," *IEEE J. Sel. Top. Quantum Electron.*, vol. 3, no. 2, pp. 353–358, 1997.

[32] K. I. Kallimani and M. J. O. Mahony, "Relative intensity noise for laser diodes with arbitrary amounts of optical feedback," *IEEE J. Quantum Electronics*, vol. 34, no. 8, pp. 1438–1446, 1998.

[33] R. Kliese, T. Taimre, A. A. A. Bakar, Y. L. Lim, K. Bertling, M. Nikolić, J. Perchoux, T. Bosch, and A. D. Rakić, "Solving self-mixing equations for arbitrary feedback levels: a concise algorithm," *Appl. Opt.*, vol. 53, no. 17, pp. 3723–3736, 2014.

[34] R. Tkach and A. Chraplyvy, "Regimes of feedback effects in 1.5-μm distributed feedback lasers," *J. Light. Technol.*, vol. 4, no. 11, pp. 1655–1661, 1986.

[35] S. Donati and R. H. Horng, "The diagram of feedback regimes revisited," *IEEE J. Sel. Top. Quantum Electron.*, vol. 19, no. 4, pp. 1500309–1500309, 2013.

[36] L. Jumpertz, M. Carras, K. Schires, and F. Grillot, "Regimes of external optical feedback in 5.6 μm distributed feedback mid-infrared quantum cascade lasers," *Appl. Phys. Lett.*, vol. 105, no. 13, pp. 131112, 2014.

[37] F. P. Mezzapesa, L. L. Columbo, M. Brambilla, M. Dabbicco, S. Borri, M. S. Vitiello, H. E. Beere, D. A. Ritchie, and G. Scamarcio, "Intrinsic stability of quantum cascade lasers against optical feedback," *Opt. Express*, vol. 21, no. 11, pp. 13748–13757, 2013.

[38] R. Kohler, A. Tredicucci, F. Beltram, H. E. Beere, E. H. Linfield, A. G. Davies, D. A. Ritchie, R. C. Iotti, and F. Rossi, "Terahertz semiconductor-heterostructure laser," *Nature*, vol. 417, pp. 156–159, 2002.

[39] J. Al Roumy, J. Perchoux, Y. L. Lim, T. Taimre, A. D. Rakić, and T. Bosch, "Effect of injection current and temperature on signal strength in a laser diode optical feedback interferometer," *Appl. Opt.*, vol. 54, no. 2, pp. 312–318, 2015.

[40] S. Barbieri, J. Alton, H. E. Beere, J. Fowler, E. H. Linfield, and D. A. Ritchie, "2.9THz quantum cascade lasers operating up to 70K in continuous wave," *Appl. Phys. Lett.*, vol. 85, no. 10, pp. 1674–1676, 2004.

[41] P. Dean, A. Valavanis, J. Keeley, K. Bertling, Y. L. Lim, R. Alhathlool, A. D. Burnett, L. H. Li, S. P. Khanna, D. Indjin, T. Taimre, A. D. Rakić, E. H. Linfield, and A. G. Davies, "Terahertz imaging using quantum cascade lasers—a review of systems and applications," *J. Phys. D: Appl. Phys.*, vol. 47, no. 37, pp. 374008, 2014.

[42] K. Bertling, T. Taimre, G. Agnew, Y. L. Lim, P. Dean, D. Indjin, S. Höfling, R. Weih, M. Kamp, M. v. Edlinger, J. Koeth, and A. D. Rakić, "Simple electrical modulation scheme for laser feedback imaging," *IEEE Sens. J.*, vol. 16, no. 7, pp. 1937–1942, 2016.

[43] G. Scalari, L. Ajili, J. Faist, H. Beere, E. Linfield, D. Ritchie, and G. Davies, "Far-infrared λ≈87 µm bound-to-continuum quantum-cascade lasers operating up to 90 K," *Appl. Phys. Lett.*, vol. 82, no. 19, pp. 3165–3167, 2003.

[44] D. Indjin, P. Harrison, R. W. Kelsall, and Z. Ikonić, "Mechanisms of temperature performance degradation in terahertz quantum-cascade lasers," *Appl. Phys. Lett.*, vol. 82, no. 9, pp. 1347–1349, 2003.

[45] S. S. Dhillon, M. S. Vitiello, E. H. Linfield, A. G. Davies, C. H. Matthias, B. John, P. Claudio, M. Gensch, P. Weightman, G. P. Williams, E. Castro-Camus, D. R. S. Cumming, F. Simoens, I. Escorcia-Carranza, J. Grant, L. Stepan, K.-G. Makoto, K. Kuniaki, K. Martin, A. S. Charles, L. C. Tyler, H. Rupert, A. G. Markelz, Z. D. Taylor, P. W. Vincent, J. A. Zeitler, S. Juraj, M. K. Timothy, B. Ellison, S. Rea, P. Goldsmith, B. C. Ken, A. Roger, D. Pardo, P. G. Huggard, V. Krozer, S. Haymen, F. Martyn, R. Cyril, S. Alwyn, S. Andreas, N. Mira, R. Nick, C. Roland, E. C. John, and B. J. Michael, "The 2017 terahertz science and technology roadmap," *J. Phys. D: Appl. Phys.*, vol. 50, no. 4, pp. 043001, 2017.

[46] M. Wienold, L. Schrottke, M. Giehler, R. Hey, W. Anders, and H. T. Grahn, "Low-voltage terahertz quantum-cascade lasers based on LO-phonon-assisted interminiband transitions," *Electron. Lett.*, vol. 45, no. 20, pp. 1030–1031, 2009.

[47] G. P. Agrawal, and N. K. Dutta, *Semiconductor Lasers*, 2nd ed.: New York: Springer, 1993.

[48] G. Agnew, A. Grier, T. Taimre, Y. L. Lim, K. Bertling, Z. Ikonić, A. Valavanis, P. Dean, J. Cooper, S. P. Khanna, M. Lachab, E. H. Linfield, A. G. Davies, P. Harrison, D. Indjin, and A. D. Rakić, "Model for a pulsed terahertz quantum cascade laser under optical feedback," *Opt. Express*, vol. 24, no. 18, pp. 20554–20570, 2016.

[49] C. Henry, "Theory of the linewidth of semiconductor lasers," *IEEE J. Quantum Electron.*, vol. 18, no. 2, pp. 259–264, 1982.

[50] I. D. Henning and J. V. Collins, "Measurements of the semiconductor laser linewidth broadening factor," *Electron. Lett.*, vol. 19, no. 22, pp. 927–929, 1983.

[51] C. Harder, K. Vahala, and A. Yariv, "Measurement of the linewidth enhancement factor α of semiconductor lasers," *Appl. Phys. Lett.*, vol. 42, no. 4, pp. 328–330, 1983.

[52] J. Faist, F. Capasso, D. L. Sivco, C. Sirtori, A. L. Hutchinson, and A. Y. Cho, "Quantum cascade laser," *Science*, vol. 264, pp. 553, 1994.

[53] M. Ishihara, T. Morimoto, S. Furuta, K. Kasahara, N. Akikusa, K. Fujita, and T. Edamura, "Linewidth enhancement factor of quantum cascade lasers with single phonon resonance-continuum depopulation structure on Peltier cooler," *Electron. Lett.*, vol. 45, no. 23, pp. 1168–1169, 2009.

[54] J. von Staden, T. Gensty, W. Elsäßer, G. Giuliani, and C. Mann, "Measurements of the α factor of a distributed-feedback quantum cascade laser by an optical feedback self-mixing technique," *Opt. Lett.*, vol. 31, no. 17, pp. 2574–2576, 2006.

[55] L. Jumpertz, F. Michel, R. Pawlus, W. Elsässer, K. Schires, M. Carras, and F. Grillot, "Measurements of the linewidth enhancement factor of mid-infrared quantum cascade lasers by different optical feedback techniques," *AIP Adv.*, vol. 6, no. 1, pp. 015212, 2016.

[56] N. Kumazaki, Y. Takagi, M. Ishihara, K. Kasahara, A. Sugiyama, N. Akikusa, and T. Edamura, "Detuning characteristics of the linewidth enhancement factor of a mid-infrared quantum cascade laser," *Appl. Phys. Lett.*, vol. 92, no. 12, pp. 121104, 2008.

[57] Y. L. Lim, P. Dean, M. Nikolić, R. Kliese, S. P. Khanna, M. Lachab, A. Valavanis, D. Indjin, Z. Ikonić, P. Harrison, E. H. Linfield, A. Giles Davies, S. J. Wilson, and A. D. Rakić, "Demonstration of a self-mixing displacement sensor based on terahertz quantum cascade lasers," *Appl. Phys. Lett.*, vol. 99, no. 8, pp. 081108, 2011.

[58] H. W. Jentink, F. F. M. de Mul, H. E. Suichies, J. G. Aarnoudse, and J. Greve, "Small laser Doppler velocimeter based on the self-mixing effect in a diode laser," *Appl. Opt.*, vol. 27, no. 2, pp. 379–385, 1988.

[59] L. Lv, H. Gui, J. Xie, T. Zhao, X. Chen, A. Wang, F. Li, D. He, J. Xu, and H. Ming, "Effect of external cavity length on self-mixing signals in a multilongitudinal-mode Fabry–Pérot laser diode," *Appl. Opt.*, vol. 44, no. 4, pp. 568–571, 2005.

[60] M. Tonouchi, "Cutting-edge terahertz technology," *Nat. Photonics*, vol. 1, pp. 97–105, 2007.

[61] D. Mittleman, *Sensing with Thz Radiation*. Berlin: Springer, 2003.

[62] J. A. Zeitler, P. F. Taday, D. A. Newnham, M. Pepper, K. C. Gordon, and T. Rades, "Terahertz pulsed spectroscopy and imaging in the pharmaceutical setting – a review," *J. Pharm. Pharmacol.*, vol. 59, pp. 209–223, 2007.

[63] A. G. Davies, A. D. Burnett, W. Fan, E. H. Linfield, and J. E. Cunningham, "Terahertz spectroscopy of explosives and drugs," *Mater. Today*, vol. 11, no. 3, pp. 18–26, 2008.

[64] V. P. Wallace, A. J. Fitzgerald, S. Shankar, N. Flanagan, R. J. Pye, J. Cluff, and D. D. Arnone, "Terahertz pulsed imaging of basal cell carcinoma ex vivo and in vivo," *Br. J. Dermatol.*, vol. 151, pp. 424–432, 2004.

[65] A. J. Fitzgerald, V. P. Wallace, M. Jimenez-Linan, L. Bobrow, R. J. Pye, A. D. Purushotham, and D. D. Arnone, "Terahertz pulsed imaging of human breast tumors," *Radiology*, vol. 239, pp. 533–540, 2006.

[66] L. H. Li, L. Chen, J. R. Freeman, M. Salih, P. Dean, A. G. Davies, and E. H. Linfield, "Multi-watt high-power THz frequency quantum cascade lasers," *Electron. Lett.*, vol. 53, no. 12, pp. 799–800, 2017.

[67] M. S. Vitiello, L. Consolino, S. Bartalini, A. Taschin, A. Tredicucci, M. Inguscio, and P. De Natale, "Quantum-limited frequency fluctuations in a terahertz laser," *Nat. Photonics*, vol. 6, pp. 525, 2012.

[68] C. W. I. Chan, Q. Hu, and J. L. Reno, "Ground state terahertz quantum cascade lasers," *Appl. Phys. Lett.*, vol. 101, no. 15, pp. 151108, 2012.

[69] Y. L. Lim, T. Taimre, K. Bertling, P. Dean, D. Indjin, A. Valavanis, S. P. Khanna, M. Lachab, H. Schaider, T. W. Prow, H. Peter Soyer, S. J. Wilson, E. H. Linfield, A. Giles Davies, and A. D. Raki, "High-contrast coherent terahertz imaging of porcine tissue

via swept-frequency feedback interferometry," *Biomed. Opt. Express*, vol. 5, no. 11, pp. 3981–3989, 2014.

[70]  Y. Ren, R. Wallis, D. S. Jessop, Degl, apos, R. Innocenti, A. Klimont, H. E. Beere, and D. A. Ritchie, "Fast terahertz imaging using a quantum cascade amplifier," *Appl. Phys. Lett.*, vol. 107, no. 1, pp. 011107, 2015.

[71]  M. Wienold, T. Hagelschuer, N. Rothbart, L. Schrottke, K. Biermann, H. T. Grahn, and H. W. Hübers, "Real-time terahertz imaging through self-mixing in a quantum-cascade laser," *Appl. Phys. Lett.*, vol. 109, no. 1, pp. 011102, 2016.

[72]  S. Barbieri, P. Gellie, G. Santarelli, L. Ding, W. Maineult, C. Sirtori, R. Colombelli, H. Beere, and D. Ritchie, "Phase-locking of a 2.7-THz quantum cascade laser to a mode-locked erbium-doped fibre laser," *Nat. Photonics*, vol. 4, pp. 636–640, 2010.

[73]  M. Ravaro, V. Jagtap, G. Santarelli, C. Sirtori, L. H. Li, S. P. Khanna, E. H. Linfield, and S. Barbieri, "Continuous-wave coherent imaging with terahertz quantum cascade lasers using electro-optic harmonic sampling," *Appl. Phys. Lett.*, vol. 102, no. 9, pp. 091107, 2013.

[74]  J. Keeley, P. Dean, A. Valavanis, K. Bertling, Y. L. Lim, R. Alhathlool, T. Taimre, L. H. Li, D. Indjin, A. D. Rakić, E. H. Linfield, and A. G. Davies, "Three-dimensional terahertz imaging using swept-frequency feedback interferometry with a quantum cascade laser," *Opt. Lett.*, vol. 40, no. 6, pp. 994–997, 2015.

[75]  P. Dean, J. Keeley, A. Valavanis, K. Bertling, Y. L. Lim, T. Taimre, R. Alhathlool, L. H. Li, D. Indjin, A. D. Rakić, E. H. Linfield, and A. G. Davies, "Active phase-nulling of the self-mixing phase in a terahertz frequency quantum cascade laser," *Opt. Lett.*, vol. 40, no. 6, pp. 950–953, 2015.

[76]  T. Hagelschuer, M. Wienold, H. Richter, L. Schrottke, K. Biermann, H. T. Grahn, and H. W. Hübers, "Terahertz gas spectroscopy through self-mixing in a quantum-cascade laser," *Appl. Phys. Lett.*, vol. 109, no. 19, pp. 191101, 2016.

[77]  A. Valavanis, P. Dean, L. Yah Leng, R. Alhathlool, M. Nikolic, R. Kliese, S. P. Khanna, D. Indjin, S. J. Wilson, A. D. Rakic, E. H. Linfield, and G. Davies, "Self-mixing interferometry with terahertz quantum cascade lasers," *IEEE Sens. J.*, vol. 13, no. 1, pp. 37–43, 2013.

[78]  A. W. M. Lee, Q. Qin, S. Kumar, B. S. Williams, Q. Hu, and J. L. Reno, "Real-time terahertz imaging over a standoff distance (>25meters)," *Appl. Phys. Lett.*, vol. 89, no. 14, pp. 141125, 2006.

[79]  H. W. Hübers, S. Pavlov, A. D. Semenov, R. Köhler, L. Mahler, A. Tredicucci, H. E. Beere, D. A. Ritchie, and E. H. Linfield, "Terahertz quantum cascade laser as local oscillator in a heterodyne receiver," *Opt. Express*, vol. 13, pp. 5890–5896, 2005.

[80]  A. Barkan, F. K. Tittel, D. M. Mittleman, R. Dengler, P. H. Siegel, G. Scalari, L. Ajili, J. Faist, H. E. Beere, E. H. Linfield, A. G. Davies, and D. A. Ritchie, "Linewidth and tuning characteristics of terahertz quantum cascade lasers," *Opt. Lett.*, vol. 29, no. 6, pp. 575–577, 2004.

[81]  S. Barbieri, J. Alton, H. E. Beere, E. H. Linfield, D. A. Ritchie, S. Withington, G. Scalari, L. Ajili, and J. Faist, "Heterodyne mixing of two far-infrared quantum cascade lasers by use of a point-contact Schottky diode," *Opt. Lett.*, vol. 29, pp. 1632–1634, 2004.

[82]  S. R. Chinn, E. A. Swanson, and J. G. Fujimoto, "Optical coherence tomography using a frequency-tunable optical source," *Opt. Lett.*, vol. 22, no. 5, pp. 340–342, 1997.

[83] T.-I. Jeon and D. Grischkowsky, "Characterization of optically dense, doped semiconductors by reflection THz time domain spectroscopy," *Appl. Phys. Lett.*, vol. 72, no. 23, pp. 3032–3034, 1998.

[84] T. Taimre, K. Bertling, Y. L. Lim, P. Dean, D. Indjin, and A. D. Rakić, "Methodology for materials analysis using swept-frequency feedback interferometry with terahertz frequency quantum cascade lasers," *Opt. Express*, vol. 22, no. 15, pp. 18633–18647, 2014.

[85] S. Han, K. Bertling, P. Dean, J. Keeley, D. A. Burnett, L. Y. Lim, P. S. Khanna, A. Valavanis, H. E. Linfield, G. A. Davies, D. Indjin, T. Taimre, and D. A. Rakić, "Laser feedback interferometry as a tool for analysis of granular materials at terahertz frequencies: towards imaging and identification of plastic explosives," *Sensors*, vol. 16, no. 3, pp. 352, 2016.

[86] M. S. Vitiello and A. Tredicucci, "Tunable emission in THz quantum cascade lasers," *IEEE Trans. Terahertz Sci. Technol.*, vol. 1, pp. 76–84, 2011.

[87] S. M. Kim, F. Hatami, J. S. Harris, A. W. Kurian, J. Ford, D. King, G. Scalari, M. Giovannini, N. Hoyler, J. Faist, and G. Harris, "Biomedical terahertz imaging with a quantum cascade laser," *Appl. Phys. Lett.*, vol. 88, no. 15, pp. 153903, 2006.

[88] P. Tewari, N. Bajwa, R. S. Singh, M. O. Culjat, W. S. Grundgest, Z. D. Taylor, C. P. Kealey, D. B. Bennett, K. S. Barnett, and A. Stojadinovic, "In vivo terahertz imaging of rat skin burns," *J. Biomed. Opt.*, vol. 17, pp. 040503, 2012.

[89] E. Betzig, J. K. Trautman, T. D. Harris, J. S. Weiner, and R. L. Kostelak, "Breaking the diffraction barrier: optical microscopy on a nanometric scale," *Science*, vol. 251, no. 5000, pp. 1468, 1991.

[90] F. Zenhausern, Y. Martin, and H. K. Wickramasinghe, "Scanning interferometric apertureless microscopy: optical imaging at 10 angstrom resolution," *Science*, vol. 269, no. 5227, pp. 1083, 1995.

[91] B. Knoll and F. Keilmann, "Near-field probing of vibrational absorption for chemical microscopy," *Nature*, vol. 399, no. 6732, pp. 134–137, 1999.

[92] F. Buersgens, R. Kersting, and H.-T. Chen, "Terahertz microscopy of charge carriers in semiconductors," *Appl. Phys. Lett.*, vol. 88, no. 11, pp. 112115, 2006.

[93] A. J. Huber, F. Keilmann, J. Wittborn, J. Aizpurua, and R. Hillenbrand, "Terahertz near-field nanoscopy of mobile carriers in single semiconductor nanodevices," *Nano Lett.*, vol. 8, no. 11, pp. 3766–3770, 2008.

[94] R. Jacob, S. Winnerl, M. Fehrenbacher, J. Bhattacharyya, H. Schneider, M. T. Wenzel, H.-G. v. Ribbeck, L. M. Eng, P. Atkinson, O. G. Schmidt, and M. Helm, "Intersublevel spectroscopy on single InAs-quantum dots by terahertz near-field microscopy," *Nano Lett.*, vol. 12, no. 8, pp. 4336–4340, 2012.

[95] M. Eisele, T. L. Cocker, M. A. Huber, M. Plankl, L. Viti, D. Ercolani, L. Sorba, M. S. Vitiello, and R. Huber, "Ultrafast multi-terahertz nano-spectroscopy with sub-cycle temporal resolution," *Nat. Photonics*, vol. 8, pp. 841, 2014.

[96] G. Acuna, S. F. Heucke, F. Kuchler, H. T. Chen, A. J. Taylor, and R. Kersting, "Surface plasmons in terahertz metamaterials," *Opt. Express*, vol. 16, no. 23, pp. 18745–18751, 2008.

[97] A. Bitzer, H. Merbold, A. Thoman, T. Feurer, H. Helm, and M. Walther, "Terahertz near-field imaging of electric and magnetic resonances of a planar metamaterial," *Opt. Express*, vol. 17, no. 5, pp. 3826–3834, 2009.

[98]   C.-M. Chiu, H.-W. Chen, Y.-R. Huang, Y.-J. Hwang, W.-J. Lee, H.-Y. Huang, and C.-K. Sun, "All-terahertz fiber-scanning near-field microscopy," *Opt. Lett.*, vol. 34, no. 7, pp. 1084–1086, 2009.

[99]   H.-T. Chen, R. Kersting, and G. C. Cho, "Terahertz imaging with nanometer resolution," *Appl. Phys. Lett.*, vol. 83, no. 15, pp. 3009–3011, 2003.

[100]  I. M. Craig, M. S. Taubman, A. S. Lea, M. C. Phillips, E. E. Josberger, and M. B. Raschke, "Infrared near-field spectroscopy of trace explosives using an external cavity quantum cascade laser," *Opt. Express*, vol. 21, no. 25, pp. 30401–30414, 2013.

[101]  R. Degl'Innocenti, R. Wallis, B. Wei, L. Xiao, S. J. Kindness, O. Mitrofanov, P. Braeuninger-Weimer, S. Hofmann, H. E. Beere, and D. A. Ritchie, "Terahertz nanoscopy of plasmonic resonances with a quantum cascade laser," *ACS Photonics*, vol. 4, no. 9, pp. 2150–2157, 2017.

[102]  R. Hillenbrand, and F. Keilmann, "Complex optical constants on a subwavelength scale," *Phys. Rev. Lett.*, vol. 85, no. 14, pp. 3029–3032, 2000.

[103]  M. B. Raschke, and C. Lienau, "Apertureless near-field optical microscopy: Tip–sample coupling in elastic light scattering," *Appl. Phys. Lett.*, vol. 83, no. 24, pp. 5089–5091, 2003.

[104]  K. Moon, H. Park, J. Kim, Y. Do, S. Lee, G. Lee, H. Kang, and H. Han, "Subsurface nanoimaging by broadband terahertz pulse near-field microscopy," *Nano Lett.*, vol. 15, no. 1, pp. 549–552, 2015.

[105]  M. C. Giordano, S. Mastel, C. Liewald, L. L. Columbo, M. Brambilla, L. Viti, A. Politano, K. Zhang, L. Li, A. G. Davies, E. H. Linfield, R. Hillenbrand, F. Keilmann, G. Scamarcio, and Miriam S Vitiello, "Phase-resolved terahertz self-detection near-field microscopy," *Opt. Express,* vol. 26, no. 14, pp. 18423–18435, 2018.

[106]  P. Rubino, J. Keeley, N. Sulollari, A. D. Burnett, A. Valavanis, I. Kundu, M. C. Rosamond, L. Li, E. H. Linfield, A. G. Davies, J. E. Cunningham, and P. Dean, "All-electronic phase-resolved THz microscopy using the self-mixing effect in a semiconductor laser," *ACS Photonics,* vol. 8, no. 4, pp. 1001–1006, 2021.

[107]  N. Sulollari, J. Keeley, S. Park, P. Rubino, A. D. Burnett, L. Li, M. C. Rosamond, E. H. Linfield, A. G. Davies, J. E. Cunningham, and P. Dean, "Coherent terahertz microscopy of modal field distributions in micro-resonators," *APL Photonics,* vol. 6, pp. 066104, 2021.

[108]  E. A. A. Pogna, L. Viti, A. Politano, M. Brambilla, G. Scamarcio, and M. S. Vitiello, "Mapping propagation of collective modes in $Bi_2Se_3$ and $Bi_2Te_{2.2}Se_{0.8}$ topological insulators by near-field terahertz nanoscopy," *Nat. Commun.,* vol. 12, pp. 6672, 2021.

# 16 Applications of Terahertz Quantum Cascade Lasers

Pierre Gellie

Lytid SAS, Paris

## 16.1 Introduction

This chapter reviews the applications of terahertz (THz) quantum cascade lasers (QCLs). THz QCLs have come a long way since their first demonstration in 2002. Although still operating at cryogenic temperatures, their applications have been multiplying steadily over the last decade, helped by the availability of compact commercial THz QCL systems and the growing adoption of the THz QCL technology in the THz scientific community. Currently, the key fields of THz QCL applications are imaging, spectroscopy, and sensing.

## 16.2 Applications of THz QCLs

Since the first demonstration of THz QCLs in 2002, a great number of scientific applications using these devices have emerged ranging from imaging [1] and high-resolution spectroscopy [2] to sensing [3]. Imaging applications have been explored for non-destructive testing (NDT) [4] and biomedical imaging [5]. Further developments in imaging and sensing have exploited self-mixing in THz QCLs for a fast phase and amplitude detection without external elements in the system [6]. High-resolution spectroscopy of gases has been achieved with several different techniques to circumvent the intrinsically low tunability of THz QCL and the lack of a simple and coherent detection scheme. For example, current sweeping [2], frequency modulation (FM) [7], or, more recently, self-mixing [8].

### 16.2.1 The Early Years (2004–2007)

In the early years, following the initial THz QCL demonstration, efforts were put into optimizing the QCL devices with little to no developments made in regards to their applications. The use of the technology was still limited to groups working on the growth and fabrication of the devices.

The first applications of THz QCLs began to emerge in 2004, taking the form of imaging and heterodyne mixing. The first imaging application was realized at 3.4 THz in 2004 by Darmo et al. [10]. The imaging system used was a raster scan imaging system where the THz radiation emitted from a QCL was focused with parabolic mirrors onto the sample, which was then moved in the focal plane of the optical system using a two-axis motorized translation stage. The transmitted or reflected light was then focused onto a liquid helium-cooled bolometer. The samples were histological slices of a rat brain about 30 μm thick. This work was thus also the first bio-medical application of THz QCLs. The following year another imaging setup was demonstrated [11] using a room-temperature Schottky diode mixer to detect the THz radiation. This work exploited the multi-mode emission spectrum of a Fabry–Pérot (FP) THz QCL with a typical mode spacing in the range of few tens of GHz that falls into the bandwidth of a fast Schottky diode mixer. The beating signal between each pair of FP modes was then acquired using a spectrum analyzer with a high signal to noise ratio of over 60 dB and a fast integration time of 100 ms. The first images obtained using this setup showed a metal razor blade concealed behind a paper sheet.

The first spectroscopy applications of THz QCL were heterodyne mixing of two THz QCLs [12] and the demonstration of a heterodyne receiver using a THz QCL as local oscillator (LO) [13, 14]. In [12], two 1 mW 3.3 THz multi-mode lasers were focused onto a point-contact Schottky diode mixer to acquire a heterodyne signal using an external amplifier. By tuning the current and temperature of the two lasers, the heterodyne signal in the range of 3–6 GHz was obtained, thus showing for the first time the potential of the THz QCL technology to be used as a LO for a room-temperature receiver based on Schottky diode mixers. Later [13, 14], heterodyne receivers using THz QCLs as LO and super-conducting NbN hot electron bolometer mixers demonstrated unprecedent sensitivity above 2 THz, establishing THz QCLs as strong contenders for stable, high-power LO sources for heterodyne instruments.

Other notable early stage spectroscopic applications include the study of cyclotron resonances in InAs and InSb quantum wells establishing THz QCLs as viable and much more compact alternative to bulky far-infrared gas lasers that have been used for similar experiments previously [15]. In [16] Dhillon et al. demonstrated the generation of THz sidebands onto a telecom wavelength carrier, which opened up a new field of potential applications of THz QCLs for telecommunications. In this work the THz QCL provided both the source and non-linear medium for the generation of the THz sidebands at 1.3 μm. A single plasmon THz QCL waveguide was modified by the insertion of a 1 μm GaAs undoped layer in the center of a QCL structure to act as a near-infrared (NIR) waveguide. In 2007, by using a double metal waveguide THz QCL with a small ridge width of only few tens of microns that enabled phase-matching for THz-NIR wave mixing at 1.55 μm, the authors were able to demonstrate sideband generation at the most relevant telecom wavelength of 1.55 μm [17].

## 16.2.2    Imaging with THz QCLs

In the late 2000s, as the power output of THz QCLs made rapid improvements with more than 100 mW of continuous-wave (CW) output [18], many imaging applications were developed to exploit the high power available with the THz QCL technology. Most of the early raster-scan imaging demonstrations utilized cooled bolometers [10], Golay cells [19], or pyro-electric detectors [20]. These are all slow thermal detectors and their use resulted in very long acquisition time, on the order of several minutes to several hours, depending on the size of the image to be taken. These detection schemes showed dynamic ranges in the range from 30 to 60 dB that were higher than most imaging systems based on pulsed THz sources at these frequencies. More recently, a significantly higher dynamic range imaging setup was demonstrated using nanowire field effect transistor as a THz detector for QCLs [21], achieving a dynamic range of over 80 dB.

To obtain microscopic THz images, a THz QCL-based confocal microscope was demonstrated in [22]. Here a 2.9 THz QCL was used in a confocal setup with two spatial filtering pinholes of 200 μm and 300 μm diameter, two parabolic 90-degrees off-axis parabolic mirrors, as well as two silicon, one polymethylpentene (TPX), and three picarin lenses. The system demonstrated 67 μm transverse resolution and 400 μm axial resolution. Another microscopy setup with high image acquisition speed was demonstrated in [23]. There, a 2.5 THz QCL was used in combination with a flat scanning mirror of 35 mm in diameter. The THz beam arrives collimated on a scanning mirror and a 76-mm-diameter high-density polyethylene (HDPE) lens focuses the beam on the sample. A Ga:Ge photoconductive detector is used to collect the transmitted beam. The system allows one to image samples as large as $40 \times 36 \, \text{mm}^2$ in 1.1 sec with 240 μs integration time per pixel and 0.5 mm spatial resolution. The dynamic range of the images was quite low, however, with only 28 dB.

Most of the previously presented imaging setups employed slow thermal detectors or other incoherent detection methods. In order to increase the dynamic range and/or image acquisition time, one can exploit the high intrinsic coherency of THz QCL emission. One coherent imaging demonstration used a THz QCLs with the stabilized output frequency [24]. The THz output from the QCL was split into three beams using beam splitters. Two beams were sent to two identical electro-optic (EO) sampling units and the third beam was focused on the sample. Each of the EO arms were composed of a ZnTe crystals followed by a quarter-wave plate, a half-wave plate, and a polarizing beam splitter with a balanced photodiode detection. These elements formed an ultra-fast near-infrared EO modulator driven by the THz field. One emission line of the 2.5 THz QCL generated two sidebands combs at $+-2.5$ THz from the frequency-comb spectrum of the fs-laser centered at 780 nm (385 THz). Since the fs-laser frequency-comb bandwidth was about 5 THz or twice the QCL frequency, the carrier comb and the two EO-generated sidebands combs overlapped and generated, on the fast, balanced photodiodes, a series of heterodyne beat notes in the range of zero Hz to half the fs-laser repetition rate of 100 MHz. Thus, these EO units formed a direct THz to MHz down-converter of the QCL signal, which was then used for frequency stabilization using fast electronics and a PLL on one EO arm [25]. The second EO arm was used for signal sampling of the THz beam reflected of the sample at normal incidence. Coherent THz imaging was realized using this setup in combination with a raster scan object displacement in reflection mode and yielded a 60 dB dynamic range using lock-in amplification with 30 ms time constant. The QCL emitted only 250 μW of incident THz power, corresponding to 2.5 pW/Hz$^{1/2}$ noise equivalent power (NEP) for the EO sampling unit.

Another elegant coherent detection scheme exploits self-mixing (SM) in THz QCLs [26], which is described in Chapter 15. SM occurs in a laser when part of the radiation is re-injected into the cavity. The back-reflected light interferes with the intra-cavity optical field and produces measurable changes to the optical and electrical characteristics of the laser. SM is intrinsically coherent and in such a scheme the THz QCL is used as an emitter as well as a coherent detector. Furthermore, SM is self-aligning and intrinsically confocal due to the fact that the light must go back into the small (wavelength-sized) laser facet. Imaging was realized with SM in [26, 27].

THz QCLs have also been used in coherent transceivers for a synthetic aperture radar (SAR) applications in [28]. SARs are commonly used in the microwave frequencies on aircrafts for remote high-resolution imaging of ground topography. In a typical SAR system, a transmitter antenna on an airborne platform moves on a horizontal plane, perpendicular to the line-of-sight to the target. The Fourier processing yields phase and amplitude information of the signal backscattered from the target, providing a two-dimensional (2D) azimuth/range image of scattering centers. The azimuth refers to the horizontal cross-range of a target, and range refers to its distance from the transmitter, measured along the line-of-sight. However, a similar principle can be used for object imaging with the antenna being fixed and the object being moved in the azimuthal direction. This approach is known as the inverse SAR or ISAR. This scheme was reported in combination with THz QCLs in [28]. The frequency stability of the radiation source in such a system is the key to achieving high resolution. Here, a portion of the radiation output from a 2.4 THz QCL was used to frequency-stabilize the QCL emission using a molecular $CO_2$ gas laser as a LO and employing Schottky diode mixers and feedback loops. The remaining part the QCL output was focused on the sample and in a quasi-monostatic configuration, the reflected signal from the sample was acquired by mixing the signal with the gas laser output using a second Schottky diode mixer. This scheme was used for stand-off imaging over a 5.5 m distance with sub-mm spatial resolution. Figures 16.1(a) and (b) show a photograph and azimuth/elevation THz image of a plastic model tank coated with aluminum. For this image the target was rotated from $-2°$ to $+3.8°$ in elevation, with a step size of $0.028°$, and from $-4.8°$ to $+4.8°$ in azimuth, with a step size of $0.02°$. This scanning range corresponds to pixel resolutions of 0.4 and 0.6 mm in these two directions. The resulting acquisition time was 40 min, and the dynamic range was estimated at $\sim105$ dB per pixel.

With the high output power available from THz QCLs, real-time high-resolution imaging using THz cameras is also possible. These cameras are usually based on microbolometer focal plane array technology derived from the mid-infrared spectral range. Although the microbolometer THz cameras offer relatively low noise equivalent powers (NEP) of tens of $pW/Hz^{1/2}$, the high number of pixels (often over 60,000) requires very high THz powers to illuminate the focal plane array to maintain a high signal to noise ratio (SNR). These cameras are also more sensitive at higher THz frequencies ($>3$ THz) due to a mismatch between pixel size (35–50 μm) and wavelength ($>100$ μm) that will quickly degrade the NEP. As a consequence, THz

**Fig. 16.1** (a) Photograph of a 1/72nd scale model T-80BV tank, (b) 2.4 THz Azimuth/elevation imagery of the scale model tank with a pixel resolution of 0.4×0.6 mm. A calibration object (dihedron) is located to the right of the image. (Reproduced from [28]).

QCLs are an ideal match to these cameras for imaging applications. The first demonstration of real-time imaging was realized in 2006 using a microbolometric camera designed for the long-wavelength infrared spectral region (LWIR, 8–12 µm wavelength) using a high-power (50 mW peak power) THz QCL emitting at 4.3 THz [29]. Detection of THz light was possible due the broadband absorbing materials used in the LWIR camera technology that also absorbed THz radiation. However, the SNR in the best possible configuration was rather low (~340) and the NEP was measured to be ~320 pW/Hz$^{1/2}$. Since then, several groups have developed customized microbolometer technology to improve the camera's sensitivity at THz frequencies, including NEC Corporation in Japan [30], National Optics Institute (INO) in Canada [31], and CEA-Leti Institute in France [32]. The state-of-art sensitivities of the THz microbolometric cameras are currently of a few tens of pW/Hz$^{1/2}$, more than an order of magnitude higher than that of the LWIR camera operated at 4.3 THz in [29]. This high sensitivity allows high-contrast, high-SNR real-time imaging using THz QCLs emitting mW level of output power and could be opening potential industrial applications for THz-QCL based real-time focal-plane-array THz imaging in the near future [33].

An example of high-resolution real-time imaging is given in Fig. 16.2. Here a 2.5 THz QCL was used in combination with a THz microbolometer camera to acquire a series of images of bubble wrap concealed in an opaque plastic bottle. The THz beam from the QCL is collimated using one large 3" diameter parabolic mirror to provide an approx. 50 mm large beam that is used to illuminate the sample. The object is rotated along its perpendicular axis on a motorized translation stage located in the

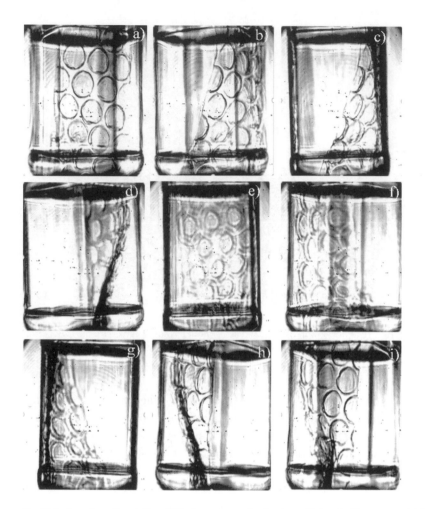

**Fig. 16.2** A series of real-time THz frames acquired using off-the-shelf commercial THz products. From (a) to (i) the sample is rotated by approximately 40 deg, starting from 0 deg (image in (a)) to 320 deg (image in (i)). A 2.5 THz QCL (TeraCascade 1000, Lytid) delivering more than 2 mW of average output power is used in combination with a THz microbolometer camera (TZCAM, I2S using CEA-LETI microbolometer technology) with a THz camera objective to acquire a set of images while rotating the sample around one axis. Several tens of images can be recorded within a few seconds. The sample is a piece of bubble wrap concealed in an opaque plastic bottle. Courtesy of CEA-TECH Aquitaine, plateforme ISO, Bordeaux, France.

middle of beam path. To image the object, a high numerical aperture THz camera objective has been developed using AR coated HRFZ-Si optics optimized for 30 cm imaging distance. A series of images were acquired at different angles that can be used to reconstruct a 3D image of the object using computed tomography software. With a full frame acquisition rate of 50 images/second (TZCam from the company I2S with CEA LETI sensor technology) a series of several tens of angular slices could be acquired within a few seconds over 180 degrees.

A multi-spectral imaging system was also demonstrated using three QCLs emitting at 2.5 THz, 3.4 THz, and 4.3 THz and a metamaterial-based focal plane array as detector [34]. Images of pellets with different chemical composition were acquired using this set-up enabling one to distinguish every pellet from each other by exploiting the difference in the chemicals absorption at different THz frequencies.

With increasing commercial availability, THz QCL technology combined with THz cameras are on the verge of breaking through to the industry-level applications. The compactness high frequency, and high output power of THz QCL sources have been exploited to speed up THz imaging to real-time, using large sensor arrays based on microbolometers [9] and opening up the prospect of wider use of THz QCLs for imaging at the industrial level for NDT applications.

A promising industrial application for THz QCLs is water-content sensing in plants. In [35], L. Baldacci et al. developed a non-invasive absolute measurement of water content in leaves and have shown very conclusive data on vine plants. There is an increasing need for new tools to assess water content in plants, in order to improve our knowledge of the use of this critical resource. The scope of this application is far-reaching, as many agricultural regions in the world now face serious threats due to global warming with water becoming a very valuable and rare resource in many parts of the world. The method described in [35] uses a combination of visible imaging and sensing using THz QCLs to measure non-invasively the water content in leaves. The method uses a linear law relating the leaf wet-mass to the product of optical depth and projected area. The measurement is performed as follow: first the THz transmission is measured using a 2.5 THz QCL operating in QCW mode and a THz sensor with lock-in amplification. The leaf thickness is then measured, and the leaf is detached to measure its fresh mass and a visible camera is used to extract the projective area. For calibration purposes the leaf is then left to dry and its dry-mass is measured after 24 hours. The coefficient of determination was found to be 0.95, thus showing that the method is robust and it can also be applied to different species.

THz imaging is considered to be a valuable NDT tool, well complementing X-ray and ultra-sound diagnostics, in particular on polymers, composite, and ceramic materials. THz QCL technology brings to the set of THz NDT tools (which are mostly based on pulsed systems) some valuable advantages: high resolution compared to sub-THz CW imaging, high dynamic range compared to pulsed system >2 THz, and real-time full field imaging using large array THz microbolometer cameras. Several proofs of concept have been shown on the use of THz QCL technology for NDT imaging applications. In [36] a THz QCL emitting at 2.5 THz was used in a raster scan imaging setup to inspect defaults in two composite materials based on carbon fiber and Kevlar. The method was able to detect delamination in Kevlar composite by using cross-polarization imaging and cracks in the carbon fiber composite. These materials are heavily used nowadays in the aerospace industry and are becoming more and more used also in the automobile industry. A reliable tool to perform NDT is therefore wished for by many industries.

Another example is using THz QCL for the inspection of injection molded polymer parts. A commercial THz QCL system emitting at six frequency bands, 2.5 THz,

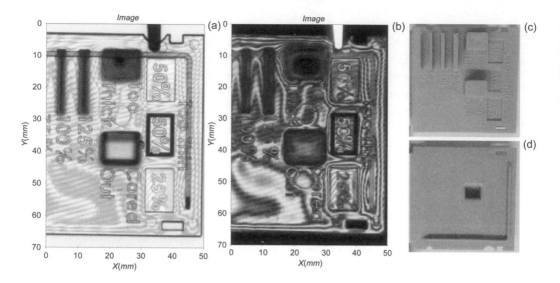

**Fig. 16.3** Simultaneous transmission and reflection imaging using THz QCLs. (a) Transmission THz image. (b) Reflection THz image. (c) Front-side photography of the injection molded PP plastic part. (d) Back-side photography of the part. The source is a commercial system (Lytid TeraCascade 1000 using the 2.5 THz band with >2 mW average output power). Raster-scan is used in combination with a beam-splitter and two pyro-electric sensors with integrated AR silicon objective (TeraPyro, Lytid). (Image acquired by A. Chopard, Lytid.)

2.9 THz, 3.3 THz, 3.8 THz, 4.2 THz, and 4.7 THz (TeraCascade 1000, Lytid), was used to acquire a set of two images simultaneously, one in transmission and one in reflection mode using a beam-splitter optical setup and two pyroelectric detectors. The sample was placed on a two-axis motorized translation stage and raster-scan imaging performed. The set of images is shown in Fig. 16.3. For this set of images, the frequency band at 2.5 THz operating in QCW mode delivering >2 mW was used to maximize transmission through the sample made of polypropylene (PP) plastic.

In this set of images, several interesting features can be seen. First, the simultaneous transmission and reflection allows to distinguish better the front and back side topology of the sample in a single-shot acquisition. Two sets of interference patterns can clearly be seen in the images due to the strong coherence of the laser. One set is approximately at an angle of 30 deg evenly spaced along the sample and is attributed to the slight tilt of the sample with regards to the normal incident radiation. This interference pattern is attributed to optical path differences between the front/back-side reflection on the part and reflections on the optics/beam-splitter. This was verified by changing the tilt angle and observing how the spatial separation of the fringes change accordingly. More interesting are the irregular patterns observed around the different topological features of the part, for example at close spatial separation in the top and right side close to the L-shape feature and at higher spatial separation in the bottom left part of the image; better contrast is obtained in the reflection image. These are attributed to FP interference between the front-side and the back-side of the part. This allows to measure with high precision the differences in sample thickness as each fringe represents one half-wavelength optical path difference. This could be used to characterize coating homogeneity or surface planarity in industrial plastic, ceramic, or composite parts.

## 16.2.3    Spectroscopy with THz QCLs

THz QCLs have very appealing properties for spectroscopic applications as they provide continuously tunable narrow linewidth high-power optical emission at frequencies above 2 THz where many important molecules have spectral fingerprints. One the first uses of THz QCLs technology for high-resolution spectroscopy was the measurement of the pressure broadening and the pressure shift of a rotational transition of methanol at 2.52 THz [37]. Here a THz spectrometer was built using a distributed feedback (DFB) QCL emitting near 2.5 THz. The QCL frequency was tuned by changing its operating current with the experimentally measured tuning coefficient of 10 MHz/mA. To monitor the THz QCL frequency in real time, the QCL output was mixed with that of an optically pumped gas laser using a Schottky point-contact diode in a corner cube reflector.

The realization of a THz quartz-enhanced photoacoustic (QEPAS) sensor for $H_2S$ trace gas detection using a 2.91 THz QCLs was reported in [38]. Details of the QEPAS technique are discussed in Chapter 13. Figure 16.4 shows the schematic diagram of the measurement system.

The quartz tuning fork (QTF) used for the THz QEPAS measurements had a 700 μm gap between prongs and a first flexion resonance of $f_0 = 2871$ Hz at 20 Torr pressure. The THz laser beam was focused between the QTF prongs by using two 2-inches diameter, 90° off-axis parabolic aluminum reflectors. The QTF is housed in an acoustic detection module (ADM) with TPX input and output windows. The THz laser beam exiting the ADM was re-collimated onto a pyroelectric detector by means of an additional aluminum parabolic mirror. Data acquisition was performed using a lock-in amplifier at $f_0 = 2871$ Hz reference frequency. A sinusoidal modulation was applied to the QCL driver as well as a voltage ramp for laser frequency tuning. A detection sensitivity of 13 ppm of $H_2S$ concentration was found with 30 sec lock-in integration time using this THz QEPAS technique and the 97.113 cm$^{-1}$ line with 0.24 mW laser power. This compares to best values of 330 ppb the for same integration time in the mid-infrared at 1266.93 cm$^{-1}$ with 45 mW laser power. However, due to the limitation of the tuning range of the THz QCL used in the experiment ($\sim$0.03 cm$^{-1}$), a much stronger THz absorption line of $H_2S$ at 95.626 cm$^{-1}$ could not be reached. If this line were to be used for sensing, the technique was estimated to be able to reach sub-100-ppb sensitivity at the same THz laser power.

Several other spectroscopic schemes have since been demonstrated using THz QCLs. Consolino et al. [2] reported a THz QCL-based spectrometer for trace gas detection using a wavelength modulation technique that yielded a four-fold increase in sensitivity compared to a direct absorption method. A heterodyne THz spectrometer based on QCLs was demonstrated in [39] using a tunable DFB QCL at 3.5 THz in

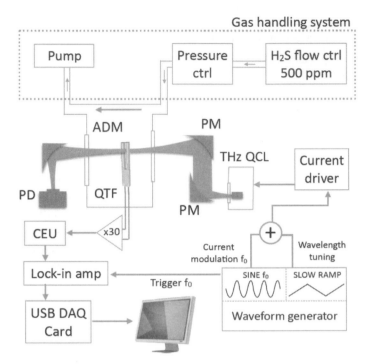

**Fig. 16.4** Schematic of a QEPAS-based trace gas sensor using a THz QCL as an excitation source. CEU – control electronic unit; PM – parabolic mirror; ADM – acoustic detection module; USB DAQ – universal serial bus data acquisition system; PD – pyroelectric detector. (Reproduced from [38]).

combination with a super-conducting NbN hot electron bolometer (HEB) bolometer for the measurement of the methanol roto-vibrational lines around 3.5 THz. Further development of this system led to a metrological grade THz spectrometer using a THz QCL [40]. Here, the QCL output is phase-locked to the tooth of a stabilized terahertz frequency comb generated via nonlinear down-conversion of the NIR frequency-comb output of a Cesium-clock-referenced Ti-Sapphire laser. The phase-locking exploited the beating between the THz comb line and a portion of the QCL output using a HEB as a mixer. The remaining part of the THz QCL beam was used for absorption spectroscopy of methane in a gas cell using a pyroelectric sensor and lock-in amplification for detection. This measurement produced THz absorption spectra of methanol in the absolute frequency scale with the absorption line center positioning uncertainty of only $10^{-9}$, ranking this technique among the most precise ever developed in the THz range.

A dual-comb spectroscopy (DCS) was also recently demonstrated with THz QCLs [41]. Applications of this technique are discussed in Chapters 7 and 14. DCS with THz QCLs allows ultra-fast and high-sensitivity spectroscopy measurement in the 2–5 THz range.

## 16.2.4 Frequency Stabilization of THz QCLs and Their Applications as Local Oscillators

THz QCLs are well suited to be used as high-frequency ($>1.5$–$2$ THz) THz LOs. They offer several orders of magnitude higher output power than electronic sources based on Schottky diode multipliers at frequencies higher than 2 THz. Compared to THz gas lasers, THz QCLs are much more compact, allowing them to be used on light airborne missions or in space. However, for the THz QCL technology to be usable as a LO in high-resolution spectrometers one needs to achieve laser frequency stabilization, good far-field beam profile, and single-mode selectivity. Control of the emission frequency and far-field profile are discussed in the other chapters of this book. Here we briefly summarize the frequency stabilization efforts. Several implementations were made to this end: stabilization to a THz gas laser using Schottky mixers, heterodyne mixing with an output from an electronic source using a HEB, stabilization using an absorption line of a gas, and phase-locking to the output of a near-infrared fs laser system.

Initially, frequency stabilization to a gas laser emission line was realized using a Schottky diode as a mixer element at 2.5 THz [37] and 3 THz [42]. To avoid using bulky THz gas lasers, a further development was to switch the reference THz sources to compact electronic sources based on Schottky diode multipliers with demonstrations at 2.7 THz [43] and 1.5 THz [44]. Due to the low available power from Schottky multiplied source (few tens of $\mu$W at frequencies above 2 THz), helium-cooled HEB mixers had to be used in [43] and [44]. In [45], a demonstration was made to stabilize a 2.5 THz QCL using the absorption line of methanol in a small gas cell, resulting in a highly compact system. However, since the tuning mechanism of THz QCLs only allows very small shifts in frequency up to a few GHz, this method can only be applied to selected DFB lasers emitting near a gas absorption line in the range 2–5 THz.

A scheme using a fs-laser and operating at room-temperature that allows to phase-lock any THz QCL was demonstrated in [25]. Here the THz QCL is phase-locked to the high harmonic of a repetition frequency of a NIR fs fiber laser using an EO sampling unit in a similar matter to that described in Section 16.2.3. This scheme can be seen as an all-optical frequency down-converter from THz to MHz frequencies. As long as the NIR comb bandwidth is larger than the QCL emission frequency, any QCL can be stabilized using this method and multiple QCL can be stabilized using a single fs-laser. The results showed that 90% of the stabilized THz QCL mode power was within a 1 Hz bandwidth [25].

THz QCL technology has proven to be an excellent candidate as LO for radio-astronomy applications using heterodyne instrumentation. Due to their high output power, multi-pixel heterodyne instruments could be imagined. Recently, several heterodyne receivers based on Schottky diode mixers dedicated to THz QCL stabilization have been developed for the 1.8–3.2 THz band [46] and are in development for the 3–5 THz band [47]. Such receivers, operating at room temperature, can be used to stabilize the frequency of THz QCLs to a well-known reference such as a 100 MHz quartz crystal. Furthermore, the LOCUS ESA project is a UK-led instrument concept for a low-cost terahertz sounder of the mesosphere and lower thermosphere (MLT) that proposes to use stabilized THz QCLs as LO for its bands 1 (4.7 THz) and 2 (3.5 THz) [48].

## 16.3    Outlook

Many industrial applications of THz technology today exploit pulsed fs-laser-based time-domain THz systems to measure layer thicknesses in multilayered materials exploiting the time-of-flight measurement approach. It is not easy to directly implement such approaches with THz QCLs because their pulses are much longer than that of pulsed fs-based THz systems. It may in principle be possible to perform thickness measurements with THz QCL frequency-comb sources or mode-locked THz QCLs, but this has yet to be demonstrated. Recent development in this field resulted in devices producing pulses of THz radiation as short as 4 ps using a dispersion compensated waveguide geometry [49]. However, the sampling of these pulses still requires complex methods based on fs-lasers [50]. The development of a simple coherent sampling scheme would open a vast field of industrial applications using THz QCLs.

For the applications where the layer thickness information is not necessary, e.g., for the identification of broad spectral features in solids/liquids, trace gas detection, or high-resolution THz imaging, THz QCLs present a strong alternative to pulsed time-domain THz systems for operation in an industrial setting as they offer compacity, stability and turn-key operation at a competitive cost. Furthermore, the much higher output powers of THz QCLs compared to time-domain systems in the frequency range from 2 to 5 THz can result in shorter measurement times and higher dynamic range of THz images.

To help improve further adoption, several developments would greatly benefit the technology. Room-temperature operation would dramatically reduce the size and cost of solutions based on THz QCLs. The current commercial products employ cryogenic cooling of THz QCLs in closed-cycle cryostats that are sufficiently compact and easy to operate. Given the current low commercial volumes for THz QCLs applications, the cost of cryocoolers impacts the THz QCL system cost only moderately. However, in the future, as volumes increase, a room-temperature operating THz QCL device will be mandatory for mass adoption of the technology.

Broad frequency tunability is also one of the most sought-after requests from users. As the power performance of THz time-domain systems and photomixers rapidly decrease above 2–3 THz, there is an overall lack of high-power THz sources in the range where THz QCLs perform best (2–5 THz). Spectroscopy is currently one the most developed application in the THz range and a tunable high-power THz QCL would be a very powerful tool. Several attempts have been made to develop broadly tunable external cavity THz QCL systems, but none have been able to provide the mode-hop free broad tuning range that users are looking for. This is partly due to the state of anti-reflection coating technology in the THz range that can currently provide, at best, few-percent residual reflectivity of a THz QCL facet. This is not enough to suppress the internal cavity modes when light is fed back into the QCL waveguide from an external grating.

Finally, an increase in the output power of THz QCL would greatly increase the dynamic range and could allow stand-off THz sensing and THz imaging. Although short-pulse THz QCL have demonstrated peak output power in the watt level, average power still remains in the range of few tens of mW at very low temperature. Given that most efficient room-temperature THz detectors are slow thermal detectors, average power is the main criteria for THz QCL usability in real-world imaging and sensing applications.

To conclude, THz QCLs have proven to be one of the most appealing sources of THz radiation for the scientific community. Many applications have been developed with THz QCLs in the 2–5 THz frequency range where they outperform competing technologies in terms of power and compactness. However, industrial applications of THz QCLs are still years behind compared to that of time-domain THz systems. The adaptation of THz QCLs for applications is nevertheless ongoing and it is clear that with the improvements in the commercial THz source technology, THz QCL systems will also find their place among industrial THz applications.

## References

[1] P. Dean et al., "Terahertz imaging using quantum cascade lasers – a review of systems and applications," *J. Phys. D: Appl. Phys.* **47**, 374008, 2014.

[2] L. Consolino et al., "THz QCL-based cryogen-free spectrometer for *in situ* trace gas sensing," *Sensors*, **13**, 3331–3340, 2013.

[3] Y Leng Lim et al., "Demonstration of a self-mixing displacement sensor based on terahertz quantum cascade lasers," *Appl. Phys. Lett.* **99**, 081108, 2011.

[4] F. Destic et al., "THz QCL-based active imaging dedicated to non-destructive testing of composite materials used in aeronautics," Proc. SPIE **7763**, *Terahertz Emitters, Receivers, and Applications*, 776304, 2010.

[5] A. D. Rakić et al., " THz QCL self-mixing interferometry for biomedical applications," Proc. SPIE *9199*, *Terahertz Emitters, Receivers, and Applications V*, 2014.

[6] P. Dean et al., "Terahertz imaging through self-mixing in a quantum cascade laser," *Opt. Lett.* **36**, 2587–2589, 2011.

[7] R. Eichholz et al., "Frequency modulation spectroscopy with a THz quantum-cascade laser," *Opt. Express* **21**, 32199–32206 2013.

[8] T. Hagelschuer et al., "Real-time gas sensing based on optical feedback in a terahertz quantum-cascade laser," *Opt. Express* **25**, 30203–30213, 2017.

[9] F. Simoens et al., "Towards industrial applications of terahertz real-time imaging," Proc. SPIE 10531, *Terahertz, RF, Millimeter, and Submillimeter-Wave Technology and Applications XI*, 2018.

[10] J. Darmo et al., "Imaging with a terahertz quantum cascade laser," *Opt. Express* **12**, 1879–1884, 2004.

[11] S. Barbieri et al., "Imaging with THz quantum cascade lasers using a Schottky diode mixer," *Opt. Express* **13**, 6497–6503, 2005.

[12]  S. Barbieri *et al.*, "Heterodyne mixing of two far-infrared quantum cascade lasers by use of a point-contact Schottky diode," *Opt. Lett.* **29**, 1632–1634, 2004.

[13]  J.R. Gao *et al.*, "Terahertz heterodyne receiver based on a quantum cascade laser and a superconducting bolometer," *Appl. Phys. Lett.* **86**, 244104, 2005.

[14]  H.-W. Hübers *et al.*, "Terahertz quantum cascade laser as local oscillator in a heterodyne receiver," *Opt. Express* **13**, 5890–5896, 2005.

[15]  D. C. Larrabee *et al.*, "Application of terahertz quantum-cascade lasers to semiconductor cyclotron resonance," *Opt. Lett.* **29**, 122–124, 2004.

[16]  S. S. Dhillon *et al.*, "THz sideband generation at telecom wavelengths in a GaAs-based quantum cascade laser," *Appl. Phys. Lett.* **87**, 071101, 2005.

[17]  S. S. Dhillon, *et al.*, "Terahertz transfer onto a telecom optical carrier," *Nature Photonics* **1**, 411–415, 2007.

[18]  B. S. Williams *et al.*, "High-power terahertz quantum-cascade lasers," *Elec. Letters* **42**, No. 2, 89-91, 2006.

[19]  K. L. Nguyen *et al.*, "Three-dimensional imaging with a terahertz quantum cascade laser," *Opt. Express* **14**, 2123–2129, 2006.

[20]  P. Dean *et al.*, "Absorption-sensitive diffuse reflection imaging of concealed powders using a terahertz quantum cascade laser," *Opt. Express* **16**, 5997–6007, 2008.

[21]  M. Ravaro *et al.*, "Detection of a 2.8 THz quantum cascade laser with a semiconductor nanowire field-effect transistor coupled to a bow-tie antenna," *Appl. Phys. Lett.* **104**, 083116, 2014.

[22]  U. S. de Cumis *et al.*, "Terahertz confocal microscopy with a quantum cascade laser source," *Opt. Express* **20**, 21924–21931, 2012.

[23]  N. Rothbart *et al.*, "Fast 2-D and 3-D terahertz imaging with a quantum-cascade laser and a scanning mirror," *IEEE Transactions on Terahertz Science and Technology* **3**, no. 5, 617–624, 2013.

[24]  M. Ravaro *et al.*, "Continuous-wave coherent imaging with terahertz quantum cascade lasers using electro-optic harmonic sampling," *Appl. Phys. Lett.* **102**, 091107, 2013.

[25]  S. Barbieri *et al.*, "Phase-locking of a 2.7-THz quantum cascade laser to a mode-locked erbium-doped fibre laser," *Nature Photonics* **4**, 636–640, 2010.

[26]  P. Dean *et al.*, "Coherent three-dimensional terahertz imaging through self-mixing in a quantum cascade laser," *Appl. Phys. Lett.* **103**, 181112, 2013.

[27]  A. Valavanis *et al.*, "Self-mixing interferometry with terahertz quantum cascade lasers," *IEEE Sensors Journal*, **13**, Issue 1, 2013.

[28]  A. A. Danylov *et al.*, "Terahertz inverse synthetic aperture radar (ISAR) imaging with a quantum cascade laser transmitter," *Opt. Express* **18**, 16264–16272, 2010.

[29]  A. W. M. Lee *et al.*, "Real-time imaging using a 4.3-THz quantum cascade laser and a 320x240 microbolometer focal-plane array," *IEEE Photonics Technology Letters* **18**, 13, pp. 1415–1417, 2006.

[30]  N. Oda *et al.*, "Development of bolometer-type uncooled THz-QVGA sensor and camera," *2009 34th International Conference on Infrared, Millimeter, and Terahertz Waves*, Busan, pp. 1–2, 2009.

[31]  L. Marchese *et al.*, "A microbolometer-based THz imager," Proc. SPIE 7671, *Terahertz Physics, Devices, and Systems IV: Advanced Applications in Industry and Defense*, 2010.

[32]  F. Simoens *et al.*, "Active imaging with THz fully-customized uncooled amorphous-silicon microbolometer focal plane arrays," *2011 International Conference on Infrared, Millimeter, and Terahertz Waves*, Houston, pp. 1–2, 2011.

[33] F. Simoens *et al.*, "Towards industrial applications of terahertz real-time imaging," Proceedings Volume **10531**, *Terahertz, RF, Millimeter, and Submillimeter-Wave Technology and Applications XI*, 2018.

[34] Z. Zhou *et al.*, "Multicolor T-ray imaging using multispectral metamaterials," *Adv. Sci.* **5**, 1700982, 2018.

[35] L. Baldacci *et al.*, "Non-invasive absolute measurement of leaf water content using terahertz quantum cascade lasers," *Plant Methods* **13**, 51, 2017.

[36] F. Destic *et al.*, "THz QCL-based active imaging applied to composite materials diagnostic," *35th International Conference on Infrared, Millimeter, and Terahertz Waves*, Rome, pp. 1–2, 2010.

[37] H.-W. Hübers *et al.*, "High-resolution gas phase spectroscopy with a distributed feedback terahertz quantum cascade laser," *Appl. Phys. Lett.* **89**, 061115, 2006.

[38] V. Spagnolo *et al.*, "THz quartz-enhanced photoacoustic sensor for $H_2S$ trace gas detection," *Opt. Express* **23**, 7574–7582, 2015.

[39] Y. Ren *et al.*, "High-resolution heterodyne spectroscopy using a tunable quantum cascade laser around 3.5 THz," *Appl. Phys. Lett.* **98**, 23, pp. 3–6, 2011.

[40] S. Bartalini *et al.*, "Frequency-comb-assisted terahertz quantum cascade laser spectroscopy," *Phys. Rev. X* **4**, 021006, 2014.

[41] J. Westberg *et al.*, "Terahertz dual-comb spectroscopy using quantum cascade laser frequency combs," *Conference on Lasers and Electro-Optics*, OSA Technical Digest, paper STu4D.2, 2018.

[42] A. L. Betz *et al.*, "Frequency and phase-lock control of a 3 THz quantum cascade laser," *Opt. Lett.* **30**, 1837–1839, 2005.

[43] P. Khosropanah *et al.*, "Phase locking of a 2.7 THz quantum cascade laser to a microwave reference," *Opt. Lett.* **34**, 2958–2960, 2009.

[44] D. Rabanus *et al.*, "Phase locking of a 1.5 terahertz quantum cascade laser and use as a local oscillator in a heterodyne HEB receiver," *Opt. Express* **17**, 1159–1168, 2009.

[45] H. Richter *et al.*, "Submegahertz frequency stabilization of a terahertz quantum cascade laser to a molecular absorption line," *Appl. Phys. Lett.* **96**, 071112, 2010.

[46] B. T. Bulcha *et al.*, "Design and characterization of 1.8–3.2 THz Schottky-based harmonic mixers," in *IEEE Transactions on Terahertz Science and Technology*, **6**, 5, pp. 737–746, 2016.

[47] B. T. Bulcha *et al.*, "Development of 3–5 THz harmonic mixer," *42nd International Conference on Infrared, Millimeter, and Terahertz Waves (IRMMW-THz)*, Cancun, pp. 1–2, 2017.

[48] B. Swinyard *et al.*, "The LOw Cost Upper atmosphere Sounder: the 'elegant breadboard' programme," *8th UK, Europe, China Millimeter Waves and THz Technology Workshop (UCMMT)*, Cardiff, pp. 1–4, 2015.

[49] F. Wang *et al.*, "Short terahertz pulse generation from a dispersion compensated modelocked semiconductor laser," *Laser Photonics Rev.*, **11**, 1700013, 2017.

[50] S. Barbieri *et al.*, "Coherent sampling of active mode-locked terahertz quantum cascade lasers and frequency synthesis," *Nature Photonics* **5**, 306–313, 2011.

# Index

absorption
    free-carrier, 21, 50, 114
    multiphonon, 102
    nonresonant intersubband, 24, 48, 50
    resonant, 124
    two-phonon, 102
absorption lines, 443
active region design, *see* terahertz QCLs
AD scattering, *see* alloy-disorder scattering
AFM, *see* atomic-force microscopy
alloy-disorder scattering, 31, 59, 64
APT, *see* atom-probe tomography
area-specific thermal conductance, 71, 72
atom-probe tomography, 52, 56
atomic-force microscopy, 380–383
axial correlation length, *see* graded-interfaces IFR
    scattering

BA lasers, *see* coherent-power scaling of
    mid-infrared QCLs
basic operation, 153
BH, *see* buried-heterostructure
Bragg diffraction, 276
    first-order, 284, 290
    second-order, 292, 296
    third-order, 278
buried-heterostructure, 52, 72

carrier leakage, 42, 44–47, 51, 63
    IFR-scattering triggered, 44, 65, 115, 128
    LO-phonon-scattering triggered, 44, 63, 128
    schematic representation, 64
    shunt-type, 43
    to the continuum, 42, 63
carrier-leakage estimates
    4.5–5.0 μm-emitting QCLs, 65
    8 μm-emitting QCLs, 67
CB engineering, *see* conduction-band engineering
CoCoA, *see* dual-comb spectroscopy
coherence, 158
coherent-power scaling for mid-infrared QCLs
    broad-area lasers, 77, 78

    *see also* mid-infrared QCLs, photonic-crystal;
        phase-locked mid-infrared QCL arrays;
        tapered mid-infrared QCLs
conduction-band engineering, 45, 46, 48
continuous-wave operation, 71, 174, 175
    core-region temperature rise, 71, 74–76
    maximum wall-plug efficiency, 72
    output optical power, 74
    threshold-current density, 71
conventional mid-infrared QCLs
    band diagram and wavefunctions, 69
    internal efficiency, 47, 68
    relative leakage-current density components, 69
CW operation, *see* continuous-wave operation

DBR grating, *see* distributed Bragg reflector
    grating
DCS, *see* dual-comb spectroscopy
deep-well active-region QCL, *see* linear-tapered
    active-region QCL
density matrix, 159
density-matrix model, 53, 71, 158
DFB, *see* distributed-feedback terahertz QCLs
    *see also* long-wavelength mid-infrared QCLs;
        mid-infrared QCLs
differential resistance, 51
DIRCM, *see* directed infrared countermeasures
directed infrared countermeasures, 397–400
    QCL-technology-based systems, 398, 399
distributed Bragg reflector grating, 86
distributed-feedback terahertz QCLs, 275
    antenna-feedback scheme, 283, 288
    Bragg mode, 284
    edge-emitting single-mode, 276
    first-order, 276
    hybrid-grating scheme, 295, 297
    out-coupling efficiency, 280, 291, 296
    second-order, 291
    second-order with central $\pi$ phase shift, 292
    second-order with dual slit, 298
    second-order with graded-photonic
        heterostructure, 293